Mastering Electronics Workbench

Version 5 and Multisim Version 6

John J. Adams

McGraw-Hill
New York Chicago San Francisco Lisbon London
Madrid Mexico City Milan New Delhi San Juan
Seoul Singapore Sydney Toronto

McGraw-Hill

A Division of The **McGraw-Hill** *Companies*

1 2 3 4 5 6 7 8 9 0 DOC/DOC 0 6 5 4 3 2 1 0

P/N 0-07-134484-5
PART OF
ISBN 0-07-134483-7

The sponsoring editor for this book was Scott Grillo and the production supervisor was Pamela Pelton. It was set in Vendome by Patricia Wallenburg.

Printed and bound by R. R. Donnelley & Sons Company.

This book is printed on recycled, acid-free paper containing a minimum of 50% recycled, de-inked fiber.

DEDICATION

To the technical support staff at Electronics Workbench. They are the front lines of the company and deserve a medal for fielding the thousands of calls it took to compile this book.

CONTENTS

INTRODUCTION

NOTE

Multisim Version 6 is the next generation of software after Electronics Workbench Version 5. The maker of both software packages now calls itself Electronics Workbench, not Interactive Image Technologies (IIT).

When you first began studying electronics you probably learned basic principles the hard way. Eventually you gained an understanding of the material being studied by building circuits out of salvaged parts. Projects that required breadboarding likely followed, with plenty of swearing when loose wiring caused problems. That was the old process of learning electronics and designing custom circuits. The new way is to power up a personal computer and launch a program. With Electronics Workbench or Multisim, infinite supplies of parts are available to you. This includes the ability to place them in any arrangement along with the processing power to simulate any breadboarded project that you can dream up.

Electronics Workbench® (EWB for short) and Multisim® are *the* virtual additions to your workshop. They are powerful interactive tools that allow quick experimentation with electronic concepts. The software programs use a personal computer to simulate what would otherwise be a real electronics workbench, complete with drawings, parts, tools, and instruments. Instead of accumulating thousands of electronic components that sit in bins collecting dust, you now have an infinitely renewable supply. Merely push the PC's power button, launch the Electronics Workbench software, and place as many components on the EWB screen as you wish. No dead batteries, no expensive ICs, no midday trips for solder or resistors; and you can all but trash the calculator and the reams of circuit calculations. It's all replaced by a clicking mouse and imaginary parts that present you with complete analyses of modern electronic circuits.

Mastering Electronics Workbench, Version 5 and Multisim Version 6 is written mainly for ***hobbyists*** interested in using Electronics

Workbench Version 5 or Multisim Version 6. Current Electronics Workbench users may want to catch up on the newer Multisim features and interface. You may have seen this program in classrooms, in corporations, or a hobbyist friend may have piqued your interest in the product. EWB and Multisim are typically advertised in electronics magazines, and the ads may have caught your eye. It's possible you have even downloaded the free demo offered on the Internet at **http://www.electronicsworkbench.com/**. Beginners can use it for LAB learning, hobbyists can use it to test-run their projects (as a breadboard substitute), and engineers can use it to prototype and debug projects. Whatever your level of electronics interest, this book will help.

Most people who first come in contact with EWB give it a test drive, get stuck in some of its more complex idiosyncrasies, and give up. I myself have been a victim of this scenario. *Mastering Electronics Workbench* will take you past that "What do I do with it now?" For example, I am finding EWB helpful in cooking up delights that I have gleaned from CMOS and TTL cookbooks, because it aids me in the design stages. In this way, I will take you through the nuts and bolts of the software and get in the trenches with you as only a fellow hobbyist can.

You will learn the secrets of experts, as well as tips and suggestions from the technicians who designed the program. They are in an easy-to-understand language that makes little use of equations and terms only an engineer would know. The book also contains circuits that will help you visualize complex simulations and draw your designs with energetic efficiency. Your newly learned abilities will shine when you take your designs from rough idea to real hardware in a flash.

When I learn a new program, I tend to get frustrated trying to find simple commands that can sometimes be buried too deeply in the application. Multisim and EWB are no different. The order of chapters in *Mastering Electronics Workbench Version 5 and Multisim Version 6* is such that it helps you learn in steps. This means learning the basics (oh, that's where that command is!), to using the basics (the triumph of your first successful simulation), to discovering new techniques that will save time and create certain effects. Soon, you will become so familiar with the software that having to whip out complex designs and simulations will not phase you.

Chapter 1, "Basics of Electronics Workbench and Multisim," is a summary of the basic elements that make up Electronics Workbench V5 and Multisim 6. This includes definitions you will encounter throughout this book, as well as condensed explanations of these powerful software packages.

Chapters 2 through 7 cover Multisim 6 exclusively. Chapter 2, "Virtual Workbench: The Multisim Interface," introduces you to menus, toolbars, and the workbench. These are used to access the software and all of its computing elements. Chapter 3, "Drawing with Multisim (Schematic Capture)," details the processes of digitally etching your designs for hardcopy, storage, or circuit simulation. Chapter 4, "Multisim Building Blocks: Components," picks apart the composition of Multisim's virtual electronic components and explains their role in each function of this software. Chapter 5, "Multisim Instruments," unravels the mystery of Multisim's virtual instrumentation. Chapter 6, "Multisim: Simulation and Analysis," helps you to interpret exactly what Multisim is trying to tell you the circuit is doing and to use that data to debug designs. Chapter 7, "Twenty-five Sample Circuits in Multisim V6," spins out practical designs to encourage you to learn the functionality of this software.

Chapters 8 through 13 cover Electronics Workbench Version 5.xx, mirroring the previous chapters in content.

Chapter 14, "Unleashing Multisim/EWB," contains tricks, power techniques, clean drawing methods, and customization secrets from the experts. They have been compiled from my own experience of years with this software, tips from the Electronics Workbench crew, and from other power users around the world.

Chapter 15, "SPICE and Netlists," touches upon the cryptic languages the computer uses to simulate your designs.

Chapter 16, "Layout and other EWB Printed Circuit Board Software," fuels your curiosity about PCB products.

The Appendices cover miscellaneous information required to run EWB or Multisim, as well as Internet addresses of various resources.

How to Use This Book

You will notice the chapter outline is divided into Multisim information followed by EWB. This is actually backwards; I chose to put the most recent Multisim 6 information in the first section of this book followed by the older Electronics Workbench 5 data, in expectation that Multisim would eventually overtake EWB in the market (however, you will see they both have their uses). I recommend reading chapter 1 in detail before deciding on which product you wish to install. Once you are up and running, try to stick with that version's chapters, because even though Multisim is the next generation up from EWB 5, the programs can be quite different from each other. Chapters 14, 15, and 16 cover information common to both versions. I also suggest you have the program up and running while exploring sections of interest. Follow along by performing the short exercises as well. It's easier to understand when you can point-and-click your way through each software feature.

Features of the Book

You may be asking yourself, "Why would I need this book when I can get the documentation for free?" Have you ever actually read an "easy-to-understand" manual? (Multisim 6 comes with an electronic format manual that must be read on screen or printed—all 989 pages of it!) I have collected information from many users who spend hours a day perfecting designs with this software. I interviewed the technical support staff of Electronics Workbench, who field thousands of questions a week and pass on to you a list of common mistakes and tips. For example, do you know the most commonly asked questions are "How do I increase the accuracy of my circuit?" or "How do I get such and such a circuit to run?" Or did you know the little known trick of holding down the Shift key while wiring to keep the wire on the same X or Y axis?

Applying the software to the real world is key as well. This book will help you dream up new methods and uses for this powerful software package. A story comes to mind of a doctor who applied EWB to model how the heart produces and handles electrical signals; what an esoteric application! This advice will help you get started and keep you inspired to continue to use EWB or Multisim.

In addition to the explanations of how EWB/Multisim works, circuits are spread throughout the book. They're to help you quickly and practically learn different features of this software.

The other advantage to *Mastering Electronics Workbench Version 5 and Multisim 6* is the companion CD containing demostration versions of Electronics Workbench 5, Multisim 6, EWB Layout and Ultiboard. Circuits included with this book can be opened with this included software. You will be able to load either one of the demonstration copies right away or order the actual software with a phone call.

Further Exploring EWB, Multisim, or EDA in General

The Internet has created a huge information database on simulation and schematic capture software, as well as on electronics in general. Start with **http://www.electronicsworkbench.com**. A good resource on Electronic Design Automation (EDA) in general is **http://www.eda.org**. (See Appendix A for a list of further Internet addresses to explore.) My own website, **http://www.basicelectronics.com/**, also has a section on Electronics Workbench and Multisim information, with a multimedia air to it. McGraw-Hill and Tab Electronics also offers several websites that can be of help in the printed material arena at **http://www.books.mcgraw-hill.com/** and **http://www.books.mcgraw-hill.com/tab-electronics/tab-electronics-home.html**.

Conventions Used in this Book

Window Menu commands are used throughout this book in the following format:

File > Print > Circuit

This means to go to the **FILE** menu, choose the **PRINT** Submenu, then select the print **CIRCUIT** command.

Explanation of Icons

Tip: While compiling this book, shortcuts and other little gems have popped up. Each hint will help with a specific procedure.

Note: An important point I wish to draw to your attention.

For EWB 5 Users: In the Multisim sections, these notes tell the readers who may have previously used Electronics Workbench 5 about recent changes to Multisim.

Exercise: A small lesson to familiarize yourself with a Multisim 6 or Electronics Workbench 5 procedure.

Caution: Take care when dealing with particular commands, actions, etc.

Summary

Electronics Workbench and Multisim sparks my enthusiasm for electronics daily. Each time I use it, the hobbyist in me marvels at its features. Discovering my own new tricks also encourages me to apply this product to all my projects and learning, and that is why I wish to share this zeal for EWB with you. In this way, I hope my excitement revitalizes or reinforces your love for this hobby by introducing and helping you to master this powerful digital tool. Enjoy!

John J. Adams

ACKNOWLEDGMENTS

Many people have helped bring this book into being. To Kristy, thank you for your unconditional love and open ear; you helped me through many frustrating times and long nights at the computer and life in general. I couldn't do it without you. To my editor, Scott Grillo, I wish to thank you for your persistence and unending patience; I am deeply grateful for this. The entire staff at Electronics Workbench did their best to bring this book to light. I particularly wish to thank the Chief Executive Officer, Joe Koenig, for bringing this book concept to me and helping to mold it as I went along. Also, extra thanks go to all the EWB technical support staff, particularly Janina Dziejko, Roman Bysh, and Luis Alves. Thanks also to others at EWB that aided in the compilation of this book in one capacity or another: Scott Duncan, Roy Mular, Beatrice Noble, Dan Harris, and Michael Benedick.

Thanks also go to a few others that either directly or indirectly helped in the compilation of *Mastering Electronics Workbench*. This includes my family; in particular Pam Cassa (mom); and Jim, Jason, Jennifer, and Jordan Adams. My research super-hero and super-friend, Noëlla Sale, deserves tons of credit. She's put in many hours of help and fact finding and verifying for this book; thanks a million, Cinabar! A few other friends have listened patiently to my rants, and I wish to thank them as well: John Dashwood, Sandi Seldney, Adam Yurkiw, Dorothy McVay, and Devora Albelda.

The last person I wish to thank is you, the reader. I really enjoyed writing this book knowing you would get much more out of it than a software manual. I hope you use it in your day-to-day actions as a hobbyist.

CHAPTER 1

Basics of Electronics Workbench and Multisim

Always look for new and creative ways to make use of Electronics Workbench or Multisim. Find examples in every stage or area of your electronics interest.

Using the Electronics Workbench simulator requires some knowledge of computers as well as electronics. If you already own a personal computer, the Electronics Workbench or Multisim software will add a new tool to your arsenal, allowing you to create endless electronic simulations. These are digital equivalents of what you would pop together in real life. The interface is very simple to learn and use, so it won't be long before you are churning out complex circuits. The software is very intuitive and, as a frequent Windows or Mac user, you will learn the ABCs of EWB in no time.

In this chapter we examine why the computer is used to simulate an electronics workbench as well as learn a few terms and basic concepts about Electronics Workbench Version 5 and Multisim Version 6. You will soon see why Electronics Workbench has become a standard in electronics simulation software.

Electronics Simulation Software

Simulators have been used to train professionals for years. An aviator sits aboard a cloned cockpit, complete with forward-looking screens, and learns to fly a jet plane without ever leaving Mother Earth. Nuclear power plant technicians are placed in a room that imitates a control center; they read the mock monitors, and make crucial decisions that may someday spare a city from a nuclear disaster.

You may not learn how to save a city with an electronics simulator, but you will definitely save time and money. Your design time is reduced to a fraction of that required to sketch, prototype and test conventional electronics. Because the software itself is the only outlay of cash required, costs are extremely low (compared to buying breadboards, parts and other developmental tools).

If you're just beginning electronics, you can use the simulator to create elementary circuits that reinforce the theories being studied. A few resistors, a battery, and ammeter help you visualize Ohm's Law. A transistor and light bulb demonstrate the gain principle upon which semiconductors are based. In fact, EWB/Multisim has all but replaced the lab time that conventional electronics courses often make use of.

It's used in most high school and community college electronics courses in this capacity.

The uses of an electronics simulator are far reaching and will be discussed in detail throughout this book. For now, keep in mind it is merely a *new tool* for your workbench that saves time and money.

The EWB/Multisim Simulator Concept

Electronics Workbench is a graphical way to simulate circuits. It replaces the normal bits associated with electronics with bytes in a computer. Gone are hand-drawn circuits, breadboarding, running the results, debugging, redrawing corrections, figuring out how to make a complex PCB layout, and the countless hours of "I hope this works" finger and toe crossing that generally go with standard electronics. Instead, you place components onto the screen and run traces between them. Schematic symbols and graphically represented instruments are used to imitate a real workbench. A resistor is replaced with a squiggly line. An oscilloscope is now a pop-up window. And a copper wire is now a horizontal or vertical line across your monitor. They still perform the same theoretic function as their real-world electronic counterparts, just more cheaply!

Once your circuit is drawn, you can run endless simulations and perform countless tests. These include placing instruments into a circuit and taking readings, controlling switches or dials with the keyboard, or watching the result with lights, LEDs, and other display items. Analysis options allow you to pick apart a circuit, revealing the essential voltage, current, and waveform readings, as well as help you to optimize a design.

The EWB/Multisim All-in-one Solution

EWB/Multisim allows for quicker learning than historic breadboard and PCB methods. The time spent wiring a circuit is all but eliminated. Also, unlike a breadboard, with its containers full of components and copper wire, there is no chance for open circuits (unless simulated on purpose). This cuts time wasted hunting for faults. An engineer can take EWB and test a theory or design a circuit that he or she has been researching or designing in a fraction of the time that a traditional design cycle would take. You can go from concept to actual PCB boards in a much shorter time frame. All in all, EWB is a timesaver for all areas of electronics. It doesn't replace the practical side of putting your hands on all the electronics morsels, but it does eliminate some of the cost, time, and frustrations related to developing and testing designs.

Now that you are confident the circuit is running smoothly and all the kinks are worked out, you can export the drawing into Electronics Workbench layout software where a *printed circuit board* (PCB) will autoroute its lines with minimal help from a user's already overworked mind. Print out the PCB onto a special sheet of circuit transfer material (available from most electronics stores) and iron it onto a copper-clad board. Go through the last few chemical, drilling, soldering steps and you've completed the project in a *fraction* of the time it would have taken using old-fashioned electronic methods.

NOTE

Actually, EWB creates schematic representations of parts which do not particularly look anything like their real-world counterparts.

The Great Simulator Debate

Two basic schools of thought exist in teaching electronics. One is the hands-on approach, making use of breadboards, components, and miles of wire. This requires the student to build a circuit from various resistors, batteries, ad infinitum. Instruments are then used to monitor the results. The lab approach was used to teach technicians and engineers for years.

Then computers came along, and the ability to simulate the electron paths and various levels of current, voltage, and resistance created a whole new approach to teaching these concepts. Now a student can create a circuit in a fraction of the time and discover the results with an accuracy previously unattainable with crude experimental equipment.

Which method is better to teach electronics? Each student has individual tastes. I, myself have had great luck with computer simulators. I simply don't see the use in spending three hours building a circuit that will teach me a minor lesson. However, it is still important to give a student a hands-on feeling preparing him for future endeavors.

I've known students who simply cannot learn with the EWB program despite its availability. This, in large part, stems from the frustration of not being able to "make a certain circuit run." I've received countless complaints of this nature while researching this book. Reading this book will give you the knowledge to utilize this tool better and get those stubborn simulations fired up. It also helps to keep in mind that computers are finicky contraptions, requiring unending patience.

Mixed-mode Simulation

Electronics Workbench/Multisim are fundamentally *mixed-mode* or *mixed-signal simulators*. This means the software handles a mixture of analog and digital components placed in the same circuit. In other words, you can hook a capacitor to an AND gate or a resistor to a CMOS inverter. The combinations are endless.

Years ago, simulators would only handle one or the other. Useless! Most real-life circuits were, and still are, a composite mixture of analog and digital components. Take, for example, a simple square wave generator, as shown in Figure 1.1. A square wave can be created using two inverters, a resistor, and a capacitor. This simulation was not possible using older electronic simulators. This simple generator is used in hundreds of applications and can now be simulated with EWB Version 5 and Multisim 6.

There are too many examples of mixed-mode simulation to list. Just remember that it is now possible to use a hodgepodge of parts, so don't be afraid to use a base of ICs and sprinkle in supporting analog components here and there.

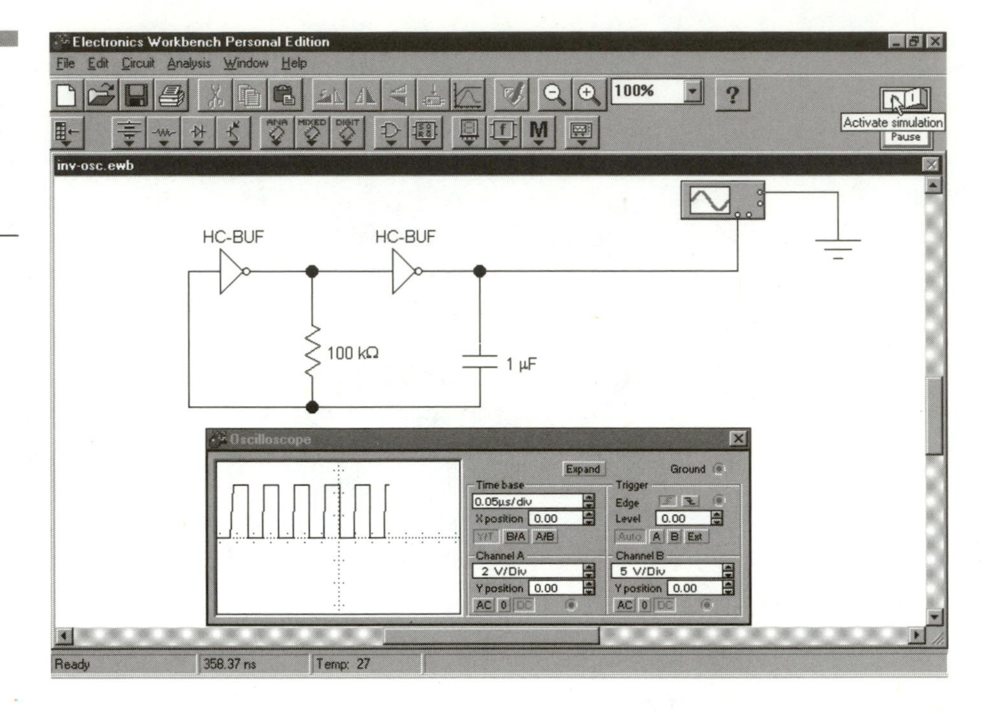

Figure 1.1
Mixing analog and digital components is now possible with EWB 5 and Multisim 6.

SPICE

Simulation Program with Integrated Circuit Emphasis (SPICE) is the computer language that makes most electronic simulations possible. The University of California at Berkeley's Electrical Engineering and Computer Sciences Department developed SPICE as an industrial standard set of algorithms for analog simulation. Here is how the prestigious institute describes SPICE:

> SPICE is a general-purpose circuit simulator with several built-in semiconductor device models. SPICE performs several analyses, including nonlinear DC, nonlinear transient, and linear AC analysis. Device types include resistors, capacitors, inductors, mutual inductors, switches, linear and nonlinear sources, lossy and lossless transmission lines, BJTs, JFETs, GaAs MESFETS, and MOSFETs.
>
> SPICE originates from the EECS Department of the University of California at Berkeley.

This quote is from the EECS Dept. Website at **http://buffy.eecs.berkeley.edu/IRO/Software/Catalog/Description/spice3f5.html**.

SPICE is a text-based language. In other words, there are no fancy drawings or mouse-clicking interfaces. Instead, a circuit is broken down into something called a Netlist (see Chapter 15). It describes each point in the circuit and what happens electrically between those locations. Do you need to learn SPICE to use Electronics Workbench/Multisim? Absolutely not! This software hides SPICE's complexity behind an easy-to-use graphical interface. The extensive learning curve typically associated with SPICE is eliminated, so educating yourself on EWB/Multisim is possible at light speed.

I bet you're thinking, "If you don't need to learn SPICE, why is it included in this book?" It sometimes helps to understand how EWB views your circuit. The data can be used to debug circuits faster or create your own circuits or components. It is not necessary to imbibe the entire SPICE language, but it helps to drink in a few gulps.

Netlists

What does a SPICE Netlist look like? Take a peek at Figure 1.2. This is an elementary circuit that displays the principle of current gain in a transistor. It's built with one battery, two resistors, one NPN transistor, and two ammeters to display the results. After exporting the file as an *.CIR file (SPICE Netlist file), you can open the Netlist with a Text Editor, such as Notepad, in Windows. Figure 1.3 shows the circuit the way your computer sees it. We explore SPICE and Netlists further in Chapter 15.

VHDL and Verilog

New to Multisim 6 are two *hardware description language* (HDL) modules called VHDL and Verilog. VHDL stands for VHSIC (Very High-Speed Integrated Circuit) Hardware Description Language. Both VHDL and Verilog are programming languages designed to emulate the behavior of advanced digital hardware circuits. These HDL-designed digital circuits can then be used in Multisim simulations. Both of these packages are not available to the Personal Edition user. You must have Multisim Professional and purchase and install the separate VHDL or VERILOG option add-ons. Power Professional includes some VHDL and/or Verilog modules, but may require you to purchase a separate add-on. For more information on VHDL and Verilog options, see **http://www.electronicsworkbench.com/html/msm_intro.html**. If you will be constructing custom integrated circuits, such as programmable logic devices, I recommend purchasing this option. Otherwise forego this lofty expense.

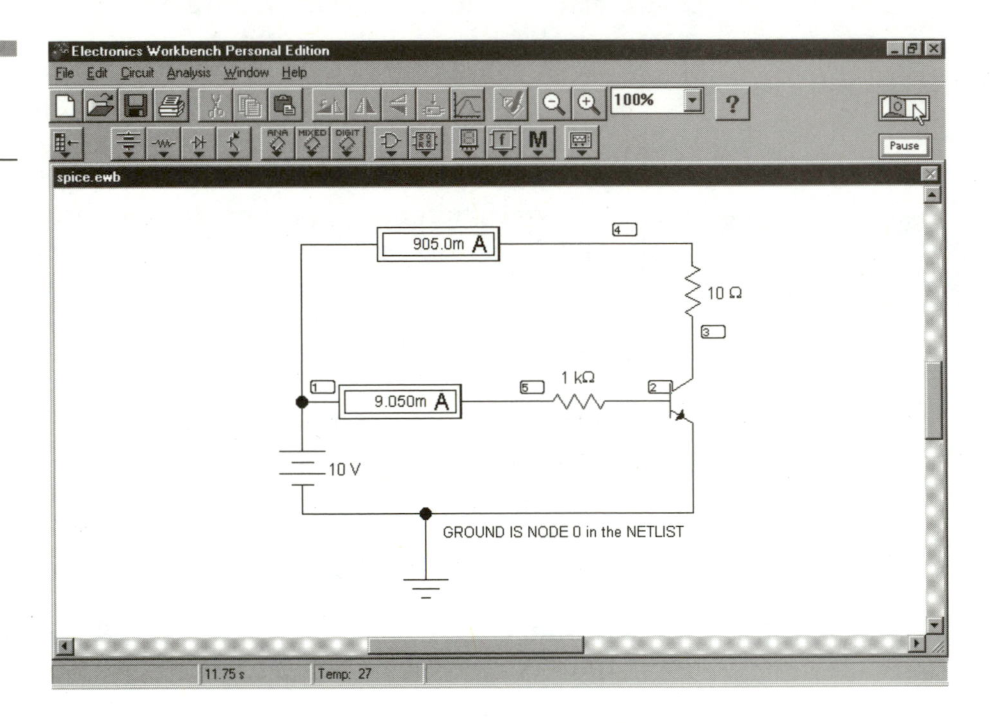

Figure 1.2
This is how we see the circuit.

Figure 1.3
This is how the computer sees the circuit.

```
**** C:\ewb5\My-Circuits\spice.ewb ***************
*  Interactive Image Technologies                 *
*                                                 *
*  This File was created by:                      *
*    Electronics Workbench to SPICE netlist       *
*    conversion DLL                               *
*                                                 *
*  Tue Nov 24 00:04:28 1999                       *
***************************************************

* Ammeter(s)
*
X_ammeter_0 1 4 ammeter_1mohm
*
X_ammeter_1 1 5 ammeter_1mohm

* Battery(s)
```

Figure 1.3 This is how the computer sees the circuit (continued).

```
*
V1 1 0 DC 10

* Resistor(s)
*
R1 5 2 1K
*
R2 3 4 10

* NPN Transistor(s)
*
Q1 3 2 0 Qnideal

* Connector(s)
* node = 0, label =
* node = 5, label =

* Misc
.OP
.SUBCKT ammeter_1mohm 1 2
    R 1 3 1mohm
    V 3 2 DC 0V
.ENDS

.SUBCKT ammeter_1mohm 1 2
    R 1 3 1mohm
    V 3 2 DC 0V
.ENDS

.MODEL Qnideal NPN(Is=1e-16 BF=100 BR=1 Rb=0 Re=0 Rc=0 Cjs=0
Cje=0 Cjc=0
+Vje=750m Vjc=750m Tf=0 Tr=0 mje=330m mjc=330m VA=1e+30 ISE=0
IKF=1e+30
+Ne=1.5 NF=1 NR=1 VAR=1e+30 IKR=1e+30 ISC=0 NC=2 IRB=1e+30 RBM=0
XTF=0
+VTF=1e+30 ITF=0 PTF=0 XCJC=1 VJS=750m MJS=0 XTB=0 EG=1.11 XTI=3
KF=0 AF=1
+FC=500m TNOM=27)

.OPTIONS ITL4=25
.END
```

Analysis

In my opinion, the most powerful tool of EWB and Multisim is their ability to analyze a circuit or component in tremendous detail. The software contains many analysis methods to dissect; create what-if scenarios; punch a circuit with any voltage, current, or signal level; or perform temperature changes. The power of analysis mode comes from its ability to troubleshoot a circuit; it allows you to break a circuit up into an easier-to-see picture than if you connected tons of instruments to it instead. Keep in mind that this feature is a learned skill. Don't let it intimidate you as a first-time user. Chapters 6 and 12 explain simulation further.

Schematic Capture Software

To create professional computer-drawn electronic circuits, a schematic capture package is required. Schematic capture software lets you automate the task of creating wiring diagrams for printouts or other use. You no longer have to spend hours of grinding pencils to their erasers or mashing your fingertips in pain. Electronics Workbench & Multisim make use of simple click-and-drag schematic editing that automates most drawing tasks. You can place a resistor symbol here, a transistor there, a digital gate to the left, an IC to the right. EWB will then, with some direction from you, neatly route all the connections into a presentable schematic drawing. The zoom-in and zoom-out feature allows you to add as many components and traces as needed to bring your design to hardcopy. This faster method makes for a much cleaner drawing than scribbles on a page. It also allows for later simulation or analysis of the circuit and the ability to use that information later to automate the task of producing a printed circuit board.

NOTE

Some editions of Electronics Workbench and Multisim limit the number of parts you can layer onto one drawing.

PCB Layout Packages

One of the biggest time-saving features of Electronics Workbench software is its ability to take your circuit designs and export them to a printed circuit board software package such as EWB Layout, Ultiboard, or Ultiroute. From there, you can convert a schematic into a usable PCB complete with multiple layers. EWB's PCB packages are extremely user-friendly, meaning you will be building circuit boards in no time.

Why You Should Use EWB Software

Electronics simulation software is actually a cost-effective replacement for most items on a real workbench. Take a look at Figure 1.4 and compare the costs. Notice that using simulation software reduces the costs of your hobby, business, or study considerably. Most people interested in modern electronics have a computer, so that expense is reduced as well.

Figure 1.4 Compare the cost of a real workbench to Electronics Workbench software.

Item	Real Workbench	EWB	Multisim
Breadboard	$2000	Included	Included
Various Components	$200.00	Included	Included
ICs/Semis	$300.00	Included	Included
Wires	$25.00	Included	Included
Multimeter	$150.00	Included	Included
Oscilloscope	$1,400.00	Included	Included
Function Generator	$500.00	Included	Included
Power Supply	$200.00	Included	Included
Paper Supplies	$20.00	Included	Included
Drafting Instruments	$100.00	Included	Included
Software Cost	NONE	$399.99	$399.99
TOTAL	$2,915.00	$399.99	$399.99

NOTE

I am not saying the real workbench is all but eliminated, but you will not need such a large cache of parts and instruments for experimenting and learning purposes.

Why purchase Electronics Workbench or Multisim software specifically? EWB is one of the most widely known mixed-mode simulators. Hundreds of thousands of students, educators, hobbyists, technicians, and electrical/electronic engineers around the world have acquired this time-saving tool. The cost, when compared to EWB's higher-end competitors, is light. Multisim and EWB's cost/feature ratio is fantastic.

Because Interactive Image Technologies has worked hard to make the interface between your mind and the computer that much simpler, in reviews it typically rates the highest for ease of use and ease of learning.

How to Make EWB Work for You

Before we go on with the rest of the book I want to stress one point:

Always look for new and creative ways to make use of Electronics Workbench or Multisim. Find examples in every stage or area of your electronics interest. This is an important point I cannot repeat enough. People, frankly, get bored with the program after whipping out a few circuits. This is because they never delve deep into the software to find the golden features that enrich the electronics experience.

I, too, fell into this trap at first. Here is how to get over that, "It's a great program but now what?" stage.

- Always keep in the back of your mind the availability of the simulator. Ask yourself, "How can I use EWB/Multisim today with this new circuit to learn a concept or to troubleshoot?"
- Make countless circuits; ones that are interesting to *you*! I use it to simulate computer parts, video game consoles, and items that are interesting to me as a hobbyist.
- Find tricks of your own. I learned most of the hints in this book, by using the programs day-to-day. I took the attitude that no matter what, I know this can be simulated somehow. A challenge will always keep a person's interest.
- Put the program through its paces and see just what it can do. Once your confidence in EWB or Multisim increases, begin using it to take your projects from napkin sketches, to design, to PCB stage.
- Be sure to read Chapter 14, Unleashing Multisim/EWB, for more usable ideas, whether you are a student, teacher, hobbyist, technician, or an engineer.
- Realize that the flow of electronics must live in our minds. The schematics *represent* the actual flow of electrons but do not appear to our senses unless hooked to a component that changes electricity into forms we can see, hear, or touch. Keep in mind that EWB is only a mind crutch for electronics; (it doesn't negate the fact you have to *learn electronics*).

Don't limit yourself to this list. I've heard many stories about uses for EWB—the list can be as endless as your imagination.

Installing EWB/Multisim

Ever tried describing an object to a person who has never actually seen it? How about outlining a procedure never performed? "OK. Did you say it was on the left or right, again?" It's time to get Electronics Workbench or Multisim installed and running so my descriptions will fall into place while I am describing it. "Oh, *that* icon! I see now."

Versions Explained

This book covers both Electronics Workbench Version 5.xx and the newer Multisim Version 6.xx. Just to clear this up, Interactive Image Technology is the creator of the Electronics Workbench software. The company now goes by the name of Electronics Workbench, and the program that was once EWB is now referred to as Multisim. So Multisim 6 is just the updated version of EWB Version 5. Wow! Confusing, but necessary before we go any further.

To daze you even more with product details, EWB and Multisim each have various editions.

- Electronics Workbench 5: Student, Personal (Hobbyist), and Professional.
- Multisim 6: Comes in three basic flavors: Personal, Professional, and Power Professional (there are other educational and student editions available as well).

Each edition of EWB 5 and Multisim 6 progressively contains more features and component models than the one before, and is more expensive. (See Figure 1.5 for a list of features that come with each.) In the book, we deal mostly with the Personal Editions.

NOTE

The 'xx' after the version number (e.g., Version 6.xx) means there may be minor updates to that version. For example, the version of Multisim at the time of writing is Version 6.11, with 6.20 on the horizon. See Figure 1.6.

Figure 1.5 Features table of Electronics Workbench 5 and Multisim 6.

ELECTRONICS WORKBENCH 5 FEATURE TABLE

Feature Table	Personal	Professional
Analyses	6	14
Analog Component	100+	100+
Digital Components	200+	200+
Device Models	4000+	8000+

MULTISIM 6 FEATURE TABLE

Feature Table	Personal	Professional	Power
Professional			
Basic schematic capture	•	•	•
Interactive simulation	•	•	•
Symbol Editor	•	•	•
SPICE analog/digital simulation	•	•	•
Component Editor	•	•	•
Complete on-line documentation	•	•	•
Industry's easiest-to-use interface	•	•	•
Export to PCB layout	•	•	•
Auto and manual wiring	•	•	•
User-configurable tool bars	•	•	•
Analysis Wizards	•	•	•
Undo and autosave	•	•	•
Multiple copies of instruments	N/A	•	•
Virtual instruments	8	9	11
Parts library	6,000	12,000	16,000
Analyses	8	15	21
Advanced schematic capture	N/A	•	•

continued on next page

Figure 1.5 Features table of Electronics Workbench 5 and Multisim 6 (continued).

Feature Table	Personal	Professional	Power
PSpice/XSPICE/BSPICE Import	N/A	•	•
Multiple workspaces	N/A	•	•
Component database	N/A	Basic	Advanced
Part search	N/A	Basic	Advanced
Bill of materials (BOM)	N/A	Basic	Advanced
Model maker (Analog and Digital)	N/A	Optional	•
HDL design/debug	N/A	Optional	•
RF design kit	N/A	Optional	•
Project/team design kit	N/A	Optional	•
Model expansion packages (I and II)	N/A	Optional	•
Post processor	N/A	N/A	•
Nested sweep analysis	N/A	N/A	•
Code modeling	N/A	N/A	•
Batched analysis	N/A	N/A	•
User defined analysis	N/A	N/A	•
VHDL Synthesis Module	N/A	Optional	Optional
Ultiboard PCB Layout	Optional	Optional	Optional
Ultiroute Advanced Autorouting and Autoplacement	Optional	Optional	Optional

Why Write About Two Versions?

When this book first began, Multisim was due for release and EWB Version 5 had already been in wide distribution for several years. This meant giving you information for both programs, because with its hefty system requirements, some of you may not be able to run Multisim and keep your sanity intact. If you own an older PC, you may want to stick with the tried and true EWB Version 5.xx. (For EWB's system requirements see Figure 1.7; Figure 1.8 shows Multisim's

EWB 5
Personal Edition

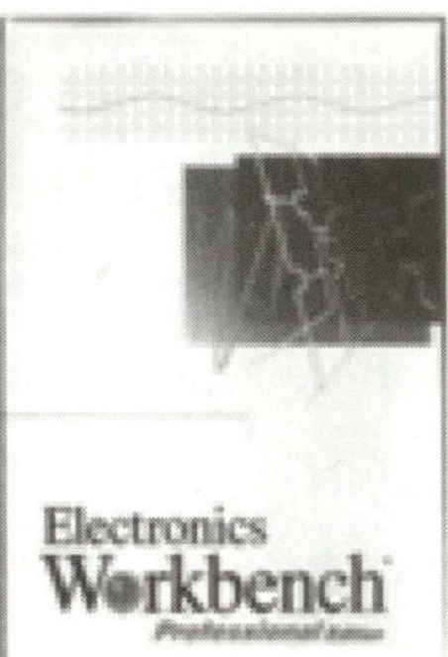

EWB 5
Professional Edition

Multisim 6
Personal Edition

Multisim 6
Professional Edition

Multisim 6
Power Professional Edition

Figure 1.6
Pictures of both versions of Workbench.

Figure 1.7
Electronics Workbench V5 system requirements.

- 486 or greater PC.
- Windows 3.11/95/NT or MAC with Emulation.
- 8-12MB RAM.
- 20MB hard disk space.
- CD-ROM drive (unless you have floppies).

Figure 1.8
Multisim V6 system requirements.

- Pentium 166 or greater PC.
- Windows 95/98/NT.
- 32MB RAM (64 recommended).
- 100 to 250MB hard disk space depending on edition.
- CD-ROM drive.
- 800X600 monitor screen resolution.

requirements. This should help you decide which version to use.) Also, Multisim has a strong "engineer-ish" feel to it. Some of the procedures are extremely complicated compared to EWB 5's friendlier hobbyist feel.

Which Version to Run

If you are purchasing software from the Electronics Workbench Company, you do not have a choice; Multisim 6.xx is the only available version.

NOTE

It may still be possible to purchase EWB Version 5.12 from the company. Contact them for further information.

If your computer does not meet the system requirements for Multisim, I highly recommend choosing a copy of EWB Version 5.12 (if you can get your hands on one). If you are planning an upgrade in the near future, you may want to forgo EWB5 and upgrade to Multisim even though you won't have blazing software speeds. But don't say I didn't warn you.

More user-friendly features, such as an Undo command, are available on Multisim. If your computer system meets or exceeds the minimum requirements, then Multisim 6 is well worth the dollars. But remember this program is much more complex than the EWB of old; despite EWB's recommended system requirements, I suggest using Multisim on nothing less than a PII-300 with 64-128MB of RAM; the more RAM the better.

I know you may want the latest and greatest product but you may be more comfortable learning EWB (with the demo even), before you purchase Multisim. I still use EWB about 50 percent of the time. This book includes the demo verision of EWB Version 5.

Machine Used to Compile this Book

Just as a note of interest, the machine used to write this book was a P-75, 32MB of RAM, 6.4 GB hard drive, 15-inch flat screen set to 800×600, Win95. EWB was also tested on a 486-66 with 12MB, 400, and

14-inch screen. My machine was eventually upgraded to a P-233 with 96MB of RAM. This sped up Multisim to a bearable walking pace.

Getting Your Hands on the EWB Program

The CD included with this book contains demo versions of the following:

- Electronics Workbench Version 5.12 with circuits from this book
- EWB Layout
- Multisim Version 6 with circuits from this book
- Ultiboard

The demos can be loaded without any passwords or keys but have limitations. See Appendix D for specific loading instructions and further data on the CD contents. Read along for more information about the demo software on the CD.

Demo EWB Version 5.12

The Demo lets you open the circuits contained in this book, as well as run a gamut of basic circuits provided by Interactive Image Technologies. You can create your own simple circuits, but unfortunately you are not able to save or print them. (See the demo's documentation for a full run-down of the demo version's features/limitations.) If you want to see how EWB works and what is possible with it, then just install this version. However, you may quickly grow tired having to remake circuits over and over again each time you start the software. The next step up is the Personal Edition.

Demo Electronics Workbench Layout

Although this product is basically the same as Ultiboard, I decided to include it as Chapter 16 uses an example. You can load Layout or Ultiboard and achieve the same results; the procedures are just slighly different. The demo offers just about every feature you can do with the retail versions of Layout, you just can't save or print the PCB files.

EWB 5.12 Personal Edition and Layout Personal Edition

You may wish to order the Personal Edition of Electronics Workbench Version 5 and Layout from the company. They are hobbyist versions and contain nearly every feature the top-end version does, but with less analysis options and modes. And unlike the demos, you can save and print files. See Appendix B for more information and costs on obtaining this software. I recommend getting these versions if you want to seriously get into schematic capture, circuit design, and PCB layout. It's a perfect get-you-going mixed-mode simulator, schematic capture, PCB layout software package to purchase if you are on a budget.

Demo Multisim Version 6.xx and Ultiboard

If you have a fast enough computer with tons of memory, load the Multisim V6 Demo for Windows 95/98/NT. You can also try out the Ultiboard Demo as well. Both demos are based on the very expensive Power Professional editions, but will certainly give you an idea of the capabilities of the Personal and Professional editions. You may note that this book is mainly written about Multisim 6.11 procedures, when in fact the demo may be a new version. There are slight differences in the component editor, but I wanted to make the newest version available to you. See Appendix D for more information.

NOTE

Ultiboard can be used in conjunction with EWB5 as well.

Multisim 6 and Ultiboard Personal Editions

These are the hobbyist versions of Multisim and Ultiboard. They contain enough features to get you going but must be ordered from Interactive Image Technologies, Of course, like everything else in life, it requires payment. See Appendix B for more information and costs on obtaining the software. I recommend getting these versions if you want to get seriously into schematic capture, circuit design, and PCB layout. It's a perfect get-you-going mixed-mode simulator, schematic capture, and PCB package to purchase if you are on a budget.

All Versions Available to You

Aside from what is contained on this CD, you can order various versions or editions from Electronics Workbench by calling 1-800-263-5552 or visiting **http://www.electronicsworkbench.com**. Newer versions and more model expansion packages are being released periodically; the easiest way to keep up with advances is to bookmark the Electronics Workbench website and frequently take a peek for updates.

Installing Your Software

Electronics Workbench 5.x can be installed onto a Windows 3.1, Windows 95/98, or Windows NT personal computer. A Macintosh with PC emulation can also run the software (at a somewhat slower speed). Multisim requires a Windows 95/98/NT system. I've included instructions on how to load the software on the disk provided with this book, as well as on how to download a demo version from the Internet. If you have a different version of EWB, EWB layout, Multisim, or Ultiboard/Ultiroute, please consult the instructions that came with it. Each version and edition has specific procedures and these seem to change constantly. When in doubt, contact tech support.

Mastering Electronics Workbench 5 and Multisim 6 Installation Interface

Refer to Appendix D for further instructions.

1. Insert the CD into CD-ROM drive.
2. If the CD interface screen does not come up automatically, run the setup.exe on the CD.
3. Follow the onscreen instructions.

Downloading Demos

You may choose to download the most recent Multisim or Ultiboard demos from the EWB Web site. However, the circuits in this book would not be included on that version. To do this, visit http://www.electronicsworkbench.com/html/demo.html and choose the appropriate version/editions. Download the file into a temporary directory and run the file. You may have to unzip the file with WINZIP. Follow the onscreen instructions.

CAUTION

Do not load more than one version of EWB into a directory. First uninstall the older version or place the new version into a different directory. For example, Install Personal into C:\EWB5 and Professional into C:\EWB5PRO\. You must install EWB5 and Multisim 6 in different directories also.

Opening and Running EWB5/Multisim 6

Once EWB V5 is installed, you can go to **Start** > **Programs** > **Electronics Workbench (Personal)** > **Electronics Workbench Personal**. This can be used with either Win95, 98, or NT.

To run Multisim, go to **Start** > **Programs** > **Multisim** > **Multisim**.

EWB and Simulator Terms

Before we embark upon learning EWB, it is best to understand a few words used in software and electronics design automation (EDA) in general. To keep verbal pace throughout the book, learn this terminology thoroughly. For a complete list, see the Glossary.

Analysis—The ability to perform specific electrical experiments on circuits or components.

EDA—Electronic Design Automation. The field of electronics that deals with using computers to automate the process of designing, testing, debugging, and analyzing electronic circuits or components.

Libraries—Also referred to as component libraries or databases; a collection of component models.

Mixed-mode Simulation—Analog and digital components simulated within the same circuit.

Models—When we refer to a component model we are speaking of the way the computer is told how that part behaves electronically. Certain variables can be set by the user or are stored on the hard drive for later access.

Schematic Capture—Drawing circuits with schematic symbols on the computer. The file can then be printed, used for simulation and analysis, or exported to a PCB program.

SPICE—Simulation Program with Integrated Circuit Emphasis. A text-based programming language used to simulate electronics circuits with a computer.

Subcircuit—A circuit within a circuit. If you use the same components in several different circuits, this automates the process further. It could save hours of repetitive design times.

These are just a few of the common words you will run into throughout this book. Others will be defined as we go along.

Summary

Electronic Workbench and Multisim have three basic functions: they operate as mixed-mode simulators, schematic capture programs, and as circuit analysis tools. Add to this their powerful layout packages and the fact that they are simple to learn and easy to operate, and you can see the advantages. What you use these products for is up to you and your imagination. This book is merely a launching point for your electronic imagination. EWB/Multisim (or any circuit simulation software) are not the be-all and end-all to learning or testing electronics. They are merely a way to enhance the learning experience, much the same way that a flight simulator teaches one to fly. Electronics Workbench does not replace the actual touch that comes with lab work, but it saves valuable time and money.

The next six chapters cover Multisim Version 6. If you are using EWB Version 5, skip to Chapter 8.

CHAPTER 2

Virtual Workbench: The Multisim Interface

The Multisim workbench can just about eliminate a real workbench.

Multisim uses a Windows-based menu and icon-driven graphical user interface (GUI). In other words, it's point-and-click software. Menu items are easily accessible with either a mouse or with swift keystroke combinations. Toolbars speed the task of executing complex commands. A workbench or project space is available to let your electronics imagination run wild. By learning these menus and toolbars at the outset, you are ensuring a deeper knowledge and understanding of Multisim and of exactly what it is capable of. This chapter also introduces the interface. Subsequent chapters further describe the interface and specific procedures. Once you know the functions of each item on the desktop, we will take a look at how they operate and interact with each other.

Overview Interface

Electronics Workbench products add new dimensions to electronics. They take the user from design stage, to development, all the way to PCB layout artwork without leaving your workstation. You are computerizing your workbench; instead of using parts and a table, you interact with the bits and bytes through a graphical interface on the computer screen. The interface is presented once the software is open, along with its menus, toolbars, toolbins, and drawing sheet. This is how you tell the program what to do.

The interface is how we interact with the computer. In the past, the only way to input information was with a keyboard, typing out long strings of complex commands. The only way to see what was being typed was with a monochromatic monitor that displayed the letters, numbers, and symbols. In today's world of visualization we use a GUI and a mouse as an input device. This method eliminates the need to memorize puzzling text commands. Now, you merely point and click the mouse or hit a few keyboard buttons to perform tasks. The Multisim interface is fashioned with easy-to-use intuitive tools, to decrease the new software learning curve. If you have ever had to learn a new program from scratch, you will appreciate the time the Electronics Workbench company has spent polishing and making this process as painless as possible. Graphic representations of components and instruments are used instead of text; parts placement is straightforward and accomplished with swift mouse work; each instrument and component is easily identifiable and closely approximates its real world counterpart.

The rest of this chapter explores the Multisim interface in detail to give you a complete understanding of the features, workings, and procedures. If you are an Electronics Workbench Version 5 user, skip ahead to Chapter 8 where you will find information on that interface.

The Multisim Screen

Take an in-depth look at Figure 2.1. Open your Multisim program by going to **Start** > **Programs** > **Multisim** > **Multisim**. See how it looks in real life? More colorful than the book, of course. The Electronics Workbench company did its homework to bring you the simplest method of using this program by making the layout very logical and

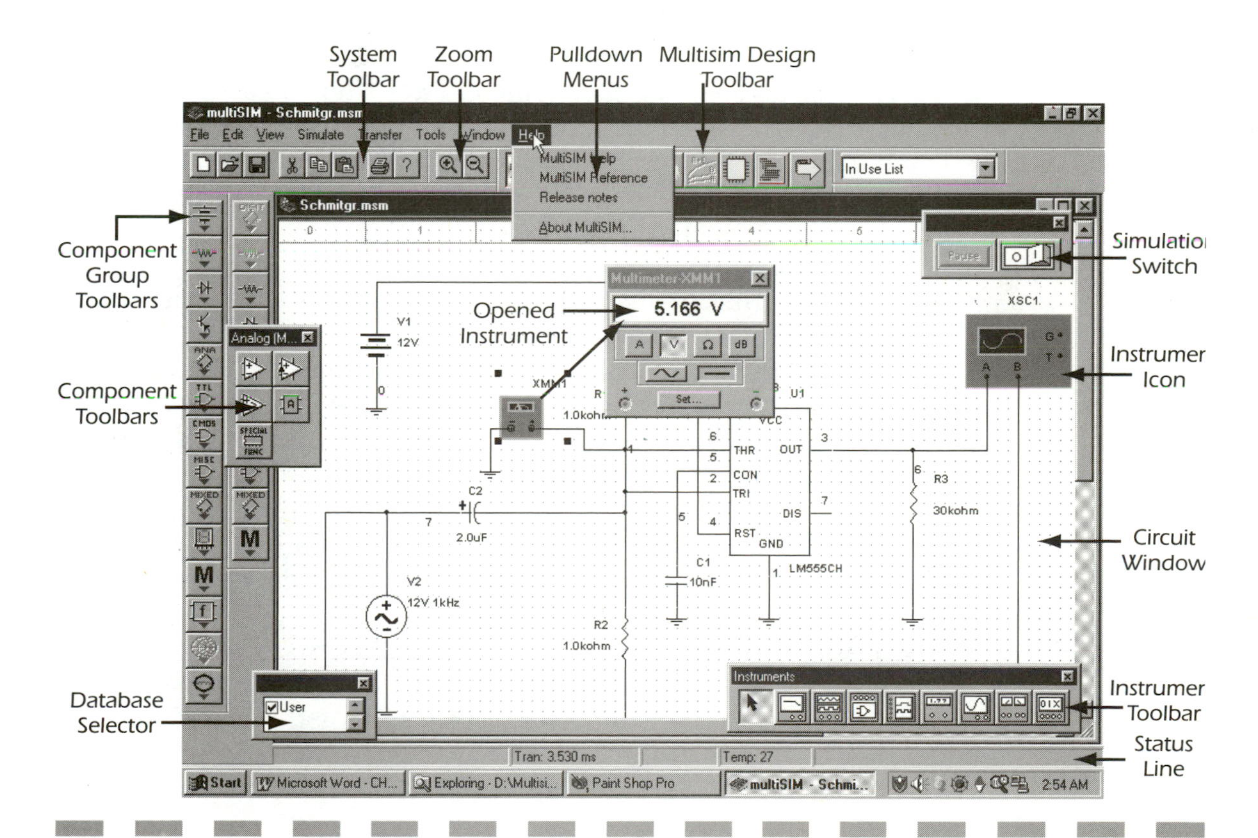

Figure 2.1 The Multisim screen.

intuitive. Here is a rundown of the areas you use in day-to-day designing:

Pulldown Menus—At the top are the familiar Windows menu items. Multisim's menu items include: File, Edit, View, Simulate, Transfer, Tools, Window, and Help. Each is broken down into logical commands or submenus.

Toolbars—A toolbar is a collection of shortcuts that, when clicked, execute a command. Think of them as one-click wonders! Multisim contains several toolbars that are grouped into common commands.

System Toolbar—This toolbar displays various System Icons that can be used as command shortcuts. If you are working away with the mouse, it is sometimes more convenient to hit an icon with your mouse pointer than go through the pulldown menus. For example, you may hit the icon that looks like a floppy disk to save your file. Otherwise you would have to go to the File Menu and choose Save. It all depends on your taste. System icons include: New, Open, Save, Cut, Copy, Paste, Print, and About.

Zoom Toolbar—Icons to control the scale of your circuit are on the Zoom bar. Zoom-in is represented by a '+' and zoom-out is a '−'.

Multisim Design Bar—Shortcuts to Component Toolbar on/off, Component Editor, Instrument toolbar on/off, Simulation switch, Analysis options, Post Processor, VHDL/Verilog (if installed), Reports, and Transfer are included on this bar.

In Use List—This presents you with a list of each component being used in the circuit. You can then select and copy that component to the circuit window.

Component Group Toolbars—This is the heart of Multisim, containing icons that represent a Parts Bin. Think of them as cabinets full of categorized electronics parts, which allow you to grab a mix and match of components or instruments to perform simulations.

Component Toolbars—These toolbars contain the individual component icons. They open when the mouse pointer is placed over a component group icon. Think of them as the little drawers within the Component Group Cabinets.

Circuit Window—This is where you lay components, run wires, and place miscellaneous text. It is essentially a drawing board or breadboard. Make sure you understand this part of the interface well, because it is used throughout the book in describing procedures. Circuit Window = Where you draw. Simple!

NOTE

This may also be called the Drawing, Workspace, or Circuit Diagram throughout this book.

NOTE

Only the Professional and Power Pro editions can have multiple Circuit Windows open at once.

Subcircuits—New to Multisim is a separate circuit window for subcircuits. It is as if the subcircuit were a circuit or breadboard of its own.

NOTE

*Use **Ctrl+Tab** on the keyboard to alternate between circuit windows.*

Instrument Toolbar—This toolbar contains instruments that can be placed onto the drawing. By default, this toolbar is deactivated. Click the Instruments icon on the Multisim Design Bar to bring it up.

Database Selector—This opens addition libraries of components and their respective toolbars. These include Multisim Master Database, Corporate Database (if supported), and the User Database.

Simulation Switch Toolbar—This toolbar starts, stops, or pauses a simulation with a mouse click. By default, this toolbar is activated. Previous Multisim releases may have this option deactivated. If so, click **View** > **Show Simulation Switch** to display it.

Status Line—This displays READY Status, Tran (Time), Temp, and Component/Key information.

NOTE

The time on the Status Line is simulated time, not elapsed time. For example, if the status line indicates 100 microseconds, 10 seconds may have gone by in real-life time. At 200 microseconds, 20 seconds may have elapsed according to your watch. Also note each circuit is different. That is to say, another simulation might display 1 second when only 1 second has gone by in real life.

Terms

It is best to learn a few of the desktop terms as used in the explanations that follow:

Desktop/Workbench—This refers to everything that is on the screen collectively: menu items, toolbars, circuit window, instruments, and so on.

Window—Any separate box that pops up with additional features or programs.

Circuit Window/Drawing/Workspace—The area in which you draw circuits.

Menu—A list of items, commands, or submenus.

Submenu—Menu within a menu.

Command—A function that the computer performs after your input. In other words, it's what you tell the computer to do. Example: The Save command stores (saves) your current circuit to the hard drive's memory.

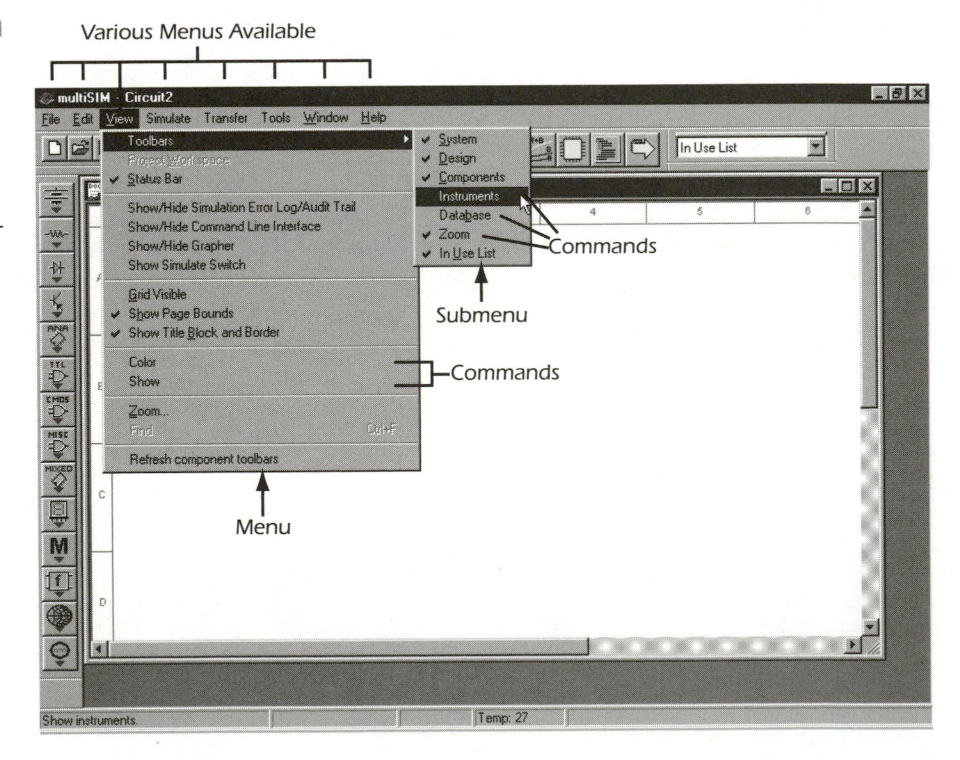

Figure 2.2
A menu can contain a list of selectable commands or submenus leading to further commands.

Hotkey or Hot Key—A fast way to access a command or feature with the keyboard. For example; Ctrl + T lets you place a block of text.

Right-Click (shortcut or Content-Specific) Menus—Windows 9x./NT has the ability to open various quick menus when you place the pointer over an area and hitting the right mouse button.

Dialog Box—A very important term: A Window that lets you add further items to a command. For example, the Component Properties Dialog Box lets you qualify a component. See Figure 2.3.

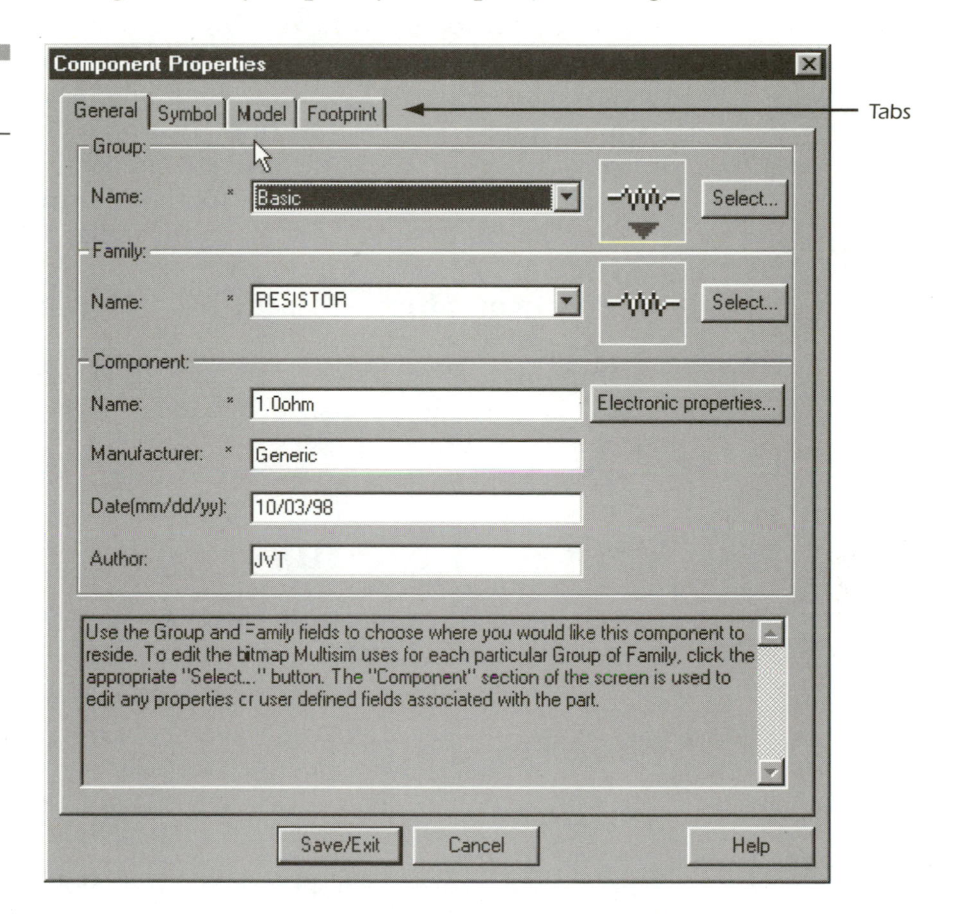

Figure 2.3
A dialog box.

Tab—In a dialog box, you will often see tabs at the top of the window that look like file folder tabs. By clicking on these, you can access other choices, items, or pages. Also see Figure 2.3.

Icon—A small picture which, when clicked, performs a command. It is a fast way of accessing common commands.

Toolbar—A group of icons. See Figure 2.4.

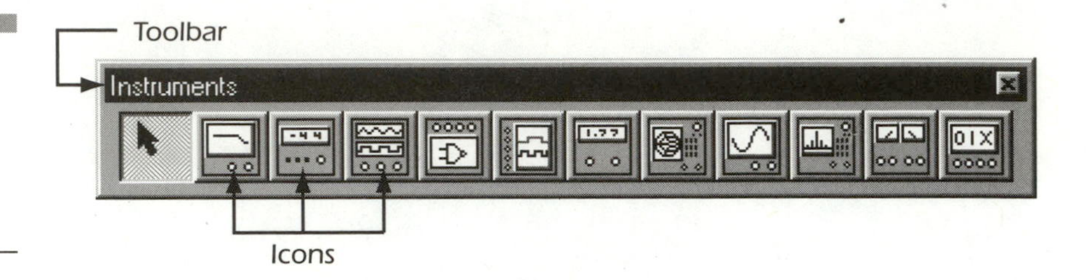

Figure 2.4 Sample of a toolbar. Notice the various icons, which, when clicked, execute a command.

Toolbox—A toolbar that contains components—otherwise known as a component toolbar. This is not an official term, but I like to use it to help you understand how Multisim mimics a real workbench.

Component—All the information that makes up a Multisim virtual electronic component. This includes the electrical properties, schematic symbols, electronic models, and the footprint or package information.

Component Browser—A dialog box that lets you search for, select, or edit components.

Component Editor—A dialog box that allows you to create or edit your own component.

Component Properties—The meat of the virtual component—these contain adjustments, such as electrical properties, symbol, model, and footprint or package. See Figure 2.3.

Multisim Menus

Some of these initial explanations are elementary in nature and can be skipped by seasoned Windows 95/98/NT users.

Menu Format

Text presented throughout the book as:

File > Save As...

directs you to access the **File** menu and choose the **Save As** command. If the line contains a submenu, it looks something like this:

View > Toolbars > (Toolbars List).

Navigating and Using Menus and Submenus

Multisim allows you to navigate menus using either a mouse or keyboard keys and hot keys.

MOUSE METHOD

To use the mouse to navigate menu items, move the mouse until the onscreen pointer is over the text of the menu item. Make sure it is not to the side, above or below the text. Click the mouse's left button and let go. This opens the menu and its list of commands. Scroll down to the command you wish to execute and click the left mouse button again. You can also click and hold the mouse button while opening the menu, then scroll down to an item, and release the button. This saves a click of the mouse. See Figure 2.5.

If a submenu exists within a menu, you see a small arrow next to the text. Scroll the mouse pointer down to that item and wait for the submenu to open. Move the pointer over to the new menu and click the appropriate command.

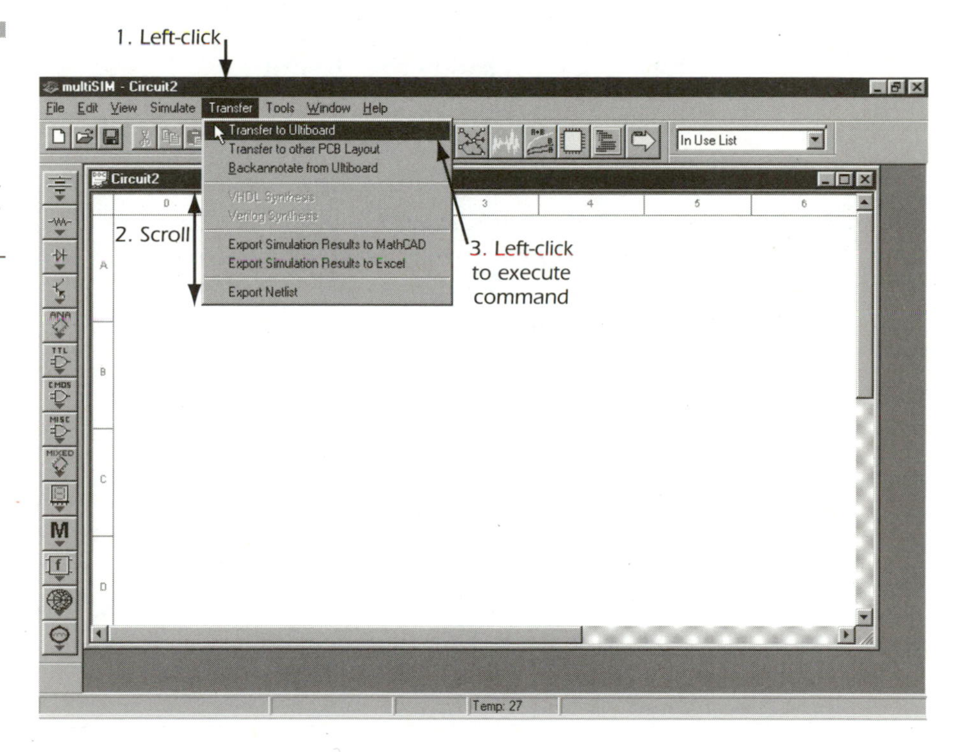

Figure 2.5
To use a menu, left-click the mouse over the text, scroll the pointer down the list, and left-click over the command.

TIP

Clicked the wrong menu? Hit the Escape (ESC) key or go to the top menu item and click the mouse again.

HOTKEYS

Most commands can be executed with a few keystrokes. For example, if you want to save a file, hit the **Ctrl+S**. Release both at the same time. Not all commands have these swift shortcuts.

You will notice that most of the menus have a letter underlined (for example, **Edit**). This means you can hold down the Alt key and the underlined letter to open that menu. Scrolling down with the arrow keys then lets you select an item. Hitting Enter executes that command. If one of the commands in the menu has a letter underlined, then you can hit the letter to execute it without having to scroll down and hit enter (for example, **Alt+F** opens File menu; and **S** saves your work.)

Which method works best; mouse or hotkey? Whichever works for you. Just learning Multisim? Try to stick with the mouse until you learn the hotkeys. I tend to use a combination of both; I use my left hand to access keyboard items and leave my right on the mouse. Whichever is handy gets the call.

TIP

WIN95/98/NT Right-clicking Shortcut Menus

To use shortcut menus instead of using the standard menus to find the command you need, use the right mouse button to click a file or folder. The menu that appears shows the most frequently used commands for that file or folder—Windows 95 Help.

If Multisim is run on Windows 95/98/NT, you have shortcut menus available to you. I like to call them context-specific menus because it's as though the program knows what you are going to do when the mouse pointer is in a specific area. For example, click the right button while the pointer is over a blank area on the Circuit Window. A menu pops up, like that in Figure 2.6. Commands can be executed by scrolling down with the pointer and hitting the left button. Many items in Multisim have these right-click menus. Browse around with the pointer and right-click here and there to discover them.

Dialog Boxes

Some commands require you to further explain what you want Multisim to do. In this case, a window opens up to let you enter or change data. See Figure 2.7. For example, if you choose **Edit** > **User Preferences,** a new window with the User Preferences dialog box opens

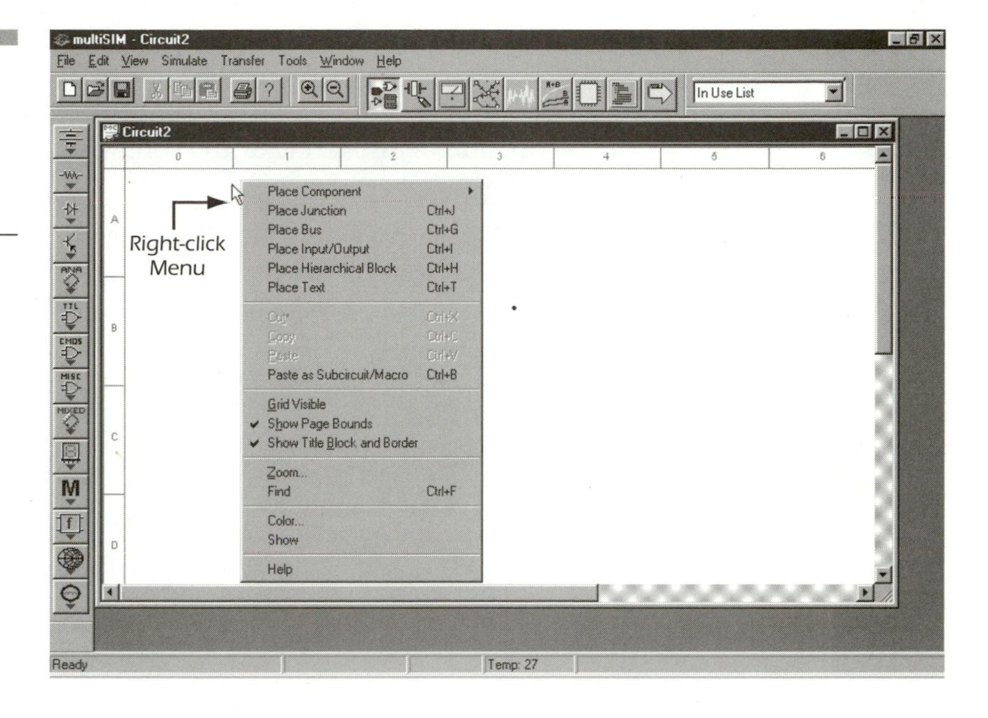

Figure 2.6 Right-click menus save having to remember where specific commands are placed.

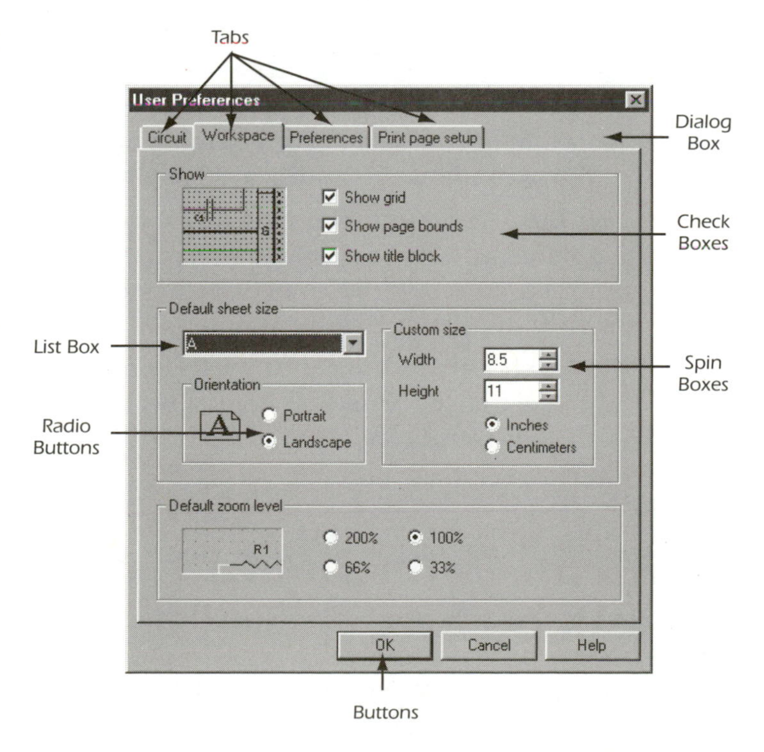

Figure 2.7 A dialog box usually is used to further explain exactly what you want Multisim to do. Several methods of entering this information are presented here.

up to allow you to modify the format of each future circuit. Memorize this term now, because it is used continuously throughout the book.

A dialog box lets you add to or modify information in the Multisim function by using one of the following:

- **Tabs**—Software uses file tabs because it can only fit so many options onto the screen at once. Flipping through these tabs can access additional pages of settings.
- **Buttons**—These execute a command, open another dialog box, or perform some operation when clicked with the mouse. For example, in Figure 2.7, the OK in the box is a button that executes the command as a whole.
- **Radio Buttons**—Used to select one of a multiple-choice list. If you select another item in the list, the last item is canceled. The term comes from the old radio preset controls. For example, under the Workspace tab, you can select Portrait or Landscape as the page orientation (not both). See Figure 2.7 again.
- **Check Box**—A check means the statement will be true or that the feature is turned on. If there is no check, then the statement is false or the feature is off. To check or uncheck, click inside the box. For example, if you do not want the Grid displayed, click the "Show grid" box until the check is gone. See Figure 2.7.
- **Spin Boxes**—Clicking the up or down arrow advances the value in the white box next to it. For example, under Custom Size, there is a spin box to adjust the Width and Height. See Figure 2.7.
- **List Box**—A list of items that you select by clicking the arrow on the right and scrolling down until you find the item you want. For example, to select the color scheme for your drawing, click the arrow in the Color List box (circuit tab).
- **Parameter Box**—Type in the setting or information by hand. For example, under the Save-As Dialog Box, you can type the filename in the File Name field.

Menu Items in Multisim Version 6.xx

Standardized selectable menu items range from File to Help. Some custom features may throw off a seasoned Windows user, but with

Multisim's easy-to-learn menus, it takes little time to acclimate. Here is an overview of each menu option and an explanation of the various ways to speed access to each. I recommend going over each item thoroughly and following along with the program. This will help later, when you just can't seem to find that command you need.

Menu Item: File

The left-most menu selection is the File menu. Refer to Figure 2.8. As with any other Windows program, it handles the computer's filing and printing tasks. Here is a rundown of each feature in this menu.

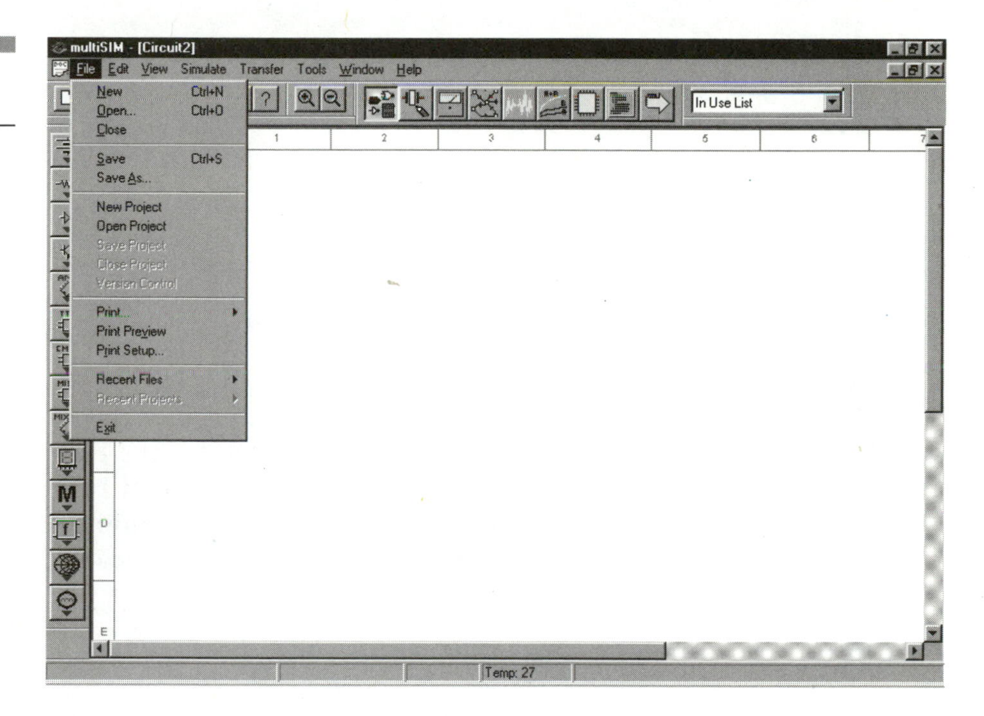

Figure 2.8
File menu.

If the item's text is grey, it means it is not selectable at that point in your session or is unavailable on your edition of Multisim.

Item: File > **New**
Hot Key: Ctrl+N
Use: Opens up a new Multisim document with the default name, Circuit1, Circuit2, etc.

NOTE

Multisim V6 has the ability to open multiple circuit windows at once. However, this is only available on the Professional and Power Pro Editions.

Item: File > **Open...**
Hot Key: Ctrl+O
Use: Launches a File Explorer to choose a file. Multisim can open either a native Multisim file (*.msm), Electronics Workbench Version 5.x file (*.ewb), Electronics Workbench Version 1-4 file (*.ca) or, by selecting All Files, you can open Netlists, which are explained further in Chapter 15.

TIP

If you don't want to risk mistakenly modifying a file just to take a peek at it, check off the 'Open as read-only' option inside the Open dialog box.

Item: File > **Close**
Hot Key: None
Use: Closes current selected circuit window. If you have not recently saved the file, Multisim asks if you wish to save before closing. Don't confuse the Close command with the Exit command.

Item: File > **Save**
Hot Key: Ctrl+S
Use: Used to save your document to the computer's hard drive or floppy disk. If it's a new file, it will switch automatically over to the **Save As** menu. Save Your Files Often! Computers tend to lock at the most inopportune time; save the heartache, frustration, and file by saving continuously.

Item: File > **Save As**
Hot Key: None
Use: Opens the Windows File Explorer to name and save your Multisim files in the native Multisim V6 format (*.msm) for later retrieval. You do not have to type in "*.msm" after your file name; this is automatically added by Multisim.

Revert to Saved is no longer available because Multisim is now able to undo certain commands. Hurrah!

The next few menu items are only available to Multisim Professional or Power Pro users, as noted by the asterisk.

Item: File > **New Project***
Hot Key: None
Use: Allows you to save all files related to a specific project into one neat and convenient location. The project's various circuits can then be added to the project folder. A toolbar that allows fast access is then added to your desktop.

Item: File > **Open Project***
Hot Key: None
Use: This allows you to open a previously saved Multisim Project File (*.msp). The folder and circuit files appear on a new toolbar on your Multisim desktop.

Item: **Save Project***
Hot Key: None
Use: Saves the project file and prompts you to save changes to any modified circuit within the project folder.

Item: File > **Version Control***
Hot Key: None
Use: Back up or restore project folders and their files.

Submenu: File > Print > (SUBMENU)

This menu item shifts to a submenu with the following commands:
Item: File > Print > **Circuit**
Hot Key: Ctrl+P
Use: Opens your computer's default Printer Dialog Box to choose print options and engage the printer. This command prints only the circuit as seen in the circuit window and nothing else.

Item: File > Print > **Bill of Materials**
Hot Key: None
Use: Opens up a new window with a list of materials used in building the current circuit. You can then choose to save this file as a text file (*.txt), print it, list additional items not included, or close this window. This is not available on the Personal Edition and is limited to basic functions on the Professional Edition.

Virtual Components are not included in the main Bill of Materials List. Hit the third icon from the right (Others) for a list of those components. They are not printable; instead, cut and paste the data into Notepad and then print.

Item: File > Print > **Database Family List**
Hot Key: None
Use: This opens a window that says, "To Access the Database Family List, select a Parts Bin and a family, then from the Browser Dialog box choose the button labeled "List report." In other words, it's a useless menu item itself.

Item: File > Print > **Component Detail Report**
Hot Key: None
Use: This opens a window that says, "To access the Component Detail Report, select a Parts Bin and a family, then from the Browser dialog box choose the button labeled Detailed Report. In other words, also a useless command itself.

Item: File > **Print Preview**
Hot Key: None
Use: Want to preview a circuit before committing to expensive toner and paper? Use this menu item. It opens a window letting you view each page Windows will print. It allows you to zoom in and out of each page. Hitting one more button takes it to hardcopy (Print).

If you're not quite sure how the circuit will look, or if it takes up multiple sheets, use this command.

Item: File > **Print Setup**
Hot Key: None
Use: This opens the Windows Printer Setup dialog box to adjust the printer. You can also hit the Page Setup... button to enter that dialog box. From there you can set margins, zoom level, and more. See the next chapter for more information on Page and Printer Setup.

Item: File > **Recent Files** > **(filenames)**
Hot Key: None
Use: Use this submenu to swiftly access one of the last four files you've worked on. It saves having to search the hard drive for recently used Multisim Files.

Item: File > **Recent Projects** > **(project names)**
Hot Key: None
Use: Quickly opens one of the last few project windows used.

Item: File > **Exit**
Hot Key: Alt+F4
Use: Shuts down the Multisim software after asking you to save the document(s) in progress.

Menu Item: Edit

Here is a list of commands (refer to Figure 2.9) that deal with manipulating the current drawing; such as placing components; cut, copy, paste; flipping, rotating; setting up your drawing; and most important, setting your preferences. Clicking the Edit menu, then scrolling down accesses commands. You can also hit **Alt+E** and scroll with the arrow keys.

Figure 2.9
Edit menu

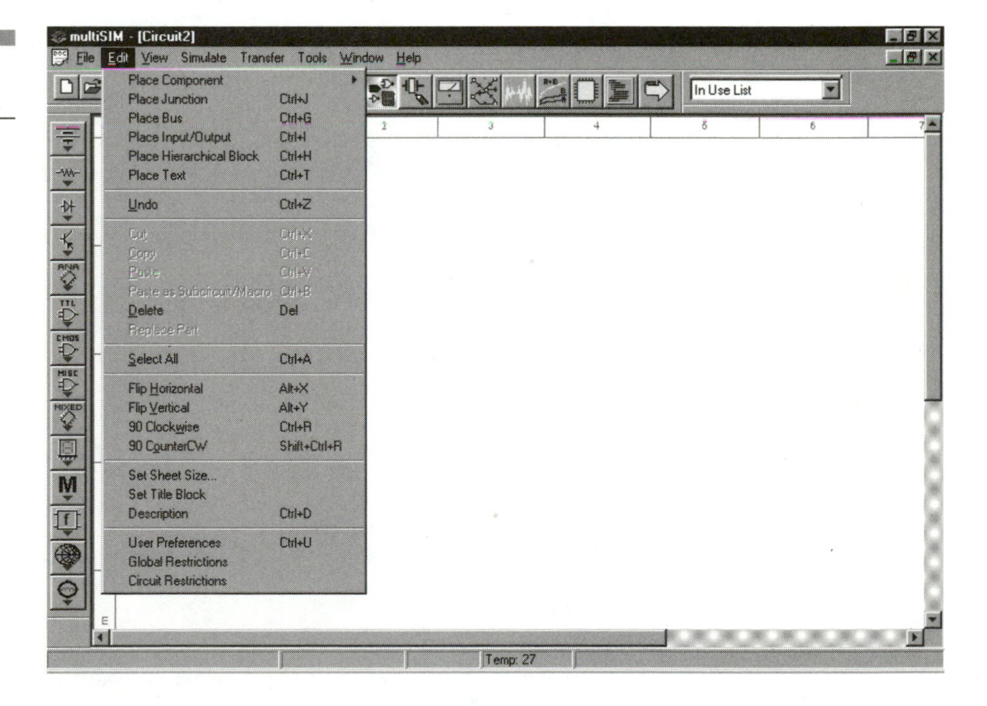

Item: Edit > **Place Component** > **(library)** > **Opens 'Browser dialog box'**
Hot Key: Ctrl+M for Multisim DB, Ctrl+L for Corporate DB, Ctrl+K for User DB (DB = Database or Library)
Use: Using this submenu, first select the part's database location (Multisim, Corporate or User). The Browser dialog box opens. Now, select a component to place into the drawing. This is explained in detail in the next two chapters.

TIP

*If you are lacking monitor real estate (and want more room for an actual circuit drawing), hide the Component Toolbar (***View** > **Toolbars** > **uncheck Components***) and use the Place Component commands instead.*

Item: Edit > **Place Junction**
Hot Key: Ctrl + J
Use: Lets you place a junction point (otherwise known as a connector).

Item: Edit > **Place Bus**
Hot Key: Ctrl + G
Use: Places a circuit bus.

Item: Edit > **Place Input/Output**
Hot Key: Ctrl + I
Use: Places an Input/Output point in the circuit.

Item: Edit > **Place Hierarchical Block**
Hot Key: Ctrl + H
Use: Allows you to place a Hierarchical Block. This feature is only available on certain editions of Multisim and has to do with Project/Team design. It is basically a subcircuit that can be edited across several circuits at once.

Item: Edit > **Place Text**
Hot Key: Ctrl+T
Use: Lets you place a block of text into the circuit. You can then change its color or position with the right-click menu.

Item: Edit > **Undo**
Hot Key: Ctrl+Z
Use: New to Multisim; the ability to go back in time and correct a mistake. This undoes the last delete action performed and reruns all

wiring previously connected. However, you cannot undo mistaken component moves, rotations, or the like; maybe this will be possible in the next version.

Item: Edit > **Cut**
Hot Key: Ctrl+X
Use: Cuts the selected item(s) but keeps a temporary copy of them on the clipboard.

CAUTION

The Clipboard is a temporary holding area to place components. But beware! There is only enough room to cut/copy one component (or set of components) to the clipboard at once. The previously stored items erase the next time you cut or copy.

Item: Edit > **Copy**
Hot Key: Ctrl+C
Use: Copies the item(s) you have currently selected in the drawing onto the clipboard. Similar to cut, except it leaves a copy of the selected item(s) in place.

Item: Edit > **Paste**
Hot Key: Ctrl+V
Use: Pastes the contents of the clipboard into the circuit. However, a copy of the item(s) remains on the clipboard to let you place as many copies as you wish. However, if another item is copied/cut to the clipboard, the previous items evaporate into the ether.

Item: Edit > **Paste as Subcircuit/Macro**
Hot Key: Ctrl+B
Use: Pastes the contents of the clipboard into a new document, creating a subcircuit for other windows. Confusing? Yes. See the next chapter for more information on subcircuits.

Item: Edit > **Delete**
Hot Key: Del
Use: Deletes the item(s) currently selected.

TIP

Selected items have four small squares around them. If multiple items are selected, each item will have four squares around it.

Item: Edit > **Select All**
Hot Key: Ctrl+A
Use: Selects every component in the circuit all at once. Use to center the circuit or copy/paste an identical circuit in a new location. You can also create a subcircuit once all items are selected.

Item: Edit > **Flip Horizontal**
Hot Key: Alt+X
Use: Flips the selected component(s) horizontally. In other words, the left side becomes the right and vice versa.

CAUTION

If you have multiple items selected and choose a flip or rotate command, each individual item is flipped or rotated in its place. In other words, the whole circuit doesn't flip or rotate.

Item: Edit > **Flip Vertical**
Hot Key: Alt+Y
Use: Flips the selected component(s) vertically. In other words, the top becomes the bottom and vice versa.

Item: Edit > **90 Clockwise**
Hot Key: Ctrl+R
Use: Rotates the selected component(s) 90 degrees clockwise.

NOTE

Many Multisim components and instruments are not able to rotate. EWB reports that this will be corrected in newer releases.

Item: Edit > **90 CounterCW**
Hot Key: Shift+Ctrl+R
Use: Rotates the selected component(s) 90 degrees counter-clockwise.

TIP

Undo does not work if you accidentally rotate or flip an item. Instead, flip or rotate the component again to counter it.

Item: Edit > **Set Sheet Size...**
Hot Key: None
Use: Opens Sheet Size dialog box. You can set such items as sheet size, orientation of the printout, and select centimeters or inches.

Item: Edit > **Set Title Block**
Hot Key: None
Use: Opens the Title Block dialog box, which lets you type in details to be printed on your circuit's title block (bottom-right corner of drawing).

Item: Edit > **User Preferences**
Hot Key: Ctrl+U
Use: Opens the User Preferences dialog box. From here, you control each circuit's properties; the changes do not take effect until you open a new drawing.

Menu Item: View

Multisim offers superior control of your digital workbench or desktop. The following menu items (refer to Figure 2.10) allow you to control the look of the interface as well as customize it to personal tastes.

Figure 2.10
View menu

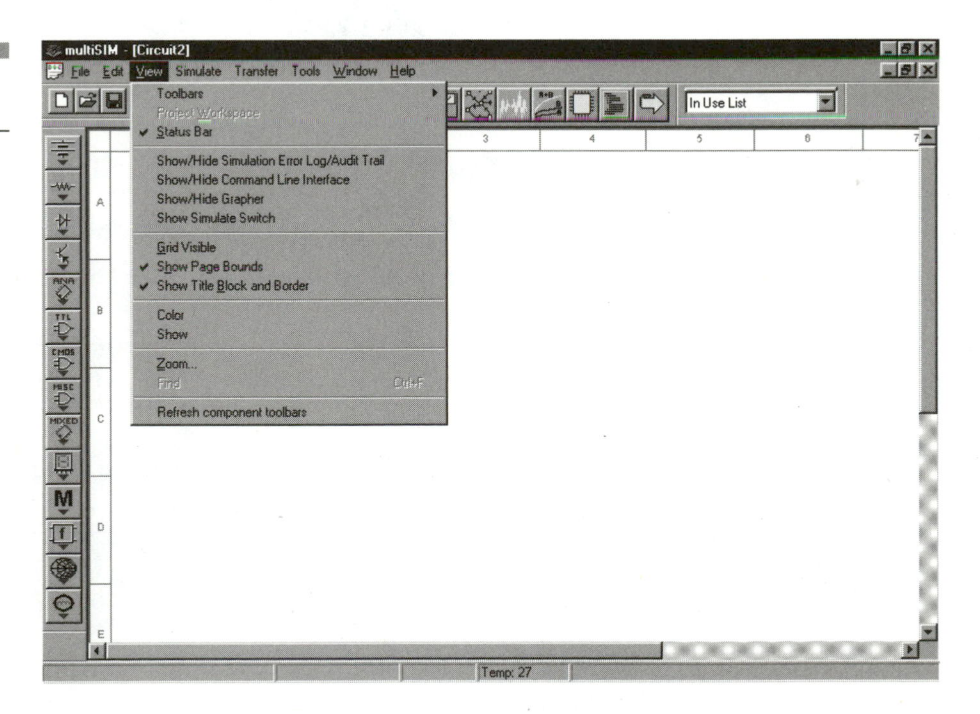

Short on drawing space? If you are using Windows 98 and have two monitors set up, move the circuit window or extraneous instruments and toolbars over to the free monitor.

Item: View > **Toolbars** > **(Toolbars Submenu)**
Hot Key: None
Use: Moves to a submenu listing all Multisim toolbars that can be displayed or hidden. A checkmark to the left indicates that the toolbar is currently displayed on your desktop. Left-click an item to deselect the checkmark, thus hiding it. Toolbars are discussed later in this chapter.

Item: View > **Project Workspace** (only on Editions with Project/Team Modules)
Hot Key: None
Use: Turns the Project Toolbar on/off. Checkmark equals on.

Item: View > **Status Bar**
Hot Key: None
Use: The grey bar at the bottom of the screen is the Status bar. It gives information such as time, date, and so forth. A checkmark equals visible, no checkmark equals invisible.

If you are short on space, hide the Status Bar.

Item: View > **Show/Hide Simulation Error Log/Audit Trail**
Hot Key: None
Use: Opens or closes the Simulation Error Log/Audit Trail Dialog Box. In earlier versions of Multisim, this box displayed each time a simulation was stopped.

Item: View > **Show/Hide Command Line Interface** (Only on Pro and Power Pro)
Hot Key: None
Use: Opens or closes the XSpice Command Line Dialog Box. Type an Xspice Command and hit return.

Item: View > **Show Grapher** or **Hide Grapher**
Hot Key: None
Use: Opens or closes the Analysis Graphs Window.

Item: View > **Show Simulation Switch** or **Hide Simulation Switch**

Hot Key: None

Use: Users of older Electronics Workbench versions will remember (with great nostalgia) the simulation switch. Multisim has it as well; it is just hidden by default (newer versions of Multisim now have this item displayed by default). Choose this command to open the ON/OFF switch and Pause button.

TIP

If you are used to the ON/OFF/PAUSE buttons, choose Show Simulation Switch and then move the toolbar to a handy location on the desktop. It can even float in the middle if that makes it easier.

Item: View > **Grid Visible**

Hot Key: None

Use: Turns the grid of the current circuit window on or off.

TIP

Leave the grid on while drawing. The visual guide helps to line up components. The grid does not reproduce in a printout.

Item: View > **Show Page Bounds**

Hot Key: None

Use: Shows or hides Page Bound. This is a dotted line that indicates where the margins of the circuit will end up on a printout.

Item: View > **Show Title Block and Border**

Hot Key: None

Use: Shows or hides the title block located in the bottom-right corner, as well as the borders. The borders are the grey lines around the drawing with A,B,C.. and 1,2,3...

NOTE

If you choose to hide the Title Block and Border with this command, they are eliminated from your printout as well.

Item: View > **Color**

Hot Key: None

Use: Opens the Color dialog box. You can choose a color scheme for the drawing on which you are working. A customization feature is also available that lets you set each circuit element's color independently.

Item: View > **Show**
Hot Key: None
Use: Opens the Show dialog box. Check the items you want to have shown. These include Component Labels, Component Reference ID, Node Name, and Component Values.

Item: View > **Zoom...**
Hot Key: None
Use: Opens the Zoom dialog box, letting you select 200, 100, 66 or 33 percent magnification.

Item: View > **Find**
Hot Key: Ctrl+F
Use: Great feature when you are working with large complex circuits. Selecting it opens the Find Component dialog box. It lists each component used in the current circuit. Click the text of the component you are looking for and hit the Select Components button. The drawing item is selected.

Item: View > **Refresh component toolbars**
Hot Key: None
Use: This command updates the Component Toolbars to show recent changes made to the user or corporate component editor.

TIP

If you have made an addition or change to the User or Corporate toolbar, you must first refresh the toolbar with ***View > Refresh component toolbars.***

MENU ITEM: SIMULATE

To see the results of a simulation, you must first run the simulation. The next commands (see Figure 2.11) deal with all simulation concerns, including analysis settings. (In my version of Multisim, the S in Simulate is not underlined. However, I can still access the menu with the hotkey Alt+S.)

Item: Simulate > **Run/Stop**
Hot Key: None
Use: Begins the simulation of the circuit. That is, the circuit is turned on (assuming everything is hooked up correctly). If you rese-

lect the item, the simulation ends and the Simulation Error Log/Audit Trail opens (newer versions do not open the Log automatically).

*Your circuit has an error if **Simulate > Run/Stop** is greyed-out. It may be something as simple as the circuit has no ground or no way to show you results—such as an ammeter/voltmeter, bulb, or instrument.*

Item: Simulate > **Pause/Resume**
Hot Key: None
Use: Used to halt a simulation temporarily. I use this to stop a circuit to have a look at the waveform or reading from the instruments before continuing.

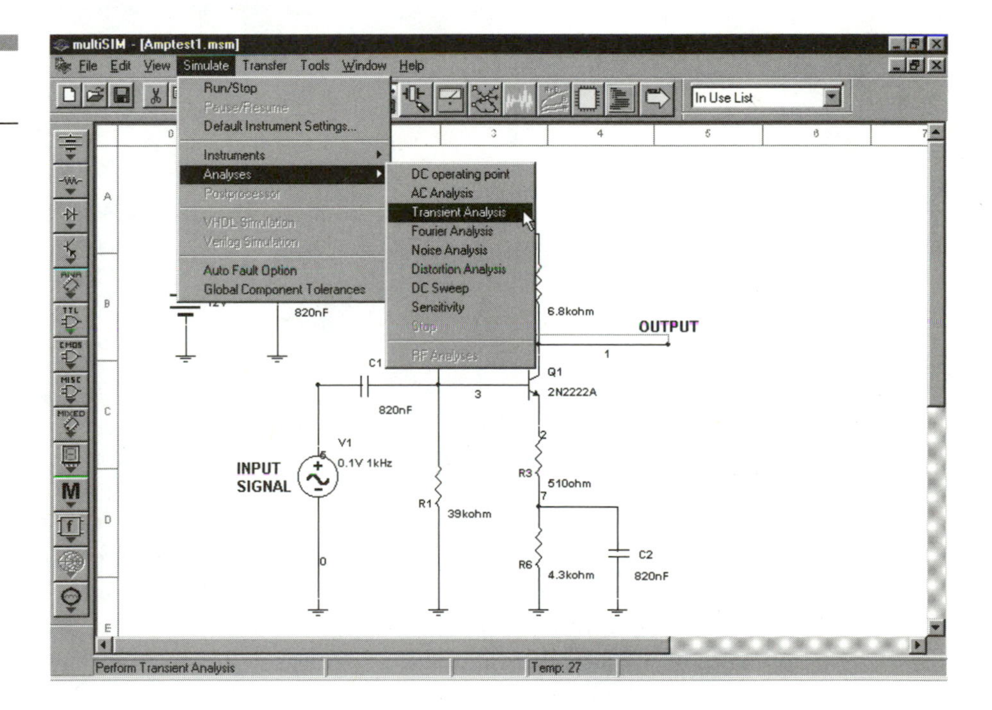

Figure 2.11 Simulate menu.

Item: Simulate > **Default Instrument Settings...**
Hot Key: None
Use: Instruments, such as the oscilloscope, spectrum analyzer, and logic analyzer, may require certain parameters to be set before a simulation. This command opens a dialog box that allows you to adjust

these simulation requirements. For a more detailed explanation see chapter 5, Multisim Instruments.

Simulate > **Instruments Submenu** opens a submenu of instruments that can be placed. Instrument submenus include multimeter, function generator, wattmeter, oscilloscope, Bode plotter, word generator, logic analyzer, logic converter, distortion analyzer, spectrum analyzer and network analyzer.

Item: Simulate > **Instruments** > **(Instruments Submenu)**
Hot Key: None
Use: Choose an instrument from the list, then place its icon onto the circuit for hookup. For more information on these instruments, see Chapter 5.

NOTE

Some of these instruments may be absent from certain Multisim editions, such as Personal and Professional.

TIP

The features table supplied by EWB says you are not able to open multiple copies of the same instrument with the Personal Edition. However, with my current 6.11 edition I am able to open multiple copies of the same instrument.

Analyses Submenu

The Analysis Submenus include DC operating point, AC analysis, transient analysis, Fourier analysis, noise analysis, distortion analysis, DC sweep, and sensitivity. (In other editions there are parameter sweep, temperature sweep, transfer function, worst case, pole zero, Monte Carlo, trace width analysis, RF analysis, batched analyses, user-defined analysis and the command to stop the current analysis. Note that not all Multisim editions include the trace width, RF, batched, and user-defined analyses options.)
Item: Simulate > **Analyses** > **(Analyses Submenu)**
Hot Key: None
Use: Multisim is packed with ways to tear a circuit to pieces and peek at just about any section of the running circuit. This submenu lets you select a method of analysis to perform. Each item will be

described in detail in the Simulation/Analysis chapter (6). For now, just see how each submenu command opens that analysis dialog box up.

Once you're more experienced with simulations, you will throw away the ON/OFF/PAUSE buttons and begin to use the Analysis menus more and more. I tend to use Analysis to troubleshoot difficult circuits faster. This is where Multisim's true power is.

The next three menu items are only available on certain editions of Multisim.

Item: Simulate > **Postprocess**
Hot Key: None
Use: Opens the Post Processor Screen. Allows you to combine several analysis results to visualize the circuit's simulation results better.

Item: Simulate > **VHDL Simulation**
Hot Key: None
Use: Opens Multisim VHDL program, if you have it installed. This is a *hardware description language* (HDL) used to design and model complex digital integrated circuits such as *programmable logic devices* (PLDs).

Item: Simulate > **Verilog Simulation**
Hot Key: None
Use: Opens Multisim Verilog program (an HDL) if installed. This is also a way to input complex digital IC information for simulation in Multisim.

Item: Simulate > **Auto Fault Option**
Hot Key: None
Use: If you are an electronics educator teaching students circuit troubleshooting, then this is the menu item to use. It opens the Auto Fault dialog box, which allows you to create random problems with the circuit. The student can then use instruments or analyses to find the fault.

Item: Simulate > **Global Component Tolerances**
Hot Key: None
Use: Multisim uses ideal components by default—there is no tolerance to them. But in the real world, resistors, capacitors, batteries, and

the like have values that are off by a large amount from what they are labeled. For example a 5 percent tolerance 100-Ohm resistor can be anywhere from 95 to 105 ohms. This command opens Global Component Tolerances, which allows you to set the tolerance for each type of circuit component.

Menu Item: Transfer

The Windows operating system creates an environment where you can transfer the data from one program to another, even though they are dissimilar. I won't get into the mechanics of this. Just realize that Multisim makes use of this application-sharing technology by including the Transfer menu items (see Figure 2.12). They let you export your circuit and simulation data to other programs, or import data into Multisim itself. For example, you can import your simulation results into Microsoft Excel™. From there it can be turned into a Web page on the Internet and read by friends or colleagues around the world. Handy! (In my version of Multisim, the T in Transfer is not underlined. However, I can still access the menu with the hotkey Alt+T.)

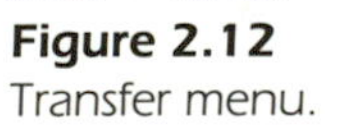

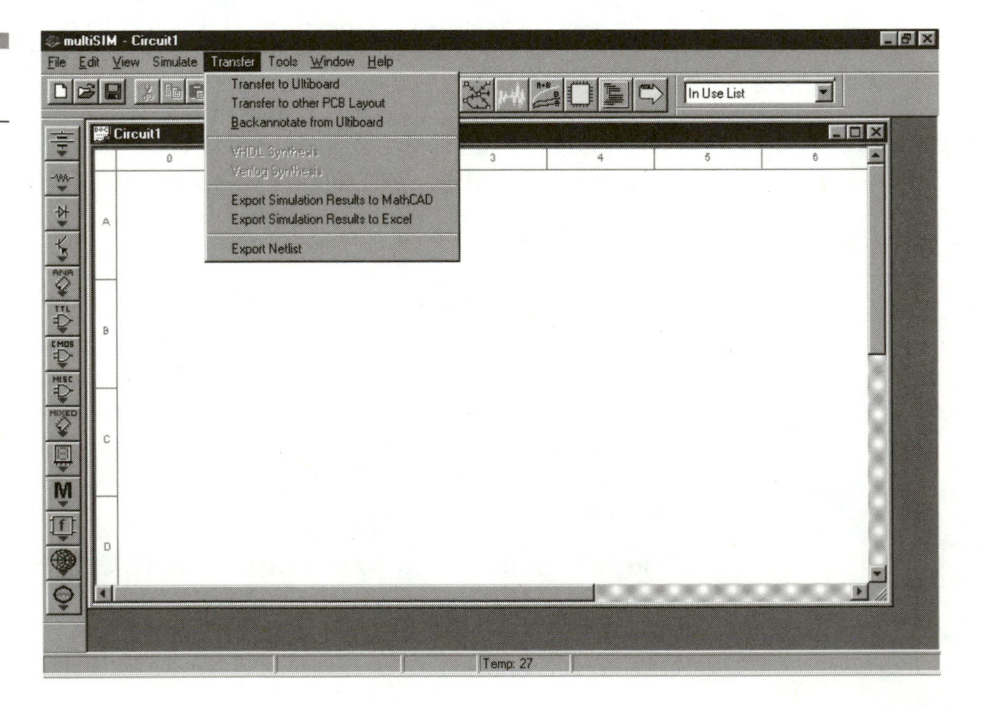

Figure 2.12
Transfer menu.

Item: Transfer > **Transfer to Ultiboard**
Hot Key: None
Use: If you have Electronics Workbench's Ultiboard Software installed, this command lets you first save the circuit as a *.plc or *.net file, then open the circuit inside EWB's Ultiboard.

Item: Transfer > **Transfer to other PCB Layout**
Hot Key: None
Use: Lets you export the current circuit design to a PCB layout package of your choice: Tango PCB, Layo1 PCB, OrCAD PCB, multiSIM PCB, Ultimate PCB, Protel PCB, or Eagle PCB.

Item: Transfer > **Backannotate from Ultiboard**
Hot Key: None
Use: Takes a file created in Ultiboard and makes a circuit out of it in Multisim. This is handy if you made changes in the PCB design and want to simulate the consequences in Multisim.

Item: Transfer > **VHDL Synthesis**
Hot Key: None
Use: If VHDL Synthesis is installed, this allows you first to save the file then open it in VHDL Synthesis.

Item: Transfer > **Verilog Synthesis**
Hot Key: None
Use: Opens file in Verilog Synthesis Program.

Item: Transfer > **Export Simulation Results to MathCAD**
Hot Key: None
Use: Lets you export the results of your simulation to MathCAD™.

Item: Transfer > **Export Simulation Results to Excel**
Hot Key: None
Use: Takes the Simulation results and puts them in a Microsoft Excel format.

Item: Transfer > **Export Netlist**
Hot Key: None
Use: Lets you save a copy of the current circuit as a SPICE Netlist File (*.cir), so you can open the file in other simulation or layout programs outside of Multisim or Electronics Workbench.

Menu Item: Tools

Only two items currently exist in this menu (Figure 2.13) but I get the feeling Electronics Workbench will be adding to it in the future.

Item: Tools > **Component Editor**
Hot Key: None
Use: Opens the Component Editor dialog box. With the introduction of Multisim comes the new Component Editor, a dialog box that lets you select, customize, search for, and store component models.

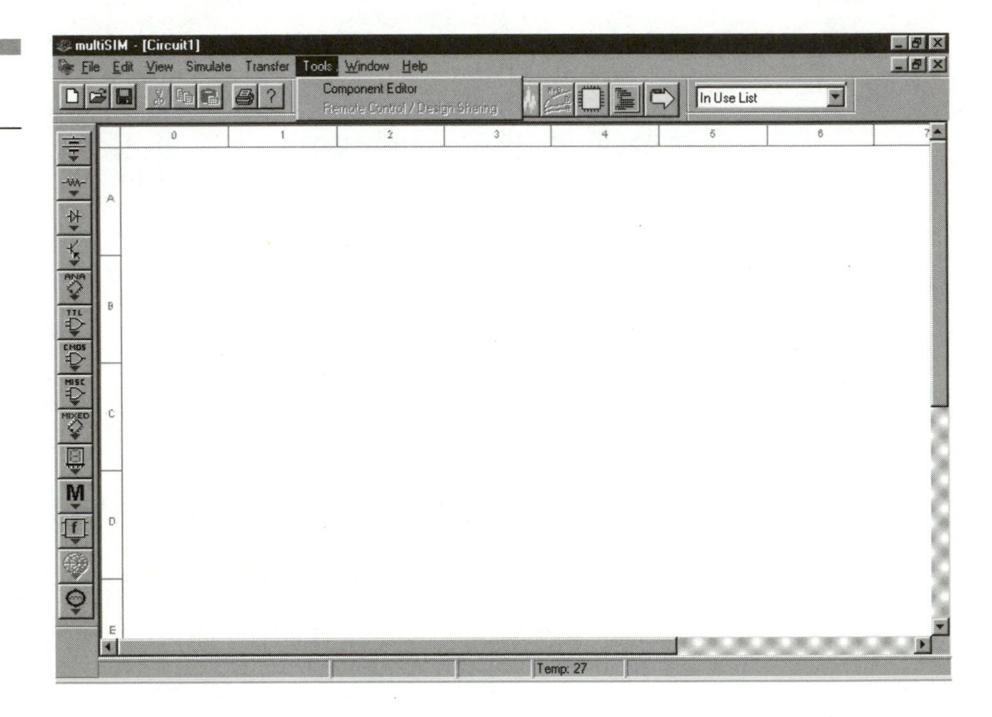

Figure 2.13
Tools menu

Item: Tools > **Remote Control/Design Sharing** (only on Pro as Option or on Power Pro)
Hot Key: None
Use: Allows you to collaborate and share designs with colleagues over a network or the Internet. Each user can control the other's copy of Multisim. Internet junkies will know this feature as *whiteboarding*. Microsoft Netmeeting is required to run it. .

NOTE

Remote Control/Design Sharing is only available on the Professional (as an add-on) and Power Pro editions of Multisim.

Menu Item: Window

The next few items control the structure of the windows in Multisim. Refer to Figure 2.14 for more information.

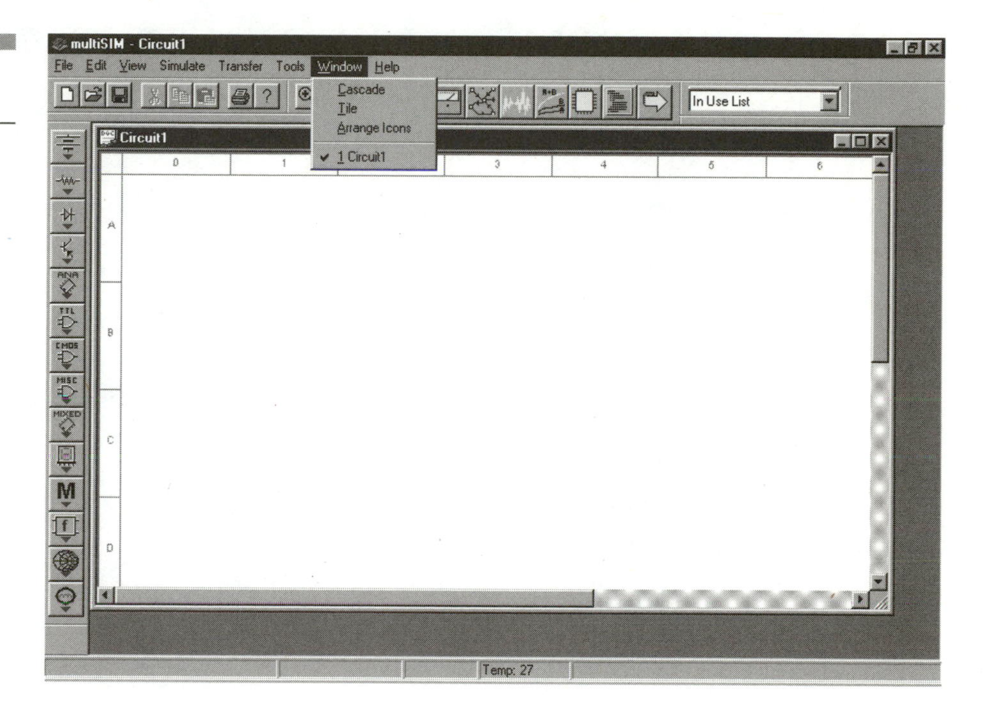

Figure 2.14 Window menu

Item: Window > **Cascade**
Hot Key: None
Use: Overlaps all open windows.

Item: Window > **Tile**
Hot Key: None
Use: Resizes each open window to fit onto one screen; like a tile floor. See Figure 2.15.

Item: Window > **Arrange Icons**
Hot Key: None
Use: If you have more than one minimized circuit, this command arranges them across the bottom of the desktop, sort of like the Windows taskbar.

Item: Window > **(CircuitNames)**
Hot Key: None
Use: Each opened circuit's name appears here. Clicking a circuit brings it forward.

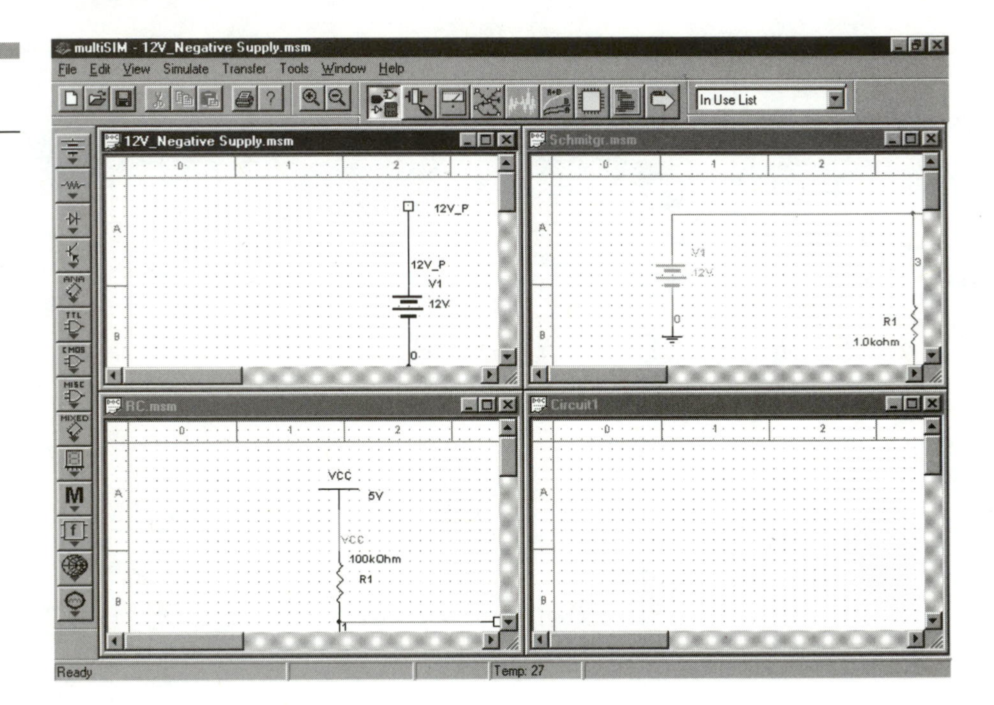

Figure 2.15
Tiling the windows

Menu Item: Help

Perplexed about how to perform a certain task, or do you require more detailed information? It is best to open up the Multisim Help menu (Figure 2.16). This lets you search for information, bookmark, or print it. It contains a cross index and search feature to answer hard-to-find information requests.

Item: Help > **MultiSIM Help**
Hot Key: F1
Use: Opens the Multisim Electronics Workbench Help Menus. If you have a component or instrument selected, that item's help box opens.

Item: Help > **MultiSIM Reference**
Hot Key: None
Use: Opens the Multisim component reference help sections. This contains the User Guide Appendices, but indexed and in a digital format.

Item: **Release Notes**
Hot Key: None
Use: Opens the Release Notes Contents Help menu to provide support on revised features and minor technical support information. It's loaded with useful information. Read it thoroughly or print it for later reference.

Item: **About MultiSIM**
Hot Key: None
Use: Displays the version and copyright information.

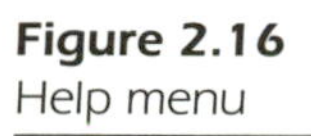

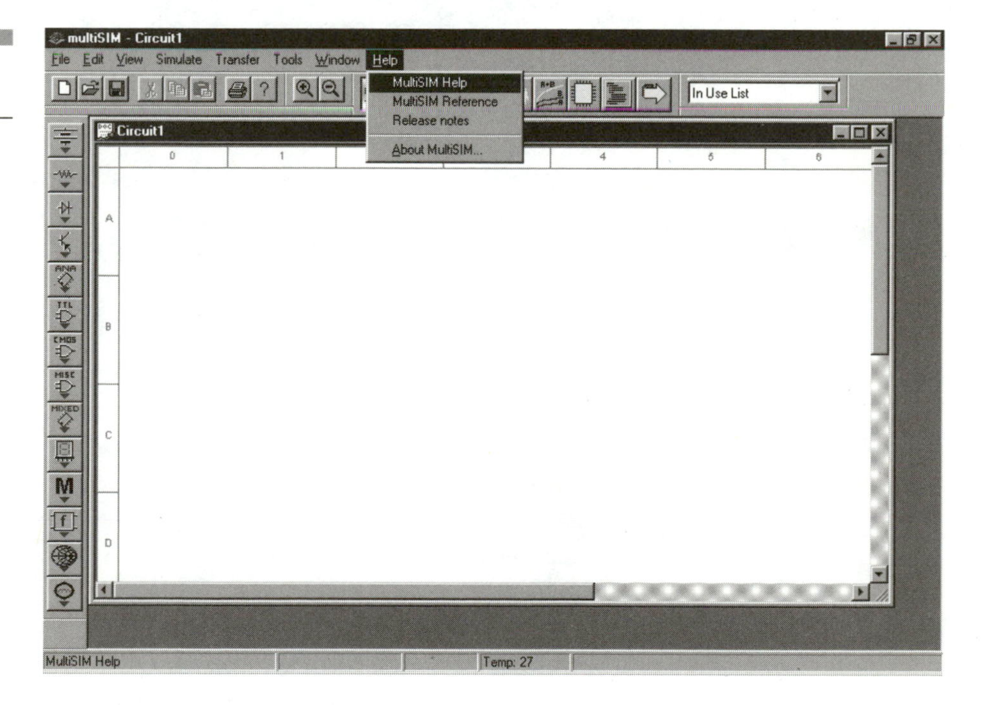

Figure 2.16
Help menu

Toolbars

Multisim makes use of standard Windows-type toolbars, to help you visualize commands and eliminate the need to memorize where menu items lie. A toolbar is a series of miniature pictures (icons) each representing a command. If you move the mouse pointer over an icon and left-click, that command is performed. For example, on the top left section of Multisim is the System toolbar (see Figure 2.17). The first icon looks like a page with a folded corner. This is the New command. To open a new document, you move the mouse pointer over the icon and click. The next icon is open. It resembles a file folder with an arrow. Most icons use some familiar image to help you figure out what they represent, but others can be quite cryptic. This section introduces you to toolbars in general and gives an overview of Multisim's toolbars.

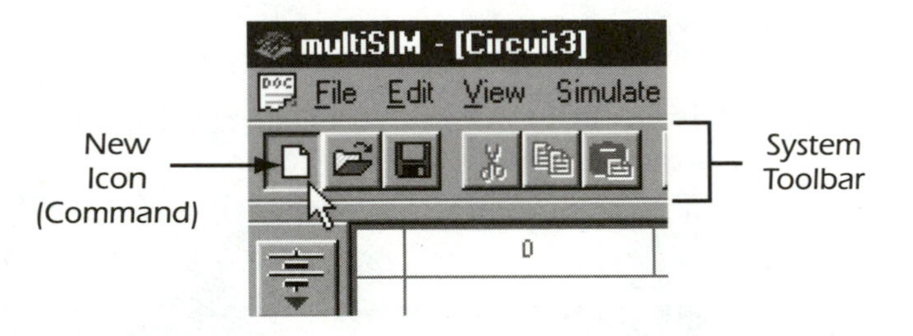

Figure 2.17
The System Toolbar is a good example of a Multisim toolbar that contains various icons.

TIP

Can't figure out what that icon is? Place the mouse pointer over it and wait a moment. A small yellow tag opens with the icon's title. See Figure 2.18.

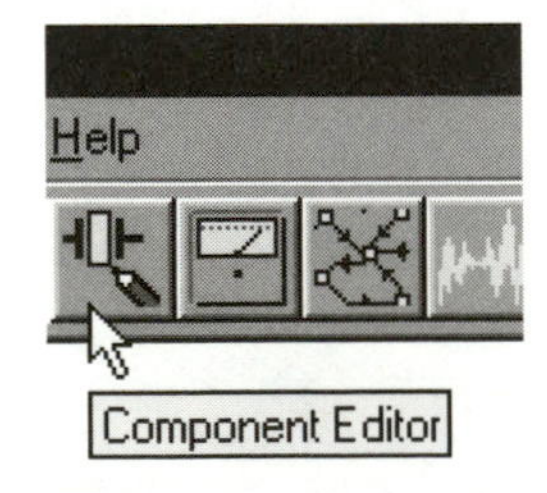

Figure 2.18
A yellow tag appears if you hold the mouse pointer over an icon for a moment.

Each toolbar contains individual icons that represent commands or opens dialog boxes that further clarify the command. You execute a command by placing the mouse pointer over the icon and left-clicking. For example, if you want to zoom in on the circuit, click the icon that looks like a magnifying glass with a '+' sign in the middle.

The Multisim desktop (or workbench) is highly customizable. You can selectively display the toolbars of your choice. They can be moved into any area of the screen or made to float, and in some cases you can control which components appear on the User or Corporate toolbars.

For more desktop customization information, see Chapter 14, Unleashing Multisim/EWB.

You will notice a grey box around each separate toolbar. Each of these toolbars can be displayed or turned off. To set which toolbars are displayed, go to **View** > **Toolbars** > **(toolbars)**. You will see a checkmark to the left of each toolbar that is currently activated. To check or uncheck a toolbar, move the mouse pointer over that item and left-click it.

Each toolbar (icons with grey box around it) can be moved to just about anywhere on the screen. This includes the top, left, right, or bottom of the desktop; it can even float anywhere in the middle. To do this, make sure the toolbar item is first displayed; then follow this procedure:

1. Place the mouse pointer between the grey line and any of the icons in that toolbar. See Figure 2.19.

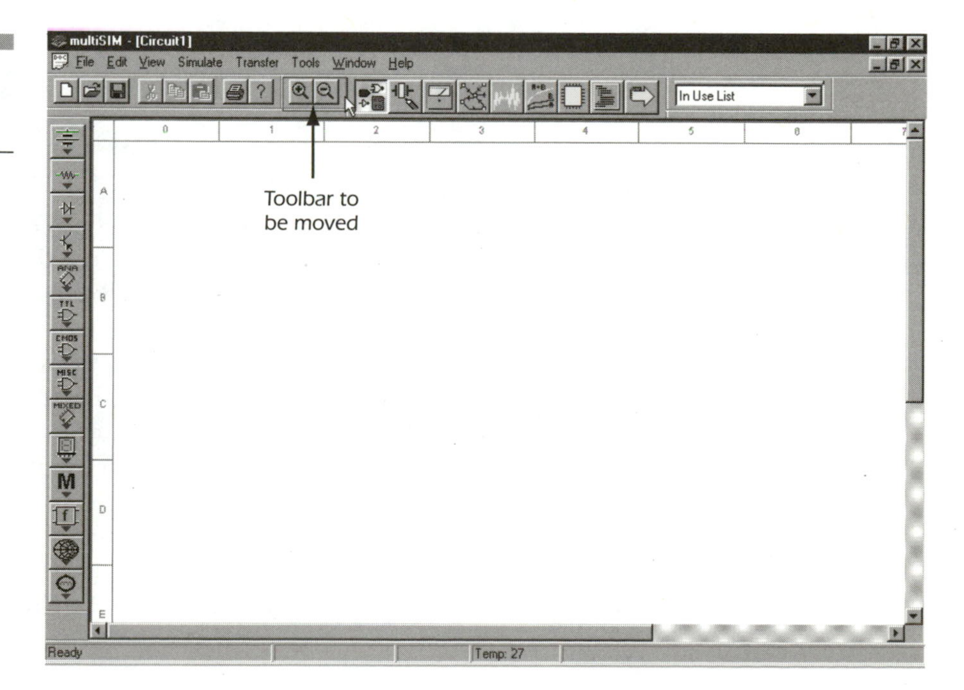

Figure 2.19 Moving toolbars–Step 1.

2. Left-click and drag the toolbar to where you wish to place it on the Multisim desktop and release the button. See Figures 2.20 and 2.21.
3. If the toolbar is close to an edge, it will create a border to lay the toolbar into. If it is in the middle of the workbench, it will float. See Figure 2.22.

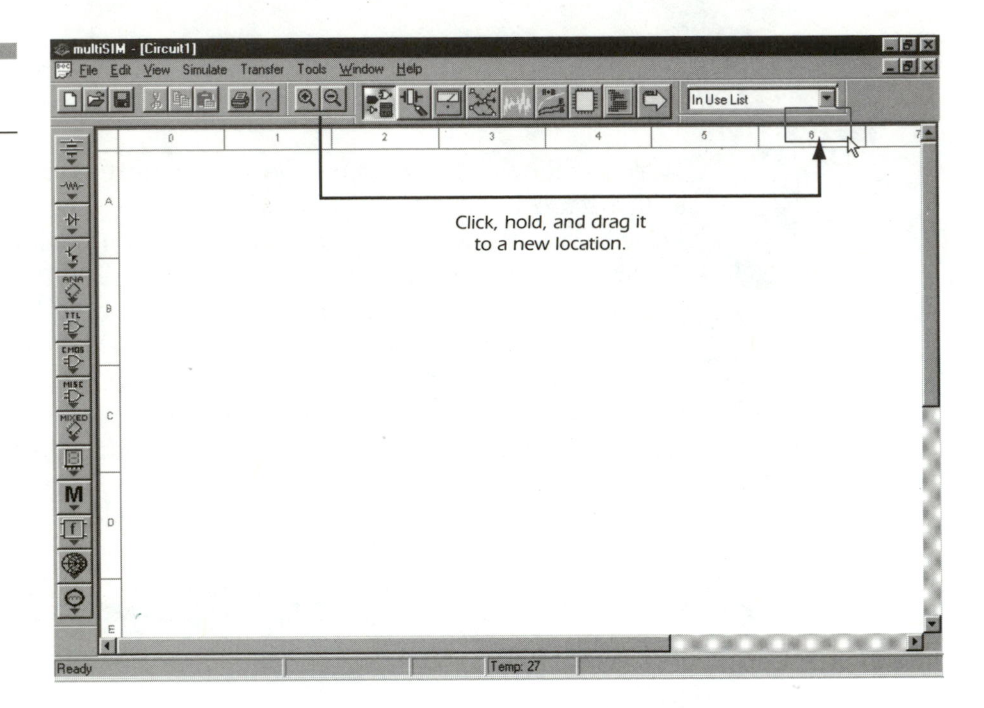

Figure 2.20
Step 2.

NOTE

Toolbar placement settings can be saved as of version 6.11. When you open the program next time, they will be in the same place you moved them. Previous versions did not save this information, requiring you to reposition them each session.

Play around with the toolbars. Try to set up a comfortable workbench just as you would in real life. Place a bunch of well-used toolbars across the top. Turn off those items you don't necessarily use that often; you can always use menu items for those unutilized commands. Doing so helps create a clean screen and speeds design time.

This section describes each toolbar item. Note that the Use descriptions are short. If you want more details, refer to the Menu Items described previously in this chapter. Refer to Figure 2.23 for more information.

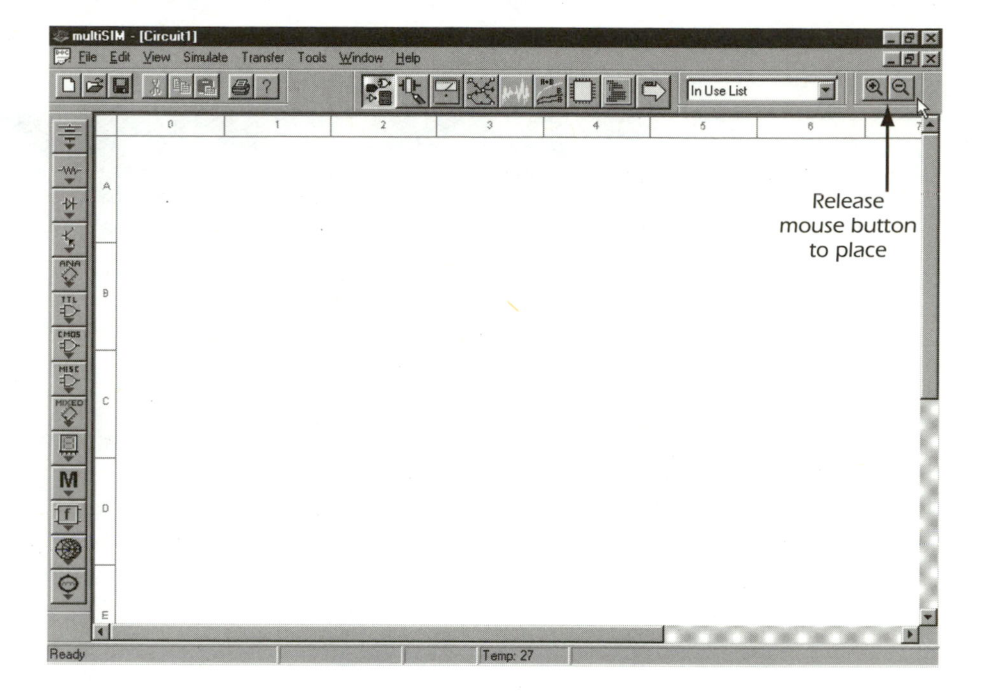

Figure 2.21
Step 2.

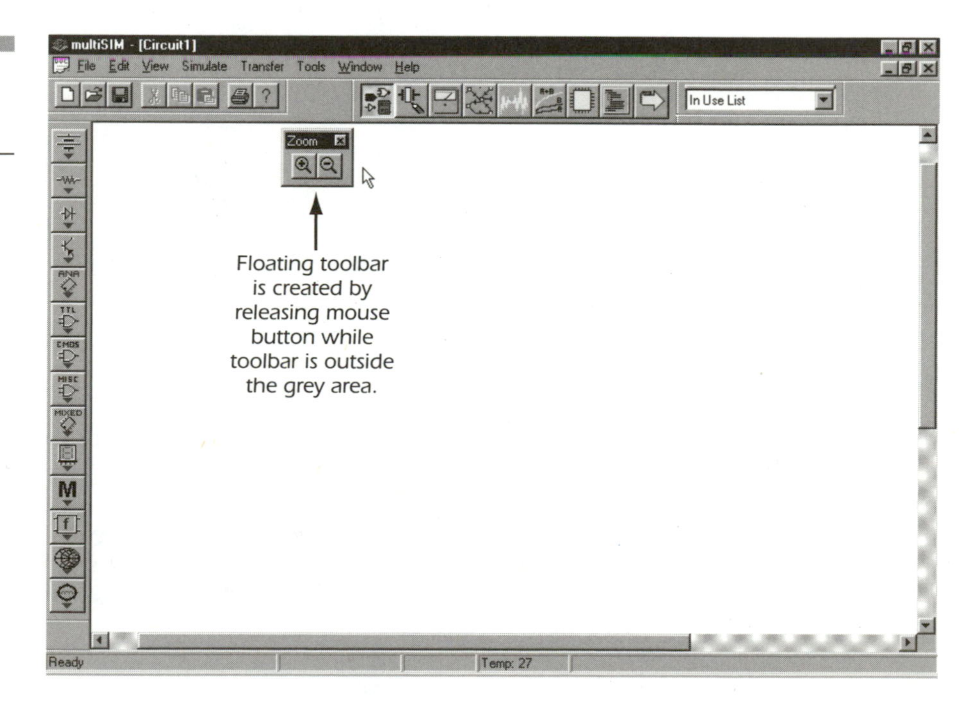

Figure 2.22
Step 3–Floating a toolbar.

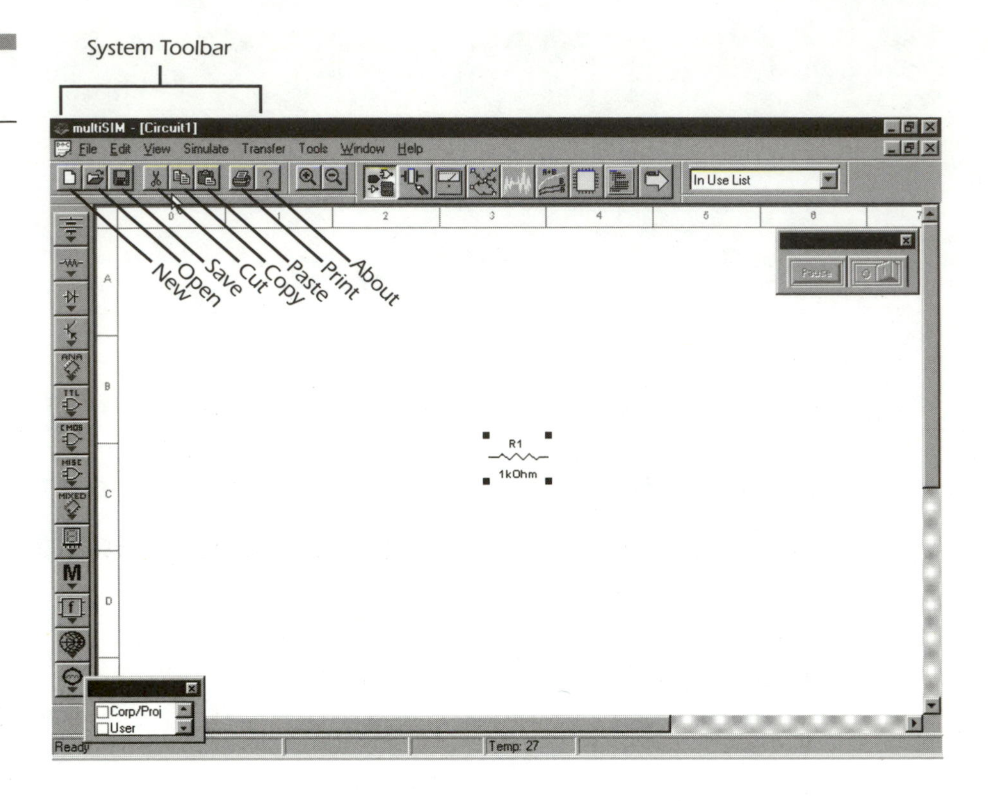

Figure 2.23
System toolbar.

System Toolbar

Icon: **New**
Use: Opens a new blank circuit window.

Icon: **Open**
Use: Brings up Windows File Explorer to open an existing circuit.

Icon: **Save**
Use: Saves current circuit.

Icon: **Cut**
Use: Cuts selected component(s) out of drawing. A copy is kept on the Windows clipboard.

Icon: **Copy**
Use: Places a copy of the selected component(s) onto the clipboard but leaves the original part(s) in place.

Icon: **Paste**
Use: Copies the contents of the clipboard into the current drawing.

Icon: **Print**
Use: Opens the Print dialog box. Prints only the circuit and nothing else.

Icon: **About**
Use: Opens dialog box containing simple information about your version of Multisim.

Zoom Toolbar

Refer to Figure 2.24 for a look at the Zoom toolbar's icons.

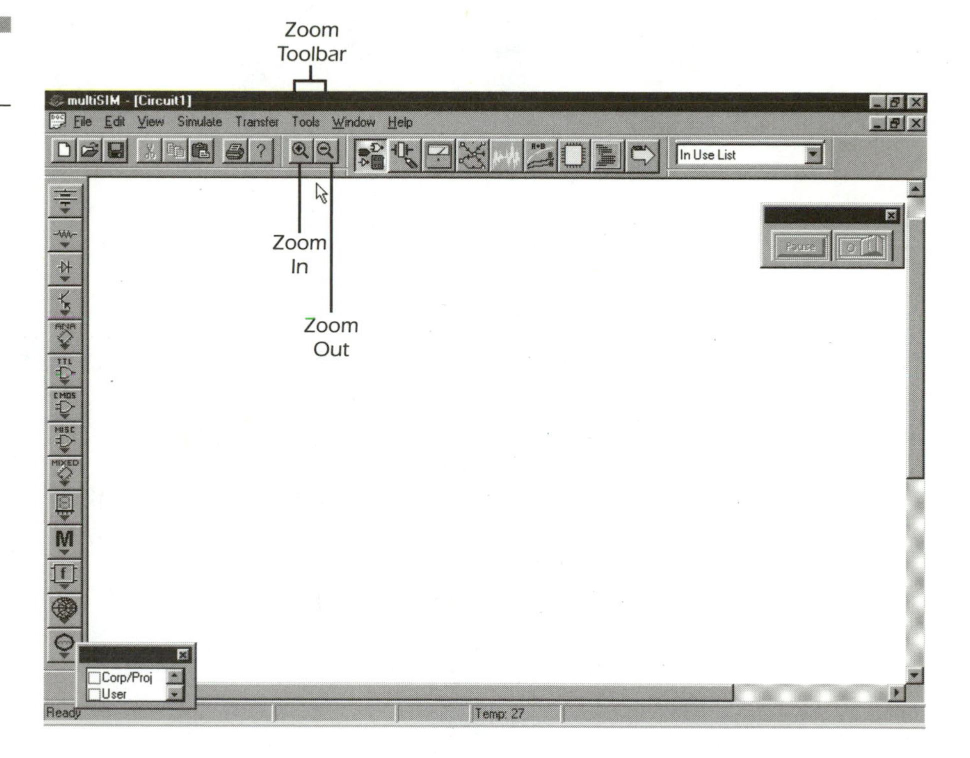

Figure 2.24
Zoom toolbar.

Icon: **Increase Zoom**
Use: Zooms into a section of the circuit.

Icon: **Decrease Zoom**
Use: Zooms out of a section of the circuit.

Multisim Design Bar

Refer to Figure 2.25 for a better look at the Multisim Design toolbar.

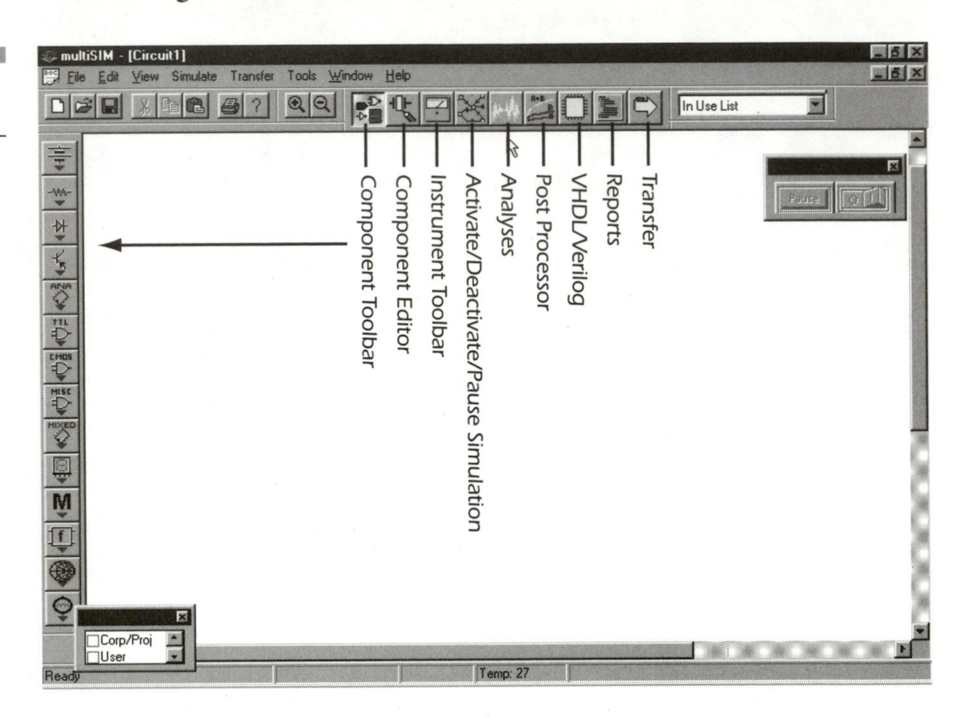

Figure 2.25
Multisim design bar.

Icon: **Component**
Use: Opens/Closes Component Group toolbar.

Icon: **Component Editor**
Use: Opens the Component Editor window.

Icon: **Instruments**
Use: Opens/Closes Instruments toolbar.

Icon: **Simulate**
Use: Gives you the simulation option of Run/Stop and Pause/Resume.

Icon: **Analysis**
Use: Opens menu to let you select a method of analysis.

Icon: **Post Processor** (only on Power Pro.)
Use: Opens the Post Processor window.

Icon: **VHDL/VERILOG** (only on editions with this feature)
Use: Presents a menu allowing you to open VHDL Simulation, Verilog Simulation, VHDL Synthesis, or Verilog Synthesis, depending on which modules you have installed.

Icon: **Reports**
Use: Presents a menu letting you open Bill of Materials, Database Family List, or Component Detail Report.

Icon: **Transfer**
Use: Presents a menu letting you select Transfer to Ultiboard, Transfer to other PCB Layout, Export Simulation Results to MathCAD, Export Simulation Results to Excel, or Remote Control/Design Sharing.

In Use List

The In Use list (Figure 2.26) is a scroll-down menu that lets you place another copy of any of the current components. Once you select which component you wish to place, move the mouse pointer onto the drawing and the part will 'stick' to the pointer. Clicking the left button again lays the component and sometimes begins running one of the wires.

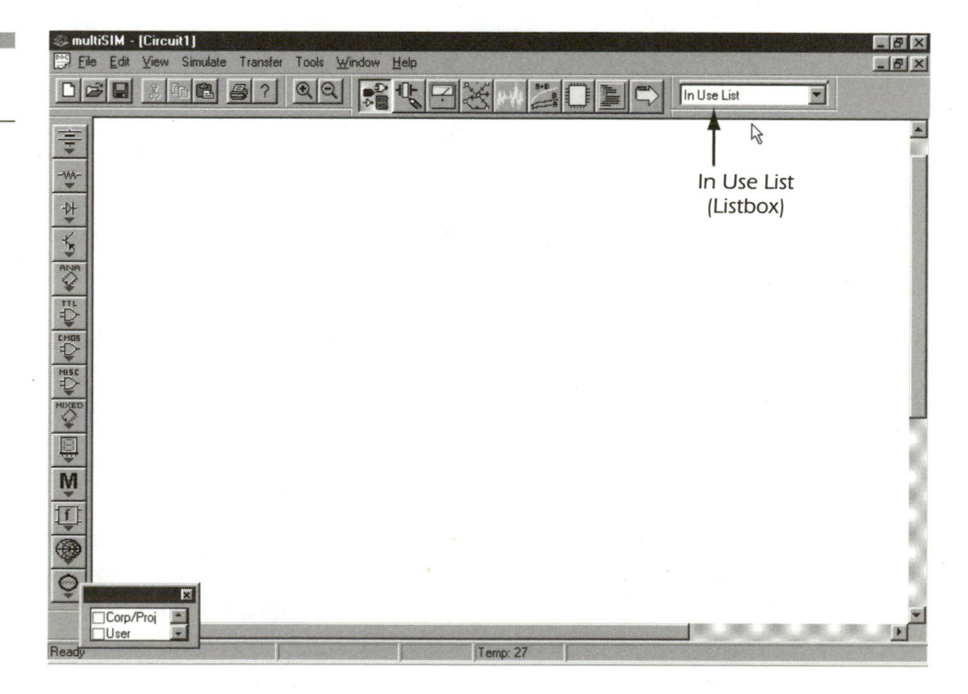

Figure 2.26
In Use list.

Component Group Toolbar(s) and Component Toolbars

The Component Group toolbars (Figure 2.27) contain icons that open the actual Component toolbars when the mouse pointer is moved over them. They are not the actual component toolbars. Think of the icons as an assortment of similar components or component subcategories. For example, the Transistor icon on the Component Group toolbar opens the Component toolbar containing all types of transistors.

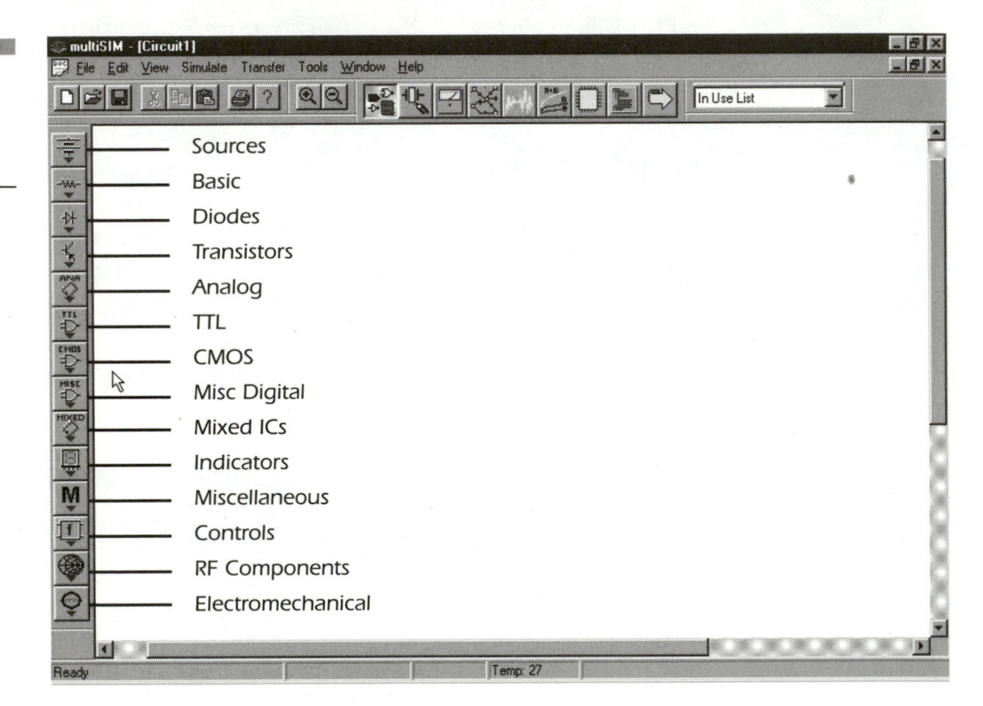

Figure 2.27 Component group and component toolbars.

The Component Group toolbars contain icons of similar components and are each listed within the Component Group Icons below. For a more in-depth explanation of each Component, see Chapter 4, Multisim Building Blocks. To see which components are included in each Component Group toolbar, refer to figures on the next page.

Don't click the icons on the Component Group toolbars. Instead, move the mouse pointer over the icon and wait a moment for the Component toolbar to open. It's annoying at first, but you'll get used to it. There is currently no option to turn this feature off. Note that as of Multisim 6.2, you can click the toolbars once to keep them open and click again to close.

COMPONENT GROUP TOOLBARS

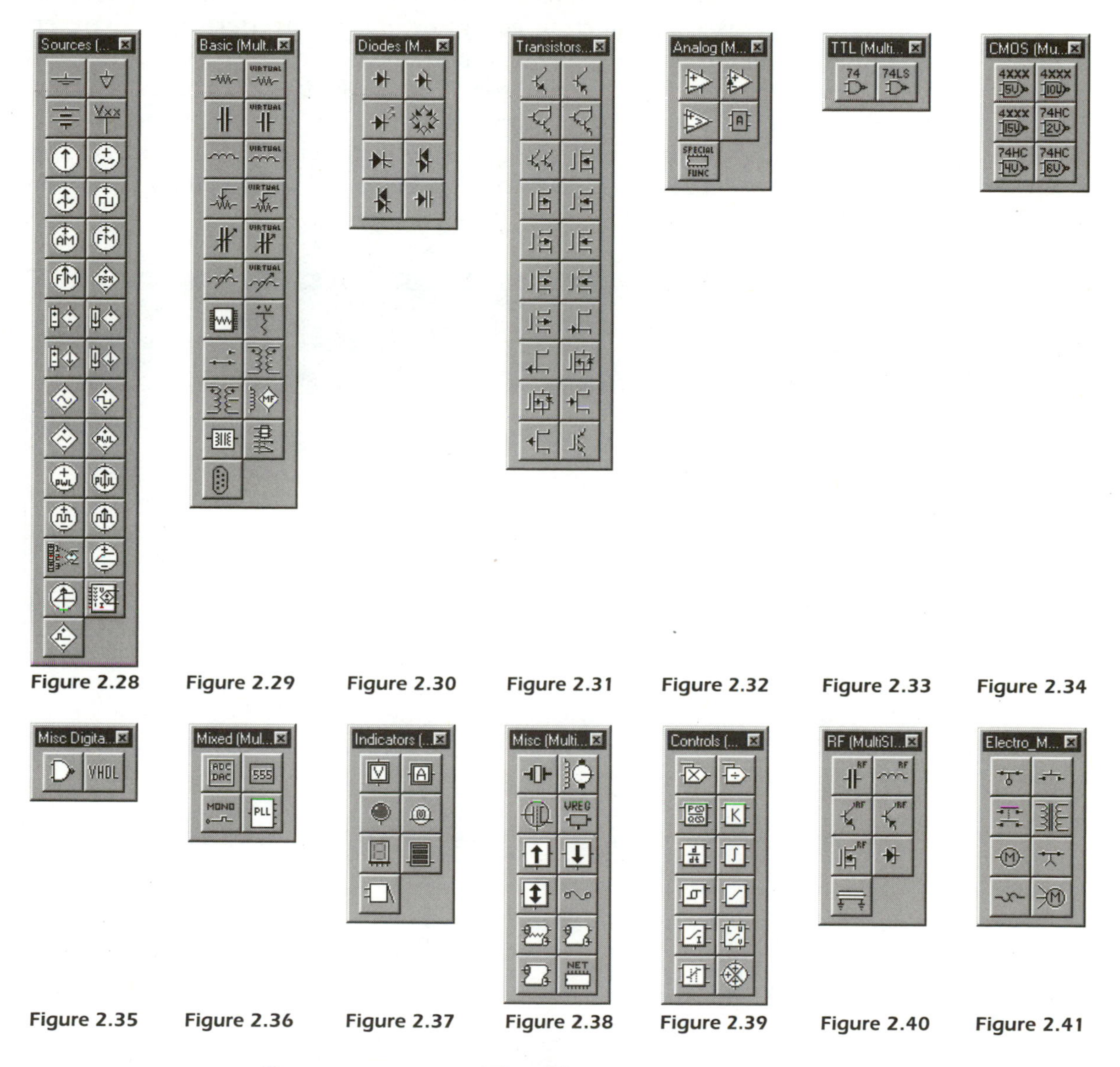

Figure 2.28 Figure 2.29 Figure 2.30 Figure 2.31 Figure 2.32 Figure 2.33 Figure 2.34

Figure 2.35 Figure 2.36 Figure 2.37 Figure 2.38 Figure 2.39 Figure 2.40 Figure 2.41

Instrument Toolbar

The Instrument toolbar contains icons for each of Multisim's virtual instruments. The Instruments toolbar is closed by default, but can be turned on by either going to **View** > **Toolbars** and checking off Instruments or by hitting the Instruments toolbar icon on the

Multisim design bar. The toolbar then floats somewhere on the desktop. The Instruments represented are (left to right) Bode plotter, distortion analyzer, function generator, logic converter, logic analyzer, multimeter, network analyzer, oscilloscope, spectrum analyzer, wattmeter, and word generator. Refer to Figure 2.42 for more information. Personal Edition users will notice that not all of these instruments are available to them; this is yet another disabled feature. More information about instruments can be found in Chapter 5, Multisim Instruments.

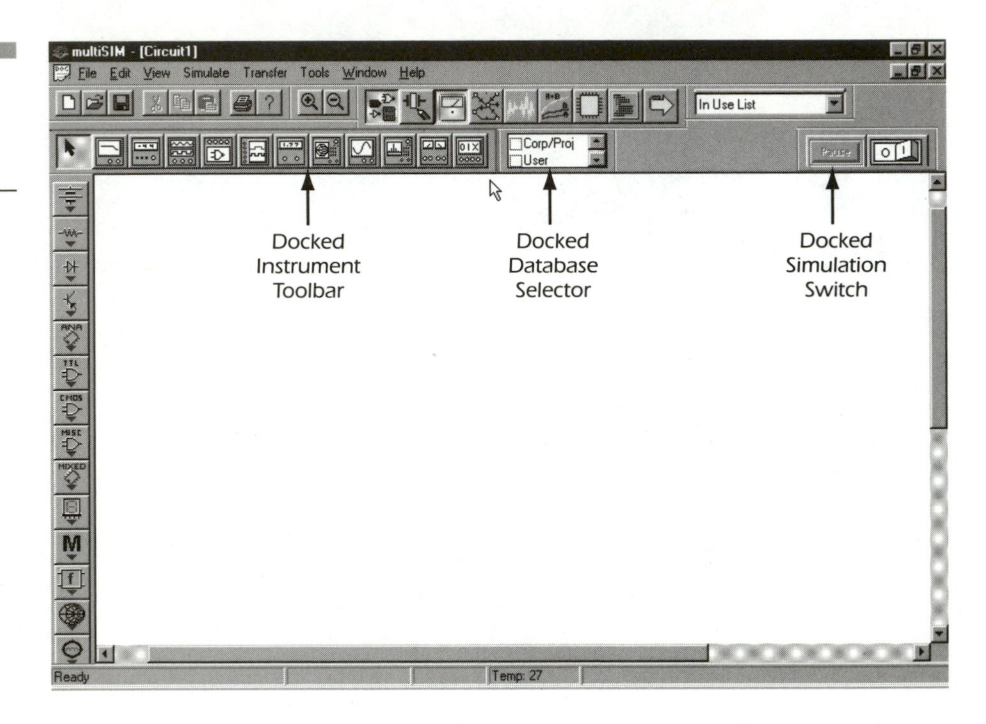

Figure 2.42
Docking toolbars maximizes drawing space.

TIP

Dock the Instrument toolbar, simulation switch, and database selector below the Multisim Design toolbar to maximize space.

Database Selector

The Database Selector toolbar item allows you to select which component group toolbars are shown on the left side of the desktop. If the item is checked, it is displayed. You may have to scroll down using the arrow keys to see the list of three databases: Multisim, Corporate, and User. (Figure 2.43). This toolbar can also be docked around any part of the desktop or allowed to float. On the Personal Edition, the Corporate Database is greyed out.

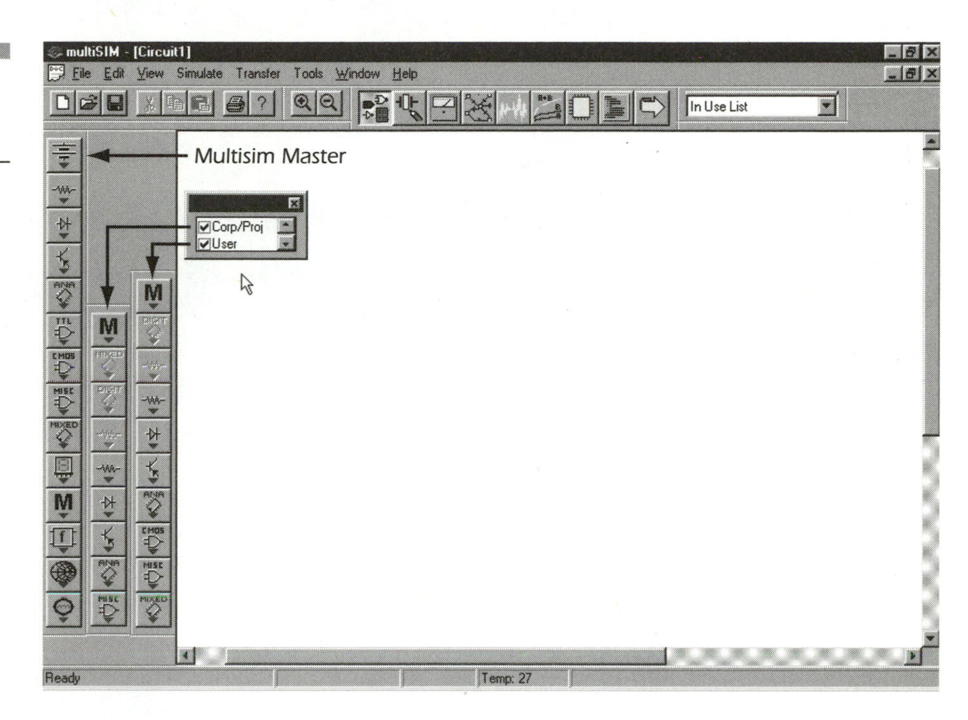

Figure 2.43 Multisim's various database toolbars.

TIP

*Close this rather useless toolbar Databae Selector) to maximize desktop real estate. Also, if you cannot see the Multisim master toolbar, go to **View** > **Toolbars** > **Database** and check on Multisim Master. Now go to **View** > **Toolbars** and check on **Components**.*

User's Guide to the Multisim Interface

Now that you've had a look at the basic components of the Multisim Interface, we will dissect and describe the different tasks of which Multisim is capable. Have a look at Figure 2.44. You will see the Interface is used to perform several functions. The first is Schematic

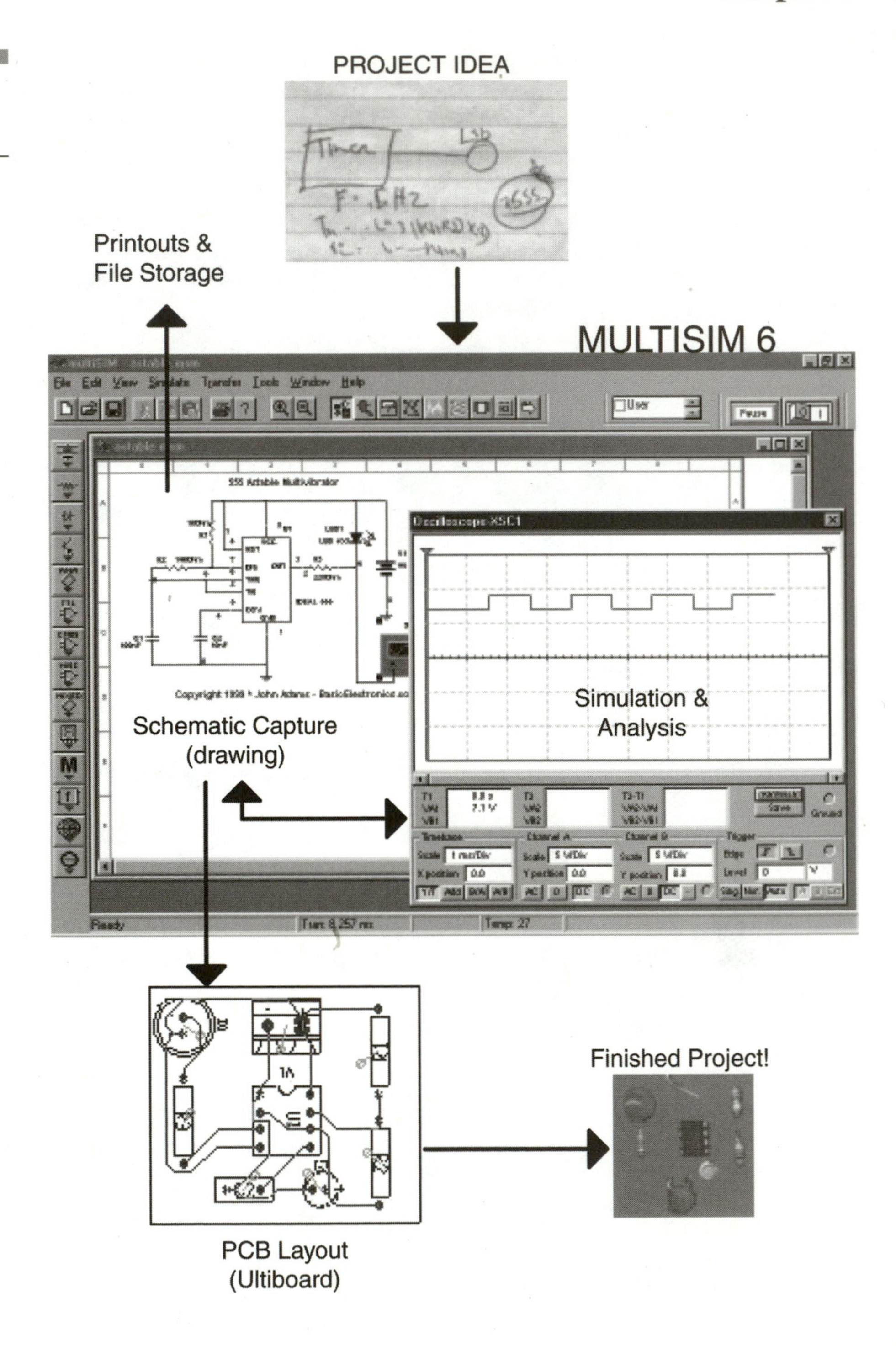

Figure 2.44
Overview of Multisim's tasks.

Capture, which is the ability to draw circuits with Multisim and save the files or output them to paper. (This is covered in Chapter 3, "Drawing with Multisim.") The second is to take the drawing and analyze or simulate that circuit (covered in Chapter 6, "Simulation and Analysis"). The Interface also gives you the ability to export your circuit designs to a PCB package. The following sections show you how to use these features and customize the interface itself to your personal tastes.

Running Multisim 6

I am assuming you have one of the Multisim versions (Demo, Personal, Professional, or Powerpro Full version) loaded and ready to open. Get those mice ready. It's time to simulate.

Let's begin by starting Multisim:

Win95/98/NT: Start > Programs > Multisim > Multisim.

A splash screen opens and, depending on the speed of your processor and the amount of RAM you have, it may take some time for the actual Multisim desktop (Interface) to appear.

TIP

You can also use the shortcut on your Windows Desktop that was likely created at the time you loaded Multisim. If it is not present, create one and point it to C:\Multisim\Multisim.exe or whichever directory the executable file is placed in.

Saving Multisim Files

To access your file in the future, you must first save it to the hard drive or to a floppy disk. Once the file is saved, you must resave and resave as changes are made. When you first begin Multisim, a new circuit document is automatically opened. It will likely be titled Circuit1.msm. Unless you want to use this generic name, I suggest renaming it at this time by saving the document. Either choose **File > Save as** or hit the Floppy icon at the top. This will open the **Save As** dialog box (see Figure 2.45). It will be opened to the directory you have set up in your User Preferences. Type in a name for the circuit in any of the formats you see in the figure (you can use up to 255 characters, including spaces). You do not need to add .msm to the end, because Multisim places it automatically. Make sure the name is easily recognizable for later retrieval. Hit the **Save** button.

Figure 2.45
Saving files to your hard drive requires you to use this dialog box.

I strongly recommend saving a circuit file every couple of minutes as you are working on it. To do this, hit the floppy icon on the Circuit Toolbar or go to **File** > **Save**. You can also just hit **Ctrl+S**. You can also set the Autosave Interval. Go to **Edit** > **User Preferences (Ctrl+U)** > **Preferences Tab** and make sure there is a checkmark in the Autosave box. Set the Autosave Interval to somewhere around five minutes or less.

Opening Multisim Files

To open Multisim circuit files you have saved or stored, you must use the Open command. **File** > **Open (Ctrl+O)** brings up the **Open** dialog box (see Figure 2.46). This opens to the directory you have set in your User Preferences. Highlight the file you wish to open. If you wish to open it as a Read-only file only, check that off now. Hit the **Open** button.

Opening Files Other Than Multisim Files

Multisim 6 has the ability to open Multisim 6 files, Electronics Workbench V5 files, and Netlist files. Use the same procedures as when opening regular Multisim files, but at the bottom of the dialog box select the new file extension with the **Files of Type** pull-down box.

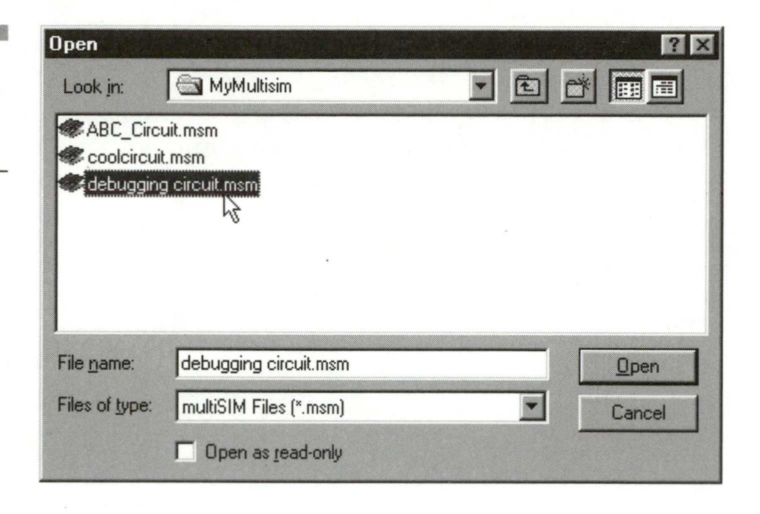

Figure 2.46
To open stored files, use the Open dialog box.

Exporting Netlist Files

If you are using a software program other than Electronics Workbench, you can typically save your files as SPICE Netlist file, which can then be imported into another EDA program. This is done by selecting **Transfer** > **Export Netlist**. Name the file in the File Name field and Multisim will add the *.cir extension to it.

Exporting to EWB Layout Products

If you have Ultiboard or one of several other PCB packages installed on your PC, you can export your Multisim design straight to it. Select **Transfer** > **Transfer to Ultiboard**. The Save As Dialog Box opens. Type in the name you wish to call the export file (use same name as the Multisim file if possible) and Multisim automatically adds the *.plc extension and creates a file with the extension *.net. These files are required for Ultiboard to work.

Sometimes, you must change a circuit while in Ultiboard. If you wish to have Multisim make the changes to your circuit schematic as well, choose **Transfer** > **Backannotate from Ultiboard**. The Open Dialog Box appears, asking for the Ultiboard File name. Find the file name and Multisim creates a schematic.

Importing Netlist Files

If you wish to open up a Netlist (SPICE) file, you must open it with the Open File dialog box. Choose Spice Netlist Files (*.cir) from the pull-down box and then select the file from the hard drive to open.

NOTE

Imported Netlists are not exactly aesthetically pleasing. You may want to rearrange the components a bit to avoid confusions.

Setting Up and Customizing Your Workbench

The first thing you notice when Multisim's Interface appears is the drastically different layout compared to EWB 5. Don't let this confuse you: the desktop is customizable. In Chapter 14, I will explain different methods to make it look similar to EWB 5.xx. You will also notice that the circuit window may be in black. Once again, this is customizable.

Accessing Menu and Toolbar Items

Menu items are placed across the top of your screen and are not customizable. They include **File**, **Edit**, **View**, **Simulate**, **Transfer**, **Tools**, **Window**, and **Help**. They are used to access commands you wish to perform. Each menu opens to a list of submenus or commands.

Toolbars are placed about the desktop. They include the System toolbar, Zoom toolbar, Multisim Design bar, In Use list, and Component Group toolbars (each opens to a subtoolbar). In addition, a few toolbars can be activated and added to your desktop. These include the Database Selector, Simulation Switch (which may be familiar to you EWB 5 users), and Instruments toolbars.

See the previous sections in this chapter for more details on menu and toolbar procedures, tips, and uses.

TIP

If you are running two monitors in Windows 98, place the toolbars on an empty screen to free up circuit drawing space on the original monitor.

Getting Help

Multisim has an outstanding Help menu that answers most common questions you may encounter. You can search it using a keyword, bookmark frequently accessed items, and print important information. The Multisim Reference is also available to help you find out more information about each of Multisim's components. Here are a few tips to save you time in using the Help menus.

OPENING THE HELP INDEX

You can either choose **Help** > **MultiSIM Help** or hit **F1**.

GETTING HELP ON A COMPONENT

Multisim lets you select a part and go directly to its corresponding help page. Some pages even display example circuits to aid in the understanding of that part. Here's how to access this:

1. Select the part.
2. Hit **F1** or choose **Help** from the top menu or right-click menu.

OPENING THE MULTISIM REFERENCE

Go to **Help** > **MultiSIM Reference**.

INDEX SEARCHES

Once you are in the Help Index window, hit the Index button. This opens up the Help Topics box as seen in Figure 2.47. Type a word in the top white box and topics appear in the box below. To select one of them, highlight it and press return or double-click it with the mouse. For example, if you were looking for information on transient analysis, start to type **T R A** and a list appears below. Double-click the selection and the Multisim User Guide box appears. See Figure 2.48.

THE FIND TAB

In this same window (Help Topics) is the Find tab. Click it (refer to Figure 2.49). Typing a keyword into the number 1 box produces a selection of suggested search words in number 2 box. You may also select a recent search word from the pull-down arrow. In the number 3 box a list of each page that contains that word or group of words as highlighted in boxes 1 and 2. For example, I typed the word "resistor" in the 1 box and the help topics suggest I look for "resistor, Resistor, resistors, and Resistors". Each is highlighted so I merely refer to box 3, which contains each page that has the word "resistor" in it. By double-clicking the words "RF Resistors" I am taken directly to that page.

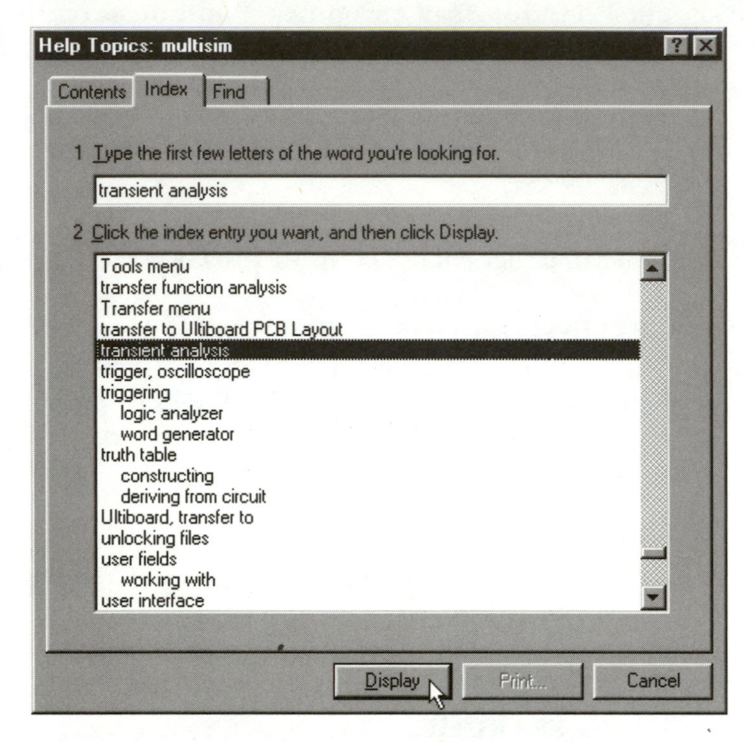

Figure 2.47
Multisim Help files are a quick way to reference the software. You can perform searches using this dialog box.

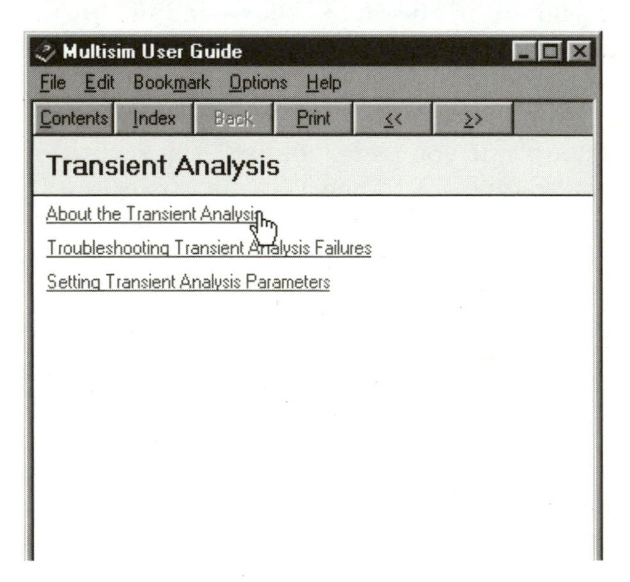

Figure 2.48
The Multisim User Guide is the 'meat' of Help. From here you can print, bookmark or cross-reference help files.

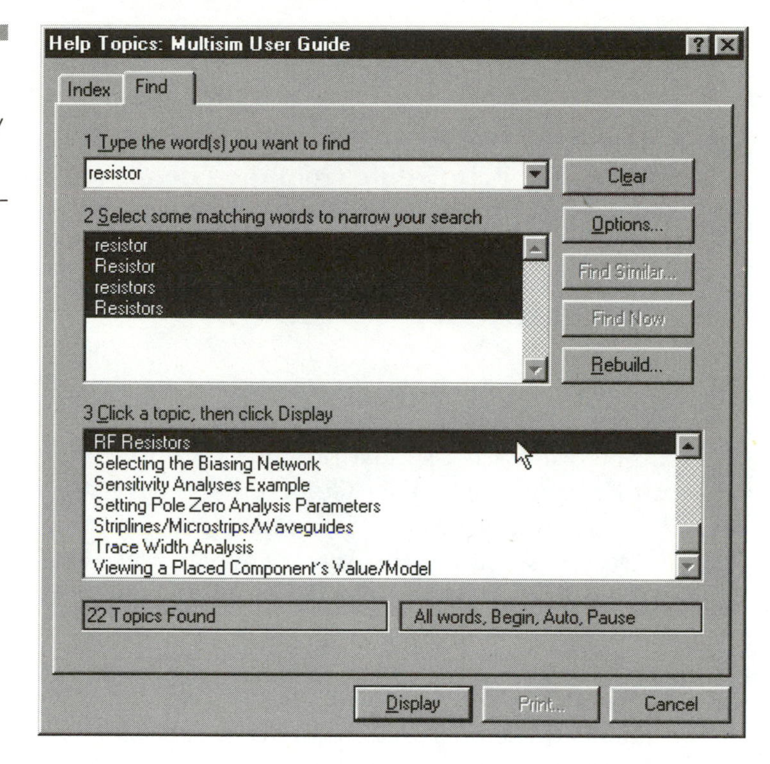

Figure 2.49
The Find tab allows you to search for any incidences of a specific word.

Adjusting the criteria can narrow searches. Hit the **Options** button and select the applicable options to fine-tune the search. Multisim has a fairly powerful Help search engine that is highly customizable using the options.

MULTISIM USER GUIDE

This is the meat of Help, where each section is described. It may contain additional links to other help files that relate to the subject. From here you can bookmark or print the files as well. (Refer to Figure 2.48).

BOOKMARKING

Any help page can be bookmarked for later reference. This works much like a web browser's bookmark feature:

1. Choose **Bookmark** > **Define**...
2. A window pops up prompting you to name the page. A title is automatically inserted into the field, but you can add whatever word(s) will help identify it later.
3. Hit **OK**.

To access these bookmarks, choose Bookmark and select the appropriate title. That page opens.

PRINTING HELP TOPICS

By choosing **Print topic** from the **File** menu, you can make a hard copy of help pages.

Multisim User Guide

Multisim now contains a paperless manual, Portable Document Format. (PDF), which is an electronic file of the Multisim manual. It can be read with a special program called Adobe Acrobat Reader or it can be printed. The User Guide itself is over 500 pages and the appendices are over 400 pages. If you have a fast printer, go ahead and print it. Otherwise, just open Acrobat when you absolutely can't find the information anywhere else. You can always print only the applicable pages.

NOTE

If you order Multisim from Electronics Workbench there may be a manual shipped with the product. At the time of this writing EWB was considering doing away with the paper version, so forgive the confusion.

HOW TO ACCESS THE MULTISIM USER GUIDE

Go to the Windows **Start button** > **Programs** > **Multisim** > **User Guide.** Acrobat opens with the document, ready to access.

TO OPEN APPENDICES

Go to **Start** > **Programs** > **Multisim** > **User Guide Appendices.**

Interface Exercises

To get used to the interface, let's try a few exercises.

1. Open Multisim.
2. Have a good long look at the screen. See what each item is, where it is placed and how it looks in relation to other items.

3. Open a few menus just to see what they contain. Do you see the submenus as well?
4. Right-click in the middle of the circuit window and have a peek at that menu.
5. Move the Zoom toolbar to make it float.
6. Turn the Instrument toolbar on. Dock it at the top.
7. Move your mouse over the various Component Group icons and let the Component toolbars open. Move the mouse away and note how quickly they disappear.
8. Try some of the hot keys, as described in the Menu section.
9. Shut some of the toolbars off and see what the screen looks like.
10. Turn the User toolbar on and then off.
11. Create a directory called **C:\MyMultisim** and save a file to it. Close the file.
12. Open some of the files in **C:\multisim\samples**.
13. Start the next chapter to learn how to draw the Multisim way.

Summary

The Multisim Interface uses simple procedures that any Windows user can master in no time. If you are in doubt about a certain procedure, read through all the chapters for further details. The help menus that are provided with EWB may shed some light on otherwise difficult tasks. Just enter the find feature in the Help Index and type a few words you think relate to what you are trying to do. If you are still having problems, you are always only a phone call away from Electronics Workbench's tech support line. See Appendix C for more information on contacting them.

The next four chapters delve more deeply into specific interface procedures, as they relate to issues such as drawing circuits, components, instruments, simulation and analysis, and building circuits.

CHAPTER 3

Drawing with Multisim: Schematic Capture

Replacing a pencil and paper with a mouse and monitor makes it harder to draw but easier to create a professional schematic.

Schematic capture is an electronic design term that describes drawing schematics with a computer. Multisim gives you the ability to create schematic representations of circuits in a high-quality format, that may include lists of parts needed to build the circuit and all other pertinent information. Circuit diagrams can then be printed or saved for archival purposes or to serve team members in the development of circuits. Schematic capture is also the starting point of simulation and analysis. Think of Multisim's Schematic Capture feature as a computer-aided drafting (CAD) program used for electronics, complete with pre-made symbols, the ability to connect their terminals together, and makes notes of circuit features.

NOTE

Also see Chapter 14, Unleashing Multisim/EWB, for plenty of valuable circuit drawing tips.

Getting Ready to Draw

To draw with Multisim it is best to adjust certain settings on your PC first. The following sections will help you:

- Set up your monitor.
- Adjust the toolbars and desktop to your liking (see Chapters 2 and 14 for tips).
- Adjust the drawing's or User's preferences in Multisim.
- Name and save the drawing.
- Draw away!

Setting Up Your Monitor

Your monitor must be set to a minimum resolution of 800×600 pixels to operate Multisim. This means there are 800 individual picture elements for the width and 600 for the height of the monitor. To do this, go to **Start** > **Settings** > **Control Panel**. Double-click the **Display** icon. Click the **Settings** tab. Under Desktop Area, move the slide until the 800×600 setting appears (see Figure 3.1). An even higher resolution is fine. You may be prompted to confirm this change; do so if it is safe (some monitors won't support higher resolutions). If you have difficulties with this procedure, refer to the Windows 95/98/NT manual or help file, the monitor's manual, or the PC video card's readme file.

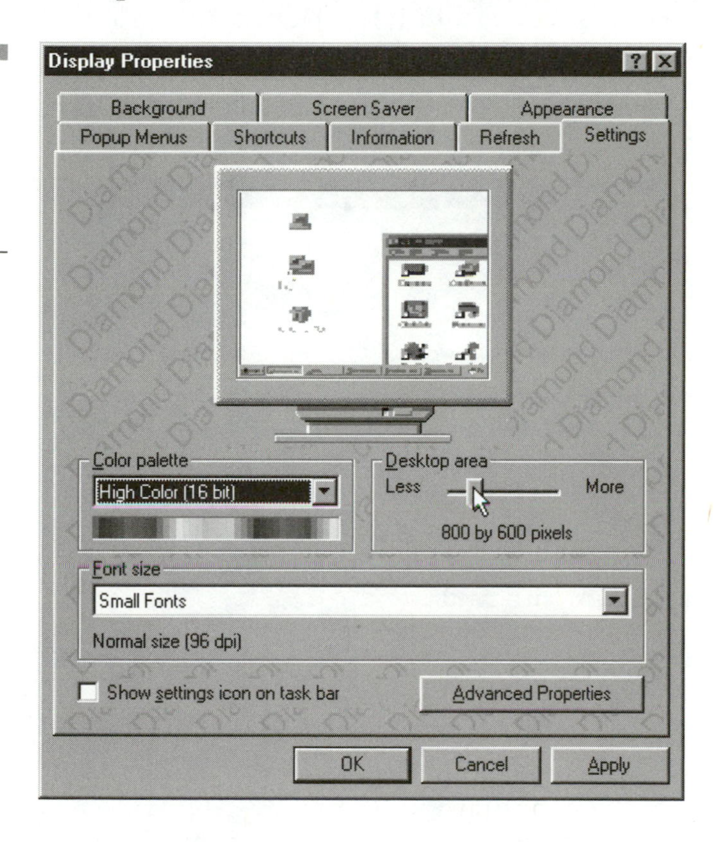

Figure 3.1 Your monitor's resolution must be set to 800×600 or greater to run Multisim.

Think of the circuit window as a sheet of paper upon which components and wiring are drawn.

Setting Up the Drawing

First things first. When you open Multisim, you must first set up your sheet of paper just as you would if you were drawing by hand.

Also required are the adjustments of certain desktop settings. You may relate this to getting your pencils and drafting tools together.

Setting User Preferences

To open up the User Preference dialog box go to **Edit** > **User Preferences** or hit **Ctrl+U**. From here you access the following pages:

- **Circuit**—Here you can adjust which schematic elements are displayed and the color scheme used for this drawing. See Figure 3.2.

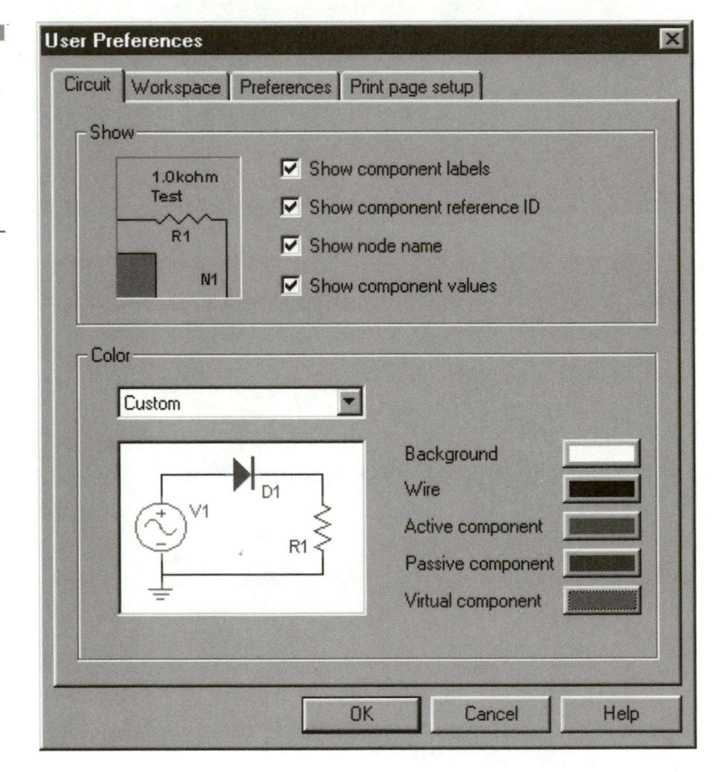

Figure 3.2
The Circuit tab under User Preferences lets you adjust what schematic items are shown.

  — **Show**—A checkmark in the appropriate box means that element will be displayed.

  — **Color**—A preset color scheme can be selected from the pull-down menu, or you can customize each drawing element's color.

- **Workspace**—These options adjust the actual drawing defaults, such as sheet size, whether to show a grid, and more. See Figure 3.3.

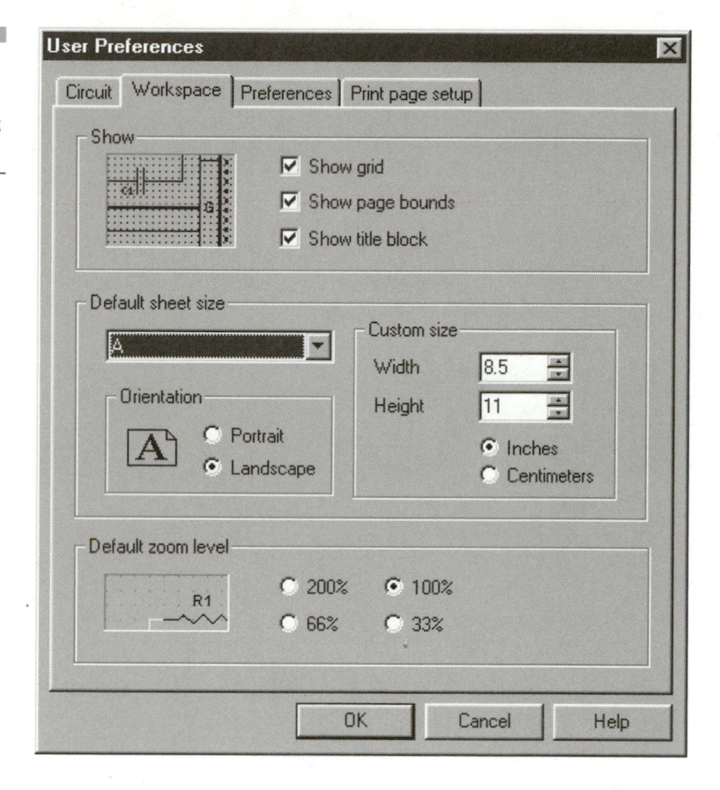

Figure 3.3
The Workspace tab sets up the drawing's sheet information.

- **Show**—A checkmark in the appropriate box means that option will be displayed.
- **Default Sheet Size**—Allows you to select a preset sheet size or customize one. Also adjusts the orientation of the drawing. Portrait place the sheet upright; Landscape places it is on its side.
- **Default Zoom Level**—Adjusts the zoom level that is used upon startup.

■ **Preferences**—These features let you adjust the software's preferences. See Figure 3.4.

- **Environment**—Lets you turn on/off the Autosave Feature and the time interval at which the computer automatically saves your. I strongly recommend leaving this feature on at all times. If you are familiar with a Windows machine, this is a given.
- **Symbols**—Allows you to select which schematic symbol set is used. ANSI stands for "American National Standards Institute" and is used mostly in North America. DIN means "Deutsche

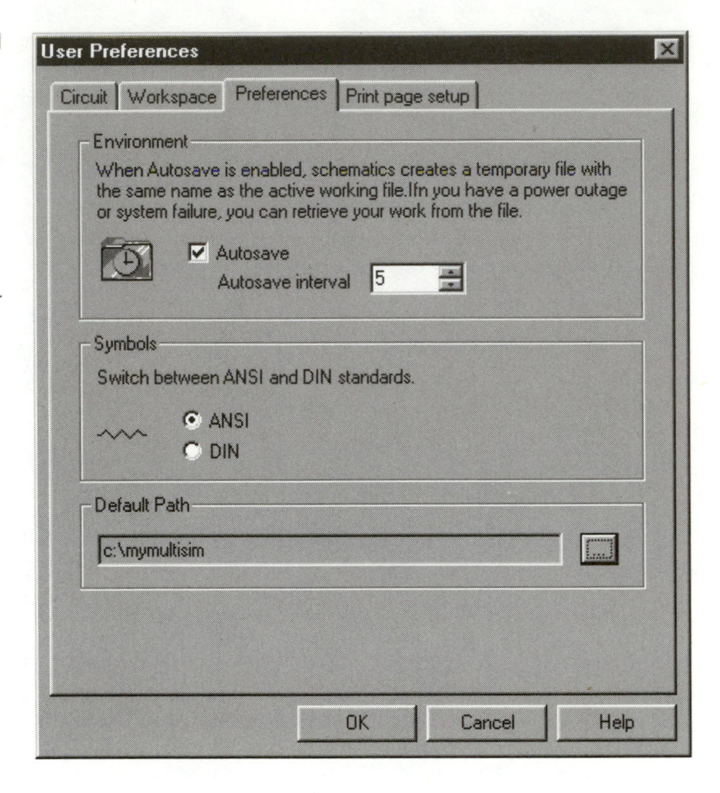

Figure 3.4
The Preferences tab lets you turn on Autosave and choose the symbol set you wish to use. You can also set the path for Multisim files.

Industrinorm (Deutsches Institut Normung)," which is based on a German standard and is often used in European countries.

- **Default Path**—Points to the path circuit files open and save into. For example, setting it to C:\MyMultisim opens and saves into your personal directory on the C drive.

- **Print page setup**—Sets up the print options as they relate to the actual printed page. See Figure 3.5.
  - **Print Mode**—Multisim does not print exactly what you see on your monitor. These two options let you adjust the output to your specific printer.
  - **Margins**—A printer can only put toner out so far to the edge of a page. With these options you can make sure Multisim does not try to color outside the lines.
  - **Magnification**—Lets you adjust the magnification of the printout. I recommend using Fit to Page unless you are looking for a specific output size.

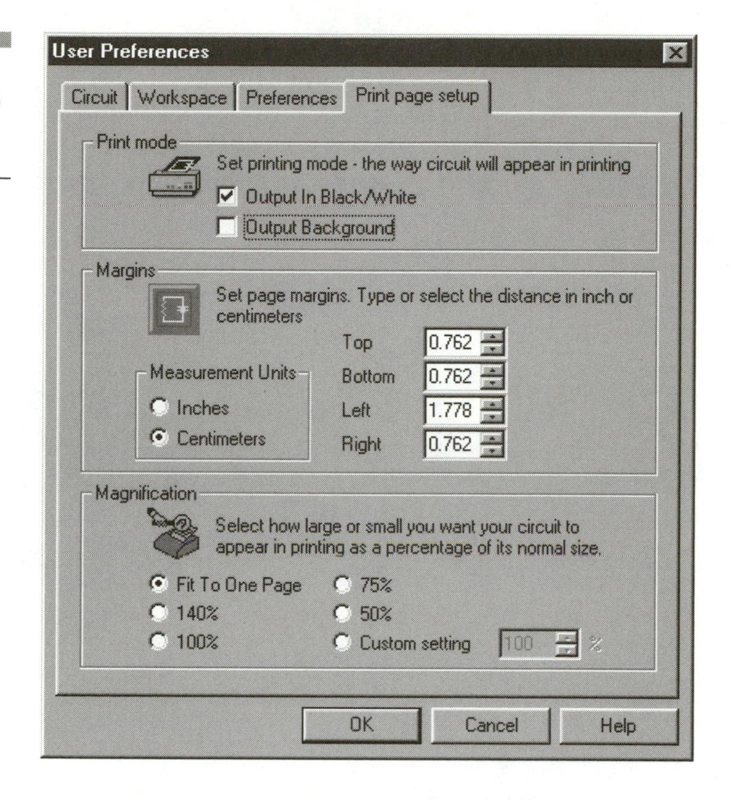

Figure 3.5
The Print Page Setup tab under User Preferences.

Open Multisim. Go to ***Edit*** *>* ***User Preferences*** *and set up the defaults to your tastes. Close the current drawing and open a new one. All future drawings will now have your preferences set.*

Print Page Setup can also be accessed by choosing ***File*** *>* ***Print Setup*** *> and hitting the* ***Page Setup*** *command button at the bottom. The format is different but the settings are the same.*

Adjusting the Current Drawing's Settings

You may have noticed that fiddling with the User Preferences has no effect on the circuit you have already open. This is because the settings can only be applied to the *next* new document you create. To edit the current drawing settings you must use various menus:

ADJUSTING COLOR AND LABELS TO BE DISPLAYED

Right-click in an empty area of the circuit drawing and choose either Color or Show to set these preferences. See Figures 3.6 and 3.7.

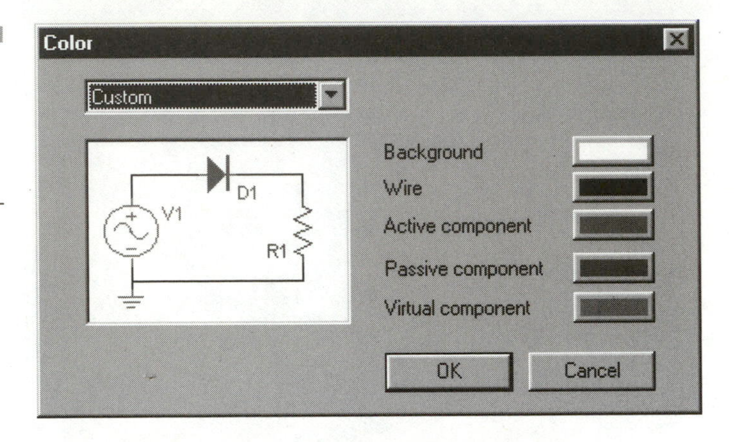

Figure 3.6 Use this dialog box to adjust the current drawing's color scheme.

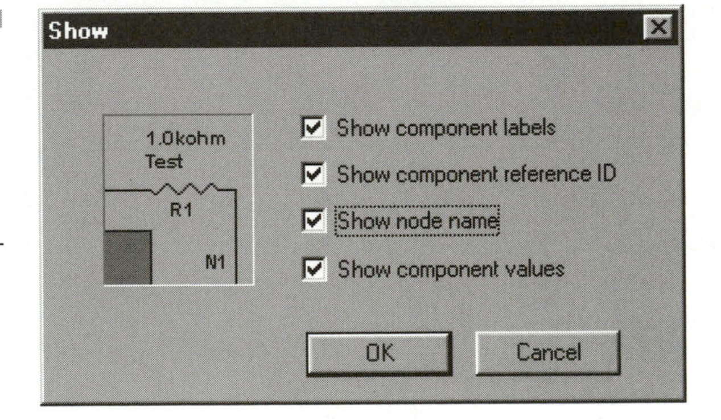

Figure 3.7 Use the Show dialog box to set what elements will be displayed on your current circuit.

ADJUSTING SHEET SIZE

Go to **Edit** > **Set Sheet Size**... See Figure 3.8.

ADJUSTING PRINT PAGE PROPERTIES

Go to **File** > **Print Setup** and choose the **Page Setup** command button. See Figure 3.9.

TIP

When you initially load Multisim, adjust the User Preferences first. Close Circuit1.msm and then open a new circuit. All documents will then be customized to your tastes.

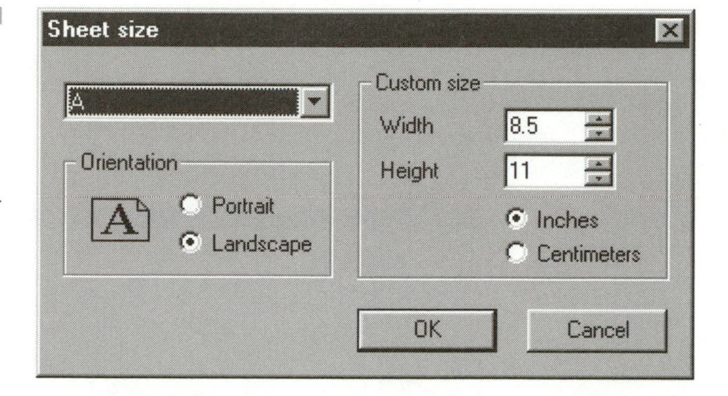

Figure 3.8 Adjusting the Sheet Size and orientation is simple with this dialog box.

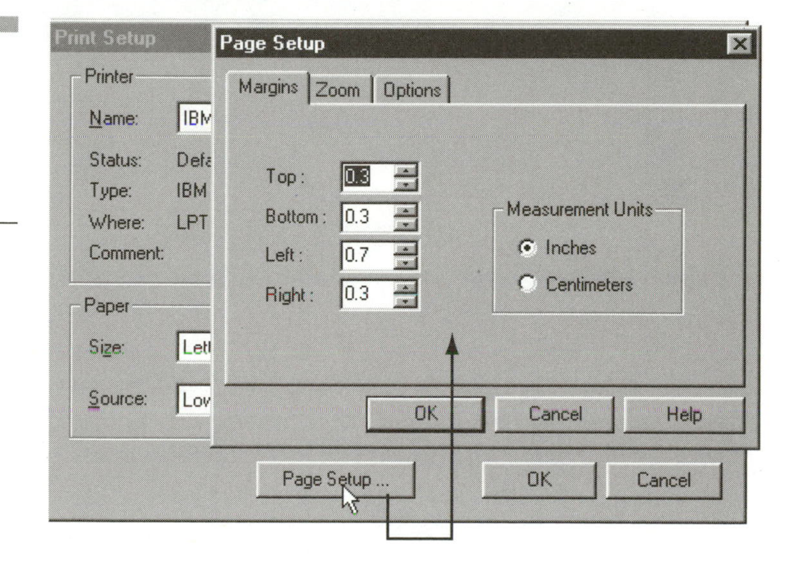

Figure 3.9 To access Page Setup, go to **File > Print Setup** and hit the **Page Setup** button.

Setting Title Block

One last item you may want to adjust before beginning the drawing is the Title Block. This helps to describe the drawing to other people or, later, helps you to remember exactly what this circuit is. To do this, go to **Edit > Set Title Block**. The Title Block dialog box opens to allow you to set each item. When you press **OK**, the information is displayed at the bottom right corner of the current drawing. See Figure 3.10.

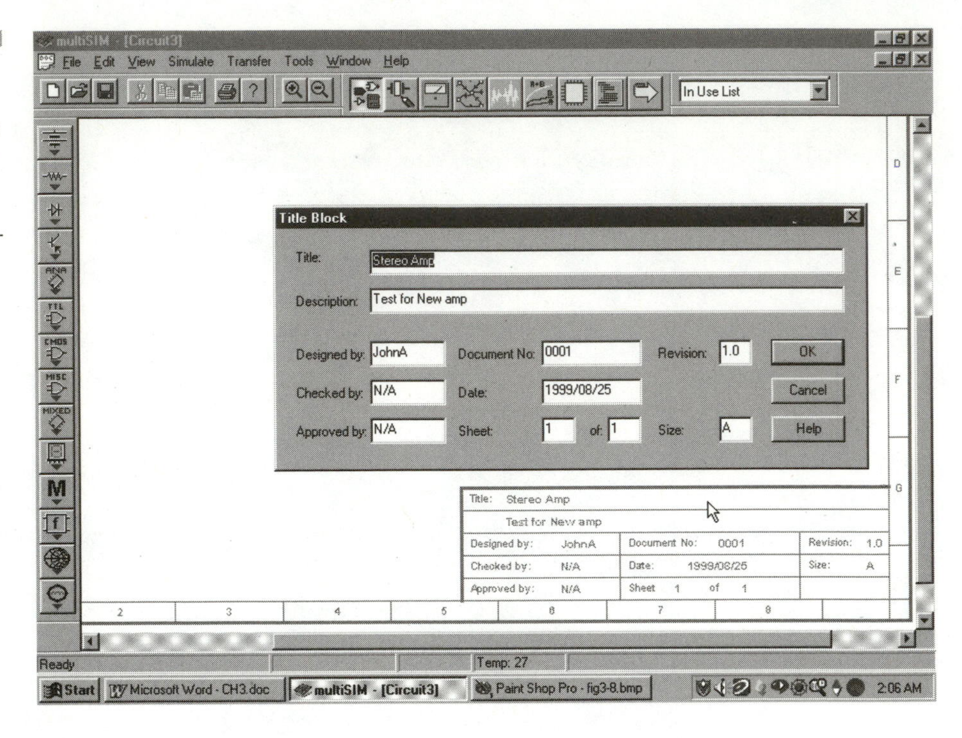

Figure 3.10
A Title Block lets you add information such as date and designer to help you later identify the drawing.

TIP

If you don't see the Title Block on the drawing, right-click inside an empty section of the circuit window and choose ***Show Title Block and Border****.*

TIP

Make a document with your Title Block partially filled in and call it main.msm. Open this file each time you start a new circuit and rename it.

EXERCISE

Begin drawing a circuit. Shut off all the labels and values. Turn off the Title Block and Border. Save the circuit and close it. Open a new circuit and begin to draw again. Does it show the border and title block?

Placing Components

Subsequent chapters go deeper into component placement, but for now, let's see how the Component Group toolbars operate. Move your mouse pointer over a specific toolbar icon, wait a moment, and another component bin appears. Do not left-click! The subtoolbar automatically opens. From there, you must move the mouse cursor directly on to the newly opened toolbar or it will disappear. The new subtoolbar contains a set of icons that represent individual components you can now choose to lay onto your drawing. Select an item by moving the mouse pointer over it and clicking the left button. See Figure 3.11.

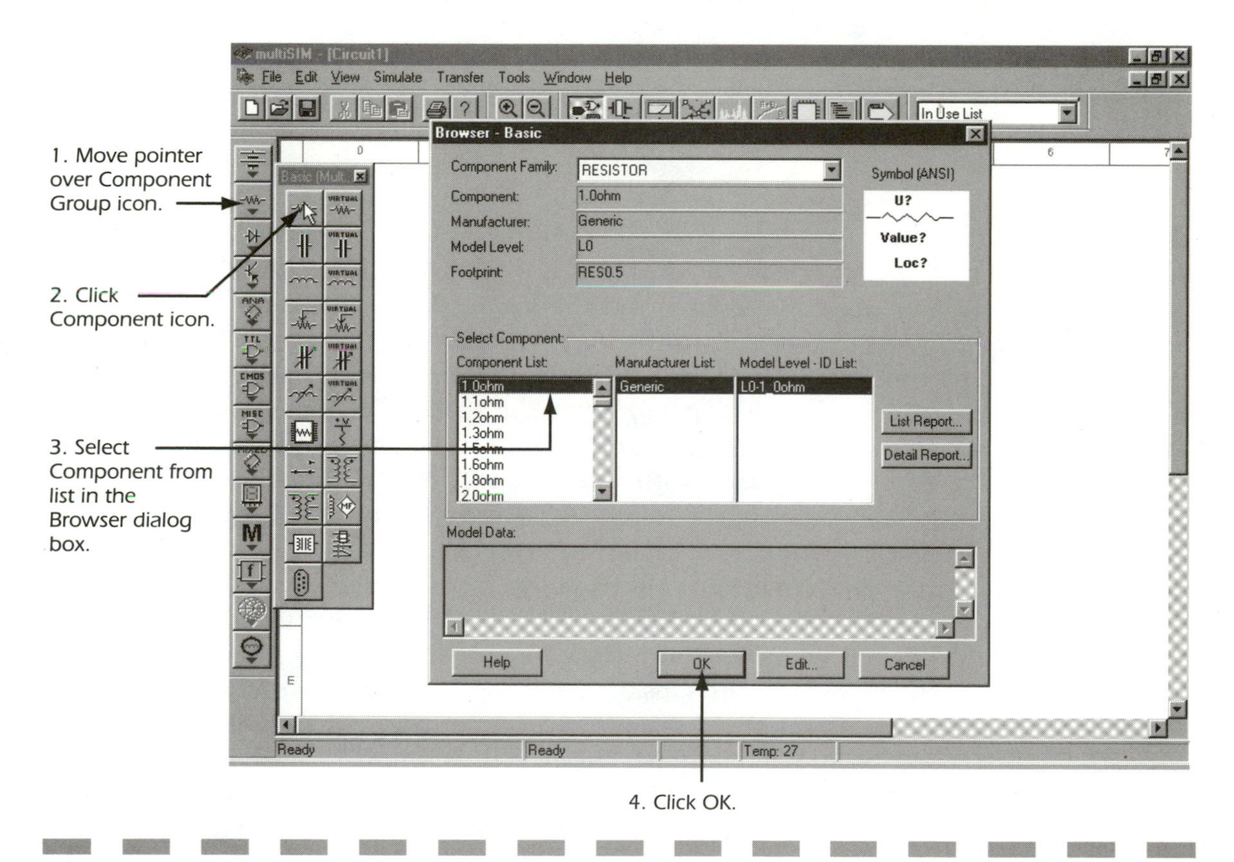

Figure 3.11 Selecting a component to place onto your circuit.

NOTE

To avoid confusion I may use two terms: Component Group toolbar and subtoolbar. The subtoolbar is a toolbar for each component group. This subtoolbar may also be called a Component toolbar.

In most cases, the Component Browser window opens (this is discussed later). After you have made a selection of which specific component you wish to use (and set its various values), you can now click the **OK** button. A different mouse pointer appears, as in Figure 3.12. Left-clicking onto the drawing (circuit window) places the component.

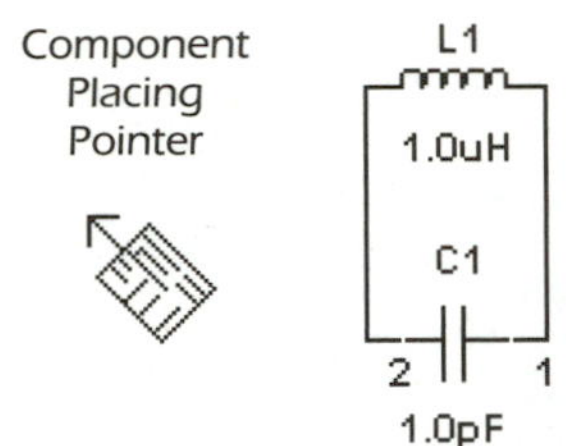

Figure 3.12 Component placing mouse pointer.

For example, if you want to place a 100-ohm resistor on the drawing, move the pointer over the Sources icon in the Component Group Toolbar and wait until the subtoolbar opens. Left-click the top left icon (resistor) and wait until the Browser Screen opens; this may take several seconds depending on the speed of your computer and the amount of RAM available. Scroll down the **Component List** and highlight **100ohm** with the mouse. Click **OK**. Now place the pointer where you want the resistor drawn and left-click. See Figure 3.13.

Some components, such as virtual resistors, caps, batteries, and others can be placed without going through the Component Browser step. To edit the properties of virtual components, double-click the left mouse button onto the component after it is placed and change the necessary values inside the dialog box. See Figure 3.14.

The Component Browser method of placing components is new to Multisim. It slows down the A-to-B techniques of schematic creation that was one of EWB 5's strengths. However, the Component Browser presents you with a more robust and accurate model when you begin simulations and analyses. The real-world modeled components incorporated into your designs will be appreciated when the circuit goes into production.

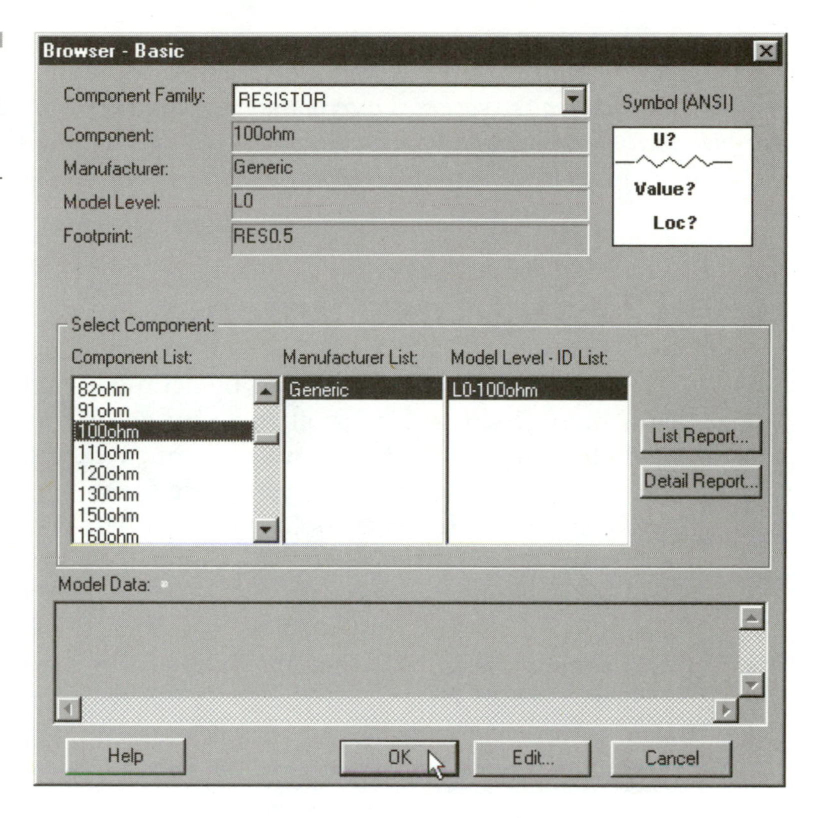

Figure 3.13 Selecting a part using the Component Browser.

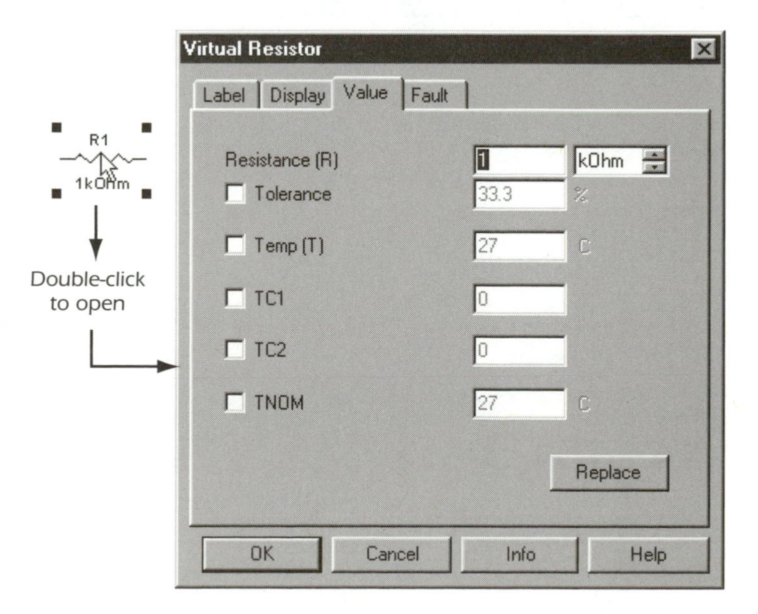

Figure 3.14 Virtual components do not use the browser. Instead double-click them to bring up this dialog box.

TIP

To speed up Multisim, install more memory (RAM) and a higher-speed processor if possible. Electronics Workbench recommends 32 to 64 MB of RAM, but if memory prices are low, I personally recommend 64 to 128 MB. The recommendation of a P-166 is low as well. Try 300 MHz or above.

Placing Instruments

The Instruments toolbar is closed by default. To choose an instrument you must first turn the toolbar on by going to **View** > **Toolbars** > and checking off **Instruments**. (It's much easier to simply click the Instruments icon on the Multisim design bar.)

Once the Instruments toolbar is open, left-click on top of one of the Instrument icons. Move the pointer over to the circuit window and left-click once more to place it. Wires can then be run to the connections on the instrument. You can open the instrument controls and options by double-clicking the instrument on the drawing. See Figures 3.15 and 3.16.

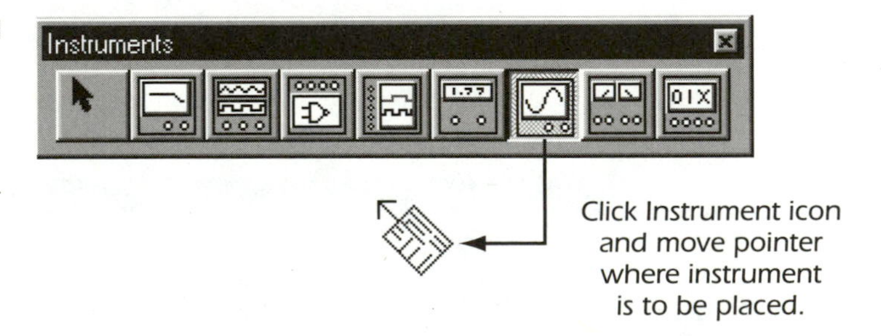

Figure 3.15 Choosing an instrument for placement.

NOTE

You can now place multiple instruments.

Quick Parts

Multisim keeps a list of parts you have placed. The In Use list on the top of the toolbar allows you to access and place the same part in a

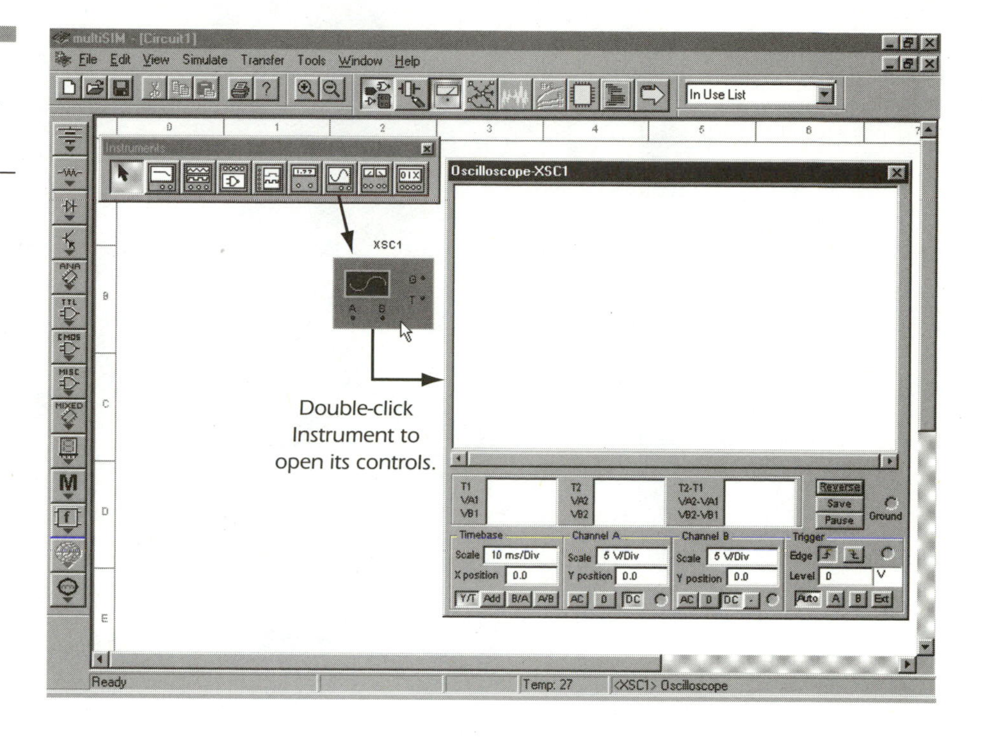

Figure 3.16 Place an instrument, then double-click its icon to open it.

different location (See Figure 3.17). For example, if you have placed a transistor, virtual resistor, ground, and battery, and want to lay another virtual resistor, this is the time-saving procedure:

1. Click the arrow on the In Use scroll-down.
2. Click the word **Resistor_Virtual** and move the mouse pointer over the circuit window. You will notice the part is stuck to the pointer.
3. Click where you want to place the virtual resistor.

NOTE

Some components may have a wire already attached to one of the connections and ready to run.

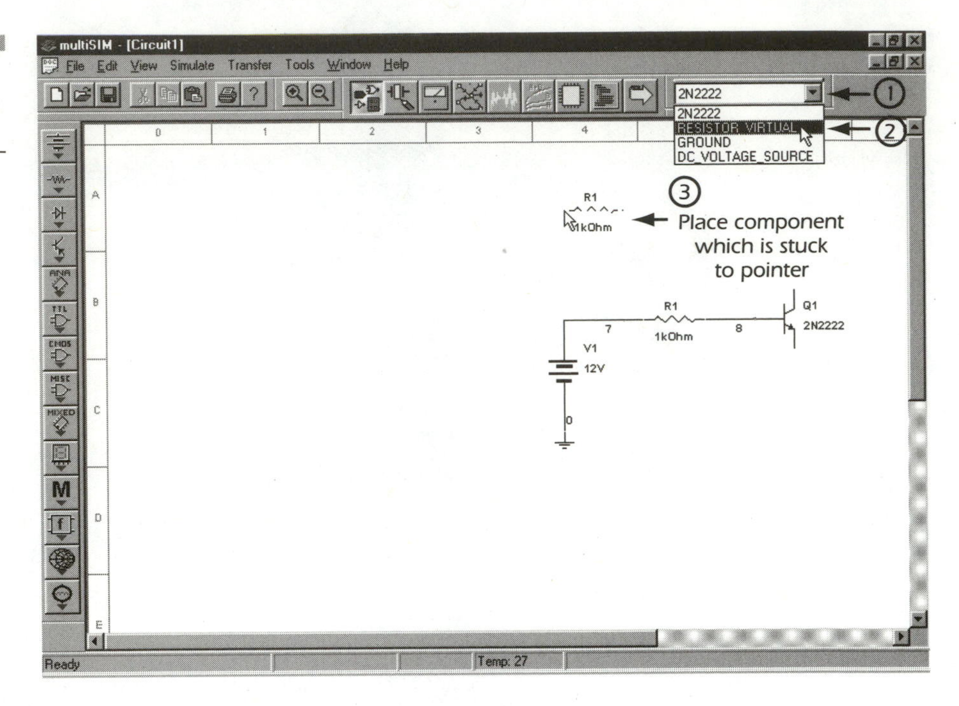

Figure 3.17
The In-Use list speeds parts placement.

Yet One More Way to Lay Components

Another way to place components better utilizes the Browser dialog box. Here is how:

1. Right-click into an empty space on the drawing.
2. Select **Place Component** > then choose which database you wish to open. This is almost always Multisim's DB. You can also just hit **Ctrl+M** instead.
3. Using the Component Family list box, choose in which family the part you want to place is contained. See Figure 3.18. For example, to lay a 2N2222 transistor, you would select **BJT_NPN**.
4. Choose **2N2222** from the Component list below and click the **OK** button.
5. Place the component onto the drawing with the subsequent pointer.

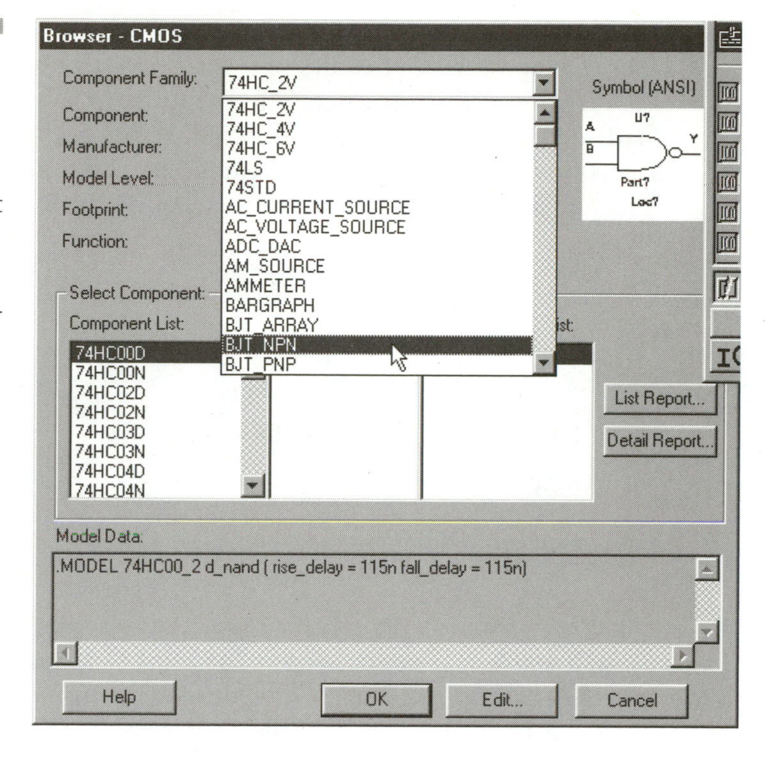

Figure 3.18
Yet another way to lay parts is to hit **Ctrl+M** then select from the Component Family scroll-down list and component list. Hitting **OK** lays that component.

Add a ground to the screen first. This makes grounds available faster via the In Use list.

Place a few components onto a blank drawing. Throw in some instruments. Get an idea how the Component Browser is used, as well as using some virtual parts. Don't worry about wiring or moving them around. Once you have too many parts on the screen and can't tell what is what, start a new drawing. Repeat until you are comfortable with placing parts and instruments.

Selecting Components, or Instruments, and Labels

To move or alter a part, instrument, or label, you must first select it. There are several ways to select one part or a group of components.

To select one component, place the mouse pointer over the schematic symbol of the part you want to select (see Figure 3.19). Click the left mouse button once. The component is surrounded by four small squares, indicating it is selected. If the mouse pointer is placed directly over the selected component its specs are displayed on the right side of the status bar, at the bottom of the screen.

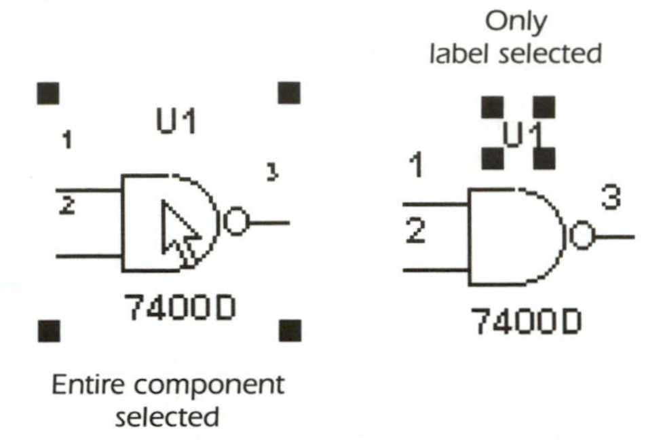

Figure 3.19
Selecting a component or label.

CAUTION

The labels of a component can be selected and moved independently from the part symbol itself. Be careful not to select the labels of a component and move those by mistake; it's a pain to get them back in place. Instead, click the component diagram itself. This selects the entire component, labels and all.

If you have selected one component and now want to select addition ones, hold the Shift key down and left-click onto the other components to add to that group (Figure 3.20). Each selected component now has the squares around it. If you mistakenly selected the wrong component and don't want to start over, just click again on the component you wish to deselect.

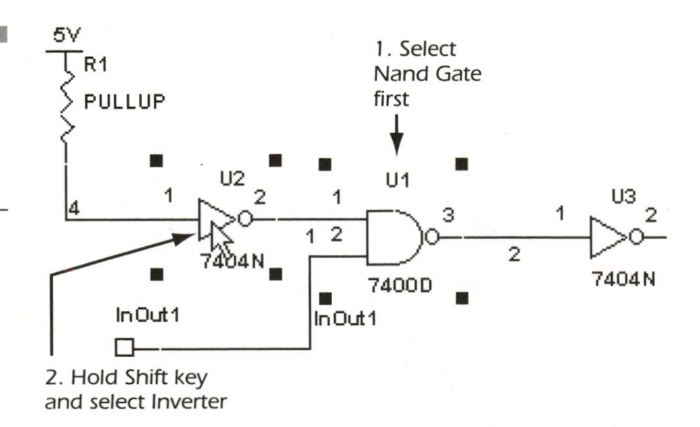

Figure 3.20
Selecting multiple components using the Shift key and mouse pointer.

Multisim also lets you surround a group of components and select them all at once. To do this, place the mouse pointer in an empty area at the top left of the section you want to select. Click and hold the left mouse button. A dotted box appears as you move the mouse. Move it to the bottom right corner of the area you are selecting and release the button. Each selected component is surrounded by four squares (see Figure 3.21). If you want to add parts to the selection, hit the Shift key and click the component you wish to add.

With EWB5, you only had to touch any section of the component while surrounding it to select it. With Multisim, you must surround the entire component, including its labels or it won't select.

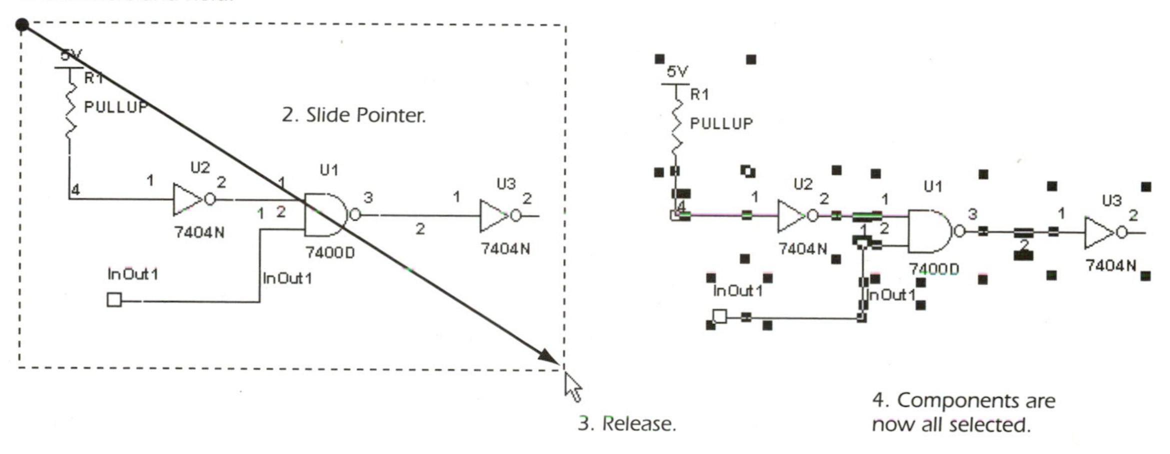

Figure 3.21 Selecting multiple components by surrounding an area.

To select the label, just place the pointer over the label's text and left-click. The four squares will now be around the label. See Figure 3.19 again. As I've said, the labels of each component can be moved independently from the parts themselves.

A minor annoyance about Multisim is that it deselects a component once placed. EWB V5 kept the component selected after placement, allowing quick changes or movement.

If you don't know what a placed component or instrument is, move the mouse pointer over it and see what it says in the Status Line at the bottom of the desktop.

Deselecting Components

Left-click a blank area in the circuit window to deselect. If you want to deselect only one component in a group of selected components, hit the **Shift** key, then left-click that component.

Moving Components and Labels

You can move a selected component or just grab the component and move it, skipping the selection step. As we said before, Multisim allows you to move component labels independently of the component diagram itself. This allows you to further customize and control the schematic.

To move a single component, place the mouse over the schematic symbol itself (not the label). Click the left mouse button and hold. Slide the part to where you wish to place it. Release the button.

To move multiple components, select components you wish to move. Place the pointer over any of the schematic symbols selected. Left-click and hold; drag and drop. See Figure 3.22.

To move a label, left-click over the text of the label and drag to new location (Figure 3.22).

You can also "bump" a selected component one grid length one way or the other. Just hit the appropriate arrow key after the component is selected.

All connections remain hooked up when parts are moved.

Just as in drawing programs, Multisim has a feature that lets you move a component along only one axis (horizontal or vertical). After you begin to move the component, hit the **Shift** key. The part can

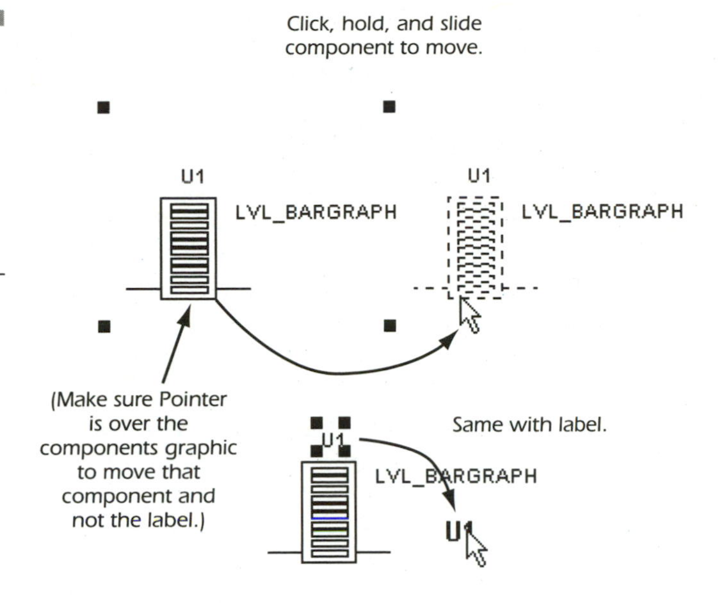

Figure 3.22 Components or labels are dragged into place with the mouse. Remember, labels move independently of the actual component.

then only be placed in the same horizontal or vertical plane. This is handy when lining up connectors for clean wiring.

EXERCISE

Place six components onto the screen. Select and move one of the components. Try moving around the labels to that part. Now select three of the six components. Move those around. Try the Shift key method of selecting an additional component.

Making Copies of Components

Sometimes it is necessary to make a component facsimile to use in another section of a drawing. Copying eliminates the hassle of laying a new component and changing the settings again and again. There are two commands used. *Cut* erases the item from the drawing but leaves a copy of the component and settings on the Window's clipboard. *Copy* merely places a carbon in the clipboard, leaving the original on the drawing. There are two ways to cut or copy a component.

NOTE

The Clipboard is a temporary storage area for components. The contents can only hold one component or one group of components at a time. When you cut or copy something else, only old part is erased.

Using the menu method, select the component(s) you wish to copy. Go to the **Edit** menu. Choose **Copy** or **Cut**. (or hit **Ctrl+X** to cut, **Ctrl+C** to copy). Choose **Edit** > **Paste** (or hit **Ctrl+V**). A copy of that part is stuck to the pointer. Move it into place and left-click. Some components automatically run a wire at this point. You can also use the icons on the toolbar that execute the cut, copy, and paste commands.

To use a quicker method, eliminate the Select step. Move the mouse pointer over the component. Right-click the component. A menu opens. Choose **Copy** or **Cut**. Place the cursor where you want the new component. Right-click and choose **Paste**. The component sticks to the pointer (moves with it). Drag the component where you wish to place and left-click.

NOTE

One handy feature about Multisim is that when you cut, copy, and paste a component, the new part's reference ID advances one number as well. So copying R1 creates R2.

TIP

You can now copy an Instrument, as well—a new feature in Multisim.

Deleting

Now that you have placed 20 parts to test the interface (four batteries, ten grounds, an oscilloscope, and parts that you have no idea what their functions are), it is time to learn how to delete them.

To delete a single component, select the component. Hit the **Delete** key. You can also go to the menu item **Edit** > **Delete**.

To delete a group of components, **Select** the chosen components and hit **Delete**.

Undeleting

Finally! Multisim lets you undelete items eliminated by mistake. Either choose **Edit** > **Undo** or hit **Ctrl+Z**. Beware: some wiring may be lost, requiring you to rerun each connection.

Place a couple of components onto the screen. Make a copy of one part and paste it somewhere else. Try cutting a component from the drawing. Now delete a few of the other components. Hit ***Ctrl+Z*** *and see if the components return to the drawing.*

Orienting a Component

Turning or repositioning a component to orient it better can make a drawing that's more aesthetically pleasing and readable. You can perform this with or without a wire connected to it (refer to Figure 3.23).

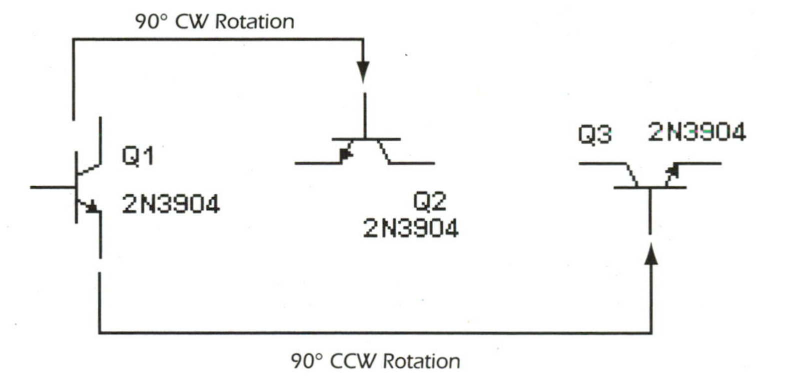

Figure 3.23 Rotating a component.

Some items, such as certain ICs and most of the indicators, switches and electromechanical devices cannot be rotated in the current version of Multisim even though the option is not grayed out. I am told rotation will be available in version 6.2.

Rotating a Part

To rotate a part select the part. Choose **Edit** > **90 Clockwise** or **Edit** > **90 CounterCW** from the menu. The part turns 90 degrees clockwise or counterclockwise. You can also hit **Ctrl+R** or **Ctrl+R+SHIFT** instead of going through the menus.

To perform a quick rotate, right-click on top of the component and choose either **90 Clockwise** or **90 CounterCW**.

Rotating a part may require repositioning a label.

Flipping a Part

Mirroring a part along the x-axis or y-axis is simple (see Figure 3.24).

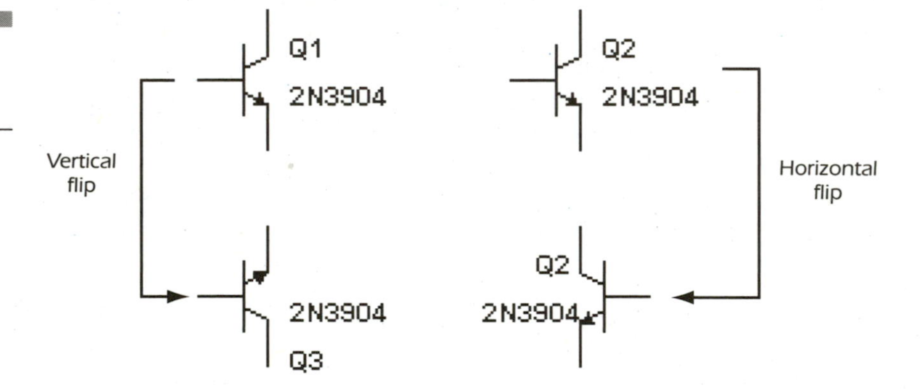

Figure 3.24 Flipping a component.

To perform a vertical flip, or mirror the part along a horizontal axis, select the part. Then use **Edit > Flip Vertical** or hit **Alt+Y**. To perform a quick vertical flip, right-click the component and choose **Flip Vertical**.

To perform a horizontal flip, or mirror the part along a vertical axis, select component, and choose **Edit > Flip Horizontal**. You can also hit **Alt+X**.

To perform a quick horizontal flip, right-click on the component and choose **Flip Horizontal**.

Place a few components onto the screen. Rotate and flip them.

Replacing a Part

When you delete a component, the wiring that connects that part into the circuit is lost. If, for example, you placed an ideal transistor into a circuit and want to replace it with a 2N3904, you can either delete the component (losing the wiring) and place a 2N3904 then rewire it into the circuit, or you can do the following (see Figure 3.25):

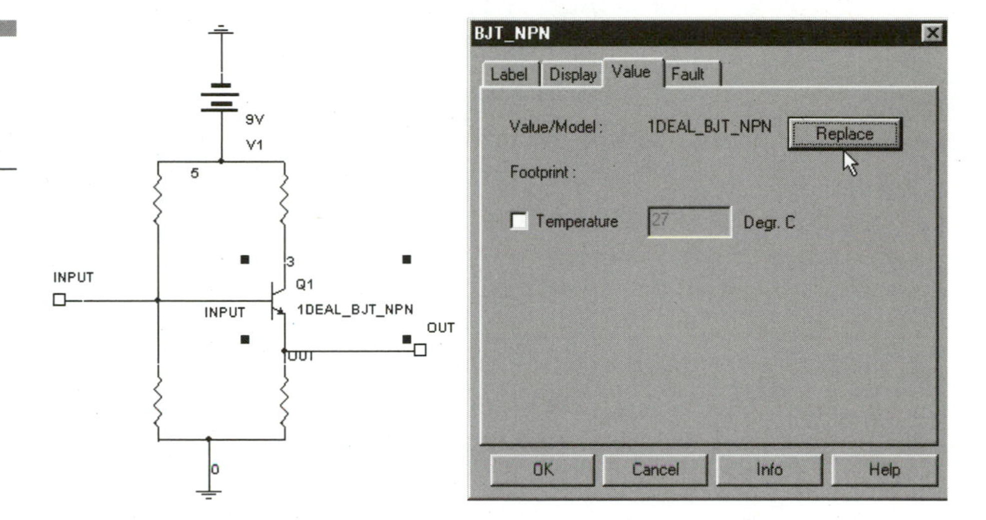

Figure 3.25 Replacing a part without having to rewire.

1. Select the ideal transistor.
2. Go to **Edit** > **Replace Part**. The browser window opens.
3. Select the **2N3904** and hit **OK**. Try it!

Another way is to double-click the component, select the **Value** tab, and hit the **Replace** button.

Multisim Wiring

Multisim's various components and instruments must be wired together to create electrical schematics and simulations. Each component contains connection(s) to which a wire can be attached and stretched to another connection. Fast wiring techniques make for speedy diagram creation. Multisim uses a more flexible wiring method than previous EWB versions, but wiring components is still a breeze.

Multisim's Manual Wiring

Previous versions of EWB had a less-than-perfect autowiring feature. Solving the spaghetti of wires in a complex circuit was, at times, trying on the eyes and mind. New to Multisim is the ability to lay wires *exactly* where you want them. You can still autowire, but now you can also draw a line with as many corners or angles as you want. You can even edit the lines after they are placed. This allows you to place the entire wire in a configuration where you wish it to go, overriding the program. Refer to Figure 3.26 and follow these steps:

1. Move the pointer over the part's terminal until a cross-hair (+) appears.
2. Click the mouse pointer then release; a dotted line appears.
3. Move the pointer in the direction you wish the wire to run and click the mouse button. The wire is still connected to the mouse pointer.
4. Move the mouse to the next point. Do this for each new corner you wish to create.
5. Left-click the button one last time when the pointer is on top of the other component's connector.

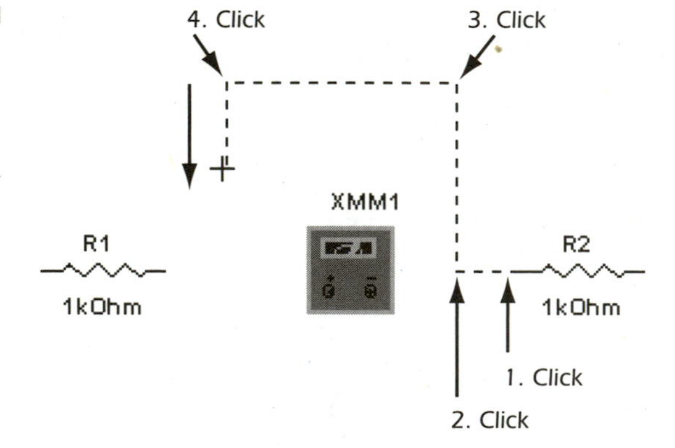

Figure 3.26 Manually wiring components together.

CAUTION

Multisim will autowire a component that has been moved even though you have manually wired it into the circuit. This requires editing the wire after Multisim does its thing.

TIP

Right-click the mouse button to release the wire from the mouse pointer if you make a mistake while running the wire. Start over.

Autowiring a Component

1. Move the pointer over the part's terminal until a cross hair (+) appears.
2. Click the left mouse button and release. A dotted line appears (see Figure 3.27).
3. Move the pointer to the other component terminal and click the left-button on top of that connection.

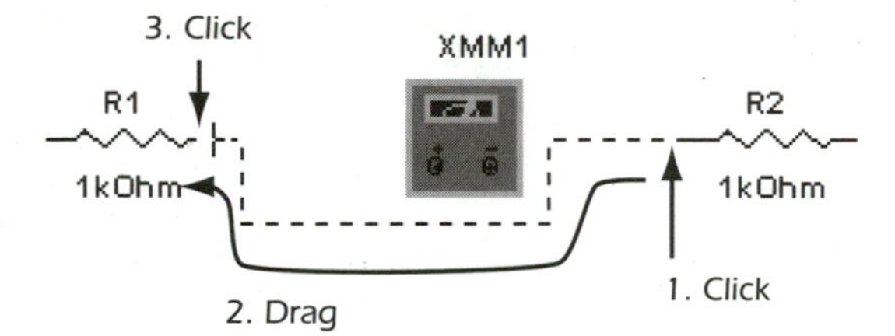

Figure 3.27 Using Autowiring to connect parts.

EXERCISE

Place two components onto the screen. Make a connection manually between two of the connectors with a few corners added in. Delete that wire. Now try the autowiring feature, running the same two connections.

Adding/Subtracting Wire Points

Multisim lets you perform *post operations* on the wire; you can add or delete any point on the selected line. Refer to Figure 3.28.

To delete points, select the wire by left-clicking anywhere along its length. Place the pointer over the corner you wish to delete. When you hit the **Ctrl** key, a box with a minus sign appears. Left-click to delete that point.

To add points, select the wire by left-clicking anywhere along its length. Place the mouse over the point in the wire where you wish to add a new corner. When you hit the **Ctrl** key, a box with a plus sign appears. Left-click to add that point.

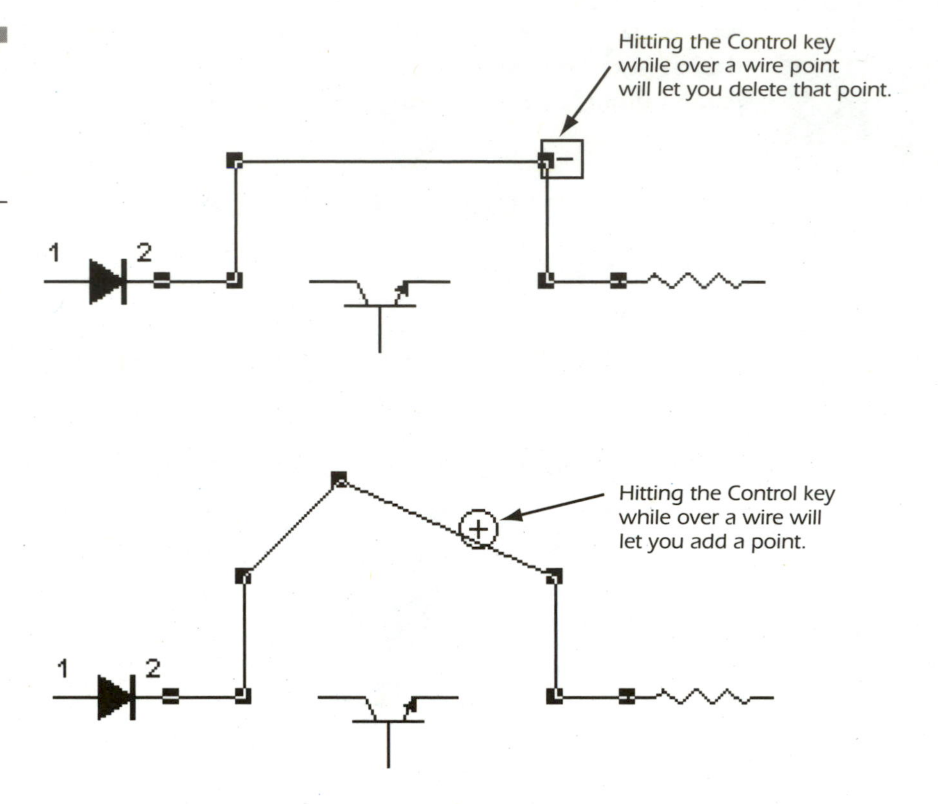

Figure 3.28
Use the Ctrl key and mouse pointer to add or delete points from a wire.

Changing the Wire's Shape or Color

You can also move any point on the line independently from the other points to create angled lines instead of those boring 90-degree scratches. You can also change the color of the wire for instrumentation signal identification or color coding.

To move any point on the line, select the line. Move the pointer over the corner you wish to move. Left-click, hold, and move the point to a new location. Release the button. See Figure 3.29.

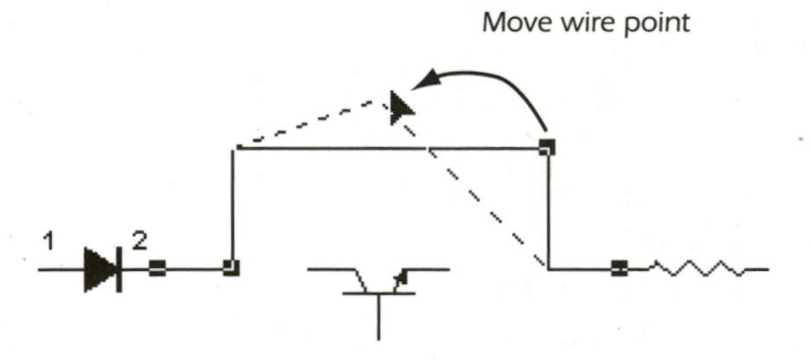

Figure 3.29
Moving line points.

To change wire color, right-click the pointer over the line and choose **Color...** The color selection window opens. Left-click on the color you wish to use and hit **OK**. See Figure 3.30.

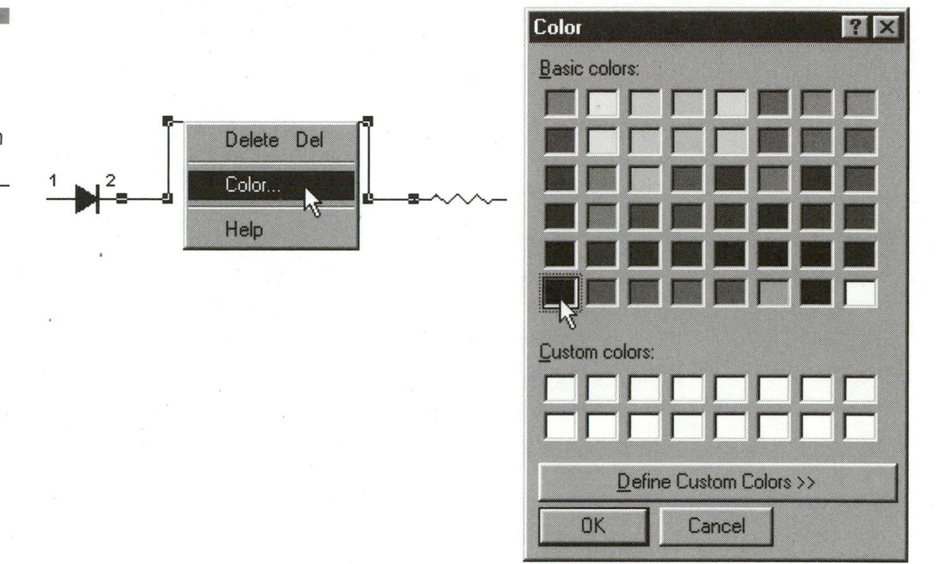

Figure 3.30 Changing the color of a wire can help identify signals on an instrument.

The wires connecting to instrument connections can be colored. This color is the same as the waveform on the instrument's graph. It helps to identify which signal is coming from which wire; for example, if you color a signal wire red, the waveform seen on the oscilloscope appears as red.

Deleting and Undeleting a Wire

If a trace is incorrect, you can delete it. Either select the wire then choose **Edit** > **Delete** or hit the **Delete** key. A quicker way is just to right-click over the wire and choose **Delete**. See Figure 3.31.

Place a few components onto the drawing board. Run a manual wire with three or four corners along its length. Delete one of the corners. Undo the delete. Now add a point in a different location. Move the point (corner) around with the mouse and see the shapes that can be created. Color the wire red.

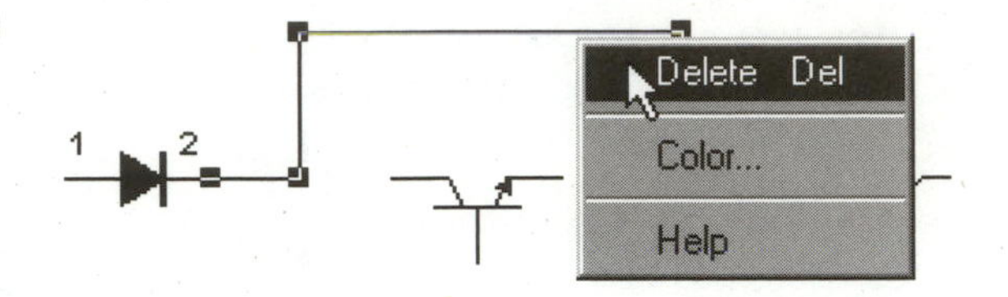

Figure 3.31
To erase a wire, select it and hit the Delete key or right-click over the wire and choose **Delete**.

Inserting a Component into a Prelaid Wire

It is possible to place some components, such as resistors and capacitors, in the middle of a wire and make the terminals connect automatically. Just select the part and drag it over the wire. Make sure it is oriented along the same axis. If not, rotate it and Multisim will automatically run the wiring. See Figure 3.32.

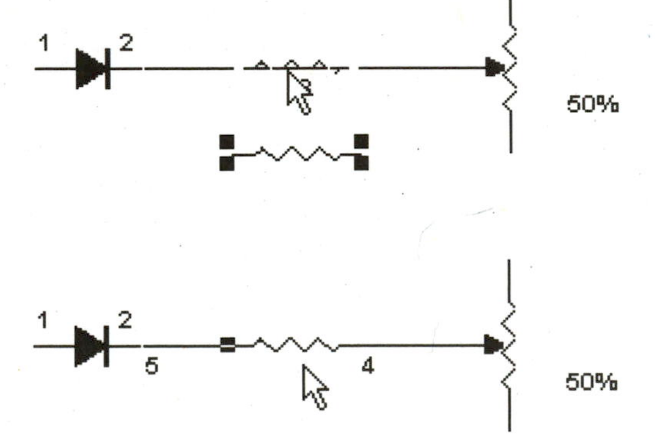

Figure 3.32
Forcing Multisim to automatically wire a component into a circuit.

Creating a Connector Anywhere Along a Wire

A *connector* (junction) can be created anywhere along a wire without your having to drag it from the tool bin. If you are already running a wire and want to connect it to the center of the wire, just move the tip of the mouse pointer onto the line and release. See Figure 3.33.

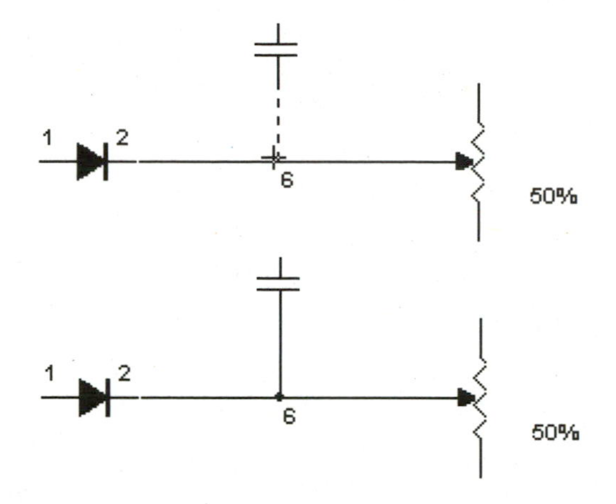

Figure 3.33 Create a connector (junction) in the middle of a wire by running another wire to it.

Notes on Nodes and Connectors

A *node* is a common reference point in the circuit. If five components are connected to each other, each wire running to the parts will have a common node (see Figure 3.34). Multisim automatically names the node with a number. However, you can change the node name by double-clicking any of the wires that make up the node and adjusting the Node Dialog Box. You will see later (in Chapter 6, "Simulation and Analysis"), why these nodes are essential to your drawing.

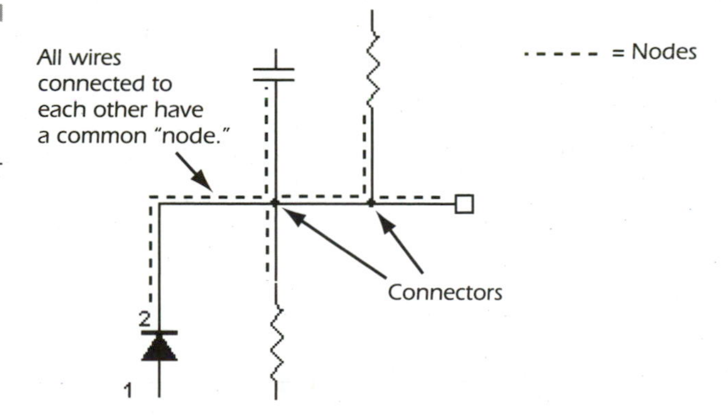

Figure 3.34 Don't confuse a connector for a node.

Connectors versus Junctions

In writing this book, I use several words to describe a connector. Multisim's official name is *junction,* which is contrary to EWB 5's *con-*

nector. Do not mistake a node for a junction, because a node can contain many junctions. A junction is merely a drawing feature to let you connect up to four wires for clean connections.

Labels, Values, Node Names, and Reference IDs

Multisim can display each component, node, or instrument's label, value, or reference ID. The *label* is the name you wish to give that part. The *value* is the component's electrical value, and the alphanumeric *reference ID* names the part for schematic reference. The node name is the name or number given to that conductor in the circuit. See Figure 3.35.

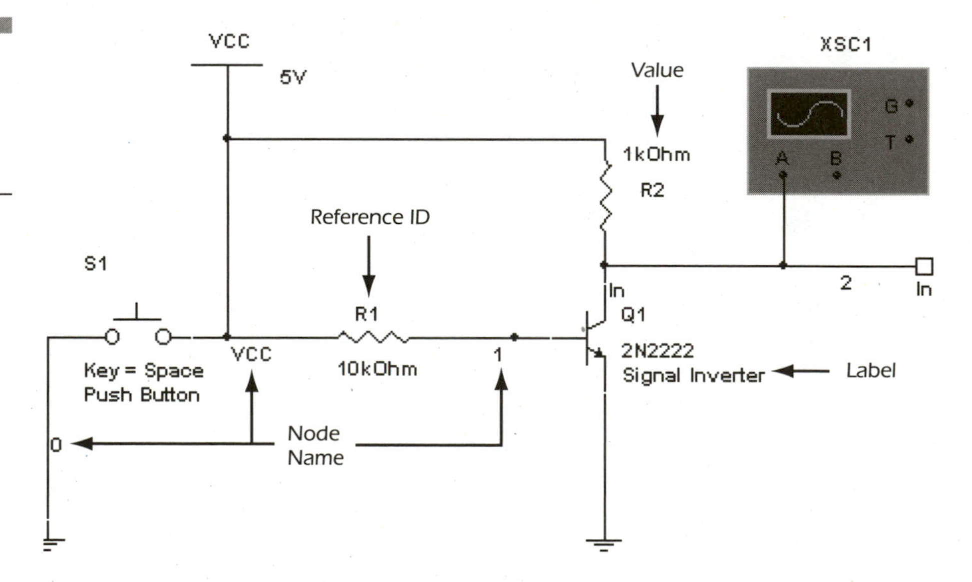

Figure 3.35
Vcc, 1, 2, and 0 (GND) are examples of nodes in this simple circuit.

Change a Component Label

You can label the component with any name you wish. Double-click the component and select the Label tab. Type in the name you wish to give the component inside the Label box. Hit **OK** (see Figure 3.36).

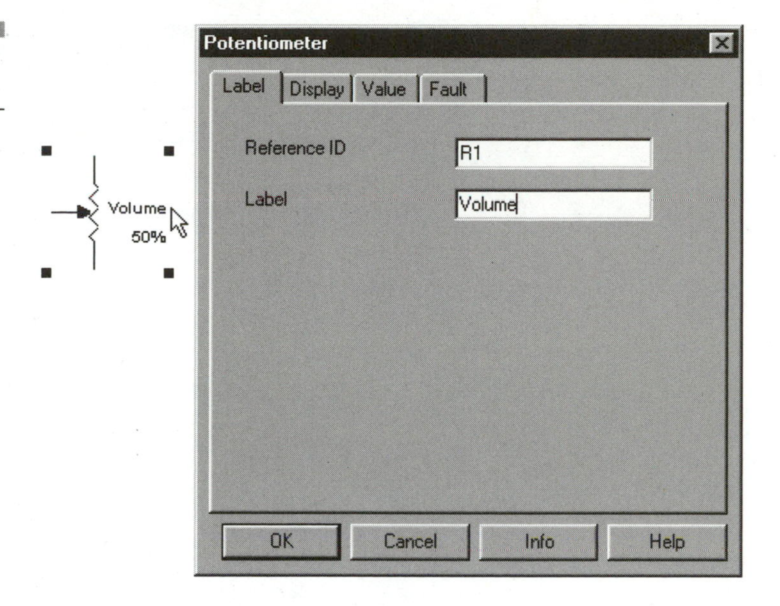

Figure 3.36
Changing Labels.

Change Value

If you have placed a virtual component, you can change the electrical value by double-clicking it and choosing the Value tab. Type the value into the box and hit **OK**. If the component was inserted from a database, you only have the option of replacing it in the circuit with a new one. Hit the **Replace** button, choose the new component, and hit **OK**. See Figure 3.37.

Change Reference ID

The reference ID gives a generic alphanumeric name to each component (see Figure 3.38). For example, if there are five resistors and two capacitors in a circuit, then the first resistor would have a reference ID of R1. The second resistor would be R2, etc. Capacitors would likely be C1, C2, etc. Multisim assigns a reference ID to each component automatically, but they can be changed. You can make this reference ID anything you want.

Double-click the component and select the Label tab. Type in the new reference ID and hit **OK**. If you use an ID name that is already in the circuit, Multisim refuses the name.

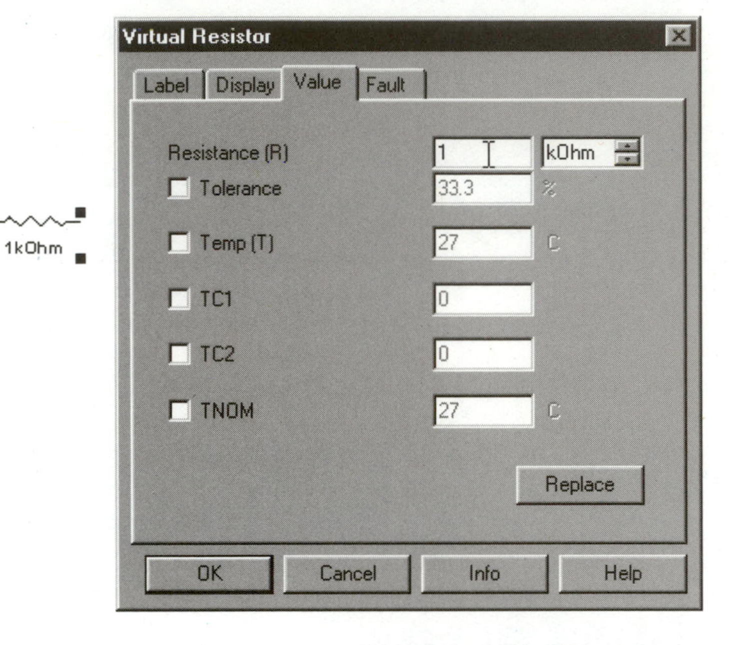

Figure 3.37
Changing the value of a component.

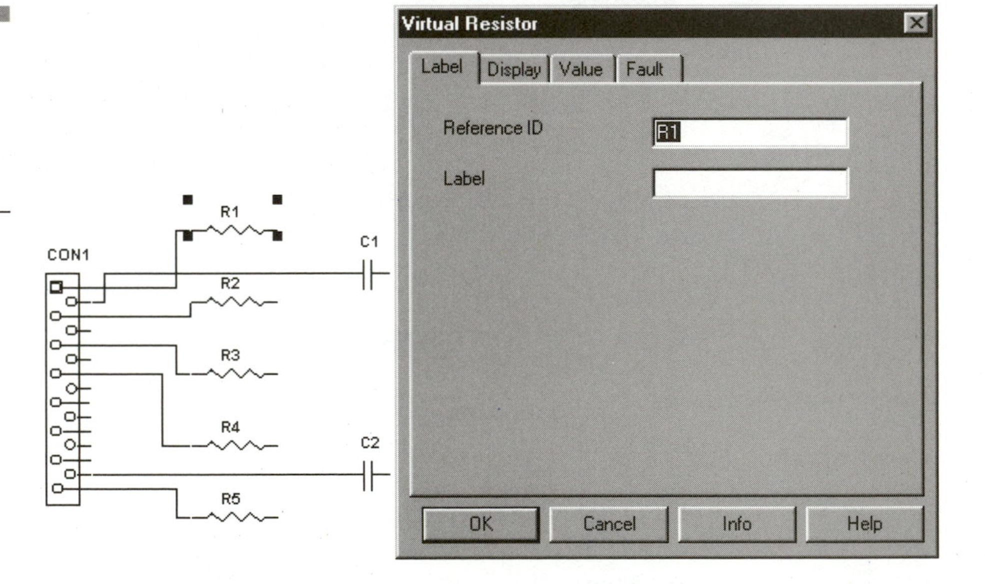

Figure 3.38
Reference IDs (R1, C2, CON1) help to further classify and identify components in your circuit.

TIP

If you want to rename C2 to C1 and C1 to C2 you must first rename C1 to something else like C-T. Then rename C2 to C1. Finally, rename C-T to C2. Whew!

Change Node Name

A name or number can identify each node. This is sometimes helpful when debugging a circuit and identifying connection points inside simulations or analyses you are running. Multisim automatically assigns node names, but these can be changed. Double-click anywhere along the wire that the node represents and the Node dialog box opens (see Figure 3.39). Type in the new name or number and hit **OK**.

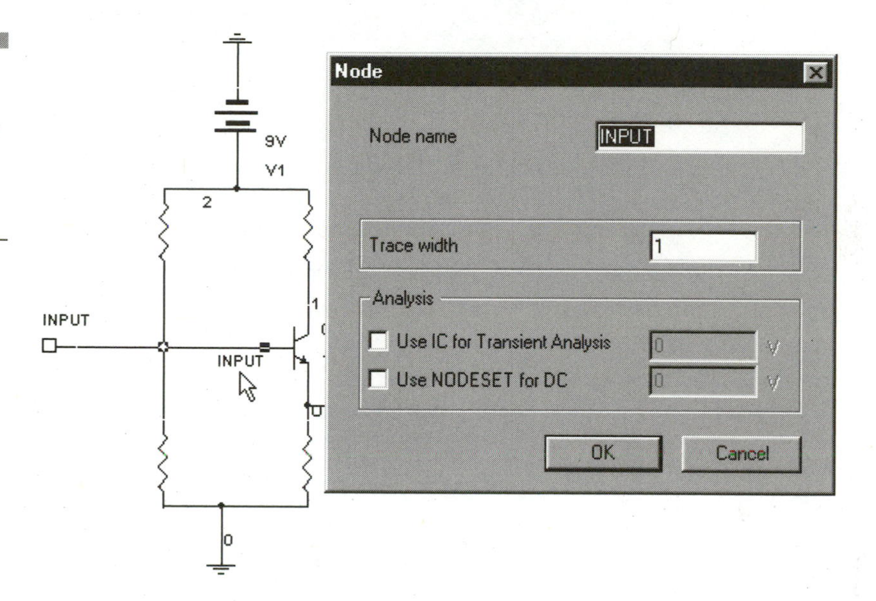

Figure 3.39 You can change a node's name instead of using what Multisim thinks it should be.

Using Global Settings

Global settings are settings that are applied to the entire drawing on which you are currently working, including which labels to show on the components. In other words, you can choose to display all the reference IDs, values, labels, and node names or any combination of them through the whole drawing. Refer to Figure 3.40.

1. Right-click a blank area of the drawing and choose **Show**.
2. Check on those items you want displayed with the components.
3. Hit **OK**. The settings are applied to each component in the circuit.

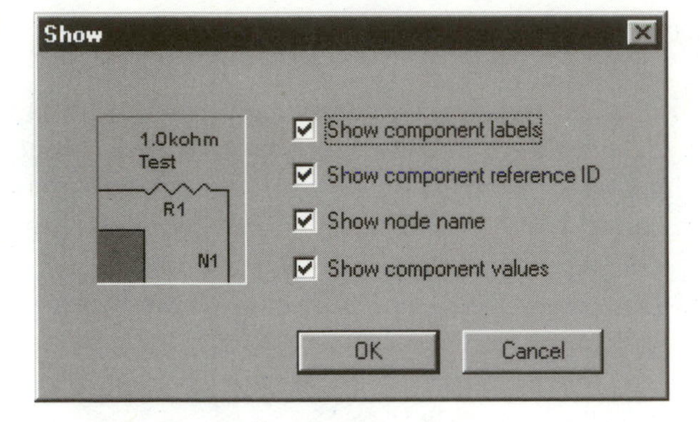

Figure 3.40
Global settings tell Multisim to place each of the chosen labels onto this drawing.

Overriding Global Settings

You can override the global settings for any separate component in the circuit to display only the information you want for that specific component while leaving the rest of the component labels intact. Refer to Figure 3.41.

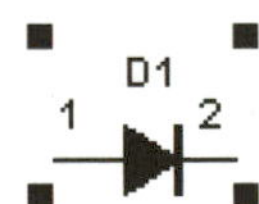

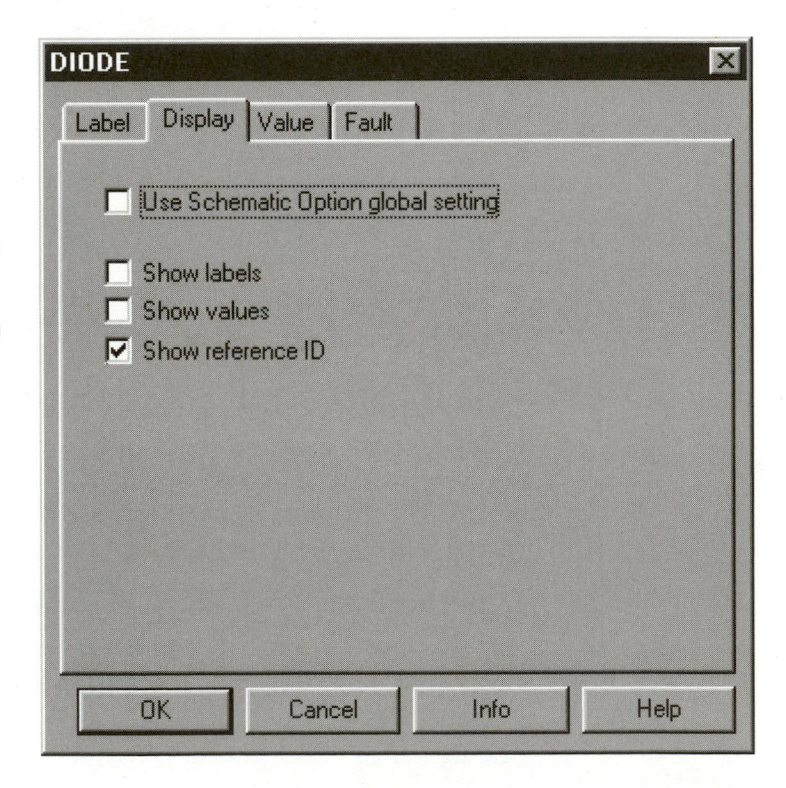

Figure 3.41
If you don't want a specific part in a circuit to use the global display settings, you can override it.

1. Double-click the component you wish to change.
2. Choose the Display tab.
3. Deselect the **Use Schematic Option global setting**.
4. Check on those elements you want displayed below that: show labels, values, reference IDs.

Text

Adding miscellaneous text is sometimes required to customize a drawing for printouts. You can pencil in personal notes, further labeling, or messages to your drawings.

Placing Text

You can position text anywhere within the drawing (Figure 3.42).

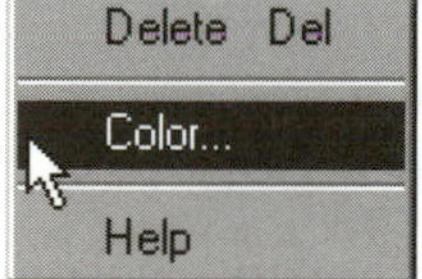

Figure 3.42 Placing text into a drawing helps to further clarify sections of the circuit or add your comments. You also have the ability to change the text color.

1. Choose either **Edit** > **Place Text** or right-click where the text is to be placed and choose **Place Text**. Hitting **Ctrl+T** also works.
2. A large capital **I** appears. Move it to where you want the text placed and left-click.
3. Type in the message or characters as you would in a word processor.
4. Click somewhere outside the text box to set the text.
5. Move text into place if required.

Changing Color of Text

To change the color of a pre-placed block of text, right-click on the text and choose Color (see Figure 3.42). Click the mouse pointer into the color box you wish to use and choose OK. You can also make a custom color by clicking into one of the empty white squares in the lower section of the dialog box and choosing **Define Custom Colors**. You can change the Red/Green/Blue or Hue/Sat/Lum balance and then hit **Add to Custom Colors** to apply the color.

As of Multisim 6.2, you can no longer adjust the text's font or size; you're stuck with what Multisim gives you (Arial 10, bold).

Advanced Schematic Capture

The next sections cover a few complex drawing procedures. Do not attempt them if you are totally unfamiliar with Multisim.

Virtual Wiring

If you are a Pro or Power Pro user you can use Multisim's new feature, virtual wiring. This lets you run a conductor across a drawing without any physical line appearing in the circuit. In other words, the wire is there, you just can't see it. This is handy if the only way around a component is straight through it. Here's how to perform this trick (refer to Figure 3.43).

1. Place the two components onto the drawing in their appropriate areas. These are the components you wish to connect virtually.
2. Place a junction (connector) slightly away from one of the component's terminals by hitting **Ctrl+J** or right-clicking and choosing **Place Junction**.
3. Connect the junction point to that component's terminal with a wire.
4. Repeat this procedure for the other component.
5. If you do not have node names selected, double-click the wire connecting the first component to its junction and note the name (usually a number). Change it if you wish.

6. Now double-click the second component's wire, to which you wish to connect virtually and call the node the same thing. Multisim will say "Node with the same name already exists. Continue?" Choose **Yes**. Components are now connected even though there is no visible wiring between them.

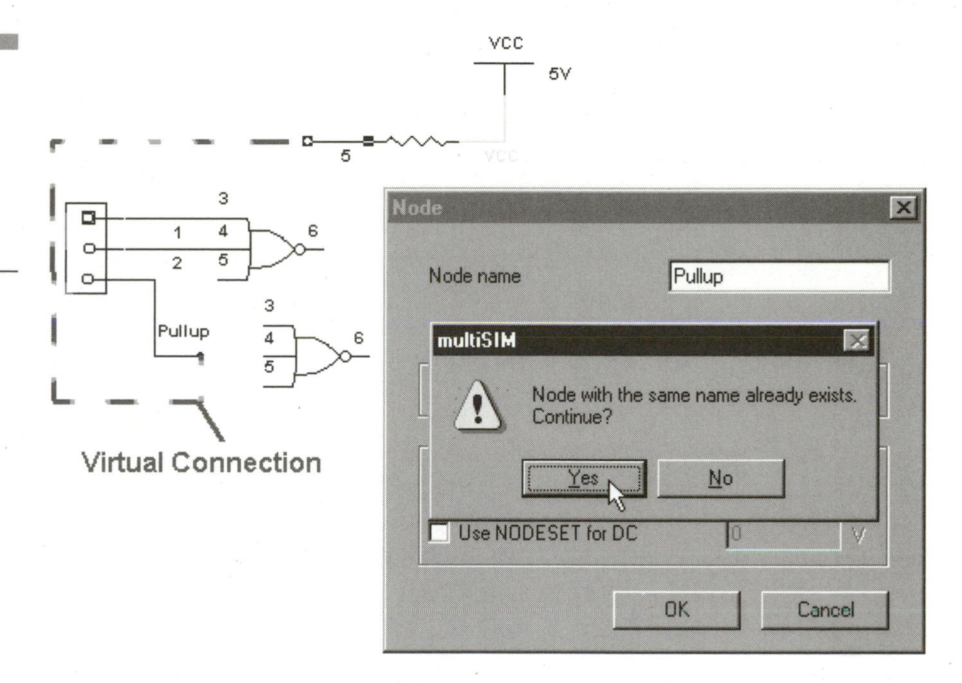

Figure 3.43 Virtual wiring can create connections without wires. However, it won't work with Personal Edition.

Bus Wiring

You can lay a *wiring bus* with Multisim, which packs a group of wires into one bold wire. You can then connect component's terminals anywhere along the bus to any line of the bus. This is handy for designing digital circuits that contain a data or signal bus containing loads of wiring (refer to Figure 3.44). Here's how to use it:

1. Choose **Edit** > **Place Bus**, right-click, and choose **Place Bus**, or hit **Ctrl+G**.
2. Cross-hairs appear. Click once over the point at which you wish to start the bus.
3. Move the cross hairs to where you want the second point and left-click again.

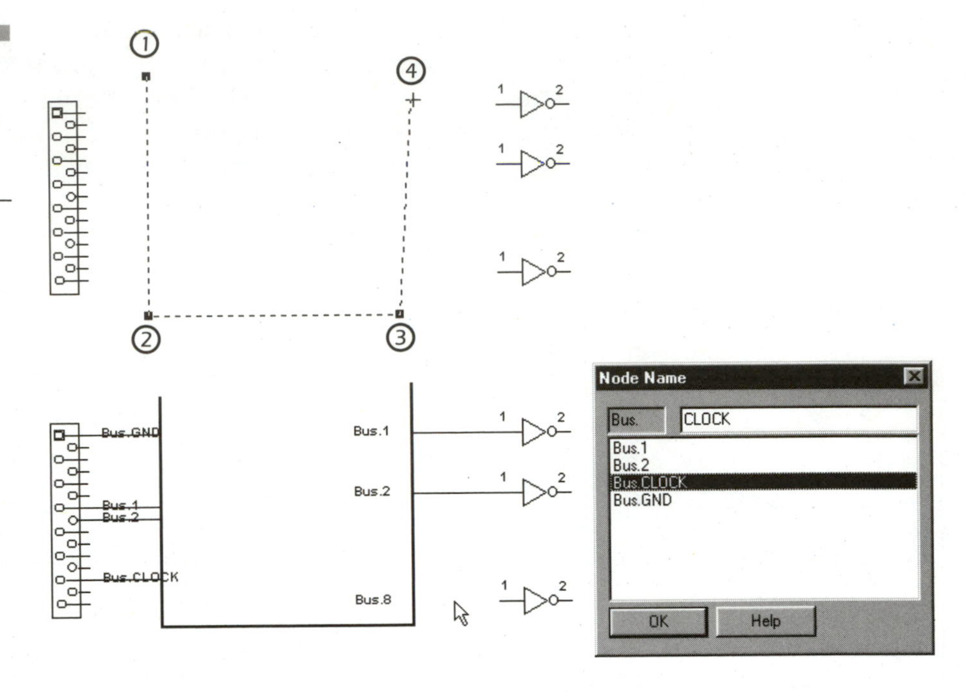

Figure 3.44 Placing a bus can save having to draw 50 lines that resemble spaghetti.

4. Place as many points on the bus as you want and, when finished, double-click on the last point.
5. To place items onto the bus, run a wire from the component's terminal to any point along the bus line and left-click. A window appears, which allows you to type in a number or name. For example, if you wanted to hook a point to the Clock signal, click the component, move and click over the bus line, and type **CLOCK**, then **OK**. If you place another wiring to the bus somewhere further along the line and name it **CLOCK** also, those two parts will be interconnected.

CAUTION

You cannot edit a bus line once it is laid down.

Subcircuits

"Circuits within circuits within..." This embodies the concept of subcircuits. They are timesavers when you must repeatedly use the same circuitry. Simply recycle saved circuits by making them into subcircuit components. Components include all the modeling, schematics, and values set in by you. You merely hook up the wiring of the new circuit and let it fly.

To create a subcircuit from an existing circuit, follow these steps (refer to Figures 3.45 and 3.46):

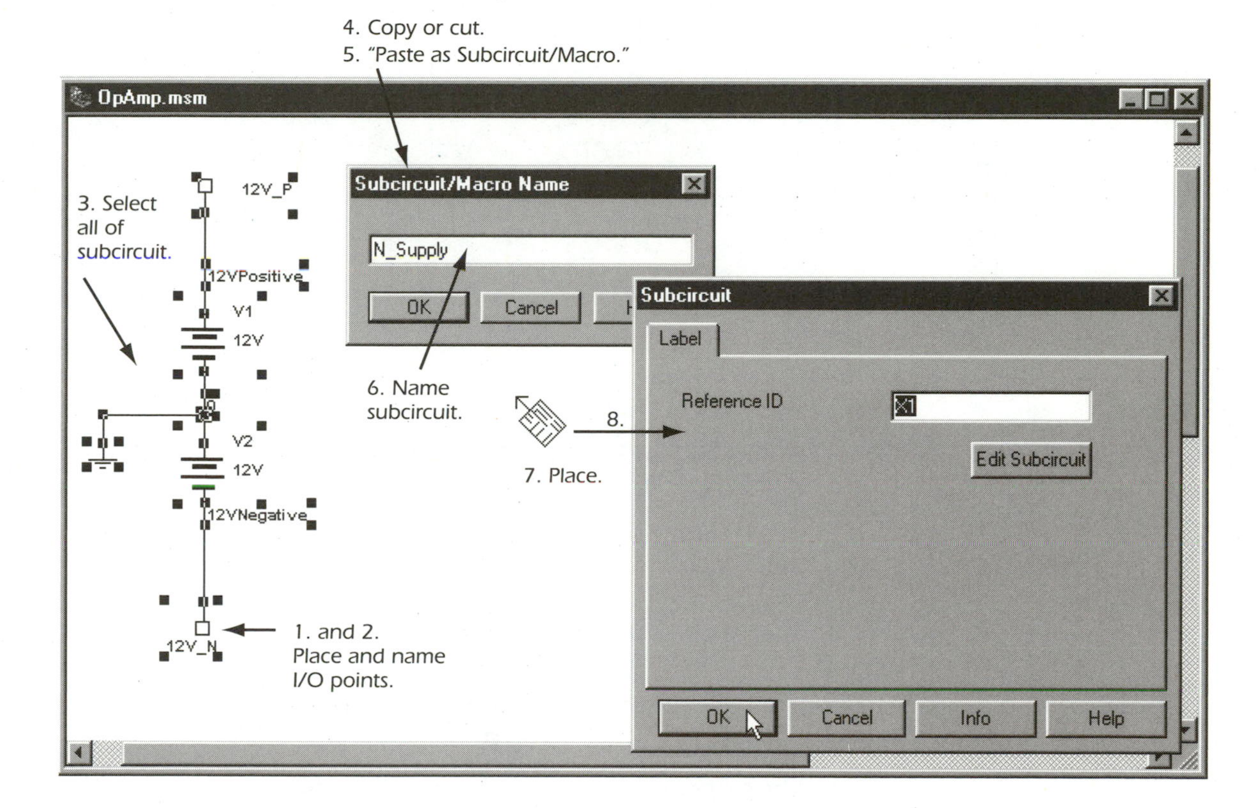

Figure 3.45 Creating a Subcircuit using Cut/Copy and Paste as Subcircuit/Macro.

1. Place input/output points at each area of the circuit that requires an external connection. (Right-click and choose **Place Input/Output** or hit **Ctrl+I**).
2. Call the I/O pins something familiar when prompted.
3. Move the mouse pointer into the upper left corner of the circuit and select the entire circuit you wish to make into a subcircuit.

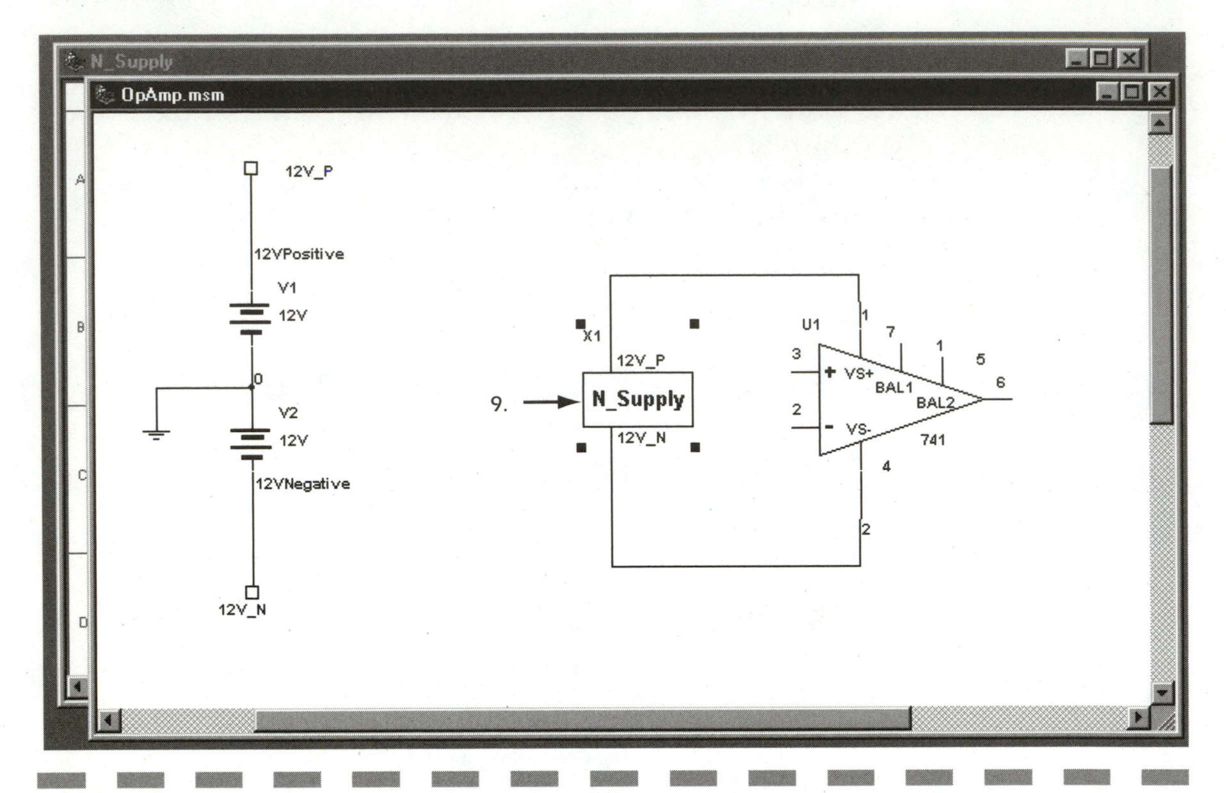

Figure 3.46 Using the subcircuit within your circuit.

4. **Cut** or **Copy** the circuit to the clipboard by using **Ctrl+X** or **Ctrl+C**.
5. Right-click onto a blank area of the circuit and choose **Paste as Subcircuit/Macro** or hit **Ctrl+B**.
6. A dialog box opens, asking you to name the subcircuit. Type in a logical name without spaces. I recommend making it as short as possible or the subcircuit takes up too much room on the drawing. Hit **OK**.
7. The mouse pointer turns to one that looks like you are placing a circuit. Click it onto the screen where you wish to place the subcircuit.
8. Another dialog box opens, asking you to set a reference ID. I usually just hit **OK**; the subcircuit is placed.
9. Make the circuit connections to the pins that you created (input/output pins).

10. If you wish to edit the subcircuit, double-click it and hit the **Edit Subcircuit** button. The subcircuit opens in a new window.

Printing Electronics Workbench Files

Have computers created a paperless society? Are you kidding? If anything, the have increased the amount of paper output from organizations. We still require printouts of circuits for fast reference and a more reliable record than that provided by the typical computer hard drive file. Multisim allows you to print schematic drawings, subcircuits, a bill of materials needed to build the circuit, a list of components in a certain family, and graphs of the instrument's readouts.

NOTE

Multisim Demo prevents you from printing circuits.

Printer Setup

When you select **File** > **Printer Setup**, the Printer Setup dialog box appears. This may be different from that shown in Figure 3.47, because each printer's setup is unique. To select a printer, choose it from the list box on the top. If you wish to adjust the printer's properties, hit the Properties button and make the appropriate settings according to your printer's manual or help file.

You can then select the size of the paper, its orientation and, in the case of my laser printer, the tray from which the circuit will print. From here, you can access the Page Setup dialog box by hitting the button at the bottom of the window.

CAUTION

You cannot print a bill of materials with the Personal Edition.

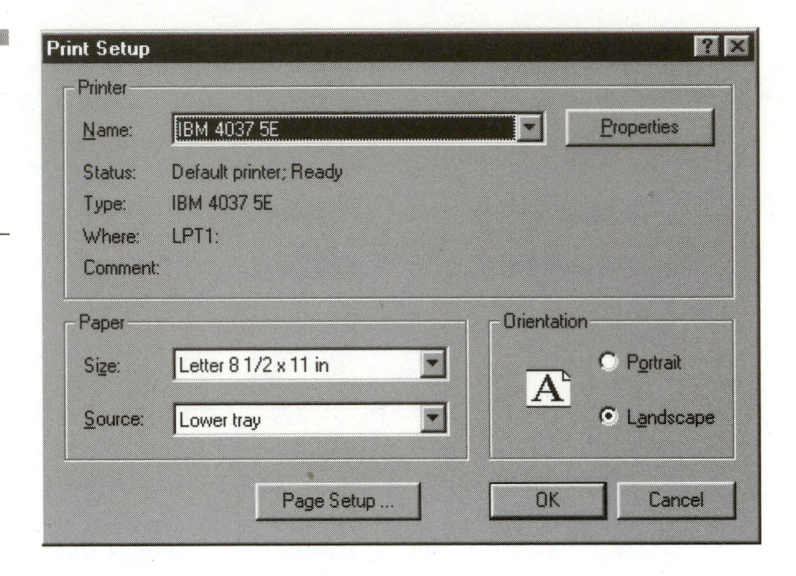

Figure 3.47 The Printer Setup dialog box tells the Windows printer exactly what to print and how it will look.

Printer Page Setup

The circuit must be set up before outputting onto paper. You may remember that the User Preferences dialog box will not affect the current circuits setting; to set up a page, you must do the following:

1. Choose **File** > **Print Setup** > **Page Setup** button. See Figure 3.48.
2. Margins are set to what Multisim thinks will be appropriate for your printer. You can edit these at this time if you wish.
3. The zoom level can be adjusted by choosing that tab and clicking the zoom level you wish. I recommend leaving it set to Fit To One Page.
4. The options tab is described in the next section.

Optional Settings

IS YOUR PRINTER COLOR OR BLACK AND WHITE?

Multisim can print a circuit in either monochromatic black and white or in a vibrantly descriptive color mode. If you have a color printer and want the printout of your circuit to be in full color:

1. Choose **File** > **Print Setup** > hit the **Page Setup** button.
2. Choose the **Options** tab.

Page Setup

Margins | Zoom | Options

Top : 0.3
Bottom : 0.3
Left : 0.7
Right : 0.3

Measurement Units
Inches
Centimeters

OK | Cancel | Help

Page Setup

Margins | Zoom | Options

Magnification
Fit To One Page
140%
100%
75%
50%
Custom settings : 100 %

OK | Cancel | Help

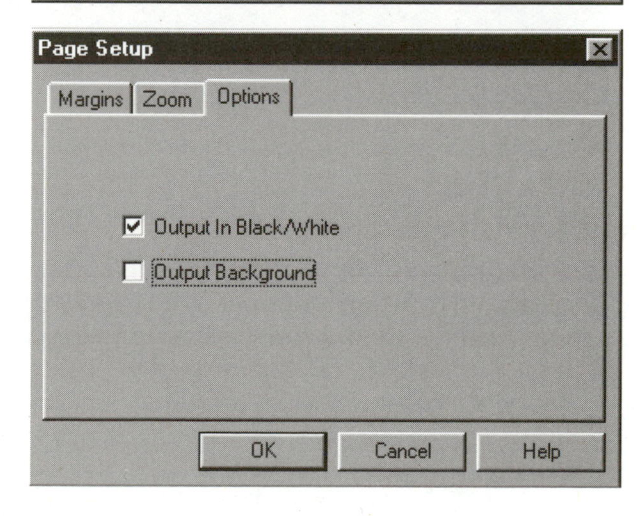

Figure 3.48
Page Setup lets you adjust Multisim's printer controls (not Windows controls).

3. Deselect the Output In Black/White tab.
4. Hit **OK** and **OK** again.
5. Go to **File** > **Print** >. Select the print out you wish to output.

NOTE

If you set the output to color and you only have a black and white printer, colored components and wires will be in greyscale.

TIP

*If you are printing in color mode, open the Options tab and make sure **Output Background** does not have a check in it, especially if you have a black background. This saves on pricey toner or ink!*

Printing a Drawing

Multisim has the ability to print out a variety of items besides the circuit itself. The following sections describe the capabilities and printing procedures.

File > **Print** > **Circuit** opens Windows' Print control box. From here, you can control which printer to use, which page(s) you want printed, and the number of copies you want. Once each item is selected, hit **OK** and the circuit is sent to the printer. You can also hit the Property button to access the selected printer's Property dialog box.

TIP

*Hit **Ctrl+P** to quickly bring up the Print box.*

PROPERTY DIALOG BOX

Each printer has a different Print Properties box. You can typically set paper size, resolution options, color options, and more. See the printer's manual for more information on these settings.

FILE > PRINT > BILL OF MATERIALS

Choosing this command opens the Bill of Materials view window (not available on Multisim Personal Edition). See Figure 3.49. You can also

Bill Of Materials View

To Printer

[illegible]	Description	Reference_ID	Package
1	LED, LED_red	LED1	CASE 279B-01
1	CAPACITOR, 330nF	C1	cap4
1	RESISTOR, 10kohm	R3	RES0.5
1	RESISTOR, 120ohm	R2	RES0.5
1	RESISTOR, 470ohm	R1	RES0.5
2	BJT_NPN, 2N2222A	Q2, Q1	TO-18
1	74LS, 74LS00D	U1	D014

Figure 3.49 A Bill of Materials is handy when shopping for parts. It is not available on the Personal Edition of Multisim.

open it by hitting the Reports icon on the toolbar and choosing Bill of Materials. From here, save the text file or print it. Hit the printer icon in the top left corner to print the file.

FILE > PRINT > DATABASE FAMILY LIST

This prints a list of every component in a specific Component family. It directs you to open the component browser, select the Component family, and hit the **List Report** button. The list opens as a Windows Notepad document. Choose Print. (Yes, it's a pain to print a list of components in that family.)

TIP

Want to know what a printout looks like before committing it to paper? Choose ***File > Print Preview****.*

FILE > PRINT > COMPONENT DETAIL REPORT

This report gives you enough details about a component to put a professor to sleep. However, like the previous report, you have to get to it through the Component Browser. Select the component from the appropriate Component toolbar to open the browser. Choose the family and component for which you want to see a report. Hit the **Detail Report** button and the Detail Report window opens. You can view the information or hit the **Print** button.

PRINTING SUBCIRCUITS

The only way to print subcircuits is first to double-click the subcircuit's symbol and choose **Edit Subcircuit**. This opens it as a new circuit window with the subcircuit inside it. Print it as you would any other circuit.

TIP

Hit the Print button on the toolbar if you wish to print only the circuit.

NOTE

See Chapter 6, Simulation and Analysis, for more information on printing graphs and simulation results.

Zooming In and Out of the Circuit Window

Sometimes it is necessary to zoom in or out of a section of your drawing. If you are making a rather large schematic, you may need to zoom out. Packing a lot of components into a small area may require a zoom in. Multisim is only able to present a maximum zoom-out of 33 or 66 percent of your drawing, or a zoom-in to 200 percent.

To zoom in or out using the menus, choose **View** > **Zoom**. A dialog box opens to allow you to choose 33, 66, 100, or 200 percent. See Figure 3.50. You can also right-click inside the drawing and choose **Zoom** to open this dialog box.

To use the Zoom toolbar, click the icon on the circuit tool bar that looks like a magnifying glass with a plus or minus sign inside it.

Summary

Drawing with Multisim can be simple for some people and difficult for others. Practice these procedures until you feel confident and are able to whip out passable designs. After that, check out Chapter 14,

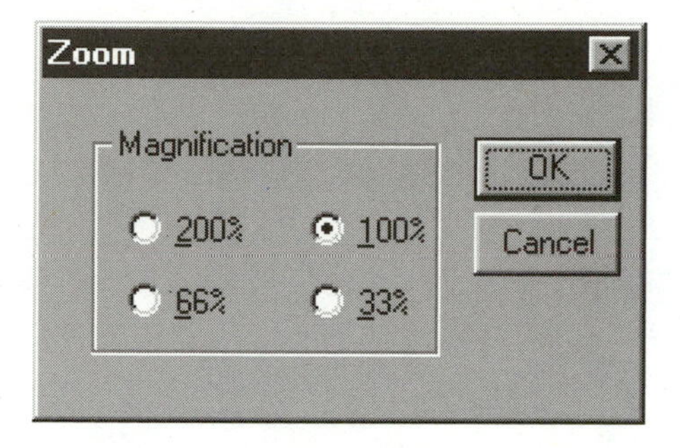

Figure 3.50
Zooming in and out of a circuit lets you see the smallest detail or back away for a birds-eye view.

Unleashing Multisim/EWB, to learn cleaner drawing techniques and tricks to speed you along. I still think drawing with a pencil and paper is faster—but then again, you can't simulate a hand-drawn graphite circuit.

CHAPTER 4

Multisim Building Blocks: Components

Components are the building blocks required to assemble our electronic toys.

Building virtual circuits requires an understanding of Multisim's building blocks. This chapter is a guide that describes the components that make a Multisim circuit. To use the software efficiently, it's important to grasp the various virtual components and how they function. This includes a working knowledge of everything from simple resistors, to MOSFETs and complex Digital ICs. Each component has values, parameters, simulation factors, electrical properties, symbols, and more, which you must comprehend to run your simulations more efficiently.

Unlike previous EWB versions, Multisim now arranges components into editable and searchable databases. Toolbar access is then available to each database of components. It's a flexible way to access modeled components or use various pre-made component libraries. It is best to understand this at the outset, because it can be quite difficult to comprehend the database structure.

This chapter also gives you the foresight to model more realistic components of your own design, so try to get an overall understanding of just how your computer sees these virtual parts and their databases. You can use this chapter as a component reference for your creations.

Virtual Components

Resistors, capacitors, inductors, transistors, and hundreds of other electrical components come together to make a modern electrical device. Each electronic component performs a set task in relation to current, voltage, or signal. For example, a resistor can be used to adjust voltage or current in a circuit. A capacitor can temporarily store a charge and later release it or be used to block DC and pass AC. Multisim contains a virtual representation of thousands of these real-world electronic components. Instead of being made of metals, plastic, carbon, ceramic, and silicon, they are represented by a calculation and schematic symbol that performs the same function inside your computer-designed circuit (see Figure 4.1). A group of virtual components wired together gives you an idea of how a real circuit would perform and the results it would produce. When I talk about components in relation to Multisim, I am referring to *all* the information that lets the computer simulate a real-world component. This includes the schematic symbol, electrical model, footprint/package, and miscellaneous electrical properties.

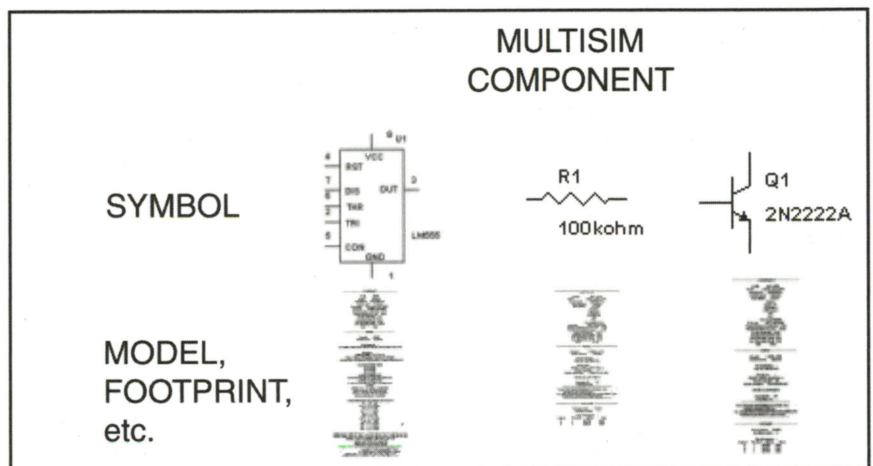

Figure 4.1 Nearly every Multisim component has a real-world equivalent.

Overview of Multisim's Components

There are three basic elements that make up a component in Multisim, the *schematic symbol*, the *electrical model*, and the *footprint/package* of the real-world component. A few other items describe the component's location in the Multisim databases, the component's family, and certain of its electrical properties. The schematic symbol is used to create the drawing and is also a visual way to tell Multisim how to connect the various components for simulation. The electrical model is a mathematical representation of a real component. It is the information the computer needs to simulate that electronic part. The footprint is the classification of how the pins are laid out. Lastly, the package is the physical form in which the chip or component comes. With this information, almost any real-world electronic component can be simulated in Multisim. The existing virtual electronic components that are shipped with Multisim can be edited, copied, created, and manipulated. The components are placed in a library (or database) that can be searched and further modified, or plugged right into your newest circuit.

NOTE

With Multisim, each of the component's elements is now fully editable and customizable, which also allows you to make your own schematic symbols.

New to Multisim

With older versions of Electronics Workbench you had to place a component, bring up its properties, and make changes to the model. Not anymore. You may have noticed while trying to place a BJT NPN transistor, for example, that it opens a dialog box before letting you place the component. This is the *Component Browser* (see Figure 4.2). It lets you select a component from a library of similar parts first. Then once the **OK** button is pressed, you are taken back into the Circuit Window to place the component. Backwards? Yes. But it gives more power and accuracy to the entire simulation and allows you to export to a PCB package with greater efficiency. It also gives you a chance to search for that exact component that fits your design.

TIP

Does the Component Browser open too slowly? Add RAM to your computer.

Databases (Libraries)

A database (DB) is a collection of data that can be accessed by a variety of methods. Let's take, for example, a database of names and addresses. You may be able to access it by searching for a certain name, or merely by selecting from a logical list of streets. In Multisim, we are deal-

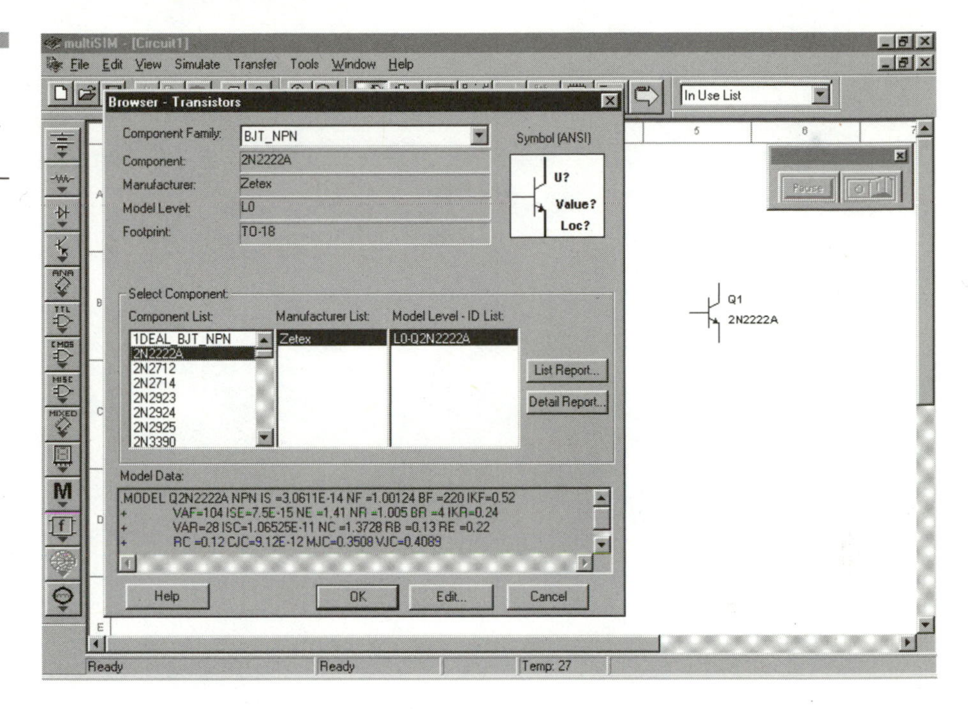

Figure 4.2
Multisim uses the Component Browser in placing parts.

ing with a database of virtual components. Multisim uses two databases for the Personal Edition and Professional Editions and three for Upgraded Professional or Power Pro. Each database contains components composed of symbol, model, and footprint and package information (data). They are categorized, labeled, and ready for you to access using Multisim's Component Toolbars. The three databases shown in Figure 4.3 are:

- **Multisim Master Component Database**—Contains all pre-made components shipped with your version of Multisim. It is not editable. You can, however, copy a Multisim component into one of the other two databases and then modify that component.
- **User**—Contains components you have created. All components are customizable. Available on all editions of Multisim.
- **Corporate Component Database**—Contains components that are available to each user on a design project or team. In this way, you can share common components your company, department, or project uses. This library is fully editable but only available on Power Pro or if you added the Project/Team Design Kit to the Professional Edition.

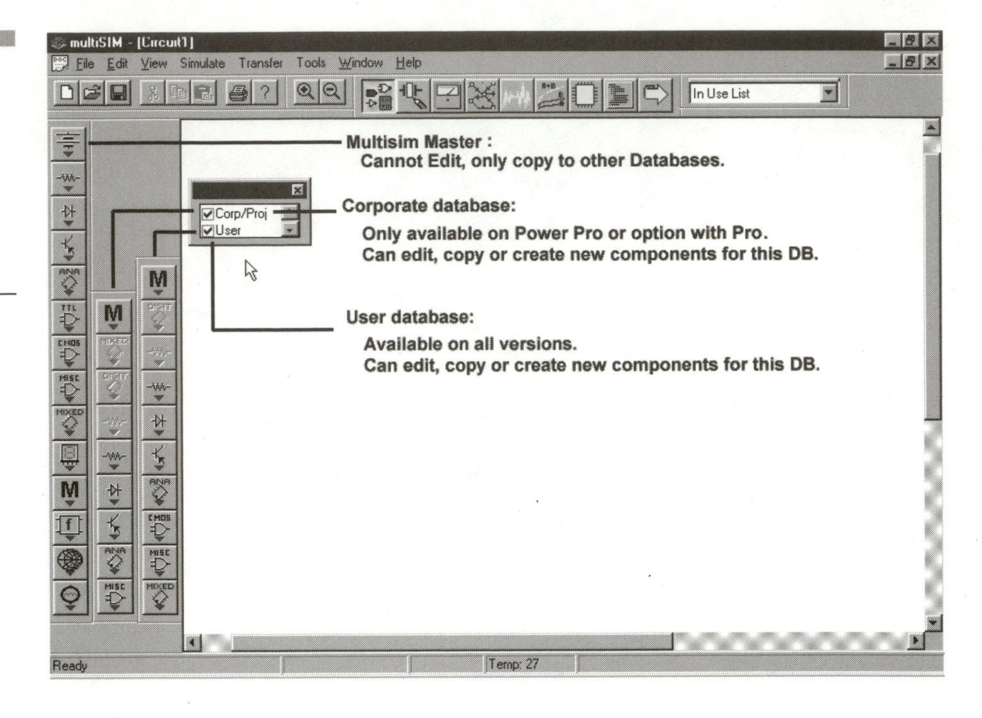

Figure 4.3 There are two or three databases of components (depending on which edition you own) included with Multisim.

NOTE

Multisim Personal ships with 6,000 parts; Multisim Professional with 12,000; and Power Pro has a whopping 16,000 usable virtual components.

EXERCISE

Turn on each database that is available on your version of Multisim with the Database Selector, as in Figure 4.4.

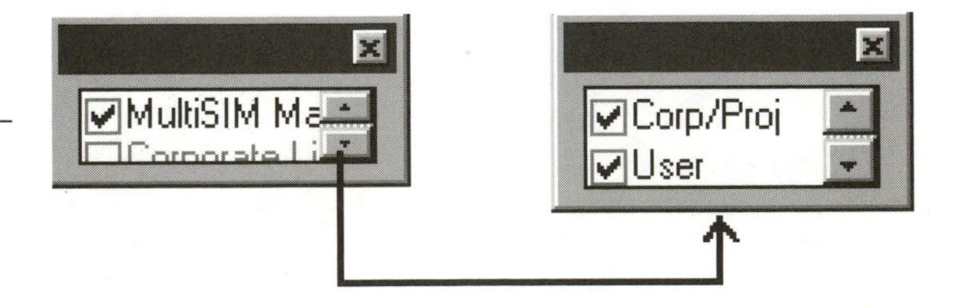

Figure 4.4 The Database Selector.

Multisim's Component Database Structure

Accessing the Multisim component databases can be extremely confusing at first, especially if you're used to older versions of EWB. Unfortunately, it is the only way to place, edit, and modify Multisim components. Take a good look at Figure 4.5; it is a chart I made to help you visualize how Multisim organizes its component data. Don't feel bad if you don't quite understand it at first. It took me two days and a huge flowchart to understand it myself. First take a look at the three divisions seen in the right column:

- **Component**—The Component (in this case a Zetex 2N2222A NPN transistor) contains all the information your circuit needs to display and simulate this component within the confines of the computer screen and simulation. It includes details about the symbol, model, electrical properties, and footprint and package of the real-world component.
- **Databases (libraries) full of components**—This is where the components are stored. It contains all information relating to the component's location in Multisim and its related database. For example, the 2N2222A is chosen from the Multisim Master Database, which contains thousands of transistor models.
- **The Browser, editor, and creator**—The most important element in this tier is the Component Browser. It's used to access each of Multisim's component libraries (databases). It also contains a search feature that lets you locate any of Multisim's components quickly, using a variety of search parameters. If you choose to modify the 2N2222A in anyway, you must copy it either to the User or Corporate library. The Component Properties dialog then comes up. From there you can modify the symbol, where it is placed on the toolbar, how it is modeled within the computer, its various electrical properties, and how it appears to a PCB package. Once the modifications are done, they are saved to your own database (User or Corporate), from which you can now quickly access it later.

By understanding the overall structure of Multisim's component and database structure you will steam through your circuit drawing and simulations. The following sections break this structure down into greater detail and outline the procedures related to each.

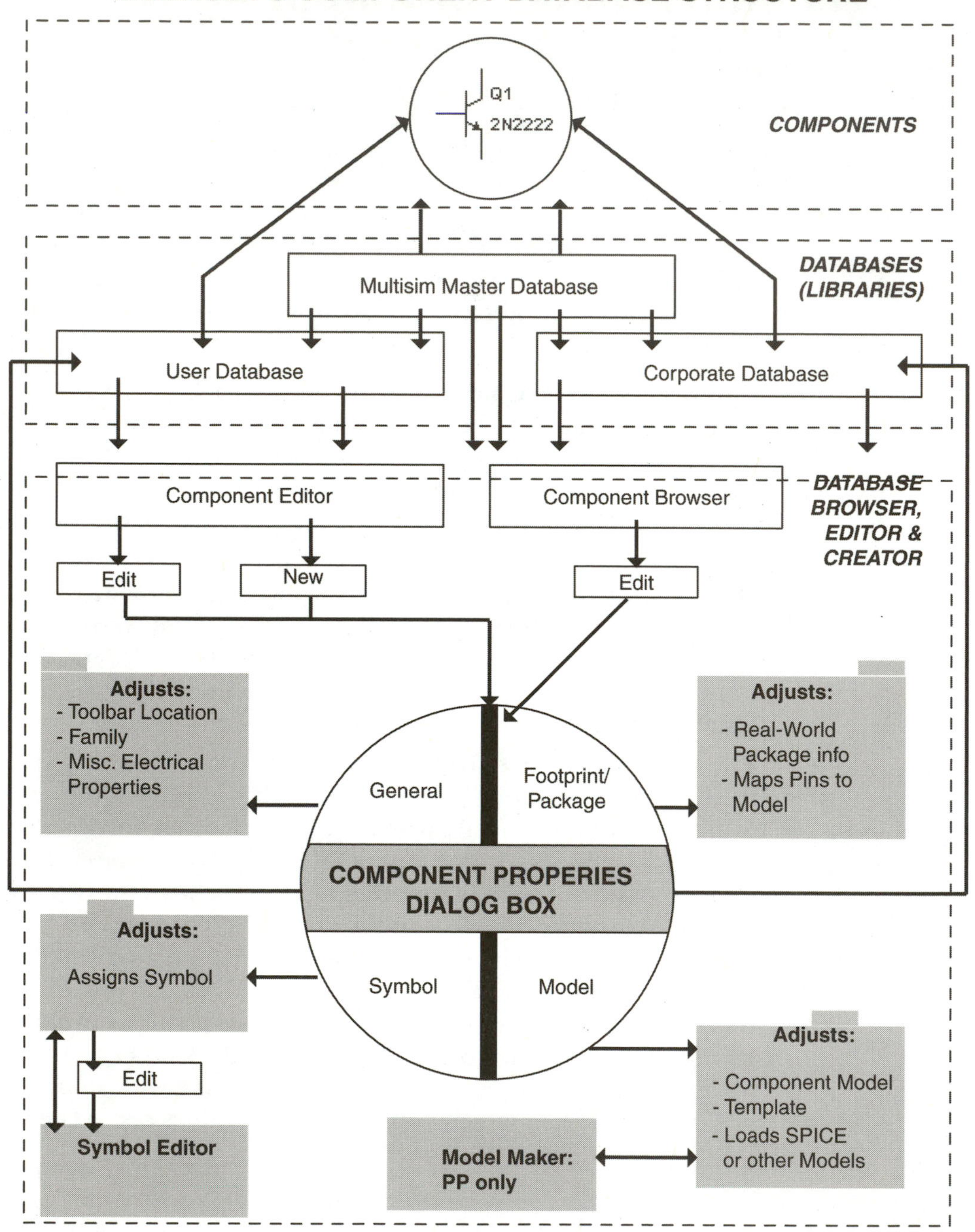

Figure 4.5 How Multisim arranges components.

Components

A Multisim component is a composite of information that includes:

- Its name, date of creation, author and copyright information.
- Its location in Multisim's databases.
- Its electrical properties.
- Its schematic symbol used in schematic capture mode.
- A computer model of the real-world component along with the model's information.
- Its footprint and the package in which the real-world component comes.

Databases (Libraries)

COMPONENT GROUP TOOLBARS

A Component Group is a collection of similar electronic devices. Examples of component groups are transistors, diodes, and indicators. The Component Group toolbar allows access to the individual Component Toolbars merely by moving the mouse pointer over the appropriate component group icon (see Figure 4.6).

COMPONENT TOOLBARS

Once you have moved the pointer over an item on the Component Group toolbar, a new Component toolbar will open. From there you can select a specific component family (see Figure 4.6 again). These families are described in detail later in this chapter. These toolbars are also called *subtoolbars*.

NOTE

Each database has its own Component Group toolbar.

Database Toolbars

For easy access, each database (Multisim Master, User, and Corporate) has its own Component Group and Component toolbars. Only the component families and components you have created or edited will

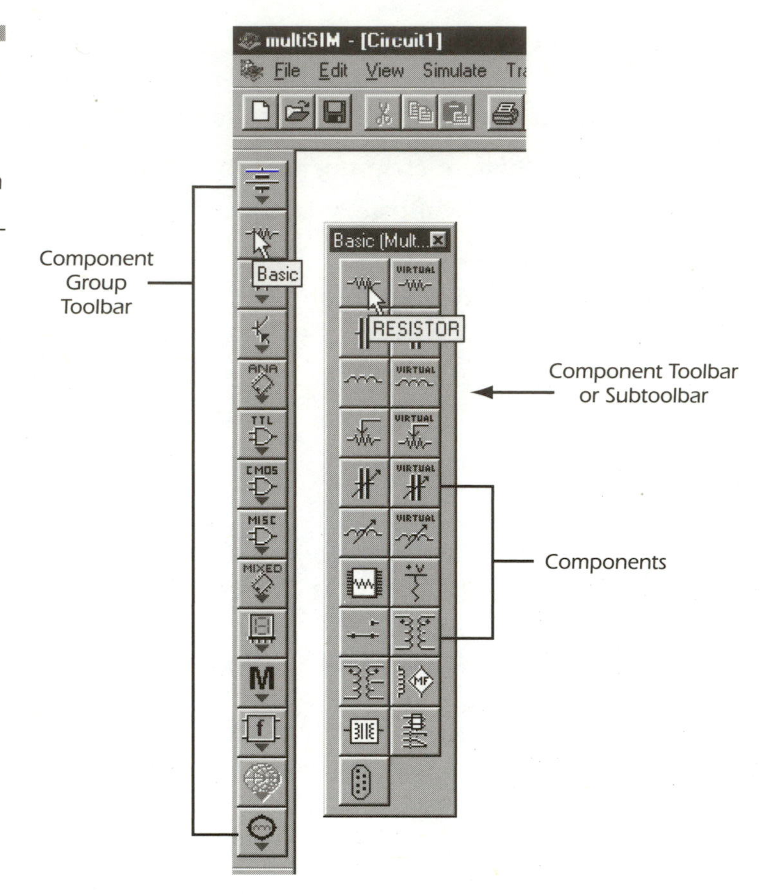

Figure 4.6 Move the mouse pointer over the Component Group toolbar icons to open the subtoolbar.

appear on the User and Corporate Toolbars. These are fully customizable and can be moved around the screen or made to float as well.

DISPLAYING AND TURNING ON/OFF DATABASE TOOLBARS

When Multisim first opens, you see a small window, as pictured in Figure 4.7. Check off which database toolbars you want displayed. You may have to scroll up or down to access the top and bottom libraries. If the Database Toolbar Selector is not activated, go to **View > Toolbars > Database**.

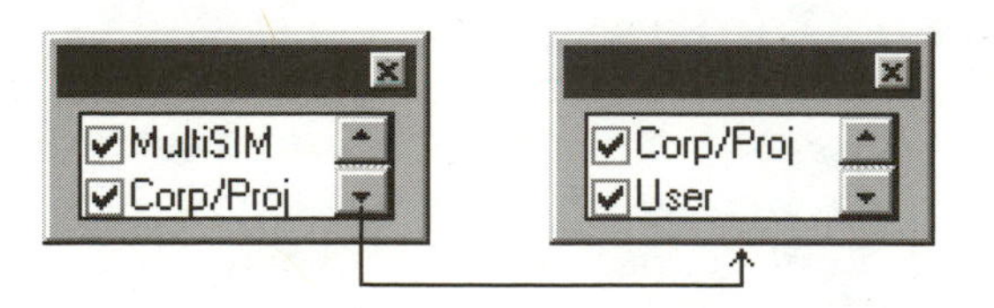

Figure 4.7 The Database Selector lets you turn the toolbars on or off.

TIP

*If your newly edited or created components are not showing up on the toolbar, go to **View > Refresh Component Toolbars**. This is just another quirk in Multisim that may eventually be fixed in later editions.*

USING USER OR CORPORATE COMPONENT DATABASE(S)
First open the appropriate toolbar with the Database Selector, if you haven't yet. Move the mouse pointer over the Component Group in which the component is placed. This opens that database's Component toolbar (subtoolbar) or immediately displays an icon of that component. If one of the Component Group toolbar items is greyed-out, it means that group has no parts in it.

EXERCISE

Open Multisim. Turn on all the databases. Move the mouse over the different Component Groups and see what each contains.

Browser, Editor/Creator, and Properties

The Component Browser

To place a component, you must first locate it from inside one of Multisim's component databases (libraries), which is why the Component Browser first opens when you go to place a component. It is the most important dialog box in Multisim, because it is the starting point of searching for a component, editing or modifying a component, and placing that component into your circuit. The next sections cover the Component Browser window and the procedures for using it.

EXERCISE

Open the Transistor toolbar and begin to place a BJT_NPN transistor; following along with the following descriptions. (See Figure 4.8).

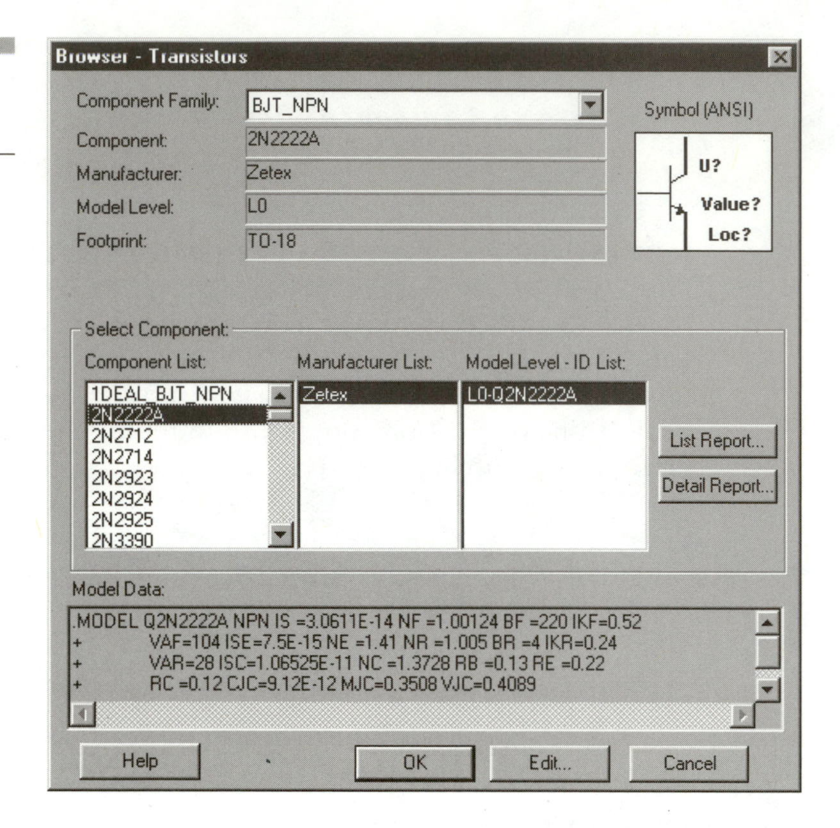

Figure 4.8 The Component Browser.

- **Component Family**—A component family is a group of specific electronics parts with specific properties. It's another way to further classify components. For example, a transistor can be from either the BJT family or the MOSFET family. A Zener Diode has a different family than, say, a Schottky diode. A Component toolbar may contain several component families. For a complete list of these, open the Component Browser and scroll through the Component Family pull-down (see Figure 4.9).

To change Component Families quickly, use the pull-down arrow and select the family. This prevents having to close the Browser box and trying to locate the new family on the Component toolbars.

- **Component**—Generic name of the current Component.
- **Manufacturer**—The supplier of the real-world component on which the virtual part is modeled.

- **Model Level**—The level of sophistication of each model. For example, regular SPICE is L0, SPICE with subcircuit is L1, etc.
- **Footprint**—The type of package the component comes in; the information is needed to export circuits to a PCB Software package.
- **Symbol**—This is the schematic symbol that is placed into the circuit window to represent the component as a whole.

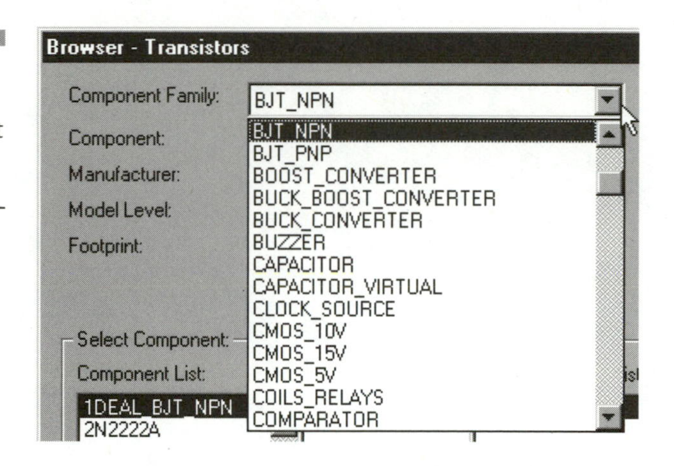

Figure 4.9 Quickly select between Component Families with this pull-down.

SELECT COMPONENT FRAME

The following items are shown within the Select Component frame.

- **Component List**—A collection of real-world part numbers from which to choose (as printed on an actual component). If you wish a printout of all the components available in that family, hit the **List Report** button.
- **Manufacturer List**—The various manufacturers that produce this part number.
- **Model Level-ID List**—Lists the model levels available for that particular component.
- **Find Button** (Not available on Personal Edition)—Opens the Search Dialog Box, giving you a powerful method to track down the exact component needed for a design. (This is described later in this chapter.)
- **List Report Button**—Opens a complete list of components in that family. When you press the button, Windows Notepad or Wordpad launches with this data. It is handy for creating a hard copy of the components available to you.

- **Detail Report Button**—Brings up an in-depth synopsis of the component properties (see Figure 4.10). You can print the data by hitting the **Print** button at the bottom or close the window by hitting **OK**.
- **User Fields Button** (not available on Personal Edition)—Lets the user add notes to a particular component that may later be used. An example would be notes on parts availability and cost.

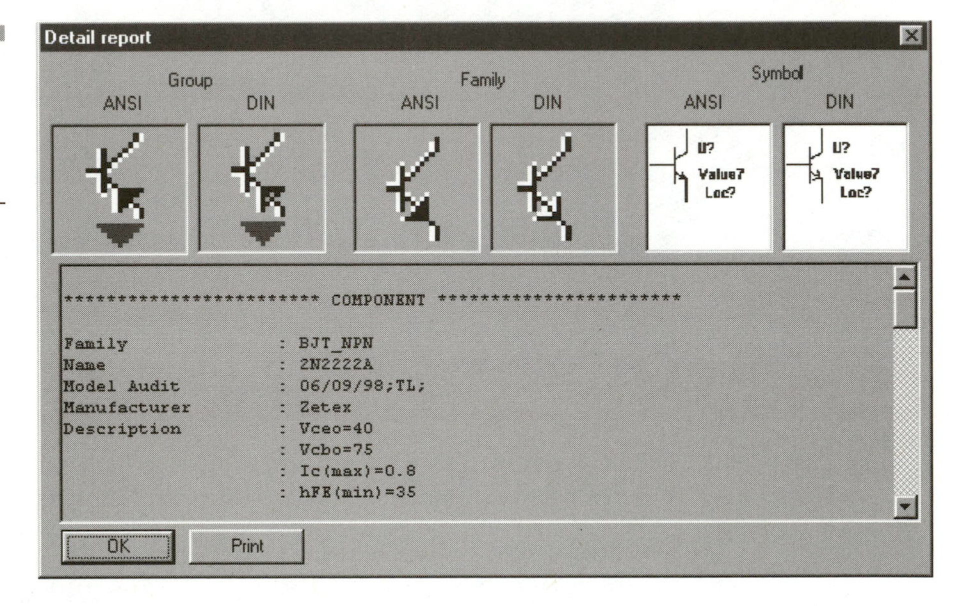

Figure 4.10 This dialog box gives a complete rundown of the component selected.

MODEL DATA FRAME

This frame displays the model information used to simulate the selected component. This is typically SPICE information (covered in Chapter 15, "SPICE and Netlists"). You will likely have to scroll the frame down to view the entire contents.

BOTTOM BUTTONS

The following items are shown in these buttons:

- **Help**—Brings up a help menu for that component, if available.
- **OK**—Executes the command with the selected parameters. Usually means Multisim is now ready to place component.
- **Edit**—Lets you make a copy of the current selected component to place in the User or Corporate library. From there the Component

Properties dialog box opens, which allows you to change the symbol, electrical properties, model, and footprint information.

- **Cancel**—Closes Browser without executing command.

EXERCISE

Open the Diodes toolbar and choose Zener Diodes. Scroll through several of them and hit Detail Report. Print a report.

Component Properties Dialog Box

The Component Properties dialog box contains the meat of each component's information. It is the nerve center of the Multisim's component databases. From here you can modify just about any electronic part's information used in the drawings or simulations or you can create a real circuit. In other words, it's the information Multisim needs to take your idea from concept to working device.

Once you have chosen a component from the Component Browser, you can hit the Edit button, thus opening the Component Properties dialog box. From here you are asked to make a copy of the component (from the Multisim Master database) to be placed in either the User (or Corporate) database; see Figure 4.11. The Component Properties dialog box now opens (Figure 4.12). After a copy of the component is created, you can then change any of the model parameters including location in the database and toolbar, the schematic symbol, the electrical model, and the footprint and package information for the real-world component. The following sections describe each tab of options available to you. These are: **General**, **Symbol**, **Model**, and **Footprint**.

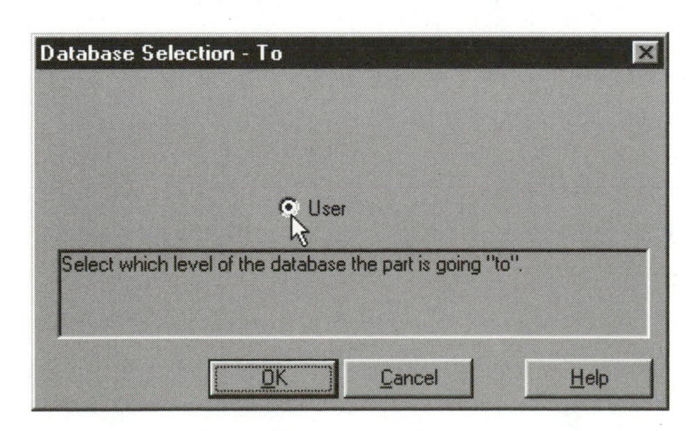

Figure 4.11
When editing a component, you are first asked to select the database to place it. It will always be User with the Personal Edition.

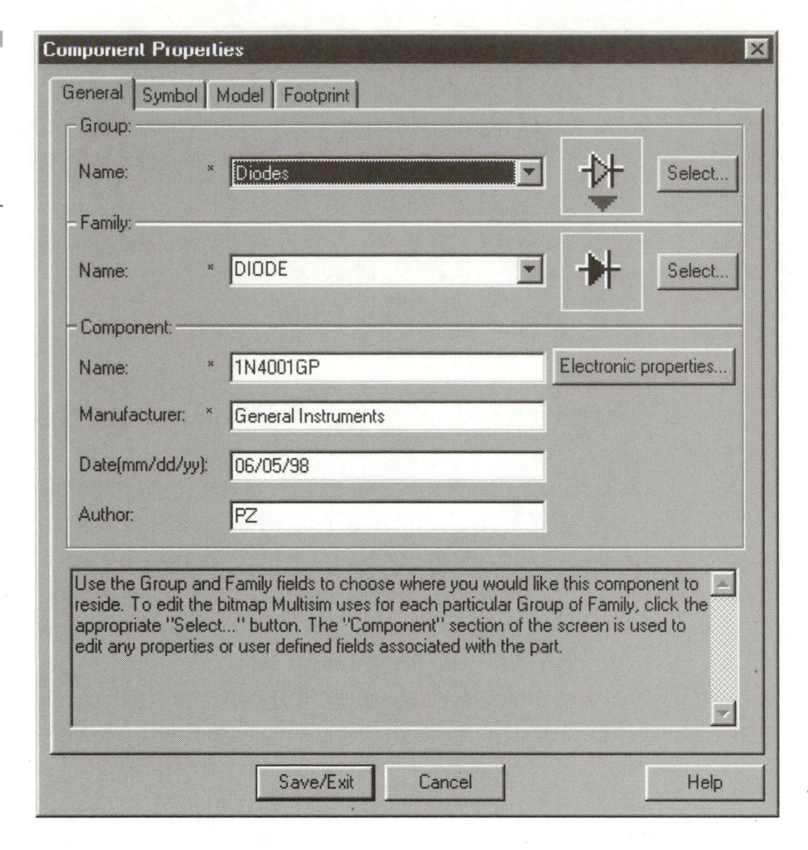

Figure 4.12 The Component Properties dialog box opens to the General tab.

*Begin to place a General Instruments 1N4001GP diode. When the Browser box opens, hit the **Edit** button. Choose to place the new component into the **User** Database.*

Additional procedures and dialog boxes that are accessed through the Component Editor, such as the Symbol Editor and Model Maker, are described later.

GENERAL TAB

For a view of the selections available under the General tab, see Figure 4.12 again.

- **Group**
  - **Name**—Sets the Component Group in which the component will be placed. It is then available to that subtoolbar.

- **Select Button**—You may want to change the Component Group icon's appearance. Hit **Select** and search for a Windows BMP file that suits your tastes. Pre-made BMPs are available in the **C:\Multisim\Buttons\Ansi\Select**... or **C:\Multisim\Buttons\Din\Select**... folders.

- **Family**
  - **Name**—Selects the Family in which your component will be placed.
  - **Select Button**—You can change the Family icon's appearance. Hit **Select** and search for a Windows BMP to suit. Pre-made BMPs are available in the **C:\Multisim\Buttons\Ansi\Select\...** or **C:\Multisim\Buttons\Din\Select**... folders.
- **Component**
  - **Name**—Customize a component's name for easy recognition.
  - **Manufacturer**—Who makes this component in real life? Type 'Generic' if you are stumped.
  - **Date(mm/dd/yy)**—When was this component model created? The date is not Y2K friendly, but I see no problem in using a four-digit year (2000 instead of 00).
  - **Author**—Creator's name or initials.
  - **Electronic Properties Button**—Opens the Electronic Properties dialog box, where you can enter specifics.
  - **User Fields Button** (not available in Personal Edition)—Lets you add user information that can later be accessed. An example is the cost of a part from a specific manufacturer.
- **Text Box**—Contains a short helpful blurb.
- **Save/Exit Button**—When all model data are entered onto each page, this button saves the data to your custom toolbar (User or Corporate) and exits the Component Properties dialog box.
- **Cancel Button**—Closes the dialog box without executing command.
- **Help Button**—Brings up a Help menu for this section.

NOTE

*Once you have completed each item, do not hit **Save/Exit**. You must now click each tab and complete each page of Properties; then hit **Save/Exit**.*

SYMBOL TAB

From this page (see Figure 4.13) you can modify, copy, or create a schematic symbol to represent your component. You can also select between symbol sets (ANSI or DIN). This is a creative new feature that lets you customize a symbol that may not be included in any of Multisim's symbol sets.

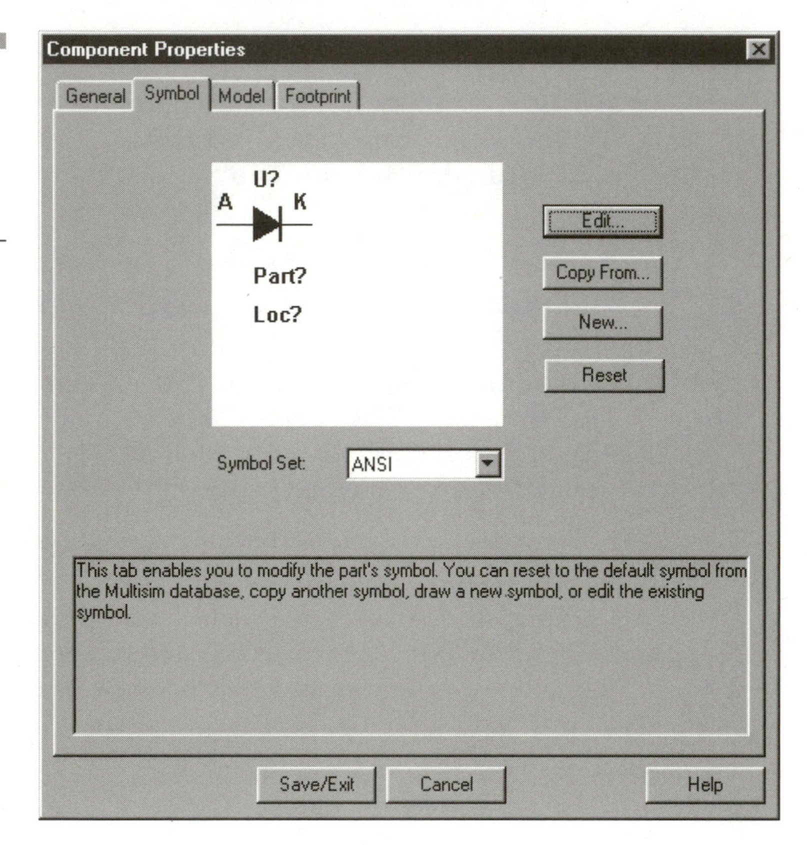

Figure 4.13 The Symbol tab contains information about the component's schematic symbol.

- **Edit Button**—Opens the Symbol Editor to modify the symbol. (More on the Symbol Editor soon.)
- **Copy From**—If you are not a computer artist or just want to copy a symbol from an existing database, you can use the Select Symbol Data dialog box, as seen in Figure 4.14. Just select the Database, Family, and Component from which you wish to copy and hit **OK**.
- **New**—Opens the Symbol Editor, where you can custom draw and label the component. (More on the Symbol Editor soon.)

- **Reset**—Hit this button if you wish to undo your recently edited or newly created symbol.
- **Symbol Set**—Select between ANSI or DIN symbol set.
- **Text Box**—Helpful information for this page.
- **Save/Exit**—Save each page's information and go back to the Multisim main program.

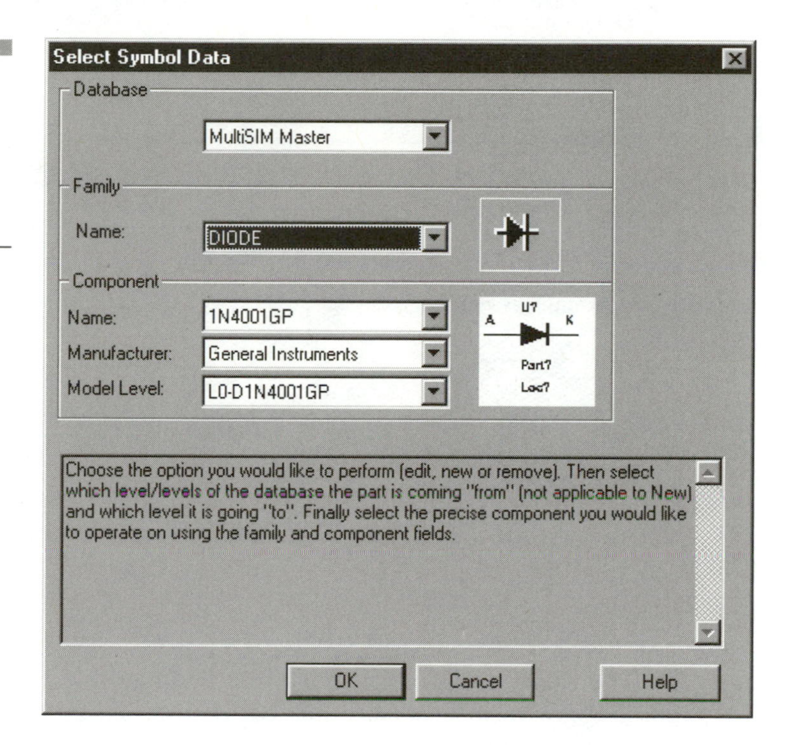

Figure 4.14 If you are not an artist, you can select a pre-made symbol from Multisim's libraries.

MODEL TAB

To create a computer simulation, the designer must model the various parts of the system. This data (see Figure 4.15) mimics a real-world counterpart with mathematics or algorithms. It is information on how electric component would react to input, what exact effect the input causes, and what is then output. A certain voltage level will produce one reaction; a change in current will cause something else to happen. In this way Multisim creates a virtual world full of components that are mirrored to the real world.

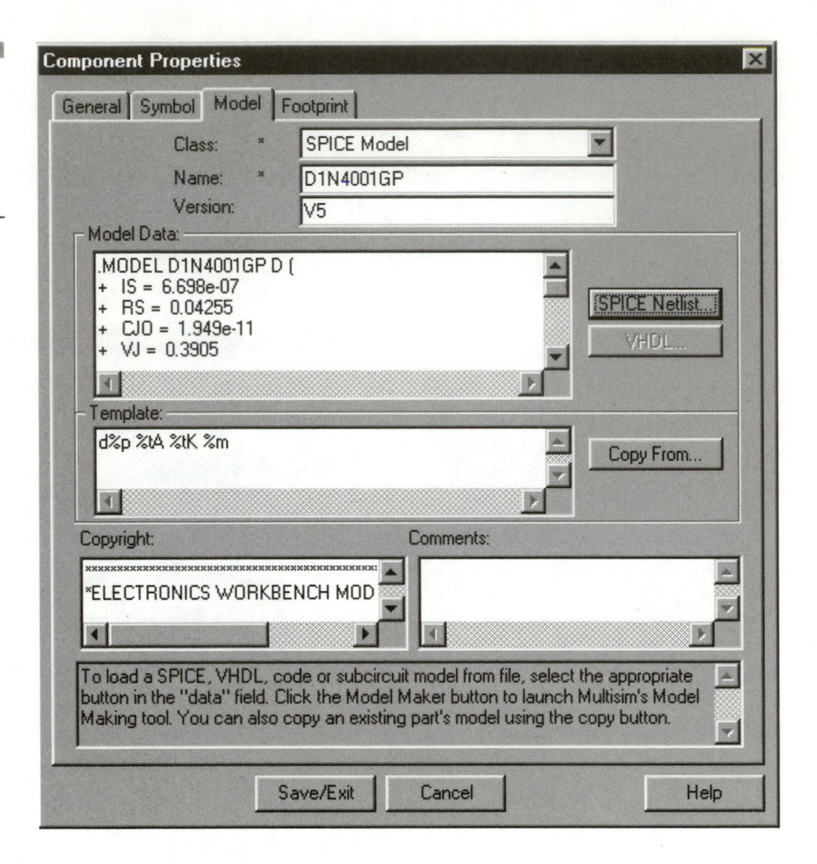

Figure 4.15 The Model tab provides simulation information about that component.

There are several ways that Multisim models these components. Each of these methods is described in greater detail throughout this book. Right now, let's just look at the procedure of creating or editing a component's model.

- **Class List Box**—Choose either Digital Table Model, SPICE Model, SPICE Subckt, VHDL, XSPICE DIGITAL MODEL, or XSPICE Subckt.
- **Name**—Name of the component's Model.
- **Version**—Current version of the component. If you are doing revisions to the component, it is handy to keep track of versions.
- **Model Data**—Actual data used to simulate the component. There are several formats:
  - **Model Maker Button** (PP only)—Opens the Model Maker program to customize your own model (described later).

- — **SPICE Netlist Button**—Lets you open a Berkley SPICE Netlist, PSPICE Netlist (not on Personal Edition), or a Digital model file from your hard drive.
- — **VHDL Button** (PP Option Only)—Allows you to open a VHDL executable file (*.vx) to use as a component model.
- — **Code Model Button**—Lets you open a Code Model DLL files from the hard drive.

TIP

Can't figure out a model's format? The button that is highlighted is the format of the model data.

- **Template**
  - — **Text Box**—The model's template field connects pins to their respective nodes in the model. You can type the information into the text box or copy the template information from another component by using the **Copy from** button.
  - — **Copy from Button**—Copies the template information from another component into the text box. It opens the Select Model Data dialog box, which allows you to first make a choice.
- **Copyright**—To protect a model as an intellectual property, a copyright statement is placed inside this window. If you are creating your own model, I advise you place a statement in this window such as, "Copyright ©2000 John Adams for McGraw-Hill/Tab. All rights reserved."
- **Comments**—If you have any additional information such as notes or further model descriptions, place it in this window.
- **Text Box**—Helpful information to complete the Model Page of parameters.
- **Save/Exit**—Saves entire Properties information to the designated Database and exits.

FOOTPRINT TAB

If you want a very accurate computer model of a component, you must include the physical characteristics of the real-world part (Figure 4.16). This includes the package it comes in, the number of pins, the assignment of logic to each pin, and the distance between pins. Once this information is assigned, you can export the circuit to a Printed Circuit Board Package much more easily.

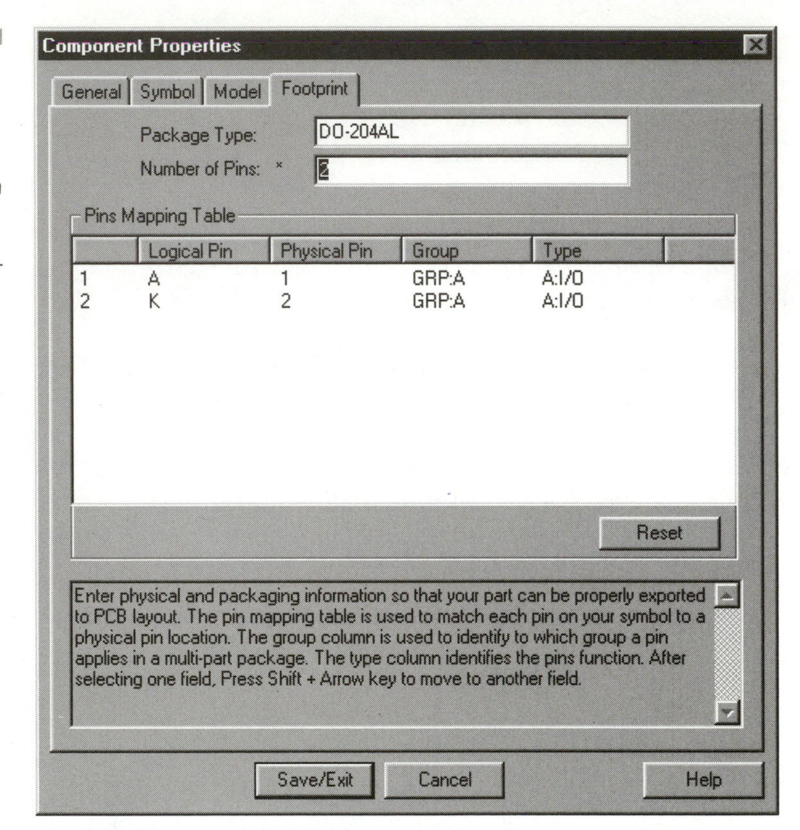

Figure 4.16
The Footprint tab cements the model and symbol to a real-world package. It can then be used in a PCB layout package.

- **Package Type**—Most components are designed in a standard package. For example, a 555 Timer IC would likely be inside a plastic eight-pin IC. The technical term for this package is DIP-8 (Dual In-line Package, eight pins). If it were a surface mount package it might be called SOIC-8.
- **Number of Pins**—Defines the number of connections the component utilizes. For example, the 555 timer would have eight pins.
- **Pins Mapping Table**—Multisim no longer makes the schematic symbol look like the real package. Instead, it places the pins in a more logical order that doesn't always match a 1, 2, 3... format. If you are going to export your designs to a PCB package, you must first describe how each pin of the component matches the model's pin. This is called *mapping*. The pin mapping table is used to match

each pin on your symbol to a physical pin location. To edit any of the fields, double-click over the text/number until a frame appears around the field.

— **ROW** #—Defines the row for each pin's information.

— **Logical Pin**—Each pin has a logical name. For example, a NAND gate may have 1A, 1B, 1Y, etc. Editing each logical pin matches the logic with the actual pin number.

— **Physical Pin**—Lists the pin numbers of the package, as defined by the manufacturer. If you are looking at the top of an IC, the upper-left pin is usually PIN 1. Pins are numbered consecutively down the left row, over to the lower-right side pin, and back up the right side to the top.

— **Group**—If there are logical pins that are grouped, they can be defined here (see Figure 4.17 for the Pin group naming conventions.) For example, a 74LS00 (Quad-NAND gate) has four gates, two power pins. You can group each gate's inputs and outputs into one group and the power pins into another.

— **Type**—Each logical pin has a specific type of input or output, voltage level, power requirements, etc. Each must be defined. For example, if you are defining the first gate's A input from a 74LS00, then it would be typed as **D:INPUT:TTL_LSRCV**. Use Figure 4.17, to define the types. (Remember **D** means Digital pin and **A** means Analog pin.)

— **Reset Button**—Undo any changes you have made to the Pins Mapping Table.

- **Text Box**—Help for Footprint Page.
- **Save/Exit Button**—Stores all component properties to the appropriate database and exits the dialog box.

Figure 4.17 Pin group naming convention.

For logical pins, use the following formats:

For:	Use:	Where:
Pins associated with one section of a component:	GRP:n	n is the section
Pins common to several sections, but not all sections:	GRP:n:m	n and m are the sections

continued on next page

Figure 4.17
PIN group naming convention (continued).

For:	Use:	Where:
Pins common to all sections:	COM	
Pins associated with voltages):	PWR:V0 orPWR: Vn	V0 is ground or n is a voltage
Unused pins (no connects):	NC	

For digital components, the pin type is used to link the I/O models to the logical core for each device. In other families, such as analog components where the simulation models are self-contained units, pin types are for information purposes only.

Use the format:

Where:

Type: Is either A (analog) or D (digital)

Mode: Is one of the following:
Input, output, I/O, 3-state, Open_drain, Open_source, Open_sink, I/O_open_drain, I/O_open_source, I/O_open_sink, Input_ECL, Output_ECL, I/O_ECL, Terminator, PowerNC

Model: Pin model name (none for analog)

Component Editor

Another way to edit, modify, or remove a component from any library is by using the Component Editor. You can access this Dialog Box by going to **Tools** > **Component Editor**, or by selecting the Component Editor on the Multisim toolbar. The next sections describe this dialog box (see Figure 4.18).

- **Operation**—Edit, New, Remove. Do you want to modify, create, or delete a component from the selected library (database)? Click one of the radio button selections to choose.
- **Database**—Depending on which operation you have chosen, this lets you set where the edited component is coming from (**FROM:**) and where it will be placed (**TO:**). If you are creating or deleting a component, it sets the appropriate database.
- **Family**—To which family does the component belong?
- **Component**
  - **Name**—What is the component's name?
  - **Manufacturer**—Who manufacturers the real world part?

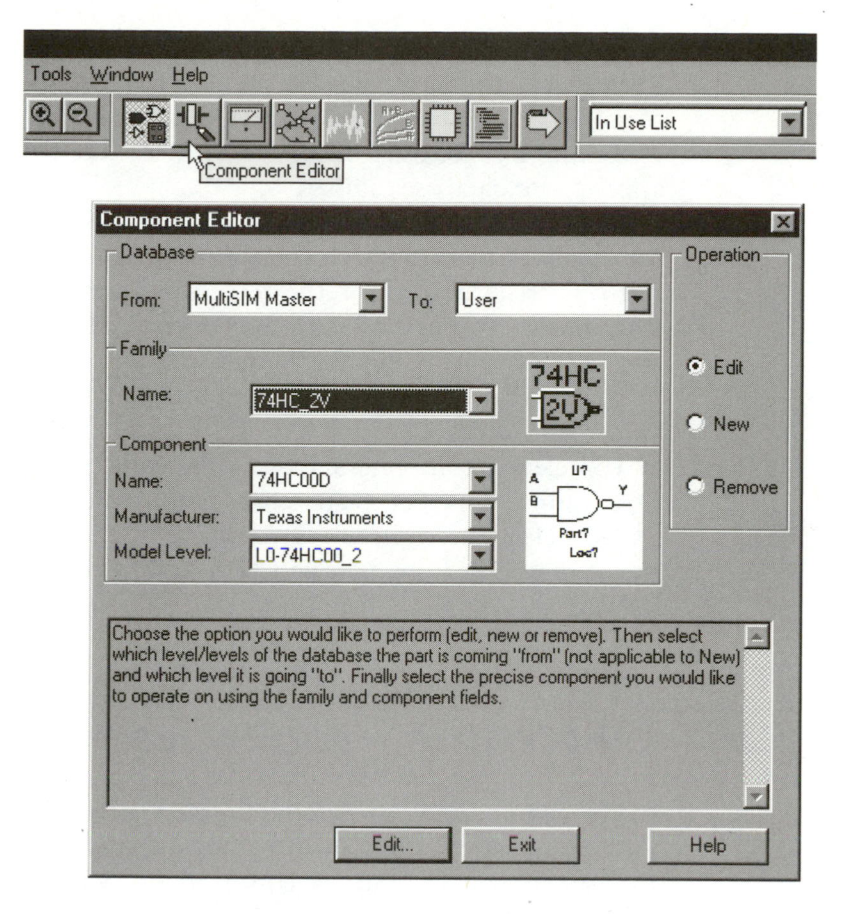

Figure 4.18 The Component Editor edits or creates new components. It is only a different way to access the component databases and not a program itself.

— **Model Level**—Sophistication level of model.

- **Text Box**—Helpful information on filling out this dialog box.
- **Buttons**—Depends on which operation you have chosen.
  - — **Edit**—Opens Component Properties dialog box with the component and database information input that allows you to edit the information.
  - — **New**—Opens Component Properties dialog box with the component information that allows you to create a new device.
  - — **Remove**—Once a component is selected, this removes it from the appropriate database.

To edit a component, select **Edit** from the Operation frame. Set which database the component is coming from and where the edited component will be going. Adjust any other information at this time.

Hit the **Edit** button at the bottom of the screen. Enter all information into the Component Properties dialog box and hit the **Save/Exit** button. Hit the **Exit** button in the Component Editor. Choose **View** > **Refresh Toolbars**. You can now use that component from the appropriate database toolbar.

To create a new component, select **New** from the Operation frame. Set the database into which the component will be saved. Hit the **New** button at the bottom of the screen. Enter all information into the Component Properties dialog box and hit the **Save/Exit** button. Hit the **Exit** button in the Component Editor. Choose **View** > **Refresh Toolbars**. You can now use that component from the appropriate database toolbar.

To delete a component from a database, select **Remove** from the Operations frame. Set the database from which the component will be erased. Select the component to delete in the Component frame. Hit the **Remove** button. You will be asked to confirm deletion; hit **OK**. Refresh the toolbar.

Electronic Properties

In the Component Browser, you may have noticed the Electronic Properties button. The button opens the dialog box seen in Figure 4.19, where you can adjust common properties as well as electrical properties that are unique to that component. Each component has the same properties in the left window but the right will vary. Let's use a Darlington NPN transistor pair as an example (as seen in Figure 4.19). You see that the left side uses the following parameters to describe the component:

- **Thermal Resistance Junction**—The thermal characteristics within the component (from the junction to the case), in watts or degrees centigrade.
- **Thermal Resistance Case**—The thermal characteristics of the whole package (component) in watts or degrees centigrade.
- **Power Dissipation**—The power dissipation of the component, in watts.
- **Derating Knee Point**—The temperature at which the power of the component or package begins to be de-rated, to operate the device in its safe operating range, in degrees centigrade.

- **Min. Operating Temperature**—The lowest ambient temperature at which the component can operate reliably, in degrees centigrade.
- **Max. Operating Temperature**—The highest ambient temperature at which the component can operate reliably, in degrees centigrade.
- **ESD Rating**—The electrostatic discharge for the component.

To change these values, enter the desired changes in the appropriate fields. To save your changes and return to the General tab, click **OK**. To cancel them, click **Cancel**.

The right side of the screen has values unique to each component. Look in Multisim's Appendices for a description of each. Our example (Figure 4.19) describes the transistor parameters of the Darlington pair.

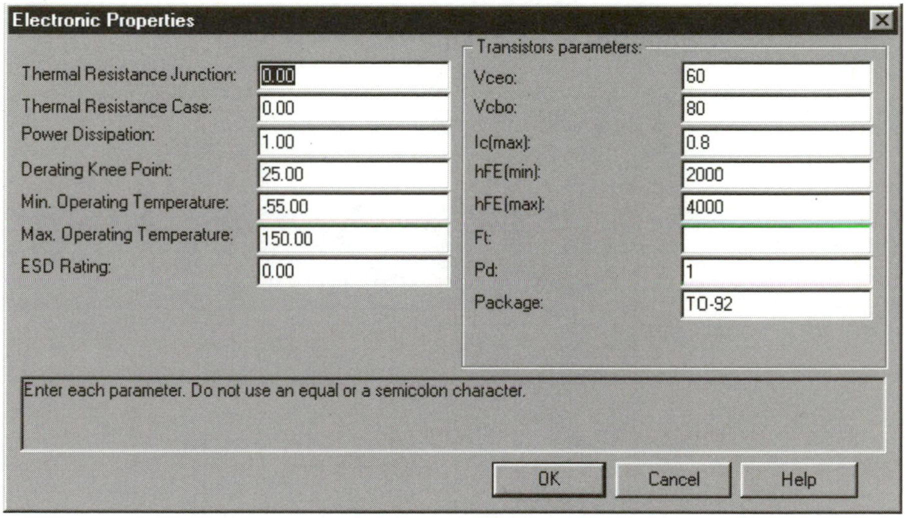

Figure 4.19 Electrical Properties button opens this dialog box.

Electronic Properties

Thermal Resistance Junction:	0.00
Thermal Resistance Case:	0.00
Power Dissipation:	1.00
Derating Knee Point:	25.00
Min. Operating Temperature:	-55.00
Max. Operating Temperature:	150.00
ESD Rating:	0.00

Transistors parameters:

Vceo:	60
Vcbo:	80
Ic(max):	0.8
hFE(min):	2000
hFE(max):	4000
Ft:	
Pd:	1
Package:	TO-92

Enter each parameter. Do not use an equal or a semicolon character.

OK Cancel Help

Symbol Editor

Finally! You can create your own symbols to use within your circuits. This is the ultimate custom feature, because you can supplement Multisim with schematic symbols not included with the software. Maybe the symbol is not quite the way you want it to appear in your schematic drawings. Change it! This includes creating or modifying its diagram, label positions, and where the logical pins are placed. The symbols can be created from scratch, or you can modify an existing drawing by adding your personal touches.

The Symbol Editor is a separate Windows application. The program works very much like any other Windows drawing program, so this section does not go into extensive detail; instead, we will use examples. No pencils required.

Opening the Schematic Symbol Editor

You must access the Symbol Editor through the Component Properties dialog box, Symbol tab. Choose either the **Edit** or **New** button and the Symbol Editor program opens.

Symbol Editor Overview

Figure 4.20 shows the Symbol Editor program. The following lists describe the Symbol Editor's functions.

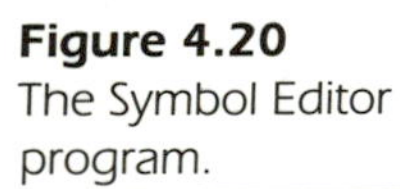

Figure 4.20 The Symbol Editor program.

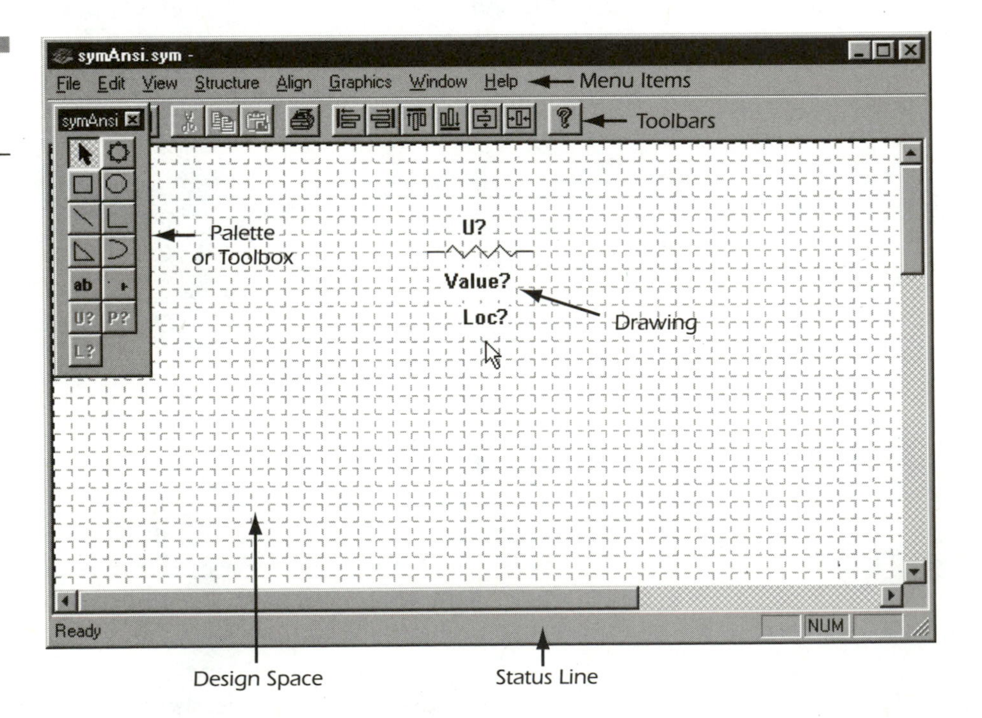

- **Menu Items**
  - **File**—New, Open, Save, Save as..., Print, Print Preview, Print Setup, Page Setup, Recent Files list, Exit. Each item is self-explanatory.
  - **Edit**—Undo, Redo, Cut, Copy, Paste, Delete, Select All, Flip Horizontal, Flip Vertical, 90 Degree Clockwise, 90 Degree CounterCW, Animation. Each item, save Animation, is self explanatory.
    - **Animation**—Want to add a bit of multimedia pizzazz to your schematic? This option lets you design an animated component, such as a light bulb that lights up.
  - **View**—Toolbar, Status Bar, Palette, Grid Visible, Snap to Grid, Change Grid, Show Page Bounds, Zoom. Each item is checked or unchecked to toggle that feature on or off. I recommend leaving the Grid visible and the Snap to Grid feature on for quicker, trouble-free drawing.
  - **Structure**—Group, Ungroup, Bring to Front, Send to Back. You can group items together to avoid small unintended changes to single items. You can also layer the items; you can place a circle behind a square or a line behind a triangle.
  - **Align**—Centers, Horiz Centers, Vert Centers, Tops, Bottoms, Left Sides, Right Sides, Abut Left, Abut Right, Abut Up, Abut Down, Align to Grid. Helps to align items with other items or with the grid.
  - **Graphics**—Fill Pattern, Pen Style, Arrowheads, Color, Font, Properties. Controls fill and line style. For example, if you want a box filled with crosshatching, you would select the item and choose **Fill Pattern** > **Diag Cross**. If you want a line to terminate with an arrow, choose **Arrowheads** > ——>.
  - **Window**—Controls the separate windows in the application.
  - **Help**—Help menus.
- **Toolbars**—In order from left to right: New, Open, Save, Cut, Copy, Paste, Print, Align Left, Align Right, Align Top, Align Bottom, Flip Vertically, Flip Horizontal, About.
- **Palette or Toolbox**—See Figure 4.21.
- **Design Space**—Space in which you draw the schematic symbol.
- **Status Line**—Information on the current selection.

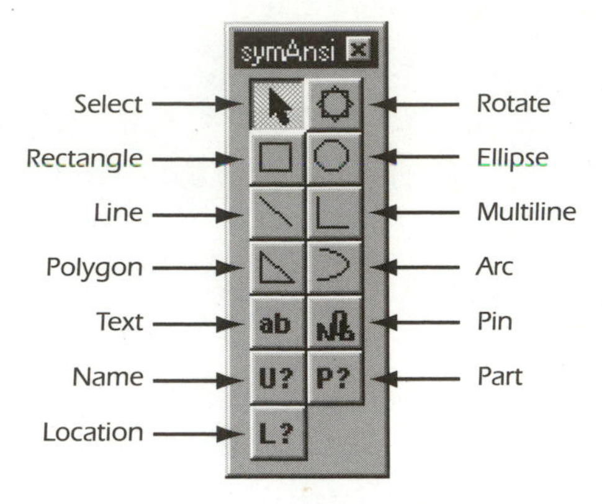

Figure 4.21
The Palette.

By choosing the **Edit** button from the Symbol tab in the Component Properties dialog box, you can select an existing drawing to modify it. Make sure you have chosen either the ANSI or DIN Symbol Set from the list box before hitting the **Edit** button. The Symbol Editor opens. When your changes are complete, hit the **Save** button and close the Symbol Editor; Multisim reopens. Save or Exit the Component Browser.

TIP

*Your new symbol may not work until you first go to **View** > **Refresh Component Toolbars**.*

If you are daring enough, you may want to whip out your own schematic symbols. You can do this by choosing **New** from the Symbol tab in the Component Properties dialog box. Make sure you have chosen either the ANSI or DIN Symbol Set from the List box before hitting the **New** button. The Symbol Editor program opens. When your changes are complete, hit the **Save** button and close the Symbol Editor; Multisim reopens. Save or exit the Component Browser.

Let's modify a resistor symbol using Symbol Editor. Say you want to create a burnt-out resistor for troubleshooting reasons. This may be used in a schematic to help a student better understand a defect in a circuit or to help a technician or circuit designer run the ramifications of a blown component. Follow these steps:

1. Begin by turning the User toolbar on with the Database Selector.
2. Start to place a 100-ohm resistor. When the Browser opens up, choose the **Edit** button and select **User Database**.

3. The Component Properties window opens. Select the **Symbol** Tab and choose **Edit**. The Symbol Editor opens.
4. Click one of the center squiggly lines of the resistor and hit **Delete** (see Figure 4.22).
5. Choose **File** > **Save** and then **File** > **Exit**.
6. The Component Properties window reopens with your newly edited symbol. Rename the component (General tab, Component, Name:) calling it **100ohm-Faulty**.
7. Hit the **Save/Exit** button. The browser opens once more. Hit the **Cancel** button. You should be back to scratch.
8. Choose **View** > **Refresh Component Toolbars**. Select your new component from the User database, Basics toolbar. The Browser opens. Place the new 100ohm-Faulty resistor.
9. Double-click the component and go to the Fault tab. Select **Open** and check on 1 and 2 terminal. Hit **OK** (see Figure 4.23). You've just created a burned-out resistor.

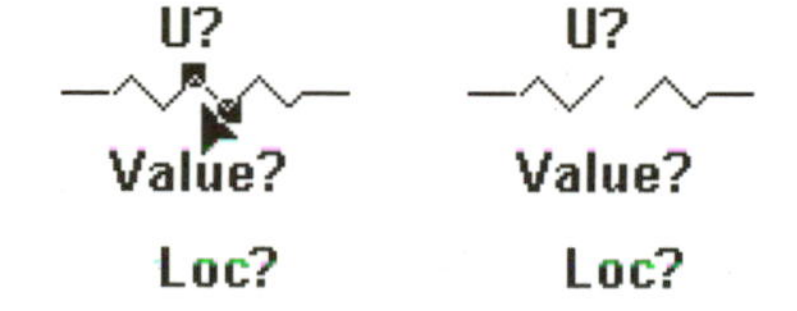

Figure 4.22 Modifying an existing symbol.

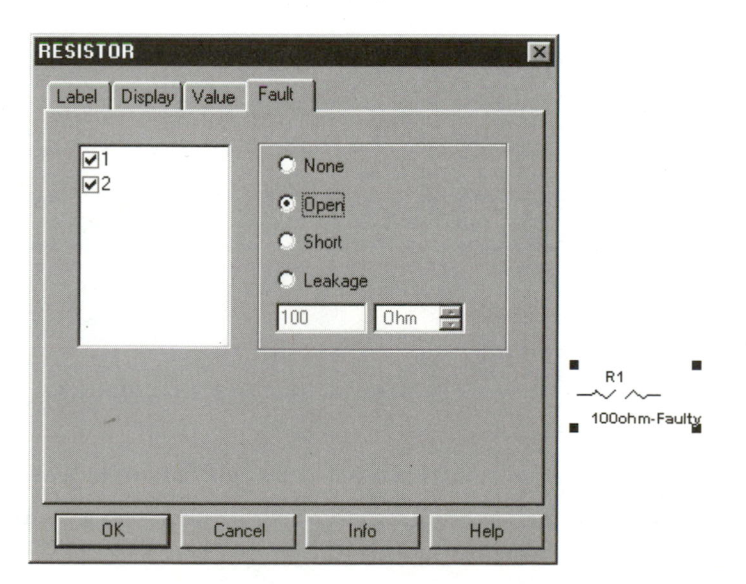

Figure 4.23 A faulty resistor created using the Symbol Editor and component Fault dialog box.

Let's say that instead you wanted to create a speaker for schematic capture and limited load simulation. Follow these steps:

1. Begin by turning on the User toolbar using the Database Selector.
2. Start to place a 16-ohm resistor from the Basic toolbar.
3. Hit the **Edit** button on the Component Browser and choose **User** database. You will copy your new component to this database. The Component Properties dialog box is open.
4. Under the General tab, name the component **16-ohmSpeaker**.
5. Choose the Symbol tab and hit the **New** button.
6. Refer to Figure 4.24. Draw a small rectangle in the middle of the screen.

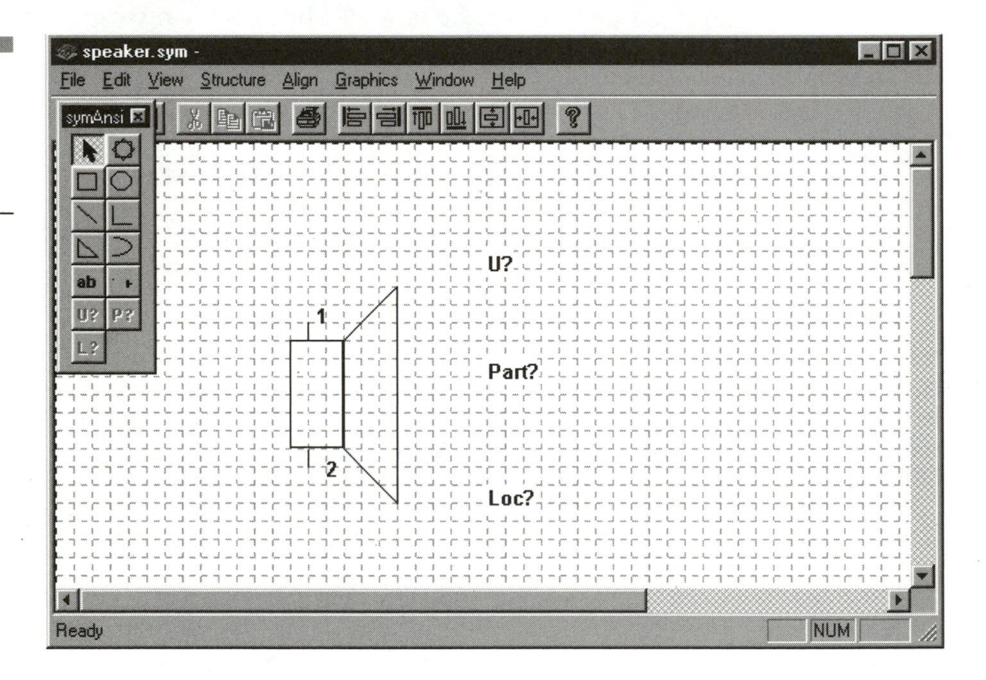

Figure 4.24
A 16-ohm speaker is used as an example of how to create a new component.

7. Use the polygon tool to draw the cone section of the speaker.
8. Place a new PIN onto the drawing, rotating it counter-clockwise 90 degrees.
9. Double-click it to open the Pin Properties window. Click **Edit**.
10. Make the settings match those in Figure 4.25. To do this, click over each row and adjust the settings at the bottom of the dialog box.

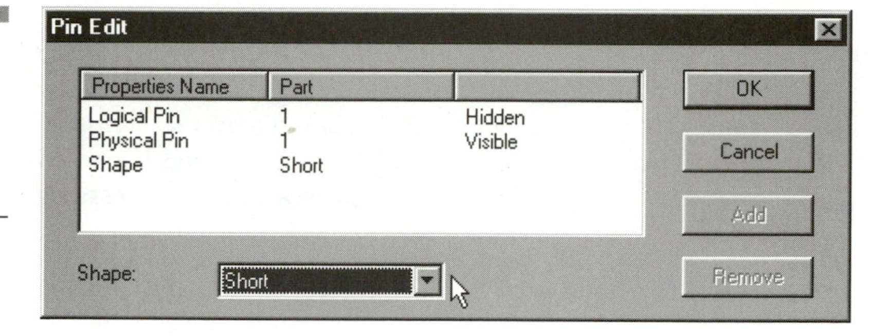

Figure 4.25 Each pin's parameters must be set with the information in this dialog box.

Hit **OK** and place the Pin at the top of the rectangle, as shown in Figure 4.24.

11. Make of copy of this Pin and place it at the bottom of the rectangle. Double-click and make the settings **2** and **2** instead of **1**. Vertical Flip the Pin.
12. Save the file as **Speaker.sym** and exit the Symbol Editor. Multisim opens once more to the **Component Properties** > **Symbol Tab.**
13. Hit the **Save/Exit** button. Hit the **Cancel** button on the Browser.
14. Refresh the toolbars and place the resistor. Select the **16-ohm Speaker** and place it onto the drawing. It will now act as a 16-ohm speaker load (see Figure 4.26).

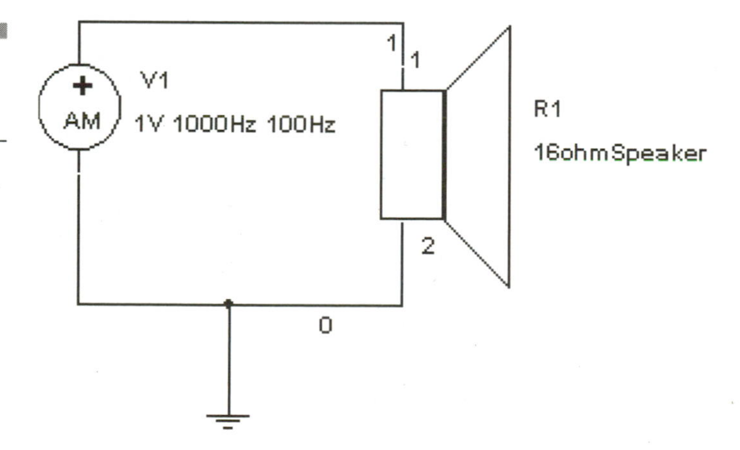

Figure 4.26 The completed Speaker component.

NOTE

You may need to add an inductor to accurately simulate the new speaker component. This exercise was merely to teach you the capabilities of Multisim. See Chapter 7 for another example of this customization.

NOTE

There is an animation feature in the Symbol Editor that allows you to create schematic items that change when a certain action happens in the simulation. An example is an LED that flashes. However, Multisim provides no documentation on its use. I have tried to make this work with no success at this time. If I can pull a procedure up from somewhere it will be posted at: ***http://www.basicelectronics.com/ewb/***.

Component Models

A component model is a computer representation of the real-world component. It's the information given to the computer to simulate that part within a circuit scenario. There are several languages or conventions in which the model can be created, and Multisim handles most of them. These include Berkeley SPICE, P SPICE, XSPICE, VHDL, and VERILOG.

How Parts Are Simulated

Place a part onto your drawing. Begin with a virtual resistor. You will notice that it has two terminals in which to connect additional components by using a wire as a conductor. The various terminals in the circuits, along with their paths to the next component, are called *nodes*.

Back to the resistor (see Figure 4.27). Let's call the left side Node 1 and the right side Node 2, and assume that electricity is flowing from Node 1 to 2. But while it goes through the resistor it experiences a change either in voltage, amperage, frequency, resistance, capacitance, or the like. In this case, a resistance is created in the flow of electricity.

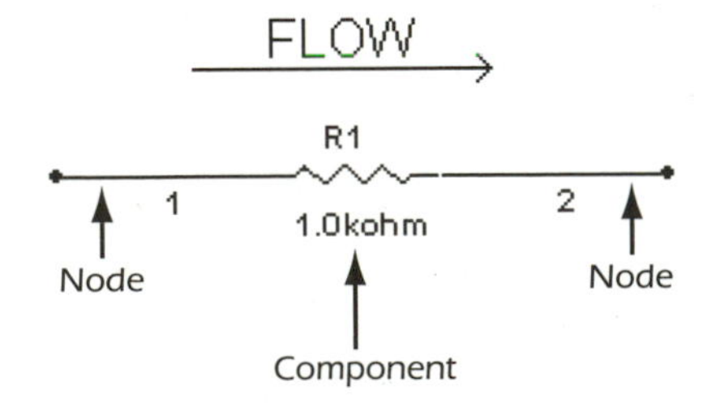

Figure 4.27 This component's information (mathematical model) tells Multisim what it is doing when placed between nodes in a circuit.

How does a Component Model tell the computer exactly what is taking place between the nodes? With a modeling language such as Berkley SPICE (which will be explained later in Chapter 15, SPICE and Netlists).

NOTE

Other modeling languages can be used, but SPICE is by far the most popular.

Each component used in Multisim uses a mathematical model, with values that can be adjusted. Input characteristics closely approximate a real component's properties and changes that may occur in the electrical flow between nodes. Multisim contains a powerful database of preset models that you can use for your own custom parts, or you can make your own.

TIP

If you own the Personal Edition of Multisim, the only way to create custom components is by creating netlists and importing them with the Component Properties window. Power Pro users can choose to use the Model Maker described below. If you still have EWB5 loaded, I recommend using it to create your SPICE netlist files and importing them instead; EWB5 had a pseudo-model—maker even with the Personal Edition.

The Model Maker

NOTE

The Model Maker is only available with Power Pro or as an option with Professional.

Multisim databases don't contain a model for every component ever manufactured. Far from it! If a component you are looking for has no existing model, you will have to create your own from a *databook*. Most companies provide a *datasheet* with the real-world component's electrical characteristics listed. Datasheets are usually downloadable for free off the Internet (on the manufacturer's web site) or offered in a databook, which contains many datasheets. These can sometimes be ordered free of charge, or purchased. Contact the manufacturer of the component for further information about datasheets.

Once you have a datasheet, you can use the Model Maker to input the information to create the component's computer model.

To open the Model Maker, with the Component Properties dialog box open to the **Model** tab, press the **Model Maker** button (if you have a version with Model Maker). You are asked to select a Model Maker and hit **Accept**. This can be BJT, Diode, MOSFET, Operational Amplifier, SCR, Zener, Waveguide, Microstrip Line, Open End Microstrip, Interdigital Capacitor, RF Spiral Inductor, Strip Line, Stripline Bend, or Lossy Line. The selected component's Model dialog box opens.

The component Model Maker opens several pages of figures that require filling in. This is the information Multisim will use to simulate the component. The program automatically inputs common figures into most of the fields, making it easier and faster to fill in the data. Once each field is filled in on a page, hit the **Next** tab. Fill out as much information as possible on each page and hit the **OK** button. You will be taken back to the Component Properties dialog box, where you will see the SPICE Information filled out in the Model Data window.

NOTE

For more information and examples of the Model Maker, see Multisim's User Guide.

Alternatives to Model Maker (Personal Edition)

Why the Electronics Workbench didn't see fit to include the Model Maker in cheaper versions is beyond me. Maybe it will be added to later versions. For now, we are stuck with having to download SPICE models or create models in the SPICE language ourselves (shiver from the thought!). To download SPICE models, do a search on semiconductor websites for *.cir files. I have found a cool way to do this using a search engine called **www.Google.com**. If you type in ".cir spice" you will get a huge list of SPICE files. If you add the part number, you will be all that more likely to find exactly what you are looking for.

To create your own models, take an existing SPICE model that is similar and make your modifications to the Netlist file. Enter the specs under the Model Data field in the Component Properties window, Model tab. You can then adjust certain electrical properties in the General tab.

If you are a Personal Edition EWB 5 user, import your SPICE models after creating them in the older simulator.

Searching for Components (Power/Pro Users)

Multisim offers a powerful search engine to find the exact component for which you are looking. Once again, however, you must own either Professional or Power Professional for this feature. To use it, go to the Browser Dialog, hit the **Find** button. This opens the Search Dialog Box, which lets you search by a variety of methods. See the Multisim *User Guide* for more details.

Multisim Component Guide

Adjustable Variables

Virtual components have values that can be adjusted by double-clicking the component and selecting the Value tab. From here you can adjust a virtual capacitor's capacitance, a virtual resistor's resistance, and so forth. If a component such as a transistor, is modeled, it can only be replaced with another model using the Replace button.

Identifying Parts

If you dragged a part onto the Circuit Window, not all the component information will show up. For instance, the part will typically only have its Model Name, neglecting the name you wish to give it. How do you discern it from all the other similar parts? By further identifying it. Take a look at Figure 4.28; which diagram is more understandable: A or B?

There are two ways to perform this task, each with multiple procedures. One is to use a label and the other is to make Multisim display all of the component's information, including Reference ID, Model Name, your new label, value, and node names.

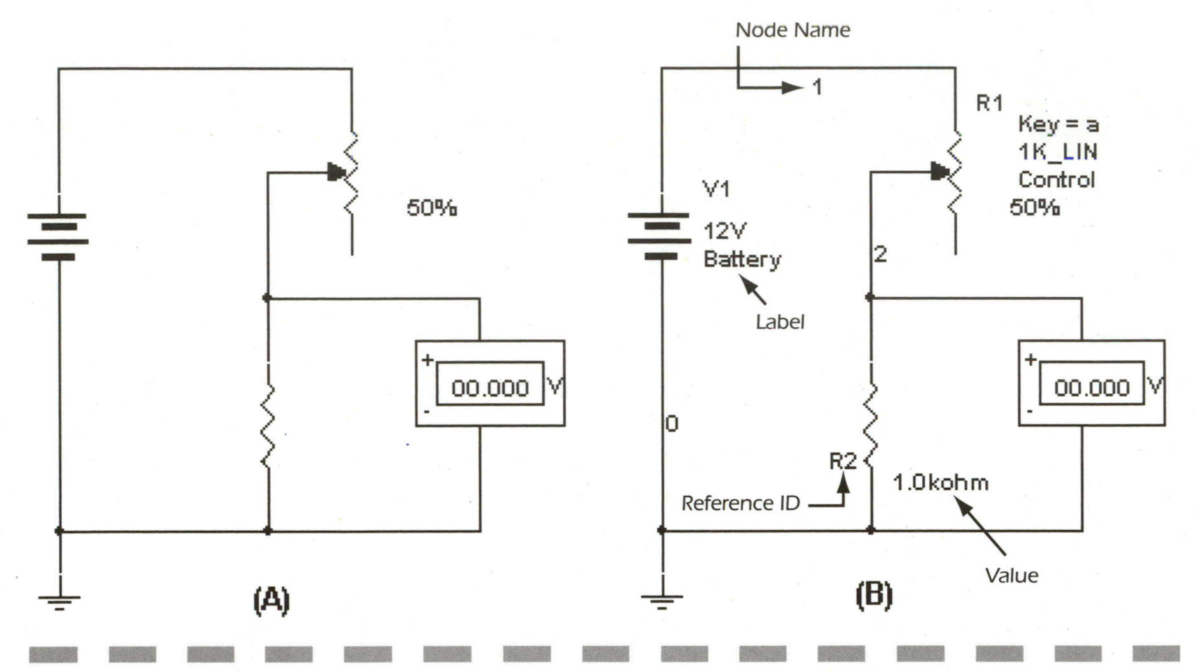

Figure 4.28 Which drawing contains more useful information? Labeling helps to identify each element of a circuit.

To use a label to identify a part, refer to Figure 4.29. If you wish to call the part by a name that suits your tastes, use the following procedure:

1. Select the component and double-click it.
2. Click the Label tab.
3. Type in the Name or Number you wish to call this component.
4. Click **OK**.

To use a reference ID to identify a part, you can let Multisim name the part for you with a generic tag (i.e., R1, C3, Q2). This is perfectly acceptable for most applications, unless you wish to further customize the information with a label. There are several methods to make Multisim display a component's Reference ID.

By using the Show/Hide Tab in Schematic Options you can tell Multisim what you want the component to display. In this case, we want the Reference ID to show automatically for every part. Here's how:

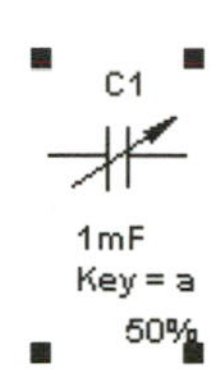

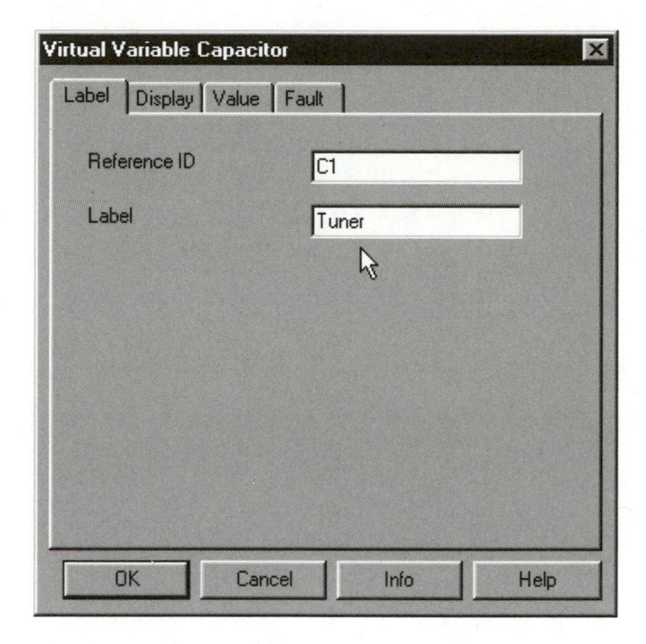

Figure 4.29 Double-click a component and choose the Label tab to access this dialog box.

1. Open the **View** menu and choose **Show** (or right-click on a blank area in the circuit window and choose **Show**).
2. Place a check in the Show Reference ID option.
3. While you are at it, check each item you want displayed globally.
4. Click **OK**. This displays the Reference ID of every part in your circuit.

If you only want Multisim to display the Reference ID of the one component on which you are working, use the following procedure:

1. Double-click the component.
2. Click the Display tab.
3. Uncheck the "Use Schematic Options global setting" box and check the appropriate items.
4. Click **OK**.

Creating a Fault in a Component

Multisim Version 6.x contains a powerful feature to deal with more realistic training and simulation. It lets you create a flaw in the circuit either to simulate those conditions for design or to help train a technician. For example, if you are simulating how a circuit would react to a deadly short, you can use this feature to see its results. Or, if there is a short in a circuit somewhere (embedded into the simulation by the instructor), you can tell a student technician to use virtual instruments to find the fault.

Under the Component Properties (double-click component) of each part is a tab titled, "Fault". This opens up the component's Fault options, which allow you to control that part's glitches. Here is a rundown of the typical faults available to you (refer to Figure 4.30).

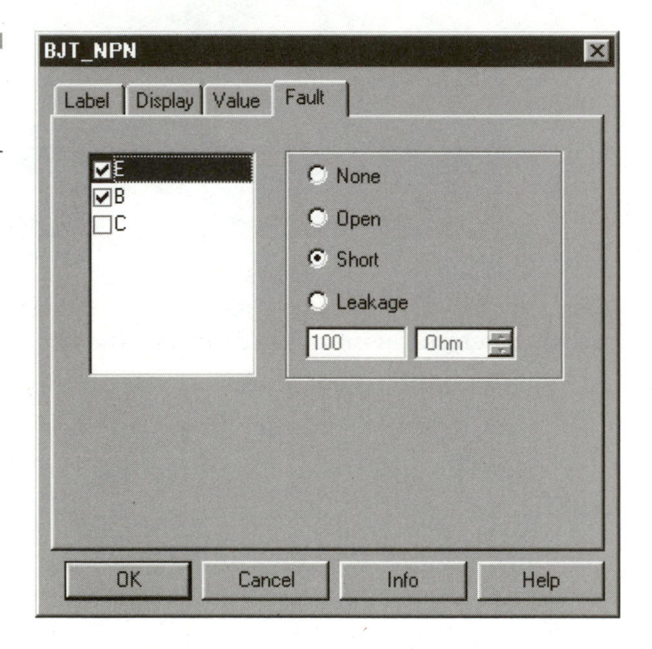

Figure 4.30 Creating a fault in a BJT transistor.

NOTE

Each of these faults can be created between any selection of the component's terminals.

- **Pin Checkboxes**—At the right are the Pinouts with a checkbox next to them. This shows between which points the fault(s) is being simulated.
- **Type of Fault**—The right box lets you select the type of fault you wish to create. The choices are:

  — **None**—(Default) No simulated fault in the component.

— **Open**—Simulates an open circuit condition between the points you have set.

— **Short**—Creates a zero-ohm short between the terminals.

— **Leakage**—Leaks a set amount of resistance between the terminals you have indicated in the spin box below.

Let's use a few examples to learn how to setup a faulty component:

1. Create the circuit in Figure 4.31.
2. Run the simulation.
3. Take note of the transistor's gain (<> = 100) and the voltage when the potentiometer is set to 50%.
4. Double-click the NPN Transistor.
5. Click the Fault Tab.
6. Select Short.
7. Check off the B and E terminal boxes. This creates a fault between the Base and Emitter of the NPN transistor.
8. Click **OK** at the bottom.
9. Run the simulation again.

What are the results? Try different types of faults and note the results.

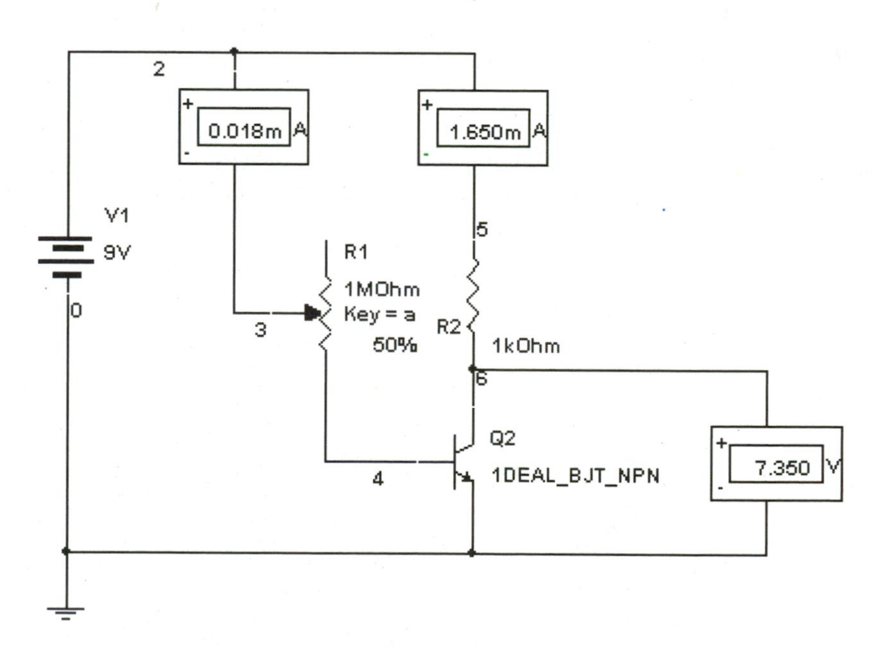

Figure 4.31
An example circuit to test the faulty transistor.

Parts Descriptions and Examples

The next sections briefly explain the use of each component. Some components also contain an example of their use for easier understanding. For detailed information on each component, go to **START** > **Programs** > **Multisim** > **User Guide Appendices**. You can also select **Component Help** from the same submenu. Another method is to place the component, select it, and hit **F1** (Help).

You will notice that some of these components have real-world equivalents (such as resistors, LEDs, etc.), but others might be considered subcircuitry (such as volt- or current-controlled voltage and current sources). If you wish to transfer your designs to a PCB layout program in the future, I suggest you only use them in test simulations. Instead, build that specific circuitry yourself, and store them as a library of subcircuits for retrieval and placement into your designs.

Sources Toolbar

For an overall view of the Sources toolbar, see Figure 4.32. Note that descriptions of all toolbar icons start at the top-left corner and go to the top-right corner, then drop one level and continue.

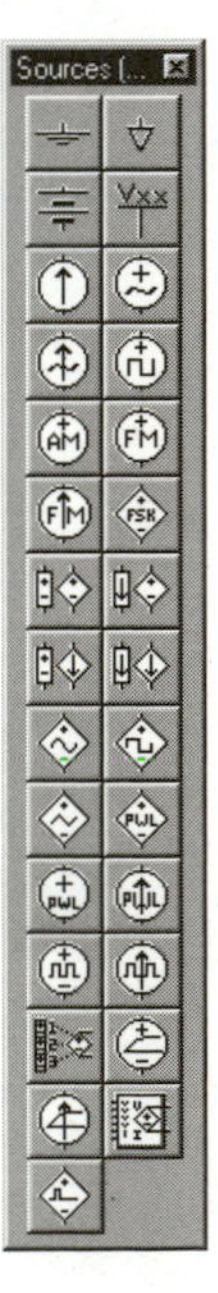

Figure 4.32
The Sources toolbar.

The **Ground** symbol represents a common point of reference for positive or negative measurements—zero volts. It is a return point in a circuit, usually hooked to the negative side of a battery or voltage source. Each circuit must contain this point of reference and thus every circuit must contain at least one ground.

The **Digital Ground** is used to connect ground to those digital components that do not have an explicit ground pin. The digital ground must be placed on the schematic, but should not be connected to any component.

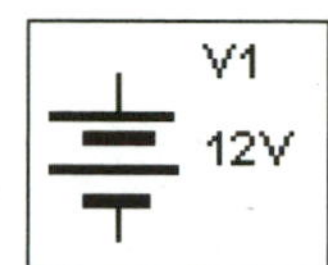

The **DC Source** or battery functions as a DC voltage source. It can operate as low as the pV range, all the way to a killer Teravolt (TV, meaning one-trillion volts) range. This lets you simulate anything from a Duracell battery to a lighting bolt.

When using a battery in a circuit, keep in mind that Multisim does not add an internal resistance, as found in a real-world battery. If you want to use a battery in parallel with another battery or switch, insert a 1 milliohm resistor in series with it.

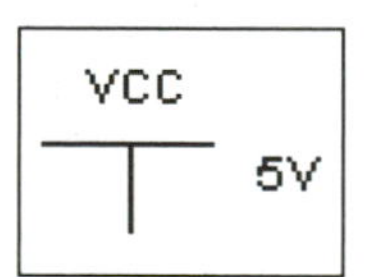

The **Vxx (Vcc or Vdd) Voltage Source** provides a voltage point of whatever value you choose. For example, if you want a +15-volts line to connect to a CMOS IC, you can place the component, double-click it, change the value to 15V, and change the Reference ID to Vdd. This is mainly used to supply voltage to the Vcc pin of an IC or connected to a gate to throw it high (logical 1).

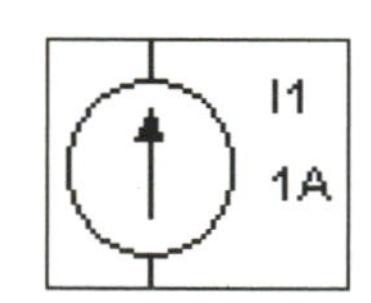

The **DC Current Source** adds a DC current to your design. It can be adjusted from a negligible pA range to a deadly house-melting TA range. This is a good component with which to learn current analysis.

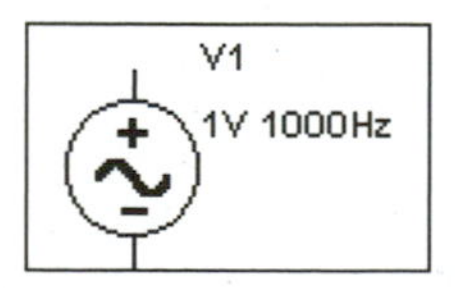

The **AC Voltage Source** injects an AC voltage sine wave signal into a circuit. This component is measured in peak voltage, not in RMS voltage as EWB 5 did. Double-click the component to set its voltage, frequency, and phase angle.

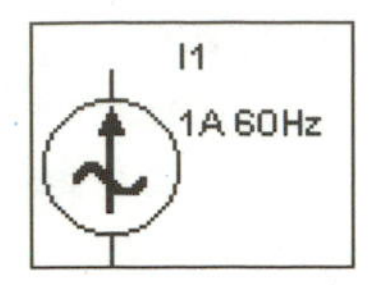

The **AC Current Source** lets you pump out an AC current. Like the AC voltage source, it is also measured in peak, not RMS, voltage. The frequency and phase angle are adjustable. Double-click the component to set these parameters.

If you are importing a design with an AC voltage or current source from EWB V5 to Multisim V6, remember that EWB reads RMS and Multisim reads peak voltage values. A 120-volt AC source coming from EWB appears as a 170-volt AC source in Multisim.

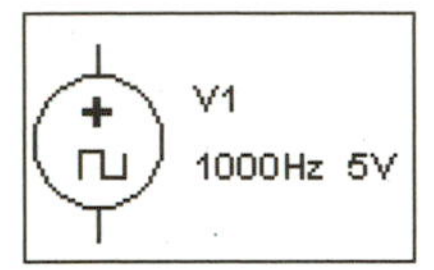

The **Clock Source** is used in many ICs that require a square wave to 'clock' their circuitry. This source lets you input an adjustable square wave into a circuit (i.e., it is a square-wave generator). You can adjust the amplitude (voltage), duty cycle (on-to-off ratio), and frequency by double-clicking the placed component. If you must run more than one square-wave clock signal in a circuit and the function generator is already used, this makes a great replacement.

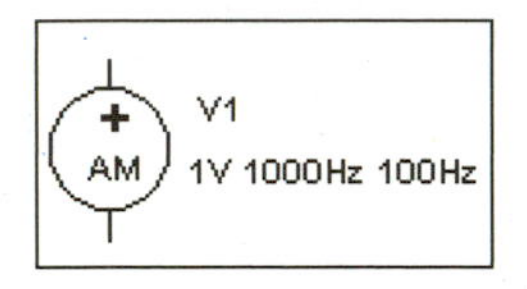

The **AM Source** (single-frequency amplitude modulation source) generates an amplitude-modulated wave. It is used to build and analyze communications circuits. Double-click the placed components to adjust carrier amplitude, carrier frequency, modulation index, and modulation frequency settings.

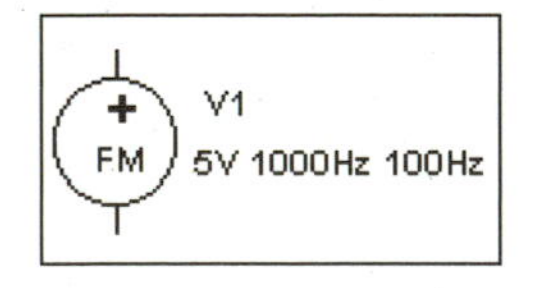

The **FM Voltage Source** (single-frequency frequency modulation source) generates a frequency-modulated wave. It is used to build and analyze communications circuits. Double-click the placed components to adjust voltage amplitude, voltage offset, carrier frequency, modulation index, and signal frequency settings.

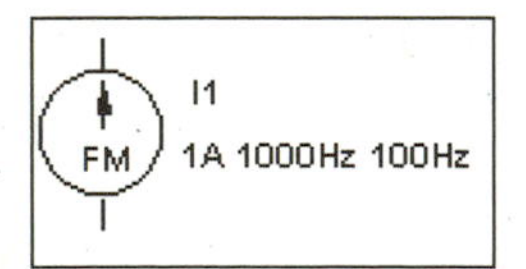

The **FM Current Source** provides a frequency-modulated current source. It is also used in communication circuits. Double-click the placed component to adjust current amplitude, current offset, carrier frequency, modulation index, and signal frequency settings.

The **Frequency-Shift Keying (FSK) Source** is used to key a transmitter for telegraph or teletype communications by shifting the carrier frequency over a range of a few hundred hertz. The frequency-shift-key—modulated source generates the mark transmission frequency, f1, when a binary 1 is sensed at the input, and the space transmission frequency, f2, when a 0 is sensed. FSK is used in low-speed modems. Figure 4.33 shows a good example of the FSK. Double-click the placed component to set the *mark* and *space* frequency as well as the amplitude of both signals.

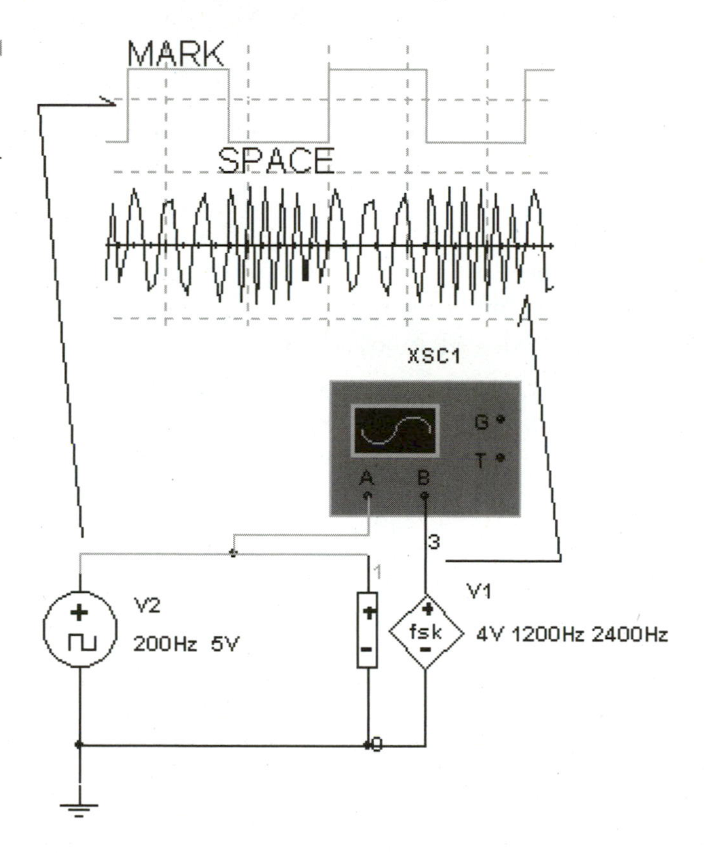

Figure 4.33 A good example of the FSK.

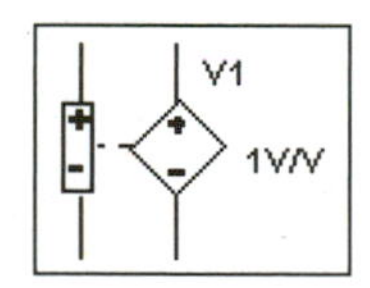

The **Voltage-Controlled Voltage Source (VCVS)** outputs a set voltage determined by the input voltage and gain (E). The input side is indicated by a rectangle with + and − terminals. Output is a diamond with + and − terminals. If, for example, you have 12 volts hooked to the input and the gain set to 1kV, then the output voltage would increase to 12 kV.

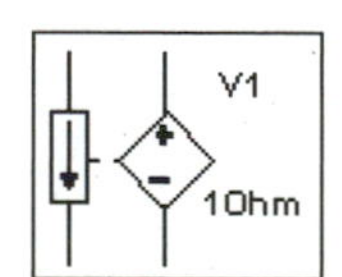

The **Current-Controlled Voltage Source (CCVS)** outputs a set voltage determined by the current at its input. The rectangle with the arrow is the current input and the triangle is the associated output voltage. Double-click the component to access the adjustments. The two are related by a parameter called *transresistance* (H), which is the ratio of the output voltage to the input current. It can have any value from mW to kW. Use the formula: H = Vout / Iin. This component is used to change a current reading into a voltage reading.

Use the CCVS to read current on the virtual oscilloscope. Just wire it into the circuit as shown in Figure 5.16 in Chapter 5.

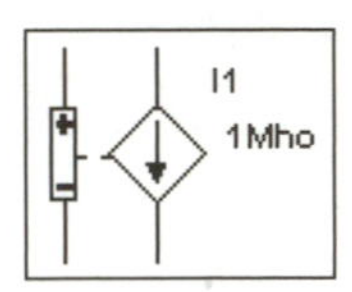

The **Voltage-Controlled Current Source (VCCS)** outputs a current value set by the inputted voltage level. Double-click the component to open its parameters. The two are related by a parameter called *transconductance* (G), which is the ratio of the output current to the input voltage. It is measured in *mhos* (also known as *seimens*) and can have any value from mmhos to kmhos. Use the formula: G = Iout / Vin. It is used to change a voltage reading into a current reading.

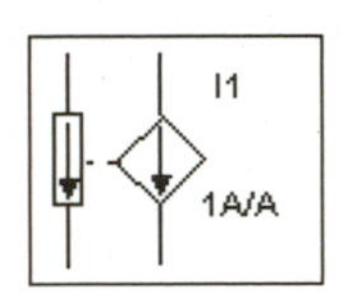

The **Current-Controlled Current Source (CCCS)** outputs a current adjusted by current input. The two are related by a parameter called *current gain* (F), which is the ratio of the output current to the input current. The current gain can have any value from pA/A to TA/A. The formula is: F = Iout / I in. It is used to change the level of a current.

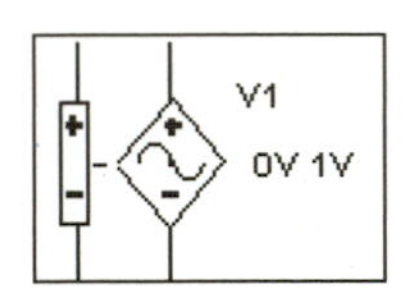

The **Voltage-Controlled Sine Wave Oscillator** may be recognized from its acronym **VCO**. It creates a waveform with a frequency determined by the level of input voltage. You must adjust the control and frequency arrays to fit your circuit by double-clicking the placed component. For example, if you want 0 volts to represent 0 Hz and 5V to represent 1,000 Hz, then adjust the settings as shown in Figure 4.34.

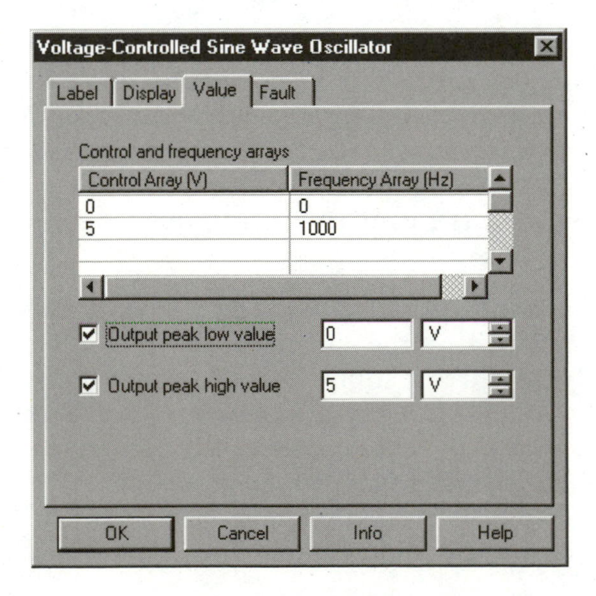

Figure 4.34
The settings in the voltage-controlled sine wave oscillator.

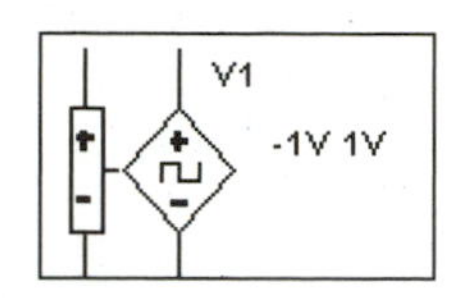

The **Voltage-Controlled Square Wave Oscillator** functions the same as the VCO but outputs a square-wave. It requires more waveform tweaking from you.

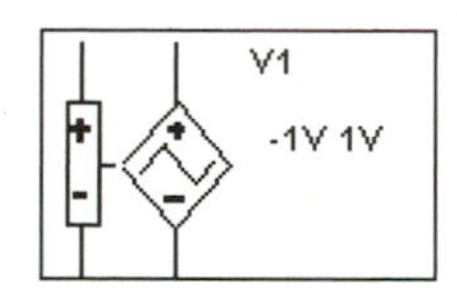

The **Voltage-Controlled Triangle Wave Oscillator** functions the same as the VCO but outputs a triangle-wave. It also requires more waveform tweaking from you.

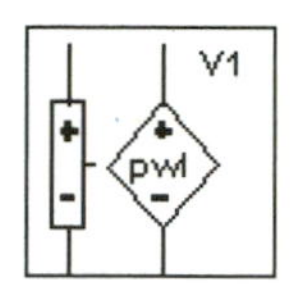

The **Voltage-Controlled Piecewise Linear Source** allows you to control the shape of the output waveform by entering up to five (input, output) pairs, which are shown in the Properties dialog box, Value tab as X, Y coordinates. The X values are input coordinate points and the associated Y values represent the outputs of those points. If you use only two pairs, the output voltage is linear. See Figure 4.35 for an example circuit.

Example

In the sample circuit shown below, a triangle waveform with uniform rise and fall slopes is modified to a parabolic waveform for which the slope increases at each reference point.

The co-ordinate pairs that perform this conversion are:

First pair	0,0	(no change)
Second pair	1,1	(same)
Third pair	2,4	(slope is increased between this pair and the last)
Fourth pair	3,9	(slope increased again)
Fifth	4,16	(even steeper slope)

Note In this example, the Y (output) is the square of the input. It is therefore an exponential.

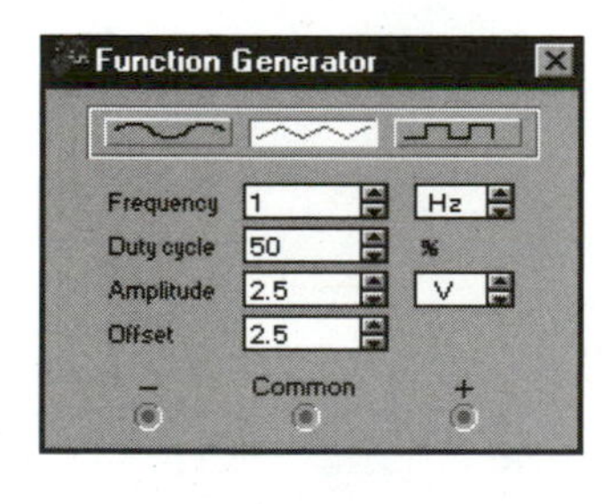

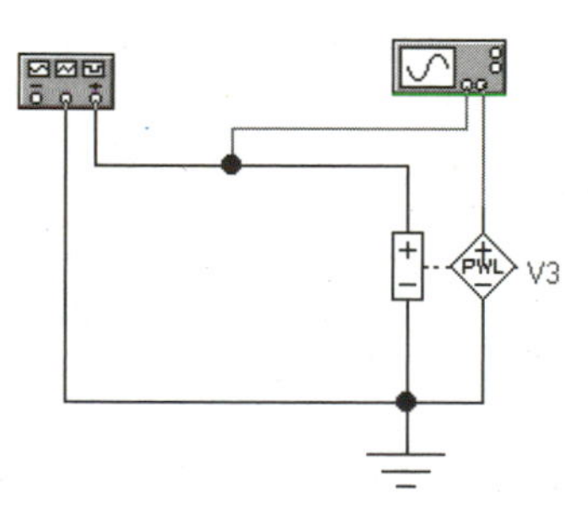

Figure 4.35 An example circuit.

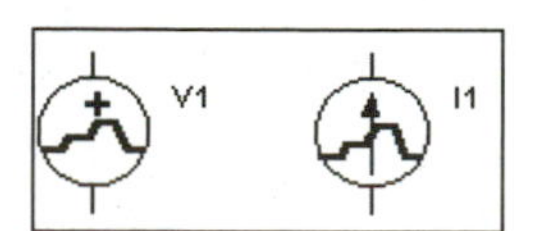

The **Piecewise Linear Voltage and Current Source** is one of Multisim's customization tools. It lets you output a voltage or current waveform that has been described in a text file. The file combines pairs of time and voltage (or current) level information about the wave to be output upon the component's terminals. See Figure 4.36. To connect and run:

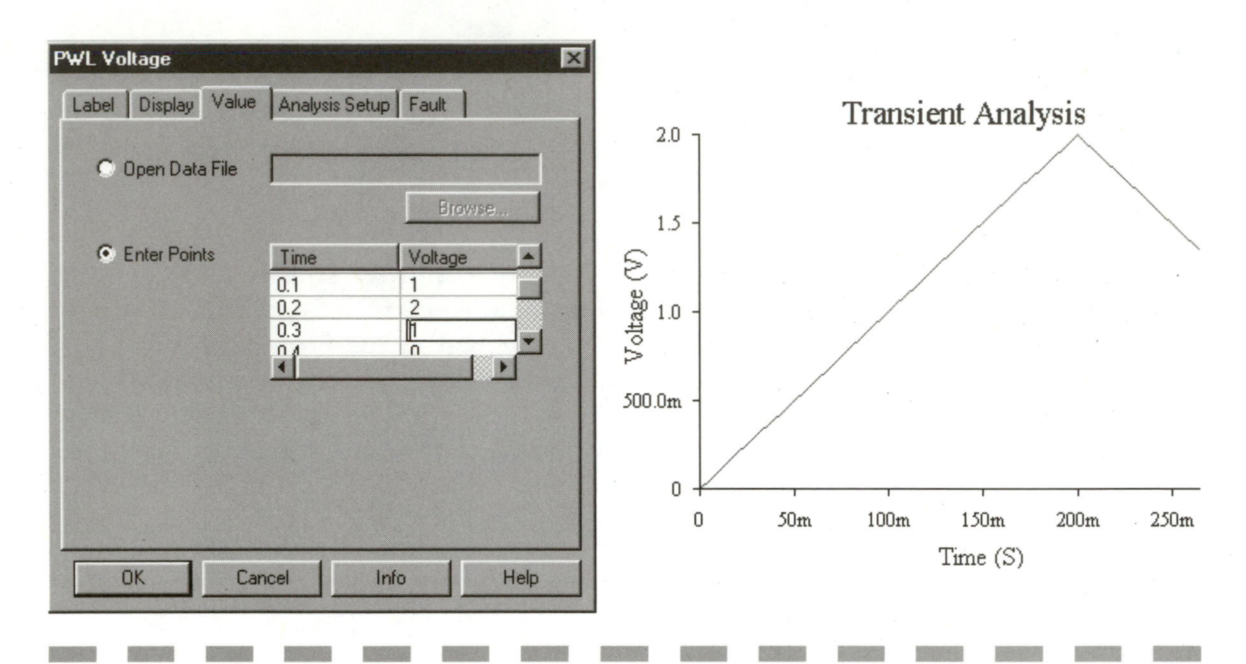

Figure 4.36 The PWL voltage source can be used to make custom voltage signals.

1. Place the component onto the screen and double-click it, choosing the Value tab.
2. Choose either Open Data File (a text file with Data Pairs) or Enter Points (input points using the Components dialog box).
3. If you chose to open a data file, hit **Browse**, find the text file that contains the wave information, and select it.
4. If you chose to input the data yourself, fill out each field on both sides.
5. Hit **OK**. Connect one of the terminals to the circuit and run. Note that the bottom terminal is 180 degrees out of phase from the top terminal and is typically connected to ground.

There are two ways to build a text file to run this circuit. The difficult one is to type the file in by hand, in two columns: left is the time points and right is the voltage level at that time (see Figure 4.37). A much simpler method is to hook the oscilloscope up to a circuit and save the data as an *.scp file. The waveform recorded with the scope is

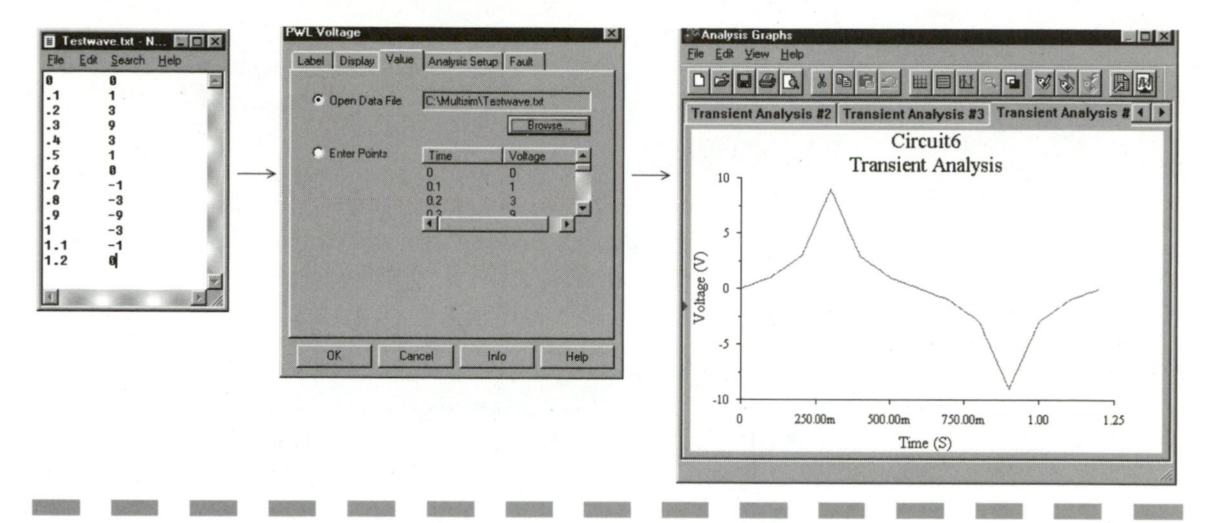

Figure 4.37 You can enter time/voltage pairs with either Windows Notepad or with the "enter points" on the Value tab. See text.

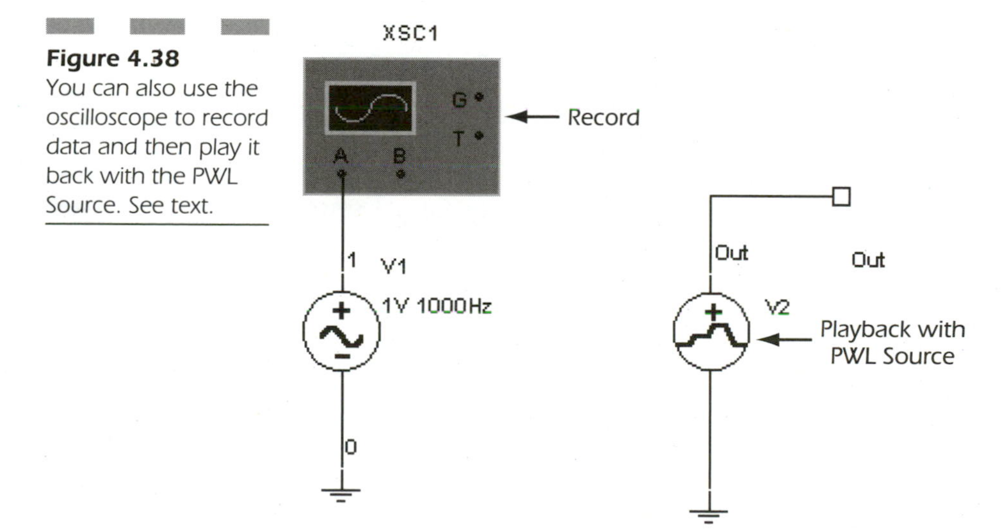

Figure 4.38 You can also use the oscilloscope to record data and then play it back with the PWL Source. See text.

now magically output by the Piecewise component into your circuit (see Figure 4.58).

Used in just about any application that needs a custom waveform, with some ingenuity, this component really releases the power of Multisim. You will see some rather exciting examples of this component's capabilities in later chapters.

*The Write component is no longer available. Instead, hook up the oscilloscope and create an *.SCP file or do a Transient Analysis and save the results as a *.txt file.*

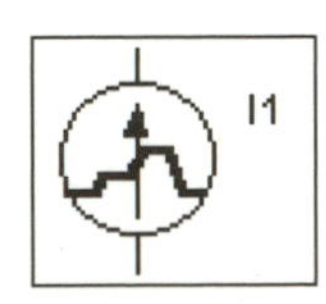

The **Piecewise Linear Current Source** is the same as the Piecewise Linear Voltage Source but produces a current wave according to your input.

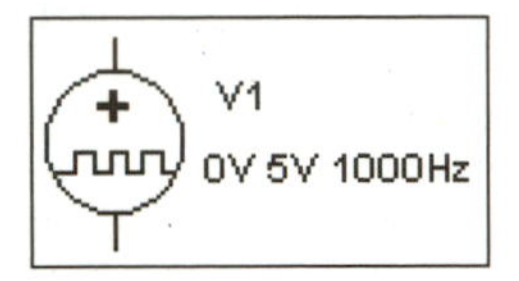

The **Pulse Voltage Source** creates a custom voltage-pulsed wave according to your settings. An example would be a heartbeat or trigger pulse. You can control the length of the pulse, the rise and fall times, and the period. At the time of this writing there was not much information available about this component; so choose the Help file for more info or see the EWB website.

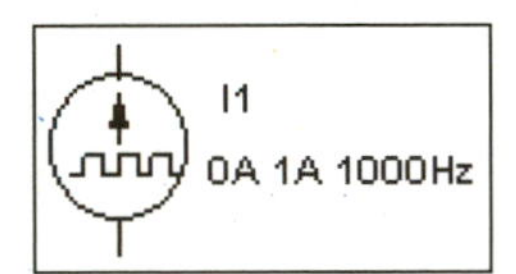

The **Pulse Current Source** is the same as the Pulse Voltage Source but creates a current pulse instead.

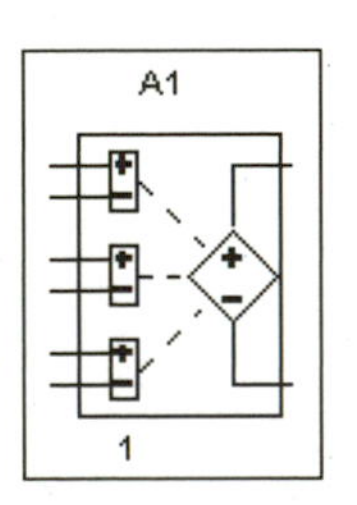

The **Polynomial Source** is a voltage-controlled voltage source that is set up using a polynomial transfer function. It is a specific case of the more general nonlinear dependent source. Use it for analog behavioral modeling. In Multisim, the polynomial source has three controlling voltage inputs: V1, V2. and V3.

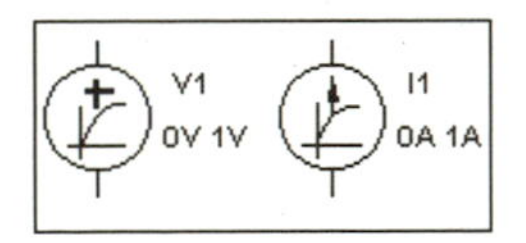

The **Exponential Current Source** and **Exponential Voltage Source** are configurable sources whose output can be set to produce an exponential signal. See the Help file for more information (as of this writing there was no additional information available).

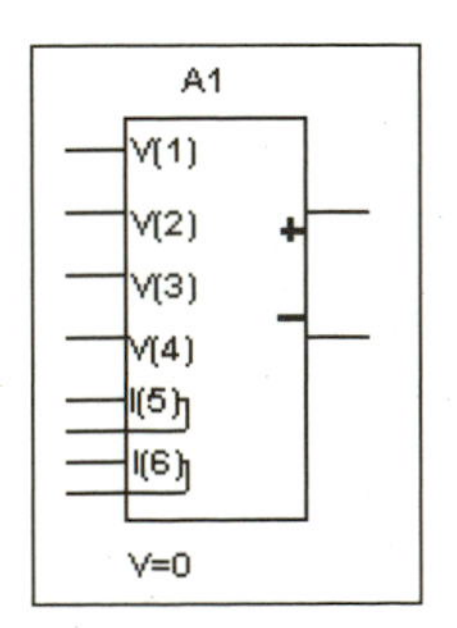

The **Nonlinear Dependent Source** is a generic source that allows you to create a sophisticated behavioral model by entering a mathematical expression. Expressions may contain the operators +, −, *, /, ^, unary-. These predefined functions, abs, asin, atanh, exp, sin, and tan are also allowed.

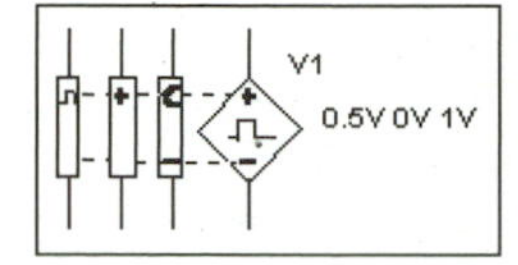

The **Controlled One-Shot** oscillator takes an AC or DC input voltage, which it uses as the independent variable in the piecewise linear curve described by the control/pulse width pairs. From the curve, a pulse width value is determined, and the oscillator outputs a pulse of that width. You can change the clock trigger value, output delay from trigger, output delay from pulse width, output rise and fall times, and output high and low values. When only two coordinate pairs are used, the oscillator outputs a linear variation of the pulse with respect to the control input. When the number of coordinate pairs is greater than two, the output is piecewise linear.

Basic Toolbar

For an overall view of the Basic toolbar, see Figure 4.39.

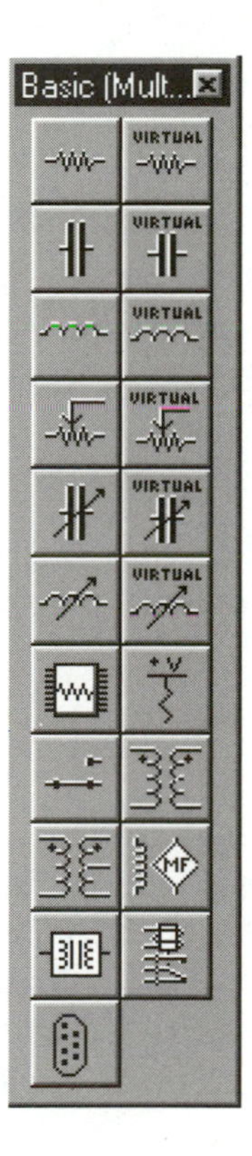

Figure 4.39
The Basic toolbar.

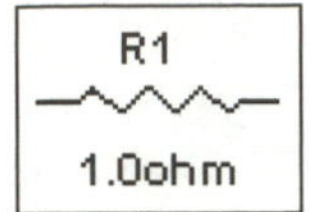

Resistors are perhaps the most versatile electronic components in existence. They can be used to control voltage and current, pull up or down a logic gate state, provide a test load, and in countless other tasks. Resistors come in a variety of sizes and power ratings depending on the application. Multisim simulates many resistor-related variables including power dissipation, tolerance, minimum operating temperature, maximum temperature, and others. Multisim makes you first open a dialog box and then pick a specific resistor, such as a 1-ohm, .5-watt unit (unless you use a virtual resistor). The only way to edit or create a resistor is by using the Component Editor and Component Properties dialog boxes.

If you simply want to place a resistive load in a circuit without having to use the Component Browser, you can choose a **Virtual Resistor**. Adjustments are made post-placement by double-clicking the component. Use it to place a resistor quickly.

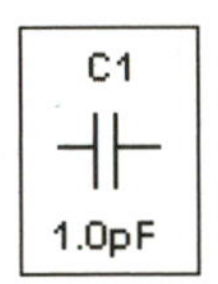

The **Capacitor** is perhaps the second most-used electronics component after the resistor. A capacitor stores energy in the form of an electrostatic field. Capacitors are typically used to filter or remove AC signals from a variety of circuits. In a DC circuit, they can be used to block the flow of direct current while at the same time allowing AC signals to pass. DC circuits also use capacitors as resistor/capacitor (RC) time constant circuits. The resistor allows the capacitor to charge at a rate set by the RC combination, then drain at a set rate. Capacitance is measured in *farads*. As with the resistors, capacitors must be placed as components with the Browser dialog box.

The **Virtual Capacitor** quickly places capacitance into a circuit.

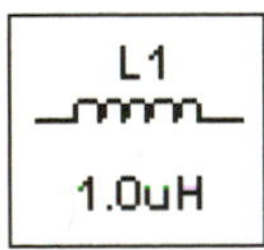

The **Inductor** is similar to a capacitor in that it stores electricity, except in the form of an electromagnetic field. It does this when the current flowing through it changes. This characteristic is useless in DC circuits, but in an AC circuit, it will oppose a change in current flow. This is called *inductance* (L), which is measured in *henrys* (H). This item uses the Inductor Browser dialog box to make a selection.

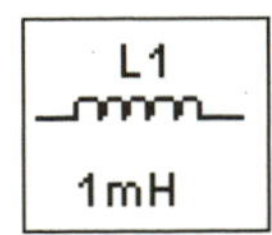

The **Virtual Inductor** quickly places an inductance source in a circuit.

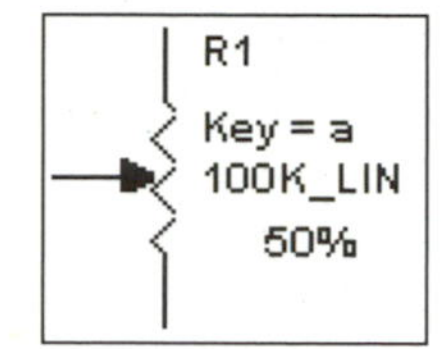

The **Potentiometer (Variable Resistor or POT)** is a user-controlled component that works in much the same way as a volume control on a stereo (most volume controls are POTs). It provides an adjustable resistance path for current to flow through. By varying the amount of resistance from one side to the other, you can control the voltage or current. This is done by first identifying a control key on the keyboard. When you hit that key the value is decreased by a preset percentage. If you hit the Shift key + the letter you chose again, the value increments. Its not a quick-turning knob-type gadget, but it will do for simulations.

You must pick the value of the POT with the Browser when you first place it. Once it is placed, double-clicking the component opens the window shown in Figure 4.40. You can set the keyboard letters you wish to use to operate the POT and also adjust the initial setting and by how much a key-press will increase or decrease the POT.

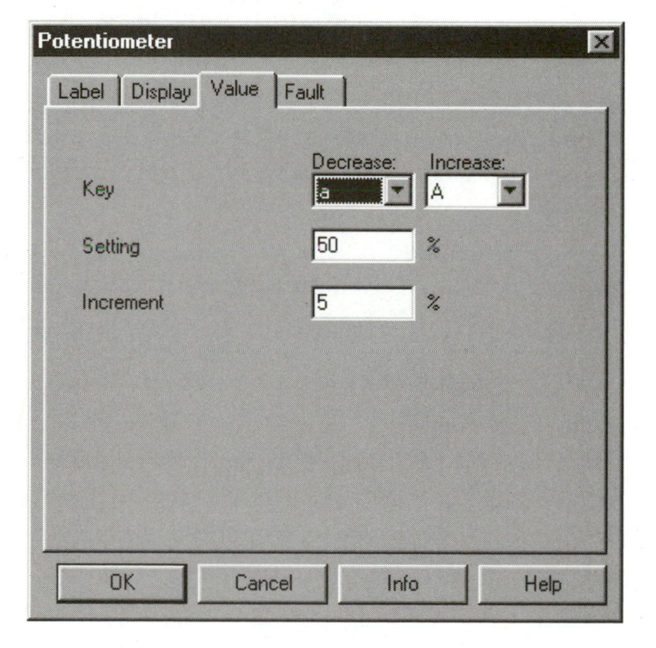

Figure 4.40 Double-click the placed POT and choose the Value tab to assign the Control key. See text.

NOTE

Potentiometers do not rotate or flip in Multisim, as of version 6.11. EWB claims this will be fixed in later versions.

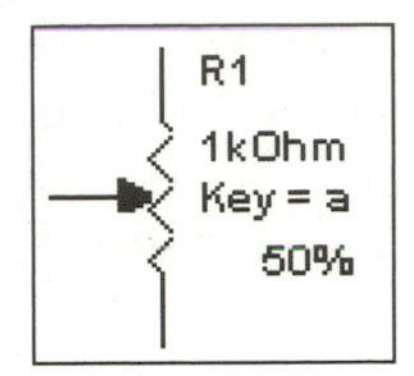

The **Virtual Potentiometer** is a quick way to place a POT. Double-click the component, and you can make it any value you wish.

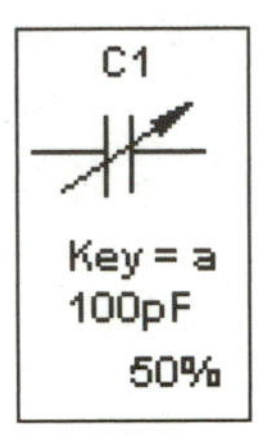

The **Variable Capacitor** allows you to adjust a capacitor's farad rating while the circuit is running. This is done by assigning a keyboard letter as a control. Pressing it lowers the value by a set percentage; pushing Shift + the same key increases the value. The Variable Capacitor operates the same way as the Potentiometer. Only three values exist for variable caps: 100pF, 30pF, and 350 pF. The only way to change this is to use a Virtual Variable Cap or to make a custom variable cap. Here's how:

1. Begin to place a Variable Capacitor. When the Browser opens, select any of the caps and hit **Edit**. Place the component into the User database. This opens the Component Properties dialog box.
2. Go to the Model tab. In the Template window, scroll until you see (* : **capacitance 100e-12**;). Change this value to the new farad value. For instance, if you want a 1μF variable cap, make the value **10e-06**.
3. Change the name of the Component and Model to 1μF. (The μ symbol is entered by holding the Alt key and typing in 0181.)
4. Hit **Save/Exit**, **Refresh User Toolbar**, and place component.

Radio frequency circuits (*tank circuits*) often require this component to dial in a certain frequency. In fact, on old radios, the frequency dial was connected to a variable air-core capacitor.

As of Multisim 6.2, you can no longer edit or copy interactive components.

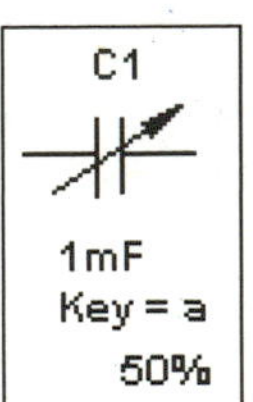

The **Virtual Variable Capacitor** allows you to make your own variable caps for simulation purposes.

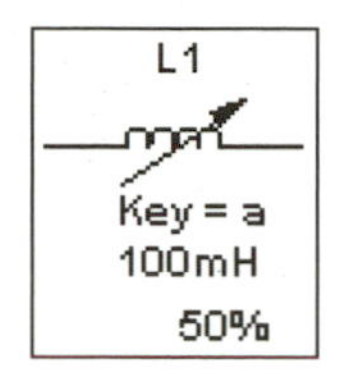

The **Variable Inductor** is the same as the Variable Capacitor, but offers changing inductance. Double-click and choose the Value tab to set the control key.

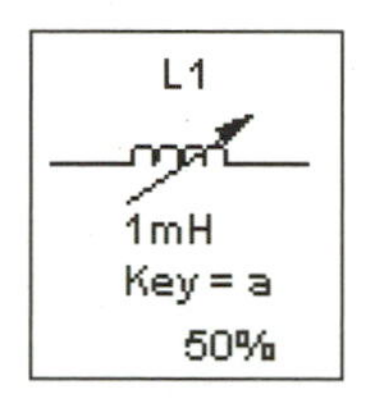

The **Virtual Variable Inductor** is a quick way to place a Variable Inductor or place a certain-valued Variable Inductor that is not covered by the previous component. Double-click and choose the Value tab to set the control key and value.

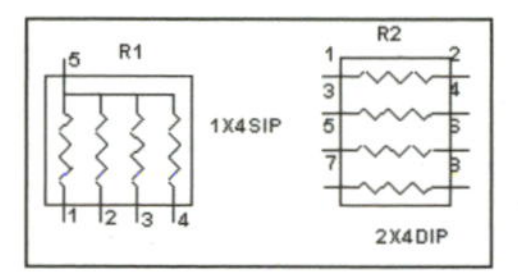

The **Resistor Pack** has several resistors within one package. Each resistor can have one terminal connected to a common pin (SIP) or have a separate pin (DIP). It's a cleaner way to design circuits using lots of resistors. When you select the component, the Browser allows you to select bused (SIP), isolated (DIP), or even a split terminator version. A typical use would be to run each line in an 8-bit bus through a 1K resistor (see Figure 4.41).

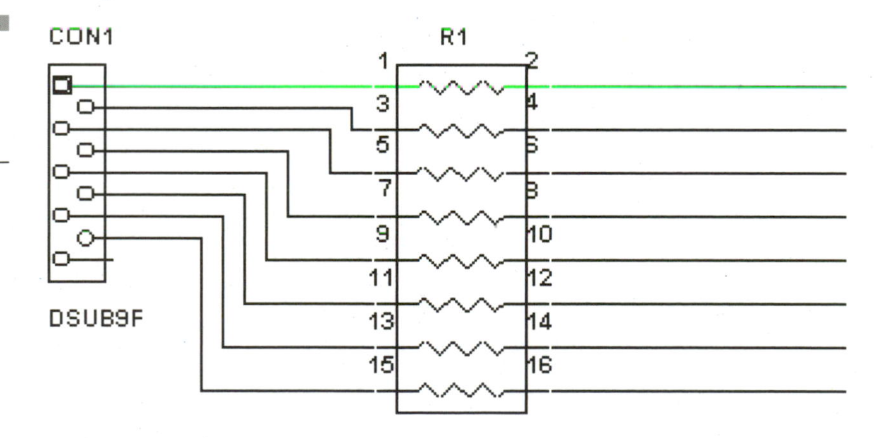

Figure 4.41
A resistor pack used inline with a data bus.

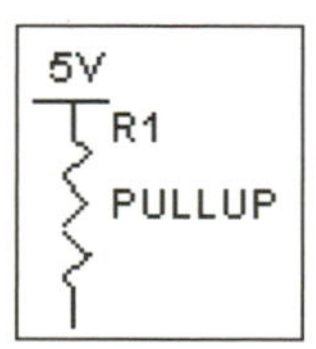

The **Pull-Up Resistor** is great when you are working on digital or mixed circuits. Sometimes one of the IC pins must be raised to 5 volts through a current-limiting resistor. This component saves having to run all sorts of irritating spaghetti lines and multiple components. The resistor has a value of 1Kohms.

TIP

To create different value pull-ups, edit an existing one by changing its model using the Component Properties window. Change the 1,000 in line 2 to the new resistance value and change the name in line 1. Be sure to change the name under the General tab as well.

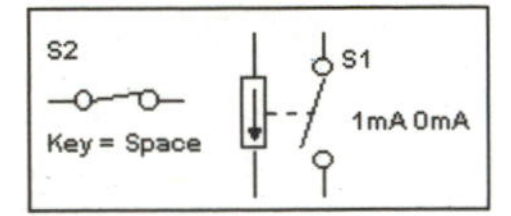

The **Switch** gives you a way to control your circuit on the fly. It conducts electricity from one terminal to the other until the appropriate key is struck, changing the flow to the other terminal. Other switches in Multisim are controlled by current, time, or voltage.

Switches are used to control a circuit while it is active, using your keyboard. Figure 4.42 is a Single-Pole, Double-Throw (SPDT) switch. Let's assume the terminal that is on the left side is number 1, the top-right terminal is A, and the bottom-right terminal is B.

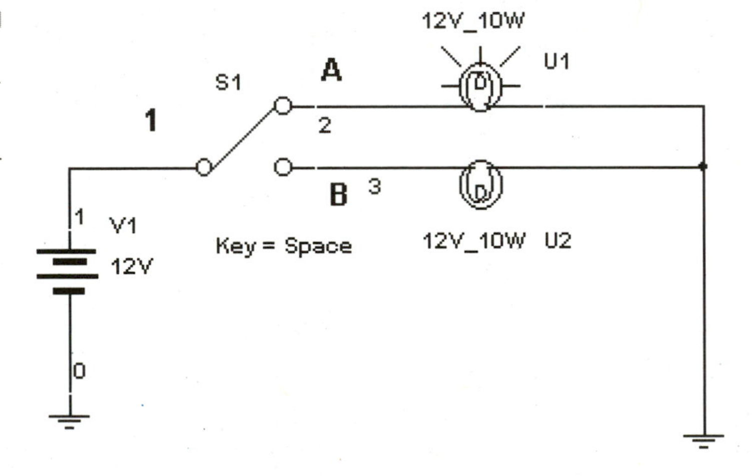

Figure 4.42
A single-pole, double-throw switch example.

- 1 to B conducts upon activation.
- Hitting the Space key makes 1 conduct to A.
- Hitting the Space key again makes 1 conduct to B once more.

Here is how to set a key to operate as a switch in your circuit (see Figure 4.43).

1. Place a switch on the drawing board and connect the appropriate terminals.
2. Double-click the switch or use the menu option to get to its properties.
3. Choose the Value tab.

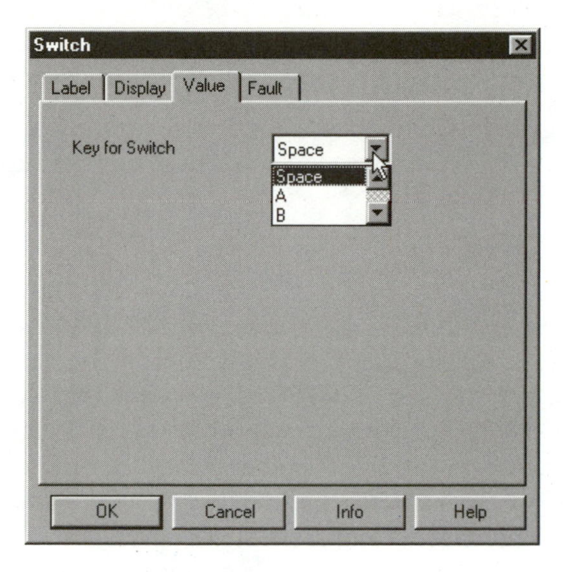

Figure 4.43 Double-click Switch and use the Value tab to assign a control key.

4. Enter the character/number used to control the switch during simulation.
5. Hit **OK**.
6. Activate the circuit.

Pushing the key once flips the switch to the other terminal. Hitting it again places it back to the original terminal.

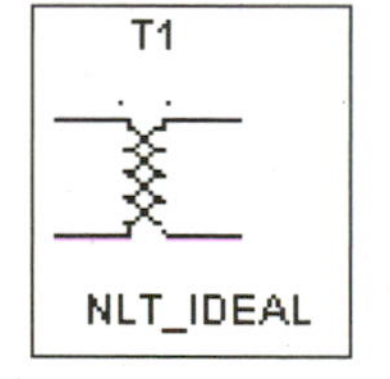

Using the **Nonlinear Transformer**, you can model physical effects such as nonlinear magnetic saturation, primary and secondary winding losses, primary and secondary leakage inductance, and core geometric size. See the Multisim Help file for detailed information on this component's use.

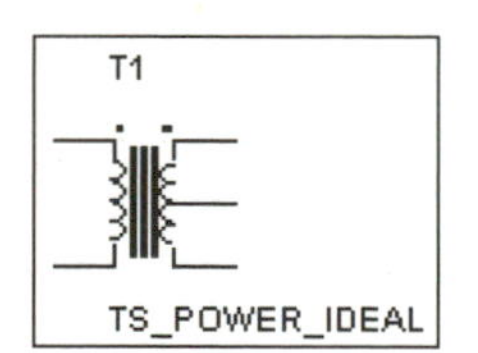

The **Transformer** is used to step down or step up voltage in an AC circuit. When you place a transformer, the Browser opens to give you a selection of pre-made units. A transformer is typically used in appliances to change step-down or step-up voltage; say 120 volts to 12 volts.

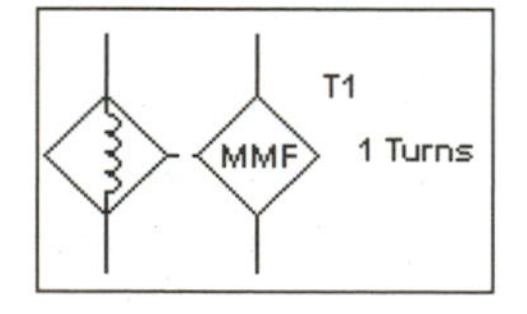

The **Coreless Coil** component is a conceptual model that you can use as a building block to create a wide variety of inductive and magnetic circuit models. Typically, you would use the coreless coil together with the magnetic core to build up systems that mimic the behavior of linear and nonlinear magnetic components. It takes a current input and produces a voltage. The output voltage behaves like a magnetomotive force in a magnetic circuit; that is, when the coreless coil is connected to the magnetic core or some other resistive device, a current flows.

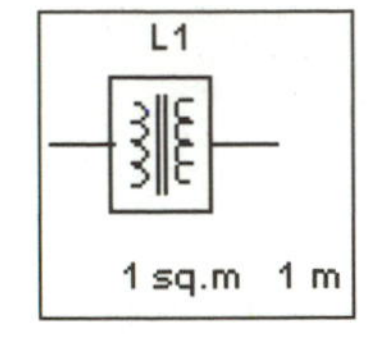

The **Magnetic Core** allows you to accurately model an inductive/ magnetic component. See the Help file for more information.

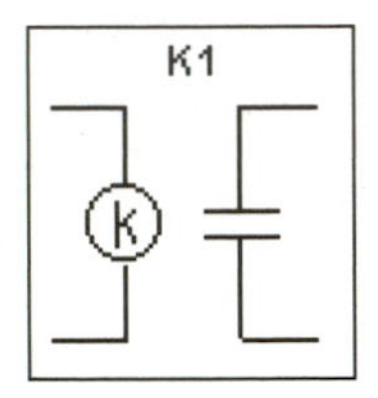

The **Relay** is modeled after a magnetic relay such as those used in automatic car door locks. A small current charges a magnetic core, causing a switch to turn on/off. The idea is that you are controlling a high-current circuit with a low-power source. Relays are often used as an interface between a low-power IC and a high-voltage high-current device. Most of the Relay components in the Browser have four connections. The two connections on the left side (circle and K) are the magnetic coil. The two terminals on right are the contacts. When the coil has a set amount of current going through it, the contacts close (or open), depending on how the component is being used. Figure 4.44 shows a low-power 5-volt clock being used to energize the coil. Two high-wattage light bulbs are hooked to the switching side of the relay.

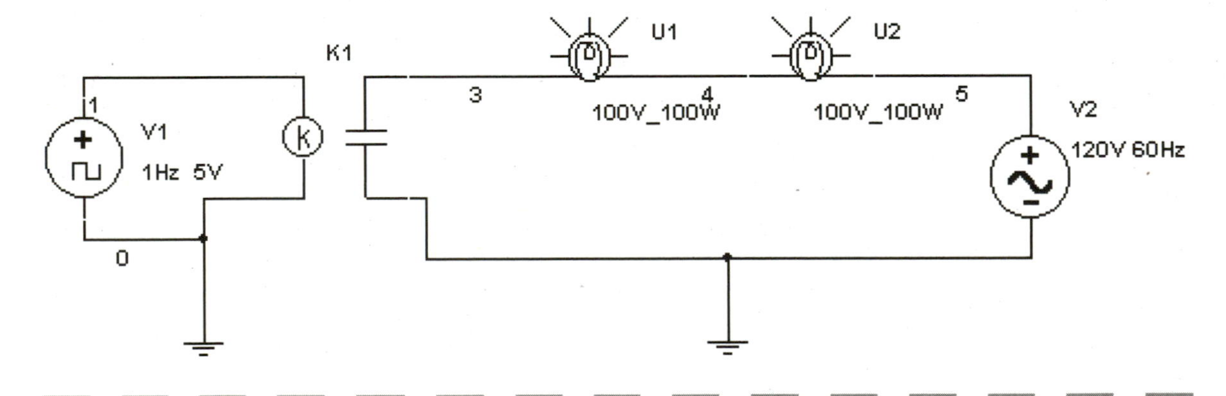

Figure 4.44 A relay example circuit.

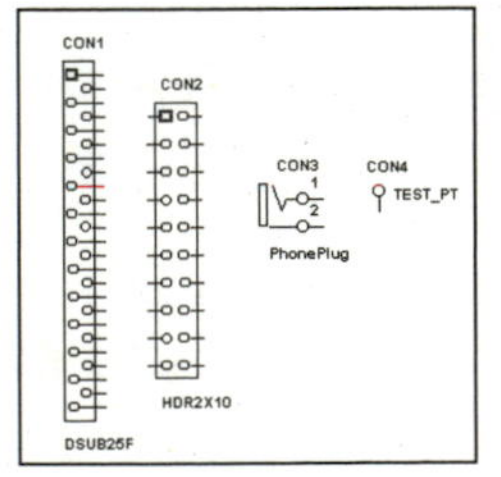

Connectors represent several real-world connectors such as DSUBs (often used on the back of computers for the parallel port connection), header strips (often used inside computers, to connect the hard drive cable), a phone plug (used in audio equipment), and a test point.

Connectors are mainly used in schematic capture and also to export this information to a PCB package.

Diodes Toolbar

For an overall view of the Diodes Toolbar, see Figure 4.45.

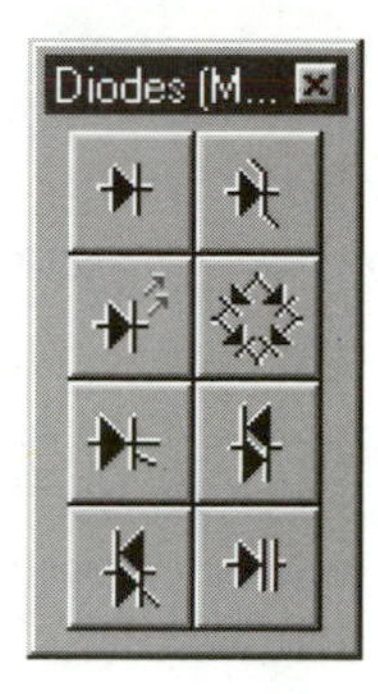

Figure 4.45
The Diodes toolbar.

Diodes are basically components that allow current to flow in only one direction. They are used in applications like simple solid-state switches in AC circuits, to help change AC into DC (rectification), and in the case of an LED, to provide a light show.

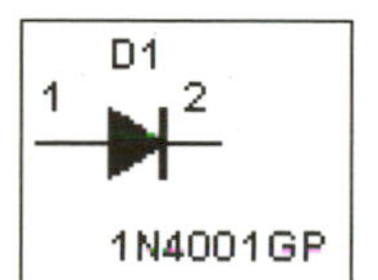

Using the **Diode** component, select from among hundreds of signal and general purpose diodes. For a detailed report of a specific diode, hit the report button in the Diode Browser.

The **Zener Diode** operates in a reverse breakdown region (called the *zener region*). It is used to provide voltage regulation. Think of it as being equal to a battery providing the amount of voltage as set in the zener voltage (Vz). See Figure 4.46.

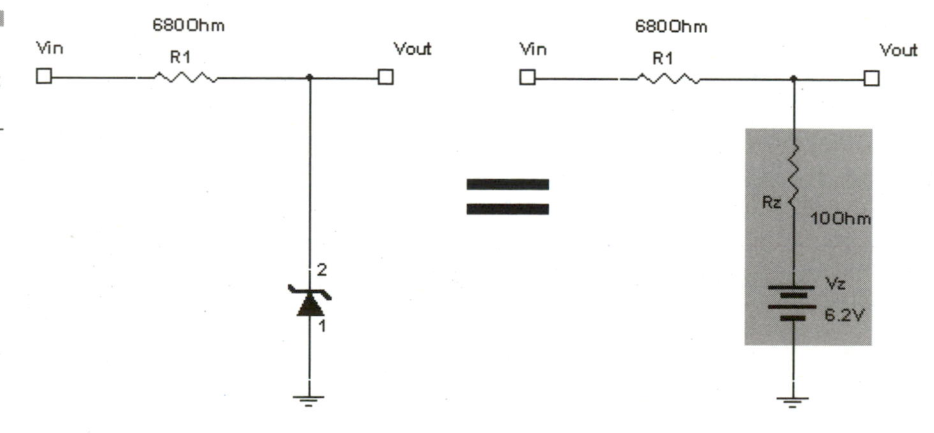

Figure 4.46
A Zener diode and its equivalent circuit.

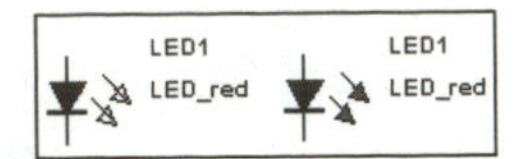

The **Light-Emitting Diode (LED)** is a special diode that emits light when current is flowing through it. Typically used as an indicator of some sort, it can, for example, signal the fact that a circuit is indeed getting current. Multisim's schematic representation of an LED is not too exciting but it does the job. When the diode is functioning, the center of the two arrows becomes filled with color when activated, indicating a lit state. Several LEDs are selectable from the Browser dialog box, including various colors as well as an infrared (IR) version.

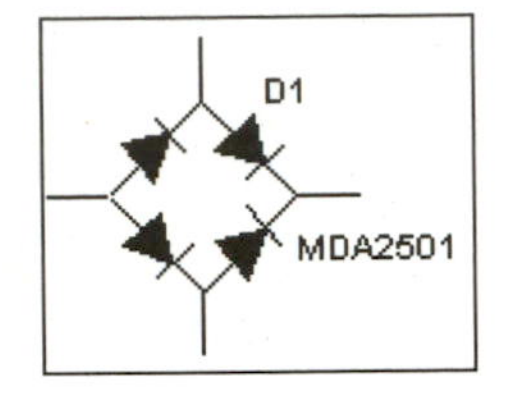

The **Full-Wave Bridge Rectifier (FWB)** is a combination of four diodes that are configured to perform full-wave rectification. It's a great tool to simulate power supplies. By using this component you can save the time that would have gone to spitting out four diodes. The other advantage is that ITT has built-in gazillions of models to choose from for accurate simulations, analysis, and export to PCB layout. (Unfortunately, Personal Version only has two, but you can download many others from a manufacturer's website.)

NOTE

Thyristors are in a family of devices containing four layers of semiconductor material. They are diodes that conduct once a certain voltage level is reached. In most cases, the cathode-to-anode current flow is controlled with a gate; When the gate is activated, there is a low resistance between the anode and cathode; when the gate is shut off, the low resistance continues until the level drops below a certain point or shuts off.

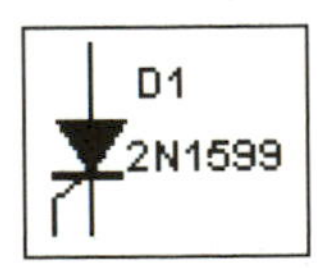

The **Silicon-Controlled Rectifier** is a thyristor that can be thought of as a one-way valve with a switch that activates the flow. It consists of three connections: the anode, cathode, and gate. When a pulse is applied to the gate (switch), the diode section conducts in one direction. When the gate is off, the diode continues to conduct until it drops below a certain level. An SCR is typically used when interfacing a low-power digital circuit to a high-voltage AC line, such as in your home. A lamp dimmer and motor control are two examples.

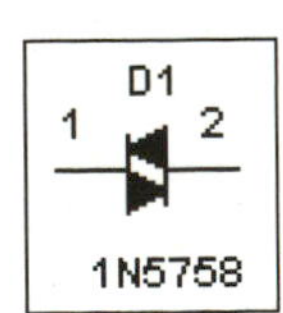

The **Diac** thyristor is a bi-directional device without a gate to switch it off/on. It conducts current in both directions, once a certain voltage is reached. There are only two terminals on this component.

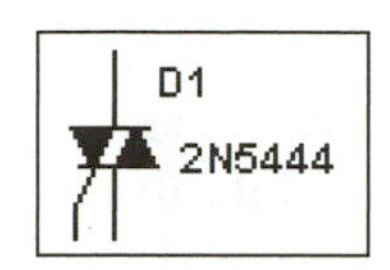

The **Triac** thyristor is a bi-directional device with a gate.

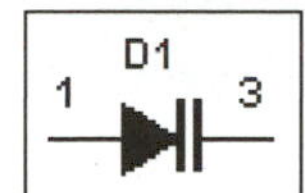

The **Varactor** diode is fundamentally equal to a voltage-variable capacitor when operated in the reverse-biased mode. The reverse-bias voltage controls the level of capacitance: raising voltage equals decreased capacitance and vice-versa. Varactors are typically used in electronic tuner circuits in TVs and other circuits.

Transistors Toolbar

The transistor is the hands-down winner for the Invention of the 20th Century Award. It has created a whole electronics industry that thrives on its ability to miniaturize. A transistor is a valve used to control electric current. It is typically used for amplification and switching in billions of circuits. A transistor usually contains three terminals. A *bipolar junction transistor* (BJT) uses a collector, base, and emitter. A *field-effect transistor* (FET) uses a drain, gate, and source. Each transistor has its own applications and form of operation. See Figure 4.47 for an overview of the Transistor toolbar.

Figure 4.47 The Transistor toolbar.

BIPOLAR JUNCTION TRANSISTORS (BJTS)

Simply stated, a small change in current flow between the base to emitter of the transistor creates a large change in current flow in the collector-to-emitter current. A BJT is made with three layers of doped semiconductor regions with two PN junctions between the layers. A BJT transistor operates on the principle that the current involves carriers of both polarities—*holes* and *electrons*.

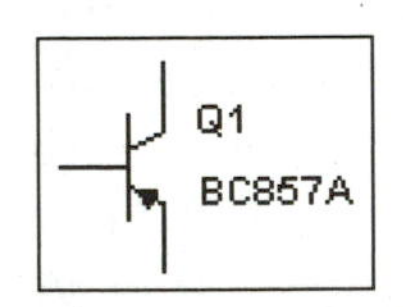

The **BJT_NPN (Bipolar Junction Transistors, Negative-Positive-Negative)** transistor consists of two negative layers separated by a positive junction layer. The NPN or negative-positive-negative transistor is the most commonly used model. It can be used for switching and amplification purposes. In an NPN the base is positive and the collector/emitter are negative. A small base-to-emitter current flow will cause a greatly amplified collector-to-emitter current flow.

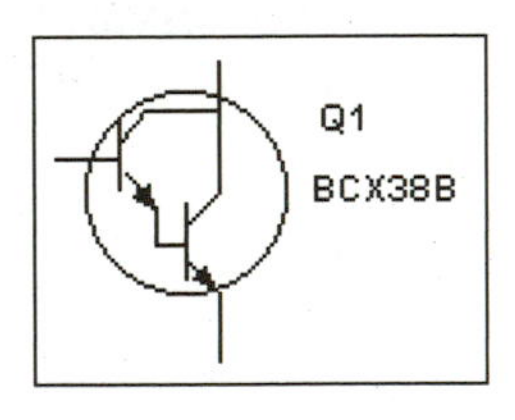

The **BJT_PNP** is the same as the BJT_NPN except that there are two positive layers and one negative layer.

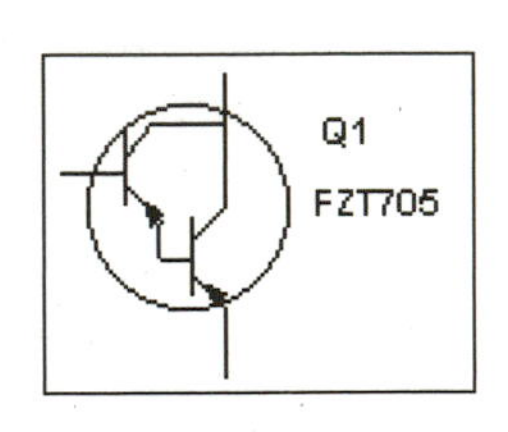

The **Darlington_NPN** is a transistor on steroids. It uses two transistors in one package to create an extremely high-gain transistor. It takes the signal coming from the first transistor and multiplies it by the gain of the next. Say, for example, a 1μA current comes into the first transistor and is amplified by 100. This gives the next transistor an input of 100μA, which is then amplified by a factor of 100 again. This gives a total current gain of 10,000 (10,000 h_{FE}). Quite a jump!

The **Darlington PNP** is the same as the Darlington NPN, but made with two BJT_ PNPs.

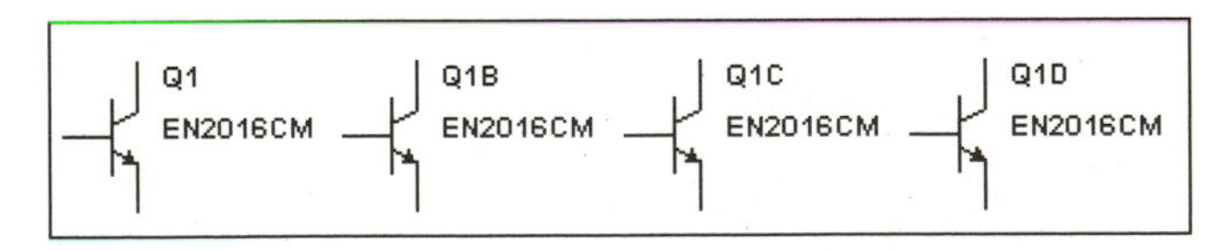

The **BJT Array** integrates several BJTs into one package. You must first select the component in the Browser window. When you go to place the component, you may be presented with the Select Section window. Choose which transistor to place from the array seen in Figure 4.48.

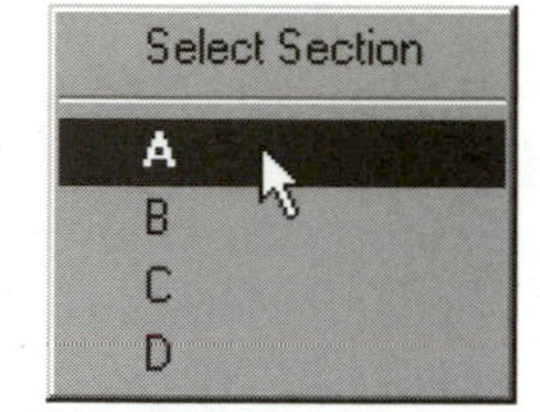

Figure 4.48 The Select Section window lets you choose which transistor to place in the array.

JUNCTION FIELD EFFECT TRANSISTORS (JFETS)

Junction field effect transistors use either a negative or positive channel with two opposite polarity gates on either side of the channel. Think of it as a garden hose with a hand wrapped around it. The top of the channel (hose) is the drain and the bottom of the hose is the source. The gate would be your hand wrapped around the hose. In an n-channel FET, if the gate-to-source junction is reverse-biased (negative voltage applied), a field is created that lowers the drain-to-source current. It's like squeezing the hose with your hand. Refer to Figure 4.51. These transistors are used in applications that require a high input impedance and high frequency response. Junction field effect transistors are shown in Figures 4.49 and 4.50.

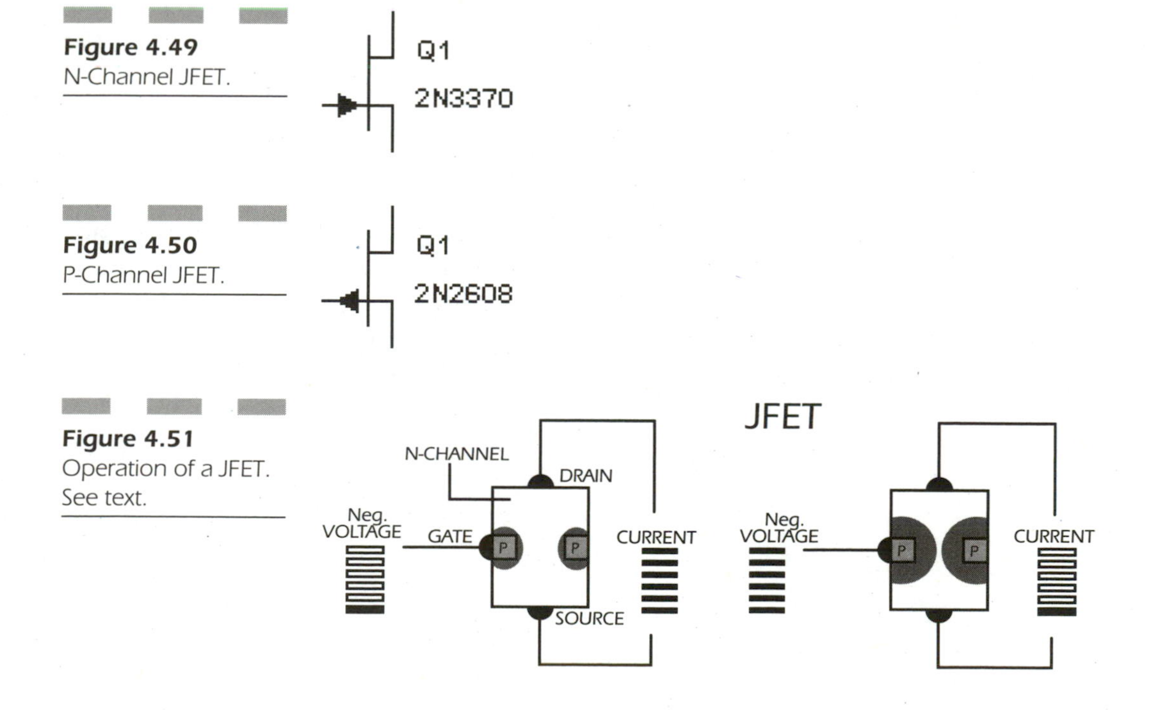

Figure 4.49 N-Channel JFET.

Figure 4.50 P-Channel JFET.

Figure 4.51 Operation of a JFET. See text.

MOSFET

Metal-Oxide-Silicon Field-Effect Transistors (MOSFETs) are a category of FET with no PN junction structure. The gate is instead insulated from the channel with a layer of silicon dioxide.

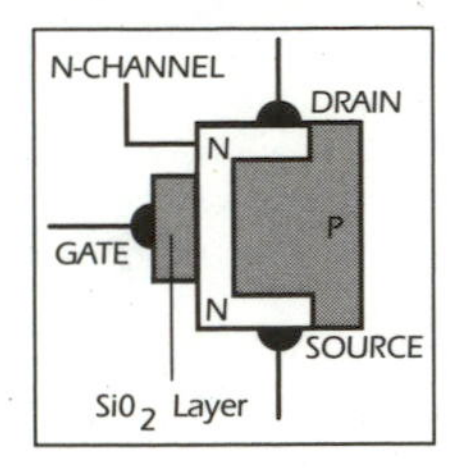

MOSFETs come in two basic types: depletion-enhancement (DE) and Enhancement only (E) mode. Each can further be broken into positive or negative channel. The DE MOSFET (Figure 4.52) can operate in either depletion mode or enhancement mode, whereas the E MOSFET can only operate in the enhancement mode. Depletion mode (in an n-channel MOSFET) occurs when the n-channel is depleted of some of its electrons, thus decreasing the channel's conductivity. A higher negative gate-to-source voltage lowers the drain-to-source's current, until ultimately it is nil. Enhanced mode (in an n-channel MOSFET) occurs when more electrons are conducted into the n-channel with a positive voltage, thus enhancing the conductivity of the channel. A greater positive gate-to-source voltage increases the channel's drain-to-source current. See Figure 4.53.

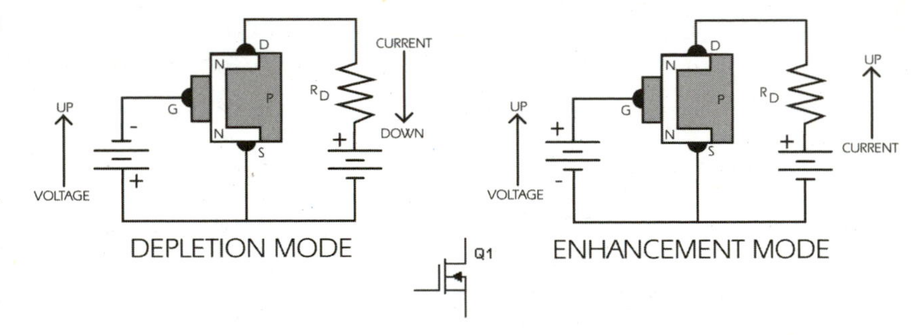

Figure 4.53
An n-channel DE MOSFET can be used in Depletion and Enhancement mode.

Enhanced MOSFETs do not have a physical channel. Instead (on an n-channel enhanced MOSFET), a positive voltage on the gate induces a channel. The higher the voltage, the greater the conductivity of the induced channel (gate-to-source). If the voltage is below a certain threshold, there is no induced channel. (See Figure 4.54.)

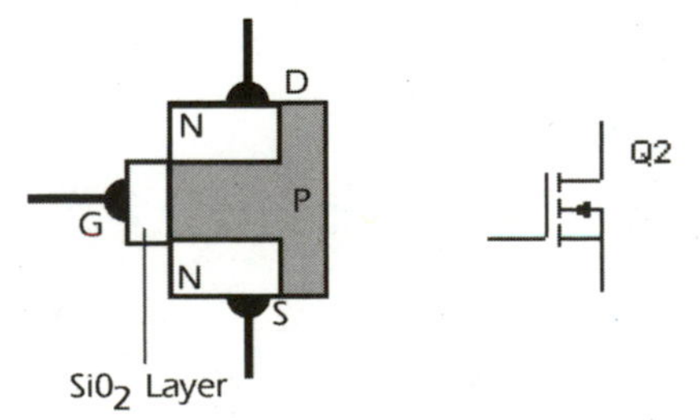

Figure 4.54
An n-channel E MOSFET only works in Enhancement mode.

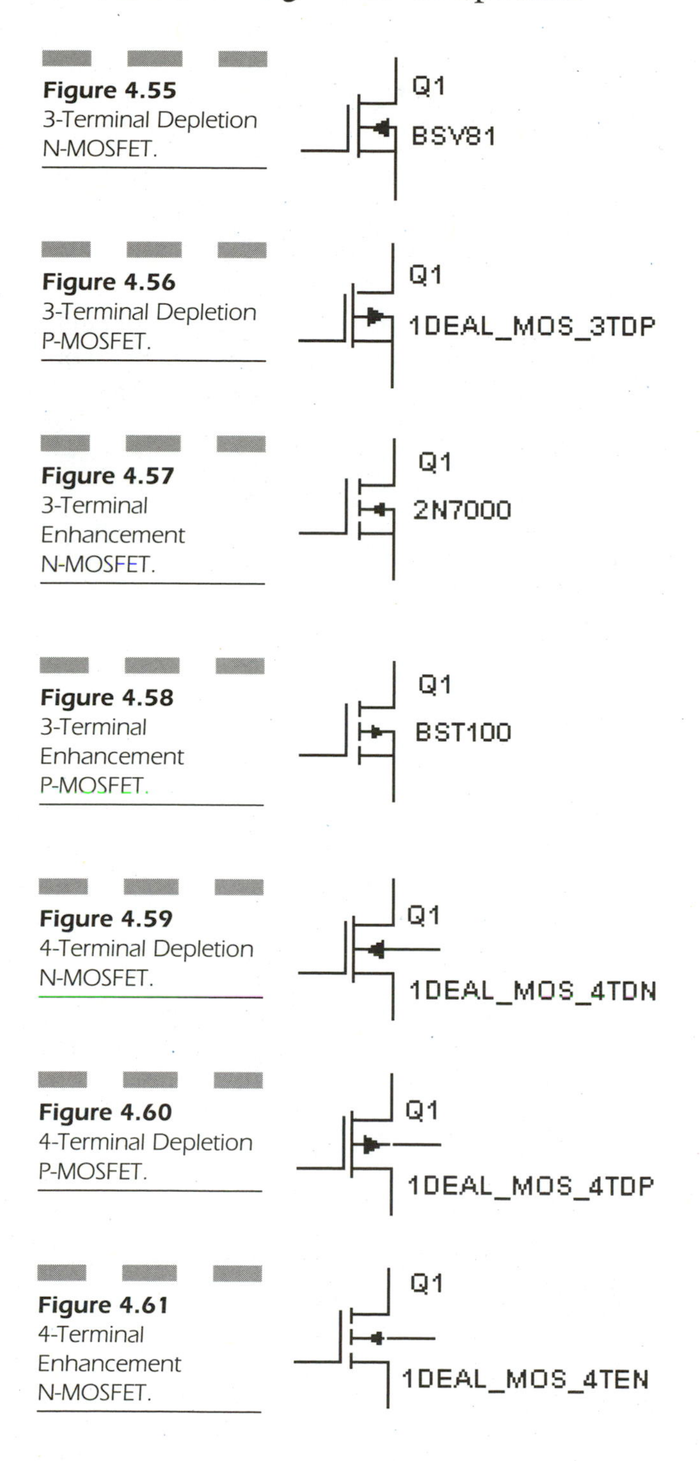

Figure 4.55 3-Terminal Depletion N-MOSFET.

Figure 4.56 3-Terminal Depletion P-MOSFET.

Figure 4.57 3-Terminal Enhancement N-MOSFET.

Figure 4.58 3-Terminal Enhancement P-MOSFET.

Figure 4.59 4-Terminal Depletion N-MOSFET.

Figure 4.60 4-Terminal Depletion P-MOSFET.

Figure 4.61 4-Terminal Enhancement N-MOSFET.

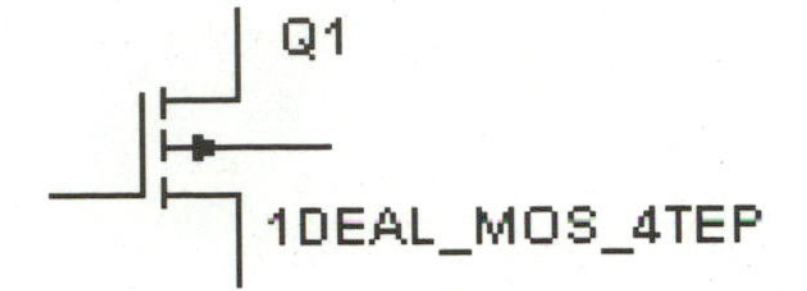

Figure 4.62 4-Terminal Enhancement P-MOSFET.

POWER MOSFETS

Power MOSFETs are used in applications that require control of higher currents and speeds of switching. A good example is a motor control circuit, which uses *pulse width modulation* (PWM) to control the speed of rotation.

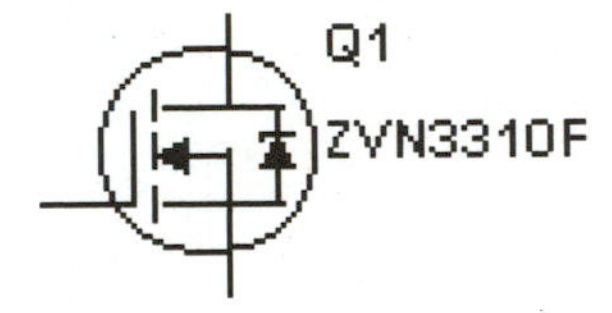

Figure 4.63 Power MOS_N (N-Channel Power MOSFET).

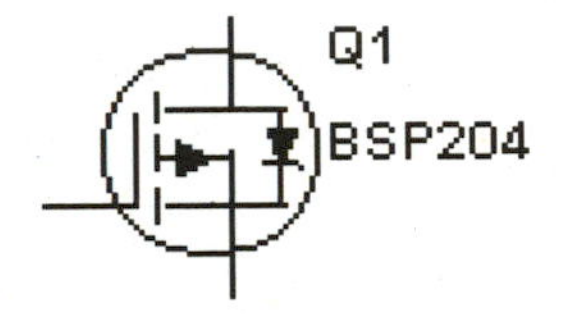

Figure 4.64 Power MOS_P (P-Channel Power MOSFET).

GALLIUM ARSENIDE FIELD-EFFECT TRANSISTORS (GAASFETS)

GaAsFETs are a high-speed field-effect transistors that use *gallium arsenide* (GaAs) as the semiconductor material, rather than silicon. They are generally used in very high frequency amplifiers (into the gigahertz range) found in satellite applications and cell phones.

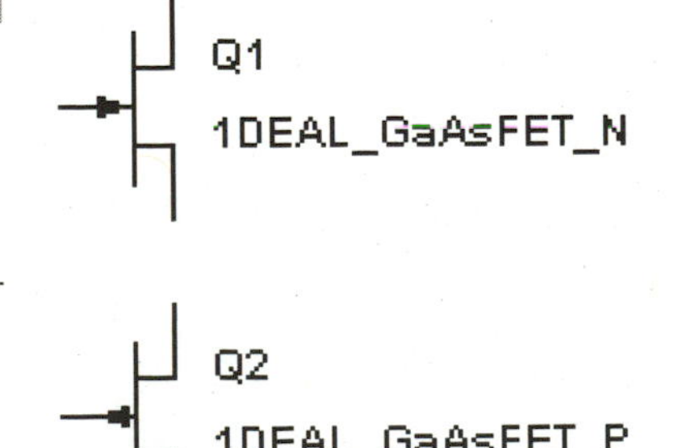

Figure 4.65 GaAsFET_N (N-CHANNEL GaAsFET) & GaAsFET_P (P-CHANNEL GaAsFET)

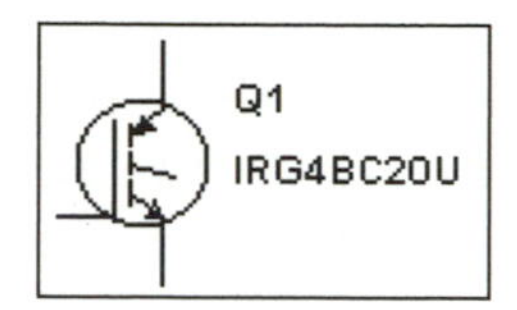

The **Insulated Gate Bipolar Transistor** (IGBT) is a MOS gate-controlled power switch with a very low on-resistance. It is similar in structure to the MOS-gated thyristor, but maintains gate control of the anode current over a wide range of operating conditions.

Analog ICs Toolbar

Figure 4.66 shows the Analog IC toolbar.

Figure 4.66
The Analog ICs toolbar.

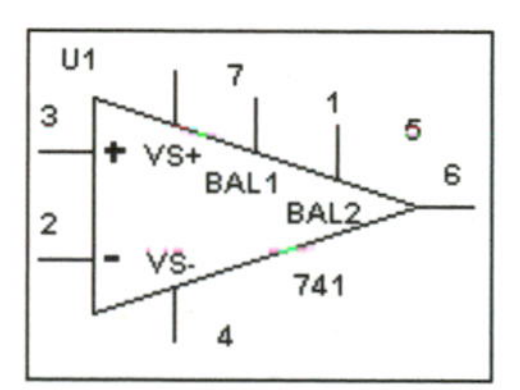

The **Opamps** or operational amplifiers are practical amplifier ICs used in millions of electronic applications. The op amp, in its simplest form, consists of three terminals: non-inverting input (A), inverting input (B), and output. It is basically a differential amplifier, meaning it will subtract the voltage of B from the voltage of A, then multiply it by the amount of gain. The results are placed at the output terminal. It can also be used as an inverting amplifier and non-inverting amplifier, such as those used in stereo circuitry.

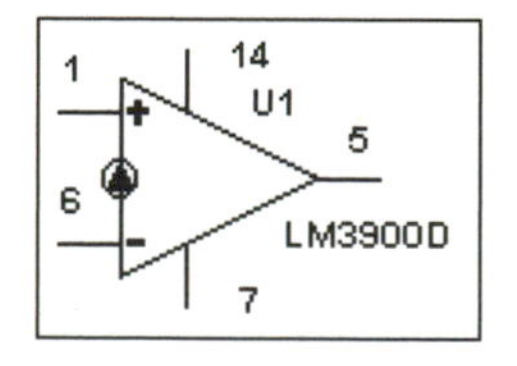

The **Norton Opamp** is a current-differencing amplifier (CDA), which is a linear device that is compatible with digital circuitry and operates from a single power supply. The Norton amplifier is similar in many ways to the comparable op amp circuit. The gain equation is identical.

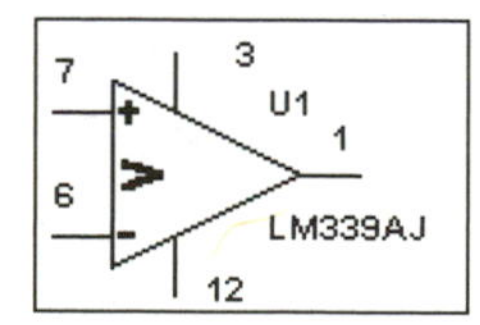

The **Comparator** component takes two voltage levels and compares them. It then sets the output to one of two states, On or Off, according to which input has a greater level. For example, if there were 5 volts on the inverted input, and 4 volts on the non-inverting input, the output would be low. However, if we change the non-inverting input to 6V, the output goes high. It's commonly used as an interface between analog and digital circuits.

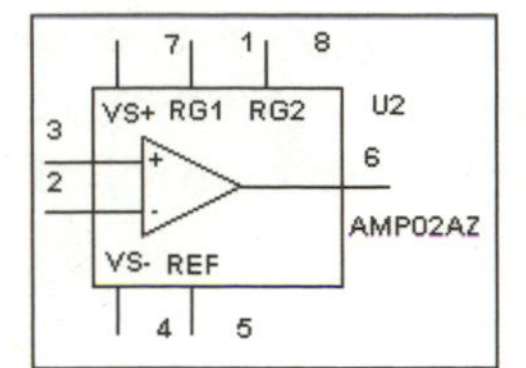

Wide Bandwidth Amplifiers have high-gain, high common-mode rejection ratio (CMRR) circuitry and are used to detect and amplify low-level signals. They are typically used in process control or measurement applications.

The **Special Function** analog circuits are not otherwise covered. Some examples are active filters, video amps, instrumentation amps, and pre-amps.

Figure 4.67
Examples of special analogy circuits..

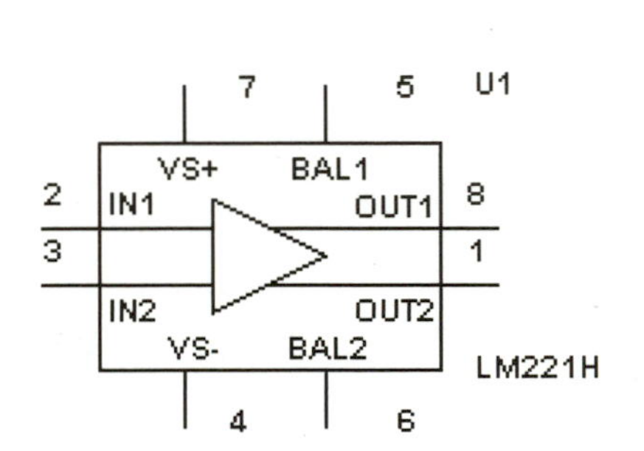

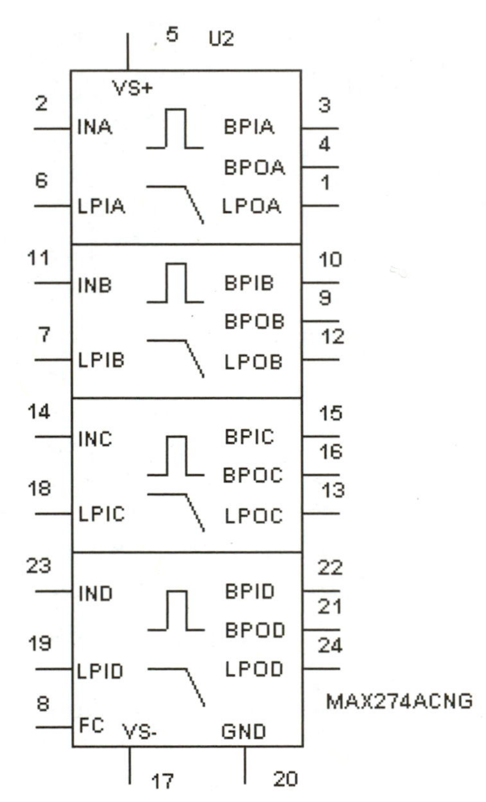

Transistor-Transistor Logic (TTL) ICs Toolbar

A TTL IC is a digital integrated circuit (such as gates, flip-flops, inverters, etc.) that operates at a greater speed than CMOS circuitry, but uses much more power. TTL ICs are typically used in digital circuits that don't require battery power such as cable boxes, televisions, computer boards, and the like. Multisim contains a large section of the 74xx series ICs (see Figure 4.68). Each section of the IC can now be separated on the drawing (i.e., A gate, B flip-flop, etc.). They are then combined when sent to a PCB Layout program.

Figure 4.68
The Transistor-Transistor Logic ICs toolbar.

The **74STD (74xx) Series** (Figure 4.69) is a standard 74xx series IC that contains a large repository of digital gates and other logic items. Choose the component you wish to place; if there are multiple sections to that IC, a window pops up, prompting you to select which section of the IC to use (see Figure 4.70). For example, a 7404 contains six inverters (A, B, C, D, E, and F).

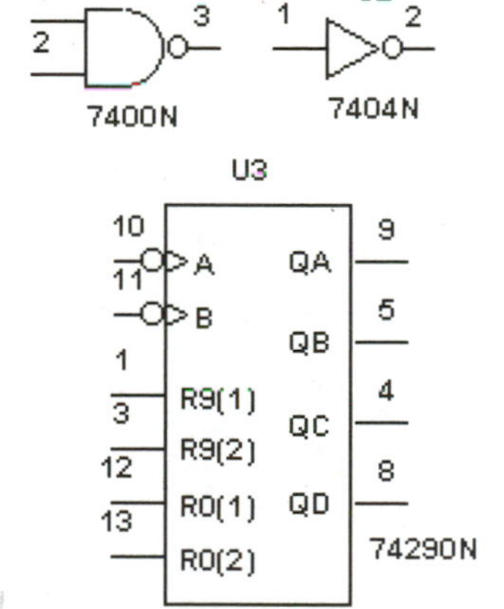

Figure 4.69
The 74xx series TTL ICs.

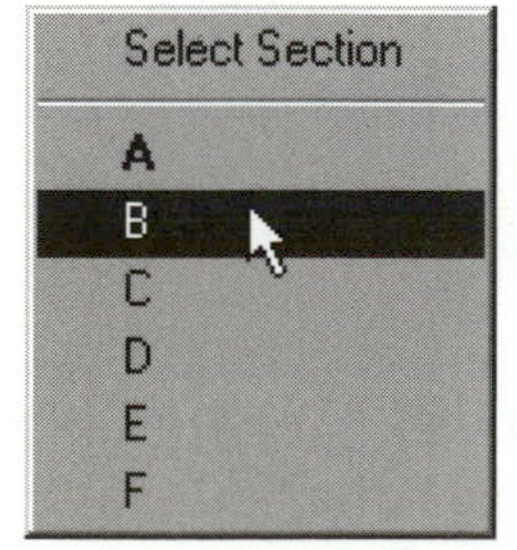

Figure 4.70
The selection window pops up letting you choose which section of the IC to place. Six inverters from a 7404 are shown.

The **74LS (74LSxx) Series** (Figure 4.71) are a Low Power Schottky (LS) series of TTL ICs that use less power than their standard 74xx counterparts but still operate at hyperspeed.

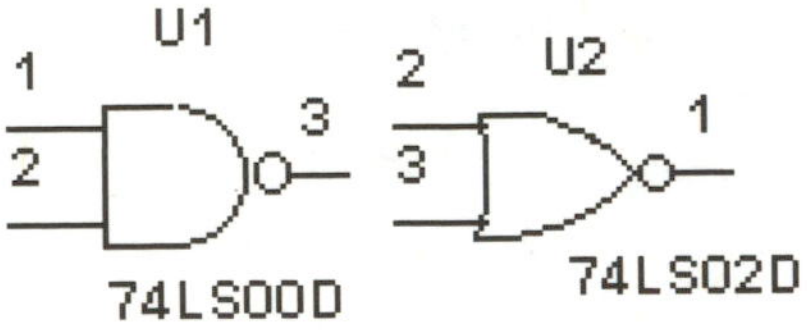

U1 74LS00D U2 74LS02D

Figure 4.71
The 74LSxx series TTL ICs.

*If you do not quite know what a TTL or CMOS IC does, place a copy of the component, select it, and hit **F1** (Help).*

Complementary Metal-Oxide Silicon (CMOS) ICs Toolbar

CMOS ICs are digital circuits (such as gates, flip-flops, inverters, etc.) that operate at a slower speed than TTL chips, but use a fraction of their power. CMOS ICs are able to operate over a wider voltage range and are based on both p- and n-channel MOSFETs. Typically, CMOS chips are used in portable applications that require battery power. An example is a laptop computer or cell phone. See Figure 4.72 for a view of the CMOS toolbar.

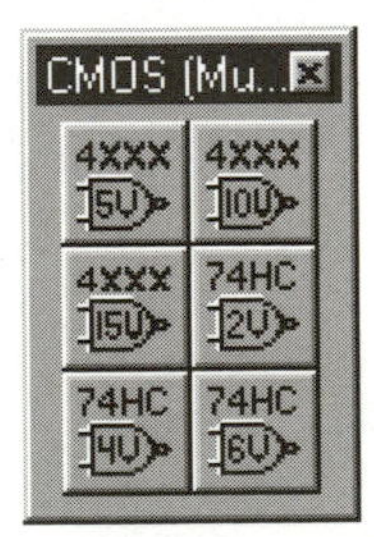

Figure 4.72
The CMOS toolbar.

The next three devices (Figures 4.73 through 4.75) are CMOS ICs that operate at their respective voltage ranges.

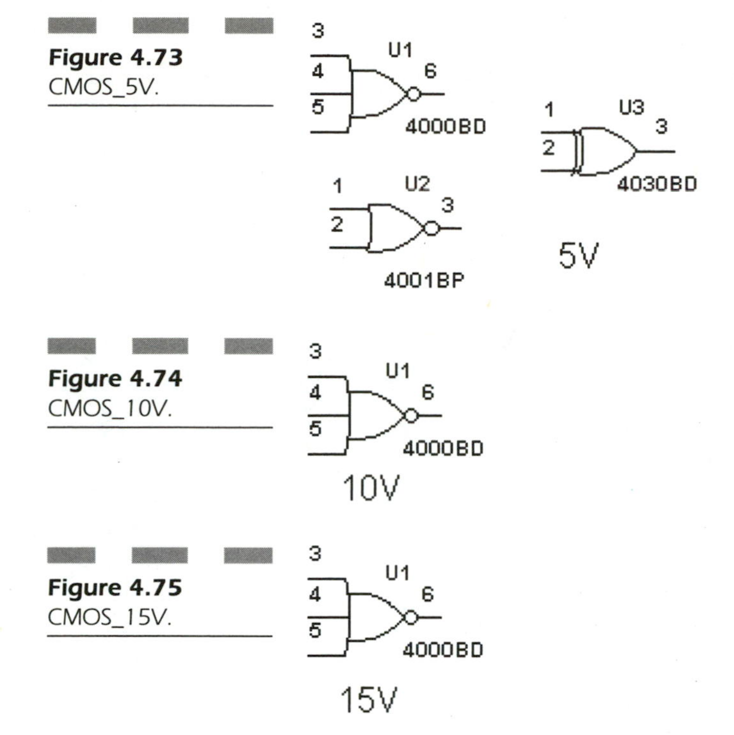

Figure 4.73
CMOS_5V.

Figure 4.74
CMOS_10V.

Figure 4.75
CMOS_15V.

The 74HCxx (Figures 4.76 through 4.78) series are CMOS ICs that are functionally equal to their TTL 74LSxx counterparts but can switch at a ten times faster rate. They can be interfaced directly to TTL ICs.

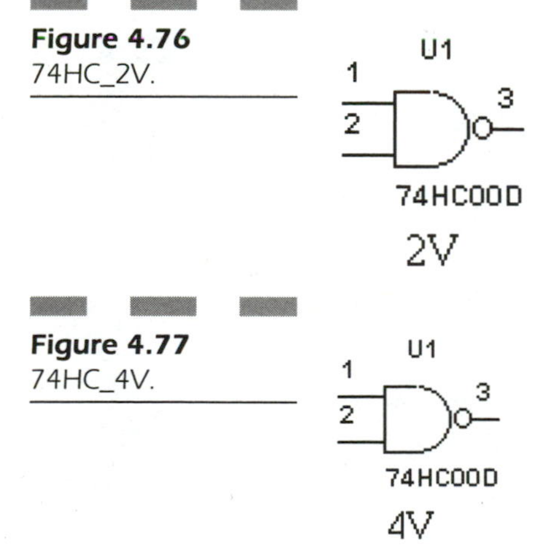

Figure 4.76
74HC_2V.

Figure 4.77
74HC_4V.

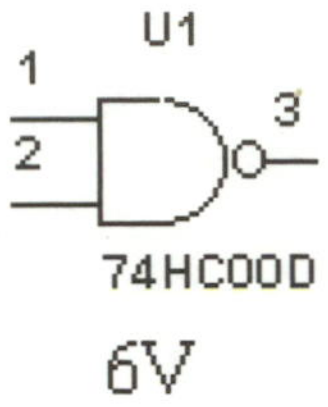

Figure 4.78
74HC_6V.

Working with TTL and CMOS IC Power Supplies

Because Multisim no longer places the whole IC symbol as one item, each digital IC needs its own power supply setup. Double-clicking any of the gates and selecting the Value tab (see Figure 4.79) accesses it. From here you can set which connection the power node and digital ground node will go. Remember that you need a digital ground on the drawing somewhere (it doesn't need to be connected to anything). This dialog box creates a virtual connection from the Vcc and Digital Ground symbols to the IC's power supply.

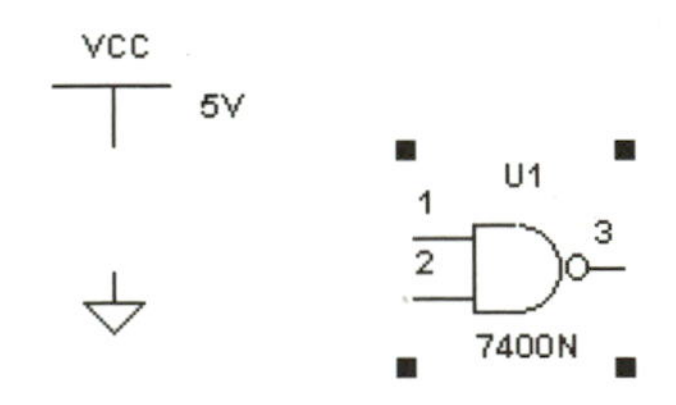

Figure 4.79
Setting up an IC's power supply.

Miscellaneous Digital Toolbar

The Miscellaneous Digital Toolbar is shown in Figure 4.80.

Figure 4.80
The Miscellaneous Digital toolbar.

The **TIL** icon is a mixture of gates, flip-flops, registers, and more. It replaces the gates and digital toolbins from the old Electronics Workbench V5. They will not export to a PCB package; instead, they offer a quick way to test digital electronic circuits.

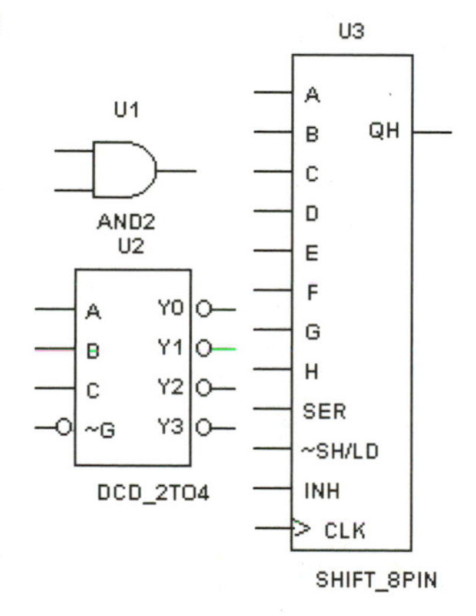

Figure 4.145
The TIL toolbin contains simple logic components.

The **VHDL** icon contains several pre-built VHDL components usable in Multisim. They include 74xxx, 75xxx, and even SRAM.

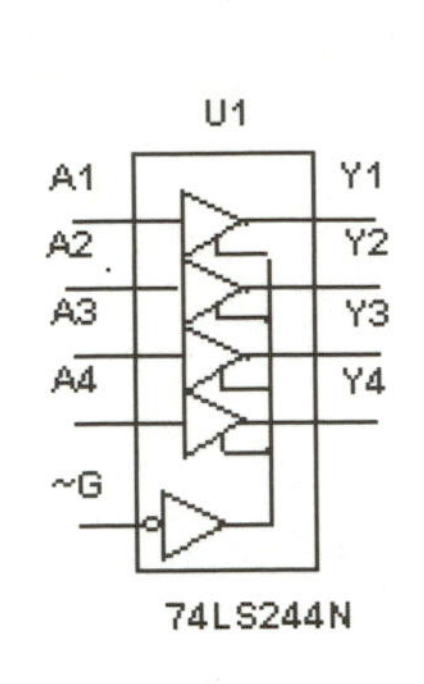

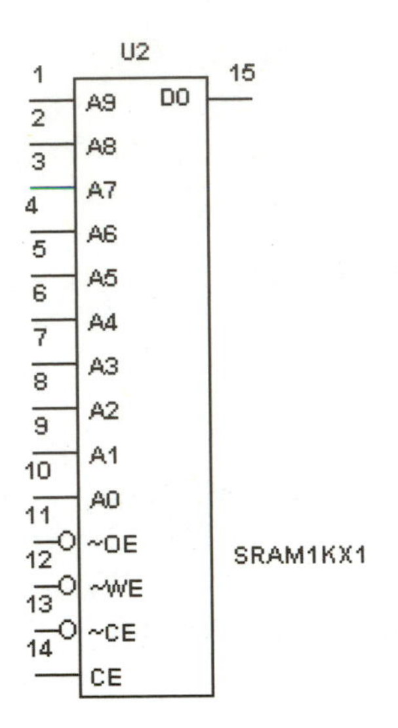

Figure 4.82 Premade VHDL components are available from this toolbin.

Mixed Chips Toolbar

See Figure 4.83 for Mixed Chips toolbar. Its components are shown in Figures 4.84 through 4.87.

Figure 4.83 The Mixed Chips toolbar.

In the **ADC_DAC (Analog-To-Digital Converters and Digital-To-Analog Converters**) an **ADC** is an encoder that converts an analog voltage into a digital word. There are five inputs and nine outputs. The **DAC** takes a digital signal and changes it to a voltage or current level. These chips are used by the millions to convert analog signals to digital, process the digital information (or even store it), and then output it through another converter as an analog signal. The finest example of ADC and DAC chip use is the modern compact disc (CD) technology. The CD recorder takes an analog signal (through a microphone) and puts it through an ADC. The digital information is then stored on the CD. The CD player then uses a DAC to turn the digital bits of a CD back into the sound waves, which are sent to a speaker.

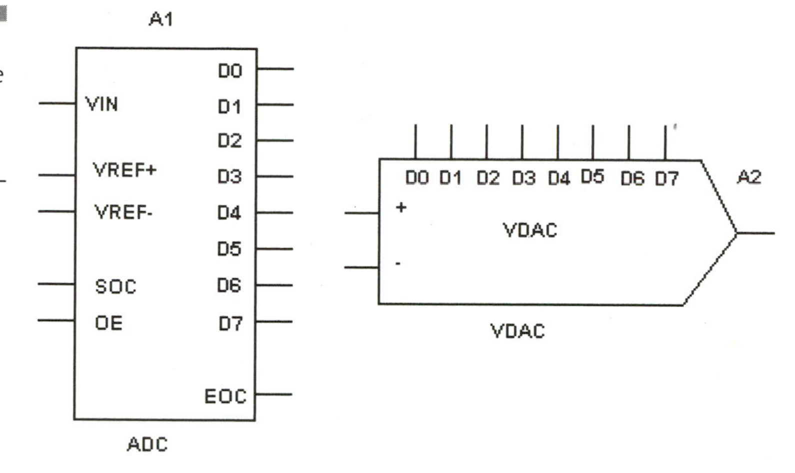

Figure 4.84 ADC and DAC ICs are used in nearly every electronics device these days.

The **Timer** is one of my all-time favorites. I can remember building hundreds of projects with 555s bought from the local electronics store. With Multisim, I now have unlimited timers. The Timer can be configured as a monostable or astable multivibrator, ADC, or in thousands of other applications, using only minimum components. It can be used for almost anything from a simple wave generator, to a short or long timer, to a missing pulse detector. A good example is an automobile's intermittent windshield wipers. Timers are even used in some solid-state ignition systems. (Chapter 7 contains several circuits that make use of this gem.)

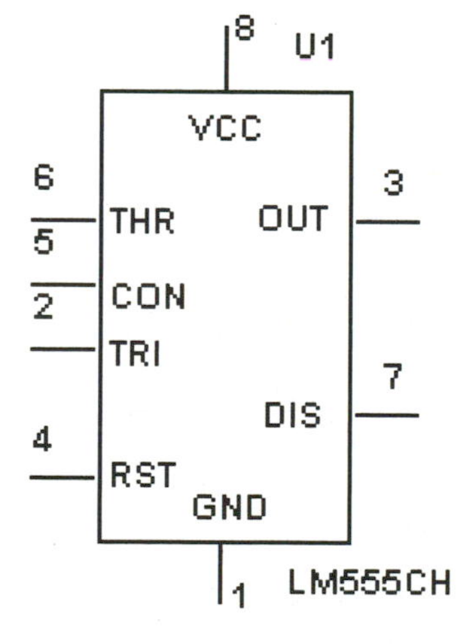

Figure 4.85 An 555 timer IC.

The **Monostable Multivibrator** produces an output pulse of fixed duration in response to an edge trigger at its input. The length of the output pulse is controlled by the timing RC circuit connected to the monostable multivibrator.

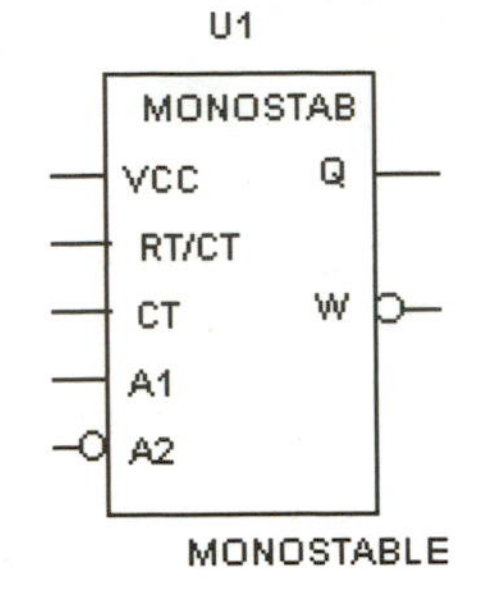

Figure 4.86
A monostable multivibrator IC.

The **PLL (Phase-Locked Loop)** IC has a built-in oscillator whose output phase and frequency are steered to keep it synchronized with an input reference signal. It's used in modern radio frequency applications to "lock" onto a certain frequency.

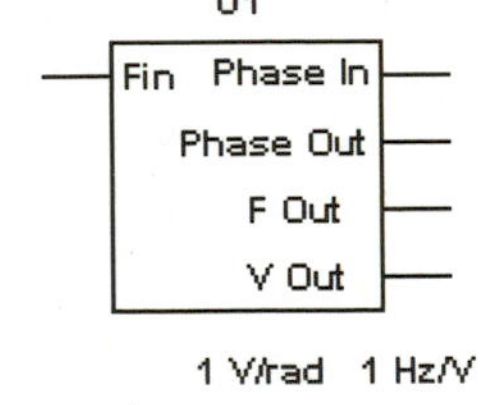

Figure 4.87
A PPL IC.

Indicators Toolbar

The Indicators toolbar is shown in Figure 4.88.

Figure 4.88
The Indicators toolbar.

The **Voltmeter** measures voltage in a circuit, and gives a visual numeric readout. If you need to place multiple indicators to show voltage in the circuit, it is easier to use many of these instead of multiple multimeters set to voltmeter. Be sure to obey the polarity and direction (because the component cannot be flipped once placed) when placing in a circuit. H means horizontal, HR means horizontal with reversed polarity, V means vertical, and VR is vertical with reversed polarity (Figure 4.89). If you are confused, look at the diagram in the upper-right corner of the Voltmeter Browser dialog box when placing it.

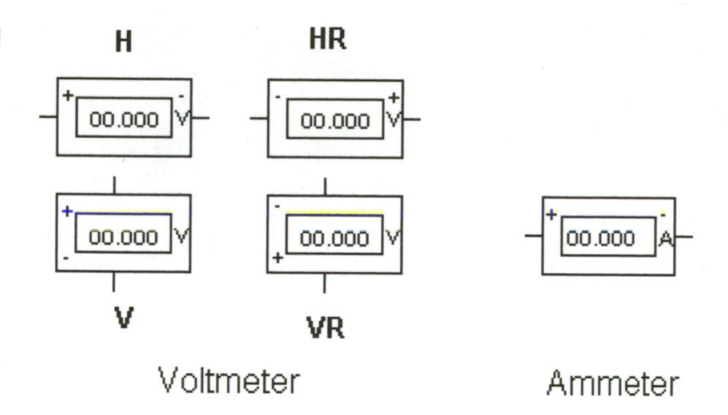

Figure 4.89
Since you cannot rotate the voltmeter or ammeter with Multisim, you will have to choose the component with the correct orientation when first placing it.

The **Ammeter** measures amperage (current) in a circuit, and gives a visual numeric readout. Obey polarity rules when placing this component. The Ammeter is used in place of multiple multimeters set to ammeter.

NOTE

You must double-click the placed Voltmeter or Ammeter and adjust it to read either DC (default) or AC (in RMS) (see Figure 4.90). You can also adjust the meter's resistance for a greater degree of simulation accuracy.

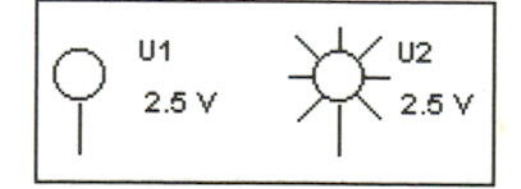

Probe is a visual indicator that will animate when a threshold voltage is reached. The circle radiates a pattern of lines when lit (threshold voltage reached). Think of it as a logic probe; it is used as a quick visual method of reading the outputs of digital circuits. You can change its color by right-clicking and choosing color. Double-click the component and using the Value tab to set the threshold or turn-on voltage.

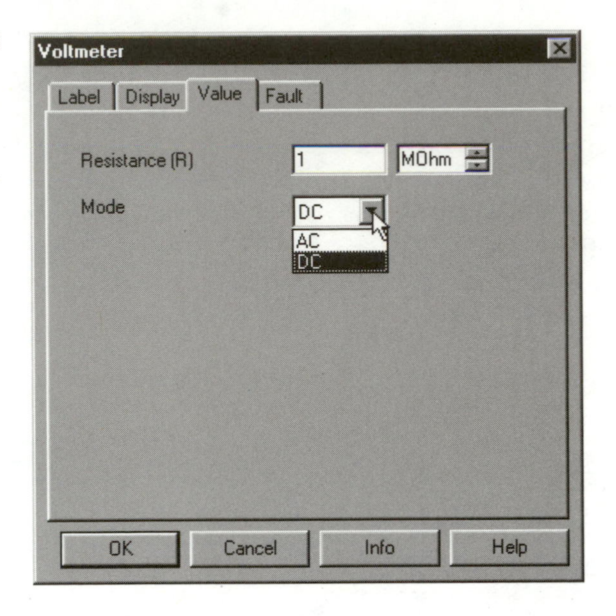

Figure 4.90
Adjust the voltmeter's or ammeter's setting by double-clicking it.

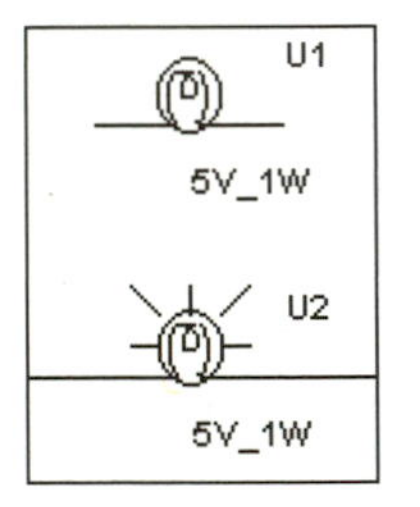

A **Lamp** is a visual indicator that animates when the correct voltage or wattage is applied. If this is exceeded, the bulb burns out. Think of it as a light bulb. You can select the wattage and voltage from the Browser before laying this component.

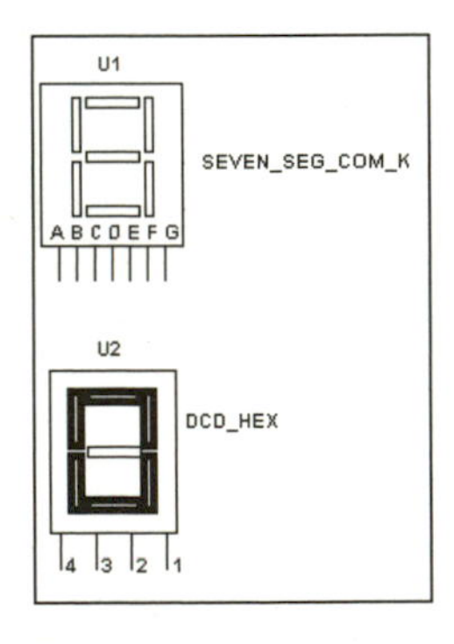

The **Hex Display** is composed of seven elongated LEDs that, when lit in a specific order, create the numbers 0 through 9 or the letters A through F. There are two types: one needs a control line for each LED (Common K); another decodes a binary signal (BCD). The component is used in any simulation that requires an alphanumeric character.

The **Bargraph** is also a series of elongated LEDs, but stacked on each other. It is typically used as a power level indicator; the more LEDs lit, the higher the level. The Decoded Bargraph has built-in voltage conversion circuitry and only requires you to hook a voltage level to it. There is also a version where each LED has its own terminal connection (Figure 4.91).

Figure 4.91 Bargraph component.

U2

U1

LVL_BARGRAPH

UNDCD_BARGRAPH

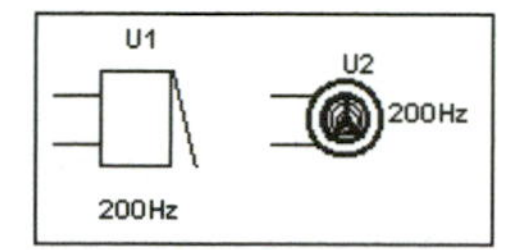

The **Buzzer** indicates when a specific current is flowing through it. Your PC speaker will "sound" at the frequency you set. Double-click the component to adjust its frequency, voltage, and turn-on amperage.

Miscellaneous Toolbar

The Miscellaneous toolbar contains various components that may be used in circuit design. The toolbar is shown in Figure 4.92.

Figure 4.92 The Miscellaneous toolbar.

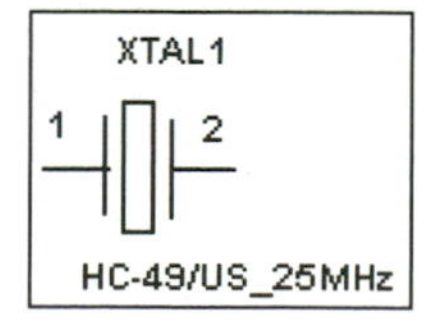

The **Crystal** is used in oscillator circuits to give a very stable and specific frequency output. It is a quartz crystal that resonates at a specific frequency, determined by its physical size. Because of its accuracy, it is used in timing applications that require an exact measurement of time. Your computer's microprocessor makes use of one such "XTAL."

The **Motor** is a virtual DC motor with connections to its stator that create a magnetic field, and terminals for its brushes, which send power to the rotor. See the Multisim Help file for more component information.

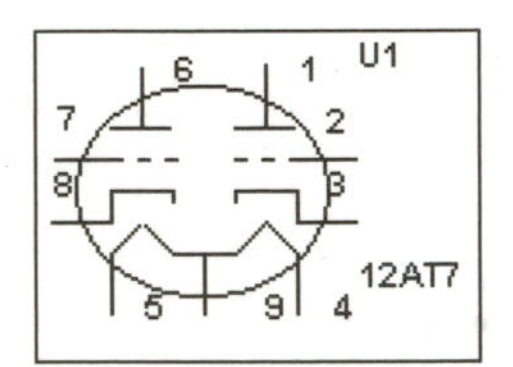

The **Vacuum Tube** is the transistor's ancestor. Even though silicon wonders have all but replaced the old tubes, some exotic stereo and musical equipment still uses them.

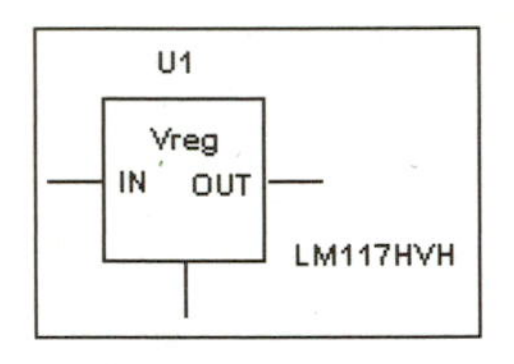

The **Voltage Regulator** takes a varying voltage source and steadies it to one continuous voltage. For example, you may have a 12-volt DC line that just went through rectification and is not a steady 12 volts. If it is fed through a 5-volt Voltage Regulator, you can then use it as a stable digital source. The Voltage Regulator is used mostly in power supplies, to give a steady voltage level.

The **Boost Converter, Buck Converter, and Buck-Boost Converters** are used to simulate DC, AC, and large-signal transient responses of switched-mode power supplies operating in both the continuous and discontinuous inductor current conduction modes (CCM and DCM, respectively). To adjust the settings, double-click the placed component.

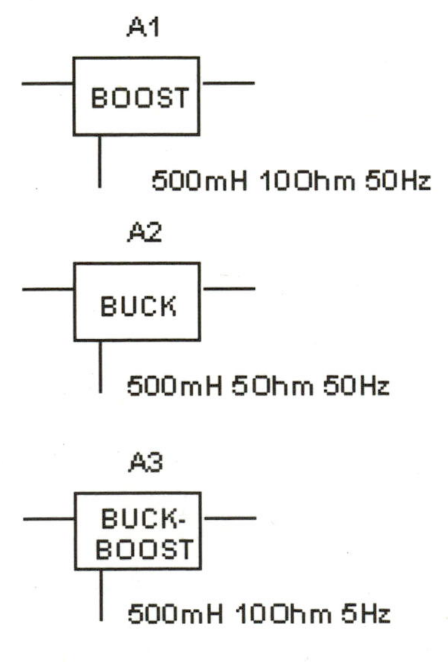

Figure 4.93
Samples of boost, buck, and buck-boost converters.

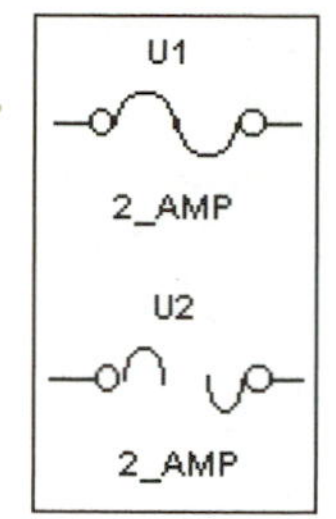

The **Fuse** is a device that protects a circuit from excessive current. You can only use one of Multisim's pre-made fuses or create a component of your own. Once the fuse is blown in the circuit, you will have to reset the simulation to replace it.

The **Lossy Transmission Line, Lossless Line Type 1, and Lossless Line Type 2** are wire components (Figure 4.94). Wire and cable are not perfect conductors. They add more and more resistance, inductance, capacitance and conductance for each meter of cable or wire. These components allow you to simulate the results of these cables and wires by inputting their characteristics. Double-click the placed component and enter each figure into Value tab. (Sometimes, these numbers are available in datasheets or catalogs from electronic distributors.) The Lossless model is an ideal cable or wire that simulates only the characteristic impedance and propagation delay properties of the transmission line. The characteristic impedance is resistive and is equal to the square root of L/C. These values are also accessed by double-clicking the placed component.

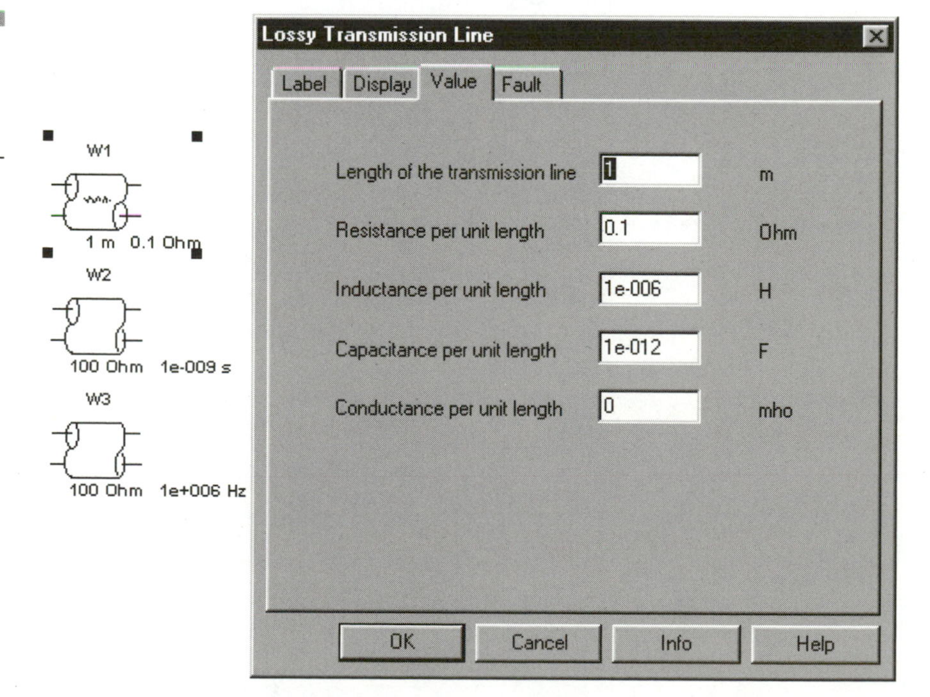

Figure 4.94 Simulate wire and cable with these components.

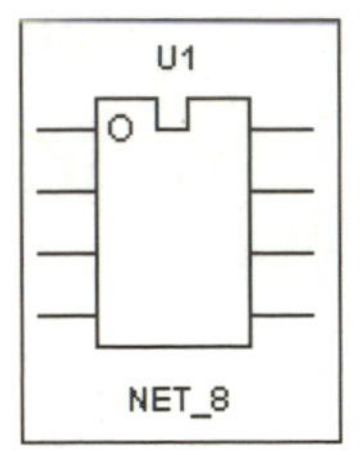

The **NET (Netlist Component)** is used when you know the SPICE language or have obtained a component model in the form of a SPICE Netlist. This component inputs that simulation into a preselected IC package template. You must first select the package (2 to 20 pins) from the Browser, then edit the component into the User database. Once inside the Component Properties dialog box, add the SPICE model to the Model tab and save or exit the design. Place the new component.

Controls Toolbar

The Controls toolbar is shown in Figure 4.95.

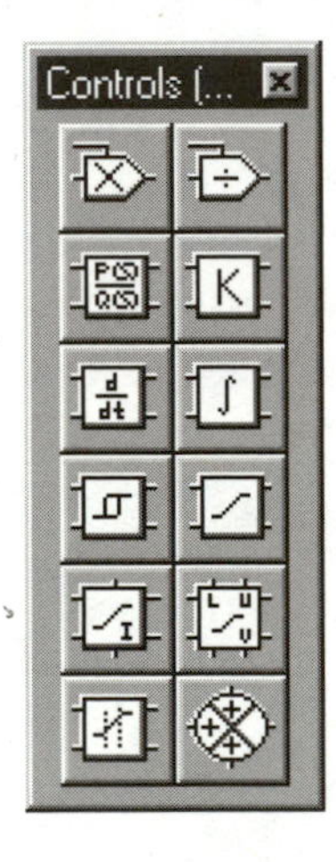

Figure 4.95
The Controls toolbar.

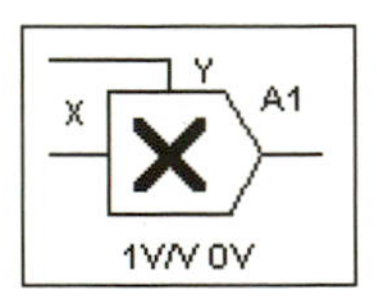

The **Multiplier** multiplies two input voltages (X and Y) and outputs the result to A. By double-clicking the placed component, you can adjust its offset and gain.

Multisim Help files contain in-depth explanations and example circuits of the Control items.

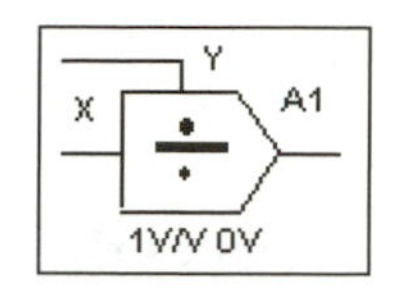

The **Divider** divides two input voltages (X and Y) and outputs the result to A. Offset, gain, lower limit, and smoothing domain are accessed by double-clicking a placed component.

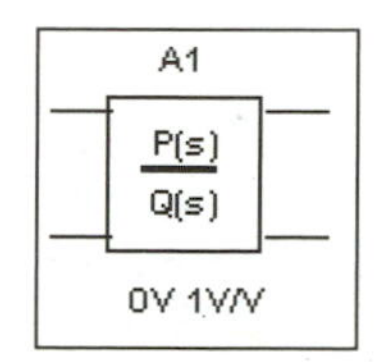

The **Transfer Function Block** models the transfer characteristic of a device, circuit, or system in the s domain. The transfer function block is specified as a fraction, with polynomial numerators and denominators. A transfer function up to the third order can be directly modeled. This component may be used in DC, AC, and transient analyses.

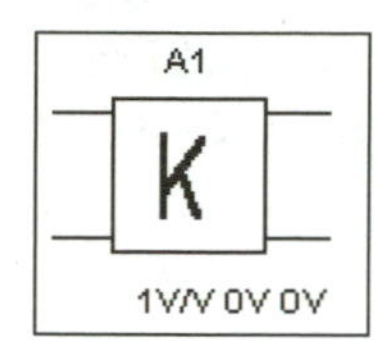

The **Voltage Gain Block** amplifies the input by a gain level (K) set by you. Double-click the placed component to adjust its settings.

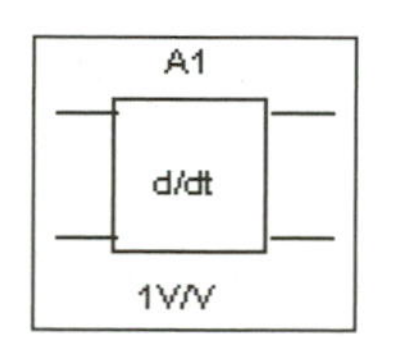

The **Voltage Differentiator** calculates the derivative of the input voltage (the transfer function, s) and delivers it to the output. It is used in control systems and analog computing applications. Differentiation may be described as a "rate of change" function and defines the slope of a curve. (Rate of change = dV/dT.) Double-click the placed component to adjust its settings.

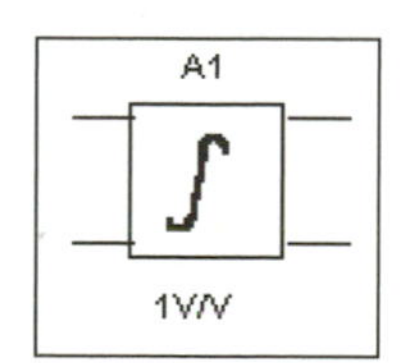

The **Voltage Integrator** calculates the integral of the input voltage (the transfer function, 1/s) and delivers it to the output. It is used in control systems and analog computing applications. Double-click the placed component to adjust its settings.

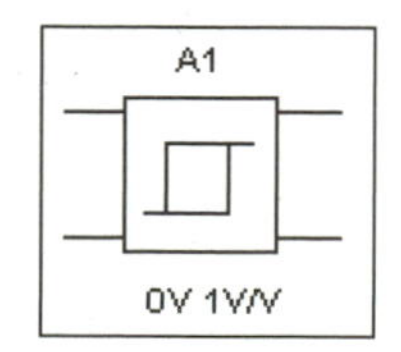

The **Voltage Hysteresis Block** is a simple buffer stage that provides hysteresis of the output with respect to the input. See the Help file for further details.

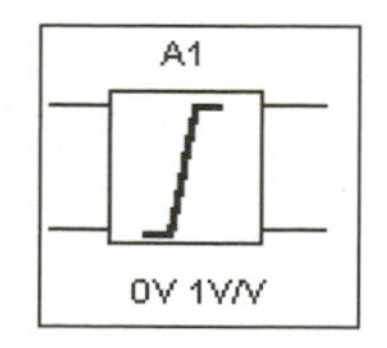

The **Voltage Limiter** is a voltage "clipper." The output voltage excursions are limited, or clipped, at predetermined upper and lower voltage levels while input signal amplitude varies widely. See the Help file for examples and further explanation.

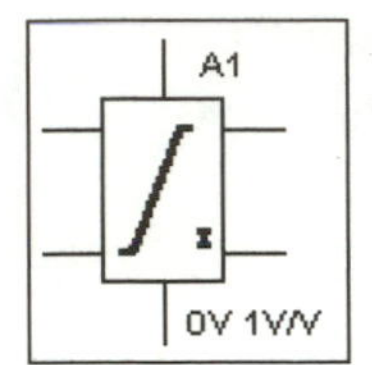

The **Current Limiter Block** models the behavior of an operational amplifier or comparator at a high level of abstraction. All of its pins act as inputs; three of them also act as outputs. The component takes as input a voltage value from the "in" connector. It then applies the offset and gain, and derives from it an equivalent internal voltage, Veq, which it limits to fall between the positive and negative power supply inputs. If Veq is greater than the output voltage seen on the "out" connector, a sourcing current flows from the output pin. Otherwise, if Veq is less than the output voltage, a sinking current flows into the output pin.

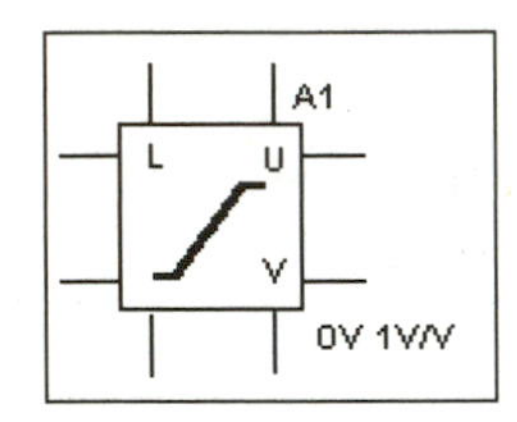

The **Voltage-Controlled Limiter** is a voltage "clippe." This component is a single input, single output function. The output is restricted to the range specified by the output lower and upper limits. Output smoothing occurs within the specified range. The Voltage-Controlled Limiter operates in DC, AC, and transient analysis modes.

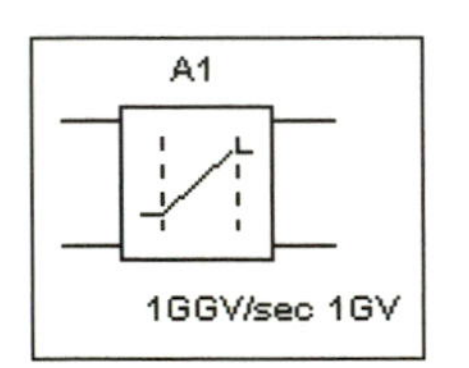

The **Voltage Slew Rate Block** limits the absolute slope of the output, with respect to time, to some maximum or value. You can accurately model actual slew rate effects of overdriving an amplifier circuit by cascading the amplifier with this component. Maximum rising and falling slope values are expressed in volts per second.

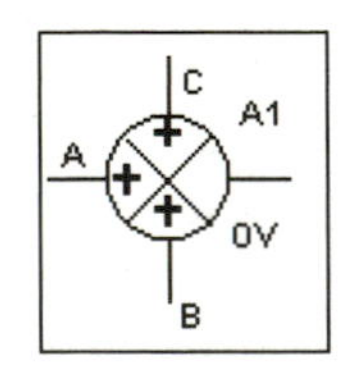

The **Voltage Summer** adds the three input voltages and outputs the results.

NOTE

The Radio Frequency (RF) Toolbar is only available to users who purchase the RF Module.

Electromechanical Toolbar

See Figure 4.96 for a view of the Electromechanical toolbar.

Electro_M...

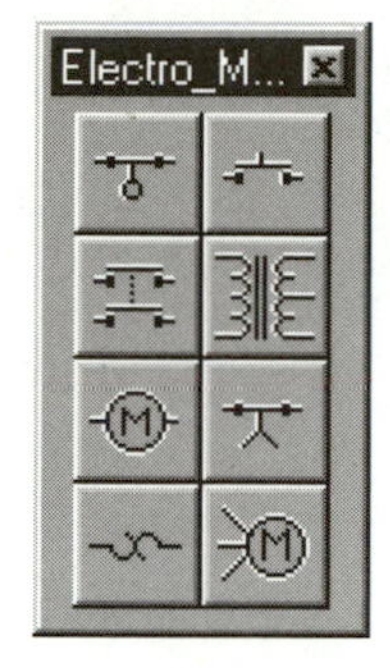

Figure 4.96 The Electromechanical toolbar.

Sensing Switches turn a circuit on or off (Figure 4.97). You can pick from a huge variety of switch symbols and footprints, but most are controlled with a keyboard letter, which is set by double-clicking the placed switch. The Value tab will open. For example, if you want the letter T to activate the switch, set the T and hit **OK**.

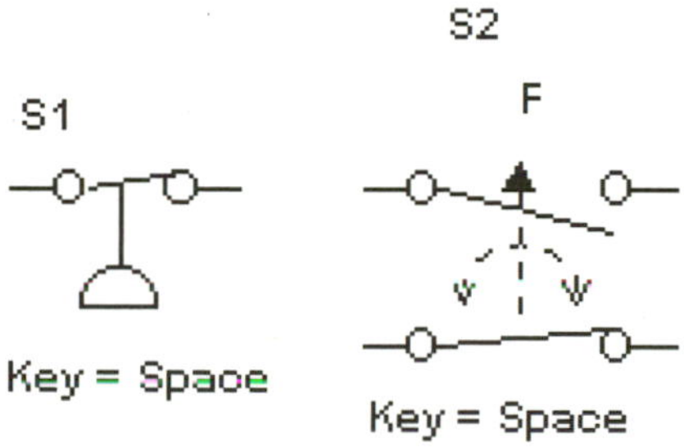

S2
S1
F
Key = Space
Key = Space

Figure 4.97 Switches.

Momentary Switches are a new component in Multisim (Figure 4.98). They mimic a real-world momentary switch with a keyboard letter (set by double-clicking the placed component). Push the button and a few moments later the switch shuts off.

S2
S1
R
Key = Space
Key = Space

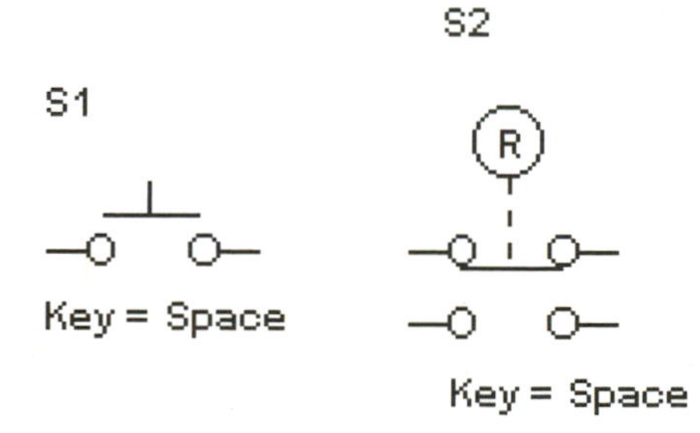

Figure 4.98 Momentary switches.

Supplementary Contacts mimic their real-world components and are controlled by a keyboard letter.

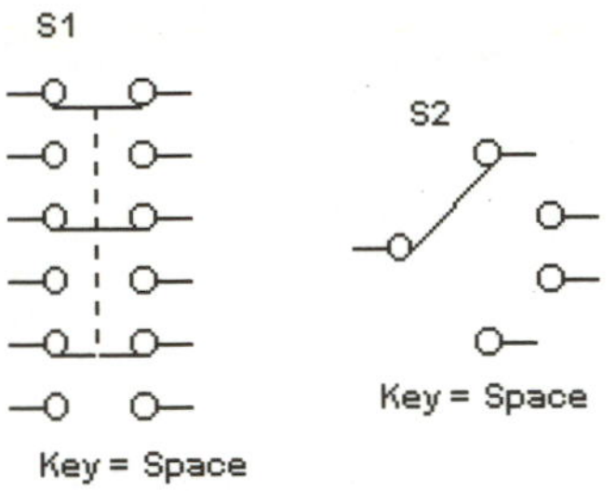

S1
Key = Space
S2
Key = Space

Figure 4.99 Supplementary contacts.

Line Transformers are simplified transformers intended for power applications, where the primary coil is connected to either 120 or 220 VAC. They perform step-up or step-down functions plus several specialized functions of voltage and current measurement.

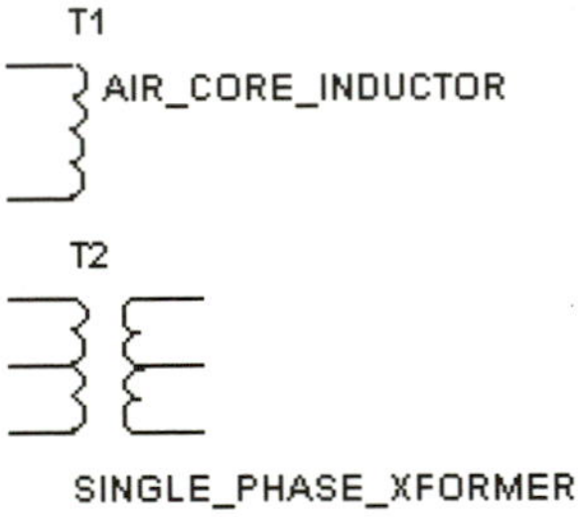

T1
AIR_CORE_INDUCTOR
T2
SINGLE_PHASE_XFORMER

Figure 4.100 Line transformer.

Coils and **Relays** can simulate a motor starter coil, forward or fast starter coil, reverse starter coil, slow starter coil, control relay, and time delay relay. See the latest Help file for more information.

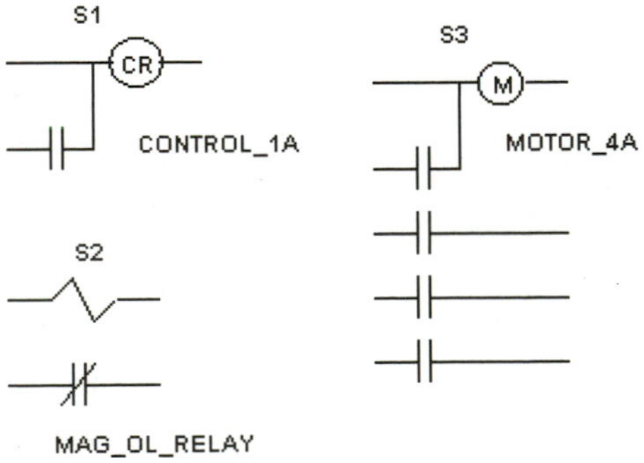

S1
CR
CONTROL_1A
S2
MAG_OL_RELAY
S3
M
MOTOR_4A

Figure 4.101 Coils and relays.

Timed Contacts include normally-open-timed-closed. See the latest Help file for more information.

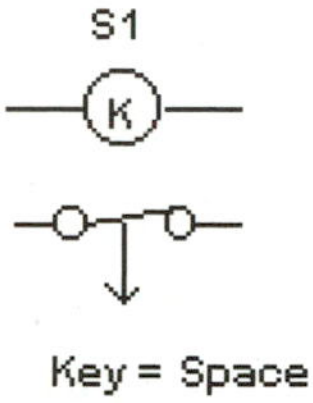

Figure 4.102
Timed contacts.

Protection Devices include fuses, overload, overload thermal, overload magnetic, and ladder logic overload protectors.

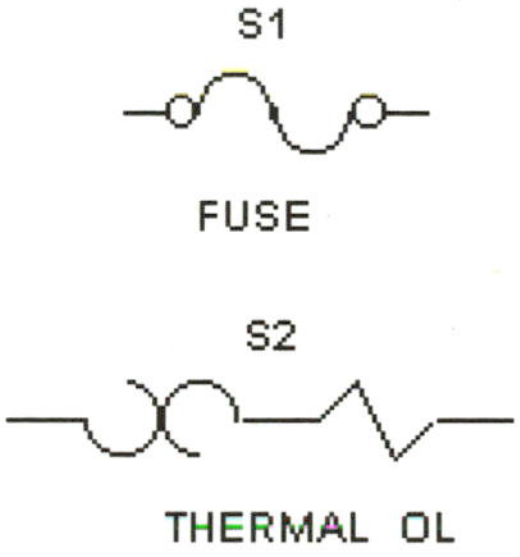

Figure 4.103
Protection devices.

Output Devices include a light indicator, motor, DC motor armature, three-phase motor, heater, LED indicator, and solenoid.

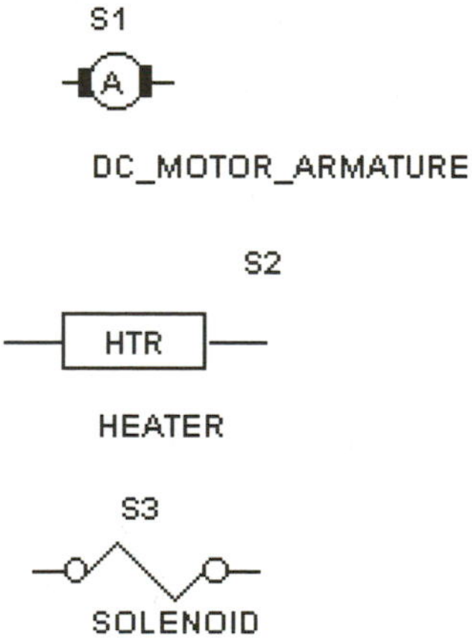

Figure 4.104
Output devices.

Getting More Details on Components

Electronics Workbench provides extensive data on each component, which include how the component is modeled, its electrical characteristics, and in some cases its applications. Under **Help** > **Multisim Reference** you will find just about the same text files as are included in the User Guide Appendices PDF document (under **Start** > **Multisim** > **User Guide Appendices**). Using the Help files saves having to print out this technical tome; simply typing in the name of the component in the Find tab of the main Help menu will also locate it. Or you can follow this procedure:

1. Place the component for which you require more information.
2. Select it by clicking over its symbol (not label).
3. Hit **F1** (Help).

There are a number of component examples on my website at **http://www.basicelectronics.com/ewb/**. If you wish to add a circuit, visit this site.

To quickly find out more information about a component, right click and choose Help.

To learn exactly what an IC contains (gates, resistors, etc.) you may want to purchase manufacturers' databooks or find datasheets on company websites.

Making Your Own Components

The information throughout this chapter should give you a good idea of what is involved in creating custom components. However, if you would like a step-by-step description, EWB puts out a great file called "Guide to Creating NEW Components with the Component Editor." It is available on the EWB website at **http://www.electronicsworkbench.com/html/create_component.pdf**. A short review of component creation follows:

1. Choose to create a new component with the Component Editor and place it into the User database.
2. The Component Properties dialog box opens and requires that each page be filled out.
3. Fill out the General tab (described earlier in this chapter), making sure to select the correct component group, family, name/date, and other parameters.
4. Fill out the Symbol tab, creating your own schematic artwork if necessary.
5. Select the model and template you wish to use from the Model tab, be it SPICE, VHDL, or another.
6. Assign a package and logical pin order to the component with the Footprint tab.
7. Choose **Save/Exit**.
8. Hit the **Exit** button in the Component Editor window.
9. Display and refresh the User toolbar if necessary.
10. Select the new component and place it.

NOTE

Multisim 6.2 uses a component wizard that aids you in the parts building process. It performs essentially the same thing as the previous versions.

TIP

*Every Multisim component is not perfect. Bugs creep into the software to wreak havoc on simulations and schematic capture. Bookmark and frequent EWB's component section, at **http://www.electronicsworkbench.com/html/msm_ts_components.html**.*

Summary

Multisim uses complex algorithms to simulate most types of real-world components, which allows you to model them accurately in your circuit designs. With the knowledge you have acquired in this chapter, you should be able to create just about any virtual component not included in the software. Using these building blocks, you can build great electronic circuits.

CHAPTER 5

Multisim Instruments

A Multisim instrument is your window to the electronic world.

You no longer have to pay thousands of dollars for electronic instrumentation and equipment—there's a cheaper alternative. One of Multisim's most powerful features is Virtual Instruments. They are just as powerful and versatile as the real thing. Look out FLUKE™! Multisim includes simple devices such as the multimeter, function generator, oscilloscope, logic analyzer, and wattmeter, and complex instruments such as the distortion analyzer, network analyzer, and spectrum analyzer. Multisim even includes instruments that have no real-world equivalent, but are nevertheless powerful tools. These include the Bode plotter, logic converter and word generator.

The primary function of these instruments is to provide an electrical signal or to read an output signal. This chapter breaks down each instrument's functions and uses.

You can now rotate/flip instruments as well as run more than one copy of an instrument (on the Professional or Power Pro Version). Infinite multimeters and scopes are possible!

Turning on the Instrument Toolbar

Multisim's Instrument Toolbar is not displayed when you start the program. Go either to the **Instrument's icon** on the Multisim toolbar or to **View** > **Toolbars** > **Instruments**. If you don't like the position of the toolbar, move it by grabbing somewhere along the perimeter of the toolbar and dragging it where you want. See Figure 5.1.

Leave the Instrument Toolbar displayed and place it right below the System Toolbar.

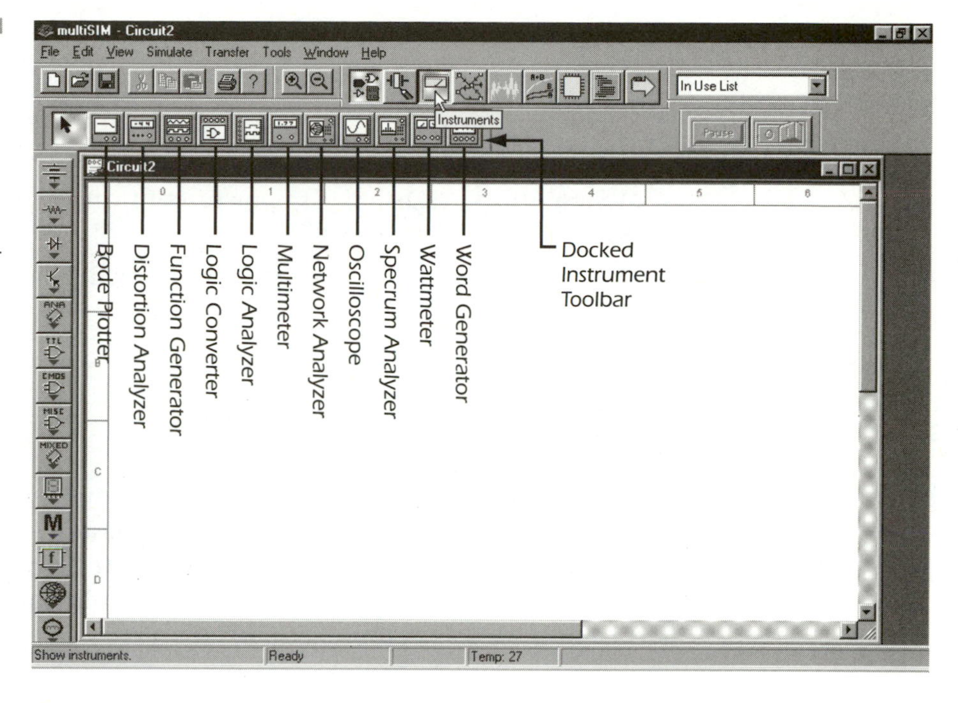

Figure 5.1
Turn the Instrument toolbar On and dock it for space savings. You can see what each instrument is on the toolbar.

Quick Overview of Instruments

This is a quick look at the instruments that Multisim offers. More detailed explanations and user guides follow. In alphabetical order the instruments are:

- **Bode Plotter**—Graphs a circuit's response to various frequencies; helpful in analyzing filter circuits.
- **Distortion Analyzer** (Pro and PP only)—Used to investigate small amounts of distortion in a circuit.
- **Function Generator**—Outputs various electrical waves into a circuit.
- **Logic Converter**—Converts a digital signal or circuit into a truth table or truth table into a digital circuit. No real-world equivalent.
- **Logic Analyzer**—Views up to 16 digital waveforms.
- **Multimeter**—Ohmmeter (measures resistance), voltmeter (measures voltage), and ammeter (measures current). Can also measure decibel loss between two points.

- **Network Analyzer** (PP only)—Used to measure RF circuit designs.
- **Oscilloscope**—Allows you to visualize an electrical wave.
- **Spectrum Analyzer** (PP Only)—Used to measure amplitude versus frequency.
- **Wattmeter**—Measures the power consumed by a circuit or component.
- **Word Generator**—Outputs a custom digital waveform into a circuit.

Multisim Version Features

The Multisim Personal Edition features the multimeter, function generator, oscilloscope, bode plotter, word generator, logic analyzer, logic converter and wattmeter. The Multisim Professional Edition has all of the above, plus a distortion analyzer. The Multisim Power Pro Edition has all of the above, plus a network analyzer and spectrum analyzer.

Placing Instruments

An instrument icon must first be placed into the circuit to attach and run the instrument. Place the mouse pointer over the instrument's icon on the toolbar and click the left mouse button once. Move onto the circuit window and place the instrument icon by clicking the left mouse button again.

Each instrument contains terminals that must be connected to the circuit. Wire each applicable terminal into the circuit. Once it is connected, double-click the instrument to open it. For example, if you are measuring the voltage across a resistor, hook up the circuit as shown in Figure 5.2.

TIP

Open up the instrument if you are having trouble seeing terminal locations. Go back to the instrument icon and move the pointer over the general area until a '+' appears. If you can't quite figure out what the terminal is, open the instrument to see the label for that point.

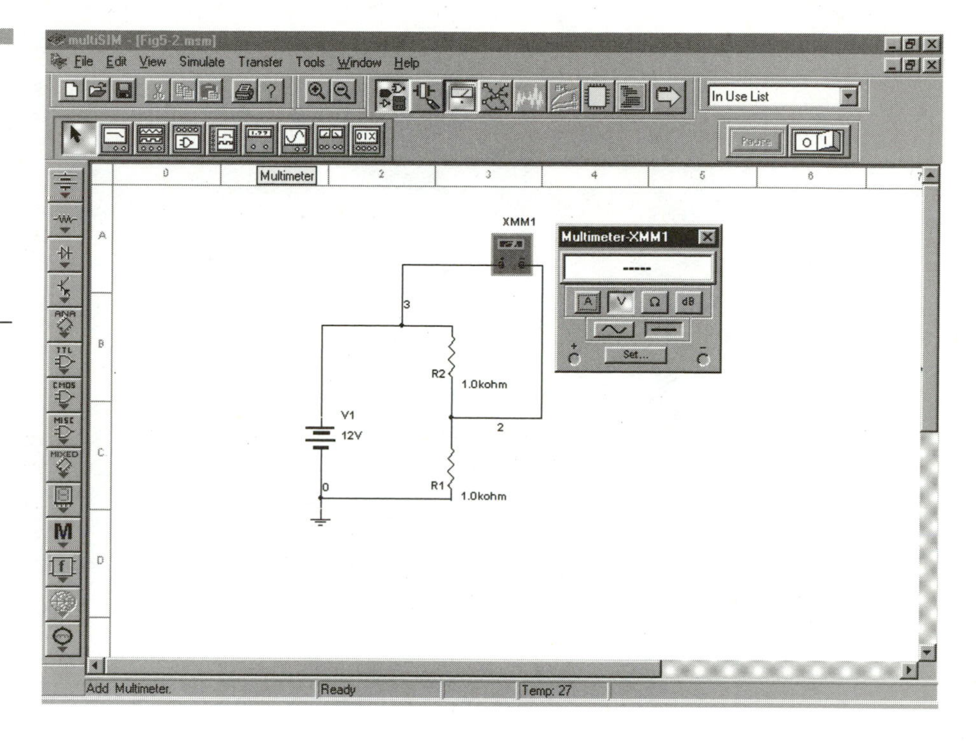

Figure 5.2 Placing, connecting, and opening instruments. In this example, we are measuring the voltage across a resistor using the multimeter.

Changing Instrument Icon Color

If you want to add some pizzazz to your circuits, color the instruments. After the icon is placed, right-click it and choose **Color**. Select which color you want from the Color Picker and hit **OK**. You can also make custom colors by pressing the **Define Custom Colors** button.

Adjusting the Instrument's Settings

The wired-in instrument must be adjusted to measure exactly what you want or to output a signal according to your specs (see Figure 5.3). Double-click the instrument's icon to open it. Each instrument has a window with various buttons, boxes, and spin controls; some of these can by adjusted while the simulation is in progress (adjusting "on-the-fly"). The controls for each instrument are described in detail later in this chapter.

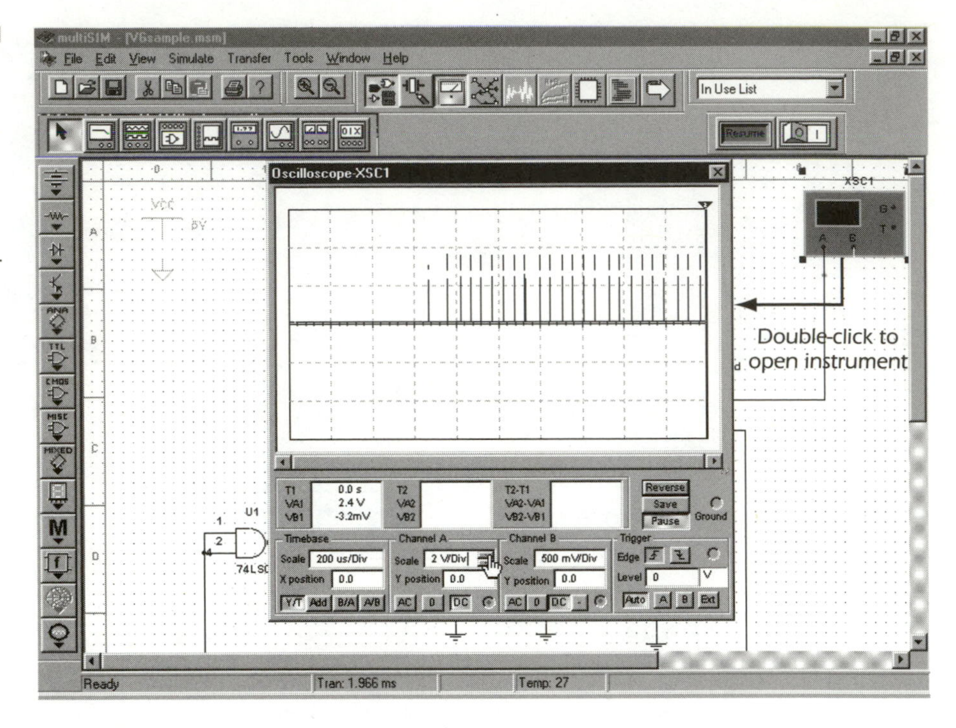

Figure 5.3
By double-clicking the instrument's icon, you can access its controls. The oscilloscope is pictured here.

Default Instrument Settings

Under the **Simulate** menu is the command to open the Default Instrument Settings dialog box. From here, you adjust instrumentation settings and control the efficiency of the simulation. The dialog box contains the following settings:

Defaults for Transient Analysis Instruments

- **Initial Conditions**—Certain conditions can be set to activate when you begin a simulation. Your options are: automatically determine initial conditions (default); set to zero; user defined (adjusted with user's parameters); and calculate DC operating point.
- **Analysis**—These options adjust exactly how the instrumentation reacts to the simulation or analysis.
  —**Start Time (TSTART)**—Start time of transient analysis; must be zero or greater.

—**End Time (TSTOP)**—End time of transient analysis; must be greater than start time.

—**Set Maximum Timestep (TMAX)**—Manually set the number of points between each cycle if this option is checked off. You can either select TMAX or choose to have Multisim generate this figure for you automatically.

—**Set Initial Timestep**—Sets the time interval for simulation output and graphing.

- **Reset to Default Button**—Puts all settings back to their default values.

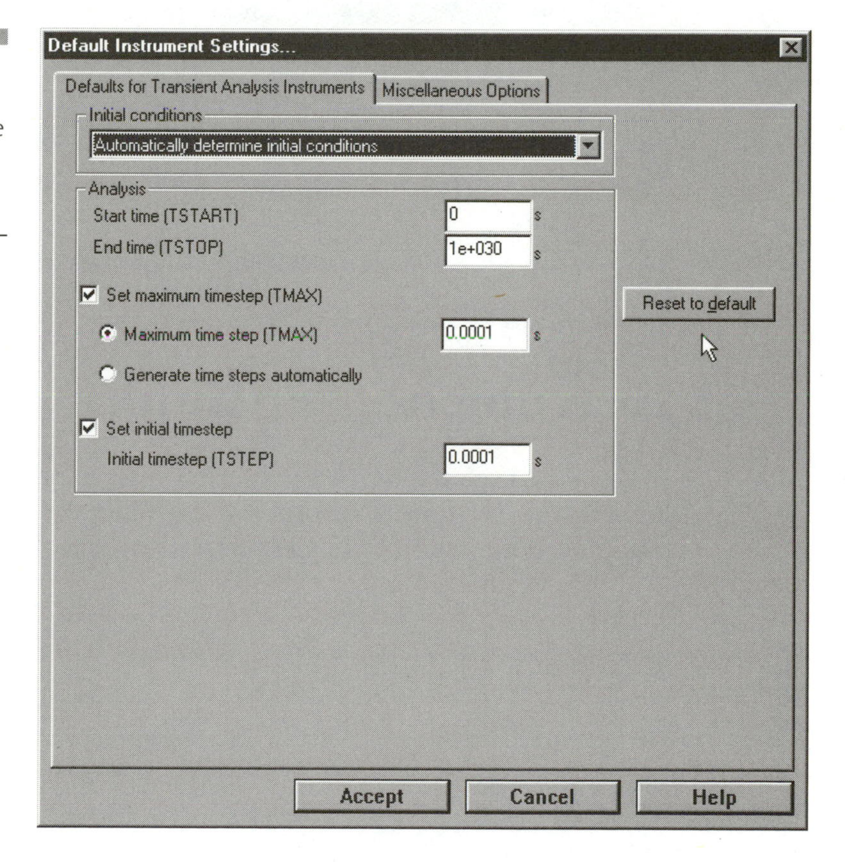

Figure 5.4 Some instruments require defaults to be set under Simulate > Default Instrument Settings.

NOTE

For information on the Miscellaneous Options tab, see the Multisim user manual.

Printing Instrument Readings and Settings

You can make a hard copy of each instrument's settings and current readout. Go to **File** > **Print** > **Instrument Maps** to open the Print Instrument Maps dialog box (see Figure 5.5). Check off the instruments for which you wish a printout and press the **PRINT** button.

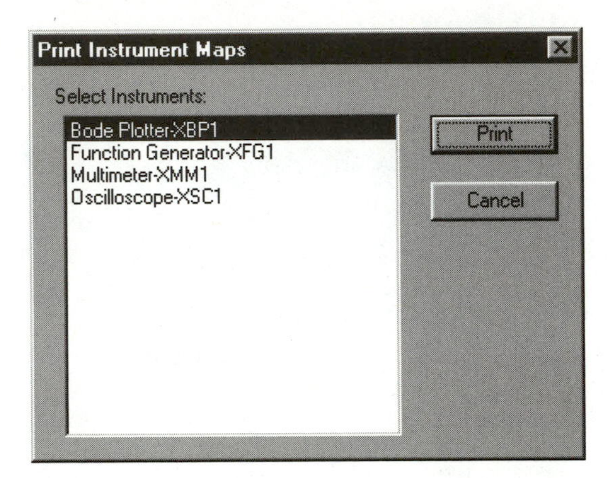

Figure 5.5 Multisim lets you print a picture of the opened instruments to later refer to its settings.

Instrument User Guide

The rest of this chapter describes each instrument. Details of how to hook up the instrument, examples, a few hints and tricks are added. The instruments are described in a logical, rather than alphabetical order.

Allow a bit of room when laying instruments for tangle free lines.

Basic Instruments

Multimeter

The multimeter is the most valuable instrument on my real workbench. It measures resistance, AC and DC voltage and current, and decibel loss. It replaces an ammeter, voltmeter, and ohmmeter. The Multisim virtual multimeter is no different. Hook up your leads, activate the circuit, and a reading appears.

Connecting the Multimeter

The positive terminal is the left connector at the bottom of the meter's icon. The negative terminal is the right connection. Think of the negative side as a black probe and the positive as a red probe. If you are reading a positive signal in reference to ground, then connect the negative to ground and the positive to a point in the circuit where a measurement is required. This can be a connector, a component's terminal, or a pin anywhere along the path of a wire (this creates a connector from the wire to the meter). The next few sections provide examples of how to hook up the multimeter.

Multimeter Operation

Once the multimeter is connected to a circuit or component group, adjust its measurement settings. Double-click the multimeter symbol. It defaults to measure DC Voltage (see Figure 5.6). The top set of buttons lets you select a mode:

- **A** for Ammeter
- **V** for Voltmeter
- Ω for Ohmmeter
- **dB** for Decibel loss between two points

The next two buttons choose between AC or DC.

- The squiggly line is AC
- The straight horizontal line is DC

The last button adjusts the meter's internal settings. The **Set** button opens the Multimeter Settings dialog box. A real multimeter is not an ideal model, because there can be induced resistance that may result in slightly inaccurate measurements. By adjusting Multisim's multimeter internal settings you can approximate these imperfections. This makes for a realistic multimeter. To change the settings, either use the spinner or type in the new value. Make sure to set the multiplier as well (n, p, M, G, etc.).

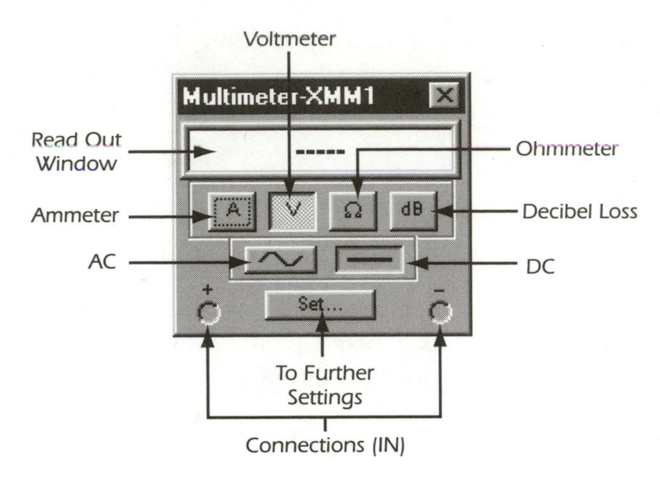

Figure 5.6 Controls and readouts on Multisim's multimeter.

NOTE

You don't have to select a range because the meter is autoranging. If you're measuring a 1V source and suddenly have to measure 24 volts, you don't have to change any range buttons or dials.

Once your preferred settings are adjusted, turn the circuit on and take the readings. Leave the multimeter open and activate your circuit. A reading should appear in the top white box. This valve may vary as time goes by.

Multimeter Test Circuits

DC READINGS

Build the circuit pictured in Figure 5.7, Measuring Voltage. Open the multimeter and activate the circuit. The settings are for a DC voltmeter, so you should see a reading of 6 volts. Press the **A** key on your keyboard and watch the results on the meter. Hit the **Shift** and **A** and note the reading now.

Modify the circuit to appear as in Figure 5.7, Measuring Amperage. Press the ammeter button (A) on the meter and reactivate the circuit. Adjust the potentiometer and note the readings.

Again, rework the circuit to appear as in Figure 5.7, Measuring Resistance. Select the ohmmeter by hitting the "Ω" symbol. Activate the circuit and vary the resistance using the potentiometer.

Figure 5.7
Measuring voltage, amperage, and resistance with the multimeter.

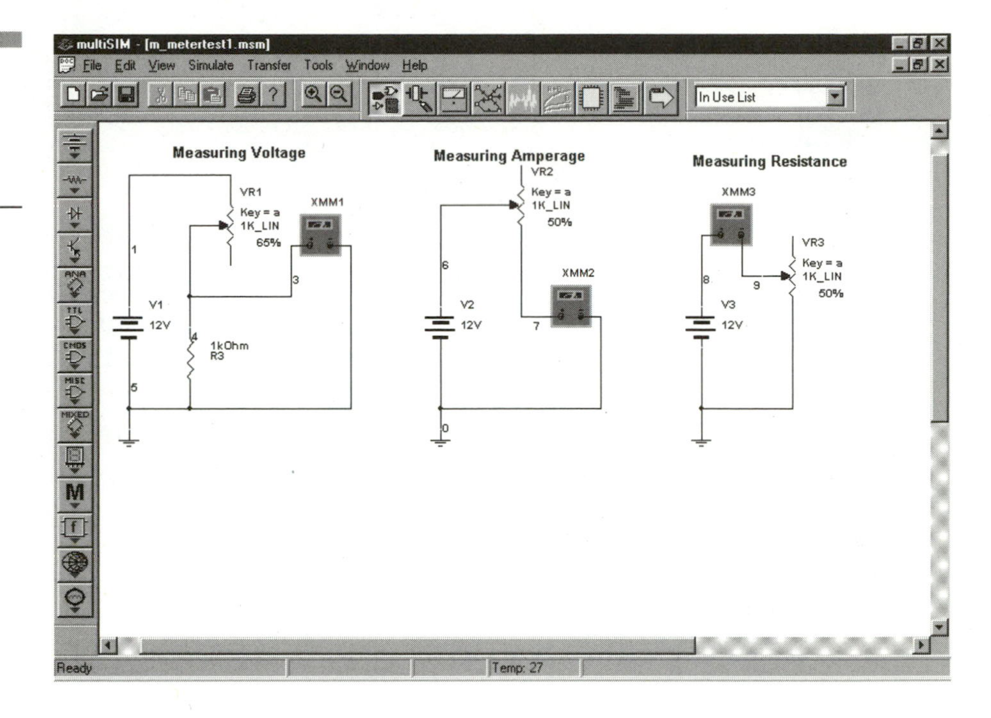

AC READINGS

Alternating current and voltage readings require that you set the meter to AC. Build the circuit pictured in Figure 5.8 and make the appropriate settings to the multimeter to read AC Voltage. Activate the circuit. What is the voltage reading? Select the AC ammeter. What are its readings?

Figure 5.8
Measuring AC with the multimeter.

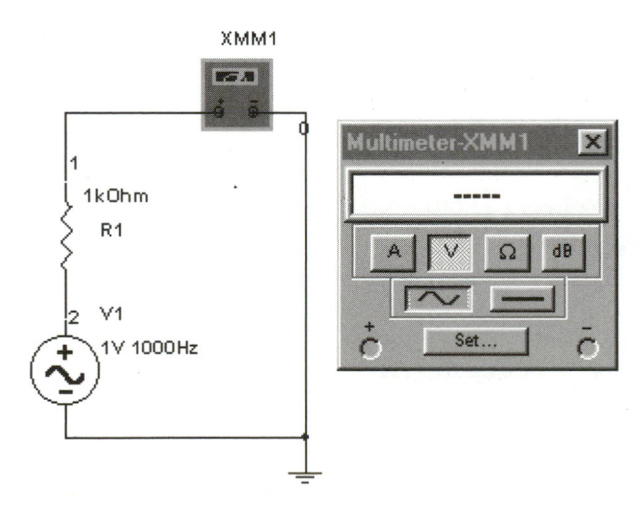

NOTE

Multisim's multimeter measures the RMS voltage/current of a circuit.

Multimeter Tips

Here are two tips to help you use the multimeter more effectively:

- **Swapping Polarity**—If the probes are on the wrong side of the meter, right-click and hit **Flip Horizontal** or hit **Alt+X** to change sides. This makes for cleaner designs.
- **Multiple Readings**—You can use more than one multimeter at a time. Just place another meter icon into the circuit and connect it. You can also use the voltmeter or ammeter components in place of the multimeter.

Wattmeter

The wattmeter is one of Multisim's new instruments. It gives a power reading according to voltage and current input. In a DC circuit this is a calculation of voltage × current. In an AC circuit, true watts = (effective voltage × effective current) × power factor. The wattmeter can also determine the power factor of a circuit, which is the cosine of the phase angle before the voltage and current. It is calculated by measuring the difference between the voltages and current, then multiplying them (between 0 and 1).

Connecting the Wattmeter

There are two sets of connections to consider when using the wattmeter. You connect the + and − of the voltage side (left set) as you would a multimeter that is measuring voltage. The current side (right set) is connected as you would connect a multimeter to read current. See Figure 5.9.

Wattmeter Operation

After the wattmeter icon is connected into the circuit, double-click it to open the Wattmeter window. That's it. Just activate the circuit and take a wattage or power factor reading.

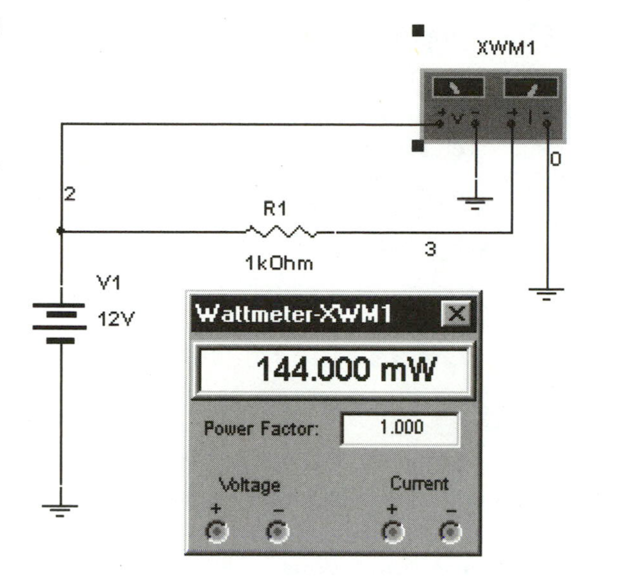

Figure 5.9 Connecting the wattmeter to read power.

Wattmeter Test Circuit

Build the circuit pictured in Figure 5.10. This circuit confirms the Maximum Power Transfer Theorem, which states, "A load will receive maximum power from a linear bilateral DC network when its total resistance value is exactly equal to the Thevenin resistance of the network, as seen by the load." Adjust the potentiometer back and forth and note that at 50% (500 Ohms), maximum power is available to the load.

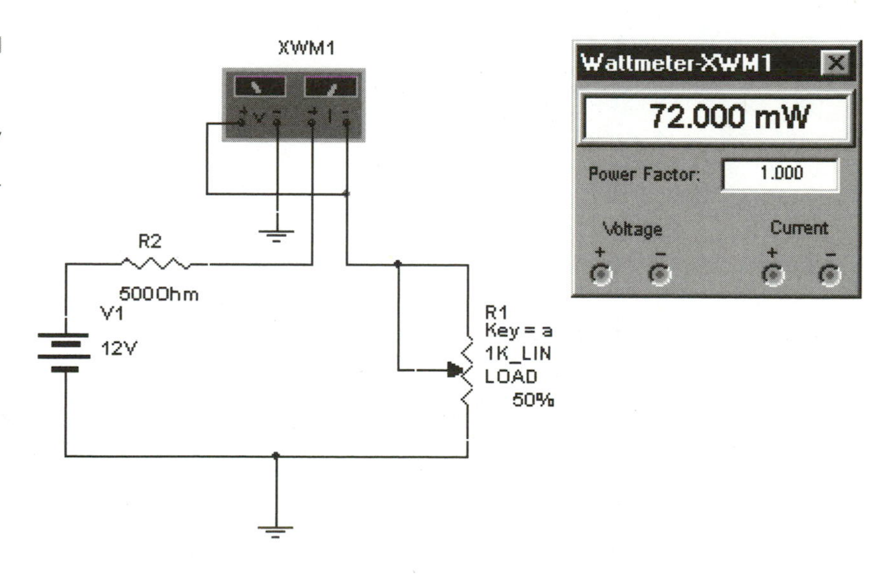

Figure 5.10 Proving maximum power transfer theory with wattmeter.

Oscilloscope

This complex instrument sometimes scares away newcomers to electronics. Don't fret. It's just another instrument to learn (and more buttons and dials to play with). An oscilloscope measures and displays the magnitude and frequency variations of electronic signals. It is used to test a functioning circuit. The "scope" gives technicians, students, and engineers a visual idea of what is happening inside a circuit. It displays waveform properties that are otherwise undetectable (because they are simply too fast). I like to think of the oscilloscope as a machine that slows time to allow us to see the intricate details (voltage, current, waveform, and frequency) of circuit operation.

Multisim provides a virtual dual-channel oscilloscope. On your real-life workbench, a scope is a major expense: aside from the hundreds or thousands of dollars required to buy a decent scope, you also need an oscillator to test your scope or learn its functions, which requires a function generator or hand-built circuits—in other words, more time and money. Multisim replaces this expense with its own savable, printable, hi-tech scope.

Connecting the Oscilloscope

Multisim's oscilloscope contains four connections (see Figure 5.11).

- Channel A (Probe)
- Channel B (Probe)
- Trigger (external)
- Ground

It is not necessary to ground the scope if the circuit being tested already contains a ground. However, I make it a practice to run a ground to it just in case I forget to ground the circuit.

If you want to 'scope out' a node in the circuit, place a wire from either the Channel A or Channel B terminal to the node you want to measure. Because the oscilloscope is a dual-channel model, you can connect another probe and read it at the same time. This is handy if you want to see what a signal is doing before and after a section of a circuit.

The external trigger is the last connection. It allows an electrical input to trigger the scope's readings; more on this in a moment.

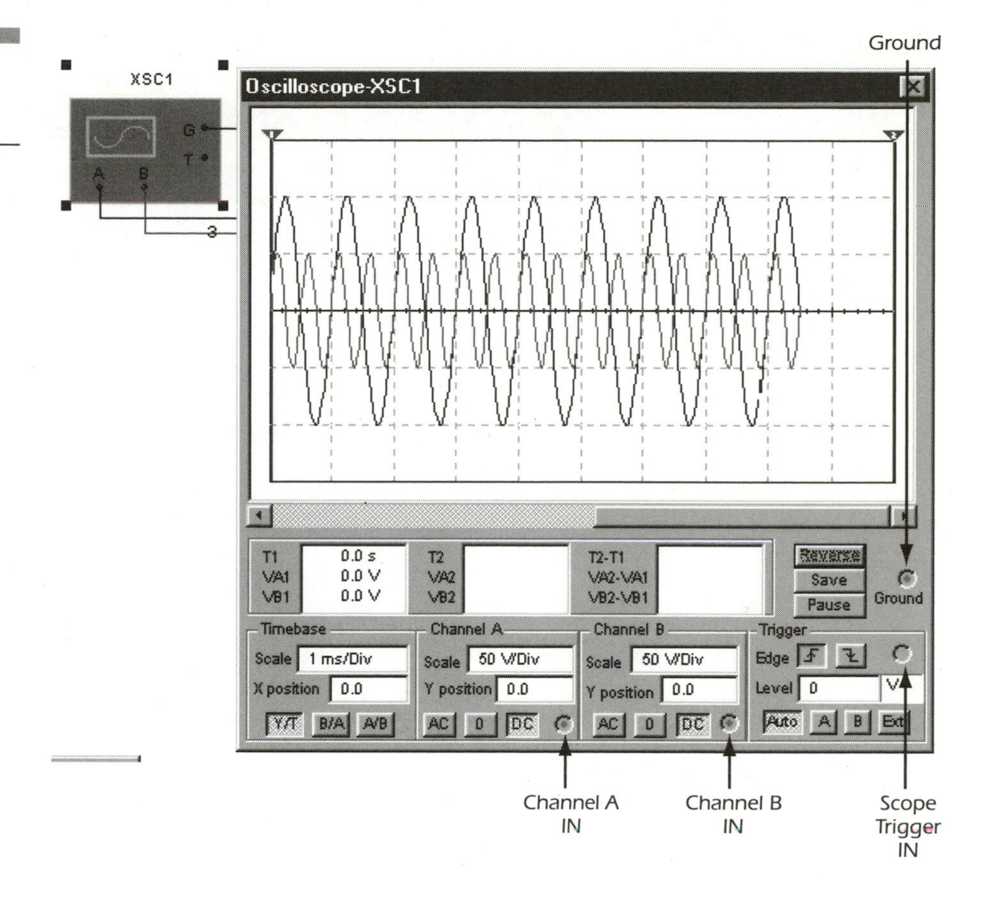

Figure 5.11 Oscilloscope connections.

Oscilloscope Operation

Once the oscilloscope is wired into a circuit, set the controls. This can be done while the circuit is in operation, but this sometimes causes all sorts of strange and quirky (but cool) effects. The controls are:

- **Graphic Display**—The readout appears here as a graphical waveform. See Figure 5.12.
- **Scrolling Bar**—Multisim records the waveform as the simulation runs. If you wish to scroll back to the beginning, use the bar below the Graphic Display window.
- **Cursor Readouts**—A set of vertical markers slides across the Graphic Display: Cursor 1 is red and Cursor 2 is blue. These can move back and forth to give you an instantaneous measurement of

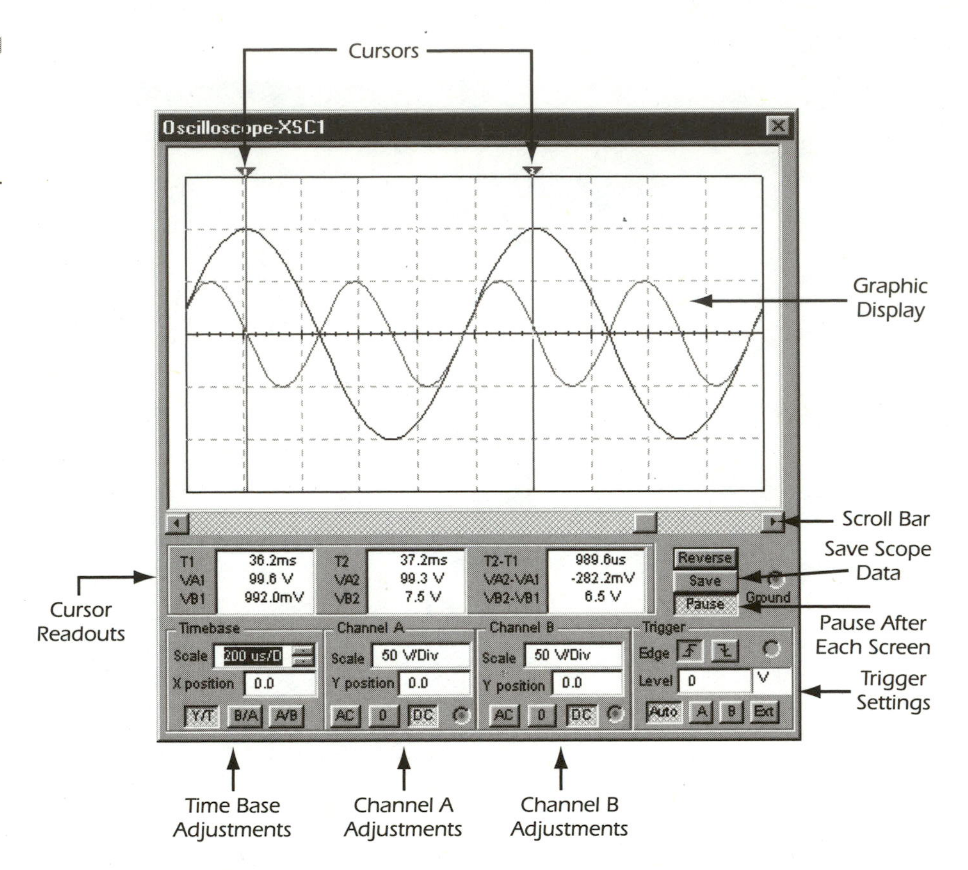

Figure 5.12 Controls and readouts of the Multisim oscilloscope.

the wave at that point in the X-axis plot. The windows below the scroll bar show these figures:

— T1 is the time position of cursor 1's point.

— VA1 is Channel A's voltage at cursor 1's point.

— VB1 is Channel B's voltage.

— T2 is the time position of cursor 2's point.

— VA2 and VB2 are voltages for cursors 1 and 2.

— T2-T1 is the time between cursor 1 and cursor 2.

— VA2-VA1 is the difference in voltage between those cursor points on Channel A.

— VB2-VB1 is the voltage difference between those cursors on Channel B.

- **Time Base**—The graphic screen of the oscilloscope has a vertical and horizontal grid. The Time Base settings control the scale of the oscilloscope's X position (vertical grid lines) when comparing the amplitude against time (Y/T).

— **Scale**—Adjusts the time units of each vertical grid line.

— **X-Position**—Controls the signal's starting point on the x-axis. When X is 0, the signal starts at the extreme left of the display. A positive value shifts the starting point to the right. A negative value shifts the starting point to the left.

— **Axes**—The axes can be switched from showing waveform magnitude against time (Y/T), to showing one input channel against the other (A/B or B/A). The latter displays frequency and phase shifts, known as Lissajous patterns, or a hysteresis loop. When comparing channel A's input against channel B's (A/B), the scale of the x-axis is determined by the volts-per-division setting for channel B (and vice versa).

NOTE

Most readings use the Y/T setting.

- **Channel A and B Settings**

— **Volts-per-Division**—Used to set the scale of the Y-Position. One horizontal grid line equals one division; if it is set to 1V, then each grid line represents one volt. The reading shoots to the top of the axis if the scale is too low. Roll the number up until you are able to view the reading. Conversely, if you can barely make out the waveform, roll the value down. In Figure 5.13, channel A is set too low and channel B is set too high.

— **Y Position**—This offsets the reading to make it clearer; if you are measuring two 0- to 5-volt square waves that are the same, offset one slightly higher or lower to better understand it. (See Figure 5.14).

— **Input Coupling**—With AC coupling, only the AC component of a signal is displayed. AC coupling has the effect of placing a capacitor in series with the oscilloscope's probe. As on a real oscilloscope using AC coupling, the first cycle displayed is inaccurate. Once the signal's DC component has been calculated and eliminated during the first cycle, the waveforms will be accurate. With DC coupling, the sum of the AC and DC components of the signal is displayed. Selecting 0 displays a flat-line reference at the point of origin set by Y.

Figure 5.13 Setting the volts/div too high or too low kills a good picture of the waveform.

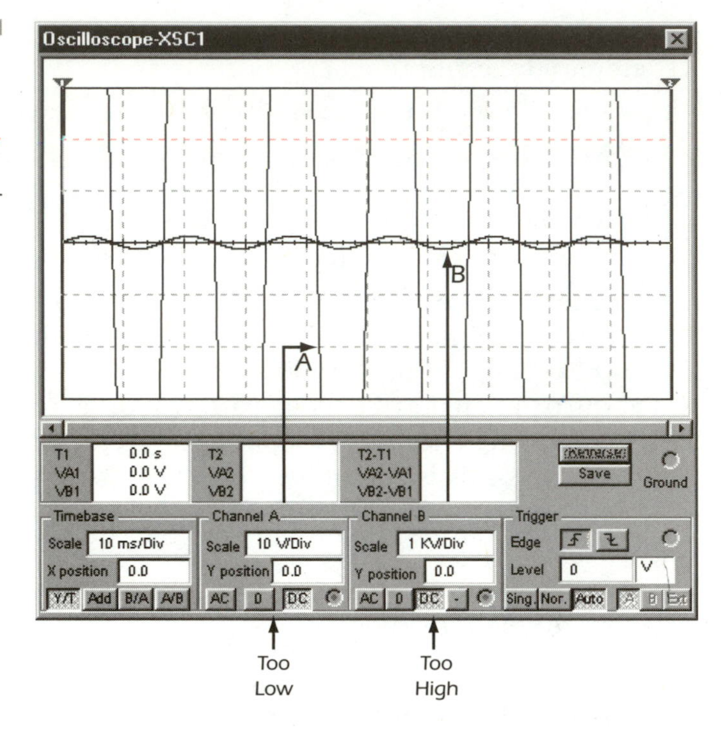

Figure 5.14 Use the Y position to view signals better if they are too similar to make out or if they are on the same line.

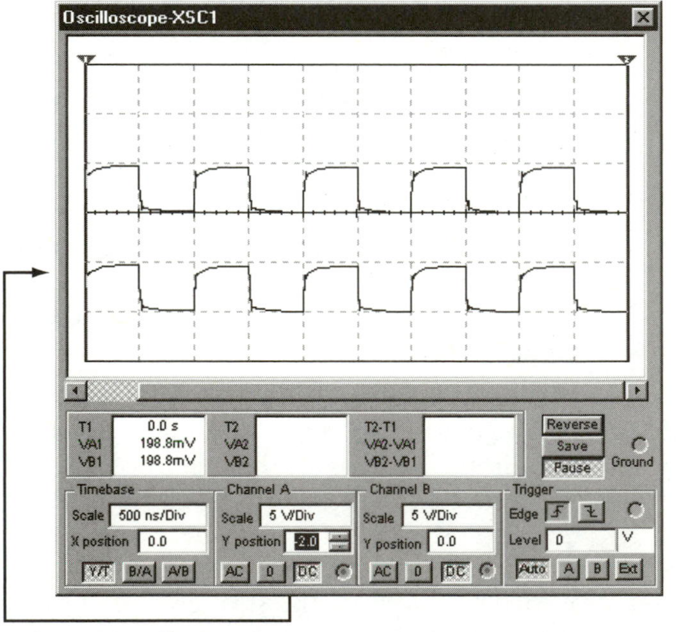

Channel A is offset to make the graph easier to read.

Do not place a coupling capacitor in series with an oscilloscope probe. The oscilloscope will not provide a path for current, and the analysis will consider the capacitor improperly connected. Instead, choose AC coupling.

- **Trigger**—These settings determine the conditions under which a waveform is first displayed on the oscilloscope.
  - **Edge**—On an ascending or descending signal.
  - **Level**—When the trigger reaches a preset level.
  - **Trigger Signal Location**—Determines from where the trigger will be received: **Auto** determines this automatically, **A** comes from channel A, **B** from channel B; and **EXT** is an external trigger signal.

The Expand button is eliminated in Multisim 6.

- **Reverse Button**—Makes a negative display (white on black instead of black on white).
- **Save Button**—Save the scope's data as a *.scp file for later use. This can be opened with Windows Wordpad or any word processor.
- **Pause Button**—Pauses the simulation at the end of each display screen to read the display. You must press the **Resume** button on the simulation switch toolbar to continue.

Readouts

Multisim's Analysis Graphs keep track of the oscilloscope's readings and converts them to a more pliable format. Go to **View** > **Show/ Hide Grapher** after the simulation stops. This opens the Analysis Graph, which has a tab for each oscilloscope reading.

Oscilloscope Test Circuits

The oscilloscope will be one of your most used instruments. Learning a few basic hookup principles makes its use easier.

BASIC VOLTAGE WAVE READING

Build the circuit as in Figure 5.15. Make sure to adjust the AC sources as shown. Open the oscilloscope and adjust its settings to match those in Figure 5.15. Activate the circuit. Hit the Pause button on the scope and examine the wave. Stop the simulation.

READING TWO WAVES AT ONCE

Referring to Figure 5.15, add a wire from R2 to Channel B on the scope. Make sure the AC source is adjusted to 50V, 2,000 Hz. Color the wire red. Activate the circuit again.

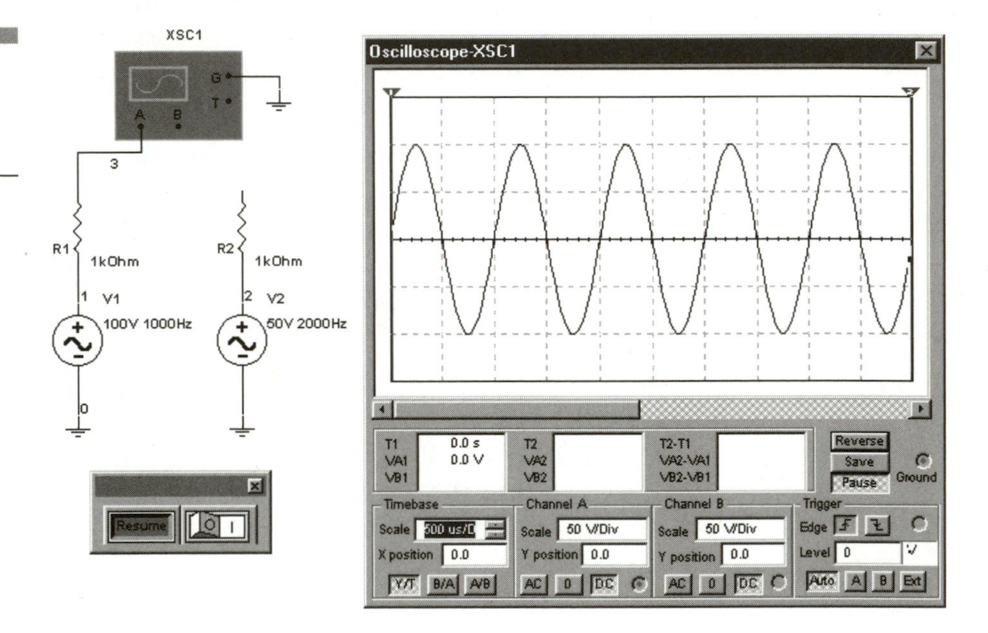

Figure 5.15 Reading a voltage wave with the oscilloscope.

Oscilloscope Tips

Here are a few tips to help you use the oscilloscope more effectively:

- **Using Colors**—Making the wires connecting two channels different colors lets you see which signal is for which channel. I typically use blue for channel A and red for channel B. I also make it a point to use channel A for before a signal and B for after. This lets me keep better track of what the circuit is doing to the signal.
- **Windows 98**—With Windows 98, you can place the instrument onto a separate screen, providing you are running a two-monitor setup.

- **Reading Current with the Scope**—The scope does not normally read a current wave, but with the addition of a current-controlled voltage source, this is possible. Hook up the circuit shown in Figure 5.16 and run it with the scope open.

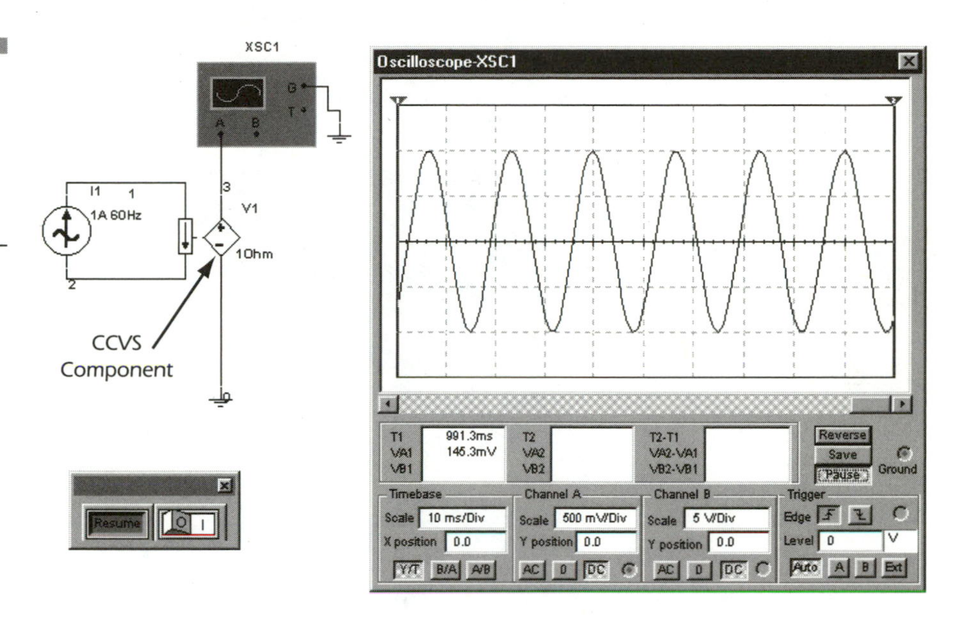

Figure 5.16
You can measure current waves with the oscilloscope by first changing the current into voltage.

Function Generator

It is sometimes necessary to inject a signal into a circuit. In Multisim, this is an optimal task for the function generator. This voltage source provides a sine-, triangle- or square-wave waveform at a frequency set by you. In reality, this is quite a costly piece of equipment, but the Multisim virtual model is included free in the price of the software.

Connecting the Function Generator

There are only three connections to worry about with this instrument: the center connection is the common connection and is usually connected to a ground; the '+' is the positive output of the function generator; and the negative (−) is the mirror of the positive output. If you want to inject a signal into a circuit, connect a ground to the common lead and the positive where the signal is going to be introduced into the circuit.

Function Generator Operation

After connecting the function generator to the circuit, double-click the icon to open the adjustments window for the instrument (refer to Figure 5.17).

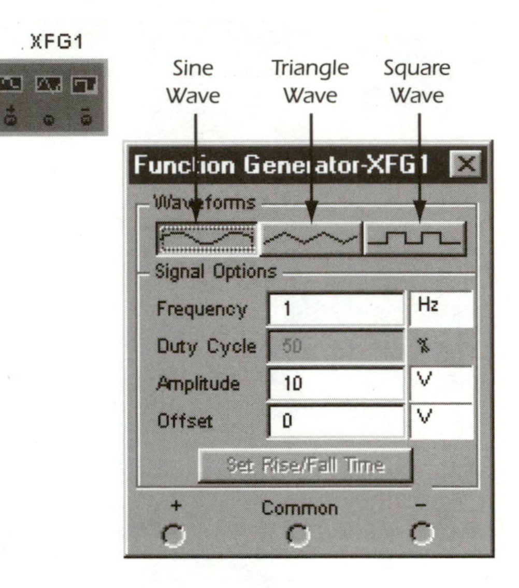

Figure 5.17 The Multisim function generator.

- **Waveforms Selection Buttons**—Select one of three waveforms to output (Figure 5.18):
  - — **Sine wave**—Gives a pure sinusoidal wave. If the amplitude's voltage setting is 10 volts, the wave swings from 0 to + 10 volts, to 0 to −10 volts, and starts over.
  - — **Triangle Wave**—This wave goes from 0 to 10 volts on a steady rise, peaks sharply, and then declines to 0 and −10 volts at a steady descent.
  - — **Square Wave**—This wave starts at 0, almost instantly rises to +10 volts and sustains that level until there is a sudden drop to −10 volts. It holds at −10 volts for a set period of time, then skyrockets back to +10 volts.
- **Frequency**—Allows you to adjust the wave's frequency (cycles-per-second) between 1 Hertz (1 Hz) and 999 Megahertz (999 MHz). Choose the variable (1 to 999) and the multiplier (Hz, kHz, MHz).
- **Duty Cycle**—You can adjust the duty cycle of the triangle or square Wave. This is the ratio of time on to time off, between 1 and 99%. See Figure 5.19: A is 20%, B is 50%, and C is 90%.

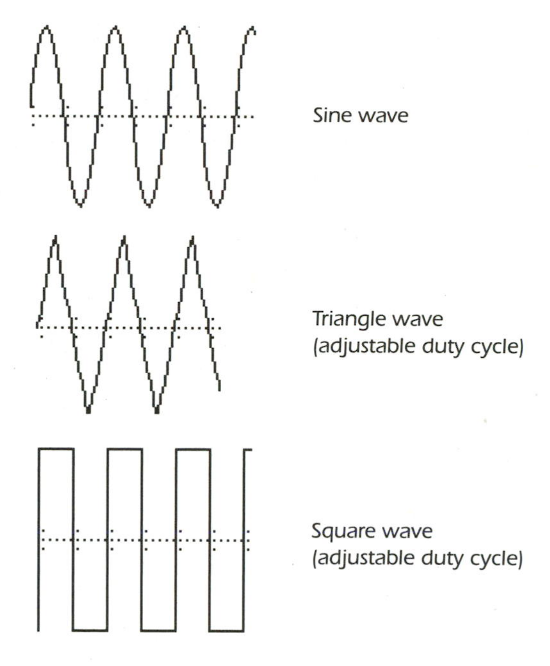

Figure 5.18 Waveforms explained.

- **Amplitude**—You can change the amplitude (height) of the wave from 1 microvolt (1uV) to 999 kilovolts (999kV). Choose the number and the multiplier.
- **Offset**—Controls the DC level at which the alternating signal varies. An offset of 0 positions the waveform along the oscilloscope's x-axis (provided its Y POS setting is 0.0). A positive value shifts the DC level upward, while a negative value shifts it downward. Offset uses the units set for amplitude.
- **Set Rise/Fall Time**—If a square wave is produced, you can add a slight delay in the time the signal takes to go high and low. This more accurately mimics true digital signals.

Function Generator Test Circuits

The fastest way to learn about the function generator is to connect it to the oscilloscope. If you are unfamiliar with the scope, review that section. Otherwise, follow these steps:

1. Build the circuit shown in Figure 5.20.

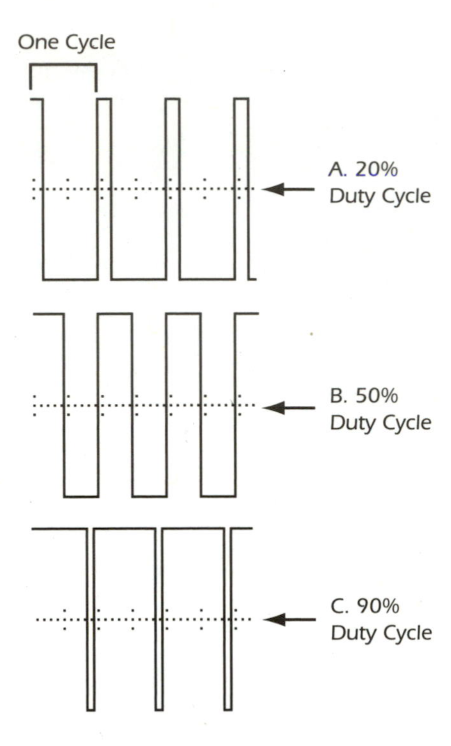

Figure 5.19
Duty cycle explained.

2. Make sure all the settings on the function generator and oscilloscope match.
3. Activate the circuit while the instruments are open.
4. Adjust a few of the settings. Change the waveforms and observe the results on the oscilloscope. Alter the frequency slightly, and then the duty cycle.
5. Try the amplitude settings. Change the offset to 5, then 10, and see where the wave positions itself on the scope.

Function Generator Tips

Here are two tips to help you use the function generator more effectively:

- **Creating a 0V to 5V Clock Signal**—The function generator's normal square wave swings from positive to zero to negative (0 to +5V to 0 to −5V to 0). However, most digital applications require a positive-only (0 to +5V to 0 to +5V to 0) clock signal. This requires that you set the amplitude to 2.5 and the offset to 2.5. See Figure 5.21.

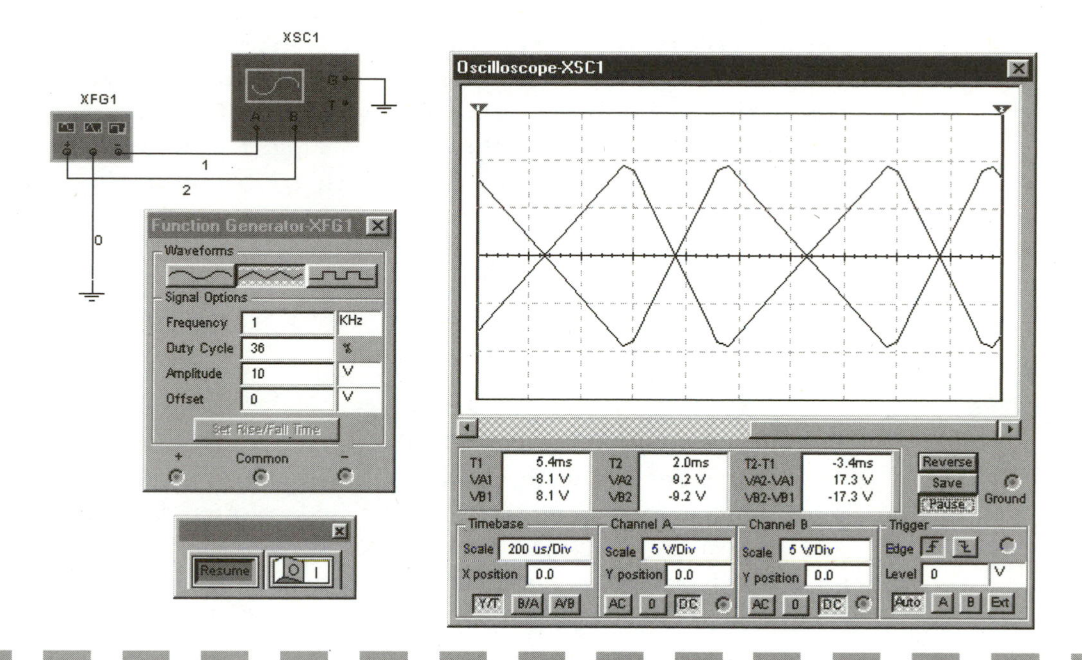

Figure 5.20 Using the oscilloscope to test the Function Generator.

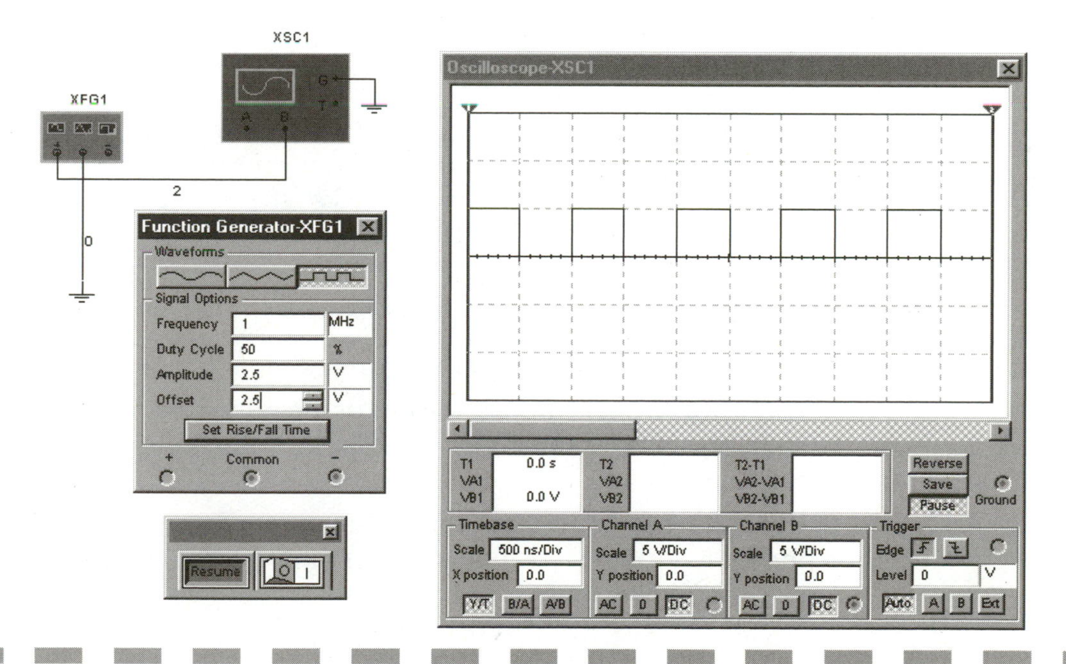

Figure 5.21 Using the Function Generator to create a clock signal.

- **More Than One Waveform Needed**—As with the multimeter, you are only able to run one copy of the function generator at once (with the hobbyist version of Multisim). If you want to inject multiple waveforms into the circuit, I suggest using an AC voltage source for a sine wave, or a clock. Both are found in the **Sources** Tool Bin.

Bode Plotter

The Bode plotter produces a graph of a circuit's frequency response and is useful for analyzing filter circuits. It is also used to measure a signal's voltage gain or phase shift. The Bode plotter performs a spectrum analysis by injecting a user-defined set of frequencies into a circuit, then charts its reaction to those numbers.

Connecting the Bode Plotter

The plotter has two sets of connections: V+In, V−In, V+Out, V−Out. The *magnitude* mode of the Bode plotter measures the ratio of magnitudes (voltage gain, in decibels) between two points, V+ and V−. *Phase* measures the phase shift (in degrees) between two points. Both gain and phase shift are plotted against frequency (in Hertz). Here's how to connect the Bode plotter in these modes, if V+ and V− are single points in a circuit (see Figure 5.22):

1. Attach the positive In terminal and the positive Out terminal to connectors at V+ and V−.
2. Attach the negative In and Out terminals to a ground.
3. If V+ (or V−) is the magnitude or phase across a component, attach both In terminals (or both Out terminals) to either side of the component.
4. In addition to connecting the Bode plotter's terminals, the circuit must contain some kind of AC source (it does not affect the operation of the plotter). In Figure 5.23, I use a function generator.

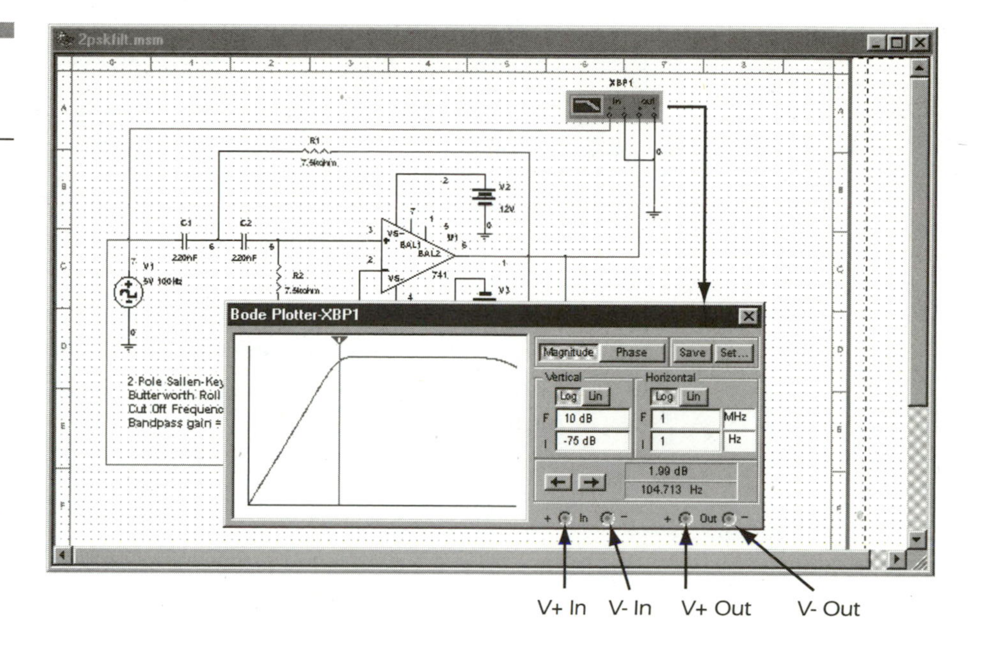

Figure 5.22 Connecting a Bode plotter.

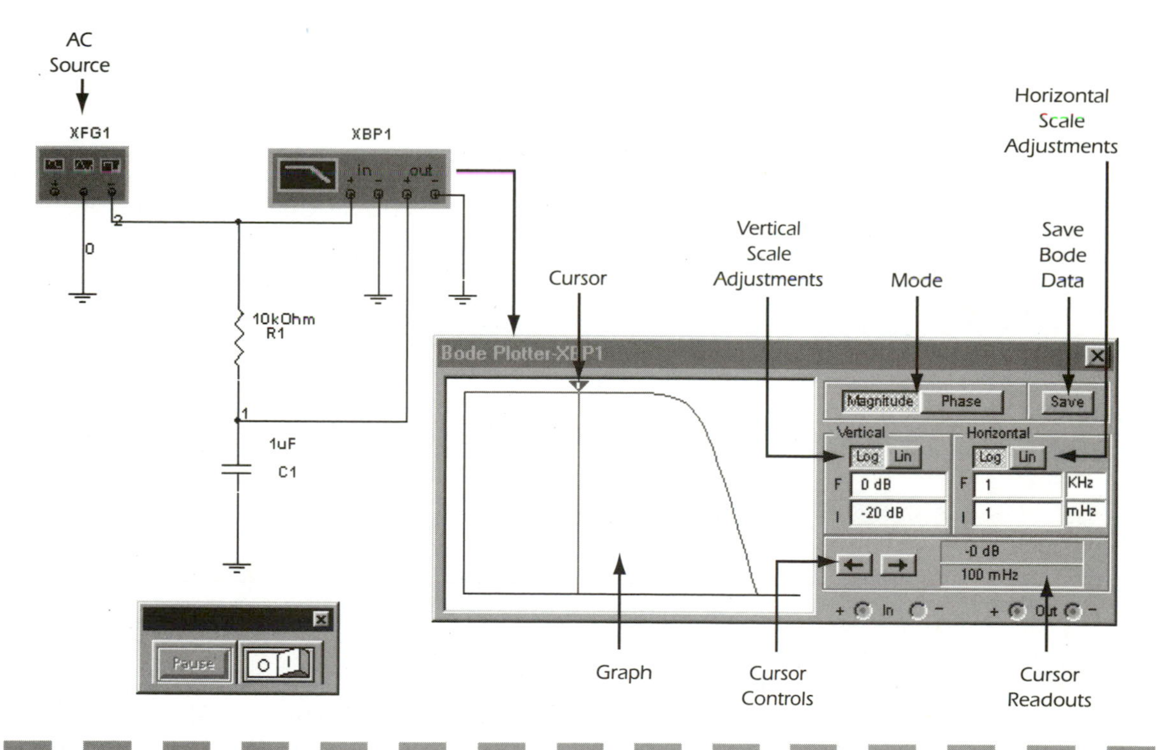

Figure 5.23 You can use the Bode plotter in filter applications to determine the cutoff frequency of this RC circuit.

Bode Plotter Operation

After placing the Bode plotter into a circuit, set it to read in the manner you choose:

- **Mode Select Buttons**—The two top left buttons mark **Magnitude** and **Phase**. Hit either to select the mode and graphs you wish to create.
- **Save Button**—The top right button saves the data to a *.bod file for later reference. This produces a text file that charts all instrument data into columns. It can later be imported into Excel for a useful hard copy.
- **Vertical Axis Settings**—Sets the plotting device's vertical axis:
  - — LOG = Logarithmic—Used when the values being compared have a large range.
  - — LIN = Linear—Graphs non-logarithmically (1,2,3,4...).
  - — F = Final value—Sets the final graphed value.
  - — I = Initial value—Sets the first graphed value.
- **Horizontal Axis Setting**—Sets the plotting device's horizontal axis. Settings are similar to vertical settings, except this adjusts the frequency ranges used for the plot.
- **Arrows**—The two arrows on the bottom left move a vertical cursor across the screen. The values are read out at that cursor position.
- **Readouts**—These two windows show the exact values at the point where the vertical cursor lies on the graph. The top field is the vertical and the bottom the horizontal readout.

Bode Plotter Test Circuit

Assemble the circuit pictured in Figure 5.23, a simple RC filter circuit. The Bode plotter will find the frequency cutoff. Adjust the settings to match the plotter as shown. Activate the circuit; the Bode plotter generates a series of frequencies starting at the initial value of 1 Megahertz and ending with the final value of 1 kilohertz. The graph is a plot of the ratio of output voltage to input voltage as a function of frequency (the Magnitude). Move the cursor to approximately −3dB. The frequency cutoff should be around 16 Hz. Try changing the value of the capacitor to 10mF and observe the new frequency cutoff.

The Bode Plotter has no real-world equivalent instrument; think of it as a function generator hooked to a circuit, with the output read by an oscilloscope.

Bode Plotter Tips

Here are two tips to help you use the Bode plotter more efficiently:

- **Saving Data**—You can press the **Save** button and store the Bode data as a *.bod file. It can be opened with a text editor or imported into Excel or Mathcad for further processing.
- **Use AC Analysis Instead**—If you want a printable graph and more control over the simulation, use AC analysis instead. Go to **Simulate** > **Analyses** > **AC Analysis** to open the dialog box. Make sure to fill out each page of information.

Digital Instruments

The next set of instruments are digitally based, meaning they either read or output a digital (0 or 1) signal. These include the logic analyzer, word generator, and logic converter.

Logic Analyzer

The logic analyzer lets you examine 16 running digital signals (waveforms) at once. It is used to visualize the output of digital logic circuits in much the same way as the oscilloscope illustrates an analog wave. If you are working on digital circuits and need to view more than two outputs (the maximum number possible with an oscilloscope) you can hook each to the logic analyzer and take a peek at up to 16 waveforms.

Logic Analyzer Connection

Each of the 16 inputs on the left side is a data input channel: the bottom left terminal is the *external clock* hook up; the bottom-center is the *clock qualifier,* and the last terminal is the *trigger qualifier.*

- **16 Data Inputs**—Any digital signal you wish to read can be hooked to these inputs directly.
- **Clock (External and Qualifier)**—Lets you use an external clocking source for the logic analyzer; also has a clock qualifier that filters the clock signal.
- **Trigger Qualifier**—An input signal that filters the triggering signal.

Logic Analyzer Operation

Open the logic analyzer (see Figure 5.24) by double-clicking its icon. Settings and functions are:

Figure 5.24
The logic analyzer.

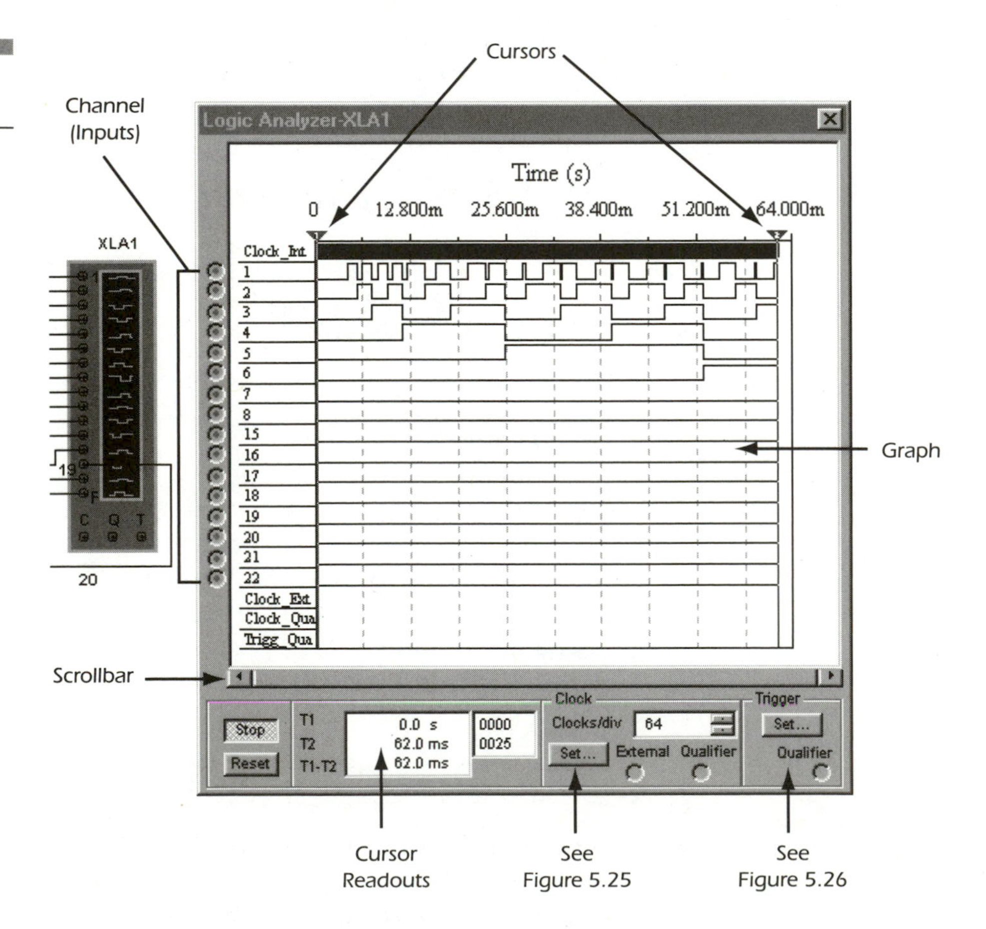

- **Graphic Display**—Shows digital waveforms, along with clock and trigger readings.
- **Scroll Bar**—Lets you scroll back and forth along the graphic window.
- **Stop Button**—Stops the instrument's reading but continues the simulation.
- **Reset**—Starts the instrument's readings from scratch.
- **Cursor Windows: T1, T2, and T1—T2**—Cursor 1 is red and cursor 2 is blue; the windows show the time position of the cursors. T1—T2 shows the time distance between cursors.
- **Clock**—To adjust the clock settings (see Figure 5.25):

  1. Click **Set** in the Clock area of the logic analyzer. The Clock Setup dialog box appears.
  2. Select external or internal clock mode.
  3. Select clock rate.
  4. Set clock qualifier, if set to external.
  5. Set the sampling settings: pre-trigger samples; post-trigger samples; threshold voltage.
  6. Click **Accept**.

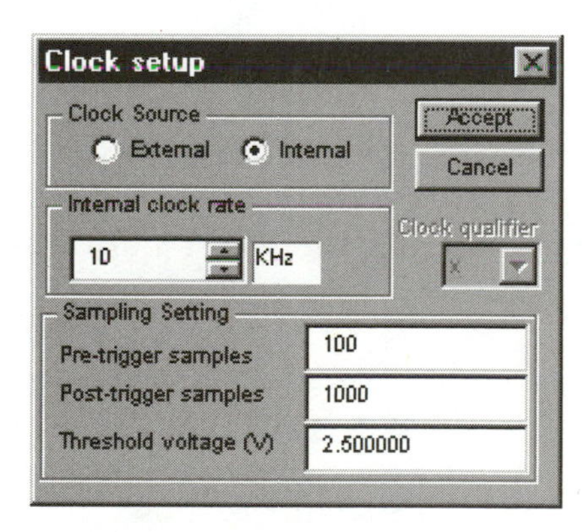

Figure 5.25 Adjusting the clock settings of the logic analyzer.

- **Clock Qualifier**—Input signal that filters the clock signal. If it is set to **x**, then the qualifier is disabled and the clock signal determines when samples are read. If it is set to 1 or 0, the samples are read only when the clock signal matches the selected qualifier signal.
- **Trigger**

  1. Hit the **Set** button. See Figure 5.26.

2. The Trigger Settings dialog box opens. Select which trigger clock edge to use.
3. Select the trigger qualifier. This is an input signal that filters the triggering signal. If it is set to **x**, then the qualifier is disabled and the trigger signal determines when the logic analyzer is triggered. If it is set to 1 or 0, the logic analyzer is triggered only when the triggering signal matches the selected trigger qualifier.
4. Set the trigger pattern (A, B, C) and its combinations. An x means the variable can be either 1 or 0. You can also enter a binary word.
5. Press the **Accept** button.

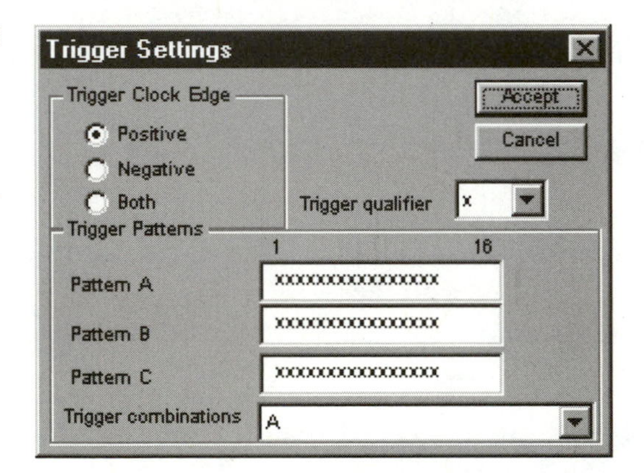

Figure 5.26 Adjusting the trigger settings of the logic analyzer.

To begin a simulation, hit the **Start** button on the simulation switch. You can stop the logic analyzer graph by hitting the **Stop** button on the logic analyzer. Hit the **Reset** button to clear the display, but continue the simulation.

Logic Analyzer Test Circuit

It's easiest to use the word generator for logic analyzer tests. Build the circuit as shown in Figure 5.27 and adjust the settings as seen in both instruments (load the up-counter in the word generator). Run the simulation and watch the readout.

Logic Analyzer Tips

These tips make using the logic analyzer easier:

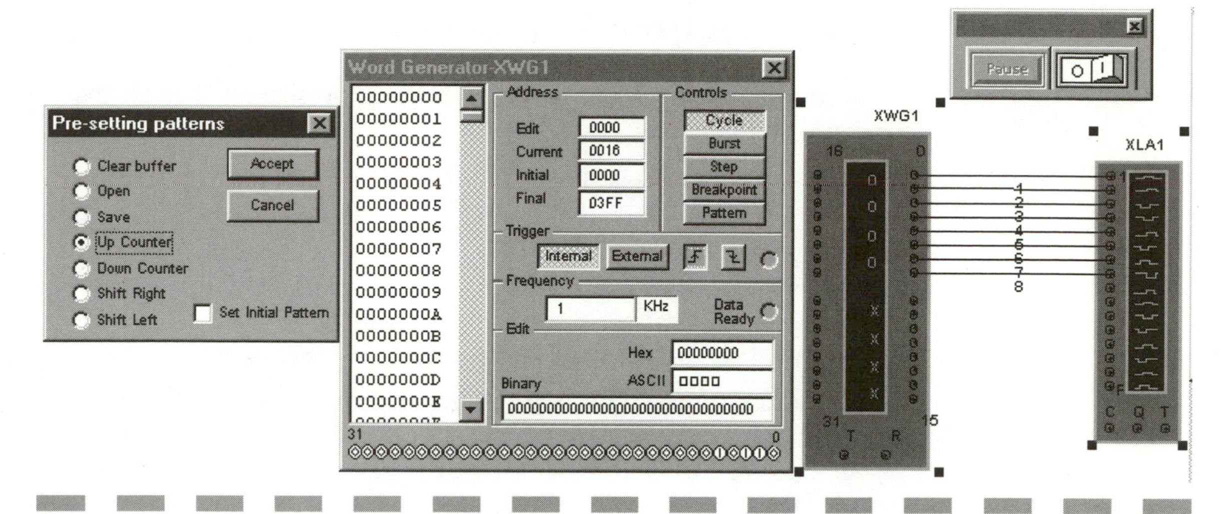

Figure 5.27 Hooking the word generator to the logic analyzer provides a quick learning circuit.

- **Use Wire Colors**—By coloring the input wires to the logic analyzer, the readout displays that color as well. This makes for a quick visual check instead of having to trace a finger across the screen for each input.
- **Use Instead of an Oscilloscope**—If you have multiple digital waves to read, use the logic analyzer instead of multiple oscilloscopes; the simulation runs much faster.

Word Generator

The word generator creates digital words or patterns of bits to feed into a digital circuit. It can be used for a variety of applications that require a specific digital input in a design. I personally use it as a shortcut to mimic a complex circuit or IC output, like a microcontroller. Think of it as a way to communicate digital words directly to a simulated device.

You can use the word patterns built into Multisim, or you can create your own and save them for later retrieval. Creating your own patterns requires setting up many parameters that are sometimes a pain to input, but there is no other way to input these digital bits into a circuit. So have patience.

Connecting the Word Generator

The word generator has 32 output channels (pins) as well as an external trigger and data-ready output.

- **Digital Output Pins**—The left row has 16 digital output channels. These output terminals send 5-volt digital signals. The top-left pin is the Least Significant Bit (LSB) and the right-bottom is the Most Significant Bit (MSB). To connect the outputs to a circuit, run wires from these terminals to the circuit's connections accordingly, keeping in mind the need to start with LSB. See Figure 5.28.
- **External Trigger**—This is an Input terminal. It allows you to control the word generator by an external means, such as a clock or other circuit. Hook up the incoming signal to the terminal and select whether you want it to cycle on a negative-going or positive-going pulse. (More on this in later sections.)
- **Data Ready**—This is an Output terminal. It lets the circuit know that data from the word generator is ready to transmit. This can be used in conjunction with the external trigger to further allow an external circuit to control the word generator.

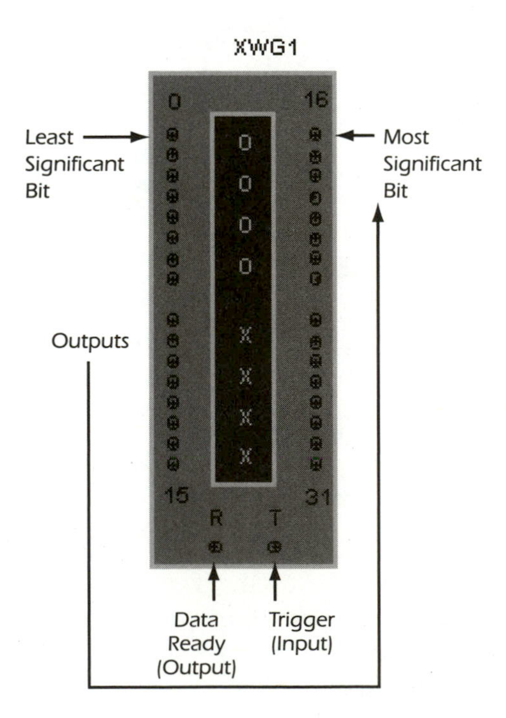

Figure 5.28
The word generator.

Word Generator Operation

To apply the word generator to a circuit window, open it by double-clicking the icon. Take a good look at the controls in Figure 5.29:

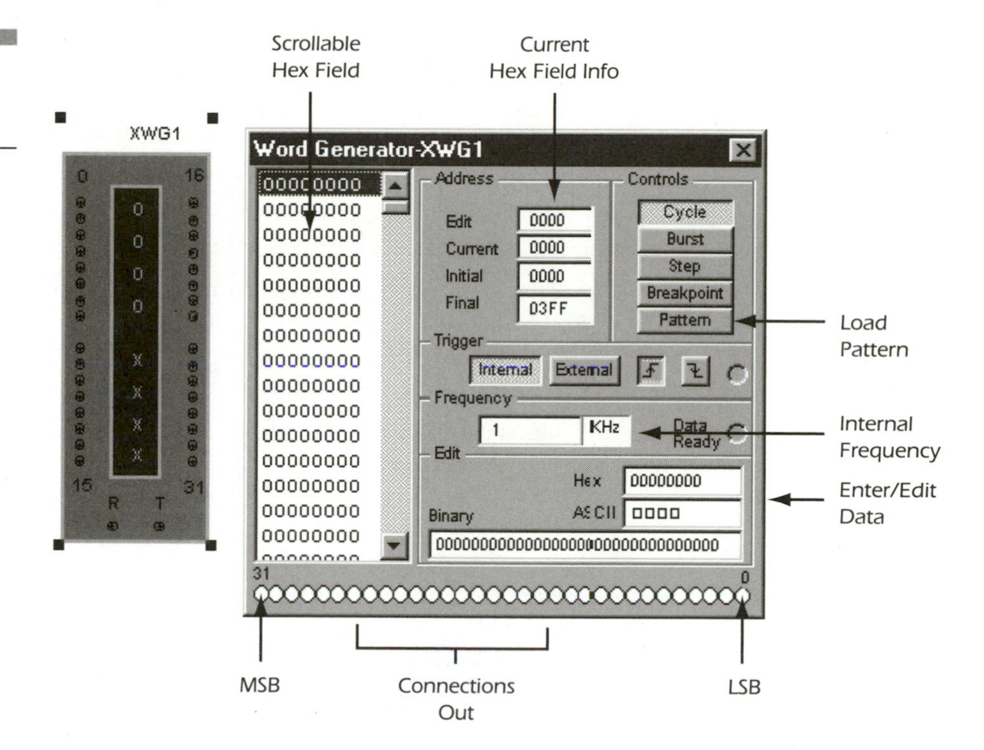

Figure 5.29 Using the word generator.

- **Scrollable Hex Field**—Controls the output of the word generator using hexadecimal numbers (0 to 9 and A to F). This is the HEX equivalent of the 32-bit binary word, as shown in the binary window to the right.
- **Address**—Sets the variables related to the HEX field addresses. Includes **Edit**, **Current**, **Initial**, and **Final**.
- **Trigger**—Select between internal or external trigger, as well as its signal trigger direction (rise or fall).
- **Frequency**—Sets the frequency at which words are sent from the generator.
- **Edit**—Lets you edit the current word in either Hex, ASCII, or Binary modes.

- **Controls**—Activates the word generator and simulation with your variable set to the outputs. These are preset ways to run the generator.
  - **Cycle**—Starts at the initiate address and goes to the end sequentially. The address is set to the beginning again and the sequence is played again.
  - **Burst**—Runs through the complete list of addresses once and then pauses the simulation.
  - **Step**—Goes forward, one address at a time.
  - **Breakpoint**—Lets you set the current address as a breakpoint. An asterisk appears next to the word at that address. When you run the simulation, it automatically pauses at that point.
  - **Pattern**—Opens the Pre-Setting Patterns dialog box (see Figure 5.30). You can select to Clear the Buffer, Open an Archived Pattern, Save a Pattern, create an Up Counter (1,2,3...), create a Down Counter (4,3,2,1), create a Shift Right (bit shifts to the right one significant bit) or a Shift Left pattern. There is also an option to set the initial pattern.

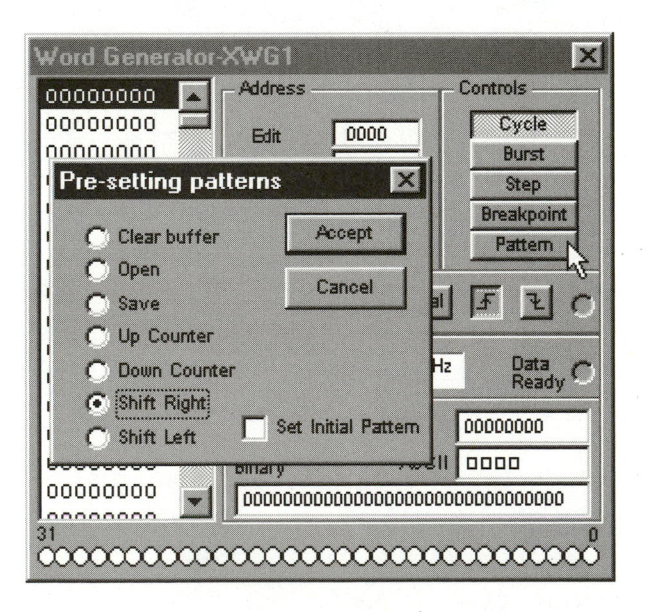

Figure 5.30 Word generator digital patterns can be loaded or saved.

- **Load**—A previously stored file can be opened with this button and stored as a *.dp file. This is a text file with eight hexadecimal characters for each entry.

- **Enter Words**—To enter words in the generator, follow this procedure:
  1. Highlight the word that exists in that address already.
  2. Enter the data into the HEX, ASCII, or Binary field.
  3. Click the next word/address you wish to edit. Repeat until complete.
  4. You can store your pattern by hitting **Pattern** > **Save** and naming the *.dp file.
- **Clear the Buffer**—To write zeros to each address, hit **Pattern** > **Clear buffer**.

Work Generator Test Circuits

Build the test circuit shown in Figure 5.31. Open the word generator. Hit the **Pattern** button and select the **Shift left** pattern. Go to line 0010 and change it to HEX 00004000 and add in each of these Hex values on each line thereafter: HEX 00002000, 00001000, 00000800, 00000400, 00000200, 00000100, 00000080, 00000040, 00000020, 00000010, 00000008, 00000004, and 00000002. Change the final address to 001D. Hit **Pattern** > **Save** and name it KITT.DP. Activate the circuit in cycle mode. Remind you of an old TV show?

The cut-and-paste feature, which allowed you to copy sections of patterns for quicker creation of complex patterns, is not available on the current version of Multisim 6 (6.11).

Word Generator Tips

Here are two tips to help you maximize your use of the word generator:

- **Make Patterns**—Create and save your own patterns. Try to build up a library of commonly used patterns and save them as *.dp files under a directory called C:\My-Multisim\Patterns\.
- **Mimic a Microcontroller's Output**—I'm not sure about you, but I use microcontrollers in many of my projects. Because Multisim currently doesn't have this component (aside from using a complex

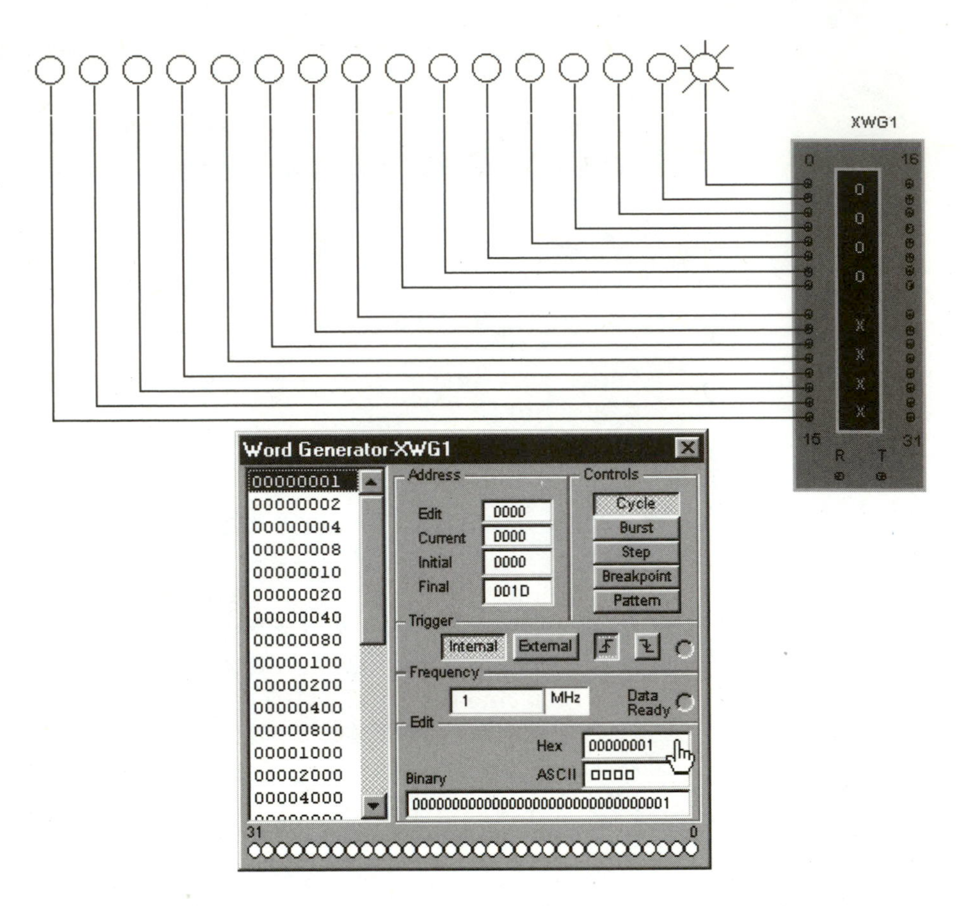

Figure 5.31 Knight Rider circuit to test the word generator.

VHDL component), I make do with a combination of instruments and parts. One is to use the word generator to emulate microcontroller outputs. For example, if I am simulating a device that turns on four relays one second apart, I adjust the generator to 1 Hz and use four of its terminals. If the controller needs an input before performing this task, I can use the external trigger on the word generator to activate the outputs.

Logic Converter

The logic generator is a timesaving digital design tool—there is no such creature in the real world of electronics. The logic converter takes a digital circuit and derives a *truth table* or *Boolean expression* from it. Another mode puts Multisim to work creating a circuit from your truth table. This is a tremendously tedious task if done by hand; the logic converter saves your brain and pencil from digital design burnout.

Connecting the Logic Converter

There are eight input connections hooked to the inputs of the digital circuit (see Figure 5.32). You can use one or all of the eight connections. The connection on the far right is the output. It reads the final outcome of the digital circuit. The following paragraphs describe the function and operation of this instrument.

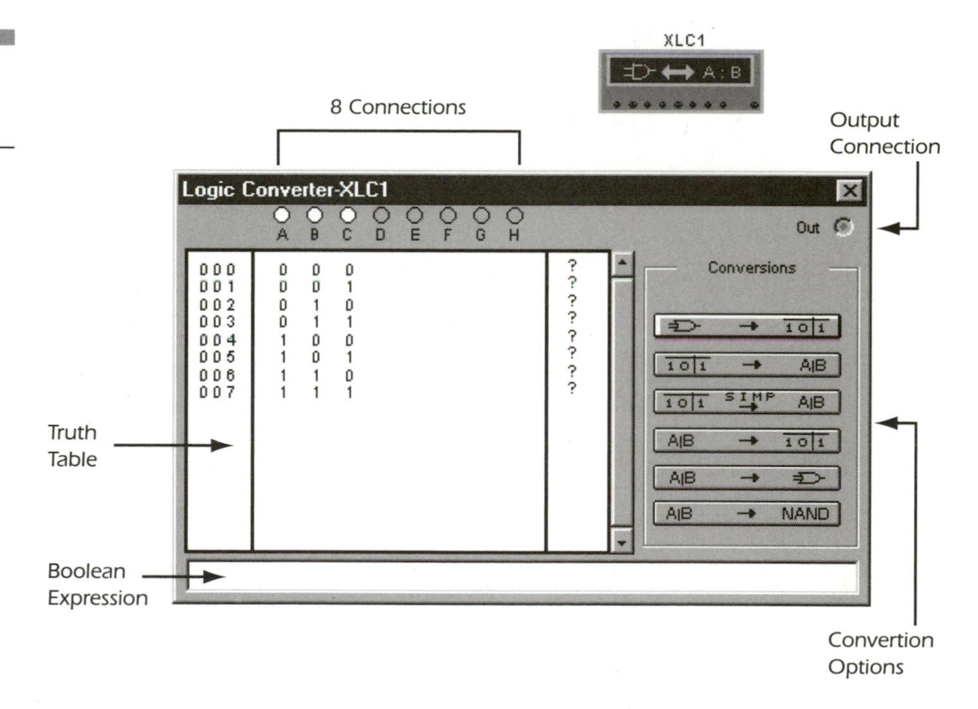

Figure 5.32 Connections on the logic converter.

DERIVING A TRUTH TABLE FROM A CIRCUIT

Figure 5.33 shows a simple digital circuit made of two NAND gates and one NOR gate. There are 16 possible input combinations for the A, B, C, and D inputs (which are connected to A, B, C, and D of the logic converter). There are also 16 possible outcomes that will be presented at the output of the circuit (hooked to the logic converter's

Output terminal). Open (double-click) the logic converter and press the **Circuit to Truth Table** button. Let the computer convert the signals into an Output table; this may take some time. You will see the results of the conversion on the far right column. As you can see, this circuit has only one combination that will activate a high output.

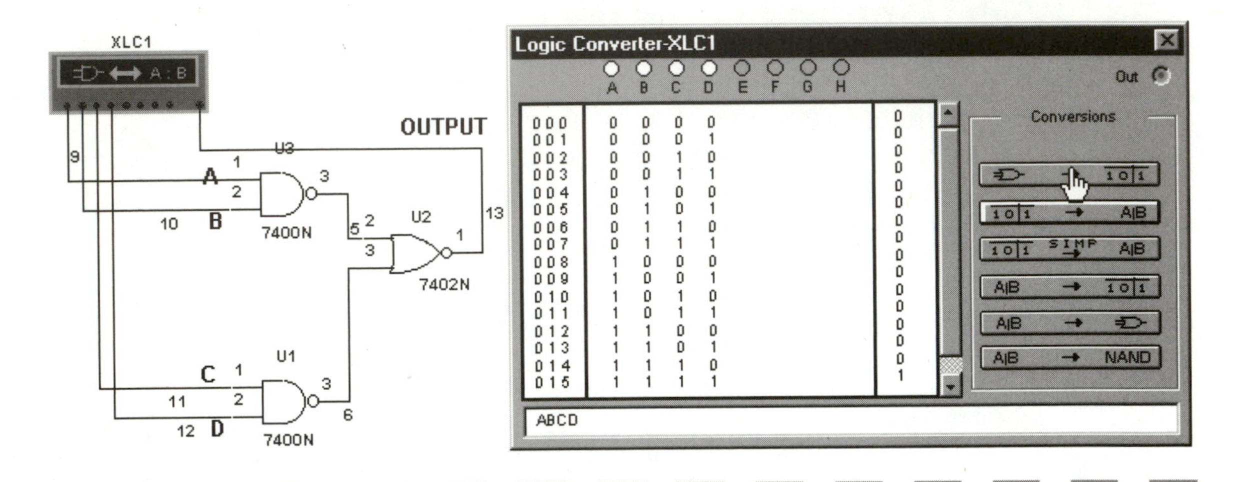

Figure 5.33 The fastest way to make a truth table!

CREATING A CIRCUIT FROM A TRUTH TABLE

This is the coolest feature of this instrument. Say, for example, that you want to build a circuit that tells when any two out of three parking spots are filled. A sensor is placed on each parking spot. These sensors must now be connected to a digital circuit to make an LED illuminate when any two spots are taken. We can use the logic converter to figure out the logic circuit. See Figure 5.34.

Here's how:

1. Open a new circuit and name it Park.msm.
2. Place a logic converter icon onto the drawing and double-click it.
3. Place the mouse pointer over the A connection (top-left terminal) and left-click it once. Do the same for the B and C connections. You should now see a series of consecutive numbers in the left column, zeros and ones in the middle column, and question marks in the right column. To create a circuit we first need to create the truth table. The question marks are the result of each row

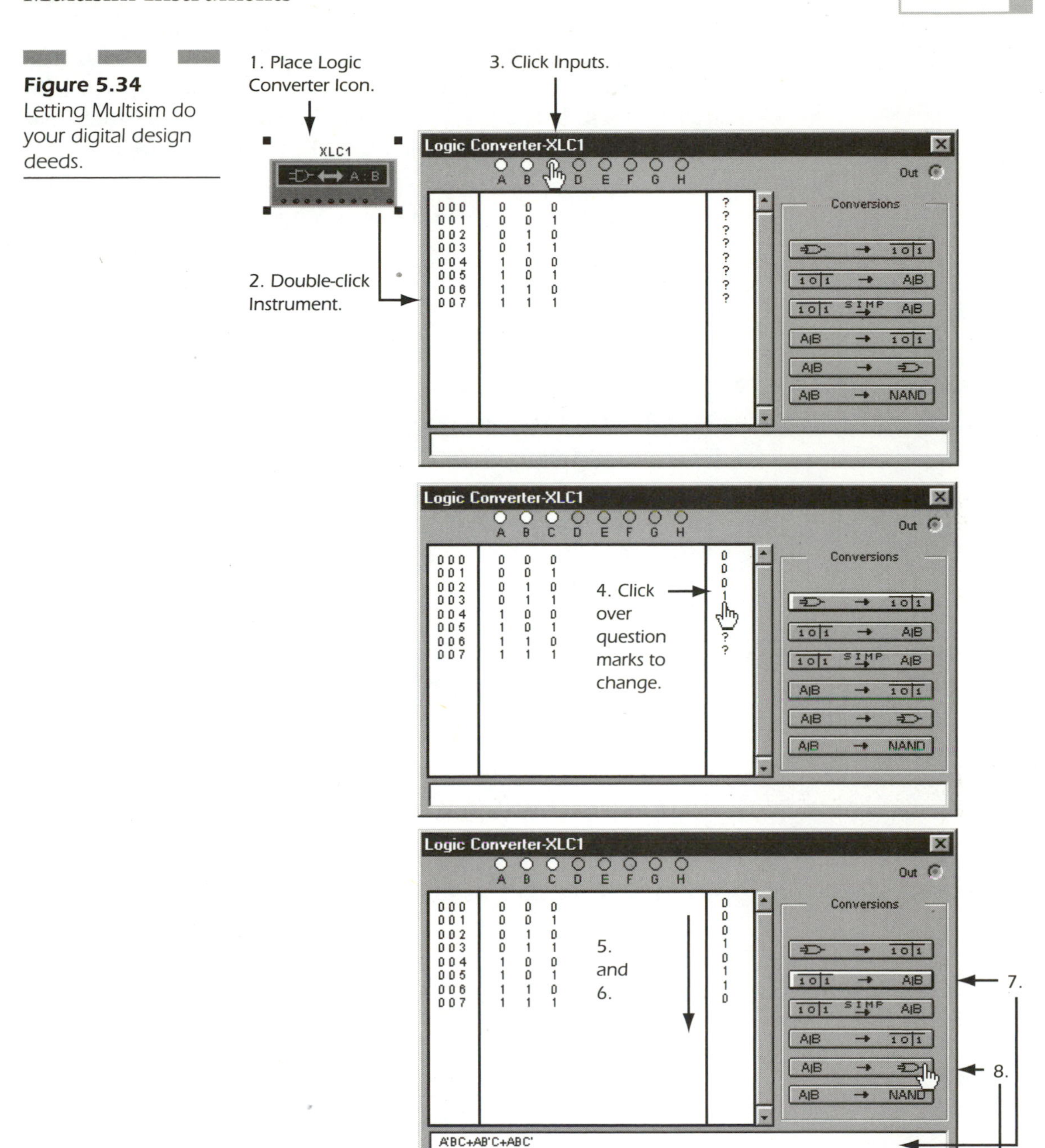

Figure 5.34
Letting Multisim do your digital design deeds.

and must be either 1, 0, or X (X means 1 or 0 is acceptable). In the case of the parking circuit, we want each row with two 1s in it to equate to a logic 1. There are three of these: 0 1 1, 1 0 1, and 1 1 0. For each of lines 003, 005, and 006 we make the result a logic 1.

4. Move the mouse's pointer over the first "?" in the right column (row 001). Click it once. It should have changed to a "0".
5. Do the same for the next two question marks as well. When you get to the third, click it twice until it is a "1".
6. Fill out the rest of the columns, making each row that contains two 1s into a 1. (There is no direct way to convert a truth table into a circuit, so we must first create Boolean expression.)
7. Press the **Truth Table to Boolean Expression** button. You see the Boolean expression in the bottom field.
8. You can convert the **Boolean Expression to a Circuit** with that button. Click it and allow it to draw the circuit for you. (See the resulting circuit in Figure 5.35.) That's it. You can now use the circuit to include in your parking circuit.

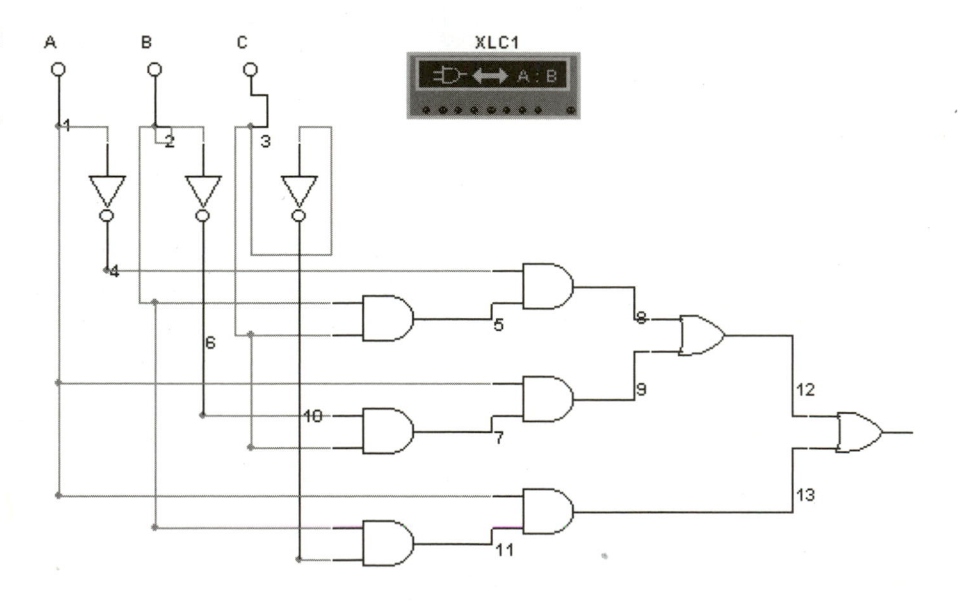

Figure 5.35
The resulting circuit.

If you wish to make a circuit entirely out of NAND gates, click the ***Boolean Expression to NAND Gate Circuit*** *button instead.*

Other Instruments

The following instruments are only available on high-end editions of Multisim. What follows is a brief description of each.

Distortion Analyzer

This new instrument is used to investigate small amounts of distortion in a circuit. It contains one input to read a signal to analyze. If you wish to see it in operation, open the file C:\Multisim\Samples\25DB_amp.msm in Multisim Pro or Power Pro.

Network Analyzer

This instrument is used to measure complex RF circuit designs. There are two inputs: Port 1 and Port 2. To see an example circuit, open C:\Multisim\Samples\Netan1.msm in the Power Pro version only.

Spectrum Analyzer

The spectrum analyzer is used to measure amplitude versus frequency, to help determine the existence of harmonics in a signal, which may interfere with other RF devices. There are several examples of this instrument under C:\Multisim\Samples\spectrum1... in the Multisim Power Pro version only.

Summary

Each instrument is complex; to learn their use requires some fiddling and experimentation. Open some of the sample circuits included in the C:\Multisim\Samples directory and locate the different instruments, then think of ways to read your own circuits using the virtual equipment. Soon you will be adjusting instrument settings as quickly and easily as you do your home theater equipment or microwave.

CHAPTER 6

Multisim: Simulation and Analysis

Investigating components/circuits and how they interact under different situations is the work of analysis and simulation.

It's a waste of resources to use a computer just to draw circuits when the same thing can be achieved with a pencil and paper. Our aim is to use circuit drawings for other rewarding purposes: in other words, simulation and analysis. It's as if you drew a circuit on paper, built the circuit in real life, and hooked up countless instruments to it. With Multisim all you have to do is draw the circuit and flip a virtual switch to get results.

NOTE

Simulation *uses the computer to simulate a real-world circuit using software.* Analysis *examines the tasks performed by a circuit, circuit segment, or component and submits them to many tests or scenarios.*

Multisim as a Simulator

Multisim simulates the electron flow through a conductor, a piece of carbon, a semiconductor, or another electronic component by pretending that real-world electronic components are strung together and active. How does it do this?

Mixed-Mode Simulation

As discussed in the first chapter of this book, Multisim is capable of simulating a mixture of analog components and digital parts. *Mixed-mode simulation* allows you to mix resistors with AND gates, capacitors with inverters, and the like.

B SPICE

Multisim's main simulation engine is based on Berkeley SPICE3F5. These algorithms describe each component in a specific circuit, as well as how they are interconnected. For example, a transistor may be described as .**MODEL 2N2222A NPN (IS=1.16e-14 BF=200 BR= 4 RB=1.69 RE=0.423 RC=0.169 CJS=0...)**. Each of the parameters describes the properties of that component. The more parameters, the more detailed the model, and thus the more realistic the part in the circuit simulation.

XSPICE

Multisim also uses Xspice, which is a set of unique enhancements made to SPICE, under contract to the U.S. Air Force. These include specialized modeling subsystems.

VHDL and Verilog

VHDL and Verilog are also used to simulate circuits in Multisim. They are both simulation-modeling languages used to describe the behavior of electronic components ranging from simple logic gates to complete microprocessors and custom chips. This is a high-end feature that could fill a book unto itself.

Using Multisim as a Simulator

You've been sweating for hours creating a perfectly drawn circuit with components, wires, labels, etc. Now what? How about activating the electrons to see if the circuit would actually work in the real world. Let's hope you hooked up some kind of indicator or instrument to see if the circuitry actually works. Let's use Figure 6.1 as an example. To simulate something, we first need a complete circuit that can operate in the real world. We also require a way to visualize the output; otherwise, how would we know what the circuit was doing? We use an indicator or instrument for this purpose.

In our example, the circuit contains a battery hooked to a switch and a light bulb (grounded). The bulb is the visual indicator needed to view the results. If we look at this from the viewpoint of a computer, all that would be seen is a list of mathematical formulas being calculated at the speed of light. That's the actual simulation.

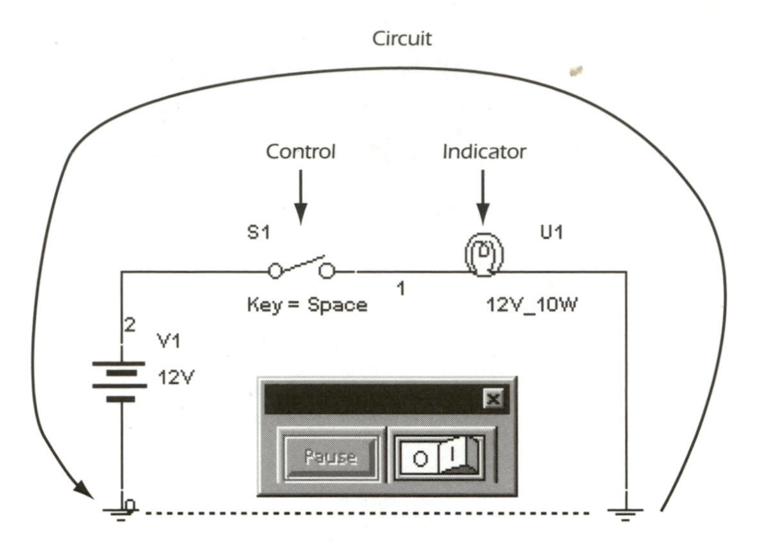

Figure 6.1 Electronics Simulation requires a "visual" to see what the circuit is doing. This can be as simple as a bulb or even a readout from an instrument.

NOTE

A Multisim simulation is basically a transient analysis *that continues to run and run until you stop it with the simulation switch. Transient analysis is explained later.*

Simulation Switch

The simulation switch activates the circuit and thus the simulation. It is located at the top right corner of the desktop when Multisim is first opened (see Figure 6.2). This toolbar can be docked or floated anywhere on the screen. '1' = activate and = '0' deactivate. A Pause button is also available to temporarily halt a simulation.

TIP

Close the simulation switch and use the Run/Stop simulation icon on the Design toolbar to save workbench space.

Simulation Error Log and Audit Trail

Multisim keeps track of each simulation from start to finish (see Figure 6.3) in a textual record similar to that of the Grapher. It logs errors, circuit information, date/time, and complex analysis information. To

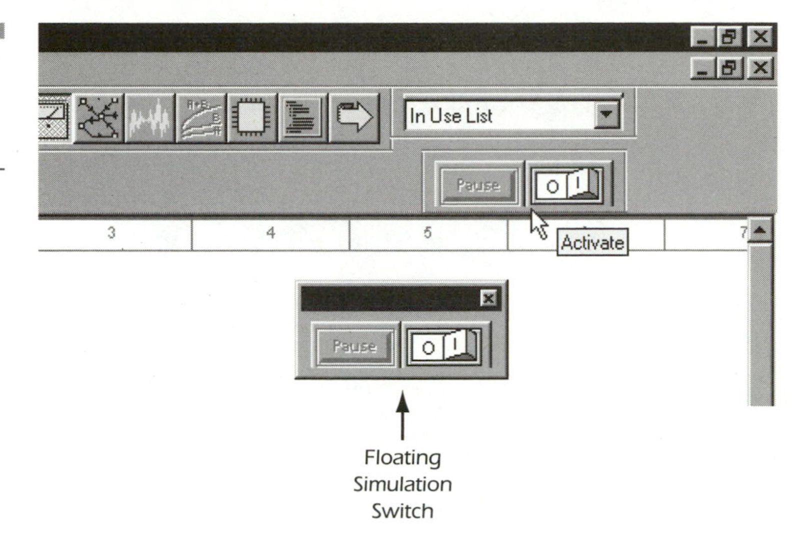

Figure 6.2
The Simulation Switch activates the circuit.

View the log or audit trail, hit **View** > **Show/Hide Simulation Error Log/Audit Trail**. To save a *.log file that can be read with any text editor, press the **Save** button and name the file. This feature is useful in diagnosing circuits that are not working quite right, usually as a last-ditch effort.

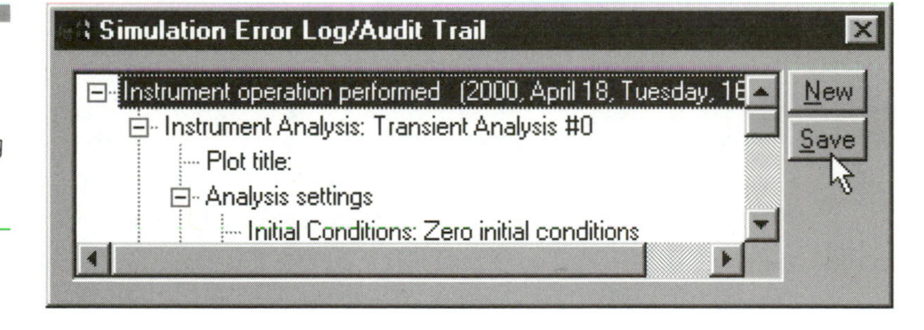

Figure 6.3
The Error Log/Audit Trail keeps track of the simulation, along with any errors that may occur.

TIP

*To make a simulation run for a preset time, choose **Simulate** > **Default Instrument Settings** and change End Time (TSTOP) to the desired time.*

VHDL Simulation

Although VHDL components can be added to a Multisim simulation, this subject is beyond the scope of this book and it is unlikely that, as a hobbyist, you would use this feature (it is a high-cost add-on to the

Professional and Personal Edition). However, it is interesting to note that if you want to simulate programmable logic devices (CPLDs and FPGAs) or complex digital chips, such as memory, CPUs, microcontrollers, and other devices that could not be reasonably modeled using SPICE, you have the ability with the high-end versions of Multisim. There is a great primer on VHDL in Multisim Appendix A. To open this, go to **Start** > **Programs** > **Multisim** > **User Guide Appendices**.

NOTE

Analyses is the plural form of analysis.

Multisim Analyses

Multisim's most powerful feature is its ability to analyze circuits inside and out. You can analyze a circuit or a component, how a component reacts to a certain circuit, how it reacts to worst case scenarios, how its reading changes with temperature—the list goes on and on.

What exactly is a Multisim analysis? The user defines areas in a circuit (nodes) from which he wishes to take readings. These can be instantaneous readings; values that change with time or frequency; or even reactions to a varying temperature, voltage, etc. Multisim calculates the appropriate variables and outputs them in the form of a table, graph, or a series of both. The results can then be recorded, printed, or sent to other programs for further processing.

Multisim performs three major types of analysis: DC, AC, and transient. Each major type can be performed in conjunction with several other analyses. DC analysis ignores all AC elements in a circuit and calculates the steady-state voltage or current for each node in the activated circuit. It is also used to determine certain transistor and diode variables to be used in other analysis. AC analysis then takes the results from the DC analysis and calculates the AC circuit response as a function of frequency. Transient analysis lets you view what a circuit is doing over time—this may become your most used feature in Multisim, because it offers greater control than if you had used Multisim instruments. Most simulations running in Multisim are of the transient variety.

TIP

Use DC analysis instead of positioning multiple multimeters around a circuit.

Why Analyze?

Why do you need such powerful analysis capabilities? Answer: the better to visualize *exactly* what a circuit is doing while active. A student can learn the characteristics of circuits and components, and engineers and designers can fine tune circuits or components they are designing. They can also test "What-if" scenarios in seconds. Technicians can simulate on the bench problems they encounter with real-world devices. The good thing about analysis is that a record can be kept of the results and graphs can be printed for later examination or presentation.

Steps of Analysis

It is easiest to breakdown the analysis processes into steps:

1. Draw a circuit, making a schematic drawing of the circuit you wish to analyze (covered in Chapter 3).
2. Set up analysis. Select the correct analysis and input the variables required.
3. Activate the analysis.
4. Read results. View the Grapher's results.
5. Make adjustments to circuit and reanalyze.
6. Save, print, or export the results. Keep your results, make a hard copy, or export them to Excel, MathCAD, or other software.

Setting Up the Analysis

I am assuming your circuit is assembled and ready. The first thing to do is to set up the options that will be used for a specific analysis. To access an Analysis dialog box, go to **Simulate** > **Analyses** > and choose the analysis method. For example, you may choose **Transient** analysis. Figure 6.4 opens with the following tabs: Analysis Parameters, Output Variables, Miscellaneous Options, and Summary. Each analysis has several pages of variables and settings to complete, but most of them are similar in function and operation.

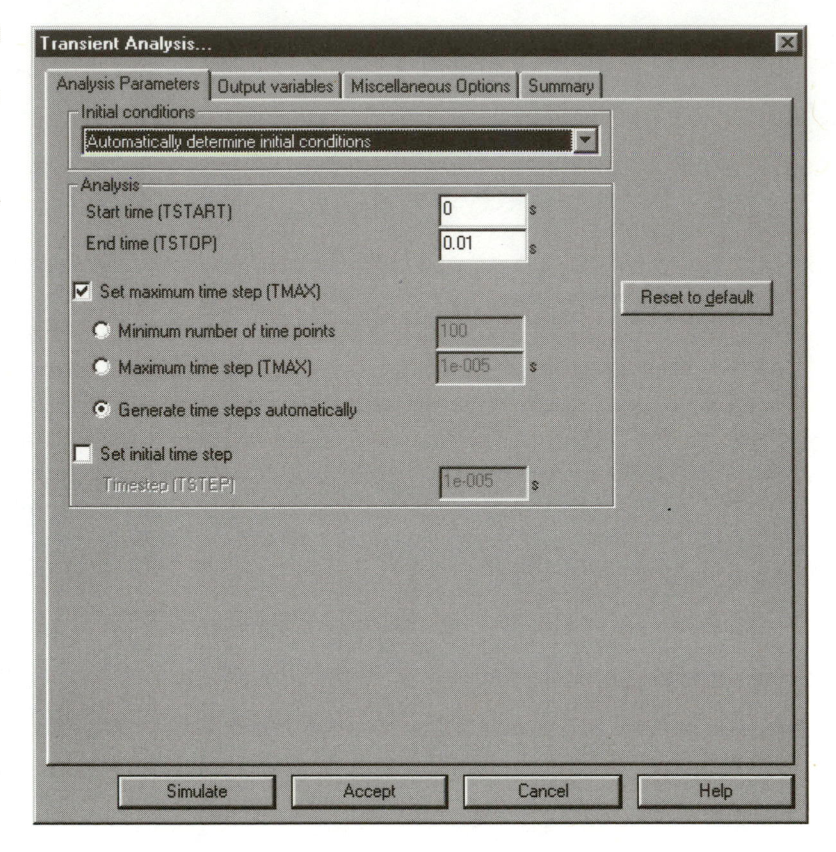

Figure 6.4 The Transient Analysis dialog box best exemplifies how an analysis is set up.

TIP

You can also hit the analysis icon on the toolbar (it looks like a yellow voice wave) to choose an analysis method quickly.

Using Analysis Dialog Boxes

Let's use examples to illustrate how to fill out some of the common Analysis pages. Go to **Simulate** > **Analyses** > **Transient Analysis**.

ANALYSIS PARAMETERS TAB

In the case of a Transient analysis, you set the initial conditions, start and stop time, and how often the simulation will take a reading. See each analysis explanation later in this chapter for setup.

OUTPUT VARIABLES

Refer to Figure 6.5; it looks complex but this is merely where you tell Multisim from where in the circuit you want to take readings. More specifically, it's where you set up which nodes/components (variables) in the circuit will display analysis results.

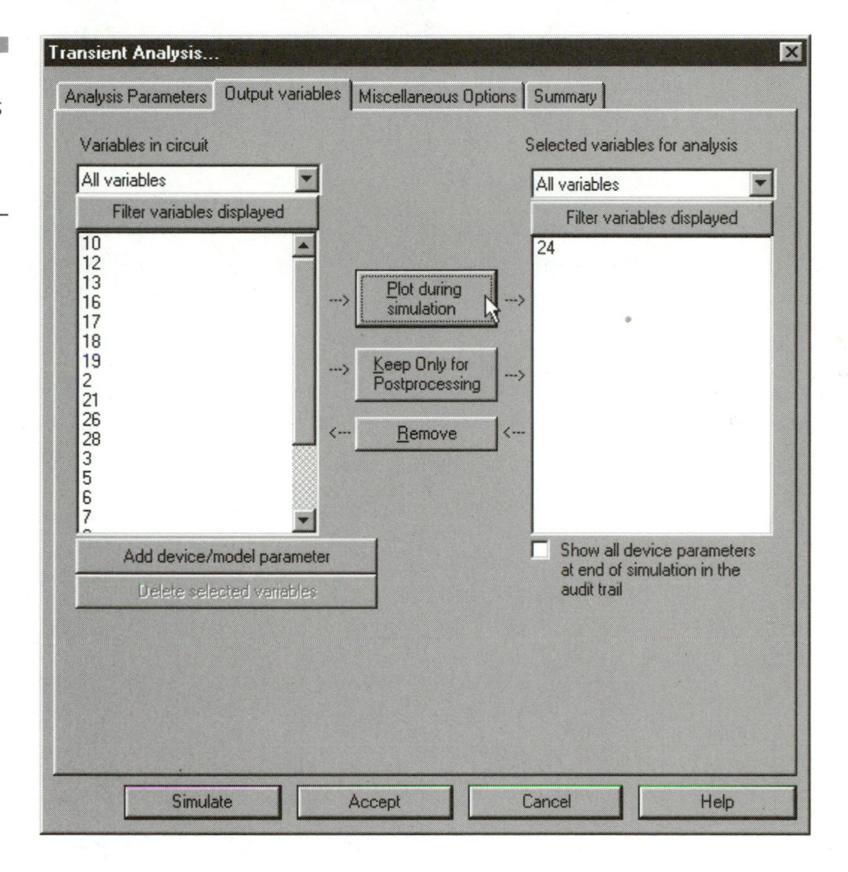

Figure 6.5 The Output Variables tab must be completed for every analysis.

- **Variable in Circuit**—This pulldown box lets you narrow the output variables selection: All Variables, Voltage and Current, Voltage, Current, and Device/Model Parameters. For example, if you just want to know which nodes in the circuit give a current reading, choose Current and a list of those components will be displayed in the Possible Output Variable Field below.
- **Filter Variable Displayed**—Opens a dialog box to let you narrow output further. Place a check in those boxes you wish displayed in the Variables Field.

- **Possible Output Variables Field**—Lists all the possible variables after you have filtered them.
- **Center Buttons**
  - — **Plot During Simulation**—Moves variables from "Variables in Circuit" list over to the "Selected Variables" list.
  - — **Keep Only for Postprocessing**—Makes variables available only in the Postprocessor, which is not a feature in the Personal Edition of Multisim.
  - — **Remove**—Moves variables from Selected list back to Variables in Circuit list.
- **Selected Variables for Analysis**—Shows the variables (nodes) displayed at the end of your analysis. You have the ability to narrow or filter variables as you did with the left field.

If you don't know which nodes you want to have graphed or simulated, go back to the circuit window, set Show Nodes, and note them on a piece of paper. Go back to the Analysis dialog box and set the variables.

- **Miscellaneous Options**—Adds a bit more customization to your analysis (see Figure 6.6).
  - — **Title for Analysis**—Choose a name that is easy to remember.
  - — **Use Custom Analysis Options** and **Analysis Options** —Lets you setup custom parameters for the analysis; I recommend leaving this alone whenever possible; more on analysis options later.
  - — **Perform Consistency Check Before Starting Analysis**—Checks to see if the circuit is able to use this analysis.

Figure 6.7 provides a detailed report of how you want this analysis performed. To expand the information, select the box containig the plus sign. To collapse the information tree, hit those boxes with minus signs.

The Art of Choosing Variables

Each analysis requires that you select one or multiple test point(s) at which the readings will take place. This requires you to fill out the Output Variables tab on the different Analysis dialog boxes. Most of

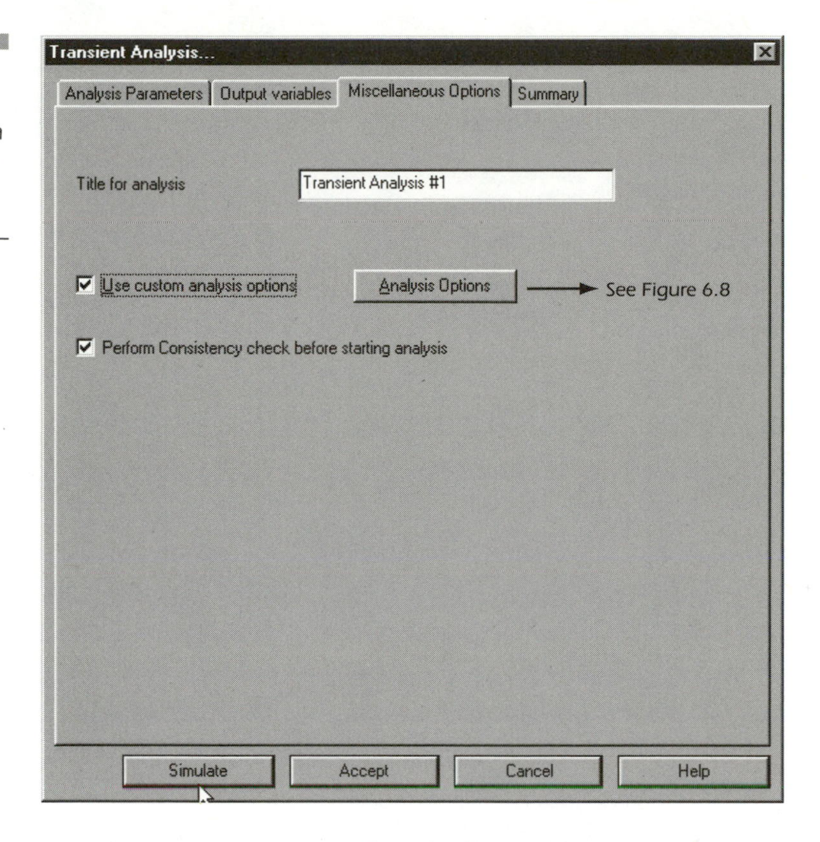

Figure 6.6
The Miscellaneous tab lets you choose a title and set advance parameters for an analysis.

the time we are comparing an input signal with an output signal; therefore the two nodes that represent these can be used. Other times you may be "sweeping" (constantly changing) the parameters of a component to see the circuit's reaction. For example, you may choose to sweep the value of a resistor as it relates to its tolerance. This requires you to set not only an output node but also the component to sweep. When you first begin analyses, you may just want to select all the variables and use **Toggle Legend** on the Grapher to sort out the information. You can then run the analysis again after you have narrowed the variables. Practice makes the selection of variables easier.

To select a series of variables, highlight one, hold the Shift key down and hit the Up or Down arrow keys. To select specific nodes, hold down the Ctrl key and click those you wish to select/deselect. Release the Ctrl key when all are selected.

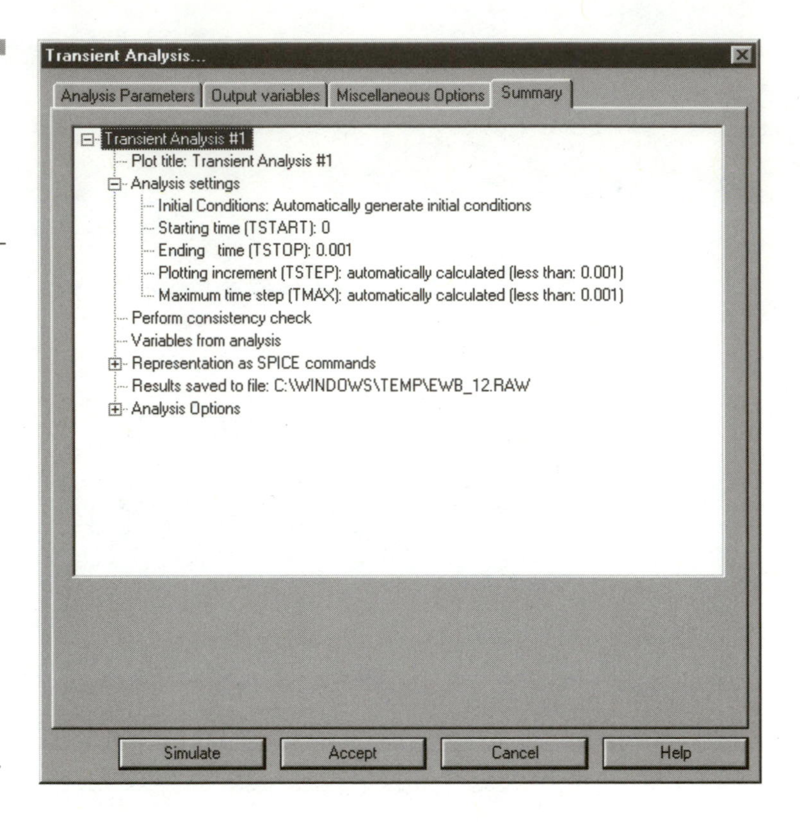

Figure 6.7
The Summary tab gives a detailed text description of the analysis about to be performed.

Adjusting Analysis Options

Customizable options give you greater control of the analyses you wish to perform. I do not recommend changing values, however, unless you absolutely know the results that will occur. Under **Simulate** > **Analyses** > **(specific analysis)** > **Miscellaneous Options** is a button to open the Analysis Options window. This is a separate dialog box that allows you to set any option. This includes specific parameters such as GMIN, ITL4, etc. If you don't know what these are, don't worry. When an item is highlighted (dark blue) you will see a description at the bottom of the screen. The Analysis Options Help file explains each item. You can access this by choosing **Help** > **Multisim Help** and typing "Analysis options" in the index field.

To change a value, highlight the row by clicking in the Parameter, Description or Value column. Go to the bottom of the dialog box and uncheck "Use default value." This lets you change its value in that field (see Figure 6.8). If you can't remember what the default value was, hit the Reset Value to Default button.

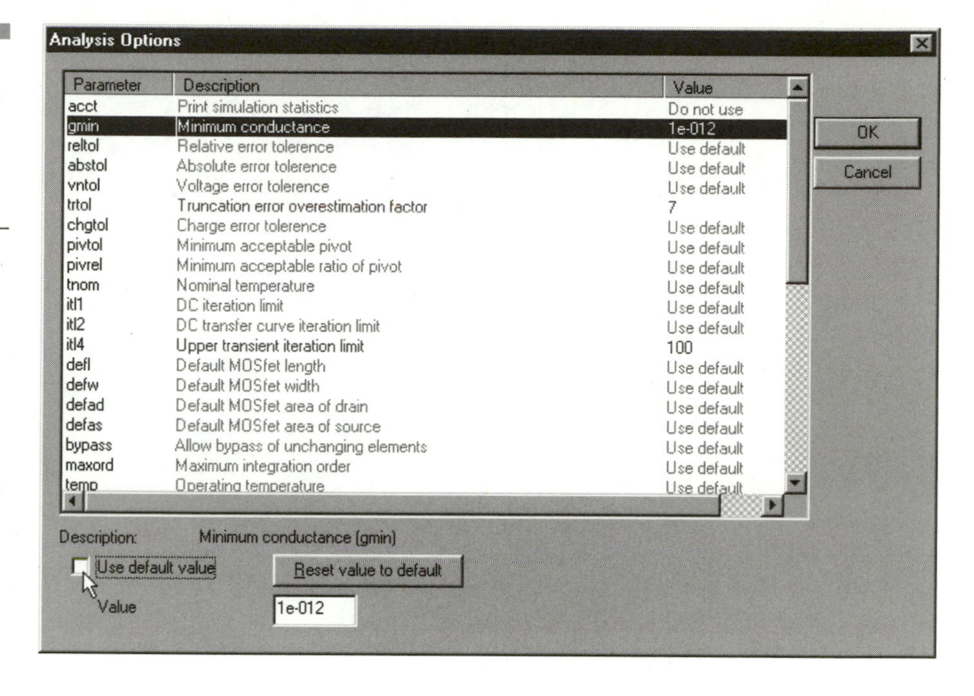

Figure 6.8
The Analysis Options dialog box can be used to enter advanced analysis parameters.

Analysis Menu

Analysis menu items can be quite complicated, so I opted for as simple an explanation as possible. For in-depth explanations of each analysis you may want to refer to the technical reference or a good circuit analysis book such as *Electronic Circuit and Simulation Methods* (McGraw-Hill), *Introductory Circuit Analysis* (Prentice Hall), or *Electronic Devices and Circuit Theory* (Prentice Hall).

NOTE

If you do not know how to use the Grapher, skip forward to that section.

DC Operating Point

Each analysis begins with a DC operating point calculation. Multisim calculates the steady-state voltage of each node in an active circuit and the current from each voltage source. It's like placing a multimeter at each node in the circuit. During the DC analysis, AC sources are nullified, capacitors are open, and inductors are shorted. Multisim determines the *quiescent operating point,* which is the steady-state operating conditions of a valve or transistor in its working circuit but in the absence of any input signal.

SETTING UP A DC OPERATING POINT ANALYSIS

Let's use two examples to explain this better. First look at Figure 6.9: it's a multiple voltage-source circuit. We will use a DC operating point to calculate the voltage at each node. The circuit consists of two batteries and three resistors.

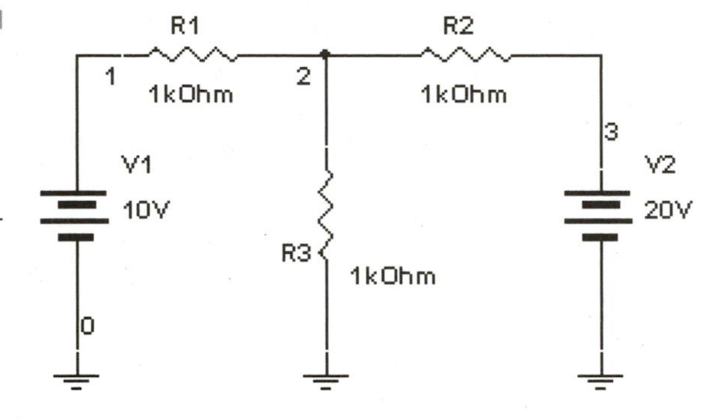

Figure 6.9 A dual DC source circuit is used to test the DC operating point analysis.

1. Build the circuit as shown in Figure 6.9 and adjust the component values to match.
2. Choose **Simulate** > **Analyses** > **DC Operating Point**. The DC Operating Point Dialog Box will open.
3. Under Output variables, highlight every variable (node) on the left field and move them to the right field by hitting the **Plot during simulation** button (see Figure 6.10).
4. Press the **Simulate** button at the bottom.
5. The Grapher opens and indicates the voltage level of each node. Note the value for node 2. (See Figure 6.11.) Close the Grapher and save the circuit as DC_OP1.MSM.

NOTE

Nodes are numbered in the order in which you connect their terminals.

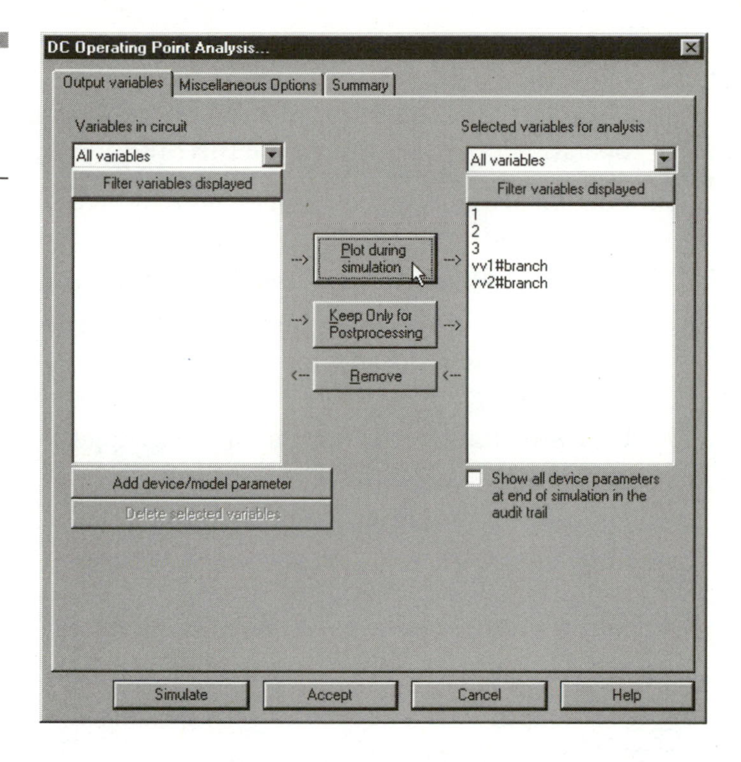

Figure 6.10 Setup the Output Variables tab as shown.

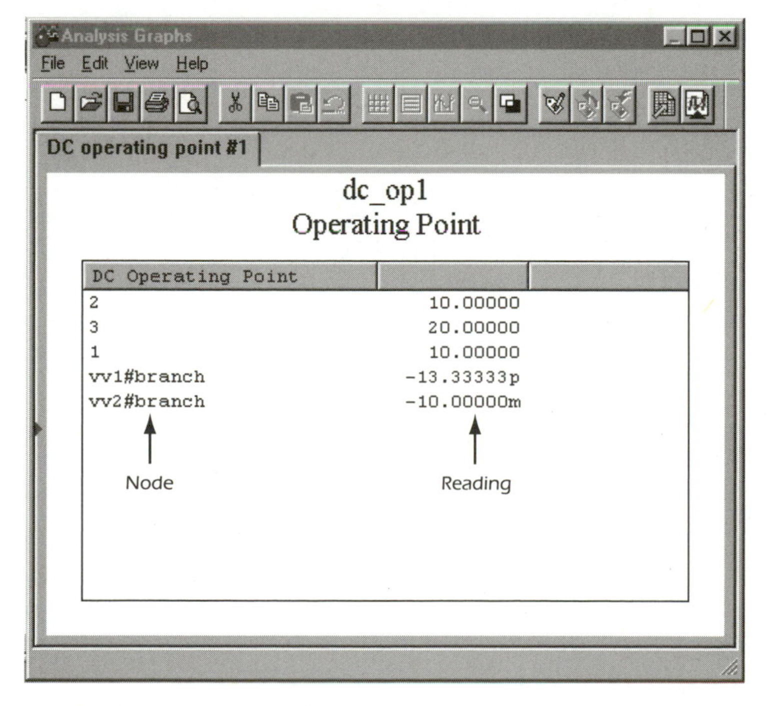

Figure 6.11 The Grapher displays the results of the DC OP analysis.

Figure 6.12 is a bit more complicated. It is a DC-biased transistor used as a simple amplifier. Say, for example, that you want to determine the DC bias point at the transistor's collector (node 1). By using DC operating point, we can figure this out in seconds:

1. Build the circuit shown in Figure 6.12 and save it as AMPTEST1.msm for later use.
2. Choose **Simulate** > **Analyses** > **DC operating point.**
3. Under Output variables, highlight the **Output** node (in this case, 1) on the left field and move it to the right field by hitting the **Plot during simulation** button.
4. Press **Simulate** at the bottom of the dialog box.
5. The Grapher pops up the voltage at node 1. This is the DC Bias Point of the amplifier. Save the circuit again.

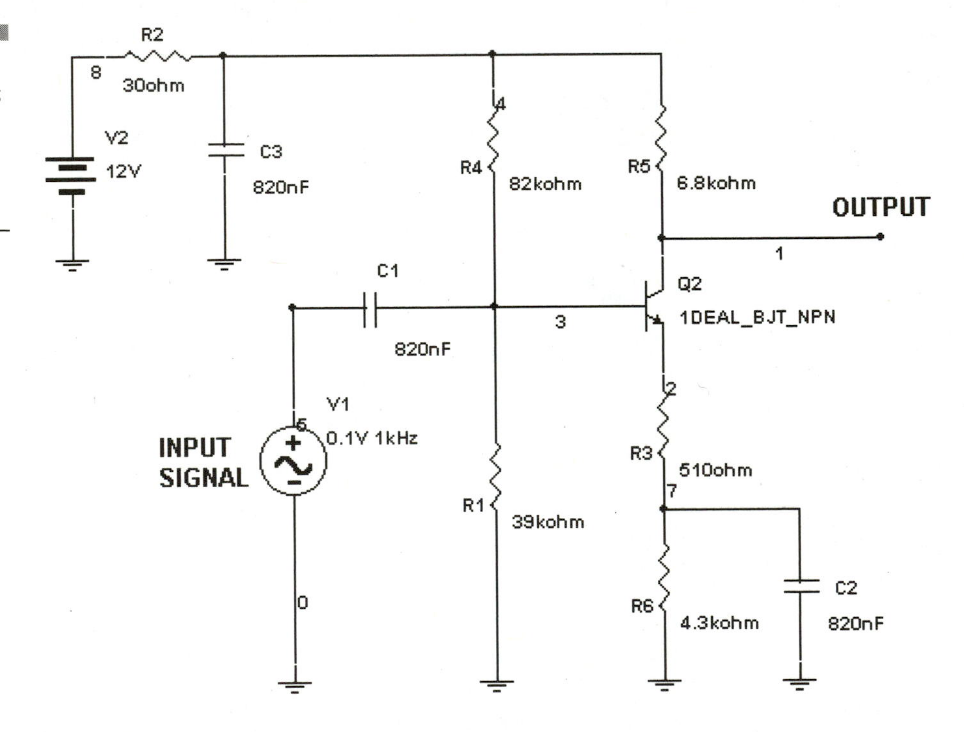

Figure 6.12 A Class 'A' amplifier is used to test many of the analyses, including the DC Operating Point.

The only readout sent to the Grapher after the simulation is a list of the voltage or current values of each node you selected. There are two columns: on the left is the variable (node), and on the right is the reading in voltage or amperage.

If you need to know the voltage readings of each point in a DC circuit fast, the DC operating point analysis is much quicker than placing a multimeter at several locations and running the simulation. It is also handy in finding the bias point of amplifiers or any other component that requires complex voltage calculations.

AC Analysis

AC analysis serves the same function as the Bode plotter: it plots the circuit's response to a series of frequencies. However, AC analysis offers more flexibility. All DC sources are given zero values, and AC components, such as capacitors and inductors, are modeled for the analysis. Remember, AC sources do not particularly matter, because Multisim inputs its own sinusoidal waveform, graphing the circuit node responses. Also, as a note, digital components are treated as large resistances to ground.

SETTING UP AN AC ANALYSIS

Examine the two filter circuits in Figure 6.13. The top shows a band-pass filter that will only let an exact frequency band through the filter; the bottom circuit shows a band-stop filter that will inhibit a certain frequency band. To perform this analysis:

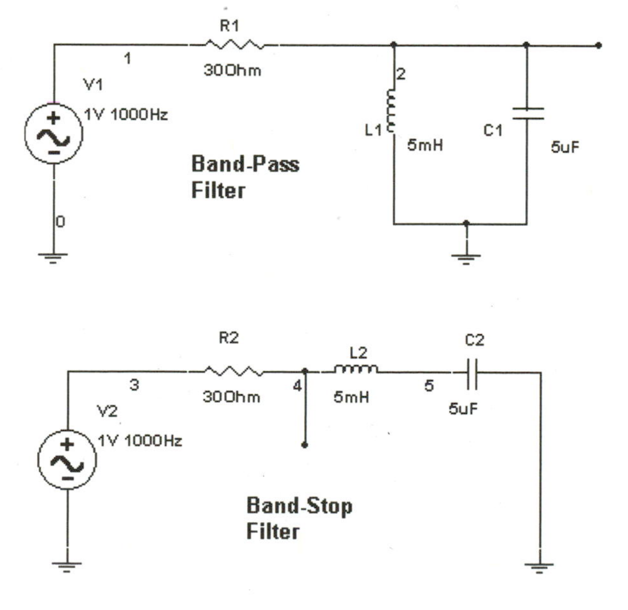

Figure 6.13 The band-pass and band-stop filters make a perfect example for an AC analysis.

1. Build both circuits and save the file as ACA_TEST.msm.
2. Choose **Simulate** > **Analyses** > **AC Analysis.** The AC Analysis dialog box opens.
3. Adjust the Frequency Parameters as seen in Figure 6.14. Make sure to set the vertical scale to linear (this makes the information easier to read).
4. Choose the two Output nodes from the Output Variables tab. In this case, they are 2 for the band-pass filter and 4 for the band-stop filter.
5. Press the **Simulate** button. The Grapher opens and begins plotting the solution.
6. If you want to check the bandwidth, turn the cursors on, position them until y1 and y2 are around 707.0000m; the bandwidth is shown in the cursor readout window (see Figure 6.15).

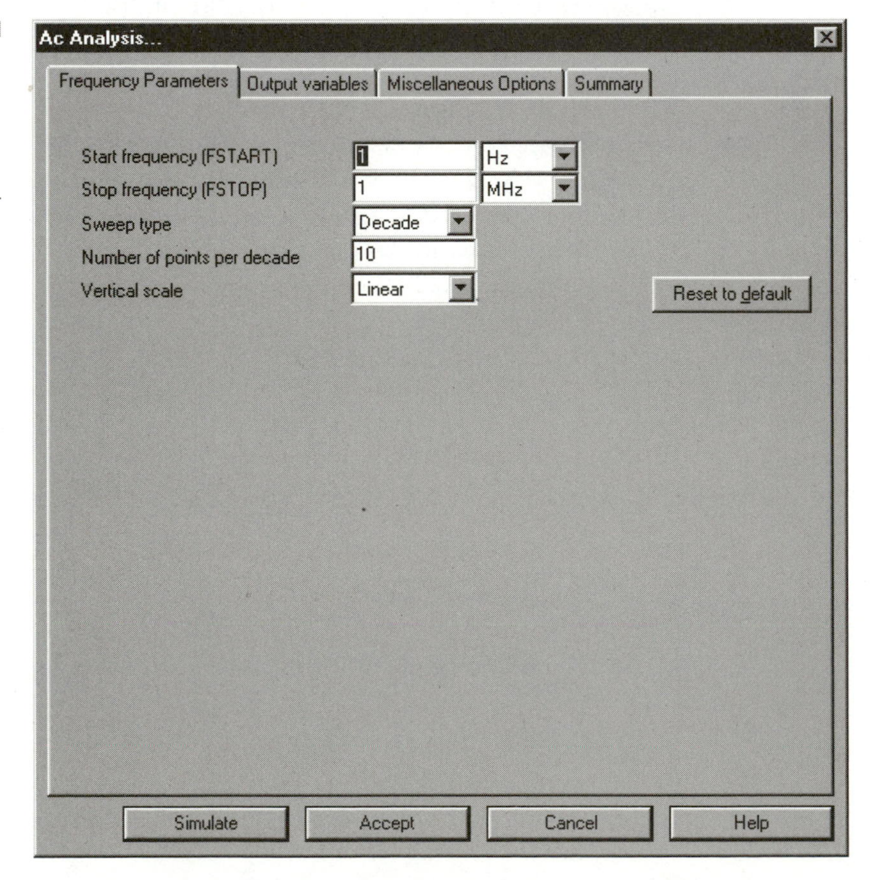

Figure 6.14
Set up the Frequency Parameters tab as shown for this AC analysis.

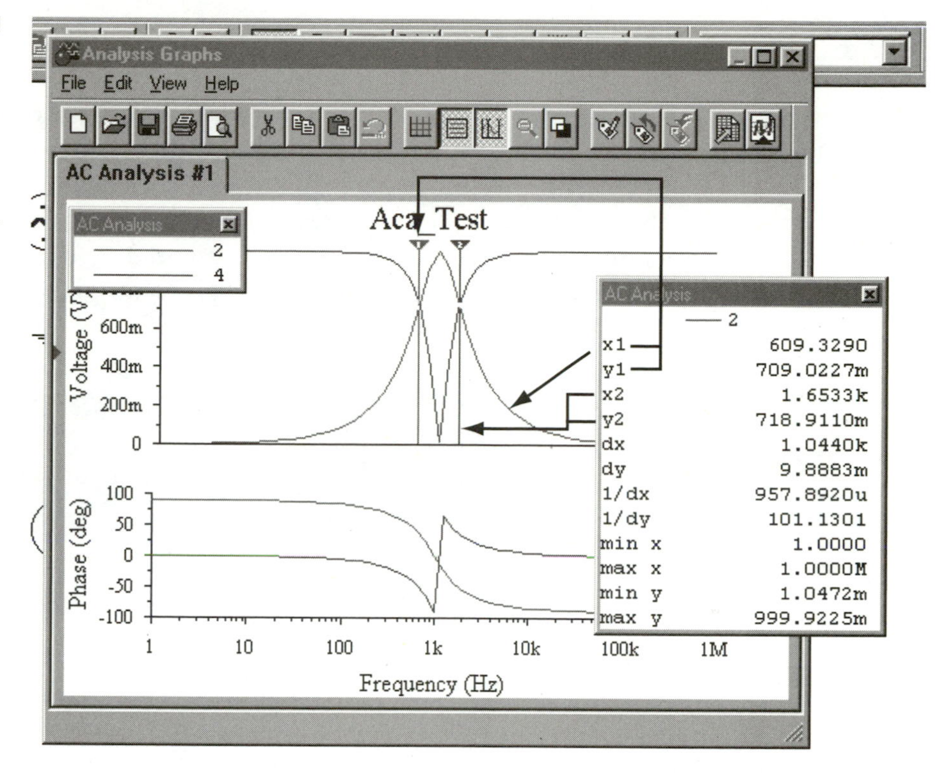

Figure 6.15
Once the Grapher appears, you can use the cursors to determine the bandwidth.

TIP

To select another graphed trace for cursor readouts, left-click anywhere along the trace. The currently selected line will be on the top of the cursor readout window (see Figure 6.16).

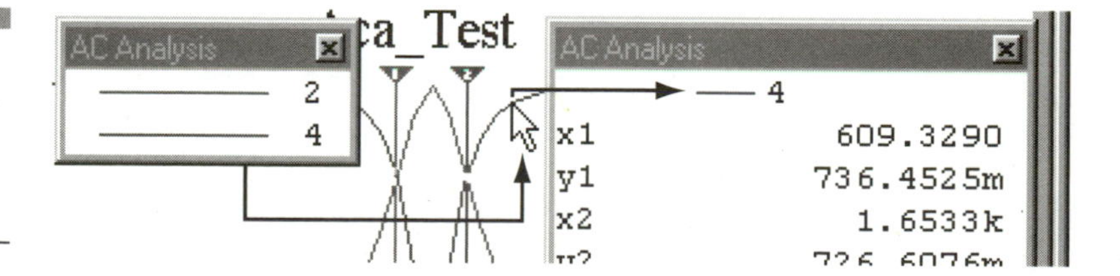

Figure 6.16
Click anywhere on a trace (waveform) to select it for the readout window.

Multisim has a Voltage VS Frequency graph, as well as a Phase VS Frequency graph. I recommend turning on the grid when using the cursors.

AC analysis is handy when designing filter circuits. It's much faster to set up than using the Bode plotter and certainly faster than pencil, paper, calculator marathons.

Transient Analysis

Transient analysis adds a time factor to your circuit's readings. DC and AC analysis are basically steady, "one moment in time" analyses. Transient analysis breaks time down into segments and figures out the voltage and current levels for each given interval. For example, a node may be 1 volt at 1 millisecond and 2 volts at 2 milliseconds, etc. Multisim strings them together to give you graphs that analyze ever-changing circuit operation. See Figure 6.17.

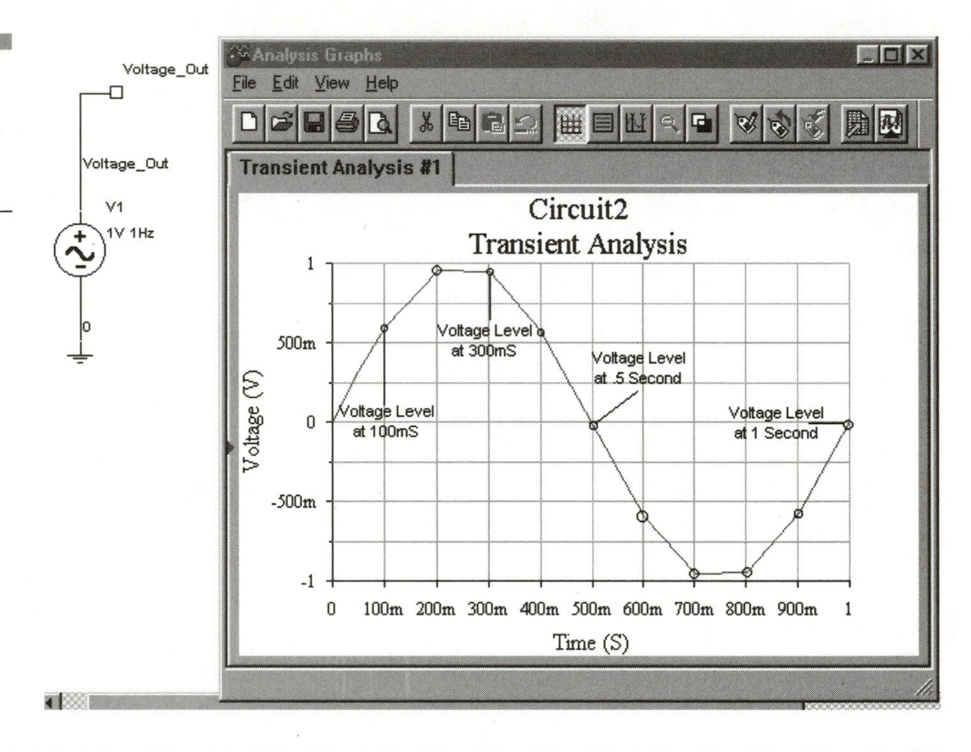

Figure 6.17
A transient analysis adds a factor of time to a circuit to create waveforms.

NOTE

Transient analysis is also called time-domain transient analysis.

Setting Up Transient Analysis

Go back to the transistor amplifier circuit we used in the DC Operating Point Analysis example (see Figure 6.18). Set up the circuit again or open the AMPTEST1.MSM file you saved.

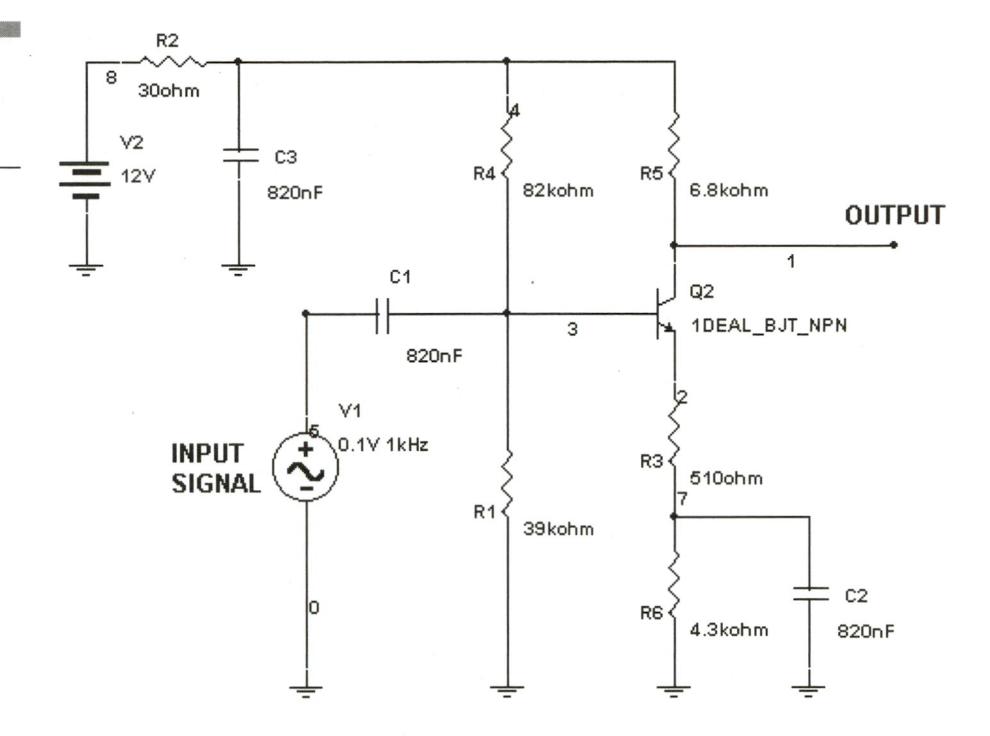

Figure 6.18 Reopen the Class A amplifier to test the transient analysis.

1. Once the circuit is opened, save the file as AMPTEST2.MSM.
2. Go to **Simulate** > **Analyses** > **Transient Analysis.** The dialog box will opens.
3. Adjust the Analysis Parameters to match Figure 6.19.
4. Select the **Output** node (in this case, **1**) and the **Input** node (**5** on this example) as the Output Variables under that tab.
5. Press the **Simulate** button.
6. The Grapher opens as shown in Figure 6.20. The blue signal is the Input AC signal; the red is the amplified Output signal.
7. Resave the circuit.

The Grapher plots a waveform for each node you have selected. The x-axis is time and the y-axis is voltage/current. Each waveform is color-coded; if you can't identify waveforms, toggle the **Legend** on. You can also use the cursors to determine specific readings along the wave.

Transient Analysis can be used for just about any circuit that requires you to view the voltage/current level over time. It replaces having to set up twenty oscilloscopes.

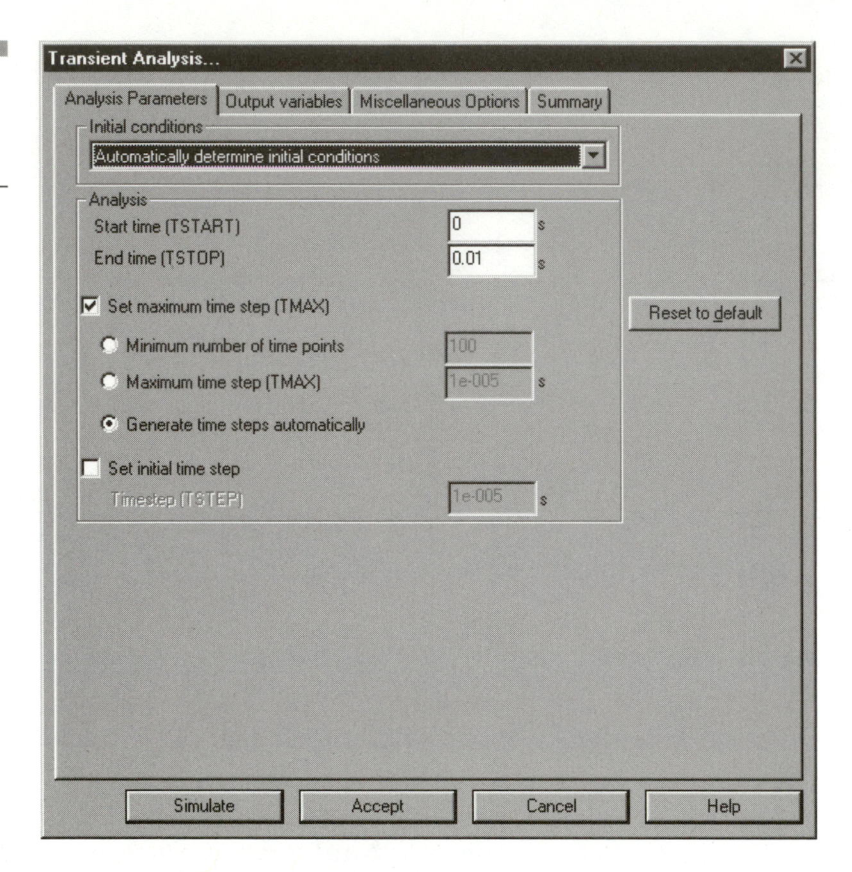

Figure 6.19 Adjust the Analysis Parameters as seen here.

Fourier Analysis

Fourier analysis breaks a complex waveform down into its component frequencies and amplitudes. Any complex waveform consists of fundamental frequencies combined with harmonics of those frequencies. For example, a complex voice wave may consist of only three fundamental frequencies along with many harmonics. By using this analysis, you can get a picture of exactly which frequencies and harmonics are involved and their relative strength. The mathematics of how this works would fill a chapter, so I simply give you the results in the following example.

Setting Up Fourier Analysis

See Figure 6.21: the three AC sources are added together to produce a complex waveform, as seen in Figure 6.22. But what elements make up this waveform?

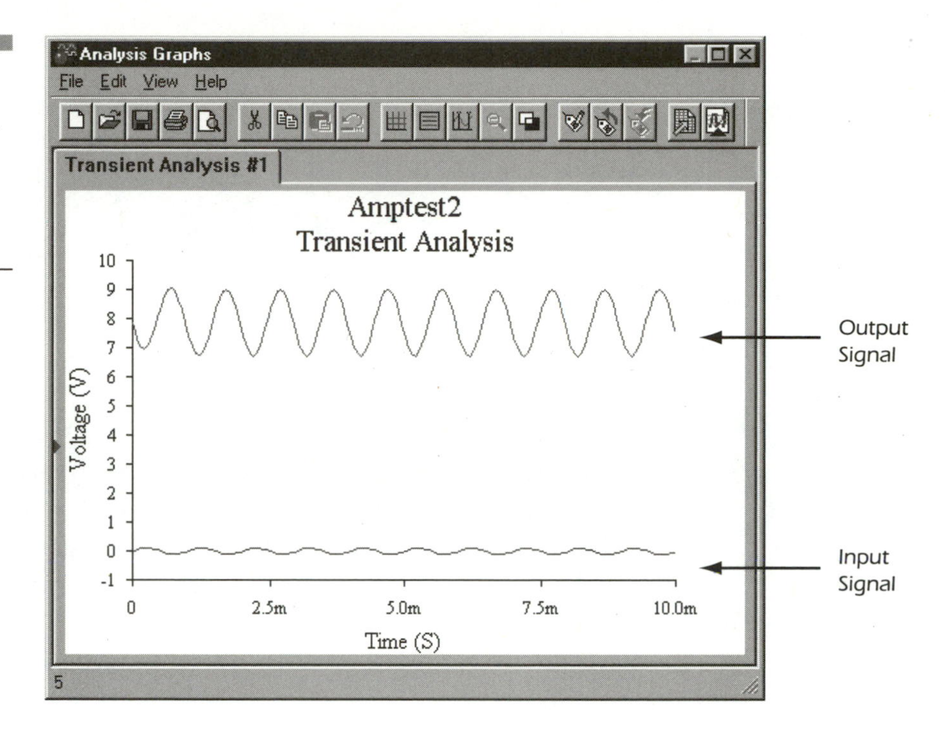

Figure 6.20
The Grapher opens with a waveform for the input (bottom trace) and output (top trace) of the amplifier.

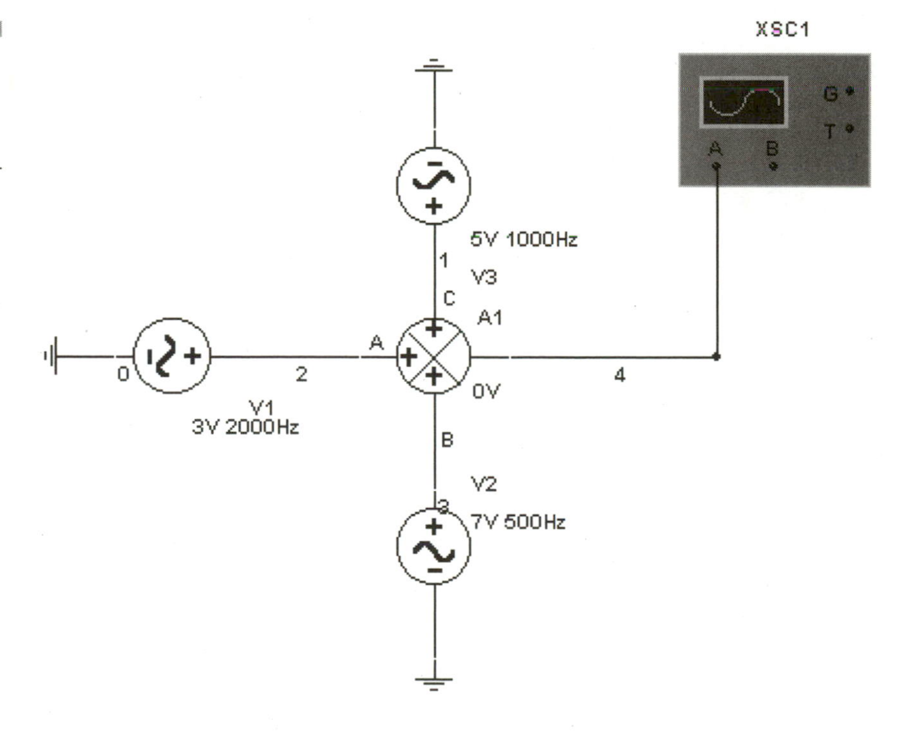

Figure 6.21
Build this three-wave adder circuit to test the Fourier analysis.

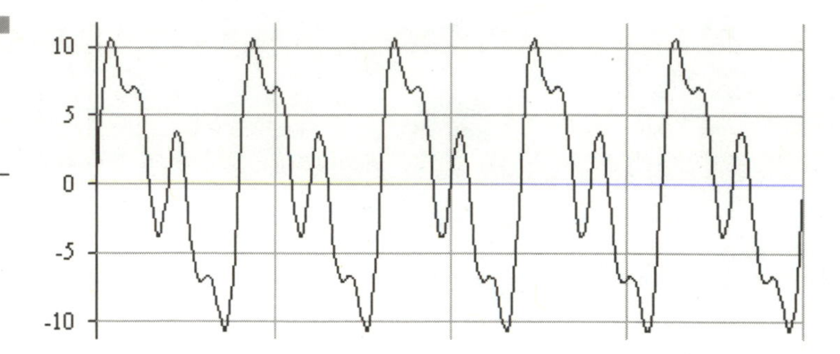

Figure 6.22 The circuit creates this non-sinusoidal waveform.

1. Build the circuit as seen in Figure 6.21 and save it as Fourier.MSM.
2. Select **Simulate** > **Analyses** > **Fourier Analysis**.
3. Choose the Analysis Parameters tab. Hit the **Estimate** button on the right. This estimates the fundamental frequency of the waveform.
4. Under **Transient analysis options**, set the **TSTOP** value to .1 second.
5. Choose the Output Variables tab and select the **Output** node (in this case, **4**).
6. Press the **Simulate** button. The Grapher for this simulation may take a moment to appear while Multisim is calculating.
7. Figure 6.23 appears. The bottom graph shows the frequencies and relative strengths (amplitude) of those frequencies. The top section lists detailed information, including the *total harmonic distortion* (THD).

Fourier Analysis can be used to analyze any complex waveform circuitry. I personally am studying speech recognition technology with it.

Noise Analysis

Noise can be defined as any unwanted voltage or current that ends up in a circuit's output and dilutes the originally intended signal. For example, a crackle may creep into an audio amplifier, causing annoyance to your ears and muddying a recording's playback. Noise-prone components can cause snow on a television picture tube. Each resistor and semiconductor in a circuit contributes to this noise, due to a combination of *thermal noise* (noise due to temperature factors) in resis-

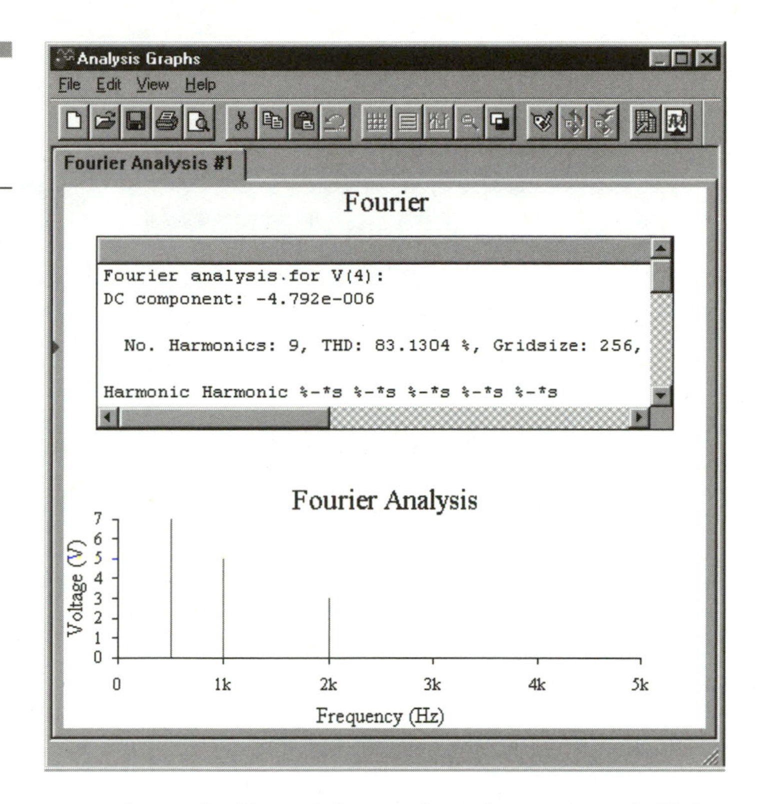

Figure 6.23 The readouts for a typical Fourier analysis.

tors, *shot noise* (found in semiconductors), and *flicker noise* (*pink noise* in BJTs and FETS operating below 1kHz).

Multisim performs an AC-like analysis on the entire circuit, using special noise models for each component. These analyses sum the magnitude of noise power from the input to output of the complete circuit; Multisim calculates the total noise produced by the circuit's components or calculates the noise of individual sections or components in the circuit. It will display the results in the form of a graph: Output Noise Power versus Frequency.

Setting Up Noise Analysis

Let's open up the AMPTEST2.msm again and follow along:

1. Open or build the circuit, setting each value.
2. Replace the transistor with a 2N3904 by double-clicking the ideal one and choosing **Replace**.
3. Go to **Simulate** > **Analyses** > **Noise Analysis**. The dialog box opens.

4. Set up the analysis parameters, as seen in Figure 6.24, making sure to check on the Set Points per Summary; leave the value at 1.

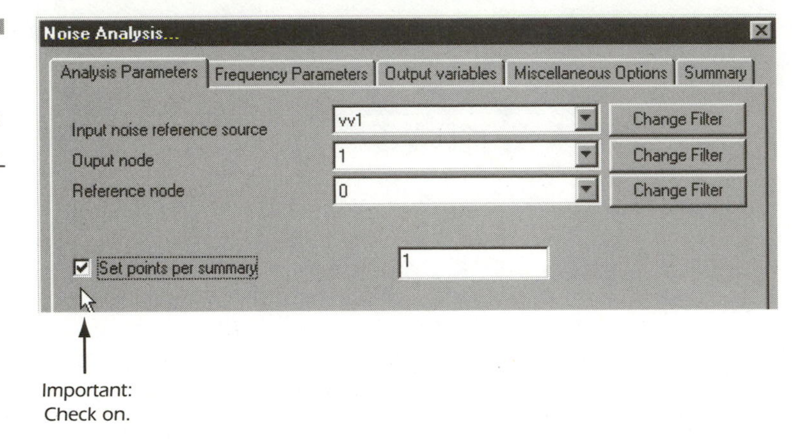

Figure 6.24 Set up the Analysis Parameters as shown for the noise analysis.

5. Choose the Output Variables tab and select inoise_spectrum and onoise_spectrum (see Figure 6.25).

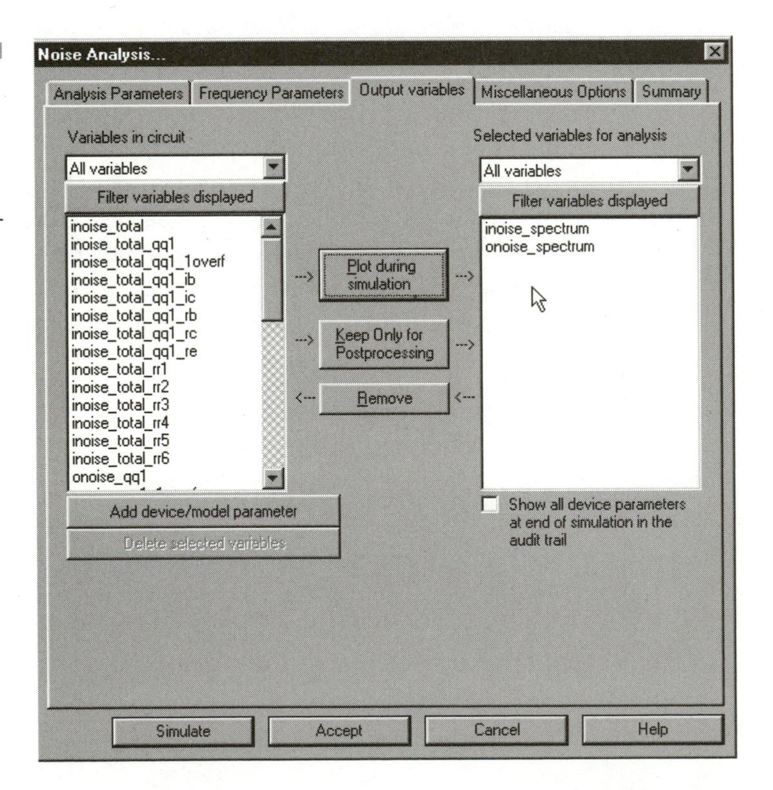

Figure 6.25 Select the inoise_spectrum and onoise_spectrum as the output variables.

6. Run the simulation by hitting the **Simulate** button at the bottom.
7. The Grapher appears, showing a graph of input-to-output noise on the top. The blue trace is input and the red is output noise spectrum (see Figure 6.26). The top graph is a comparison of input noise to output noise or the level of noise for a specific component at a varying frequency. The bottom graph shows *integrated noise*—a text version listing the noise level of each individual component or node you selected to read.
8. Save the circuit as NoiseTest.MSM.

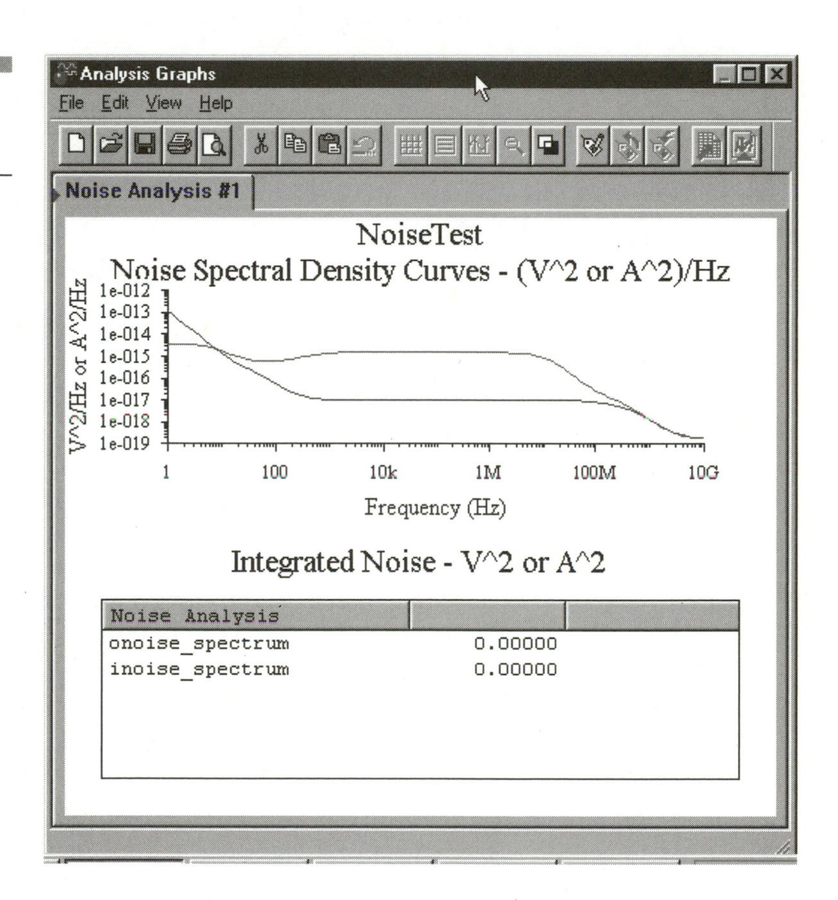

Figure 6.26 The results of the noise analysis.

Noise analysis can be used if you are trying to create a circuit with the lowest noise possible. It is often used in telecommunications applications for this purpose.

Distortion Analysis

Distortion is any change in the original waveform (see Figure 6.27 for a few examples). Distortion analysis is used to locate small amounts of distortion that are not being picked up by transient analysis. These include *harmonic distortion* (created by harmonic energy) and *intermodulation* (IM) *distortion* (created by mixing signals at different frequencies). Multisim does a small-signal distortion analysis of the circuit, sweeping one or two frequencies. The results are graphed depending on the number of frequency sweeps used.

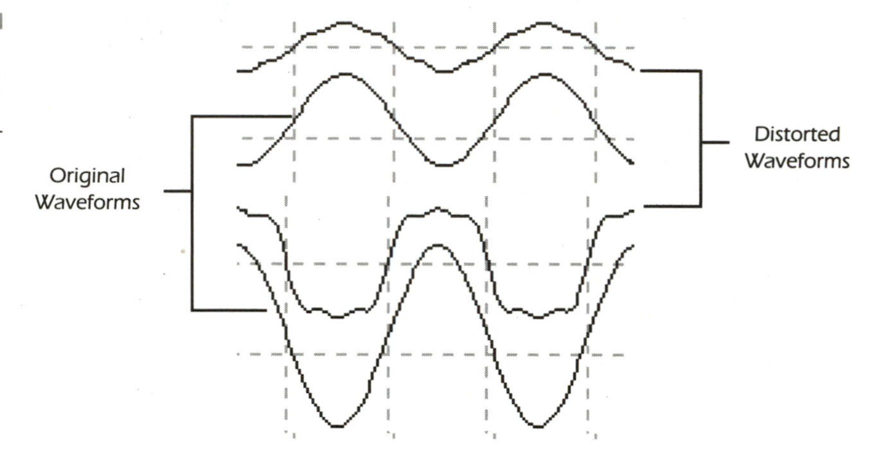

Figure 6.27 Examples of distorted waveforms.

NOTE

If F2/F1 Ratio is checked on, then an IM distortion analysis is performed; otherwise a harmonic distortion analysis is done.

Setting Up Harmonic Distortion Analysis

If you are designing audio or video circuits that must have as little distortion as possible, use this analysis:

1. Open the Noisetest.MSM file and save it as DistortionTest.MSM.
2. Choose **Simulate** > **Analyses** > **Distortion Analysis**.
3. Leave the Analysis Parameters as they are. Go to the Output Variables tab and select node **1** as the variable.
4. Hit the **Simulate** button.

5. Two pages of graphs (Figure 6.28) opens; toggle between the two. Four graphs are created: a voltage/frequency graph and a phase/frequency graph for both the second and third harmonics.
6. Leave the circuit open.

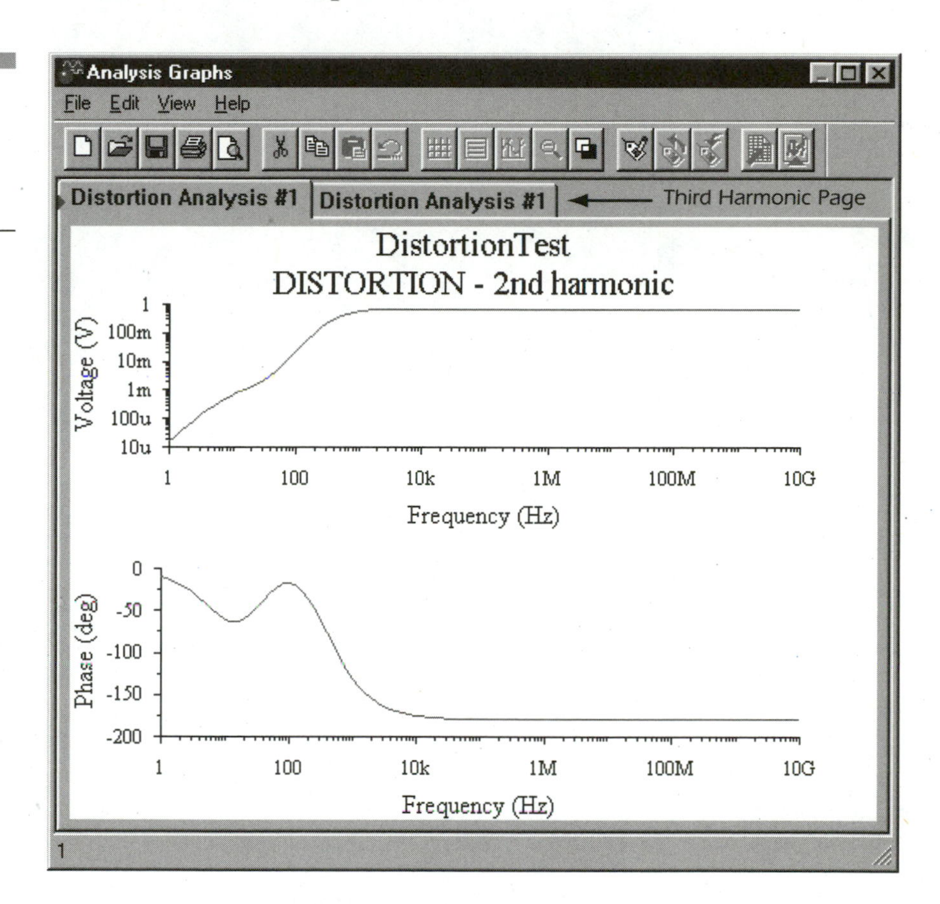

Figure 6.28 The harmonic distortion analysis creates four graphs (two per page).

Setting Up an IM Distortion Analysis

1. With DistortionTest.MSM still open, choose **Simulate** > **Analyses** > **Distortion Analysis** once again.
2. This time check on the F2/F1 ratio and hit the **Simulate** button.
3. Three pages of graphs are created: Frequency(F) 1 + Frequency(F) 2, F1—F2, and 2(F1)—F2 (see Figure 6.29).

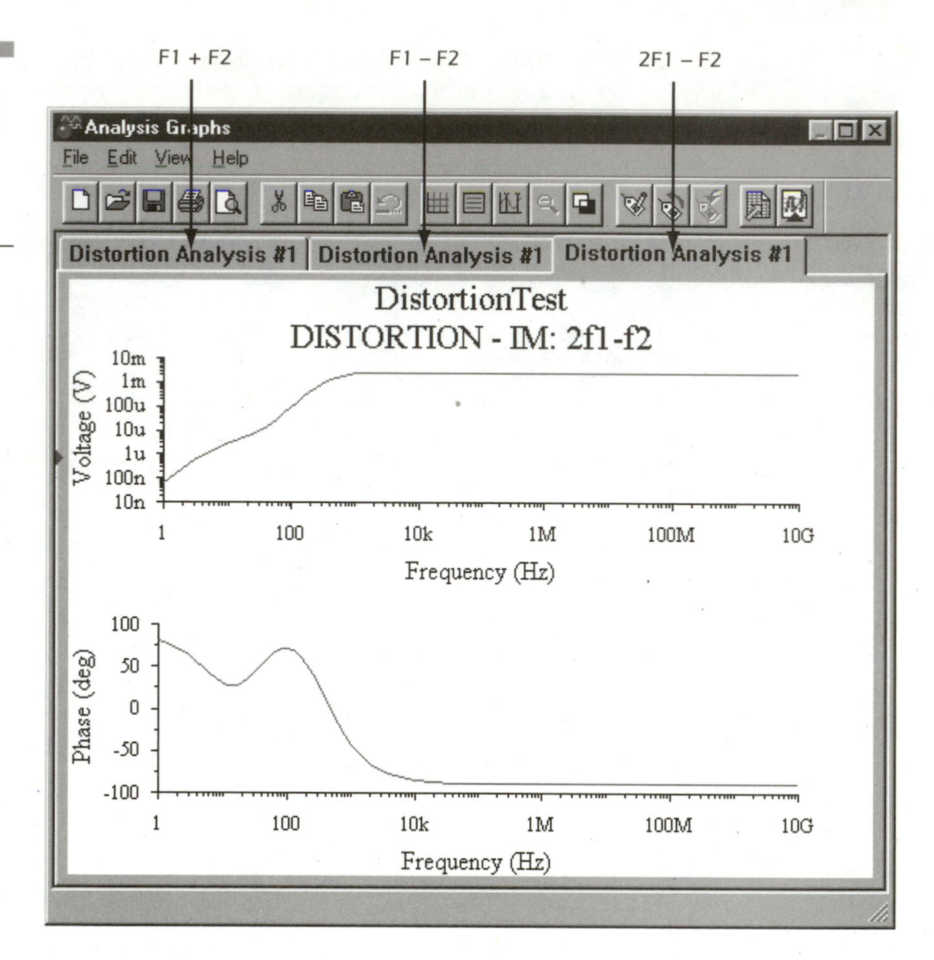

Figure 6.29 The Intermodulation (IM) Distortion analysis creates six graphs (two per page).

DC Sweep

Havae you ever wanted to see how a circuit responds to varying DC voltages? DC sweep automatically adjusts (sweeps) the DC voltage of one to two voltage sources in the circuit (according to your settings). It uses this information to graph the circuit's response. Technically, it is performing a DC operating point calculation for each voltage level(s), which saves you having to perform a DC operating point calculation, change the DC voltage levels, and do another DC Operating point calculation. It's also a great analysis to visualize non-linear components, such as transistors and diodes, or just to see how a DC circuit responds to various DC voltage levels.

Setting Up a Single-Source DC Sweep

DC Sweep can be used for any circuit for which you wish to map the results of varying DC voltages. It's also a fast way to make transistor curve graphs. We will use a simple Zener diode to test this circuit.

1. Build the circuit as seen in Figure 6.30 and save it as DCSWEEPTEST.MSM.

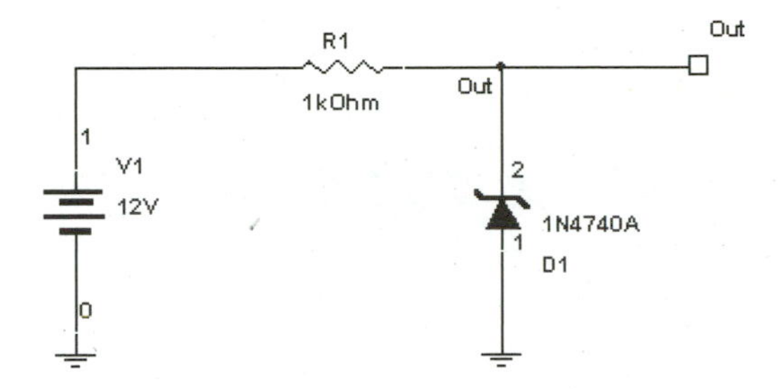

Figure 6.30
A simple Zener-control voltage regulator circuit is used to test the DC sweep analysis.

2. Choose **Simulate** > **Analyses** > **DC Sweep**.
3. Adjust the Analysis Parameters page as seen in Figure 6.31.

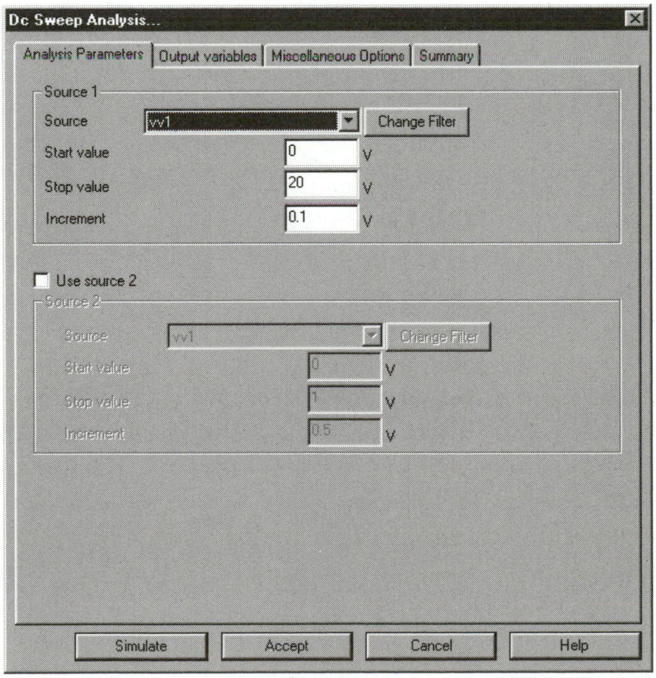

Figure 6.31
Adjust the DC Sweep Analysis Parameters as shown.

4. Choose **Out** and **1** as the Output variables.
5. Hit **Simulate**.
6. The graph shows the output voltage (y-axis) as it relates to the sweeping DC input voltage (x-axis). The voltage steadies at around 10 volts (see figure 6.32). *Note: I have offset the sweeping voltage reading (Node 1) slightly for clarification.* If you are only reading one voltage source, then the X-Axis is varying voltage source and Y-Axis is the output voltage. If you are using two voltage sources, the second voltage source will be represented incrementally by alternating color lines. To see what each line is, toggle the Legend.

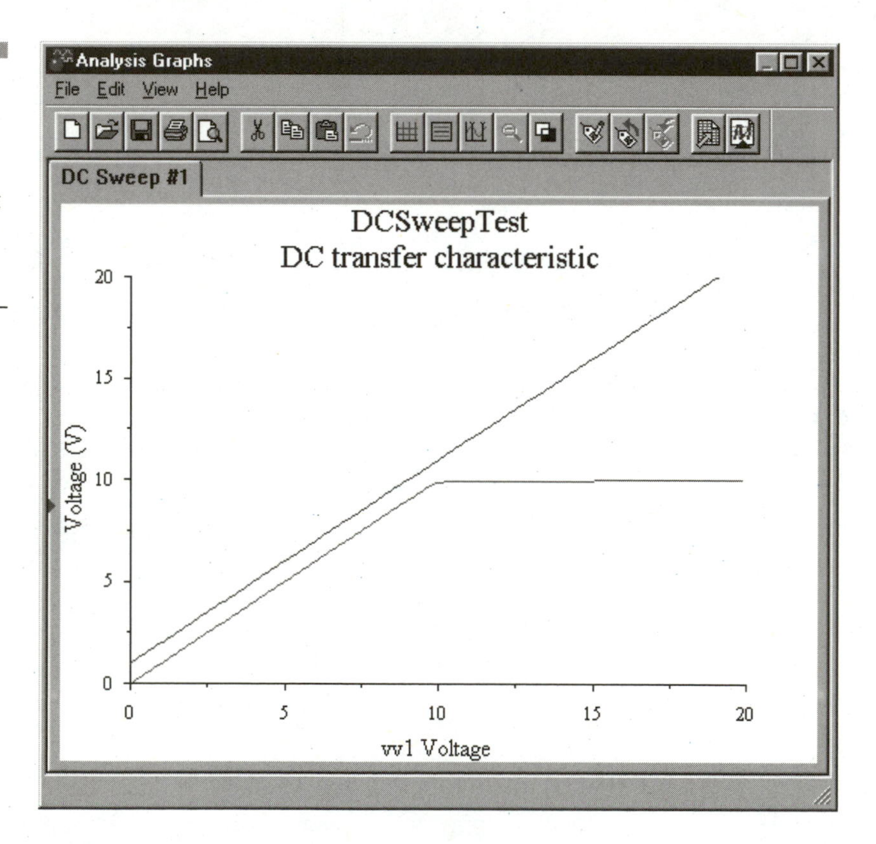

Figure 6.32
The resulting graph shows the output voltage will climb with the input until it reaches a level of 10 volts, at which point it steadies.

NOTE

The next chapter gives an example of a dual DC sweep.

Sensitivity

Depending on the application for a circuit, real-world components must be selected accordingly. For example, if you were building a complex sensitive instrument that required a very exacting reading, you wouldn't use a cheap transistor that might not be highly accurate. Conversely, if you were building a simple circuit just to help demonstrate the principle of a transistor, you wouldn't need an ultra-expensive precision unit. DC sensitivity takes the guesswork out of determining which components need to be high-end and which you can select from the bargain bin at Joe's Electronics Shop. It determines which components will most affect the DC bias point. In this way, you can see which components are most sensitive to variation, and you can refine your design. Two sensitivity analyses can be performed on a circuit: DC Sensitivity and AC Sensitivity. DC performs a DC operating point analysis, then figures out the sensitivity of each output for all of the device values and model parameters. This produces a DC Report, which tells you how much each variance in a component's parameter will affect the output of the circuit by 1 volt.

The AC Sensitivity is a small-signal analysis which measures the sensitivity of the voltage or current with respect to one or more specific parameters of a component; it tells you the effect of a small variation in each of that component's parameters on the overall output.

Setting Up DC Sensitivity Analysis

If you are designing circuits and need to know exactly where to put the money or where to skimp, this is the most valuable analysis you can use.

1. Re-open DISTORIONTEST.MSM.
2. Choose **Simulate** > **Analyses** > **Sensitivity Analysis**.
3. Set the Analysis Parameters as seen in Figure 6.33.
4. Select all the variables.
5. Hit the **Simulate** button.
6. You see the DC Report appear with parameters on the left, and the variation on the output voltage on the right (see Figure 6.34).
7. Save the circuit as SENSITIVITY.MSM.

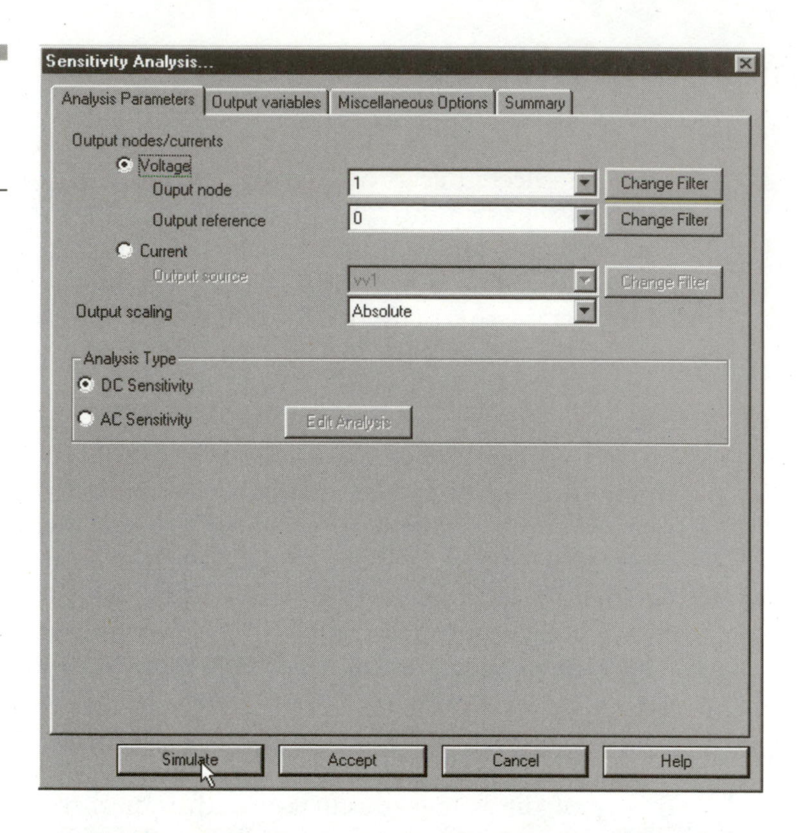

Figure 6.33 Set up the DC Sensitivity analysis as shown.

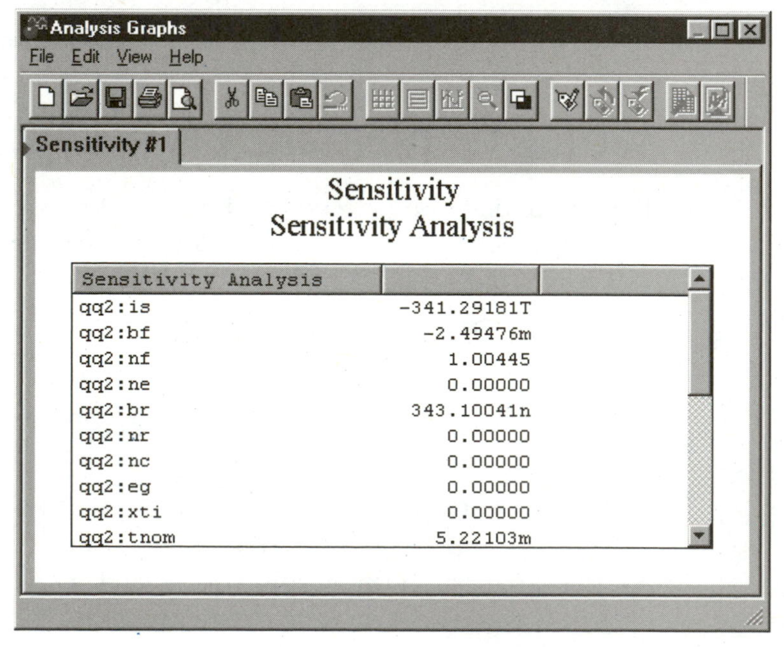

Sensitivity Analysis		
qq2:is	-341.29181T	
qq2:bf	-2.49476m	
qq2:nf	1.00445	
qq2:ne	0.00000	
qq2:br	343.10041n	
qq2:nr	0.00000	
qq2:nc	0.00000	
qq2:eg	0.00000	
qq2:xti	0.00000	
qq2:tnom	5.22103m	

Figure 6.34 A DC Sensitivity Analysis creates a DC report with the parameter on the left and the reading on the right.

Setting Up AC Sensitivity Analysis

1. Open SENSITIVITY.MSM.
2. Choose **Simulate** > **Analyses** > **Sensitivity Analysis.**
3. Adjust the Analysis Parameters to read AC Sensitivity, as seen in Figure 6.35.
4. Go to the Output Variables tab and select all the variables to be graphed.
5. Hit the **Simulate** button.
6. A Voltage/Frequency graph appears with each parameter on its own line. Below is a Phase/Frequency graph showing the same values (Figure 6.36). AC Sensitivity graphs voltage versus frequency and phase versus frequency of each selected component or model parameter when swept with an AC signal.

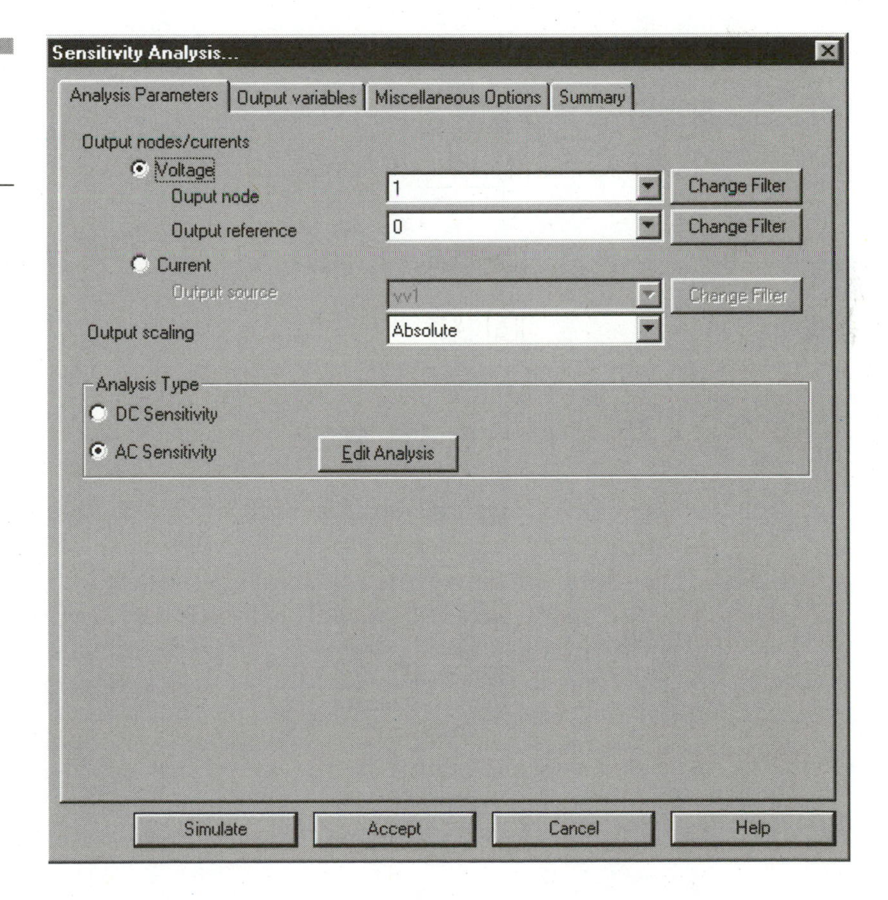

Figure 6.35
Set up the AC Sensitivity Analysis as shown.

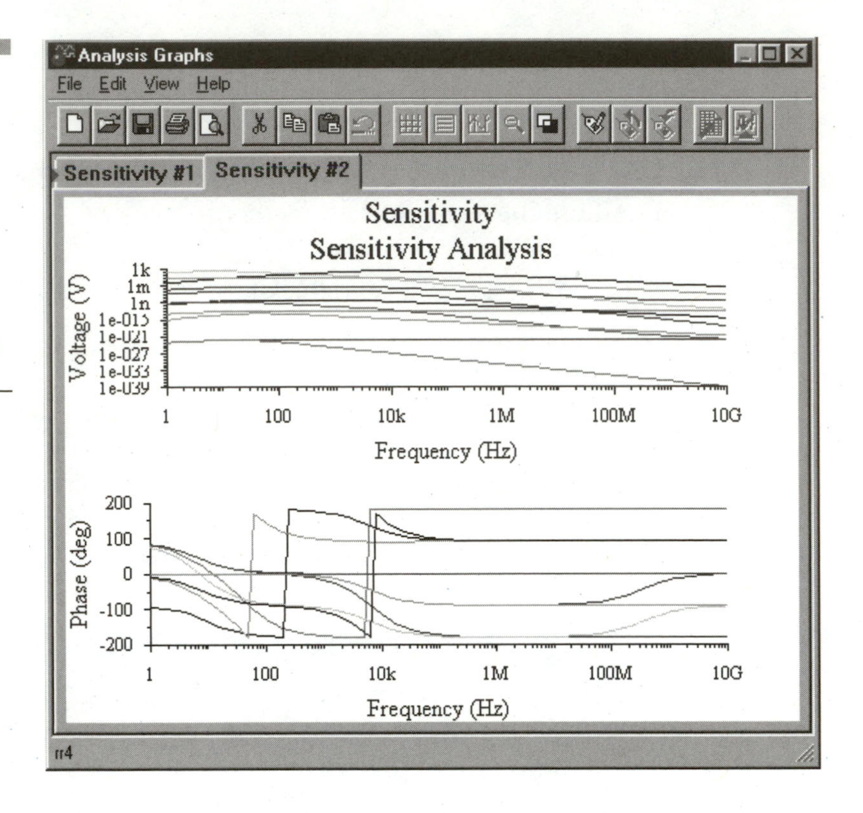

Figure 6.36 Two graphs result from the AC Sensitivity Analysis: Voltage/Frequency and Phase/Frequency. You can see a colored trace for each parameter that was swept with an AC signal.

Right-click on a graphed line to determine the label.

Other Analyses

The remaining analyses are only available in the Professional and Power Pro editions. I've provided a brief explanation for each, but see the digital version of the Multisim *Users Guide* for further information.

Multisim Professional can perform all of the above described simulations with the addition of parameter sweep, temperature sweep, pole zero, transfer function, worst-case, Monte Carlo, and trace-width analysis.

- **Parameter Sweep**—This analysis replaces having to run a simulation, change a component value/temperature/model parameter, and then run the simulation again and again. It is a quick way of fine-tuning a component value or other circuit parameter.

- **Temperature Sweep**—Each component responds differently to varying temperatures. This analysis simulates varying temperatures and how the circuit would react; the results are output according to your settings.
- **Pole Zero**—This analysis helps to determine the stability of a circuit. It calculates whether an output signal will remain bounded or increase indefinitely, which could possibly damage components.
- **Transfer Function**—This calculates the DC small-signal transfer function between an input source and two output nodes (for voltage) or an output variable (for current). It also calculates input and output resistances.
- **Worst Case**—This is a statistical analysis that gives you an idea of the circuit's response to the worse possible changes in certain component parameters. For example, if you were designing a power supply and wanted to know how certain parameters, when taken to the edge, would affect power output, you would run this analysis.
- **Monte Carlo**—Want to "roll the dice" with your circuit? Monte Carlo randomly changes component variables and notes the circuit's response.
- **Trace-width Analysis**—The conductive traces on a printed circuit board must be a certain width to handle the circuit's current; this prevents traces burning up. The trace-width analysis helps to determine each trace's width in order to pass it to the Ultiboard program.

Multisim Power Pro—performs all of the above described analyses with the addition of batched analysis, nested sweep, user-defined analysis, noise figure analysis, RF analysis and matched network designer. (See the Multisim Help file for a detailed explanation of each.)

- **Batched Analysis**—Allows the user to perform multiple analyses at once, thus saving time.
- **Nested Sweep**—With this analysis, you can reuse results from one temperature sweep, parameter sweep, or DC sweep to another. In other words, you can nest one sweep analysis into the next into the next.
- **User-defined Analysis**—Multisim presents you with a window that lets you type in SPICE analysis commands. This feature requires a thorough knowledge of the SPICE language.
- **RF Analyses**—RF analyses, noise figure, and matching networks analyses are performed through the Network Analyzer instrument.

Each of these analyses is part of Multisim's RF Design module (standard in the Power Professional version, optional in the Professional version). They are used to analyze complex Radio Frequency (RF) circuits, such as those used in modern communications (cell phones, satellites, and others).

Grapher

Now that you know how to start and run an analysis it's time to figure out how to interpret exactly what Multisim is trying to tell you about the results. Grapher is a visual way to read results. After an analysis has been run, open the Grapher by going to **View** > **Show/Hide Grapher**. (The Analysis Graphs window automatically appears at the conclusion of an analysis.)

The Grapher window remains open and visible until you either close the window or choose **View** > **Show/Hide Grapher** from the menu again (this removes the check from the menu item).

If you are running dual monitors in Windows 98, place the Grapher on a second screen.

Analysis Graphs Window

The popup window containing each analysis graph is called Grapher (see Figure 6.37). There are standard menu items, toolbar icons, and pages of graphs, which in turn may contain one or more graphs per page. First, lets look at the menu items:

When in Grapher, Multisim will not display an icon's name when the pointer is placed above it.

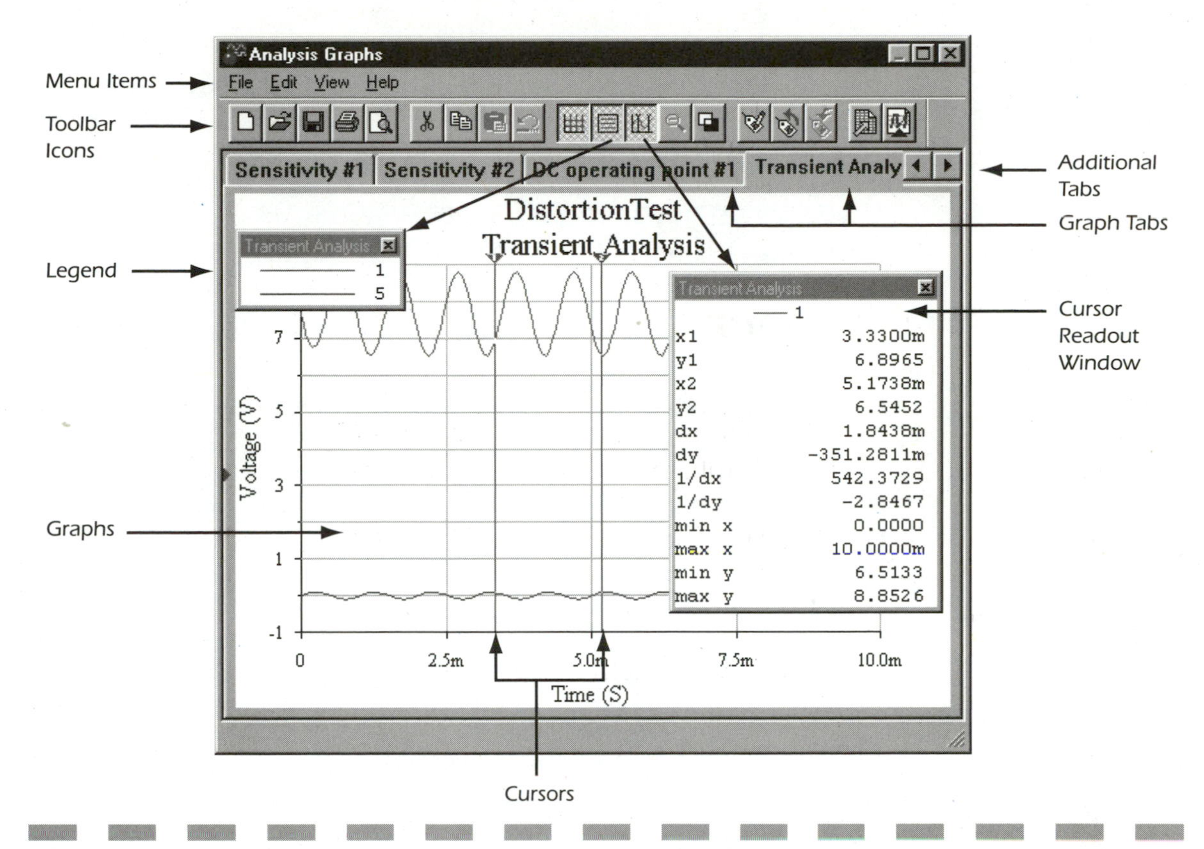

Figure 6.37 The Grapher displays graphs and readouts for each analysis performed. They can then be saved or printed.

There are many right-click menus within Grapher to speed actions. Move the mouse around and check them out.

Grapher Menu Items

The various menu items found in Grapher are shown in Figure 6.38.

Figure 6.38 Grapher menu items.

- **File**
  - — **New**—Creates a new page or lets you delete all existing pages.
  - — **Open**—Opens an existing Grapher file (*.gra), Scope data (*.scp), or Bode plotter result (*.bod).
  - — **Save**—Stores the current Grapher files.
  - — **Save as**—Name and store the current Grapher file as a *.gra file or as a text file.
  - — **Print**—Opens the Print dialog box to make a hard copy of the selected graphs.
  - — **Print Preview**—Preview each page before printing.
  - — **Print Setup**—Opens the Print Setup dialog box.
  - — **Exit**—Closes Grapher.
- **Edit**
  - — **Undo**—Undoes the last action.
  - — **Cut**—Cuts the current selected graph or page.
  - — **Copy**—Copies a selection from the current page.
  - — **Paste**—Pastes a selection into the current page.
  - — **Clear Pages**—Select which pages to delete.
  - — **Properties**—Allows you to modify page or graph properties.
  - — **Copy Properties**—Copies page properties.
  - — **Paste Properties**—Pastes page properties to other page.
- **View**
  - — **Toolbar**—Turns toolbar off/on.
  - — **Status Bar**—Turns bottom status bar off/on.

NOTE

Help on Grapher is only accessible from Multisim's main help menu. The Help menu item inside Grapher itself is useless.

Toolbar Icons

At the top of the window you will see a row of icons for Grapher's toolbar (see Figure 6.39). From left to right they are:

New, **Open**, **Save**, **Print**, **Print Preview**, **Cut**, **Copy**, **Paste** and **Undo**. Each of these items has been described in the menu section. Additional icons are:

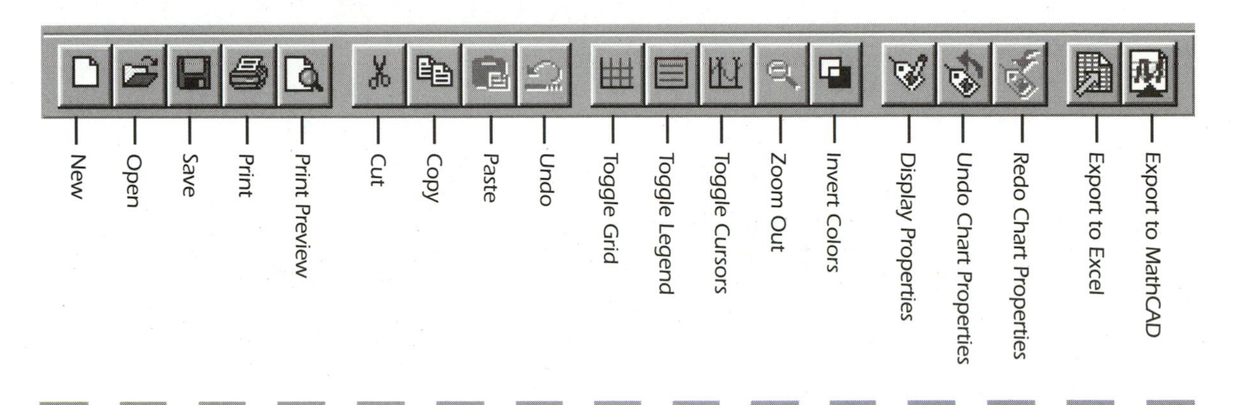

Figure 6.39 Grapher icons.

- **Toggle Grid**—Applies a grid to a graph for easier reading.
- **Toggle Legend**—Turns on a chart that shows what each color line (trace) represents.
- **Toggle Cursors**—Turns on the cursors and also opens a readout window.
- **Invert Colors**—Changes white to black, black to white, etc.
- **Display Properties**—Opens a dialog box to adjust the selected graph's properties.
- **Undo Chart Properties**—Undoes the last action taken on the Properties Dialog Box.
- **Redo Chart Properties**—Redoes the actions after you used the Undo Chart Properties command.
- **Export to Excel**—Turns the current graph into a Microsoft Excel file.
- **Export to MathCAD**—Turns the current graph into a MathCAD file.

Pages and Graphs

Each analysis is shown on a separate page. These pages are named and numbered and accessible by clicking the tabs at the top (see Figure 6.40).

You will notice a red arrow to the left of the leftmost tab while clicking the tabs; this shows the pages selected. If you were to hit the Properties icon on the toolbar, it would bring up that page's properties. Each page may also contain more than one graph (see Figure 6.41). To select a graph, left-click anywhere on that graph and the red arrow shifts to the graph, meaning it is selected. When you hit the Properties icon, the graph's properties are accessible, instead of the page's properties.

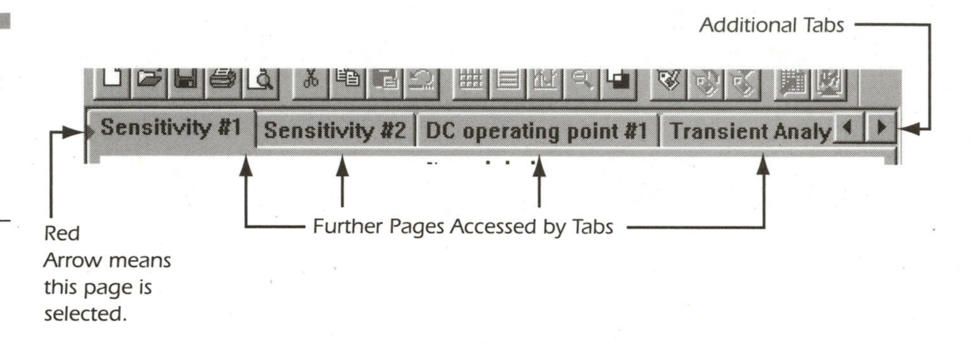

Figure 6.40 One or more graphs are placed on a page. Flip between Grapher pages with the tabs as shown.

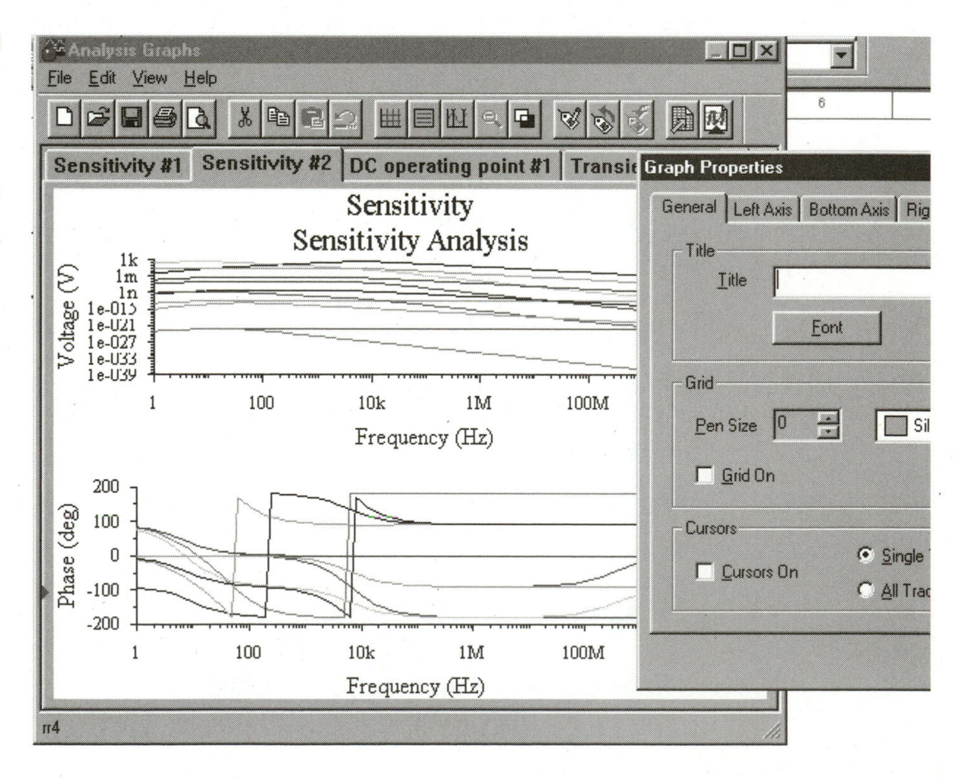

Figure 6.41 There may be more than one graph on a page. Select a graph by clicking the mouse pointer on it.

Using Grapher

Here is a list of steps and routines to squeeze the power out of the analysis graphs feature:

WORKING WITH PAGES

Once a simulation has run, Grapher opens with a page for each analysis (in some cases, multiple pages). Each page may contain one or more graphs or tables. If you have previously run other analyses, they will be available as well. You must use the tabs at the top to change between pages. If there are too many pages to list with tabs, then Multisim creates two arrows to the right of the tabs; hitting them lets you scroll to view more tabs (to the right or left) (see Figure 6.40).

NOTE

*Multisim keeps a record of the graphs of each analysis you have performed since you started the current session. The only way to delete them is to select that page/graph and hit the Cut icon. You can also press **File > New > Delete All Pages** to get rid of them all at once.*

SELECTING A GRAPH

Each analysis you select to run has a page. Choose the page you want to view by clicking the tab until it appears. The graph(s) should now be visible. Click anywhere on the graph with the left mouse button to make a little red triangle appear on the right-hand side. See it next to the mouse pointer on Figure 6.42? That means that graph is selected.

NOTE

If more than one graph exists on a page, just click the mouse button over the one you want to select.

ZOOMING IN AND OUT OF A GRAPH

This is a feature that lets you really get a close look at a section of the selected graph. It lets you zoom into any area, which you specify with a box (see Figure 6.43). To place the zoom box:

1. Select the graph you want to zoom in on.
2. Click and hold the left mouse button.

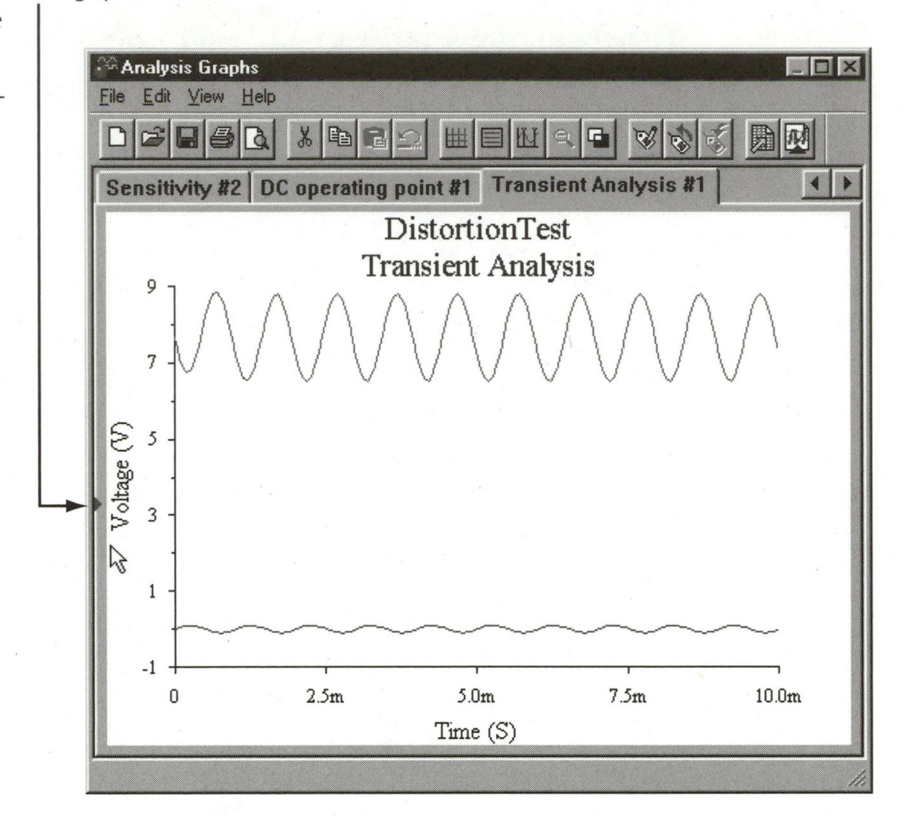

Figure 6.42
The red arrow on the left indicates that graph is selected.

3. Drag the mouse to enclose the area you want to surround. A dotted box appears. Manipulate the box until it surrounds the area of the graph you wish to zoom into.
4. Release the mouse button. A zoomed view of the graph appears.

If you want to zoom in further, repeat steps 2 through 4. When you want to zoom back out of the graph (unless you want to leave it zoomed in) hit the Restore Graph icon or right-click and choose the same.

CHANGING A PAGE'S PROPERTIES

You can open the Properties window of a page by clicking that page's tab and then hitting the Properties icon. Figure 6.44 shows the result. You can choose the tab name, title (and font), and color of the page. There is also a setting to show/hide specific graphs on the page. Make changes and close the window by choosing the **OK** button.

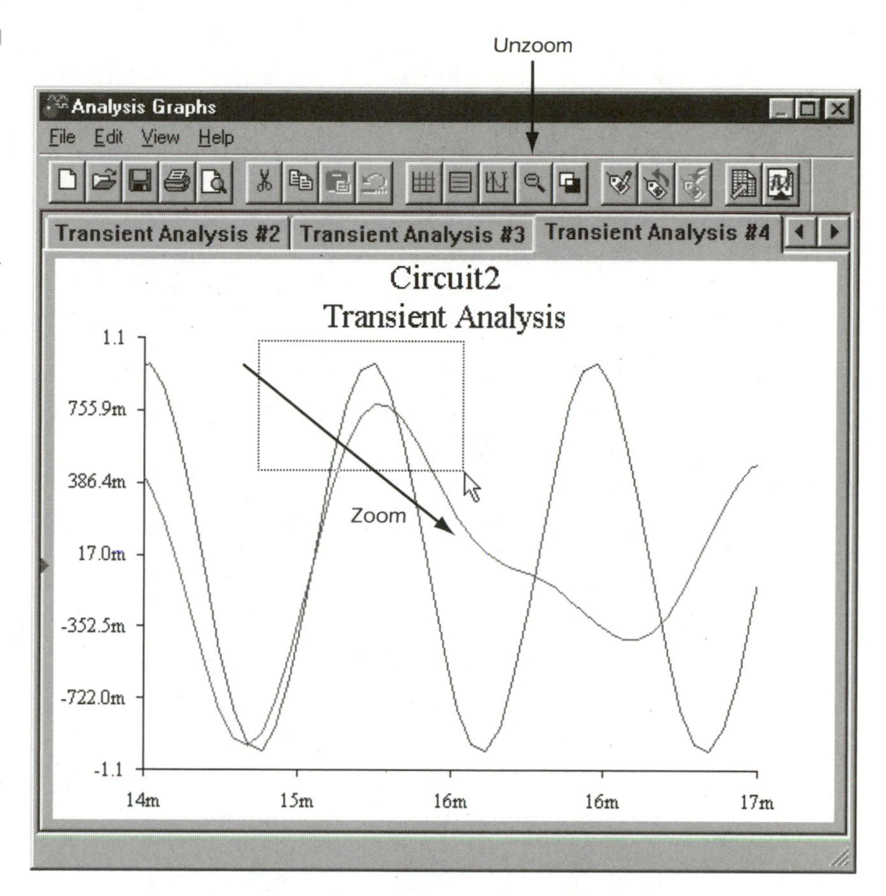

Figure 6.43
Zoom into a section of a graph by clicking and dragging the mouse pointer, surrounding the area you wish to enlarge.

Figure 6.44
The Page Properties dialog box.

INVERTING THE PAGE'S COLORS

Want to make the page black and the traces and text white? Hit the Invert Colors Icon on the toolbar. If the pages are in color try to use a color combination that will not make your data disappear.

CHANGING A GRAPH'S PROPERTIES

Each item on each tab page can be adjusted to customize the currently selected graph. See Figure 6.45 through 6.50.

- **General Page**—Title is the name of the graph, grid adjusts the grid color and line width, trace legend turns on the legend, cursors can be turned on/off or used to adjust which traces you want read out (see Figure 6.45).

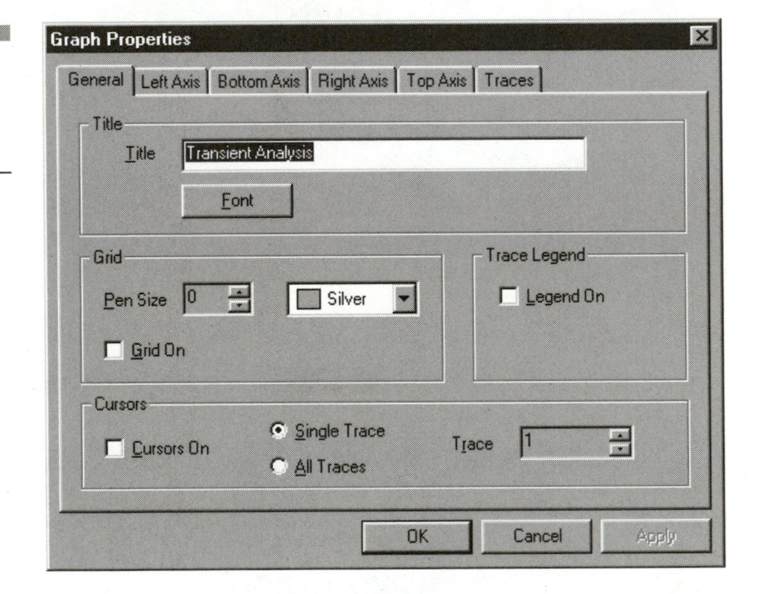

Figure 6.45 General Page of the Graph Properties dialog box.

- **Left Axis, Bottom Axis, Right Axis, and Top Axis**—Figures 6.46, 6.47, 6.48, 6.49 show the adjustments possible for each axis: the axis label and font, axis pen Size, color, the scale of that axis, the range of that axis, and the divisions (ticks per axis). The most important thing to remember is that you can adjust the area you wish to have displayed on the graph. Say, for example, you wanted to print out the wave from .017ms to .023ms. Adjusting the bottom axis range's minimum and maximum readjusts the graph. You can use the same method to shrink the amplitude with the left axis. Play with each setting until you are happy with how the graph looks for printout or storage.

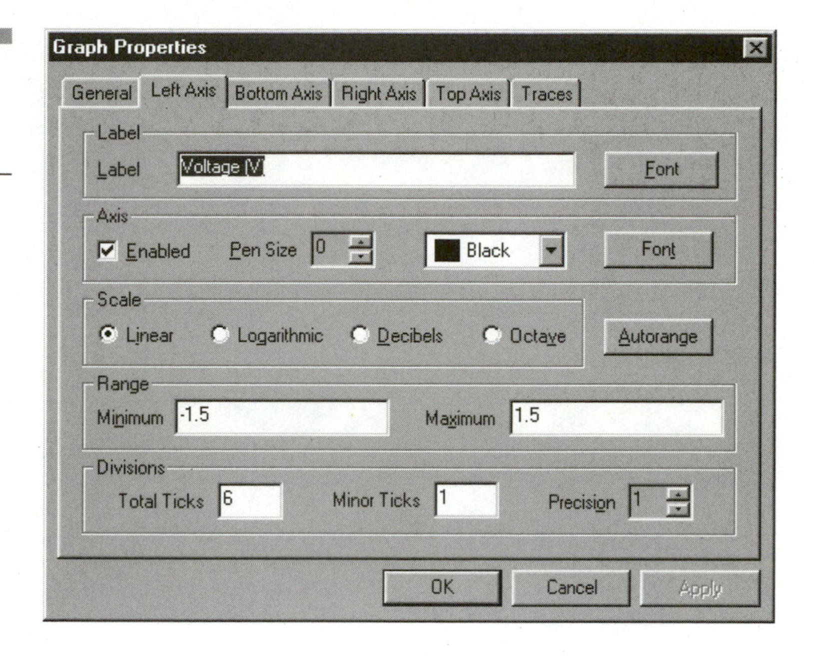

Figure 6.46
Left Axis Page of the Graph Properties dialog box.

Graph Properties
General
Left Axis
Bottom Axis
Right Axis
Top Axis
Traces
Label
Label
Voltage (V)
Font
Axis
Enabled
Pen Size
0
Black
Font
Scale
Linear
Logarithmic
Decibels
Octave
Autorange
Range
Minimum
-1.5
Maximum
1.5
Divisions
Total Ticks
6
Minor Ticks
1
Precision
1
OK
Cancel
Apply

Figure 6.47
Bottom Axis Page of the Graph Properties dialog box.

Graph Properties
General
Left Axis
Bottom Axis
Right Axis
Top Axis
Traces
Label
Label
Time (S)
Font
Axis
Enabled
Pen Size
0
Black
Font
Scale
Linear
Logarithmic
Decibels
Octave
Autorange
Range
Minimum
0
Maximum
0.1
Divisions
Total Ticks
4
Minor Ticks
1
Precision
0
OK
Cancel
Apply

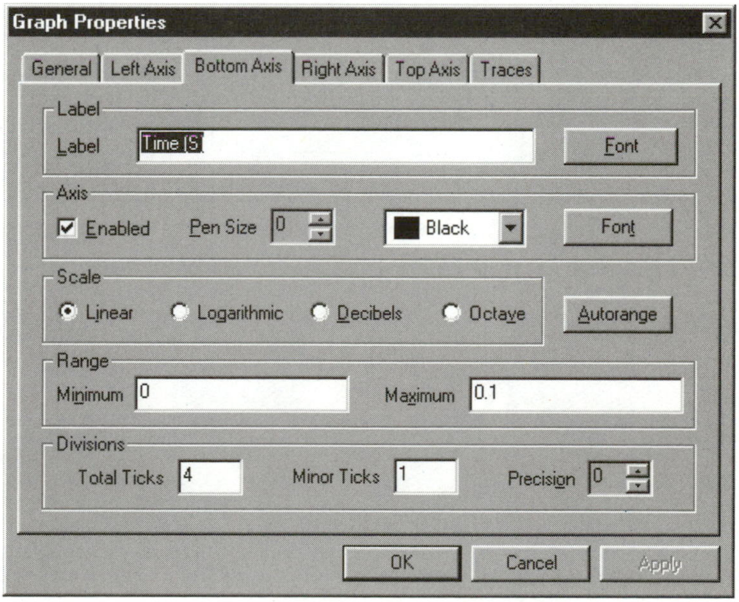

Graph Properties
General
Left Axis
Bottom Axis
Right Axis
Top Axis
Traces
Label
Label
Font
Axis
Enabled
Pen Size
0
Black
Font
Scale
Linear
Logarithmic
Decibels
Octave
Autorange
Range
Minimum
-1
Maximum
1
Divisions
Total Ticks
0
Minor Ticks
0
Precision
0
OK
Cancel
Apply

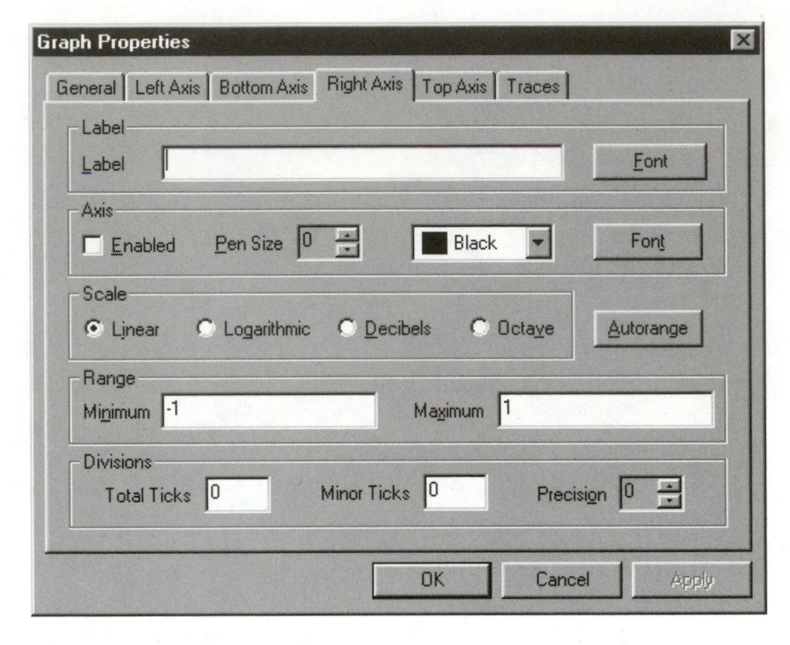

Figure 6.48
Right Axis Page of the Graph Properties dialog box.

Graph Properties
General
Left Axis
Bottom Axis
Right Axis
Top Axis
Traces
Label
Label
Font
Axis
Enabled
Pen Size
0
Black
Font
Scale
Linear
Logarithmic
Decibels
Octave
Autorange
Range
Minimum
-1
Maximum
1
Divisions
Total Ticks
0
Minor Ticks
0
Precision
0
OK
Cancel
Apply

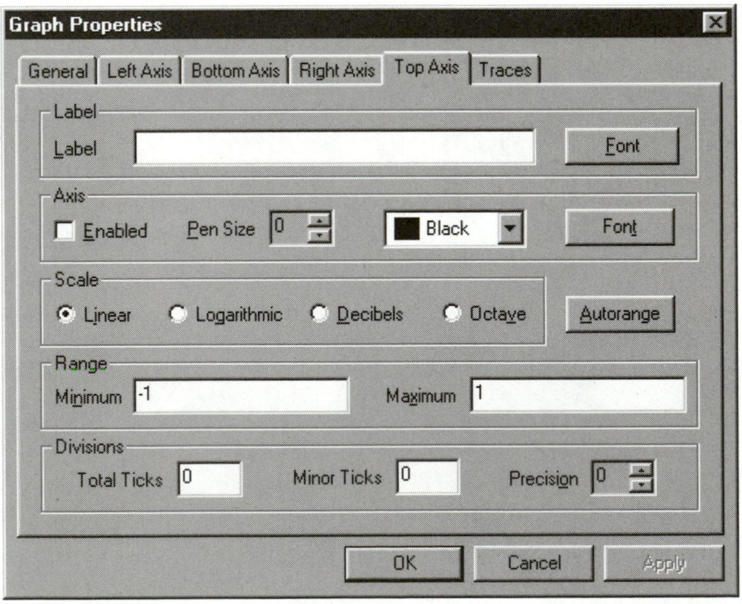

Figure 6.49
Top Axis Page of the Graph Properties dialog box.

- **Traces**—You can adjust the name of the trace, the width, the color, the X and Y range, and you can also offset the trace on an x or y axis (see Figure 6.50).

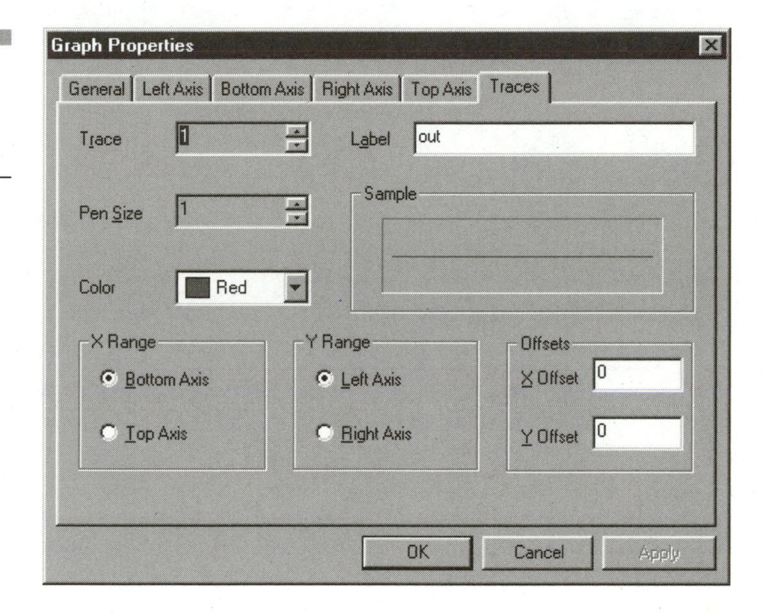

Figure 6.50 Traces Page of the Graph Properties dialog box.

CREATING A NEW PAGE

To create a new page on which to paste other graphs, hit the New icon or choose New Page from the File menu. A window pops up, asking you to name the tab. Type in a name to create a new document. To make a title on the page, click once onto that tab. Hit the Properties icon. Type a name into the title box and select your font. Pick the color you want for the page, also. Hit **OK**.

CUTTING/COPYING AND PASTING GRAPHS OR PAGES

You can cut and copy a graph (or a zoomed-in section of a graph) and paste it onto another page. By creating a new page, you can cut and copy existing graphs then paste them into the new page (see Figure 6.51).

- **Cut**—Select the graph you want to cut. Make sure the red triangle is on that graph. Hit the Cut icon or choose **Edit** > **Cut (Ctrl+X)**.
- **Copy**—Select the graph you want to copy. Make sure the red triangle is on that graph. Hit the Copy icon or choose **Edit** > **Copy (Ctrl+C)**.
- **Paste**—Choose the page to which you want to paste by hitting its tab. Click the Paste icon or choose **Edit** > **Paste (Ctrl+V)**.

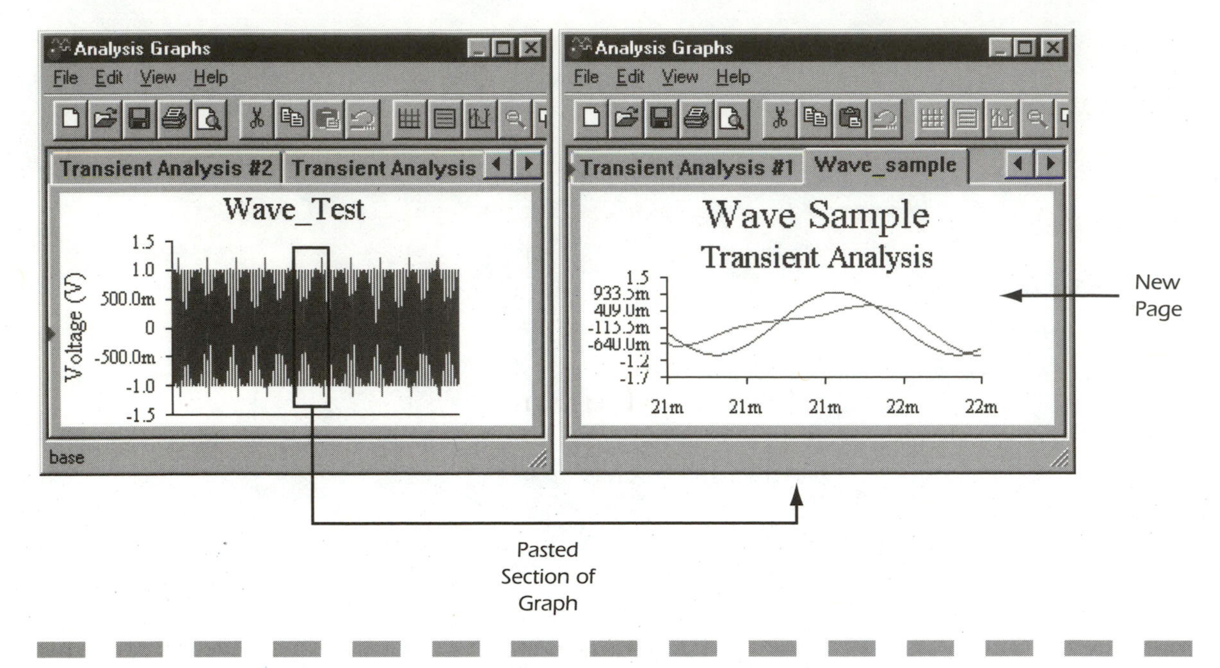

Figure 6.51 Create a new page to paste graphs (or sections of graphs) into.

PRINTING

You can send one or all graphs to a printer for hard copy. If you only want one graph printed, select the page in Print Preview and only print that page with the printer control. Hit **File** > **Print** or **Ctrl+P**. The Printer dialog box opens. Adjust the settings and choose **OK** to print.

RIGHT-CLICKING

Some menu items are available by placing the mouse over an area and hitting the right mouse button; this brings up a *hot menu*. Experiment with this.

PHASE GRAPHS

On a phase graph, you may have noticed a steep line in the middle of a waveform (see Figure 6.52). This means the graph is going past a specified amount of degrees (y-axis) and is starting off on the opposite end of the graph. Multisim chose to do this instead of enlarging a graph to enormous proportions. I try to picture in my mind the graph being one continuous waveform.

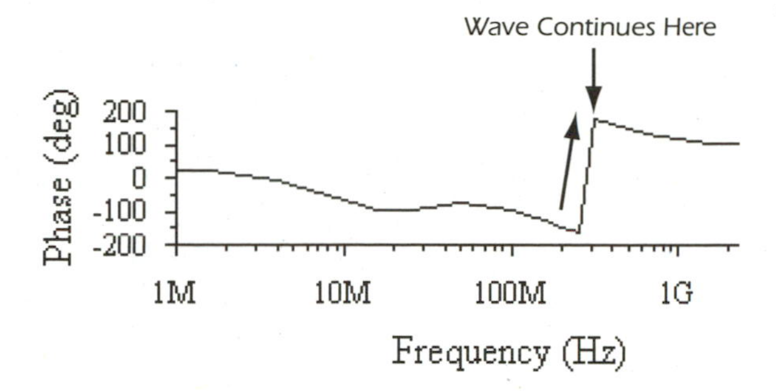

Figure 6.52
A phase graph.

Working with Cursors

When a graph is selected, a small window opens when you hit the Toggle Cursor icon on the toolbar (as pictured in Figure 6.53). Let's call this the Cursor Readout window. On the graph, you will notice the two cursors running top to bottom. Cursor 1 is red and cursor 2 is blue. You can move various cursors back and forth across the graph and take readings at those points by placing the mouse pointer over the triangle at the top (see Figure 6.53 again), left-clicking, and holding it. Drag the mouse left or right to move the cursor.

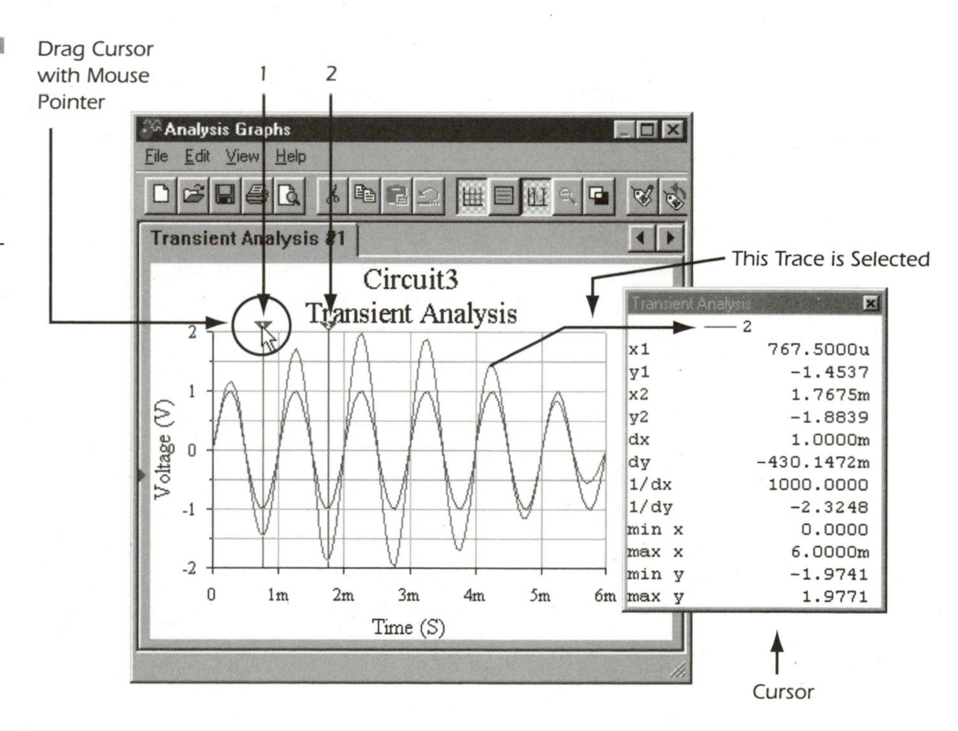

Figure 6.53
The cursors and cursor readout window are used to take numeric readings of a graph.

To see the readouts of each cursor's position, you must first select a trace. Just click over the trace you wish to view and the readings appear in the Cursor Readout Window. To view the numbers on the other traces, alternately click the pointer over those. Here is a list of what each item in the window means:

- **x1 and y1** = coordinates for cursor 1 (red).
- **x2 and y2** = coordinates for cursor 2 (blue).
- **dx** = x-axis delta between the two cursors (i.e., difference between cursor readout 1 and 2 on the x-axis).
- **dy** = y-axis delta between the two cursors. (i.e., difference between cursor readout 1 and 2 on the y-axis).
- **1/dx** = reciprocal of the x-axis delta.
- **1/dy** = reciprocal of the y-axis delta.
- **min x, min y** = x and y minima within the graph ranges (i.e., the lowest reading on the x or y axis).
- **max x, max y** = x and y maxima within the graph ranges (i.e., the highest reading on the x or y axis).

Say, for example, you want to know the exact voltage of the waveform at 1ms (see Figure 6.54). Grab the red cursor (1) and move it until **x1** reads approximately 1.0000m. What does y1 show as the voltage? Another cool thing about the cursors is that you can figure out the peak voltage in a split second. Do you see it as **min/max y**?

TIP

***1/dx** can be used as a simple frequency counter. Place cursor 1 at the start of a cycle and cursor 2 at the end of a cycle and take the reading. In our example in Figure 6.55, the frequency is 1,000 Hz or 1KHz.*

EXPORTING THE RESULTS

Multisim lets you export the results of a graph to either Microsoft Excel or MathSoft MathCAD (see Figure 6.56). Select a graph and click either the Excel or MathCAD button in the top-right corner. Multisim informs you of the creation of a new Excel or MathCAD document and prompts you to hit **OK**. Doing so opens either Excel or MathCAD, with the graph's information inputted. You can then work with the data or save/print as desired.

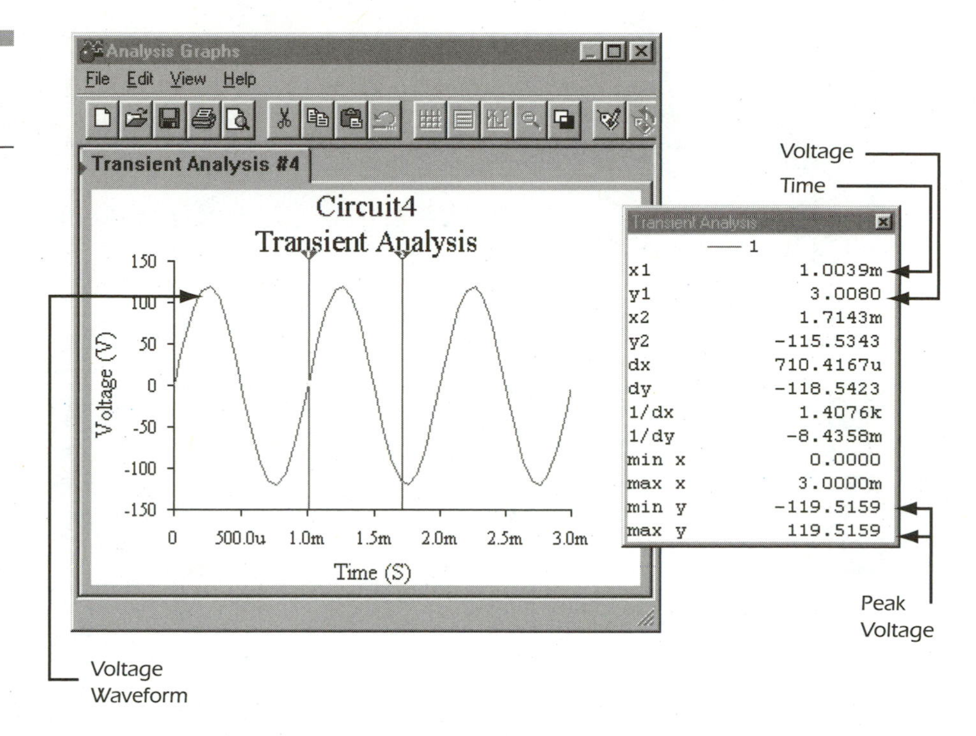

Figure 6.54
An example of how to use cursors.

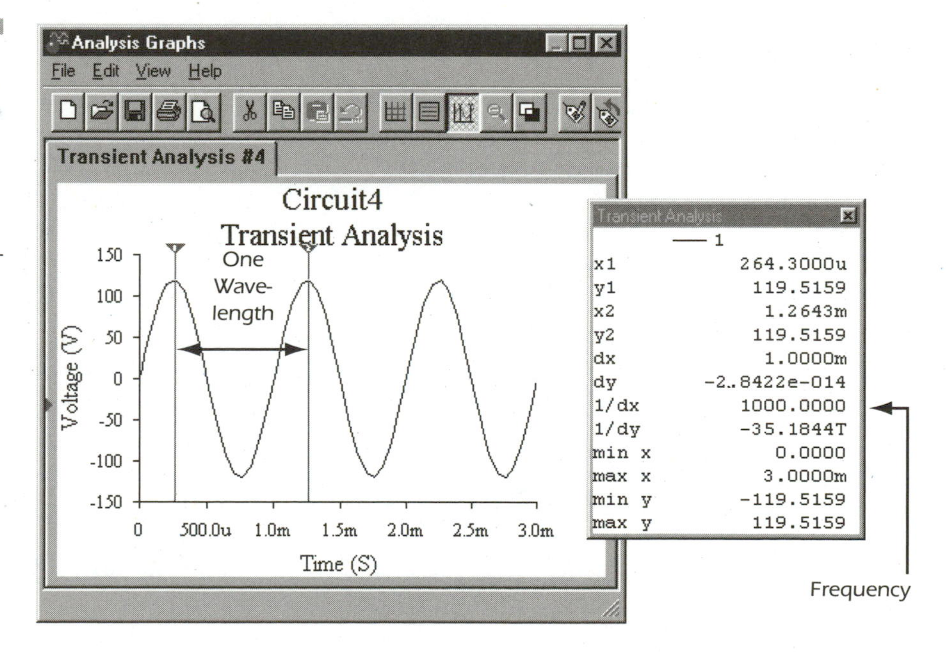

Figure 6.55
Place the cursors one wavelength away from each other and use the 1/dx reading as a frequency counter.

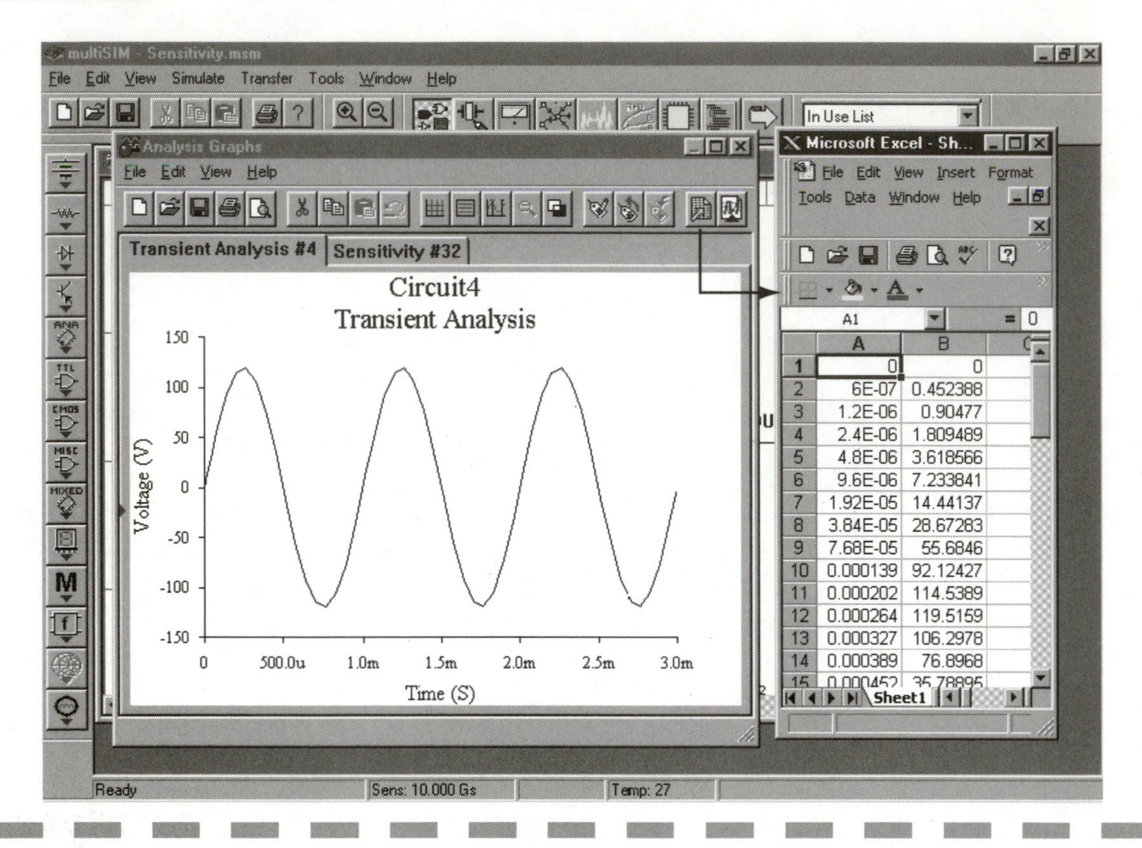

Figure 6.56 You can export the graph data into Microsoft Excel (as shown) or MathSoft MathCad for further analysis or interpretation.

Post Processing

Just as a note, there is an option available on the Power Professional version of Multisim that lets you further process the information you collected with Analysis. Detailed descriptions can be found in the *Users Manual.* Just remember that Multisim is capable of complex mathematical manipulation of each Analysis result. For more information, see Chapter 9 in the Multisim *User Guide.*

Auto Fault Option

Multisim can create a series of faults in a circuit to teach a student or technician how to troubleshoot. When you choose **Simulate** > **Auto**

Fault Option a dialog box opens up, as pictured in Figure 6.57. Set the number of faults of each type you want created in the circuit. The Any option randomly selects the faults. Hit **OK**.

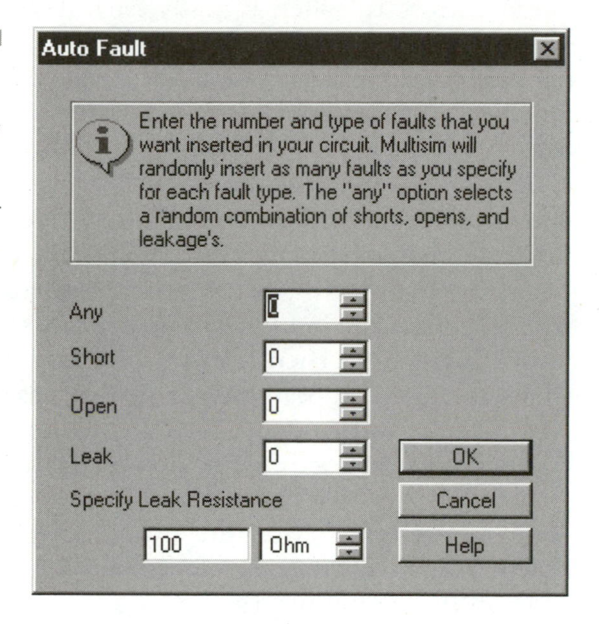

Figure 6.57
The Auto-Fault feature is a great way to learn circuit troubleshooting.

Tweaking Simulations and Analyses

GETTING A DIFFICULT SIMULATION TO RUN

Multisim can be a "bang-the-keyboard" program at times. Usually it's because you cannot get a simulation or analysis to work the way you think it should, which often results in annoying cryptic error messages. I would estimate that 80 percent of the simulations I create don't work the first time the switch is flipped. In fact, when working on any mildly complex simulation, I've come to expect virtual electrical sparks to fly. This is not always due to poor design but rather a matter of telling Multisim exactly what is happening or tweaking the simulation's parameters until all is working correctly. The following are a few observations and hints I have formulated. (See the "Debugging Circuits" section in Chapter 14 for more suggestions.)

GROUND AND DIGITAL GROUND

Make sure a ground is placed in the circuit and, if using a digital circuit, be sure a digital ground is present somewhere on the drawing.

CONNECT ALL LINES

Make sure all lines which are supposed to be connected actually are connected. Sometimes Multisim's autorouting feature makes it appear that a line is connected when it really isn't. Also, if you have been editing connectors, an entire line may have disappeared without your even knowing it. Go over each wire, remembering that a junction (connector) is present only when an intersecting line is actually connected.

FAILURE TO CONVERGE IN DC OPERATING POINT

Multisim uses a method to analyze circuits called *Newton-Raphson.* I won't go into the gory details, but at its simplest this algorithm guesses at what the solution to a circuit can be according to all its voltage and current information and then tries to solve the circuit. If it doesn't quite come out right, it repeats this procedure again and again (iteration). If the voltages and currents finally come together within a certain level, convergence occurs. If it doesn't match the user-defined convergence parameters within a set number of repetitions, then the circuit "Fails to Converge." Multisim usually displays a cryptic error message such as "Singular matrix," "Gmin stepping failed," "Source stepping failed," or "Iteration limit reached."

To correct these errors, Multisim recommends the following:

1. Check the circuit topology and connectivity. Make sure that:
   - The circuit is correctly wired, and includes no dangling nodes or stray parts.
   - You haven't confused zeros with the letter O.
   - Your circuit has a ground node and every node in the circuit has a DC path to ground. Make sure no section of your circuit is completely isolated from ground by transformers, capacitors, etc.
   - Capacitors and voltage sources are not in parallel.
   - Inductors and current sources are not in series.
   - All devices and sources are set to their proper values.
   - All dependent source gains are correct.
   - Your models or subcircuits have been correctly entered.
2. Increase operating point analysis iteration limit to 200—300. This allows the analysis to go through more iterations before giving up.
3. Reduce the RSHUNT value by a factor of 100.

4. Increase the Gmin minimum conductance by a factor of 10.
5. Enable the option Use zero initial conditions.

TRANSIENT ANALYSIS TROUBLES

"Timestep too small." This is the most common error message I get, and here is what Multisim recommends:

"If transient analysis is being performed (time is being stepped) and the simulator cannot converge on a solution using the initial time step, the time step is automatically reduced, and the cycle is repeated. If the time step is reduced too far, an error message ("Timestep too small") is generated and the simulation is aborted. If this occurs, try one or more of the following:

1. Check the circuit topology and connectivity. (See step 1 of Troubleshooting DC Operating Point Analysis Failures.)
2. Set relative error tolerance to 0.01. By increasing the tolerance from 0.001 (0.1 percent accuracy), fewer iterations are required to converge on a solution and the simulation finishes much more quickly.
3. Increase transient time point iterations to 100. This allows the transient analysis to go through more iterations for each time step before giving up.
4. Reduce the absolute current tolerance, if current levels allow. Your particular circuit may not require resolutions down to 1 mV or 1 pA. You should allow at least an order of magnitude below the lowest expected voltage or current levels of your circuit.
5. Realistically model your circuit. Add realistic parasitics, especially junction capacitances. Use RC snubbers around diodes. Replace device models with subcircuits, especially for RF and power devices.
6. If you have a controlled one-shot source in your circuit, increase its rise and fall times.
7. Change the integration method to Gear. *Gear integration* requires longer simulation time, but is generally more stable than the trapezoid method."

BREAKING THE CIRCUIT UP INTO WORKABLE PIECES

If a rather complex circuit refuses to run, begin by dividing it into subcircuits and chart their results. You can cut out the subcircuit and create a new document until that section runs correctly. Then paste each of the working sections into a new document.

RESTARTING MULTISIM

Multisim will often outright refuse to perform a procedure. If all else fails, try restarting the program. When certain memory errors occur, I also recommend saving all your work and resetting the entire system.

NOTE

Visit EWB's website at ***http://www.electronicsworkbench.com/****. It contains a Technical Support section and a list of common simulation and analysis problems. Look under the Analysis and Simulation sections. Bookmark that page and visit it often for new information.*

BUILD CIRCUIT IN USER-FRIENDLY EWB5 FIRST

If you have used EWB Version 5 and then moved to Multisim 6, you will know what I mean by EWB5 being more user-friendly. If you just can't get a circuit to run in Multisim, try it in EWB5. I can't count how many times I've built a circuit in EWB5, switch it on, and it ran without a hitch; when I import or rebuild the circuit in Multisim 6, it simply won't run. After some debugging I can eventually get it to run in the newer Multisim.

MAKE SURE YOU HAVE THE LATEST 6.XX VERSION

No software program is perfect. Minor improvements need to be added and bugs squashed. EWB periodically sends an updated version to registered customers. For example, Multisim V6.20 is due for release as of this writing. Contact EWB for the most current version number.

NOTE

E-mail EWB's Tech Support as a last resort. You can have an EWB tech staff member look at your circuit. Give them time to answer your e-mail because they get busy at times. There is an e-mail form at ***http://www.electronicsworkbench.com/html/support_form_product.html****.*

Making the Most of Your Analyses

Multisim's Analysis is only as useful as you make it. Learning the different analyses is simple; just open circuits you have made or found in books and practice on them. EWB's website provides primers and test circuits to help with this. I post test circuits as well, at **http://www.basicelectronics.com/ewb/**.

Which Analysis Should I Use?

Most newcomers to EDA are intimidated by the prospect of actually using the Analysis features in Multisim. It's no wonder, with the complex engineering explanations the Electronics Workbench company and other sources give. The main analyses (DC OP, AC, Transient, and DC Sweep) are actually quite easy to use once you get the hang of them. After learning a bit about them, it's simply a matter of choosing which real-world instruments you would have hooked into the circuit if it were actually built and replacing them with the analyses. This makes for quicker, more accurate, customizable results. Here are a few replacements:

- **Multimeter**—Use DC Operating Point to determine steady voltage levels as well as current.
- **Oscilloscope**—A transient analysis replaces an infinite number of scopes when set up correctly. This allows more traces to be plotted, and for you to narrow the readings down to only sections you *really* need.
- **Bode Plotter**—Although the Bode plotter is a great addition to the Multisim software, it can be a pain to set up. Running an AC analysis instead speeds the process.

Analyses can also replace certain components or subcircuits. For example, DC sweep can be used in place of a potentiometer, battery, and resistors in a variable voltage source configuration.

Which analysis do I personally make the most use of and why? By far, the winner is transient analysis, because it is much faster than hooking in an oscilloscope and futzing with its settings. I also use DC operating point, DC sweep, and AC analysis a lot.

Summary

Analysis can be one of the most complex sections of Multisim to learn. However, once you figure it out, the Analysis icon on the toolbar will become the most-used button. The next chapter makes use of everything you have learned up to this point by helping you to build some test circuits.

CHAPTER 7

Twenty-five Sample Circuits in Multisim V6

Building a circuit piece by piece reinforces your understanding of the whole.

There is no substitute for hands-on knowledge. Constructing an object with your own hands (be it with tools or a mouse and monitor) helps to seat concepts in the mind. Building the circuits in this chapter gives you a progressively better understanding of the workings and capabilities of Multisim Version 6. The circuits are designed to take you from simple "get used to the software" projects, to more complex simulations. Follow the text as best as you can but don't be afraid to experiment with your own alterations, tests, or designs. Most of the following circuits come from my twisted mechanical mind but some are compiled from various other sources (credits are attached to each). Open up Multisim and let's build some imaginary gadgetry.

Building Them Yourself

Instead of just opening these circuits from the CD, I recommend building each circuit as you see it in the diagram; then run it. The whole point of this chapter is to learn the workings of the program. Each circuit contains custom settings that also must be adjusted or certain analyses that must be performed (see the attachments to each circuit for details). If you are having trouble making out a certain feature of the circuit, open it up from the CD and take a peek, then go back and build it yourself.

CD with Circuits

The included CD contains all of Chapter 7's circuits. They can be opened from the CH7 folder once the Multisim 6 demo is loaded. See the CD's readme.txt file under CD_DRIVE_LETTER:\readme.txt for further information or any changes in the curent edition. When loading the CD, there will be further instructions as well.

Where to Get More Circuits to Build

Just about any circuit that can be built in real life can be constructed in a digital domain. Multisim can be used to test the circuits you may find in books or magazines, so that you can carry them over into your own projects; I recommend subscribing to a few electronic periodicals and collecting circuits that way. Also, Tab Electronics puts out several great books series that contain circuit cookbooks: the *Encyclopedia of Electronic Circuits* (Volume 1 through 6), *The Master Handbook of IC Circuit Applications*, the *McGraw-Hill Circuit Encyclopedia and Troubleshooting Guide* (Volume 1 through 4), and many more. You can also search the Internet, using the keyword *circuits*.

For more Multisim circuits, visit the new Circuit Knowledge Center at ***http://www.electronicsworkbench.com***

Ohm's Law Test Circuits (OHMS1.MSM and OHMS2.MSM)

For the first circuit, I have chosen a build-it-in-two-minutes lesson. Walking through this first circuit helps you to bring a simple drawing together:

1. Open Multisim and follow along. I hope you know how to place components and wire them together, as described in the previous chapters on Multisim.
2. Set up the document to display every label and node available by right-clicking the empty drawing and choosing **Show**.
3. Place a DC voltage source (battery) in the middle of the screen.
4. Place a ground below the battery, a few grid segments below.
5. Place a 1.0-Ohm resistor above and to the right of the battery.
6. Place an AMMETER_V to the right of and parallel to the battery. All the components are placed (see Figure 7.1). Now it is time to connect them.
7. Connect the negative terminal of the battery to the ground.
8. Connect the positive terminal to the left terminal of the resistor.
9. Connect the right terminal of the resistor to the positive terminal of the ammeter.

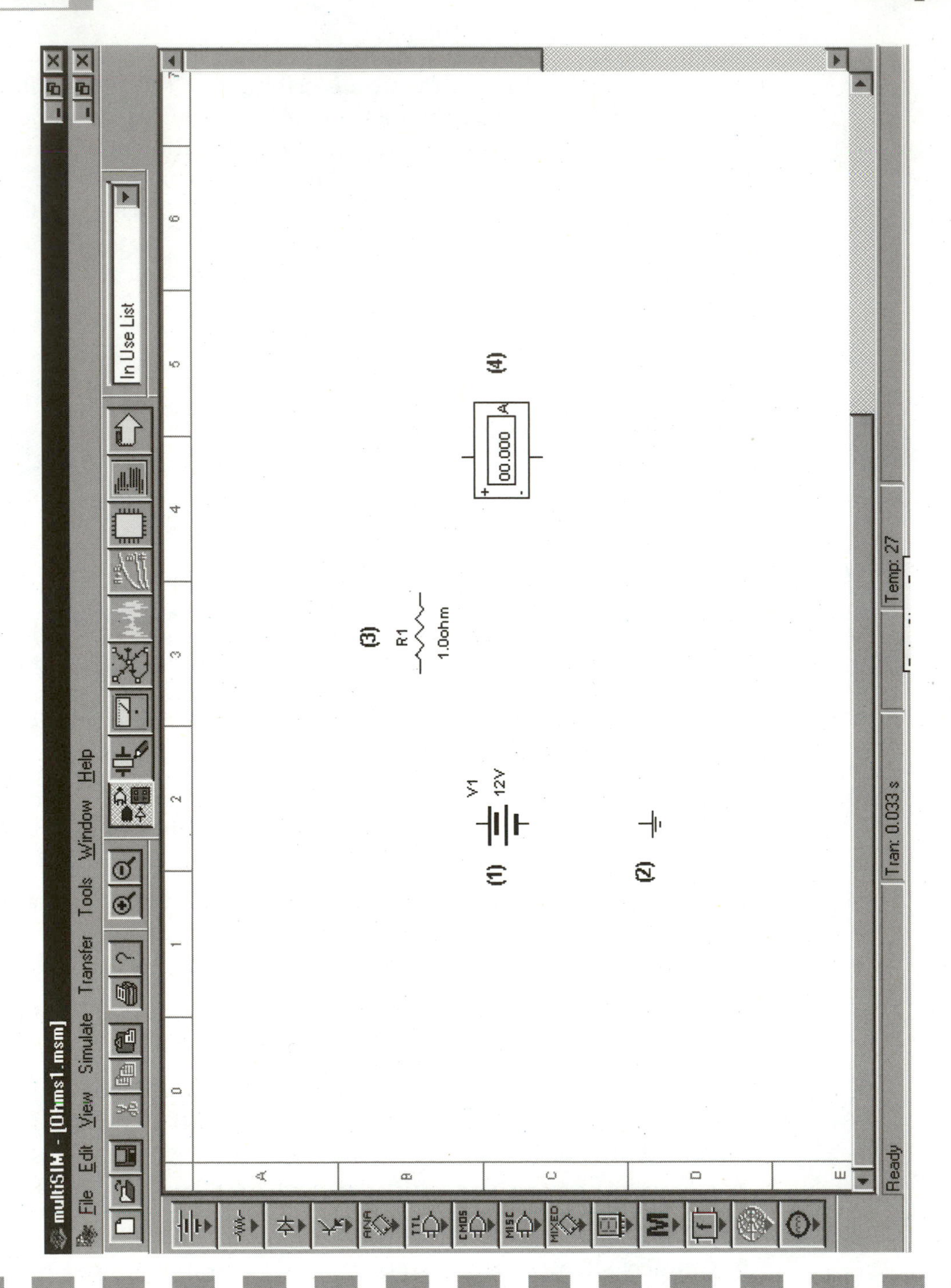

Figure 7.1 Building OHMS1.MSM.

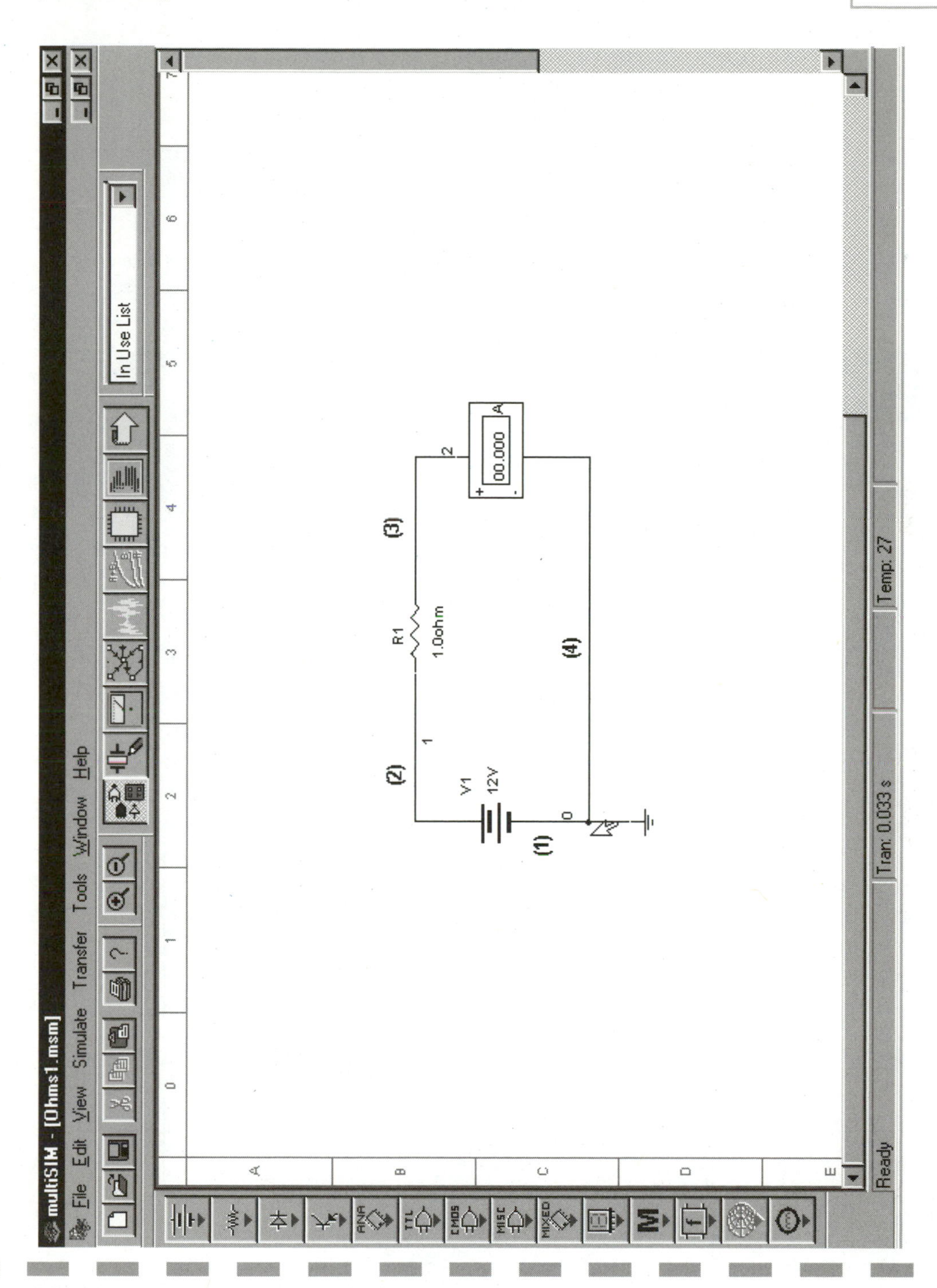

Figure 7.2 OHM'S LAW, PART I—OHMS1.MSM.

10. Connect the negative terminal of the ammeter to the center of the ground wire, thus creating a connector there. The circuit should now look like Figure 7.2.
11. Activate the circuit and note the amperage going through the circuit. Is this a correct reading? Remember I = V / R. Deactivate the circuit.
12. Save the circuit as Ohms1.msm.

NOTE

Notice that the nodes were named in the order in which you wired the circuit.

13. Now let's change the resistor in the circuit to a potentiometer. Double-click the resistor and choose **Replace** from the Value tab.
14. When the browser opens, change the Component Family to potentiometer and select a 1K_LIN Pot. See Figure 7.3.
15. Activate the circuit and note the ammeter reading.
16. Hit the 'a' key on the keyboard once and watch the readings. Hit it twice more, and note the readings. Hit **Shift+A** now.
17. Deactivate the circuit and save it as Ohms2.msm.

That's it. Simple, huh?

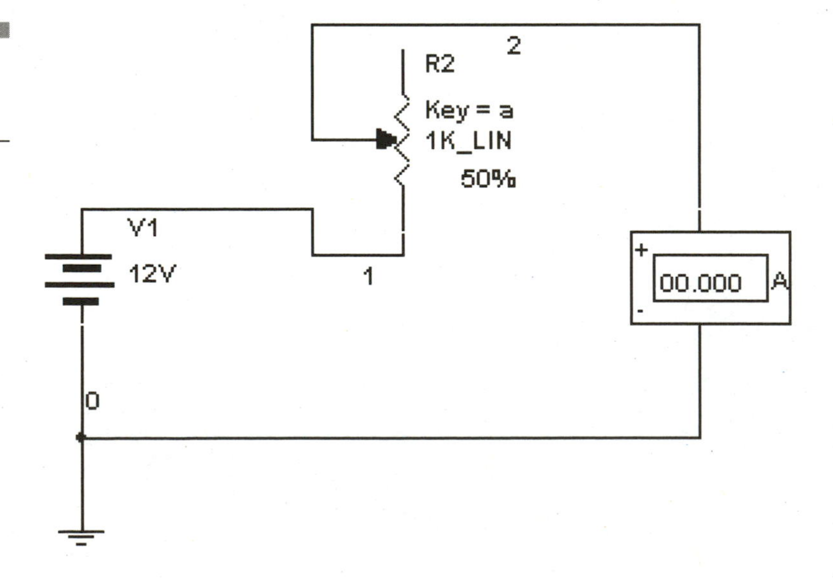

Figure 7.3
OHM'S LAW, PART II—OHMS2.MSM

Voltage Divider (Volt_Div.MSM and Volt_div2.MSM)

Resistors can be used to derive various voltages from a single source. If the ground is offset, you can even produce a negative voltage. In the circuit in Figure 7.4, we are able to create four voltage levels using three resistors in series. The circuit in Figure 7.5 uses two resistors to produce a +6V and −6V level. This next circuit's instructions are more condensed now that you know the simple procedure of building and activating a circuit. Let's start with the circuit in Figure 7.4:

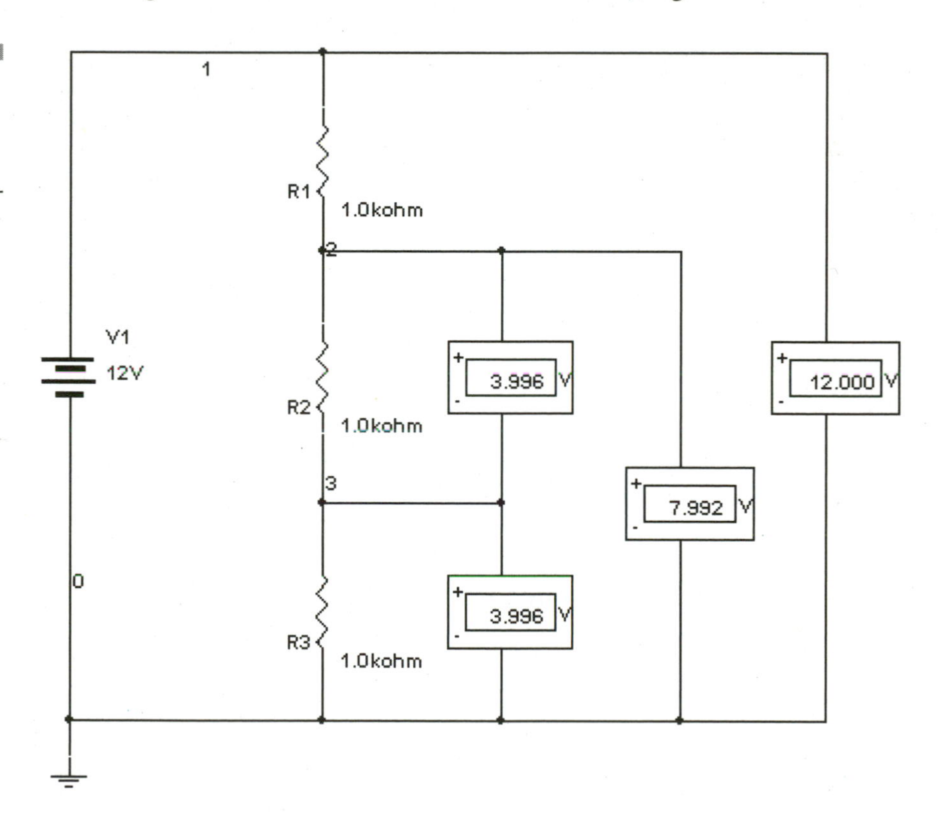

Figure 7.4
VOLTAGE DIVIDER, PART I—VOLT_DIV.MSM

1. Place all components as in Figure 7.4.
2. Wire the components together as shown, in the order that the nodes are in on the schematic. Remember Node 0 is ground.
3. Activate the circuit and note the readings on the four Voltmeters.

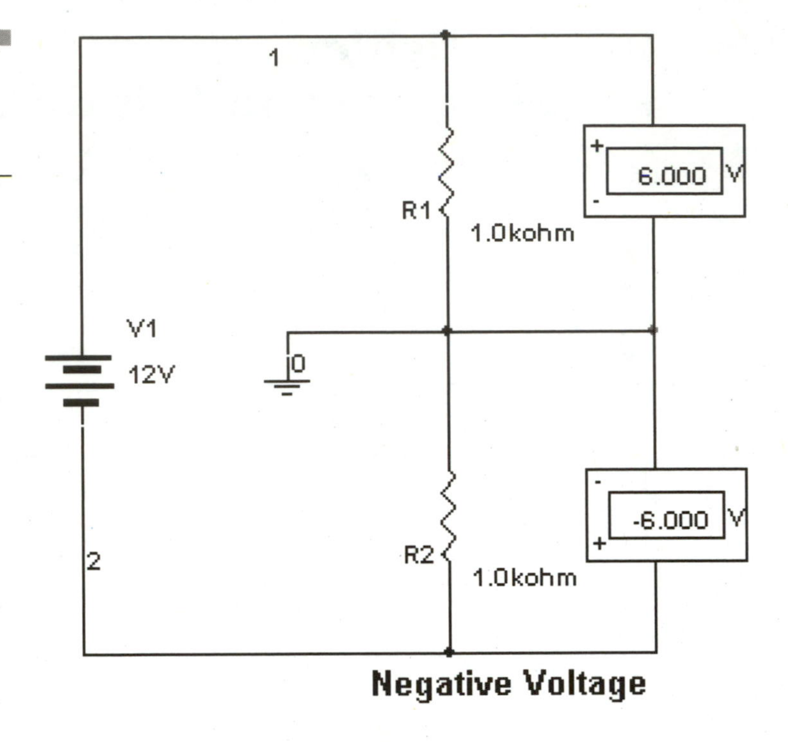

Figure 7.5 VOLTAGE DIVIDER, PART II—VOLT_DIV2.MSM.

4. Deactivate the circuit and save it as Volt_Div.msm.
5. Perform a DC analysis by selecting all of the nodes in the circuit. Do the voltages match the ammeter readings?

Now build the circuit as seen in Figure 7.5 and again run the simulation and DC analysis again.

- **Custom Settings**—Make sure Show Nodes is set in the Schematic options.
- **Analysis**—Perform a DC Operating Point Analysis with all nodes and voltage sources selected.

Kirchhoff's Current Law (Kirchhoff1.MSM)

The remaining circuits are described in detail, but their construction is left up to you. Now that you are familiar with Multisim's basic operation, you should have no trouble building these test circuits.

Kirchhoff's Current Law states "The sum of all the currents flowing to a point in a circuit must be equal to the sum of all the currents flowing away from that point." We have input a 1-ampere current source into three resistors in parallel (Figure 7.6). Each branch's ammeters give you the amount of current flowing through each path. When they are recombined, another Ammeter is used to prove this law.

- **Custom Settings**—None.
- **Analysis**—None.

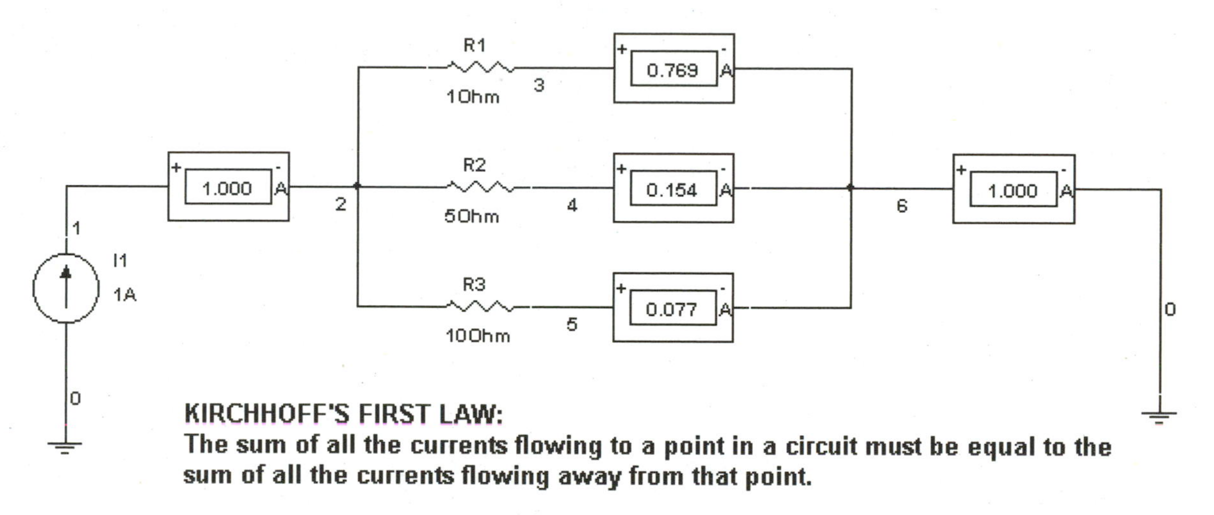

Figure 7.6 KIRCHHOFF'S CURRENT LAW—KIRCHHOFF1.MSM.

Kirchhoff's Voltage Law (Kirchhoff2.MSM)

Kirchhoff's Voltage Law states "The sum of all the voltage sources acting in a complete circuit must be equal to the sum of all voltage drops in that circuit." We have provided a 10-volt battery as voltage source. It's connected to four resistors in series and the voltage drops are recorded with voltmeters hooked to each. Do the voltage drops add up to the voltage sources? See the circuit in Figure 7.7.

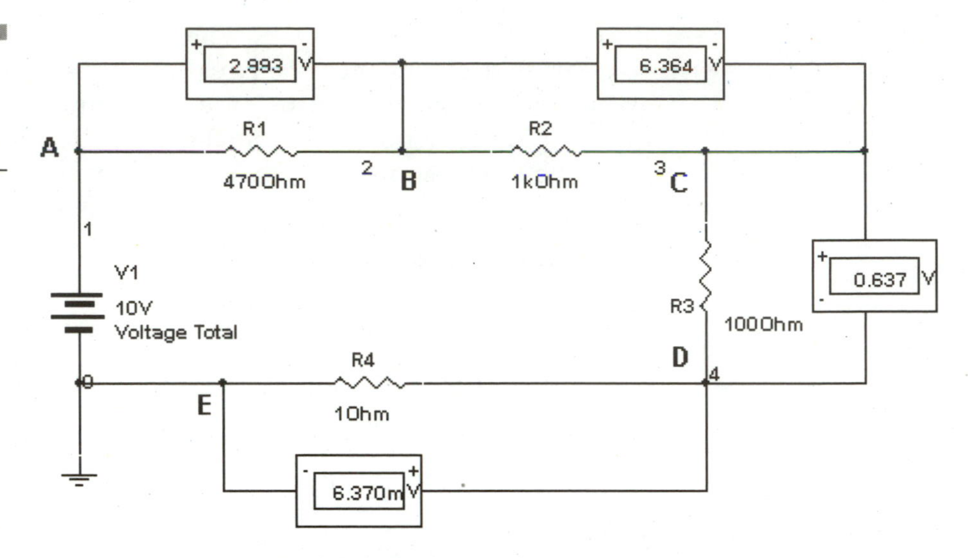

Figure 7.7 KIRCHHOFF'S VOLTAGE LAW—KIRCHHOFF2.MSM.

KIRCHHOFF'S SECOND LAW:
The sum of all the voltage sources acting in a complete circuit must equal the sum of all voltage drops in that circuit.

- **Custom Settings**—Label the nodes A, B, C, D, E, F, or place text in proximity.
- **Analysis**—Perform a DC operating point analysis to confirm readings.

Wheatstone Bridge (Wheatstone.MSM)

Build the circuit pictured in Figure 7.8. This circuit consists of three resistors and a potentiometer in a bridge configuration. The Wheatstone Bridge is often used to find an unknown resistance. When the bridge is balanced (V1/V3 = V2/V4) then the voltage on the multimeter should read near zero volts. If R1 = R3 * (R2/R4), at what percentage does R1 need to be set to balance the bridge? Remember each 5 percent change of the Pot is 50 ohms. Figure it out and set it to that figure using the 'r' or 'R' key. Does the multimeter read near zero volts now?

- **Custom Settings**—Set the Pot to use the R key. Adjust the multimeter to read DC voltage. The voltmeter is placed across the leads of the multimeter merely as a faster way to take readings. Try both the multimeter and voltmeter.
- **Analysis**—None.

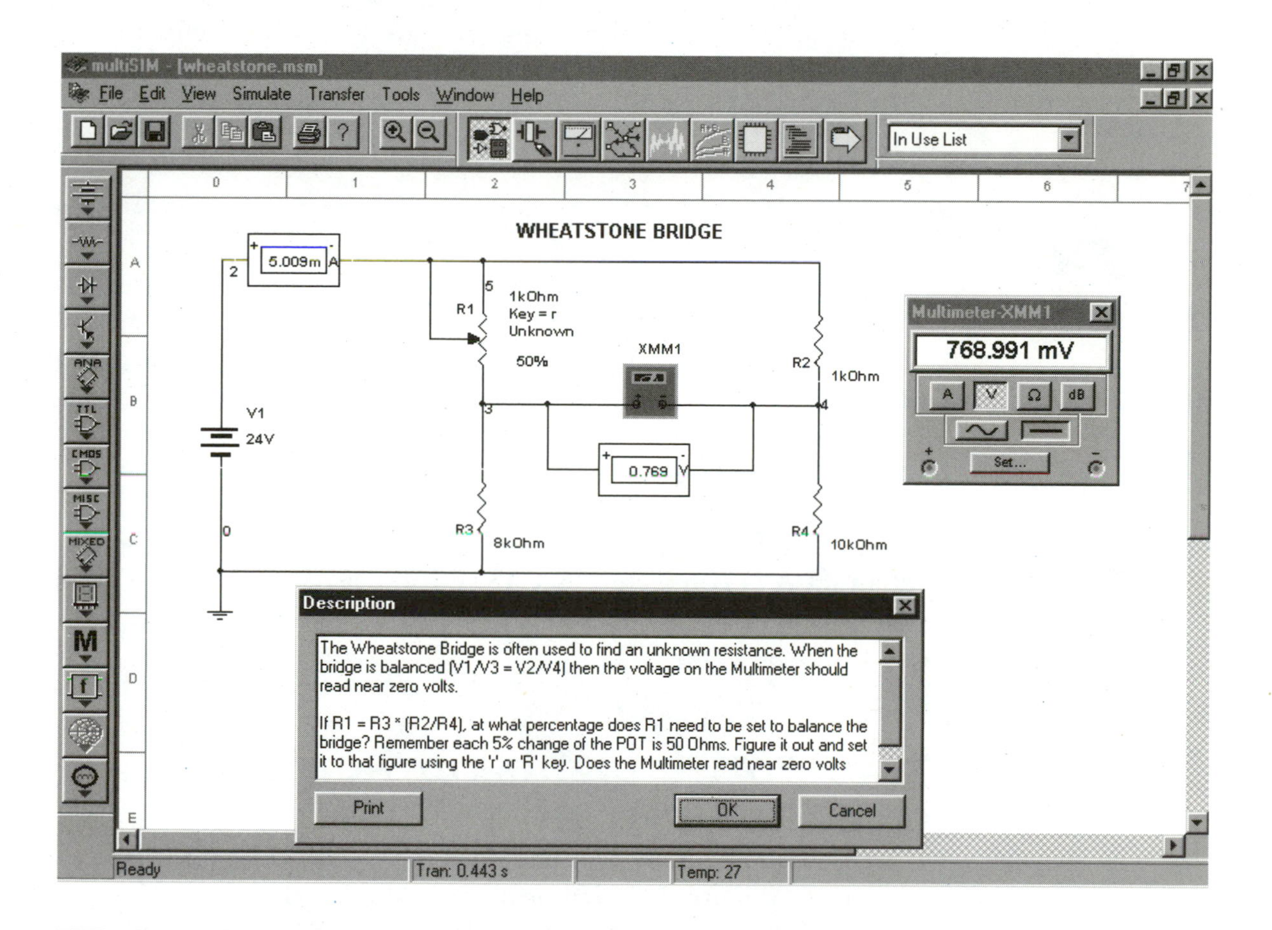

Figure 7.8 WHEATSTONE BRIDGE— WHEATSTONE.MSM.

Resistor/Capacitor (RC.MSM)

Build the circuit shown in Figure 7.9. A resistor and capacitor are often used in conjunction with each other in a DC circuit as a timing device. When the circuit is built, open the oscilloscope, hit the **Pause** button, and activate the circuit with the Simulation switch. You will see the 10V battery begin to charge the capacitor through R1. The scope will pause at the end of the screen. Click the pointer into the circuit window somewhere and hit the space bar; this switches the RC circuit to a discharge cycle. Click the **Resume** button on the simulation switch and watch the results of the capacitor discharging through R2.

NOTE

Multisim 6.2 no longer has a pause button on the oscilloscope.

- **Custom Settings**—Under **Simulate > Default Instrument Settings**; set initial conditions to zero. Adjust the oscilloscope's settings to match the diagram, making sure to hit the **Pause** button before activating the simulation.
- **Analysis**—Run a transient analysis, setting initial conditions to zero and selecting node 5 (or the same node to which the scope's channel A is connected). Set TSTOP to .005seconds. Simulate it. Toggle the switch and run the same transient analysis.

Automotive Lights-on Reminder (Reminder.MSM)

This circuit (Figure 7.10) demonstrates uses for the virtual buzzer, bulb, diodes, and switches by modeling an automobile lights-on reminder. If your headlights are turned on and the ignition key switch is turned off, a warning buzzer sounds to alert you that if you walk away from the car, a dead battery will be waiting in the morning.

The circuit uses two diodes. When both the ignition key switch (Letter K) and the headlights (Letter H) are on, the buzzer draws no

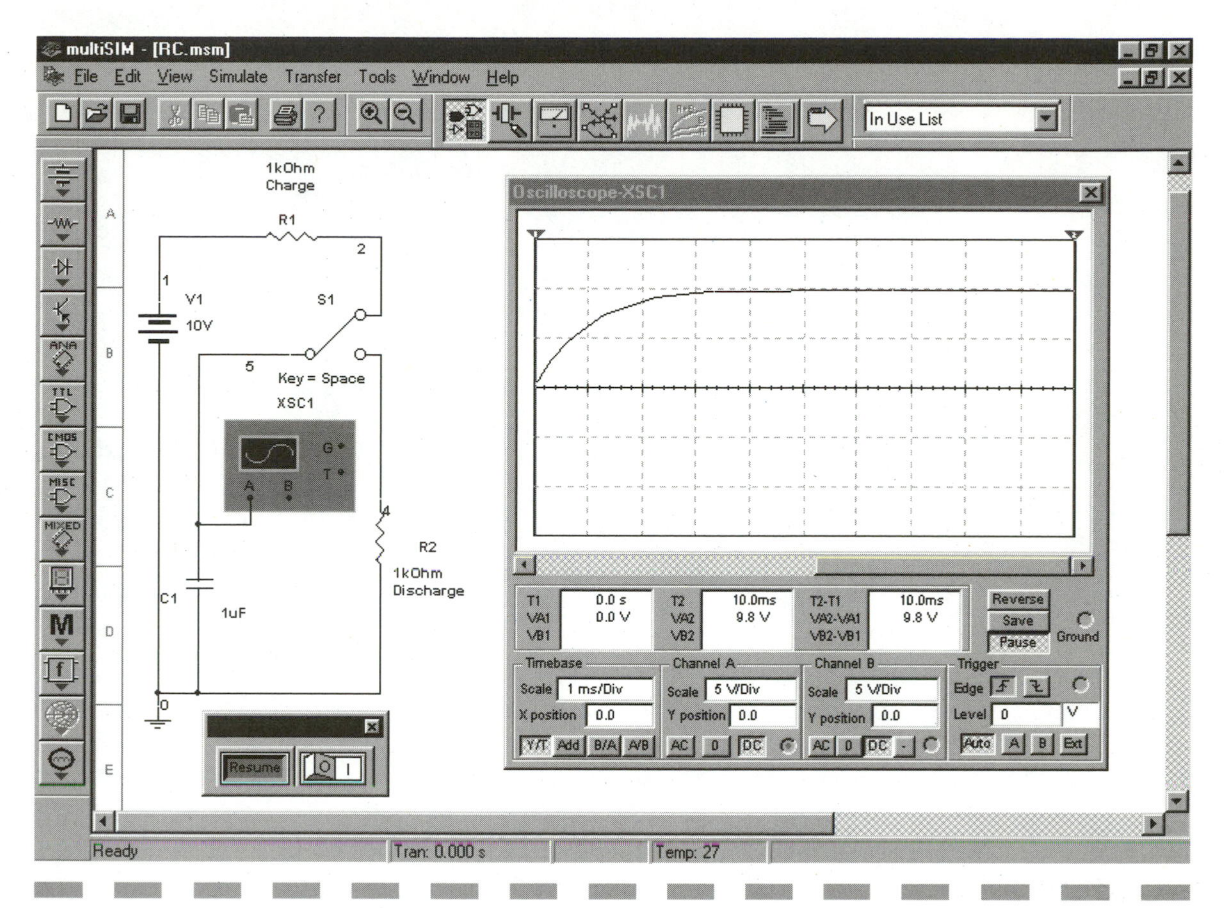

Figure 7.9 RESISTOR/CAPACITOR CIRCUIT—RC.MSM.

current and remains silent. With only the key switch on, diode 1 (D1) is reverse-bias and prevents current from flowing through the buzzer. When the headlights are on and the key switch is off, the buzzer energizes and alerts the driver to turn the headlights off. With the key switch off and lights on, D2 is reverse-bias, stopping current from flowing to the keyswitch. The 1K resistor prevents a short circuit when the key switch is on.

To use the circuit, activate it and use the K key to control the ignition key switch and the H key to control the lights. Turn the ignition on and off with the headlight in the off position (switch in the up position is off). Now turn the headlights on, making sure the key switch is also on. Now turn off the headlights. What happens?

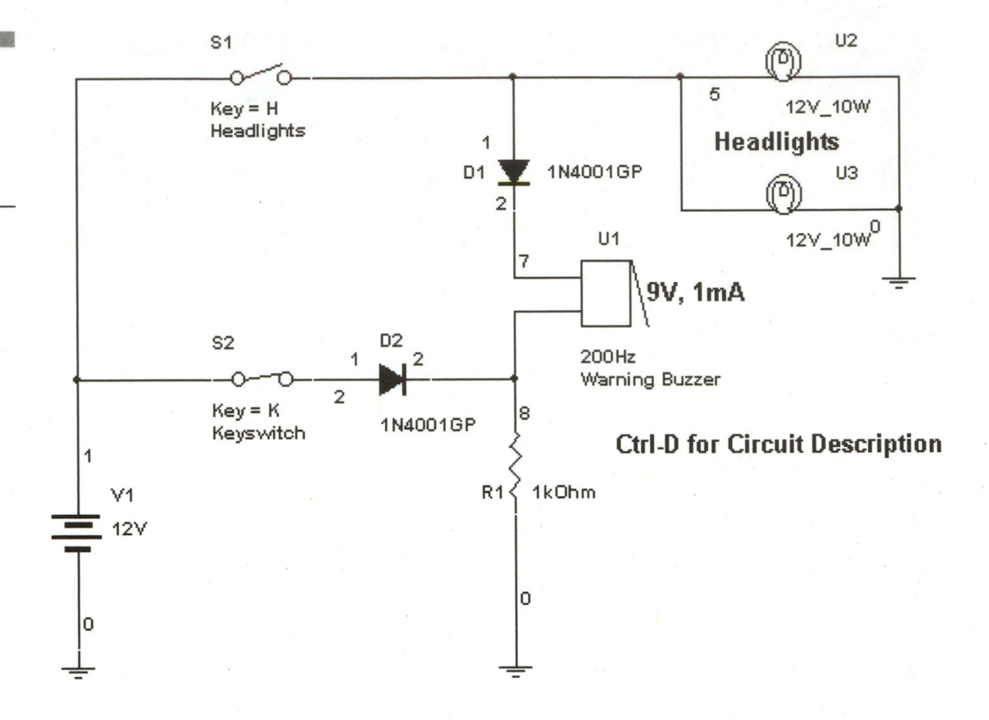

Figure 7.10 AUTOMOTIVE LIGHTS-ON REMINDER—REMINDER.MSM.

CAUTION

Because of a bug in Multisim V6.2 or lower, you must build the Reminder circuit, save it, close it, and then reopen it to run correctly. Go figure!

- **Custom Settings**—Assign the headlight switch to the H key and the keyswitch to the K key. Set the buzzer to 9V, 1mA, 200 Hz.
- **Analysis**—None.

This circuit originally appeared in *Popular Electronics.*

Two-transistor Multivibrator (MV.MSM)

This is a demonstration of a classic *astable multivibrator* (oscillator) circuit built with two transistors, two capacitors, and four resistors (see Figure 7.11).

Build the circuit and adjust the component values as shown. For the oscilloscope:

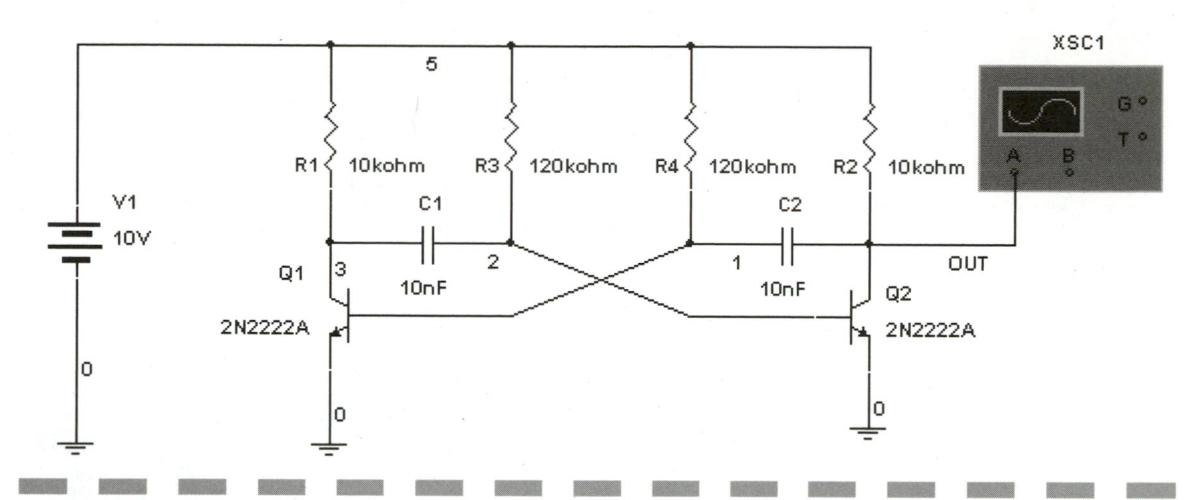

Figure 7.11 TWO-TRANSISTOR MULTIVIBRATOR—MV.MSM.

1. Open the oscilloscope and set the timebase to 1ms/Div and channel A to 5 volts.
2. Activate the simulation.

You can also use transient analysis to run this circuit (preferable):

1. Go to **Simulate** > **Analyses** > **Transient Analysis**.
2. Set the TSTOP = .01sec; check on the Maximum time step and set TMAX to 1e-05 seconds (this disables automatic time steps, which will not work for this simulation).
3. Choose the Out node as the output variable.
4. Hit the **Simulate** button. The multivibrator takes a moment to stabilize. Use the cursors and determine the frequency (this is the 1/dx reading when the cursors span one wavelength).

Simple Push-pull Amplifier (PUSHPULL.MSM)

This circuit (Figure 7.12) uses two BJT transistors to create a simple Class B push-pull amplifier: an npn on the top and pnp on the bottom. The function generator simulates signal input. The 1K-ohm resistor (RL) simulates a load. A resistance is first used between the bases of the transistors. However, disconnect node 2 and connect it to the two diodes; as you will see, they prevent crossover distortion. Figure 7.13 shows the two resulting waveforms. Open the scope and run the simulation, noting the crossover distortion. Connect the diodes and watch the distortion disappear.

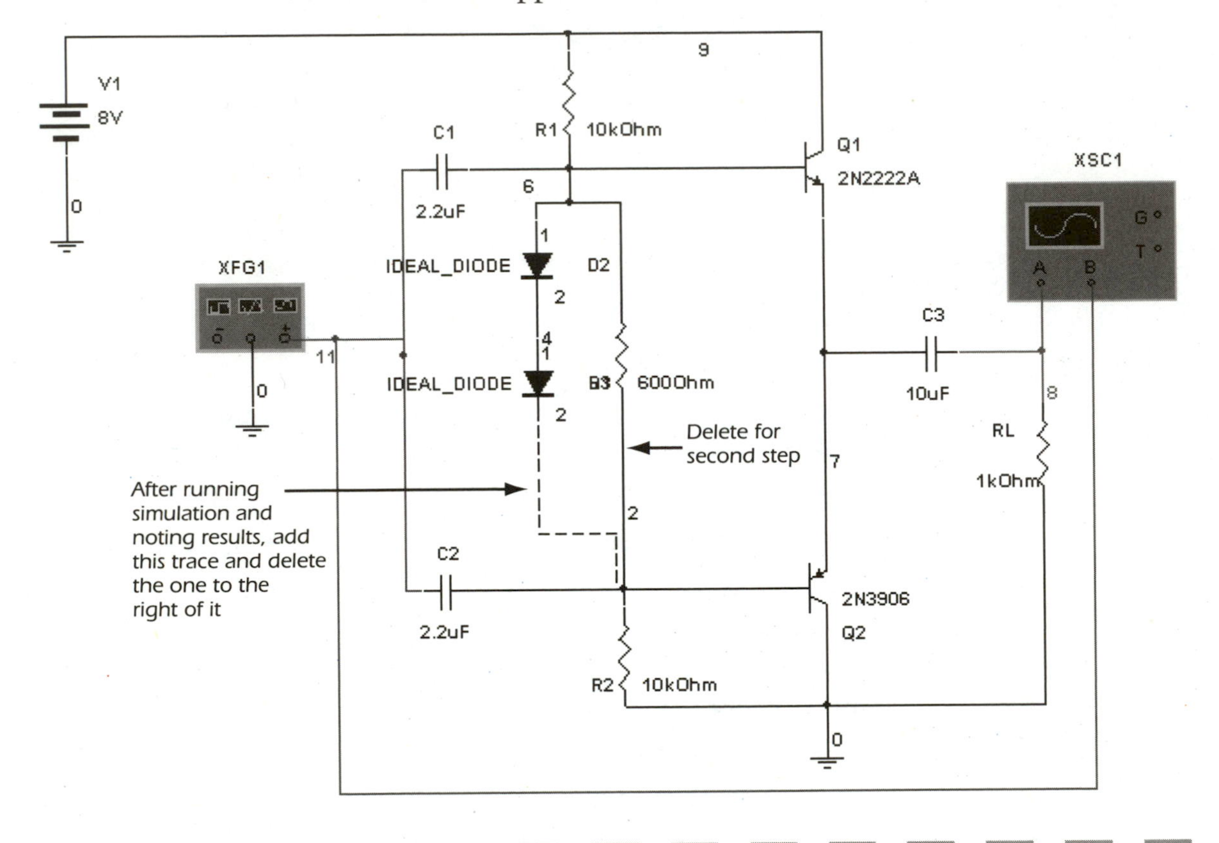

Figure 7.12 SIMPLE PUSH-PULL AMPLIFIER— PUSHPULL.MSM

- **Custom Settings**—Set the oscilloscope as shown in Figure 7.13. Set the function generator to 1kHz, 4V, sine-wave.
- **Analysis**—None.

This circuitis based on a circuit from *150 Circuits.* 1993, Interactive Image Technologies Ltd.

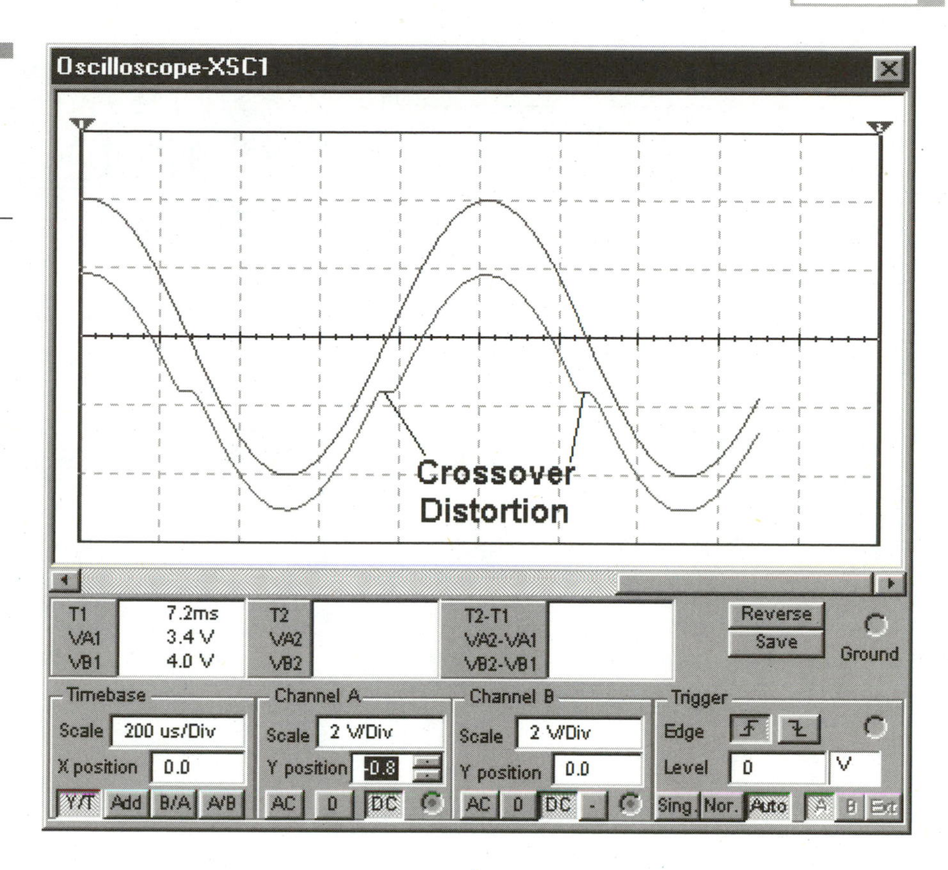

Figure 7.13 Distortion is corrected using two diodes in the push-pull amplifier.

FET Curves (FETCURVE.MSM)

Ever wanted to create current—voltage curves (drain—source characteristic curves) to visualize a certain transistor's properties? This could take hours using a pencil and calculator (see Figure 7.14). With this circuit, you can do it in seconds. It is an excellent example of a two-voltage source DC sweep. You essentially turn V1 (voltage supply) into a current reading output with the current controlled voltage source (CCVS), then sweep it. Build the circuit as shown in Figure 7.14. Open the DC sweep analysis and set the parameters described below under Analysis. Simulate. The Grapher opens, and each colored line represents a half-volt increase on the gate; this may take a minute or so. Replace the MOSFET with a similar model and rerun the DC sweep. Notice the characteristic changes? Remember, you can save the results as graph data.

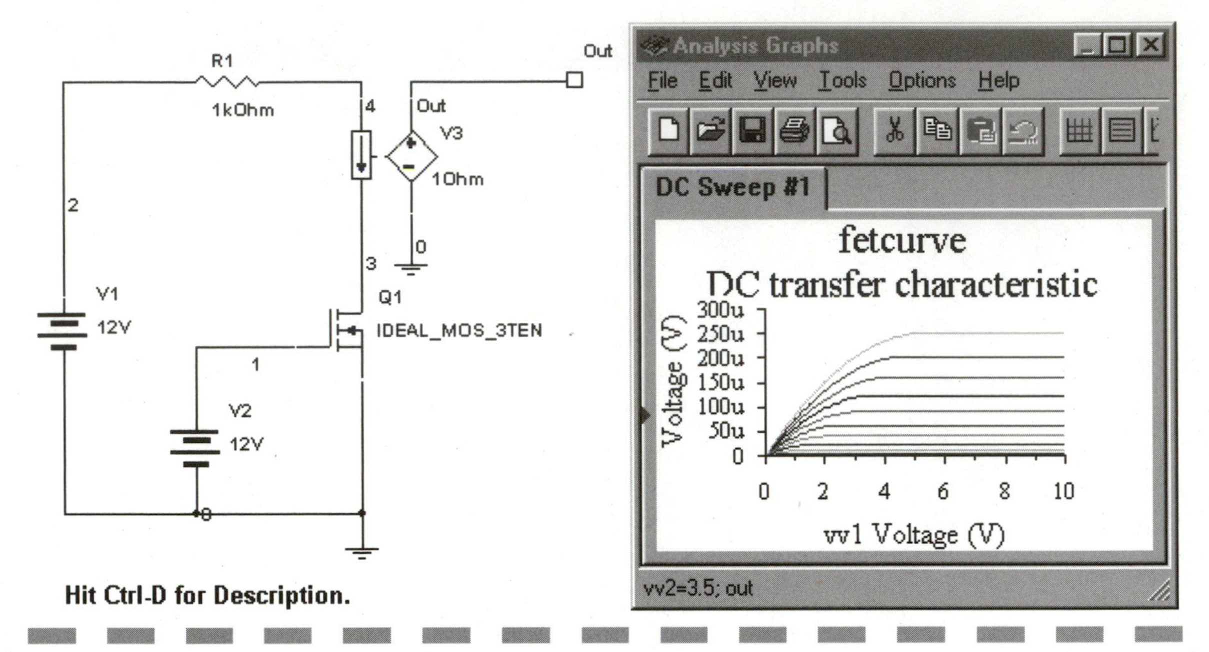

Figure 7.14 FET CURVES— FETCURVE.MSM.

- **Custom Settings**—None.
- **Analysis**—Under DC sweep, set source 1 as V1, start value as 0V, stop value as 10V, and increment as 0.1V. Check on **Use Source 2** and set source 2 as V2, start value as 0V, stop value as 5V and increment as .5V. Under the Output variable tab, choose Out as the node to read (or whichever node is the output).

This circuit originally appeared on the Interactive Image Technologies, Ltd. (EWB company) FTP site (**ftp.interactive.com**).

12-Volt Simple Power Supply (POWERSUP.MSM)

Figure 7.15 is an easy-to-build power supply, which will convert 120-volt AC into approximately 12-volt DC. The alternating current is stepped down with the transformer, put through a bridge rectifier (four diodes), and then filtered with capacitors. The DC voltage output settles at around 16.5 volts. A simple Zener diode steadies the resulting voltage to just below 12V. Beware! This is by no means a design for a real power supply; it is merely to give you an idea of how to use the components involved.

- **Custom Settings**—None.
- **Analysis**—None.

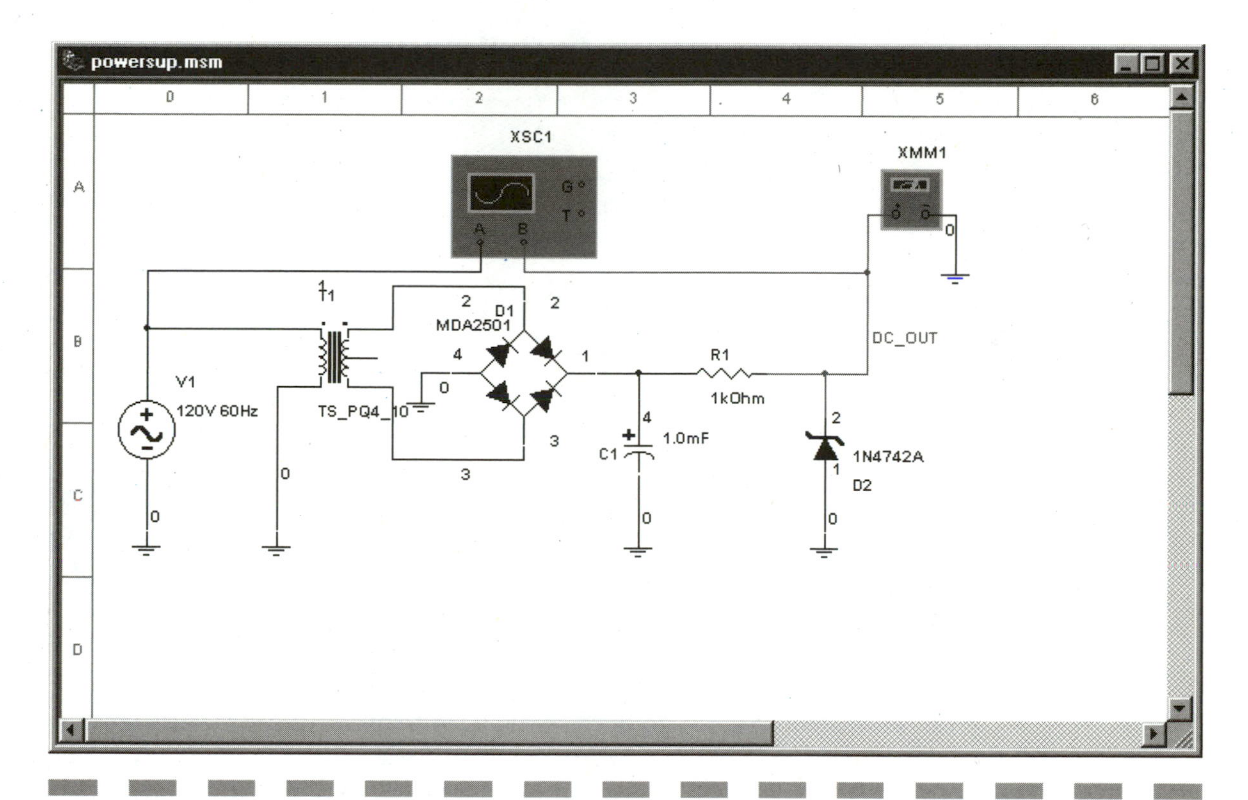

Figure 7.15 12-VOLT SIMPLE POWER SUPPLY— POWERSUP.MSM.

Astable Multivibrator (ASTABLE.MSM)

The circuit in Figure 7.16 demonstrates one of the thousands of uses of the 555 timer. In this case, it is used as a basic oscillator operating at a higher frequency. The design uses a 500 Hz oscillator to light an

LED. R2 and C1 control how long the LED will be lit and R1 + R2 and C1 controls the time at which the LED will be extinguished. Time high (LED OFF) = 0.693 × C1 (R1 + R2). Time low (LED ON) = .693 × C1 × R2. Build the circuit as shown and activate it. Try changing the values for R1, R2, and C1 and note the results. You may need to hook up the oscilloscope if the frequency becomes too high to view with the LED.

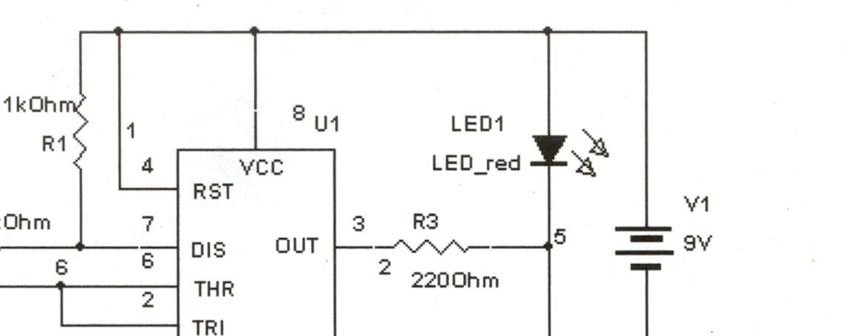

Figure 7.16 ASTABLE MULTIVIRATOR—ASTABLE.MSM.

I decided to use a higher frequency in this example circuit because Multisim is much too slow to view results in a lower frequency range. In other words, if I used the .5 Hz example from Chapter 13, it would take Multisim a few minutes to change the LED from on to off. In a real-world circuit, the LED would flash way too fast to see. This compromise is used in many circuits throughout this chapter.

- **Custom Settings**—None.
- **Analysis**—None.

This circuit is available at my website (**http://www.basicelectronics.com**). For more information on 555 timers in general, see the website.

Multisim V6.11 and previous versions need a fix to the 555 timer. According to the help file from the EWB website: "555 timer does not function properly: The 555 timer requires a little change in its simulation model only. The parameters under .MODEL comparator climit, in-offset and gain, must be changed to in-offset=0.07 gain= 100. To apply the change, start up Component Editor. Choose FAMILY: TIMER, Component: LM555CH>Edit>Model Tab>MODEL DATA. Scroll through the window until you see the statement line: MODEL comparator climit (in_offset=0.0 gain=10). Type in 0.07 for in_offset and 100 for gain. Follow by Save/Exit. This revised 555 Timer model will be saved in either User or Corporate Database. Select View followed by Refresh component toolbars. Your 555 timer will be now shown in one of the above-mentioned library. As well, before running simulation, please make sure to determine initial conditions for simulation to be SET TO ZERO. To do so, please choose Simulate>Default Instruments Settings>Initial Conditions>SET TO ZERO."

Car Battery Level Indicator (BAT_LEV.MSM)

Have you ever had a dead car battery in the middle of winter, in 30 below weather, while dressed in light clothes? Let me tell you, it's not a pleasant experience. The circuit in Figure 7.17 demonstrates several components that simulate this condition. The car battery is a subcircuit of sorts that simply uses a potentiometer to control voltage from 12V to 0V (remember the voltage divider circuit earlier?). This is connected to a series of comparators. Another voltage divider is used to derive reference voltages for the other side of each comparator. R8 is used to offset the voltage ladder so any voltage under 7 volts will not register on the bar graph.

Run the simulation and hit the letter **b**, watching the voltage come down on the voltmeter. The LEDs on the bar graph will alternately extinguish as the voltage hits a preset level.

- **Custom Settings**—Set the potentiometer to 100 percent to begin the simulation and assign the letter B to control it.
- **Analysis**—Select the nodes of each point on the voltage reference resistor ladder and perform a DC operating point analysis.

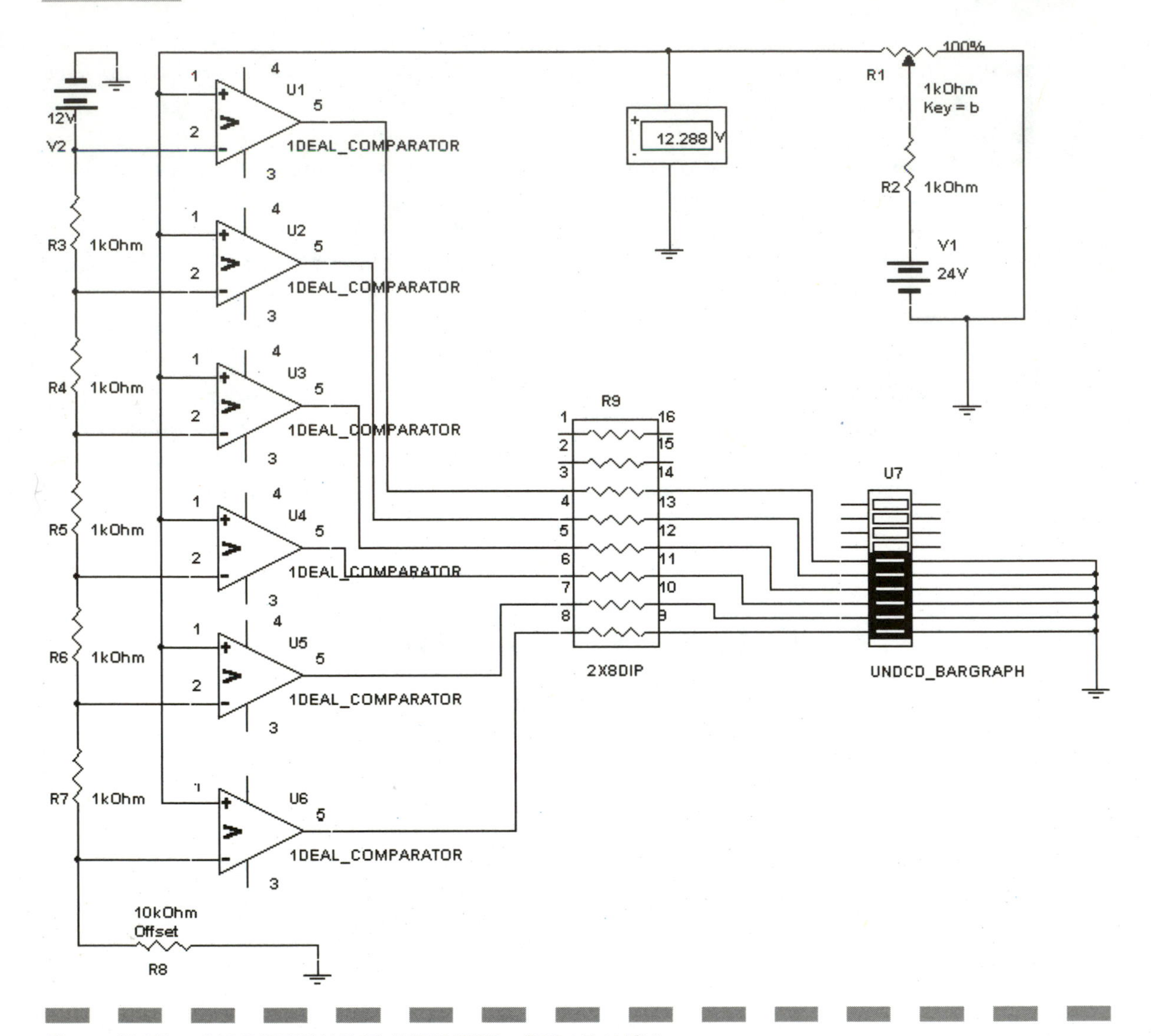

Figure 7.17 CAR BATTERY LEVEL INDICATOR—BAT_LEV.MSM.

Inverting Operational Amplifier (PGM_OPAMP.MSM)

The op amp is one of the most widely used ICs in electronics. The most basic configuration used is the *inverting amplifier*. The circuit in Figure 7.18 controls the gain by altering the resistance on the input. By hitting combinations of A, S, D, and F keys you can control the gain of the amplifier from −15 to 0. Open the scope and run the simulation. Change over to the main screen and begin hitting combinations of A, S, D, and F; note the results on the scope. Remember, each resistor added in parallel lowers the total resistance and thus raises the gain on the amplifier (Gain = −R5/Rtotal of R1, R2, R3, and R4).

- **Custom Settings**—Switches set to A, S, D, and F. Set the oscilloscope to a timebase of 1.0ms/Div and each channel to 500mV/Div.
- **Analysis**—None.

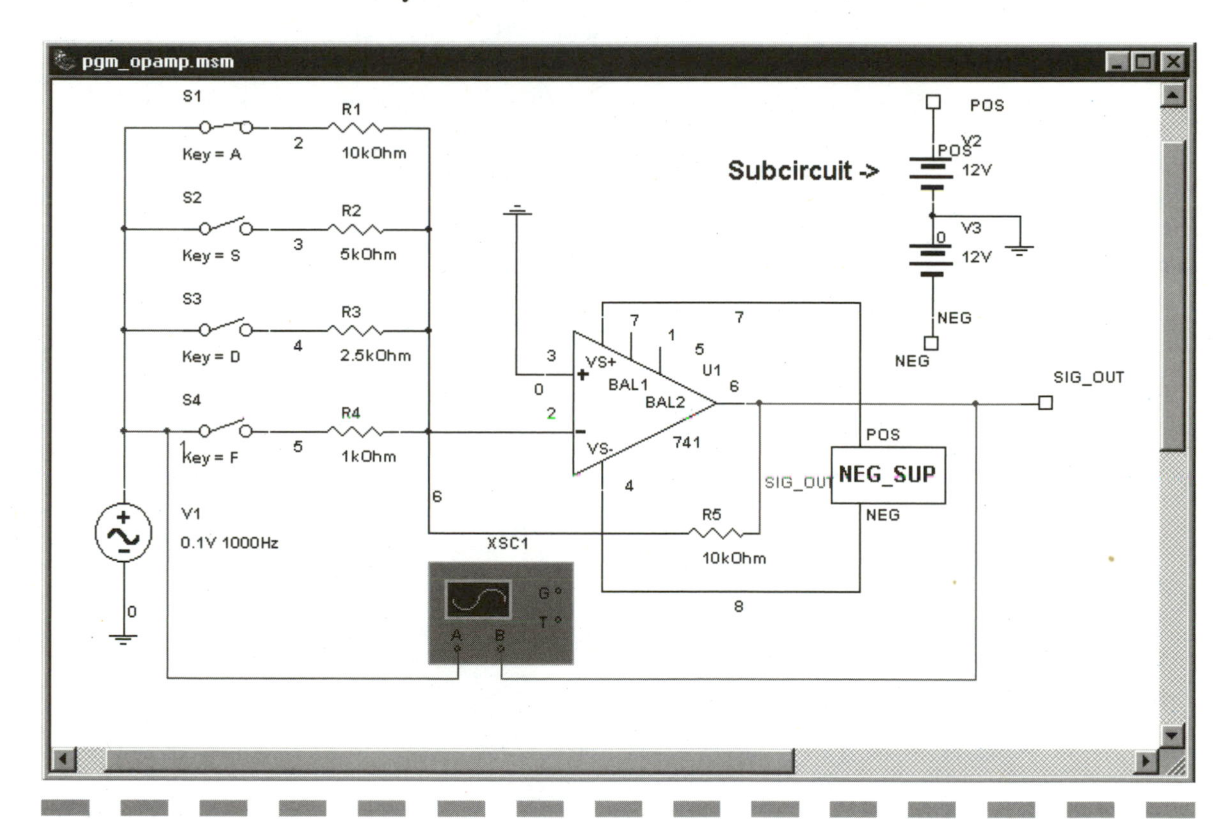

Figure 7.18 INVERTING OPERATIONAL AMPLIFIER—PGM_OPAMP.MSM.

The NEG_SUP subcircuit in the op amp example can be used to supply negative voltage to the appropriate terminals on many ICs. Use it with other circuits to shorten design time.

Analog-to-digital Converter (ADC_DAC.MSM)

Analog-to-digital converter and digital-to-analog converter ICs are finding their way into millions of modern applications. These two components are in heavy use because the world is generally analog and most processing is done in the digital domain, then sent out as an analog signal again. A great example of this is compact disc (CD) technology; it uses an analog-to-digital conversion (ADC) technology to change analog sound waves into digital words that are stored as digital bits onto a plastic disc. When the disc is placed into a player, it converts the digital bits back into analog waves with a digital-to-analog converter (DAC).

In our example circuit in Figure 7.19, we take an analog wave, send it through an ADC, and then back to a DAC. Hook up the circuit as seen and open the scope. Run the simulation and notice the stepped wave output. Note that I have flipped the DAC for clarity; the numbers are reversed. Modern DACs will smooth this wave out before it is used as an output signal.

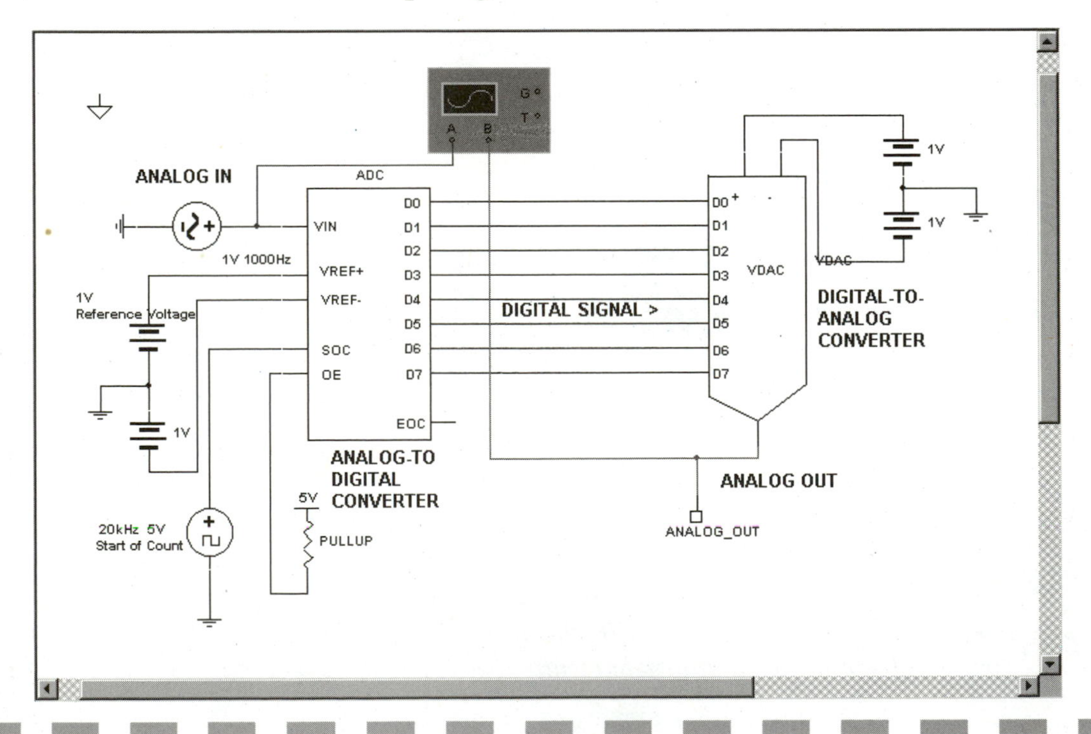

Figure 7.19 ANALOG-TO-DIGITAL CONVERTER— ADC_DAC.MSM.

This is an extremely slow simulation for Multisim. Have patience (or a fast computer!).

- **Custom Settings**—Once you have run the circuit as shown, change the start-of-count to 40 kHz and note the results. This setting changes how many voltage samples are taken per second.
- **Analysis**—None.

Simple Digital Circuit (ADDR_DEC.MSM)

Figure 7.20 shows a simple decoder that uses four NOT gates and four AND gates to turn a two-bit binary number into one of four decimal numbers. Please note that I am using four of the six NOT gates from a 7404 and an entire 7408 IC, both from the TTL toolbar. The truth table is pictured in Figure 7.20 as well. Build the circuit as shown; Activate and use the A and B key to display the four possible outputs.

- **Custom Settings**—Make the top switch A and the bottom switch B.
- **Analysis**—None.

Decimal-to-BCD Seven-segment Display (DECTOBCD.MSM)

The circuit in Figure 7.21 helps you learn the function of digital gates as well as the seven-segment display. It's a rather complex looking circuit but builds fast by using cut and paste for those sections such as the switches. Each pushbutton switch is hooked to ground through a 10K pull-down resistor and a Vcc component. Unfortunately, you can no longer assign numbers to mechanical components with Multisim, so I had to use letters. Q is 1, W is 2, etc. For example, when you hit the letter 'Q' (1), the signal wire is temporarily connected to the Vcc (High) then returns to ground (Low). This network of switches gives a signal to the logic gates (consisting of six NOR gates, two NAND gates, one NOR gate being used as a NOT gate, and a triple-input NAND

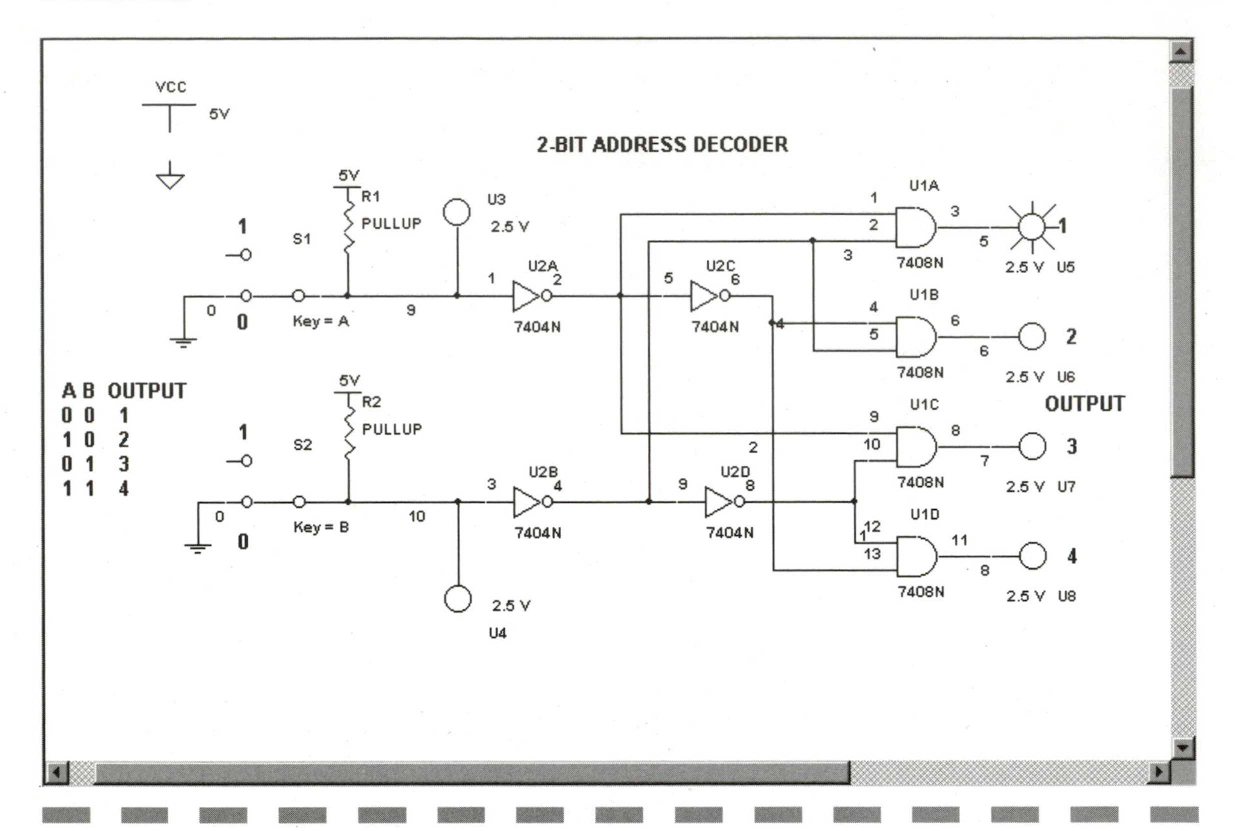

Figure 7.20 SIMPLE DIGITAL CIRCUIT— ADDR_DEC.MSM

gate [7410]). When the circuit is activated, the logic decodes the decimal and turns it into a four-bit binary number (BCD). The BCD signal is fed to the seven-segment decoder, which changes it into a visual digit between 0 and 9. If you hit the letter R (4) on the keyboard, 4 is changed into 0100 and fed to the display, which shows the number 4.

- **Custom Settings**—Build it as you see it. Set the switches to Q(1), W(2), etc.
- **Analysis**—None.

Flip-flop Circuit (FLIPFLOP.MSM)

I decided to use a game to demonstrate the use of the RS flip-flop as well as the D flip-flop (see Figure 7.22). D flip-flops are used as a count-

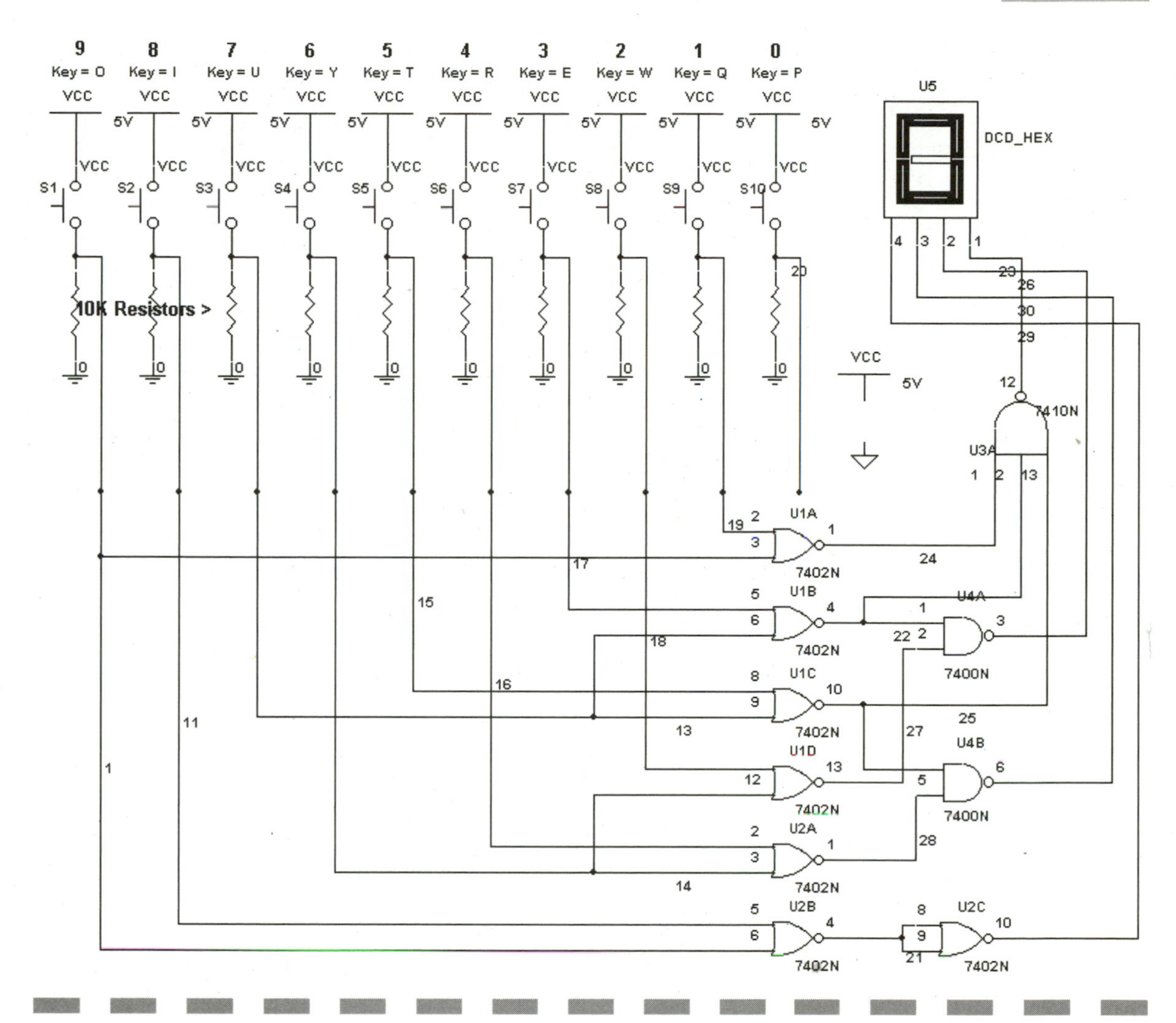

Figure 7.21 DECIMAL-TO-BCD-7 SEGMENT DISPLAY—DECTOBCD.MSM

er activated by a High signal on the D terminal and then clocked with a clock source. This in turn lights up four lights, one after the other. The inverted output of the "Go" D flip-flop and an AND gate are used to prevent players from hitting their keys too soon; it will only make a ground signal available to the RS flip-flop when the countdown is done. The RS flip-flop basically determines which signal was switched over to ground (Low) first. The results of the contest are sent to either the Q or inverted Q outputs (Q with line over it means "Not Q" or simply the opposite of Q).

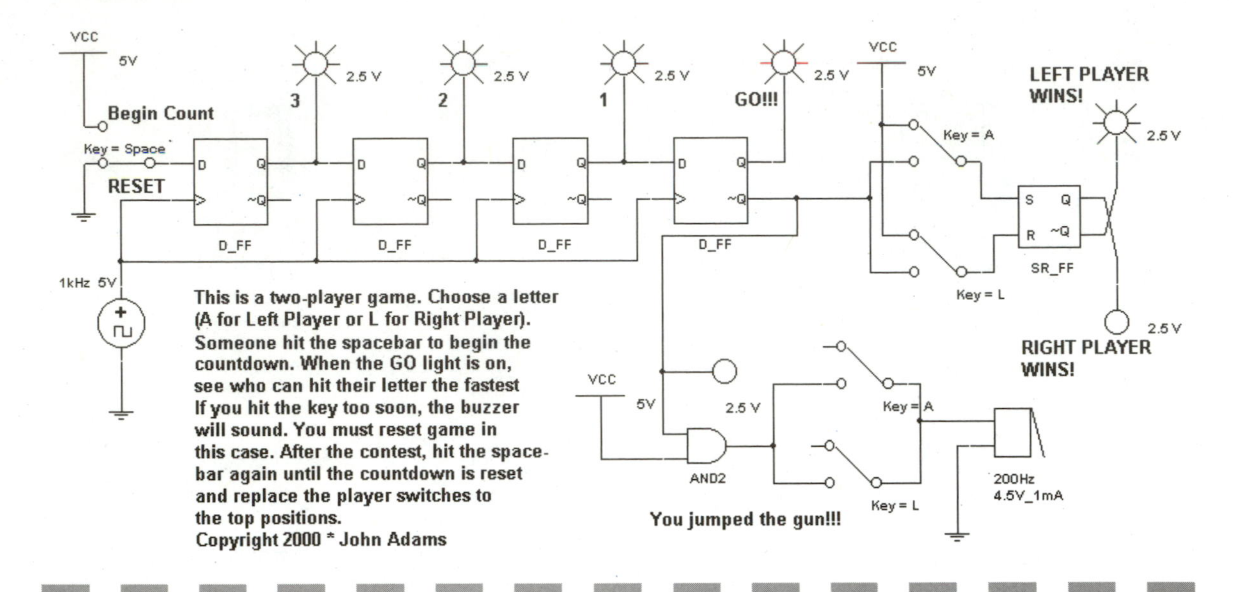

Figure 7.22 FLIP-FLOP CIRCUIT—FLIPFLOP.MSM.

"This is a two-player game. Choose a letter (A for Left player) and (L for Right-Player). Someone hit the spacebar to begin the countdown. When the GO light is on, see who can hit their letter the fastest. If you hit the key too soon, the high-pitch buzzer will sound. You must reset game in this case. After the contest, hit the spacebar again until the countdown is reset and replace the player switches to the top positions."

- **Custom Settings**—Set the **Begin Count/Reset** switch to the spacebar. Make the top switch (left player) the A key and the bottom switch (right player) the L key. Set buzzer to 4.5V, 1mA, and set the frequency as shown in Figure 7.22.
- **Analysis**—None. Have fun instead!

NOTE

The original circuit used buzzers to indicate the winner. However, Multisim currently has a bug when running multiple buzzers on the same drawing. This has not been addressed as of Version 6.2.

Parallel-to-serial-to-parallel Circuit (PAR-SER.MSM)

This rather complex circuit (Figure 7.23) demonstrates many digital components, including the use of ICs. Make sure to adjust the two clock sources as shown, paying attention to the duty cycle of the left one. If you are having trouble with the circuit, open it from the CD and try to figure it out; print it if necessary. When running the circuit, give it time, because this is a slow simulation (minutes per count on my slow 233 Mhz MMX PL). Also, be sure to use the 74LS04D, 74LS165D, 74LS164D, and 74LS273DW ICs and not the N varieties or you will get an error. You may also want to experiment with ITL4 in the Analysis Options to speed the simulation slightly.

This is a simplified parallel-to-serial back to parallel converter. The word generator parallel loads a binary number into the 74LS165 IC (parallel-load 8-bit shift register), which converts the signal into a serial output. The 74LS164 (8-bit shift register serial in/parallel out) IC then converts the serial signal back to a parallel output. The 74LS273 is an octal D-type flip-flop used to clock the output to two decoded seven-segment displays. *Note*: The purple lines are parallel signals and the red is the serial data line. Have patience. The simulation can be extremely slow.

- **Custom Settings**—The left clock source must be adjusted to 1 kHz and 87 percent duty cycle. The main clock is simply set to 8 kHz. The word generator should be loaded with the up counter pattern and set to an external negative-going trigger.
- **Analysis**—None.

Control Circuit (CONVERT_TEMP.MSM)

Moving from the United States to Canada, I was forced to learn the metric method of reading temperature. I still have difficulty (along with most Canadians) in converting the weather report's Celsius readings into the familiar Fahrenheit scale. The circuit in Figure 7.24 uses a few of the control components to do just that.

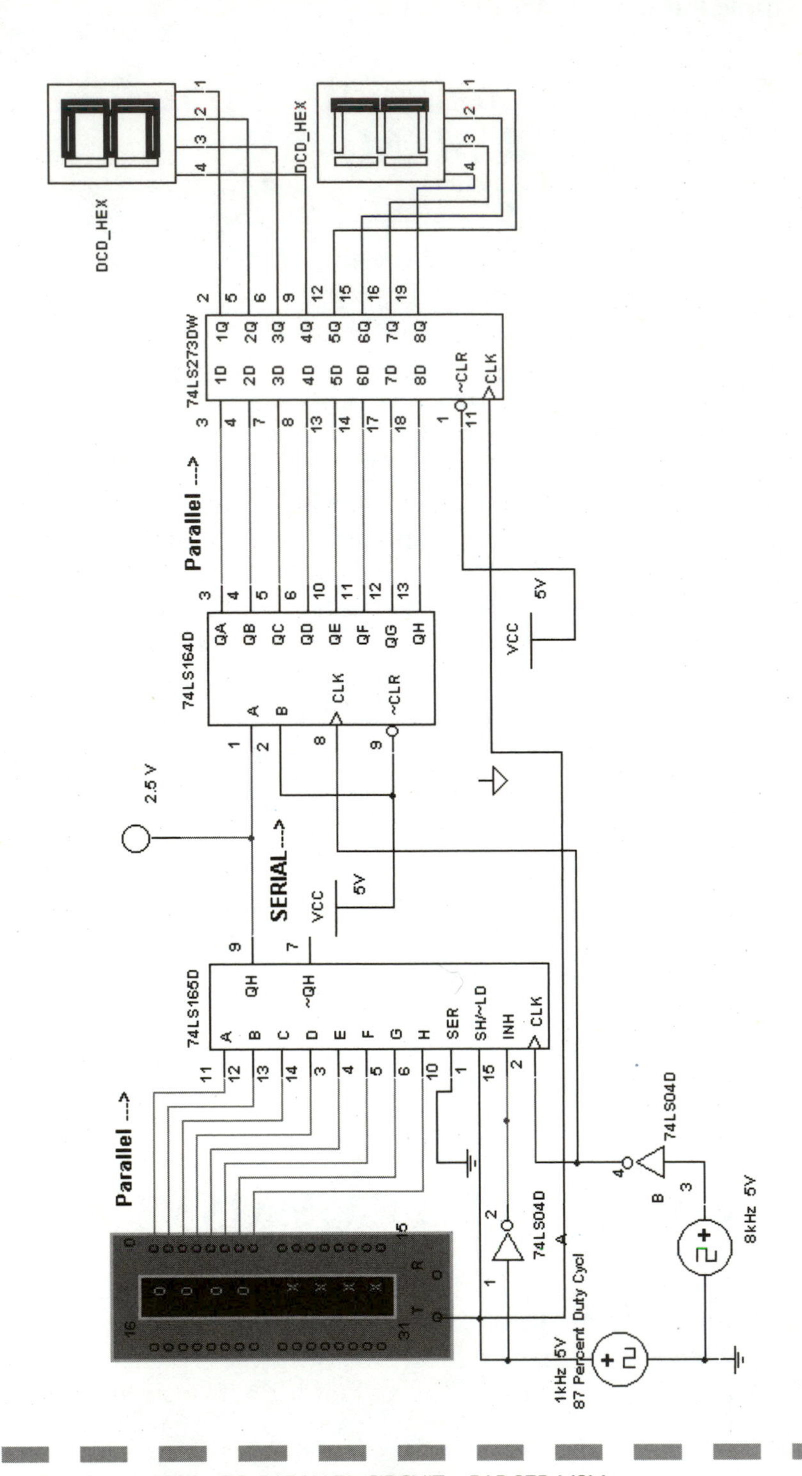

Figure 7.23 PARALLEL-TO-SERIAL-TO-PARALLEL CIRCUIT—PAR-SER.MSM.

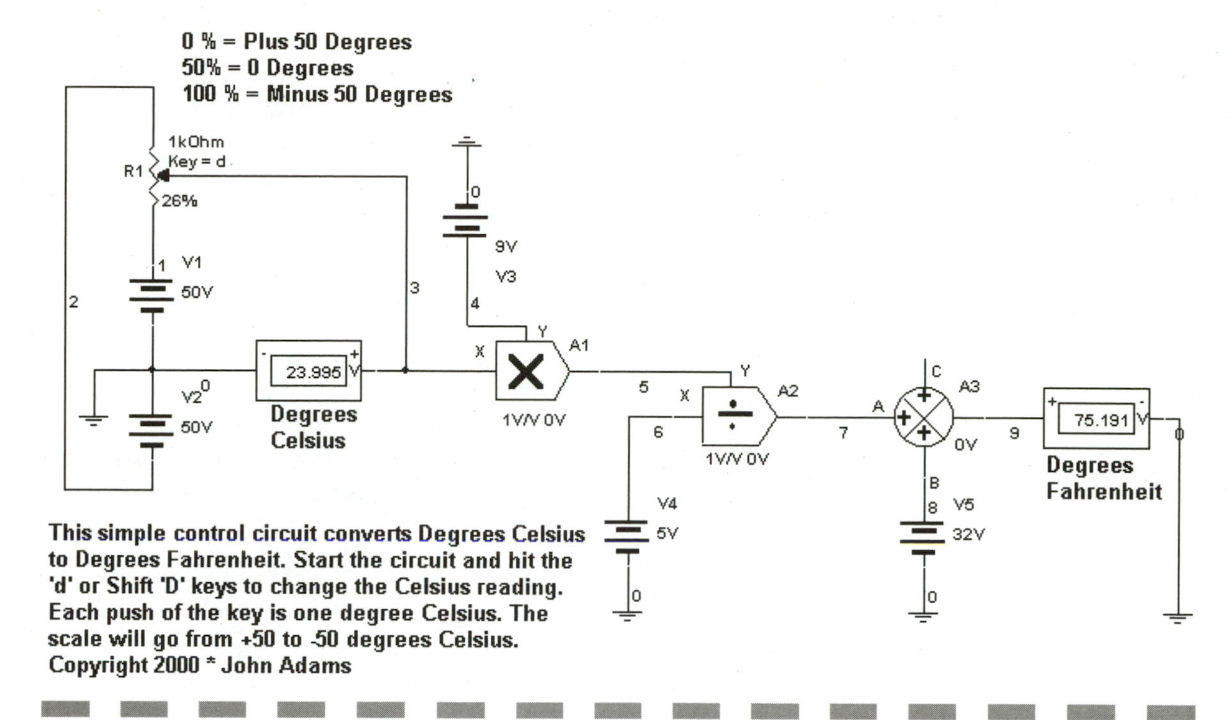

Figure 7.24 CONTROL CIRCUIT—CONVERT_TEMP.MSM

This simple control circuit converts degrees Celsius to degrees Fahrenheit. Start the circuit and hit the **d** or Shift **D** keys to change the Celsius reading. Each push of the key is one degree Celsius. The scale goes from +50 to −50 degrees Celsius.

- **Custom Settings**—Set the potentiometer key to D and make its increments 1 percent.
- **Analysis**—None.

Netlist Example (NETLIST.MSM)

Building a component from a SPICE model you downloaded from the Internet can be time consuming and complex. However, infinite expansion of the User library awaits the patient hobbyist; this next rather long example shows you how.

SPICE Netlists and models of real-world components can be downloaded from semiconductor websites. This lets you add parts not included with Multisim 6. SPICE Netlist files use the extension *.cir and can be downloaded into a temporary directory or one assigned to SPICE Netlists. Personally, I use **C:\MULTISIM\SPICE_MODELS**. In this example, I visited **http://www.national.com/models/spice/ComLinear/** and downloaded the **clc449.cir** file to this directory (1.1-GHz ultra-wideband op amp; note that the website lists this as a 1.2-GHz model). Once the SPICE Macromodel Netlist is in the directory, follow the following instructions. (Version 6.2 procedures might be slightly different but the data is the same).

1. Start Multisim and save the drawing as NETLIST.MSM.
2. Turn on the User toolbar with the database selector.
3. Begin to place a 'DIV' op amp. When the Component Browser opens, hit the **Edit** button and choose to place it in the User database. You can also use the Component Editor to access this same dialog box.
4. The Component Properties dialog box opens to the General tab. Change the name of the part to 'CLC449_OPAMP'. The manufacturer is National Semiconductor and the date/author is unimportant.
5. Change to the Model tab (see Figure 7.25). Name the model CLC449. Version is unimportant.

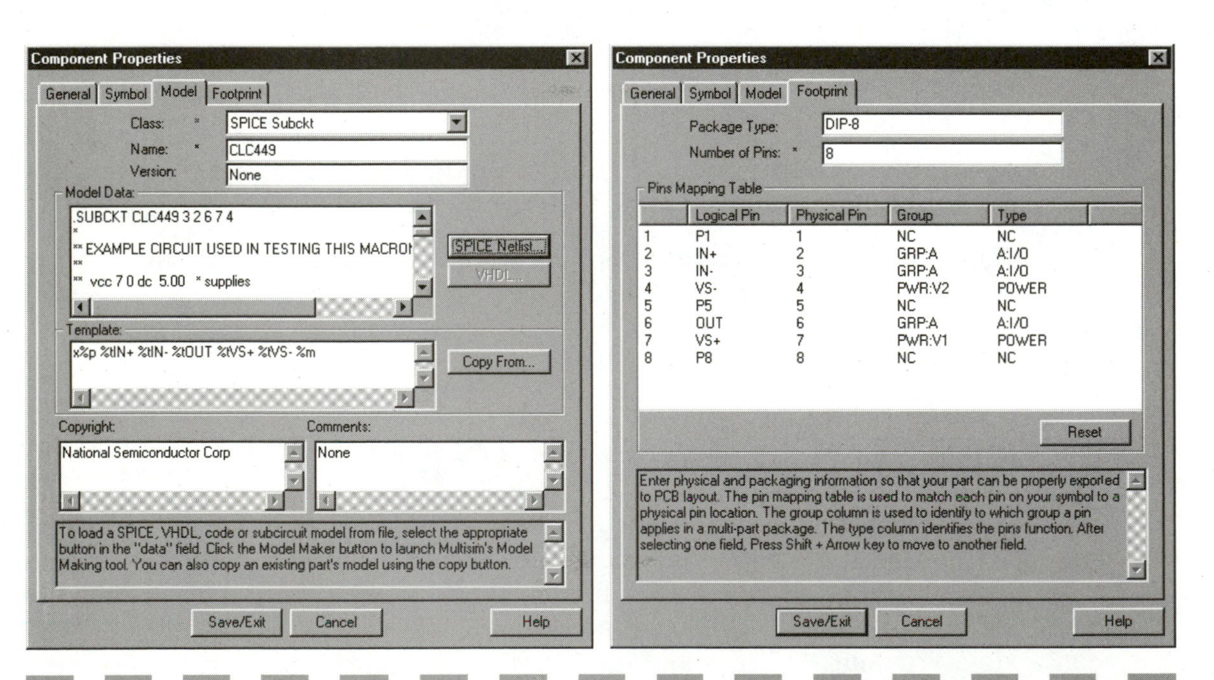

Figure 7.25 Changing a DIV OPAMP into a National Semiconductor CLC449. Make the dialog boxes match the diagram.

6. Hit the SPICE Netlist button and navigate to the CLC449.CIR file. Open it.
7. The Macromodel loads into the model data window. You need to delete each comment from the top section until the first line reads **.SUBCKT CLC449 3 2 6 7 4**.
8. Edit the template to read **x%p %tIN+ %tIN- %tOUT %tVS+ %tVS- %m**.
9. Change to the Footprint tab (see Figure 7.25). Change the package type to **DIP-8**, number of pins to **8**.
10. Match the Pins Mapping Table to that seen in Figure 7.25.
11. Hit the **Save/Exit** button.
12. The Component Browser opens once more. Hit **cancel**.
13. Go to **View** > **Refresh component toolbars.** The Analog family icon should now appear on the User toolbar.
14. Place your new component from the User toolbar onto the drawing and build the circuit in Figure 7.26.
15. Adjust the instruments as noted below in custom settings and run the circuit.
16. Run the AC analysis.

*Opening the *.cir file into a blank circuit page can give an exact idea of a Netlist subcircuit. Beware! Some of these can be enormous and take minutes to convert.*

- **Custom Settings**—Set the function generator to 1kHz, 1V. Adjust the oscilloscope's timebase to 500us/Div and channel A and B to 2V/Div. Display the Out node.
- **Analysis**—Go to **Simulate** > **Analyses** > **AC Analysis**. Set FSTART to 1Hz, FSTOP to 10GHz, decade sweep, 10 points, log scale, and select the Out node as the node for analysis. Hit the **Simulate** button. Toggle the cursors and see if the amplifier's bandwidth does indeed drop at around 1.1 GHz.

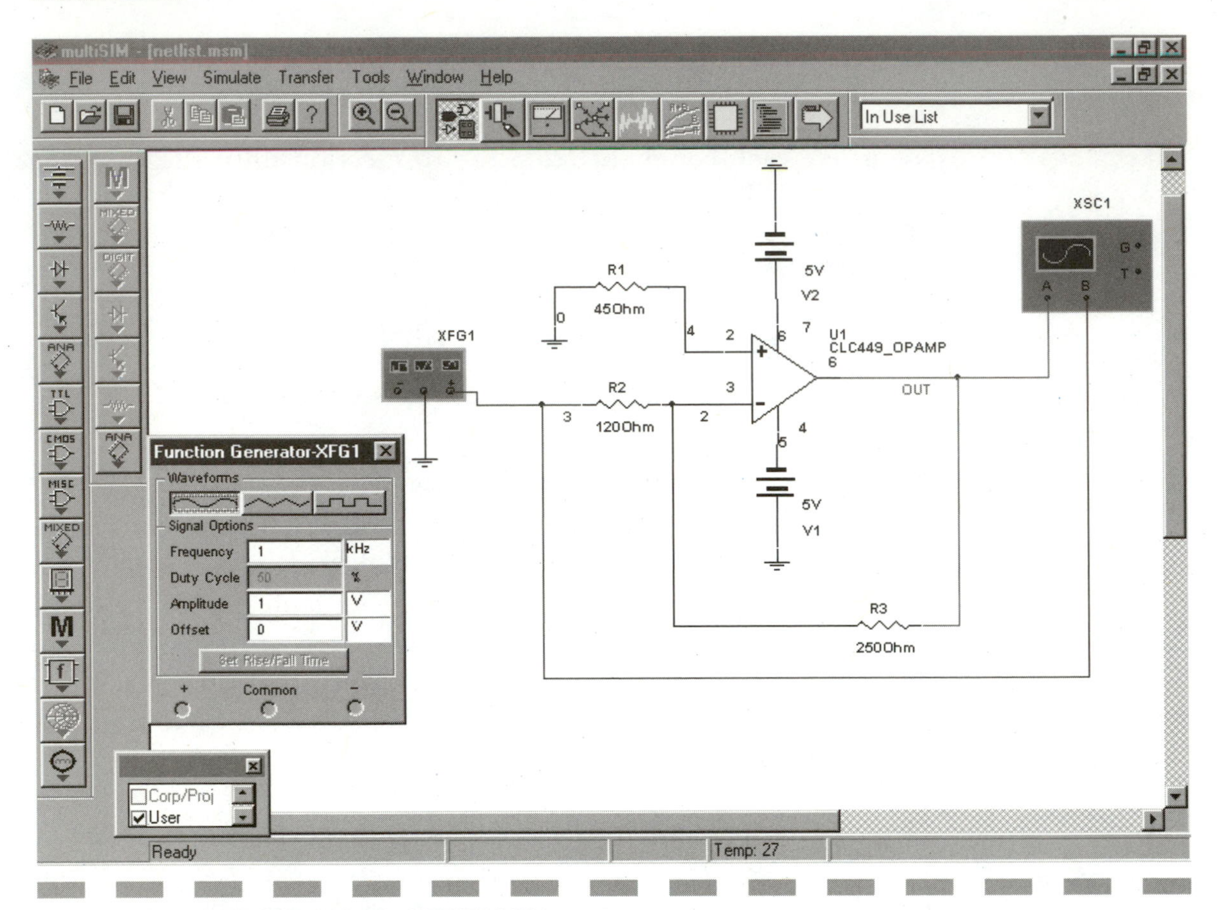

Figure 7.26 NETLIST EXAMPLE—NETLIST.MSM.

NOTE

You can typically download a datasheet for a SPICE model component and get an idea of its pinouts and applications.

PWM Voltage Controller (CHEAP_PWM.MSM)

By varying the duty cycle of a square wave, you can control the voltage level of many DC-based devices such as LEDs, transistors, DC motors, etc. In this quick-and-dirty pulse-width modulation circuit, I used a 555 timer to control the On to Off time of a 1 kHz square wave. The resulting signal is put through a capacitor to even out the voltage level. If you were using this to control the brightness of an LED, the capacitor could be eliminated because the human eye cannot 'see' the LED going on and off at even lower frequencies. Build the circuit in Figure 7.27 and adjust the custom settings. Open the oscilloscope and run the circuit. Hit the **w** or **Shift+W** keys to control the duty cycle of the output and thus the voltage level.

- **Custom Settings**—Assign the W key to the potentiometer. On the oscilloscope, set the timebase to 0.50ms/Div, and channel A and B to 5V/Div. Be sure to use the LM555H timer or the simulation will not be stable.
- **Analysis**—None.

NOTE

Cheap_PWM.MSM was built in Multisim 6.2. If you are using 6.11 or earlier you must use a corrected version of the timer, as described under the astable multivibrator circuit earlier in this chapter.

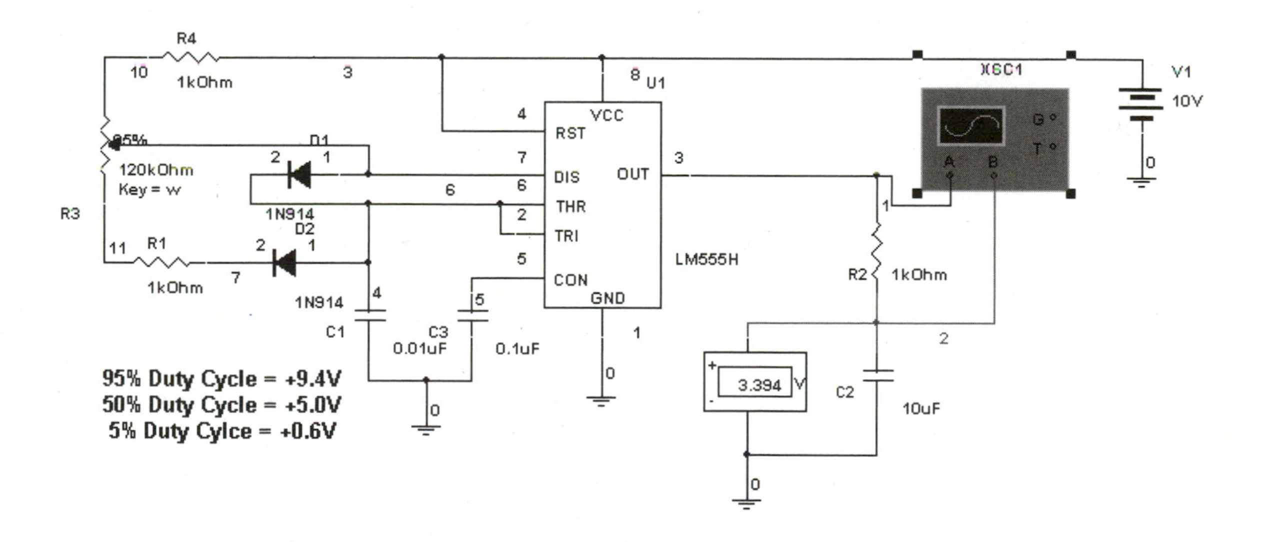

Figure 7.27 PULSE WIDTH MODULATION VOLTAGE CONTROLLER—CHEAP_PWM.MSM.

DC Motor (DCMOTOR.MSM)

The simple circuit in Figure 7.28 demonstrates the use of the DC motor component. You can investigate its speed characteristics by adjusting the current in the stator and armature. This circuit is from EWB's FTP site. You may want to experiment with a feedback system that adjusts each field to receive the optimum speed of the rotor.

In the DC shunt motor, motor RPM (from a tachometer output) is proportional to voltage (1 RPM per volt). Speed characteristics may be investigated by varying the field or armature supply voltages. Speed may be varied by inserting a variable resistance in series with the field. This method only allows an increase of RPM as field current decreases. Speed may also be varied by inserting a variable resistance in series with the armature. This method only allows for a decrease of RPM as resistance is increased. In the example circuit the series resistance in both field and armature is minimal when the percentage indicates 0 percent. Field resistance is increased by pressing **Shift+F** and for the armature circuit by pressing **Shift+A**.

- **Custom Settings**—Assign the Field potentiometer to F and the Armature to A.
- **Analysis**—None.

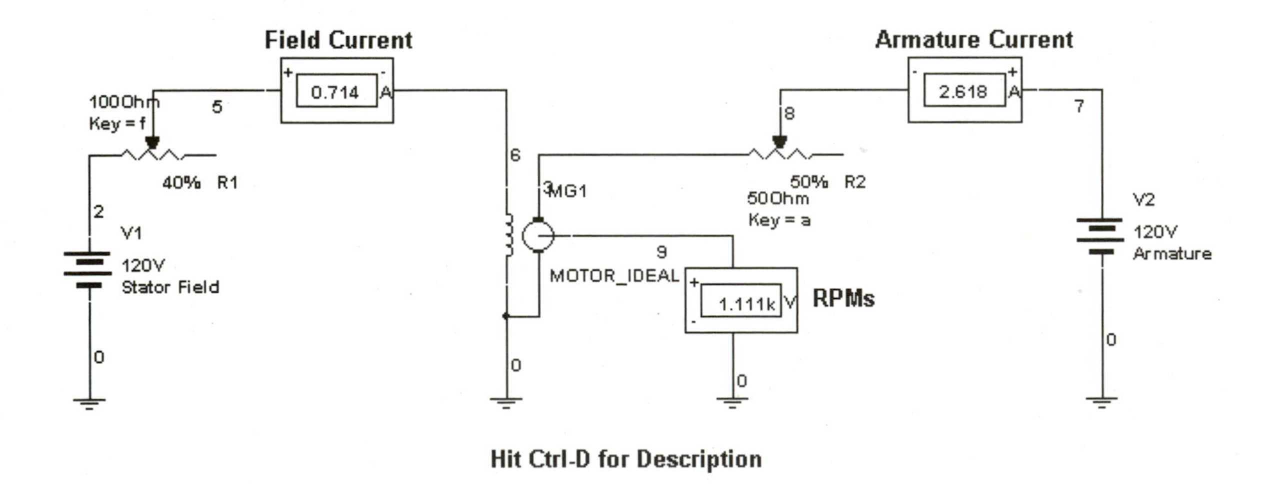

Figure 7.28 DC MOTOR—DCMOTOR.MSM.

This circuit originally appeared on the Interactive Image Technologies, Ltd. FTP site.

Phototransistor Test (PHOTOTRANS.MSM)

This circuit helps reinforce two concepts: emulation and the use of Multisim's schematic capture (symbol editor) to your advantage. With the circuit in Figure 7.29 you will realize that even though Multisim has its limitations, you can still bypass these to emulate what you are trying to simulate. In other words, if it's not built into Multisim, build it in! In this case we are using an ideal BJT as a pseudo-phototransistor. We are creating a phototransistor symbol (schematic capture) with the Symbol Editor.

This demonstration circuit emulates a phototransistor that is being used to count objects passed between the IR LED and the phototransistor. What's the secret to this circuit? The "Object" is actually a subcircuit with a 90 percent duty cycle clock source and a 4.7M resistor. This is connected to the transistor. The *base pin* for the transistor has to be placed a few grid items above the rest of the symbol and changed to an input wedge using the Symbol Editor. There is still a connection to the actual IDEAL_BJT. The circuit is not a real-world circuit but can easily be converted to such. RA would be replaced by a potentiometer to be able to adjust the sensitivity of the phototransistor. The NOT gate may need to be replaced with a transistor or Schmitt-triggered device as well.

Here is a condensed version of how to build this circuit. For further information on the Symbol Editor and editing components, see Chapter 4:

1. Open Multisim and turn on the User database.
2. Begin to place an IDEAL_BJT_NPN and choose to edit it from the Component Browser. Direct it to the User database.
3. The Component Properties dialog box should now be open to the General tab. Name the component **Phototransistor**.
4. Flip to the Symbol tab and hit the **Edit** button. The Symbol Editor should now be open.

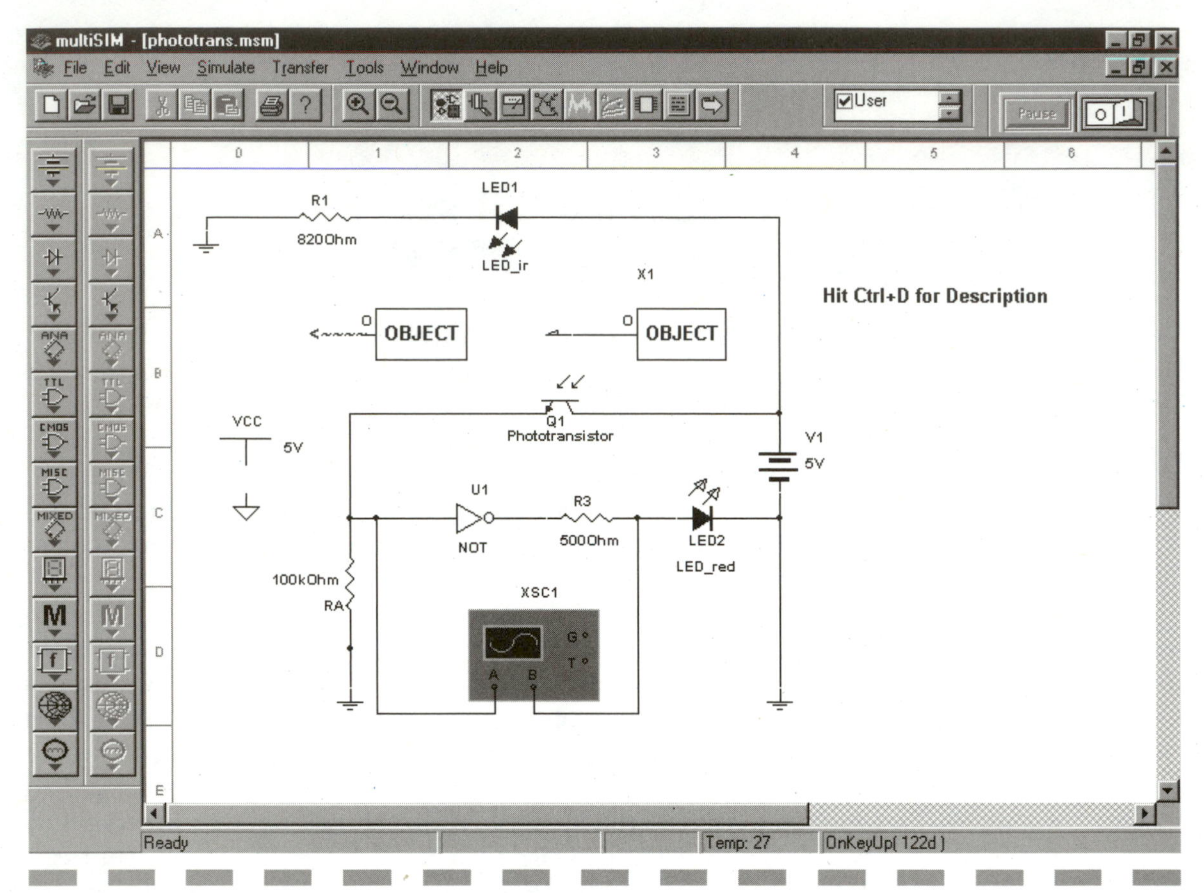

Figure 7.29 PHOTOTRANSISTOR TEST—PHOTOTRANS.MSM.

5. Move the base pin five grid lines left, then double-click it. Hit **Edit**. Single-click the word **Line** and use the list box to select **Input Wedge** and rotate it clockwise. Hit **OK**, **OK**.
6. Go to **View** and take check off the **Snap to Grid**.
7. Go back to the drawing and use the draw line tool to create one of the LED light arrows. Use cut and paste for the second.
8. Exit the Symbol Editor, answering **Yes** when asked to save changes.
9. Select the **Save/Exit** button on the Component Properties dialog box.
10. Hit the **Cancel** button on the Browser.
11. Go to **View** > **Refresh component toolbars**.

12. Place the new phototransistor from the User database under the transistor icon, BJT_NPN sub-icon. Rotate the component clockwise. The pin for the base is the arrow; use it to connect the Object subcircuit.

Figure 7.30 shows the Object subcircuit as well as the actual phototransistor with its base disconnected.

- **Custom Settings**—As described above. Set the Object subcircuit's clock source to 5V, 1 kHz, 90% duty cycle. Be sure to use a **TIL NOT gate** and not a TTL or CMOS variety.
- **Analysis**—None.

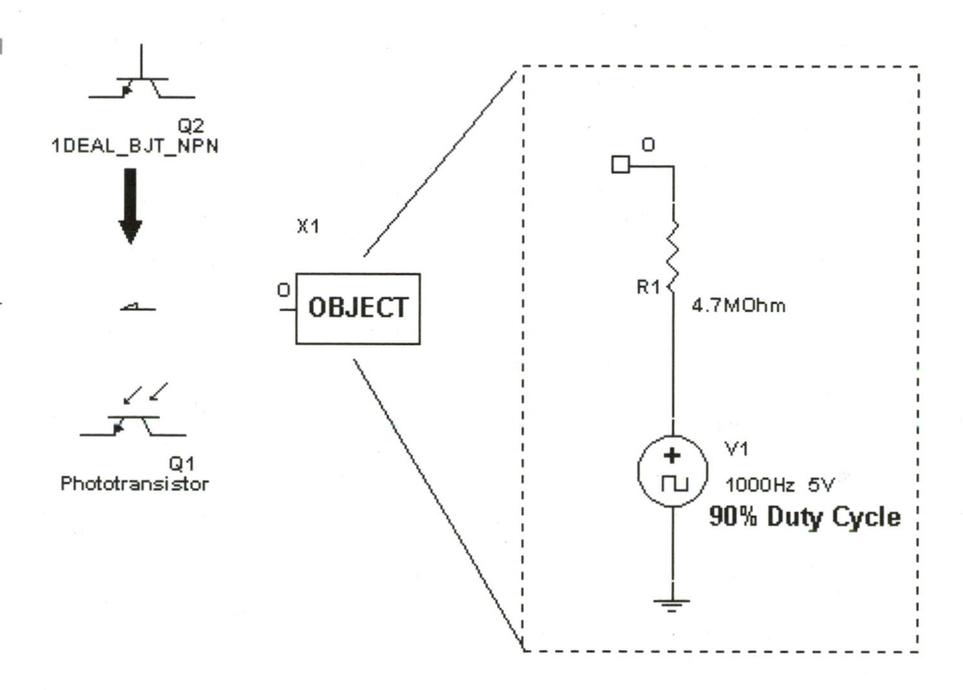

Figure 7.30 Editing a BJT_NPN to create the Phototransistor. Note that the small arrow above the Phototransistor is the actual BASE PIN.

Piecewise Source (PIECEWISE1.EWB, PIECEWISE2.MSM)

In my opinion, the piecewise linear source component and the SCOPE data are the most valuable features included with Multisim 6. They let you record a waveform (digital or analog) onto your hard drive (as a text file, *.scp in this case), then play it back through a different circuit. The simple circuits in Figures 7.31 and 7.32 demonstrate this ability. In piecewise1.msm (Figure 7.31), three waveforms are mixed to create a simulated voice wave and recorded to piecewise1.scp (Figure 7.33). Build this file and activate it for a minute or so. Use the Save button on the oscilloscope and save the file as piecewise1.scp into any directory. In piecewise2.msm (Figure 7.32) we are playing back this file through two inverting amplifiers. Build the circuit and assign the piecewise1.scp file to the Piecewise Linear Source. Open the oscilloscope and activate the circuit.

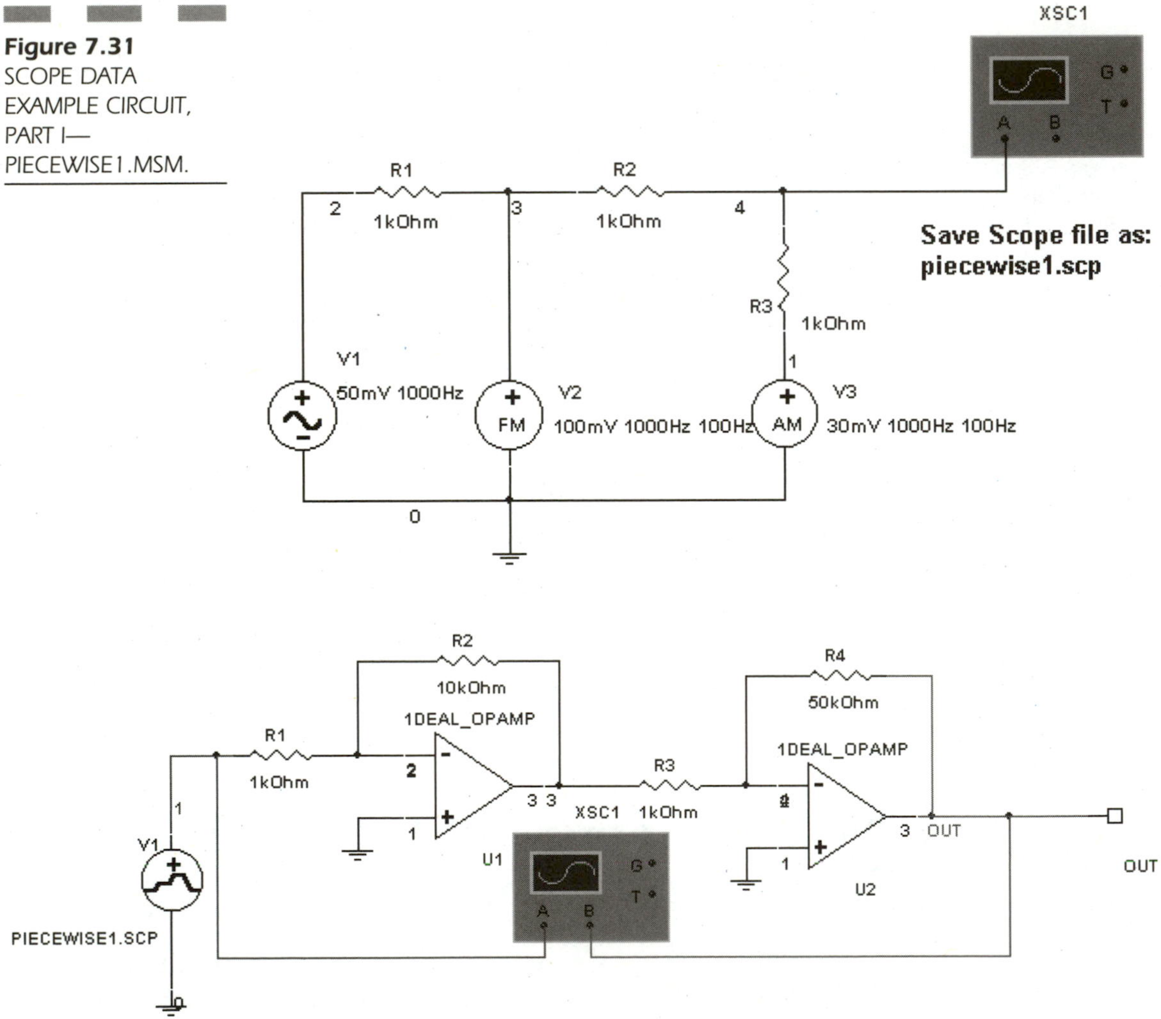

Figure 7.31 SCOPE DATA EXAMPLE CIRCUIT, PART I—PIECEWISE1.MSM.

Figure 7.32 USING THE PIECEWISE FILE WITH THE PWL SOURCE COMPONENT, PART II—PIECEWISE2.MSM

- **Custom Settings**—Save the piecewise file to **C:\MULTISIM\PIECEWISE1.SCP** or into any other directory. Assign the file to the PWL component in the second file.
- **Analysis**—None.

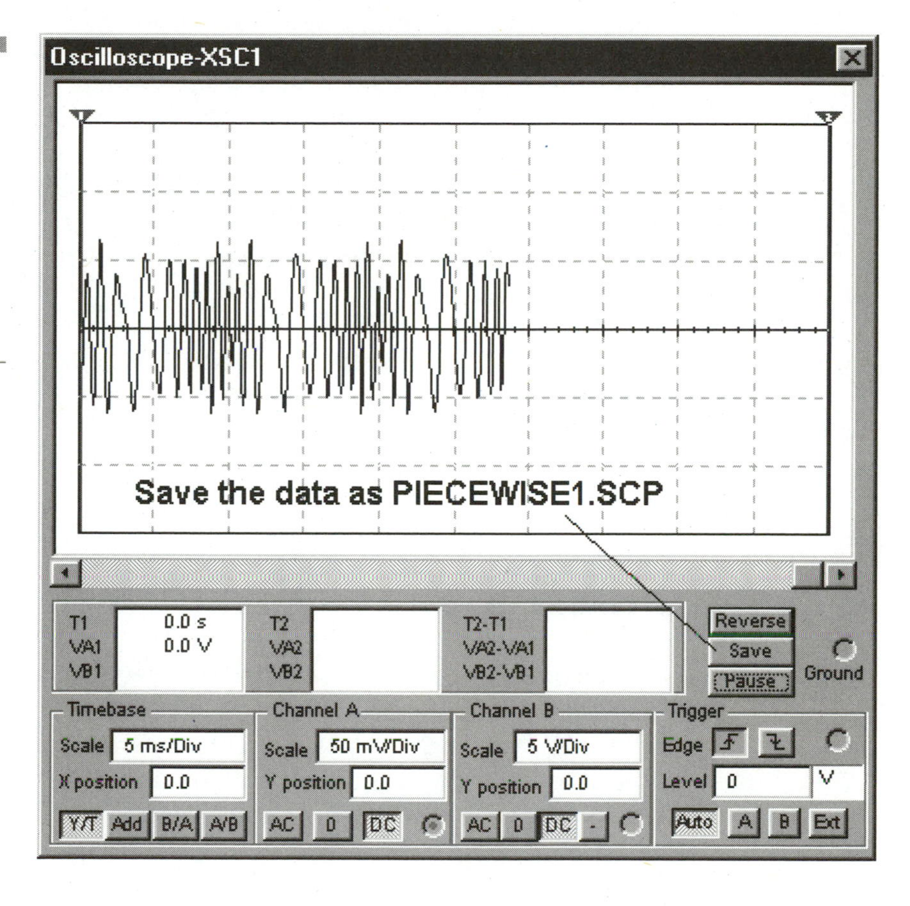

Figure 7.33
The results of the PIECEWISE1.MSM circuit. Save the SCP data to use with piecewise2.msm. Remember, a bug in Multisim 6.2 or less requires you then to edit the file. See text.

In Multisim 6.2 and lower, there is a bug that requires you first to delete each line in the SCP file before "0.000000000000e+000 0.0000e+000". Do this in Windows Wordpad and save the file in text format. Reopen the file by double-clicking the PWL source and loading it. The data appears in the window when this is done correctly. I've noticed inconsistent results with the PWL source but decided to include this simulation in hopes EWB will correct the inconsistencies in later versions. If you are getting errors with this simulation, restart Multisim. You will also notice that it takes some time to load the SCP file. Multisim is not locked; it is just thinking.

*The Write Data component was eliminated from Multisim 6 (and will be missed). The oscilloscope's Save button (SCP data) replaces it. You can also use the Grapher and save the file as a *.txt grapher text file if it is the only analysis open.*

I am developing advanced circuits and special utilities that will let you take these Multisim waveform text files and convert them into sound files (to play back through your sound card). A more advanced version will also be developed that lets you connect a digital voltmeter to a real-world circuit and read data to a Multisim piecewise file. This can then be used in your Multisim simulations and output to a black-box device connected to your computer. Watch my website (**basic-electronics.com**) for further details; this is going to be one awesome hobbyist application.

Break a large/slow circuit, such as that in PAR-SER.MSM, into smaller circuits and use the piecewise system to transfer data between them.

Making Your Own Circuits

This chapter was merely an introduction to teach you the various Multisim 6 components and their functions. The software is actually capable of extremely detailed simulations that can help your hobby or education immensely. (Just think of the ways we overcame the limitations of Multisim in the phototransistor circuit—which, by the way, I developed to use with an HO slot car lap counter.) Find projects that interest you and simulate them. Once they are developed, you have the option of transferring them to Ultiboard and creating PC boards for your own use. There's nothing like holding that PCB in your hand knowing it was created from scratch.

CHAPTER 8

Virtual Workbench: The EWB Interface

The Electronics Workbench can just about eliminate a real workbench.

Electronics Workbench uses a Windows-based menu and icon-driven graphical user interface. In other words, it's point-and-click software. Menu items are easily accessible with either a slide and click of the mouse or with swift keystroke combinations. Toolbars speed the task of executing complex commands. A workbench or project-space is available to let your electronics imagination run wild. By learning menus and toolbars at the outset, you are ensuring a deeper knowledge and understanding of EWB and its capabilities. Once you know the functions of each item on the desktop, we will take a look at how they operate and interact with each other.

Overview Interface

Electronics Workbench products add new dimensions to electronics. They take the user from design stage, to development, all the way to PCB layout artwork without leaving your workstation. You are computerizing your workbench; instead of using parts and a table, you interact with the bits and bytes of Electronics Workbench through a graphical interface on the computer screen. The interface is presented once the software is opened, along with its menus, toolbars, toolbins, and drawing sheet. These items allow us to tell the program what to do.

The interface is how we interact with the computer. In the past, the only way to input information was with a keyboard, typing out long strings of complex commands. The only way to see what was being typed was with a monochromatic monitor that displayed letters, numbers, and symbols. In today's world of visualization we use a graphical user interface (GUI) and a mouse as an input device. This method eliminates the need to memorize puzzling text commands; you merely point and click the mouse or hit a few keys to perform tasks. The Electronics Workbench interface is fashioned with easy-to-use, intuitive tools that decrease the new software learning curve. If you have ever had to learn a new program from scratch, you will appreciate the time the Electronics Workbench Company has spent polishing and making learning Electronics Workbench as painless as possible. Graphic representations of components and instruments are used instead of tons of text; parts placement is straightforward and accomplished with swift mouse work; each instrument and component is easily identifiable and closely approximates its real-world counterpart.

The rest of this chapter explores the Electronics Workbench Version 5 interface in detail to give you a complete understanding of the features, workings, and procedures. If you are a Multisim Version 6 user, go back to Chapter 2 where you will find information on that interface.

The Electronics Workbench Screen

Take an in-depth look at Figure 8.1. Open your EWB Version 5 program by going to **Start** > **Programs** > **Electronics Workbench 5.12** > **Electronics Workbench**. The Electronics Workbench company did their homework to bring you the simplest method of using this program with a layout that is very logical and intuitive. Here is a rundown of the areas you will be using in day-to-day designing:

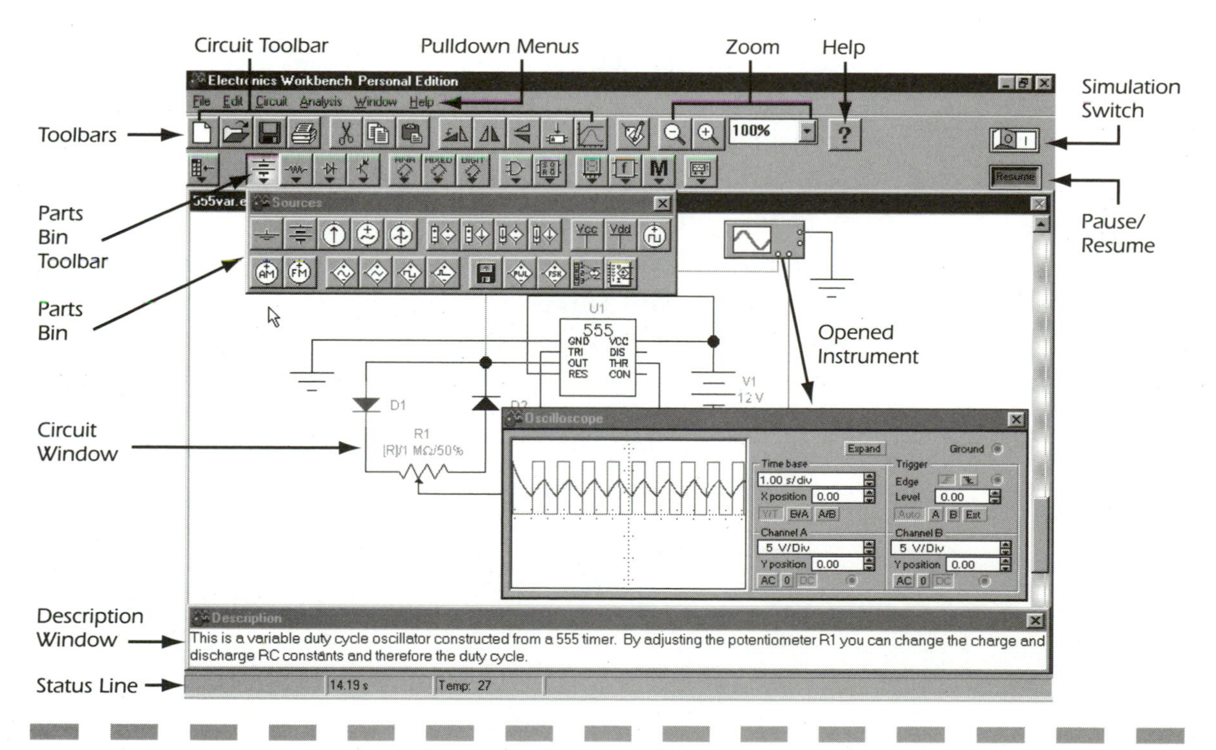

Figure 8.1 The Electronics Workbench V5 screen.

- **Pulldown Menus**—At the top are the familiar Windows menu Items. EWB's menu items include: **File**, **Edit**, **Circuit**, **Analysis**, **Window**, and **Help**. Each is broken down into logical commands.
- **Toolbars**—A toolbar is a collection of shortcuts that, when clicked, execute a command. Think of them as one-click wonders! Electronics Workbench Version 5 contains several toolbars that are grouped into common commands.
- **Circuit Toolbar**—Displays various icons that can be used as command shortcuts. If you are working away with the mouse, it is sometimes more convenient to hit an icon with your mouse than go through a pull-down menu. For example, you may hit the icon that looks like a floppy disk to save your file, rather than go to the **File** menu and choose **Save**. It all depends on your tastes. Circuit icons take up almost the entire horizontal surface and include New, Open, Save, Print, Cut, Copy, Paste, Rotate, Flip Horizontal, Flip Vertical, Create Subcircuit, Display Graphs, and Component Properties.
- **Zoom Toolbar**—Controls the scale of your circuit. Zoom-in is represented by a '+' and zoom-out is a '–'. There is also a pulldown list that lets you quickly adjust the zoom view.
- **Help Button**—This icon quickly opens a help file on the selected component.
- **Circuit Activate Switch**—Activates or turns off the circuit you have created in the Circuit Window.
- **Pause/Resume**—Temporarily halts or restarts the simulation.
- **Parts Bin Toolbar**—This is the heart of Electronics Workbench; it contains icons that represent a parts bin. Think of it as a cabinets full of categorized electronics parts that allow you to grab a mix-and-match components or instruments to perform simulations.
- **Parts Bins**—Contains the individual component icons. They open when you click over the Parts Bin Toolbar icons. Think of them as the little drawers within the part bin cabinets.
- **Circuit Window**—Here you lay components, run wires, and place miscellaneous text. It is essentially a drawing board or breadboard. Make sure you understand this part of the interface well because it is used throughout the book in describing procedures. Circuit Window = where you draw. Simple!

NOTE

This may also be called the "drawing," "workspace," "project space," or "circuit diagram" throughout this book.

- **Description Window**—An area where you can type in miscellaneous notes regarding the circuit; it is a separate window that can be opened/closed.
- **Status Line**—Displays READY Status, Tran (Time), Temp, and Component/Key information.

NOTE

The time on the Status Line is simulated time, not elapsed time. For example, if the status line indicates 100 microseconds, 10 seconds may have gone by in real-life time. At 200 microseconds, 20 seconds may have elapsed according to your watch. Also note that each circuit is different. That is to say, another simulation might display 1 second and only 1 second has gone by in real life.

Terms

It is best to learn a few desktop terms so as not to misunderstand them in the explanations that follow:

- **Desktop/Workbench**—This refers to everything that is on the screen collectively: menu items, toolbars, circuit window, instruments, etc.
- **Window**—Any separate box that pops up with additional features or programs.
- **Circuit Window/Drawing/Workspace**—The area in which you draw circuits.
- **Menu**—A list of items or commands.
- **Command**—A function that the computer performs after your input. In other words, it's what you tell the computer to do. For example, the Save command stores your current circuit to the hard drive's memory (see Figure 8.2).
- **Hotkey or Hot Key**—A fast way to access a command or feature with the keyboard.

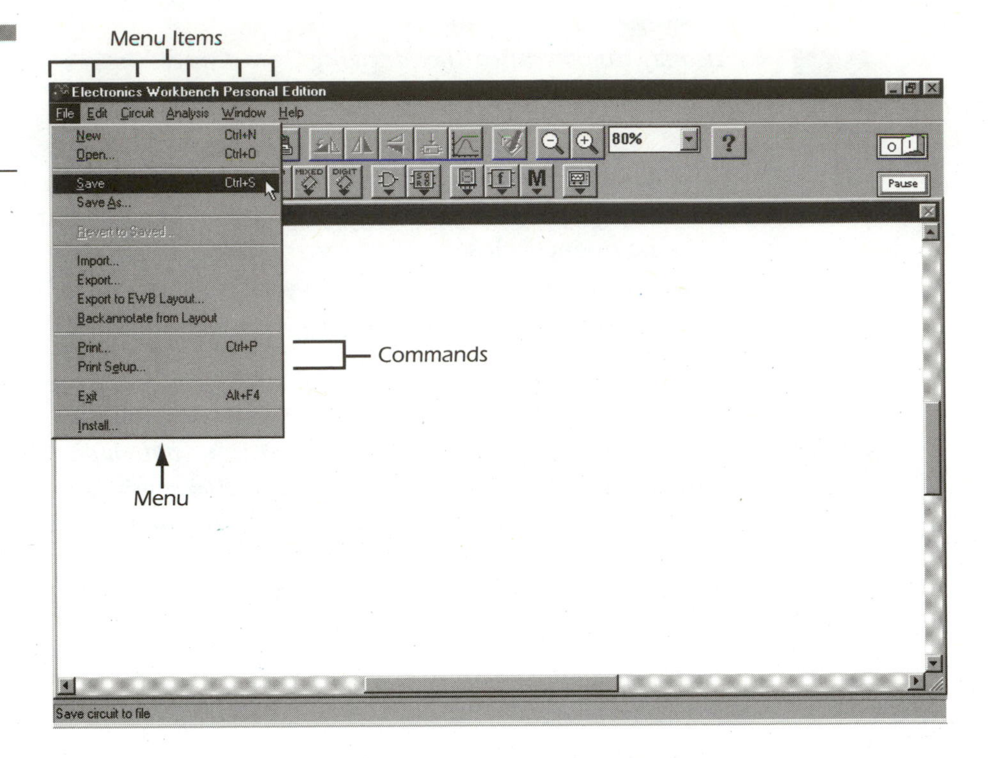

Figure 8.2 A menu contains a list of selectable commands.

- **Right-Click** (shortcut or content specific) **Menus**—Windows 9x./NT has the ability to open various menus by placing the pointer over an area and hitting the right mouse button.
- **Dialog Box**—Very important term. This is a Window that lets you add further items to a command. For example, see Figure 8.3.
- **Tab**—In a dialog box, you often see tabs at the top of the window that look like file folder tabs. By clicking each, you can access other choices/items. Also see Figure 8.3.
- **Icon**—A small picture that, when clicked, performs a command. This is a fast way of accessing common commands.
- **Toolbar**—A group of icons (see Figure 8.4).
- **Toolbin**—A toolbar that contains components, otherwise known as a *component toolbar.* It helps to use this term to understand how Electronics Workbench mimics a real workbench.

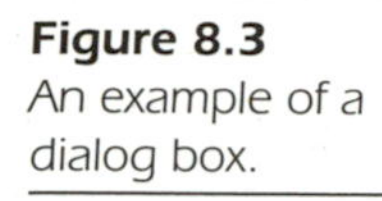

Figure 8.3
An example of a dialog box.

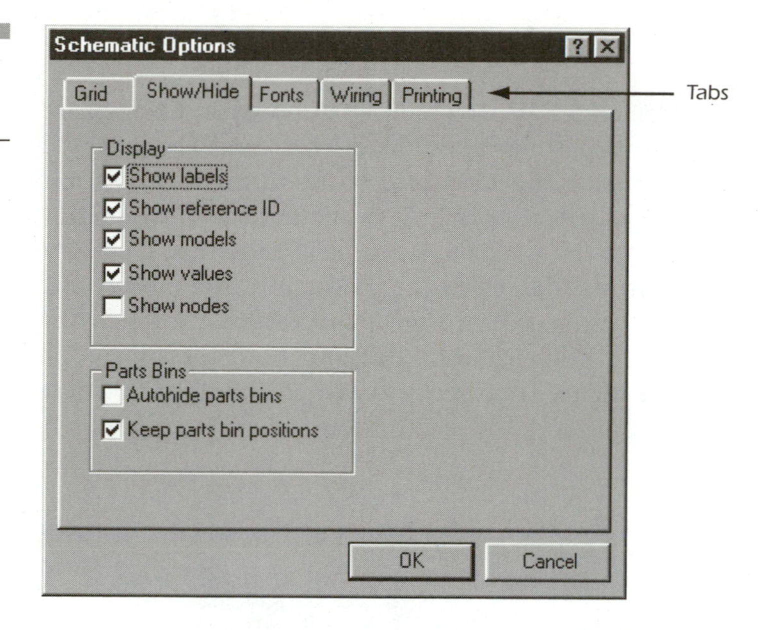

Figure 8.4
Sample of a toolbar. Notice the various icons, which, when clicked, execute a command.

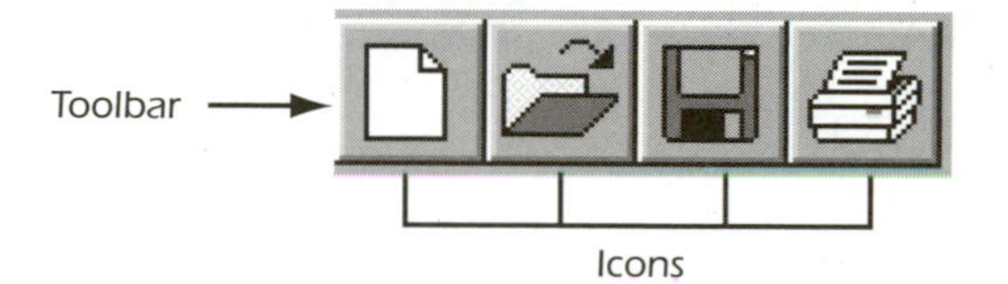

Electronics Workbench 5 Menus

Some of the initial explanations are elementary in nature and can be skipped by seasoned Windows 95/98/NT users.

Menu Format

Text that appears as:

File > Save As...

directs you to access the **File** menu and choose the **Save As** command.

Navigating and Using Menus

Electronics Workbench V5 allows you to navigate menus using either a mouse or with keyboard letters and hotkeys.

To use the mouse to navigate menu items, move the mouse until the onscreen pointer is over the text of the menu item. Make sure it is not to the side, above, or below the text. Click the mouse's left mouse button. This opens the menu and lists each command. Scroll down to the command you wish to execute and click the left mouse button again. You can also click and hold the mouse button while opening the menu, then scroll down to the item you wish to use and release the button. This saves a mouse click (see Figure 8.5).

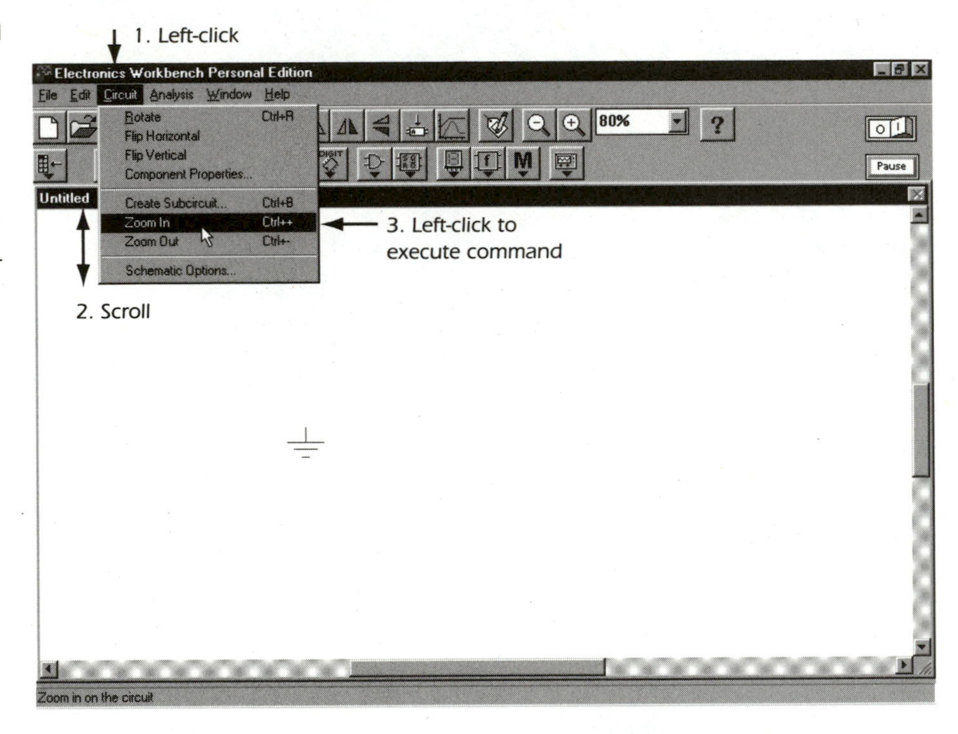

Figure 8.5
To use a menu, left-click the mouse over the text, scroll the pointer down the list and left-click over the command.

Clicked the wrong menu? Hit the Escape (Esc) key or go to the top menu item and click the mouse again.

Most commands can be executed with a few keystrokes. For example, if you want to save a file, hit the **Ctrl** key and then the letter **S**. Release both at the same time. But not all commands have a shortcut.

You will notice that most of the menu items have a letter underlined (**Edit**). This means you can hit the Alt key, then the underlined letter to open that menu. Scrolling down with the arrow keys lets you select an item. Hitting **Enter** executes that command. If one of the commands in the menu has a letter underlined, then you can hit the letter to execute it without having to scroll down and hit enter (**Alt+F** opens File menu and pressing letter **S** saves).

The choice of using the mouse or hotkeys is up to you. Just learning Electronics Workbench? Try to stick with the mouse until you learn the hotkeys. Myself, I tend to use a combination of both. I use my left hand to access keyboard items and leave my right on the mouse. Whichever is most handy gets the call.

WIN95/98/NT Right-clicking Shortcut Menus

"To use shortcut menus instead of using the standard menus to find the command you need, use the right mouse button to click a file or folder. The menu that appears shows the most frequently used commands for that file or folder."—*Windows 95 Help*

If Electronics Workbench is run on Windows 95/98/NT, you have shortcut menus available to you. I like to call them *context specific menus* because it's as though the program knows what you are going to do when the mouse pointer is in a specific area. For example, click the right button while the pointer is over a blank area on the Circuit Window. A menu will pops up resembling that in Figure 8.6. Commands can be executed by scrolling down with the pointer and hitting the left button. Many items in Electronics Workbench have these right-click menus. Browse around with the pointer and right-click here and there to discover them.

Dialog Boxes

Some commands require you to explain further what you want Electronics Workbench to do. In this case, a window opens to let you enter or change data (see Figure 8.7). For example, if you choose

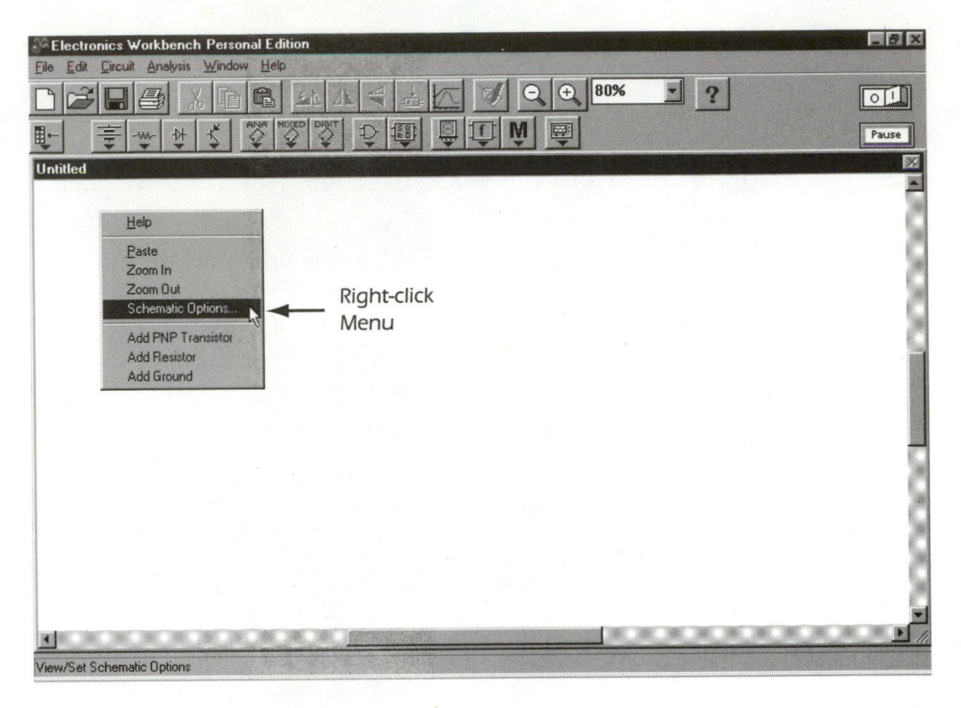

Figure 8.6 Right-click menus save having to remember where specific commands are placed.

Analysis > Analysis Options a new window with the Analysis Options dialog box opens up to allow you to modify the five pages of analysis settings. Memorize this term now, because it is used continuously throughout the book.

A dialog box lets you add or modify information to EWB function by using one of the following (see Figure 8.7):

- **Tabs**—Software uses file tabs because it can only fit so many options onto the screen at once. Flipping through these tabs can access additional pages of settings.
- **Buttons**—Execute a command, open another dialog box, or perform some operation when clicked with the mouse. For example, in the Analysis Options, the **OK** button executes the complete command with your settings.
- **Radio Buttons**—You can select one from a multiple choice list. If you select another item in the list the last item is canceled. The term comes from the old radio preset controls. For example, under the Analysis Options, Instruments tab, you can select among three options for how to set the initial conditions of the simulation.
- **Check Box**—A check means the statement is true or that the feature is turned on. If there is no check then the statement is false or

Figure 8.7
A dialog box usually is used to further explain exactly what you want Electronics Workbench to do. Several methods of entering this information are presented here.

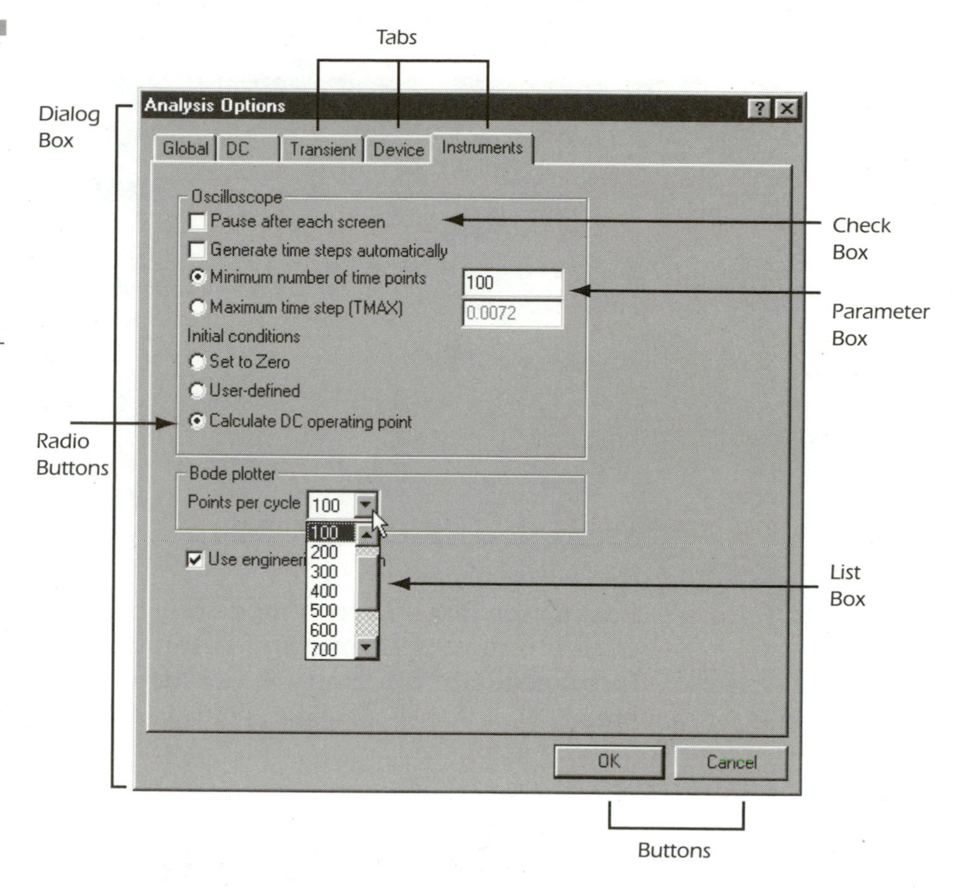

the feature is off. To check or uncheck, click inside the box. For example, on the Instruments page of the Analysis Options box, you can check off "Pause after each screen" to make it halt at the end of a scope screen.

- **Spin Boxes**—Clicking the up or down arrow advances the value in the white box next to it. For example, if you double-click a placed resistor and choose the Value tab, you can choose Ω, kΩ, or MΩ. See Figure 8.8.
- **List Box**—A list of items that you select by clicking the arrow on the right and scrolling down until you find the item. In the Analysis Options dialog box on the Instruments tab, you can change the Bode plotter's points per cycle with a pulldown.

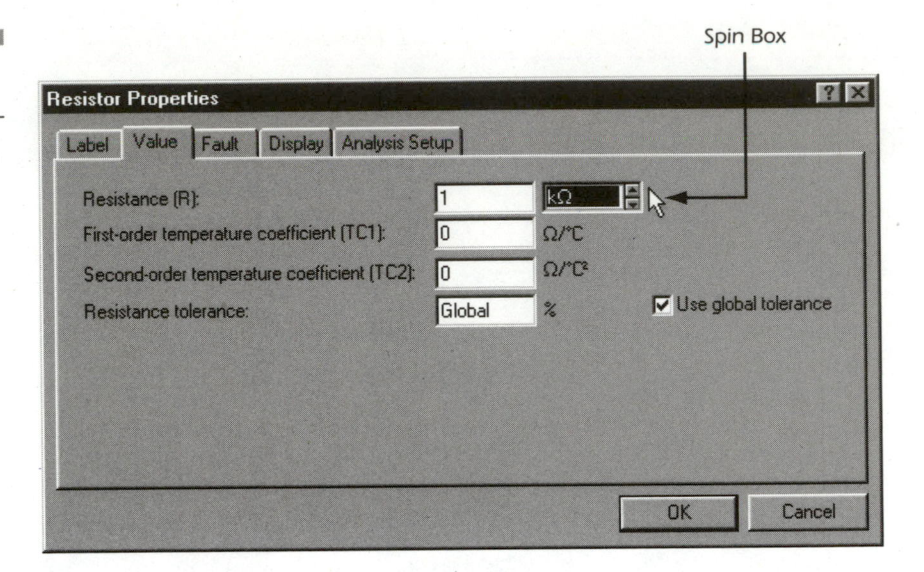

Figure 8.8
Spin boxes.

- **Parameter Box**—Type in the setting or information by hand. For example, under "Minimum number of time points" in the Instrument tab, you can type a numeric value into that parameter box.

Menu Options in EWB V5

Standardized selectable menu items range from File to Help. Some custom features may throw off the seasoned Windows user, but with Electronics Workbench's easy to learn menus, it takes little time to acclimate. What follows is an overview of each menu option and an explanation of the various ways to speed access to each. I recommend going over each item thoroughly and following along with the program. This helps later when you just can't seem to find that command you need.

Menu Item: File

The left-most menu selection is the **File** menu (Figure 8.9). As in any other Windows program it handles computer filing and printing tasks. Here is a rundown of each feature in this menu:

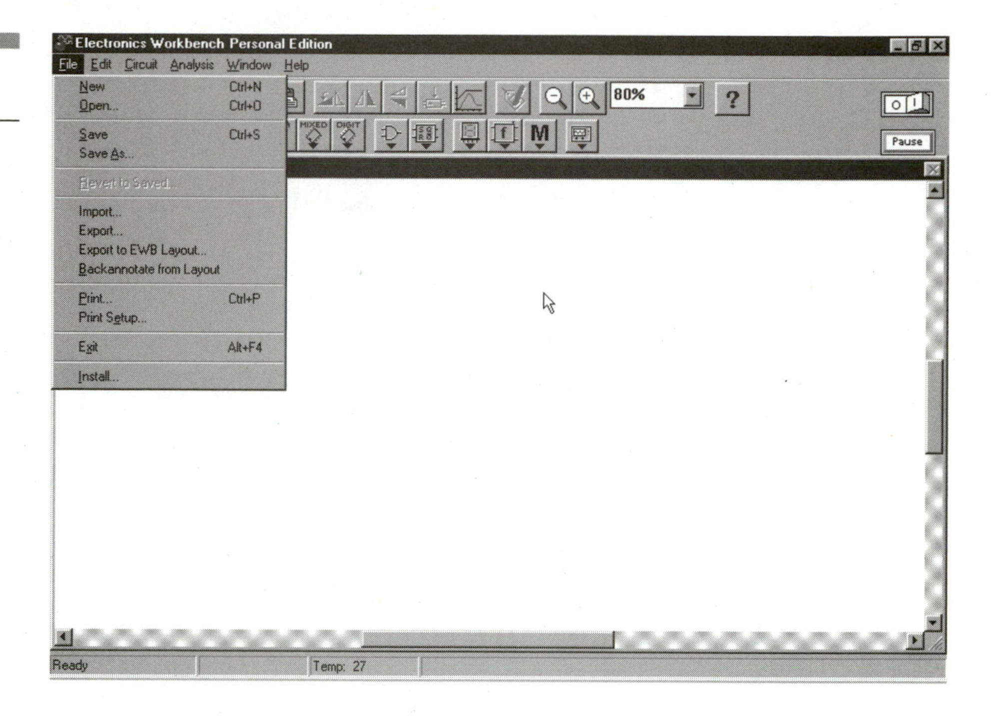

Figure 8.9
File menu.

NOTE

If the item's text is grey, it means it is not selectable at that point in your session, or not available on your edition of EWB.

Item: File > **New**
Hot Key: Ctrl+N
Use: Opens up a new Electronics Workbench document with the default name, 'Untitled'.

Item: File > **Open...**
Hot Key: Ctrl+O
Use: Launches a File Explorer to choose an analog (*.CA), digital (*.DA), or Electronics Workbench (*.EWB) file to open.

Item: File > **Save**
Hot Key: Ctrl+S
Use: Saves your document to the computer's hard drive or floppy. If it is a new file, it switches automatically over to the **Save As** menu. As a note, save often! Computers tend to lock at the most inopportune time. Save the heartache, frustration, and file by saving often.

Item: File > **Save As**
Hot Key: None
Use: Opens the File Explorer to name and save your files for later retrieval.

Item: File > **Revert to Saved**
Hot Key: None
Use: Opens the last save of the current file. EWB Version 5.x has no Undo command, so this is the next best thing. If you have made a few screw-ups and don't want to have to go through fixing them, use **Revert to Saved**. Hopefully, you will have saved the file recently; I cannot stress this enough: save often.

Item: File > **Import**
Hot Key: None
Use: Imports a Netlist text file (a *.cir SPICE file). This is an important feature for compatibility issues, because a lot of archived circuits use this format. If you wish to open them in EWB, you will have to import using this menu item.

Item: File > **Export**
Hot Key: None
Use: Reverse of the import function. It saves your data as a Netlist file for use with other related software. If you have ever wondered how EWB sees your drawing, view this *.cir file with a text editor such as Windows Notepad. This file helps you learn the SPICE language.

Item: File > **Export to EWB Layout**
Hot Key: None
Use: If you ordered EWB's PCB Layout software, you can export your EWB design directly into it using this feature. Save the file as an EWB Layout (*.plc) document and this automatically opens the Layout program and imports your components and node info. This is one of EWB's most hobbyist-friendly features.

NOTE

*Save the Layout file into the directory it prompts (**c:\ewb5\layout\pcbs**). You cannot save it into an alternative personal directory or the software will not function.*

Item: File > **Backannotate from Layout**
Hot Key: None
Use: You made a circuit in EWB Layout and wish to import it back into EWB? Use this feature. It is handy if you made revisions on your PCB but not on the *.EWB file.

Item: File > **Print**
Hot Key: Ctrl+P
Use: Brings up EWB's print options box. You can select the scale and which items of your file you wish to print.

Item: File > **Print Setup**
Hot Key: None
Use: Opens up the Windows Printer Setup box to let you set up your printer to your taste.

Item: File > **Exit**
Hot Key: Alt+F4
Use: Shuts down the EWB software after asking you if you wish to save the document in progress.

Item: File > **Install**
Hot Key: None
Use: You can purchase modules that contain other libraries and circuits; use this command to integrate them into EWB.

Menu Item: Edit

The Edit menu lists those commands that deal with manipulating the current drawing, such as cut, paste, and others (see Figure 8.10).
Item: Edit > **Cut**
Hot Key: Ctrl+X
Use: Cut the item(s) selected, keeping a copy in the clipboard for later processing.

NOTE

*If you cut or copy, the current contents of the clipboard are replaced with the new components. If in doubt about what is actually in the clipboard, choose **Edit > Show Clipboard** to display the item(s).*

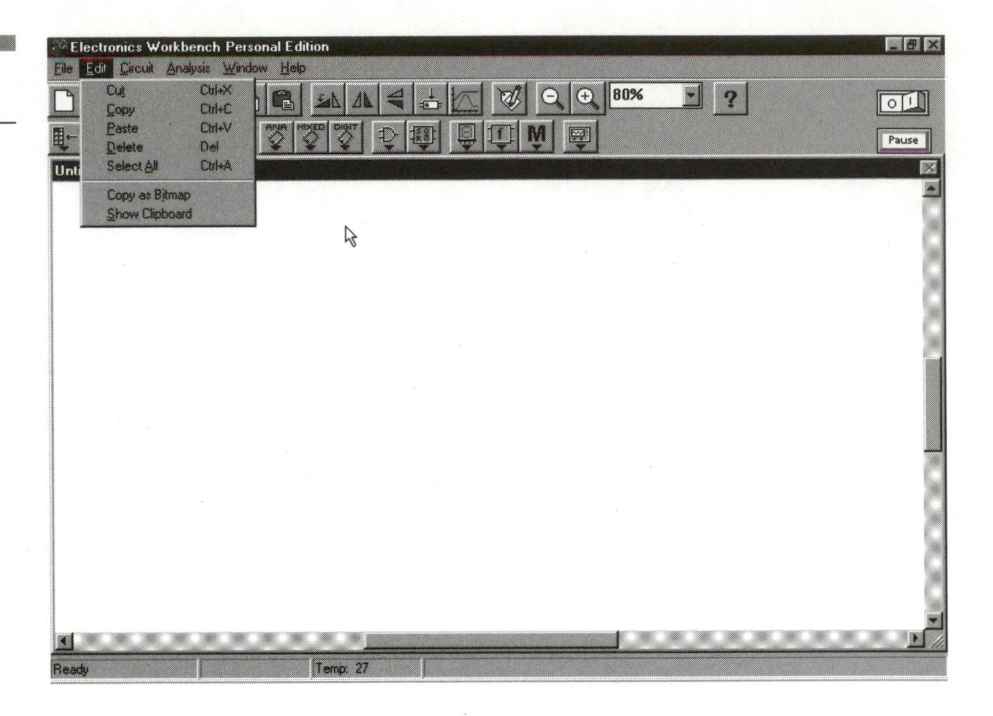

Figure 8.10
Edit menu.

Item: Edit > **Copy**
Hot Key: Ctrl+C
Use: Copies the selected item(s) into the clipboard. Similar to cut, except it leaves a copy of the selection(s) in place.

Item: Edit > **Paste**
Hot Key: Ctrl+V
Use: Pastes the contents of the clipboard into the circuit.

Item: Edit > **Delete**
Hot Key: Del
Use: Deletes selected item(s) from the circuit.

TIP

You will be prompted to confirm deletion. This is somewhat annoying, but because there is no undo feature in EWB Version 5.x, it may save sorrow. Consider using cut as an alternative. It will save a copy of the item(s) into the clipboard (just in case), allowing you to paste it back into the circuit. (All connections will be lost). It's not much consolation, but it saves having to reset all the component parameters.

Item: Edit > **Select All**
Hot Key: Ctrl+A
Use: Selects every component at once. This is handy if you have to center the circuit or want to copy and paste an identical circuit in a new location. You can also create a subcircuit once all items are selected.

Item: Edit > **Copy as Bitmap**
Hot Key: None
Use: Churns out a bitmapped copy of your circuit to the clipboard so that you can paste it into a drawing program or word processor. It's great if you are writing about a circuit and want to paste it into a document as a figure. After you choose **Copy as Bitmap** a crosshair appears if the mouse is moved onto the Circuit Window. Click and hold the left mouse button in the top left corner of the area you want to copy. Slide the mouse down and to the right until the box surrounds the area you wish to copy as a bitmap (see Figure 8.11). It will now be available to other programs, by simply using the paste feature of your WP or drawing program (see Figure 8.12).

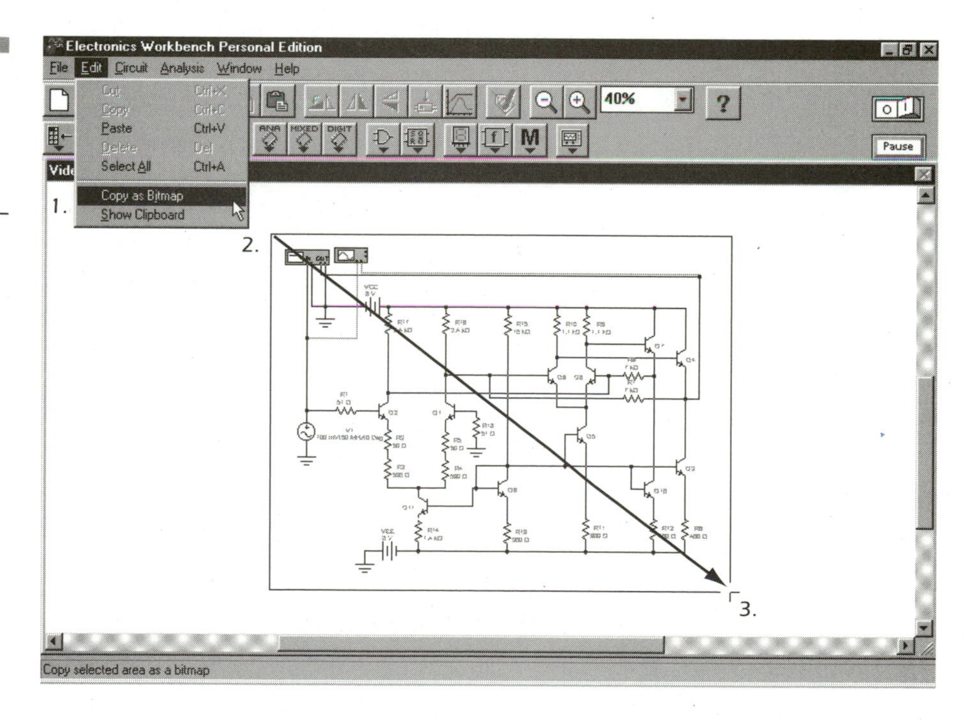

Figure 8.11
Copy as Bitmap is a great tool to import circuit diagrams into other software.

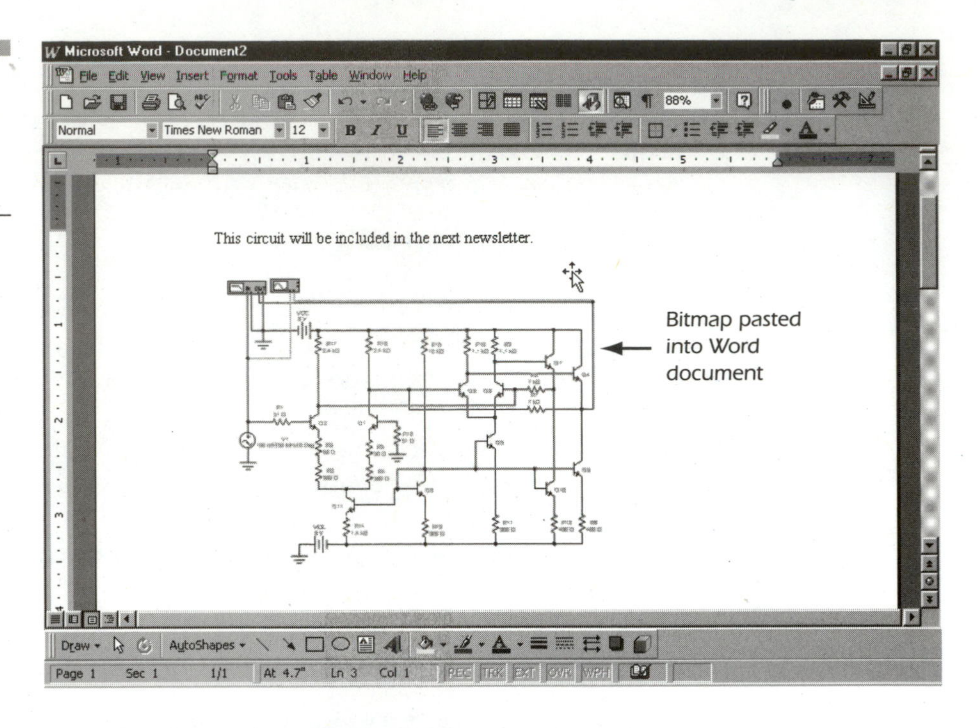

Figure 8.12 Here the Copy as Bitmap feature is used to paste a circuit into Word 97.

Item: Edit > **Show Clipboard**
Hot Key: None
Use: Opens Windows Clipboard Viewer to display the clipboard contents. Certain versions of Windows do not contain this program. If this is the case, download or find the file called clipbrd.exe and place it in the **c:\windows** directory. I've posted this at **http://www.basicelectronics.com/downloads/clipbrd.exe**.

Menu Item: Circuit

Items in the Circuit menu (Figure 8.13) manipulate circuit diagrams, zoom in and out of a drawing, and set schematic options. They save time and give you greater control of how the drawing will ultimately look.

Item: Circuit > **Rotate**
Hot Key: Ctrl+R
Use: Rotates the selected item 90 degrees counter-clockwise from its current orientation.

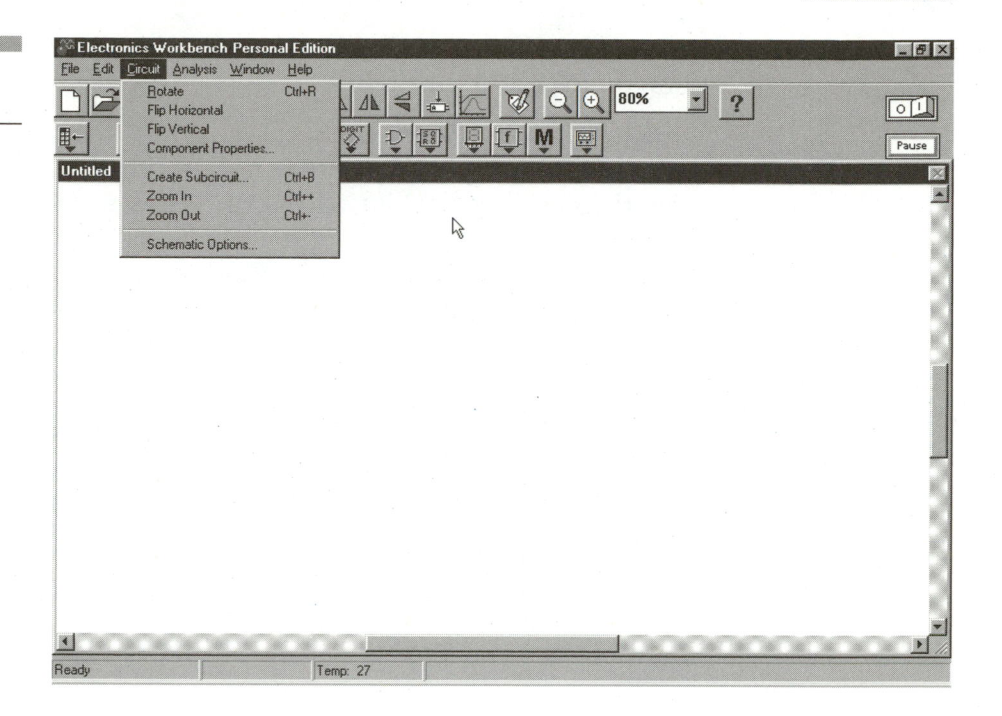

Figure 8.13
Circuit menu.

NOTE

Some components, such as ICs and some indicators, cannot be rotated or flipped.

Item: Circuit > **Flip Horizontal**
Hot Key: None
Use: Mirrors the component along a vertical plane (left side becomes right, and vice versa).

Item: Circuit > **Flip Vertical**
Hot Key: None
Use: Mirrors the component along a horizontal plane (top becomes bottom and vice versa).

Item: Circuit > **Component Properties**
Hot Key: None
Use: Opens the Component Properties box for the selected item to make revisions and choose models; same as if you double-clicked the component..

Item: Circuit > **Create Subcircuit**
Hot Key: Ctrl+B
Use: By selecting the components and choosing this command, you'll be able to create a subcircuit (circuit within a circuit) for use in

other circuits. See Chapter 9, "Drawing with EWB" for more information on EWB and subcircuits.

Item: Circuit > **Zoom In**
Hot Key: Ctrl ++. (You either have to hit Ctrl, Shift, and the Plus sign twice or you can hit Ctrl and the Plus sign on the etreme right of the keyboard twice.)
Use: Zooms into the Circuit Window at increments of 20 percent per keystroke.

Item: Circuit > **Zoom Out**
Hot Key: Ctrl +−
Use: Zooms out of the Circuit Window at increments of 20 percent per keystroke.

Item: Circuit > **Schematic Options**
Hot Key: None
Use: Opens the Schematic Options dialog box to make changes to the way documents are viewed.

Menu Item: Analysis

A simulation must first be run in order to see its results. The Analysis menu commands (Figure 8.14) describe the procedures to set up and start simulations.

Item: Analysis > **Activate**
Hot Key: Ctrl+G
Use: Begins the simulation of the circuit, the circuit is turned on (assuming everything is hooked up correctly).

Item: Analysis > **Pause**
Hot Key: F9
Use: Used to temporarily halt a simulation. Use it to stop a circuit and have a look at the instrument's waveform or reading, then continue.

Instead of using the Simulation Pause button, use the ***Analysis*** *>* ***Analysis Options*** *>* ***Instruments*** *tab and check off the* ***Pause after each screen****. Click OK. This pauses the simulation when the waveform reaches the end of the oscilloscope screen.*

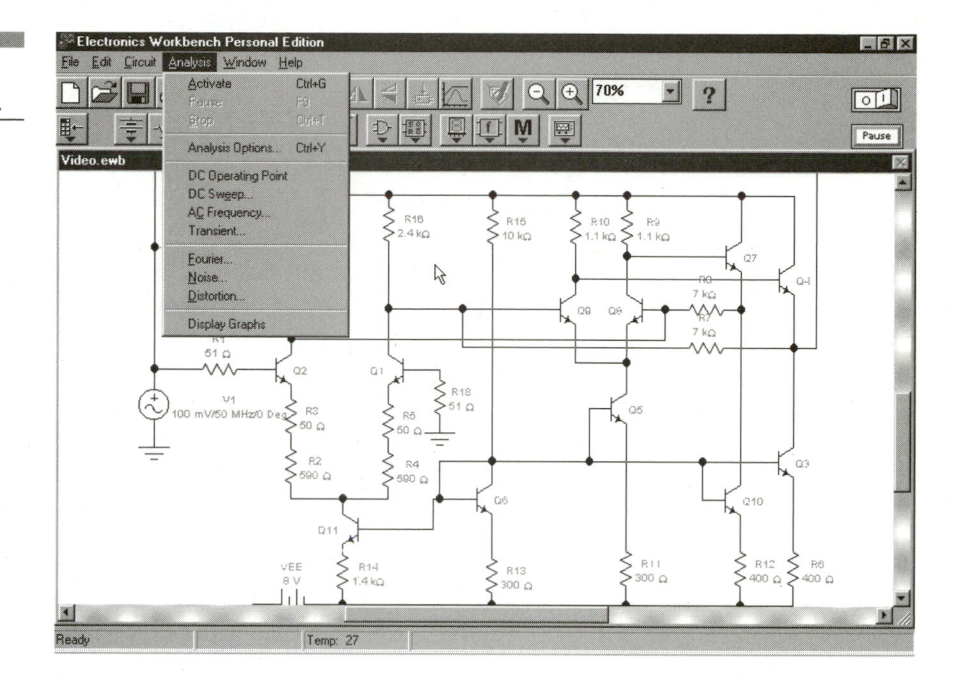

Figure 8.14
Analysis menu.

Item: Analysis > **Analysis Options...**
Hot Key: Ctrl+Y
Use: Opens the Analysis Options box to customize the analysis.

Item: Analysis > **DC Operating Point**
Hot Key: None
Use: Determines the DC operating point of a circuit. See Chapter 12 for more information.

Item: Analysis > **DC Sweep...**
Hot Key: None
Use: Computes the DC operating point of a node in the circuit for various values of one or two DC sources in the circuit. More information in Chapter 12.

Item: Analysis > **AC Frequency...**
Hot Key: None
Use: Calculates the AC circuit response as a function of frequency. More information in Chapter 12.

Item: Analysis > **Transient...**
Hot Key: None
Use: Computes the circuit's response as a function of time. Also called *time-domain transient analysis.* More information in Chapter 12.

Item: Analysis > **Fourier...**
Hot Key: None
Use: Evaluates the DC, fundamental, and harmonic components of a time-domain signal. More information in Chapter 12.

Item: Analysis > **Noise...**
Hot Key: None
Use: Used to detect the magnitude of noise power in the output of electronic circuits. See Chapter 12 for more information.

Item: Analysis > **Distortion...**
Hot Key: None
Use: Used to investigate small amounts of distortion, which are normally not resolvable in the transient analysis. See Chapter 12 for more information.

Item: Analysis > **Display Graphs**
Hot Key: None
Use: Brings up the Analysis Graphs window. Electronics Workbench V5 creates complete records as a simulation runs so you can completely analyze a circuit. See Chapter 12 for more information.

Menu Item: Window

The Window commands (Figure 8.15) control what tasks are displayed and how they will be viewed. This makes it easier to see just what is happening by showing as much information as possible.

Item: Window > **Arrange**
Hot Key: Ctrl+W
Use: Neatly organizes all windows currently open. If you are somewhat confused by staring cross-eyed at five windows, hit this command to put things in order.

Item: Window > **1 Circuit**
Hot Key: None
Use: Brings circuit window to the front.

Item: Window > **3 Description**
Hot Key: Ctrl+D
Use: Opens up the Description window. This lets you create or read notes about a specific circuit. It is great for documenting a circuit.

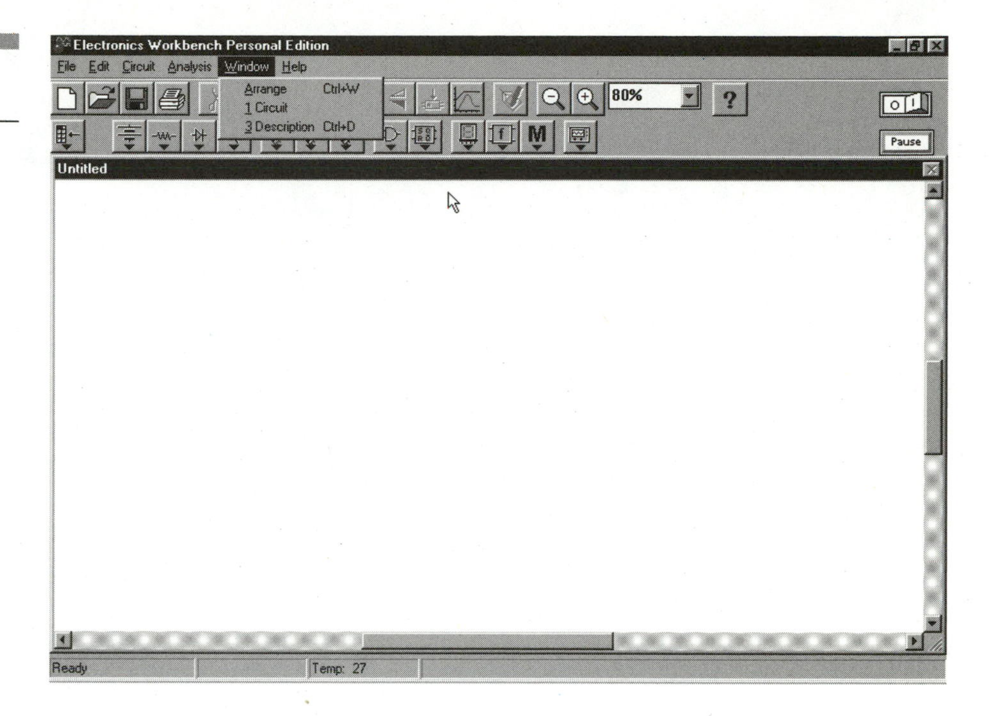

Figure 8.15
Window menu.

Menu Item: Help

Perplexed about how to perform a certain task? Want more detailed information? Open the Electronics Workbench Help menu (Figure 8.16). It lets you search for information, bookmark it, print it, and more.

Item: Help > **Help**
Hot Key: F1
Use: Opens the Electronics Workbench Help menus. If you have a component or instrument selected, that item's help box opens.

Item: Help > **Help Index...**
Hot Key: None
Use: Opens Electronics Workbench Help main content page. From there you can select a topic or hit the **Index** tab to search for items.

Item: Help > **Release Notes**
Hot Key: None
Use: Opens the Release Notes Contents Help menu. It provides support on what recent features have been revised and also provides minor technical support information. It contains loads of useful

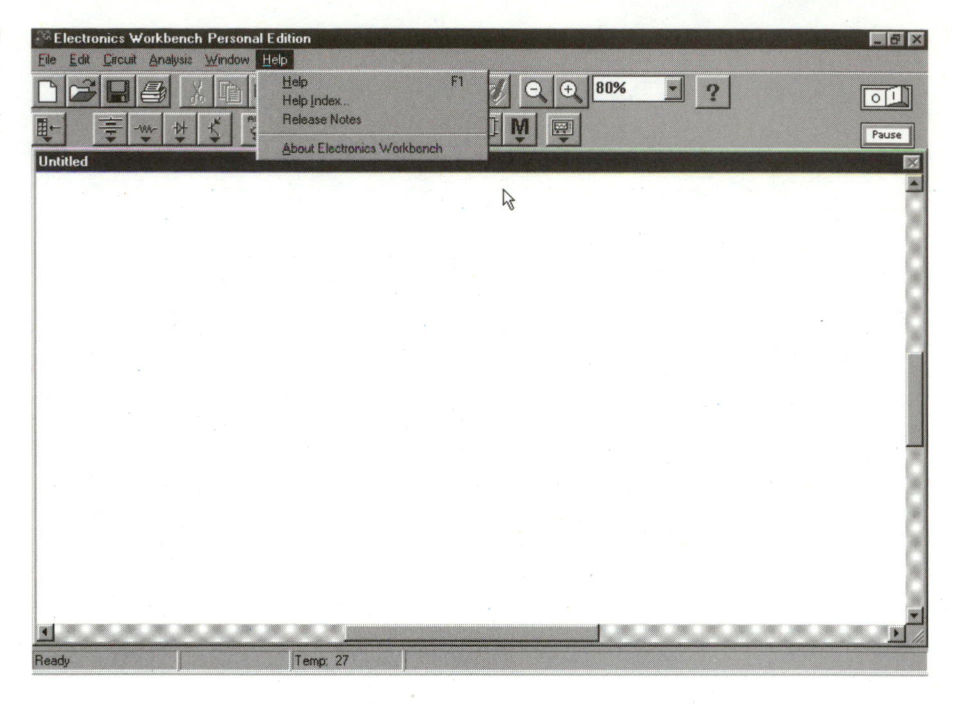

Figure 8.16
Help menu.

information, and I recommend reading it thoroughly or printing it for later reference.

Item: Help > **About Electronics Workbench**
Hot Key: None
Use: Displays version information and to whom the program is registered. Whenever calling Interactive Image Technologies technical support department, have this screen's information ready and note your serial number.

Toolbars

Electronics Workbench makes use of standard Windows-type toolbars that help to visualize commands and eliminate the need to memorize where menu items lay. A toolbar is a series of miniature pictures (icons), each of which represents a command. If you move the mouse pointer over an icon and left-click, that command is performed. For example, on the top section of Electronics Workbench V5 is the Circuit toolbar (see Figure 8.17). The first icon looks like a page with a folded corner. This is the **New** command. To open a new document, you merely move the mouse pointer over the icon and click. The next icon is **Open**. It resembles a file folder with an arrow. Most icons use some familiar image to help you figure out what they represent, but others can be quite cryptic. This section introduces you to toolbars in general and gives an overview of Electronics Workbench's toolbars.

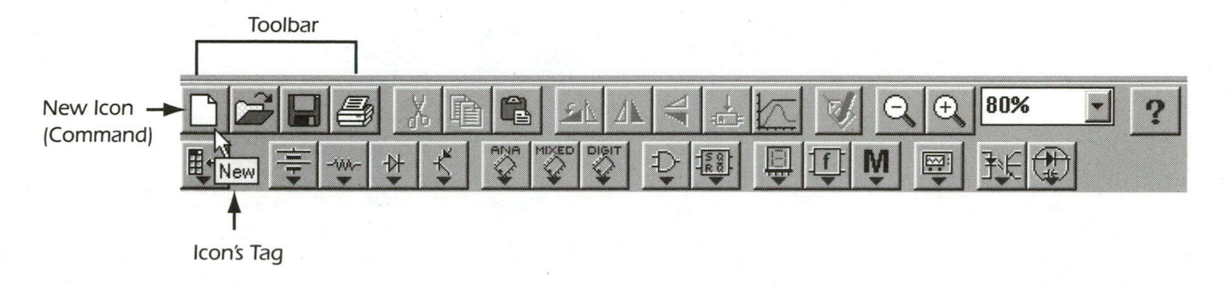

Figure 8.17 Toolbars full of icons speed access to commands. Can't figure out what an icon is? Place the mouse pointer over it and wait a moment until the yellow ID tag pops up.

Can't figure out what that icon is? Place the mouse pointer over it and wait a moment. A small yellow tag will open with the icon's title (see Figure 8.17).

Activating Commands using Icons

Each toolbar contains individual icons that represent a command or open a dialog box, which further clarifies the command. You execute the commands by placing the mouse pointer over the icon and left-clicking. For example, if you want to zoom in on the circuit, click the icon that looks like a magnifying glass with a '+' sign in the middle.

Customizing Toolbars

Electronics Workbench 5.1x features several pseudo-customization features that let you play around with the toolbars.

If you don't like the location of an icon, you can hit **Shift**, then click and drag the icon to a new location on the toolbar. The new location must be on the same horizontal axis. When you restart EWB, the icon position information is reset (see Figure 8.18).

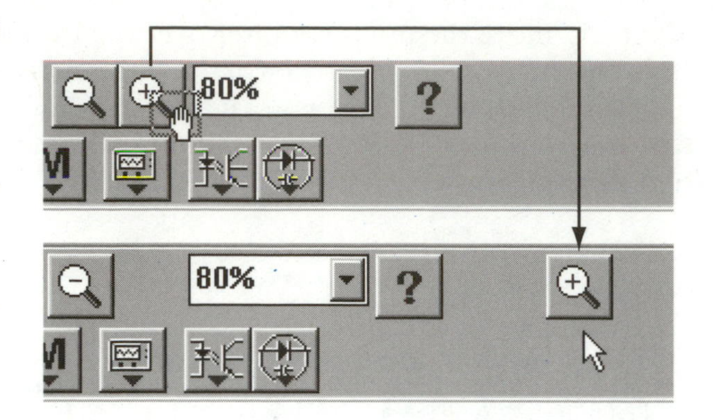

Figure 8.18
Want to customize the toolbars? Move the icons around. Hold the pointer over the icon and hit shift. Drag it to a new location.

NOTE

Toolbin icons are also customizable (movable).

If you can't stand closing the parts bins after selecting each component, hide the Parts Bin. Go to **Circuit** > **Schematic Options** > **Show/Hide** tab. Place a check in the Autohide parts bin box. Now when you select a part, the parts bin will automatically close.

If you move a parts bin from its default location, it will continue to open in that position if the Keep Part Bin Position option is checked under **Circuit** > **Schematic Options** > **Show/Hide** tab.

Electronics Workbench Toolbars and Icons

This section describes each toolbar item. The "Use" descriptions are short, but if you want more detail, look at the corresponding command in the Menu Items section.

Circuit Toolbar

The Circuit Toolbar is shown in Figure 8.19.

Icon: New
Use: Opens a new blank circuit window.

Icon: Open
Use: Brings up Windows File Explorer to open an existing circuit.

Icon: Save
Use: Saves current circuit.

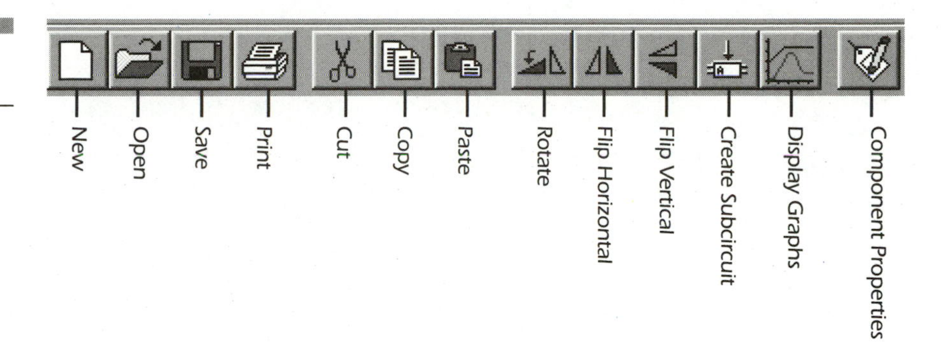

Figure 8.19
Circuit toolbar.

Icon: Print
Use: Opens the Print dialog box. Only prints the circuit and nothing else.

Icon: Cut
Use: Cuts selected component(s) out of drawing. A copy is kept on the Windows clipboard.

Icon: Copy
Use: Places a copy of the selected component(s) onto the clipboard but leaves the original part(s) in place.

Icon: Paste
Use: Copies the contents of the clipboard into the current drawing.

Icon: Rotate
Use: Swivels select component(s) 90 degrees counter-clockwise.

Icon: Flip Horizontal
Use: Mirrors the component on a vertical axis.

Icon: Flip Vertical
Use: Mirrors the component on a horizontal axis.

Icon: Create Subcircuit
Use: Makes a circuit within a circuit.

Icon: Display Graphs
Use: Opens the Analysis Graphs window.

Icon: Component Properties
Use: Opens the dialog box for the selected component's properties.

Zoom Toolbar

The Zoom Toolbar items are shown in Figure 8.20.

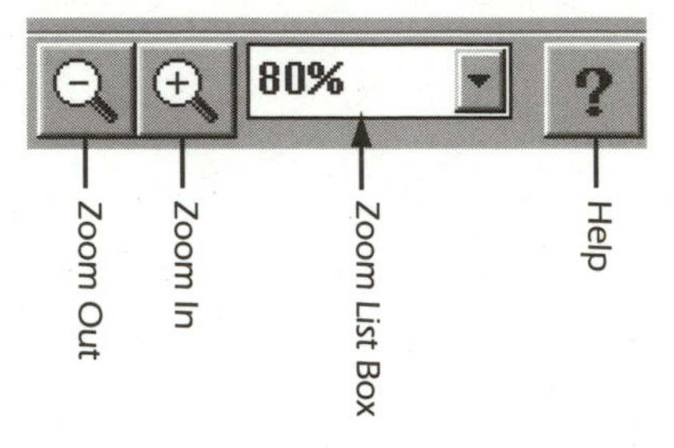

Figure 8.20
The Zoom Toolbar.

Icon: Decrease Zoom
Use: Zooms out of a section of the circuit.

Icon: Increase Zoom
Use: Zooms into a section of the circuit.

Icon: Quick Zoom
Use: Controls zoom factor of circuit.

Icon: Help
Use: Opens help file on current selected component.

Circuit Simulation Switches

The Circuit Simulation Switches Toolbar is shown in Figure 8.21.

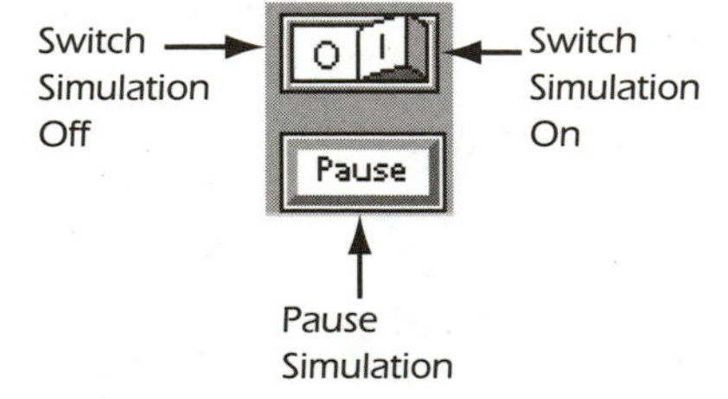

Figure 8.21
Circuit Simulation Switches.

Icon: Activate Simulation/Stop Simulation
Use: Quick way to run or stop the current simulation.

Icon: Pause Simulation/Resume Simulation
Use: Quick way to pause or restart the current running simulation.

Parts Bin Toolbar

The Parts Bin Toolbar is shown in Figure 8.22.

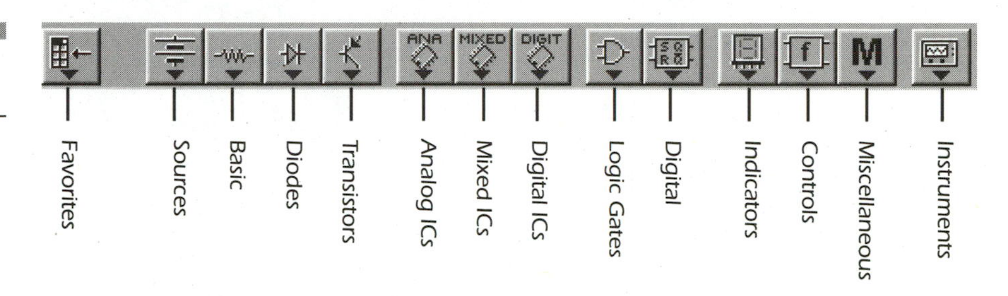

Figure 8.22
Parts Bin toolbar.

Icon: Favorites

Use: This is not an actual component but merely a list of all the subcircuits you have created. You can also add commonly used parts for quicker access. It is a fast way to place a pre-defined circuit into the current drawing. See the subcircuit section in the next chapter for more information.

TIP

To add components to the Favorites toolbin, right-click onto the opened toolbin icon and choose ***Add to Favorites.***

The Parts Bin Toolbar (PBT) contains icons that open each individual parts bin. When you left-click one of these icons, a separate toolbar opens containing several component choices. Think of the PBT as an assortment of similar components or component subcategories. For example, the Transistor icon on the PBT opens the Transistor parts bin, which contains all types of transistors. (See toolbars on next two pages.)

For a more in-depth explanation of each Component, see Chapter 10, "EWB Building Blocks."

INDIVIDUAL PARTS BINS

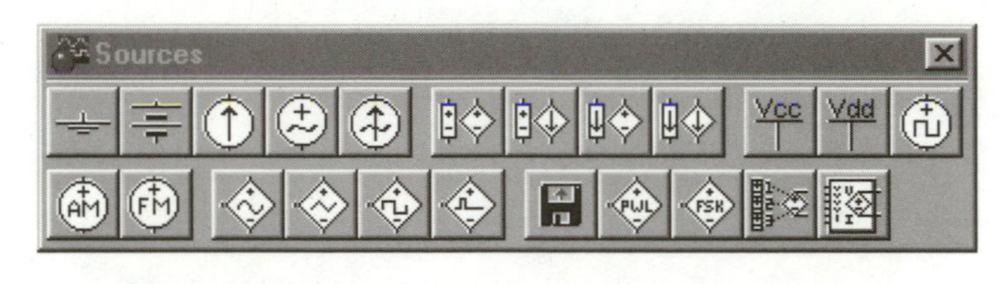

Figure 8.23

Figure 8.24

Figure 8.25

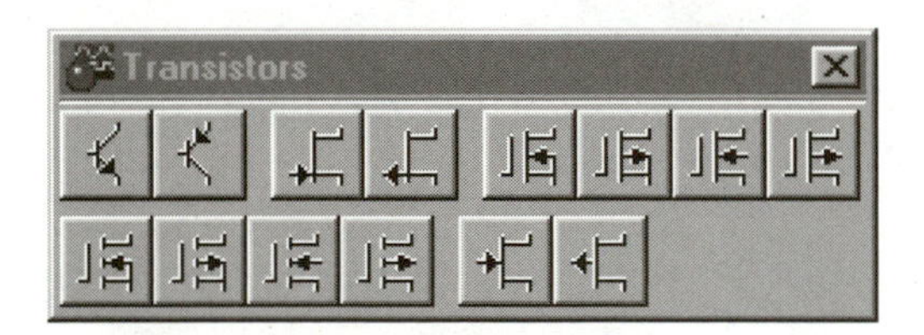

Figure 8.26

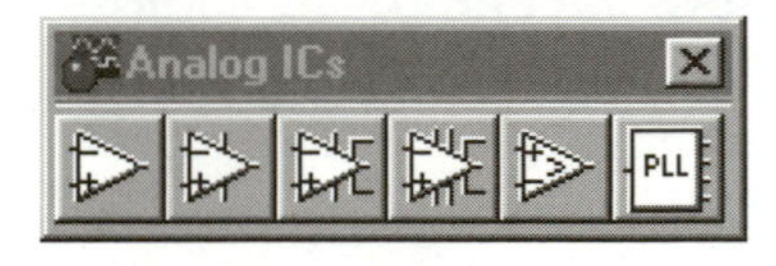

Figure 8.27

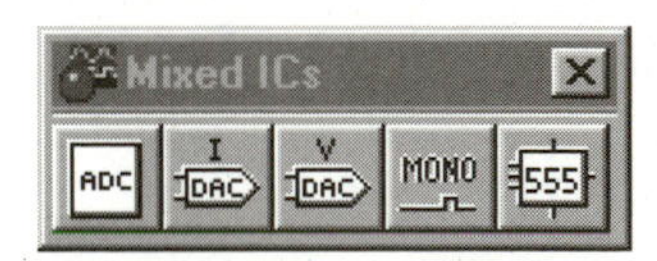

Figure 8.28

Figure 8.29

INDIVIDUAL PARTS BINS

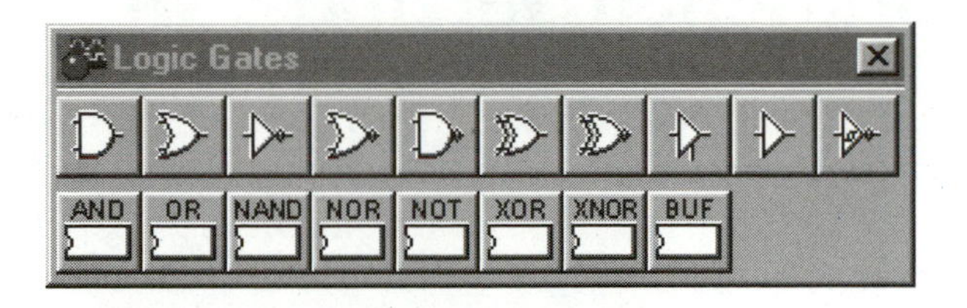

Figure 8.30

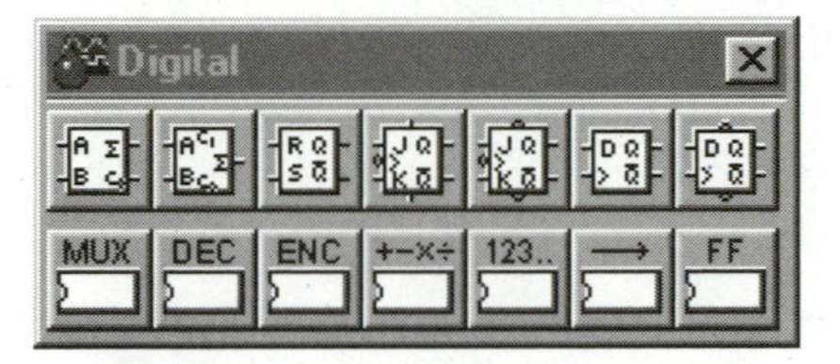

Figure 8.31

Figure 8.32

Figure 8.33

Figure 8.34

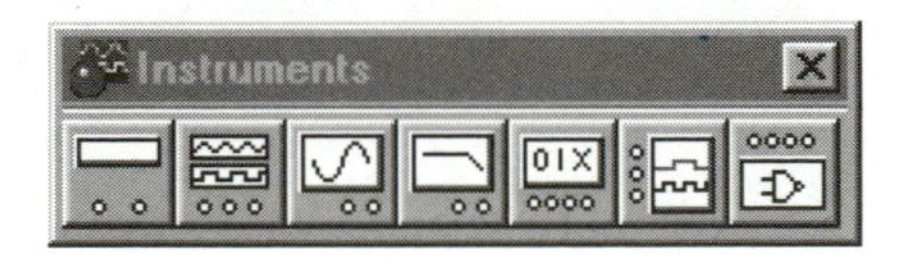

Figure 8.35

User's Guide to the Electronics Workbench Version 5 Interface

The EWB Interface is used to perform several functions (see Figure 8.36.) The first is *schematic capture,* which is the ability to draw circuits with EWB 5 and save the file or output it to paper. (This is covered in Chapter 9, "Drawing with EWB.") The second is the ability to take the drawing and analyze or simulate that circuit; this is covered in Chapter 12, Simulation and Analysis. We may add to this list the ability to export your circuit designs to a PCB package. The remaining functions allow you to perform various tasks and customize the interface itself to your personal tastes.

Running Electronics Workbench V5

I am assuming you have one of the EWB versions (Demo, Personal, or Professional) loaded and ready to open. Get those mice ready. Time to simulate.

Let's begin by starting Electronics Workbench:

Win95/98/NT: Start > **Programs** > **Electronics Workbench 5.xx** > **Electronics Workbench**

A splash screen opens and, depending on the speed of your processor and the amount of RAM, you have available, it may take some time for the actual EWB desktop (interface) to appear before you.

You can also use the shortcut on your Windows Desktop that likely was created at the time you loaded EWB. If it is not present, create one and point it to ***c:\ewb5\Wewb32.exe*** *or to whichever directory the executable file is placed.*

Saving EWB Files

To access your file in the future, it must first be saved to the hard drive or to a floppy disk. Once the file is saved, you will want to resave as changes are made.

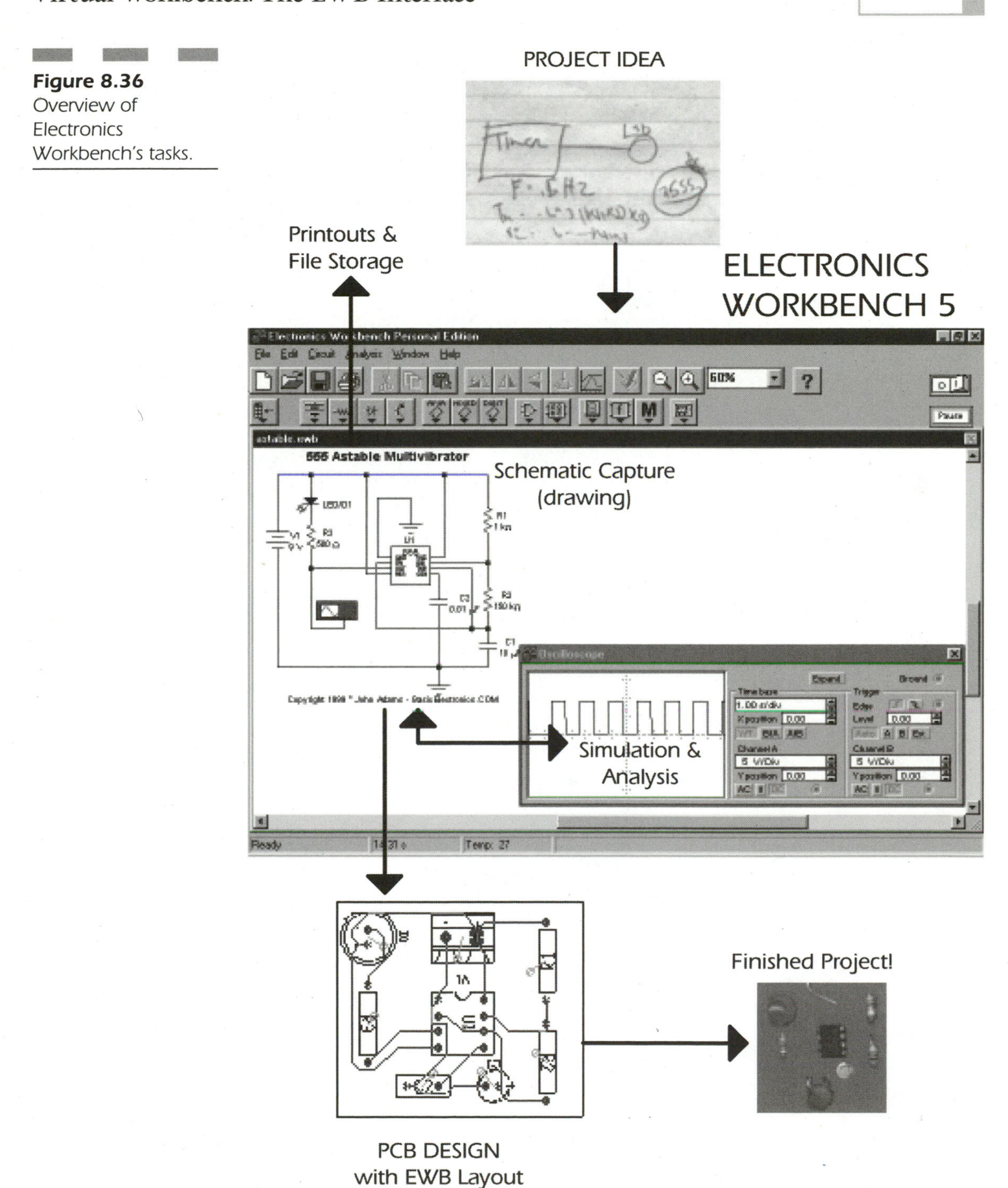

Figure 8.36
Overview of Electronics Workbench's tasks.

NOTE

*EWB files are relatively small in byte size and Zip into a much smaller space for archiving. Just use a Zip program such as WinZip (**www.winzip.com**).*

To save a new file, go to **File** > **Save As** or hit the floppy disk icon on the toolbar. The Save Circuit File dialog box opens (see Figure 8.37). The top box (next to the Save in) is the directory into which the file will be saved. You can use the next icon (the folder with the arrow) to go up one directory or double-click a folder in the main field to go into a directory. If you want to create a new directory, hit the icon that looks like a folder with a star on it. Type in the name of your file in the File Name field. Don't worry about adding .ewb to the end; this is done automatically. If you wish to change the file type, do so now with the arrow. Hit the **Save** button and the file will be saved to the hard drive with an .ewb extension, meaning it is an EWB5.1 File.

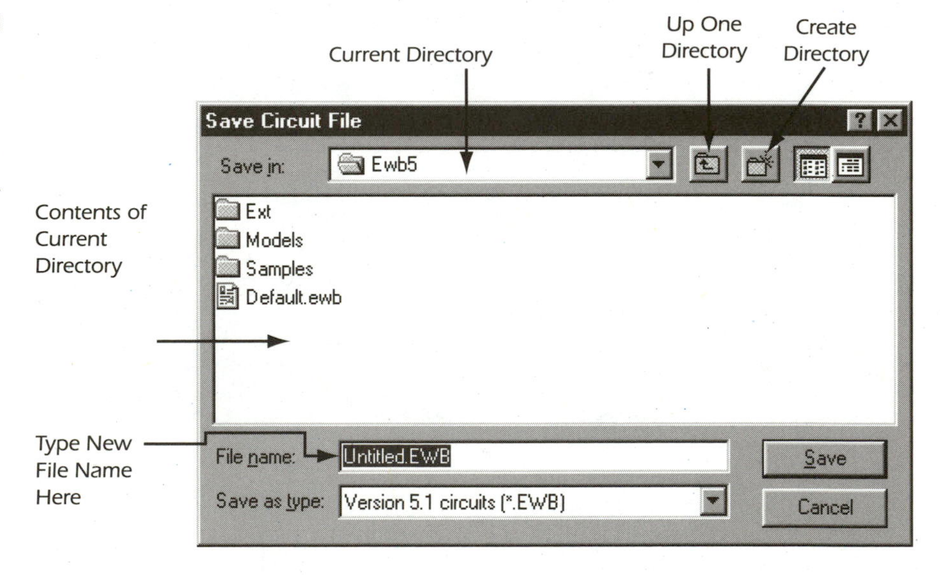

Figure 8.37 Saving files to your hard drive requires you to use this dialog box.

TIP

Think of the PC's hard drive as a file cabinet that stores your circuits.

I strongly recommend saving a circuit file every couple of minutes as you are working on it. To do this, hit the floppy icon on the Circuit Toolbar or go to **File** > **Save.** You can also just hit **Ctrl+S**.

Opening EWB Files

To open Electronics Workbench circuit files you have saved or stored, you must use the Open command. **File** > **Open** (**Ctrl+O**) brings up the Open dialog box (see Figure 8.38). This opens to the **C:\EWB5\SAMPLES** directory. If you wish to read from a personal directory, navigate there and open the *.ewb file. Hit the **Open** button or double-click the file name.

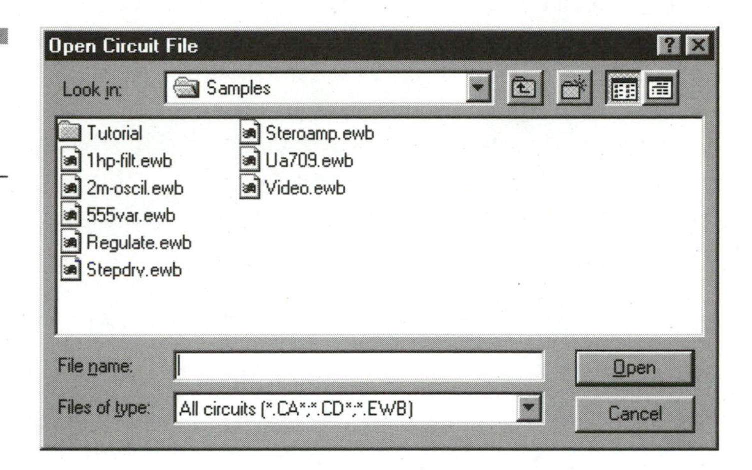

Figure 8.38
To open stored files, use the Open dialog box.

Importing Netlist Files

If you wish to open up a Netlist (SPICE) file, you must import it. Choose **File** > **Import** and the Import Dialog Box opens. Locate the *.cir file and hit **Open**.

NOTE

Imported Netlists are not exactly aesthetically pleasing. You may want to rearrange the components to avoid confusion.

Exporting Netlist Files

If you are using a software program other than Electronics Workbench, you can typically save your files as SPICE Netlist files. They can then be imported into another program. This is done by selecting **File** > **Export.** Name the file and Electronics Workbench adds the *.CIR extension to it.

Exporting to EWB Layout

If you have EWB Layout installed on your PC, you can export your EWB design straight to it. Select **File** > **Export To EWB Layout**. The Export to EWB Layout dialog box opens. Type in the name you wish to use (use same name as EWB file if possible) and EWB automatically adds the *.plc extension. *Note*: Do not change directories, because Layout needs the file to be in that directory. Hit the **Save** button and EWB Layout opens automatically with your file ready to prepare.

If you have to change a circuit while in Layout, and you wish to have EWB make the changes to your circuit schematic as well, choose **File** > **Backannotate from Layout**. The Backannotate dialog box opens, asking for the Layout file name. Select from the pull-down field which file you wish to bring into EWB.

Accessing Menu and Toolbar Items

Menu items are placed across the top of your screen and are not customizable. They include **File**, **Edit**, **Circuit**, **Analysis**, **Window** and **Help**. They are used to access commands you wish to perform. Each menu opens to a list of commands.

Toolbars are placed across the top of the desktop. They include the Circuit toolbar, Zoom toolbar, Parts Bin toolbar, Favorites icon, and Simulation controls

See earlier sections of this chapter for more details on menu and toolbar procedures, tips, and uses.

Getting Help

Electronics Workbench Version 5.xx has an outstanding Help menu that will answer most common questions you may encounter. You can search with a keyword, bookmark frequently accessed items and print important information. To open the Help Index choose **Help** > **Help Index** or hit **F1** when no component is selected.

EWB lets you select a part and go directly to its corresponding help page. Some pages even display example circuits to aid in the understanding of that part. To access component-sensitive help:

1. Select the part.
2. Hit either **F1**, **Help** > **Help**, or choose **Help** from the top menu or right-click menu.

Once you are in the Help Index window, you can press the **Index** button. This opens the Help Topics box, as seen in Figure 8.39. Type a word in the top white box and topics appear in the box below. To select one of them, highlight it and press return or double-click it with mouse. For example, if you were looking for information on transient analysis, start to type **T R A** and a list appears below. Double-click the selection and the Electronics Workbench Help window appears (see Figure 8.40).

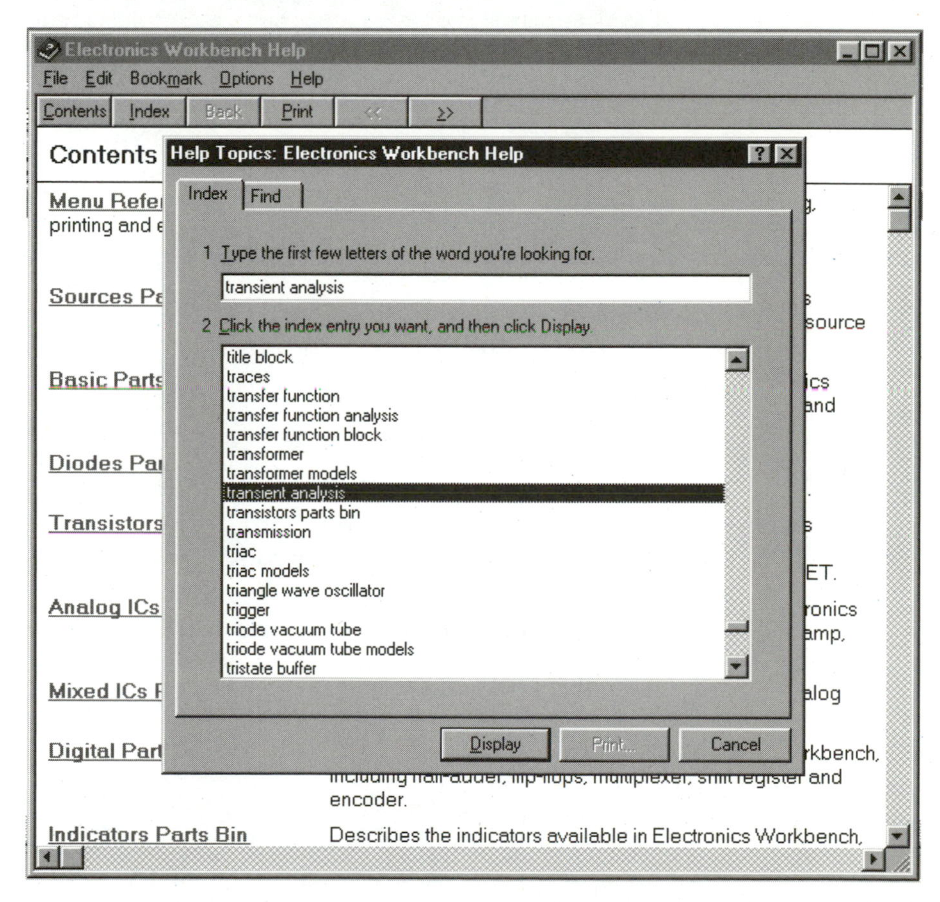

Figure 8.39 Electronics Workbench Help files are a quick way to reference the software. You can perform searches using this dialog box.

In this same window (Help Topics) is the Find tab. Click it and refer to Figure 8.41. Typing a keyword into the number 1 box causes a selec-

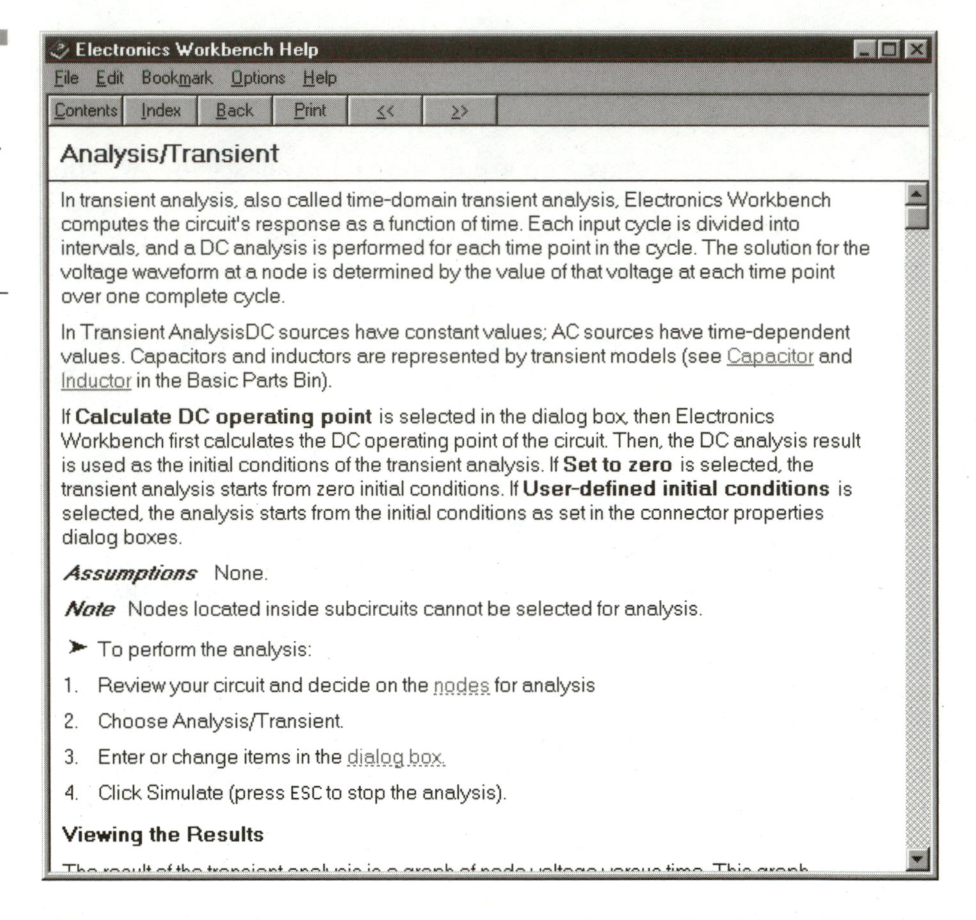

Figure 8.40 Electronics Workbench Help Window is the 'meat' of Help. From here you can print, bookmark, or cross-reference help files.

tion of suggested search words to appear in number 2 box. You may also select a recent search word from the pull-down arrow. In the number 3 box is a list of each page which contains that word or group of words as highlighted in box 1 and 2. For example, I have typed the word "resistor" in the 1 box and the help topics suggest I look for "resistor, Resistor, resistor's, resistors, and Resistors." Each is highlighted so I merely refer to box 3, which contains each page that uses the word "resistor." By double-clicking the words "resistor pack" I am taken directly to that page.

Adjusting the criteria can narrow searches. Hit the **Options** button and select the applicable options to fine-tune the search. EWB has a fairly powerful Help search engine that is highly customizable using these options.

Any help page can be *bookmarked* for later reference. This works similar to a Web browser's bookmark feature:

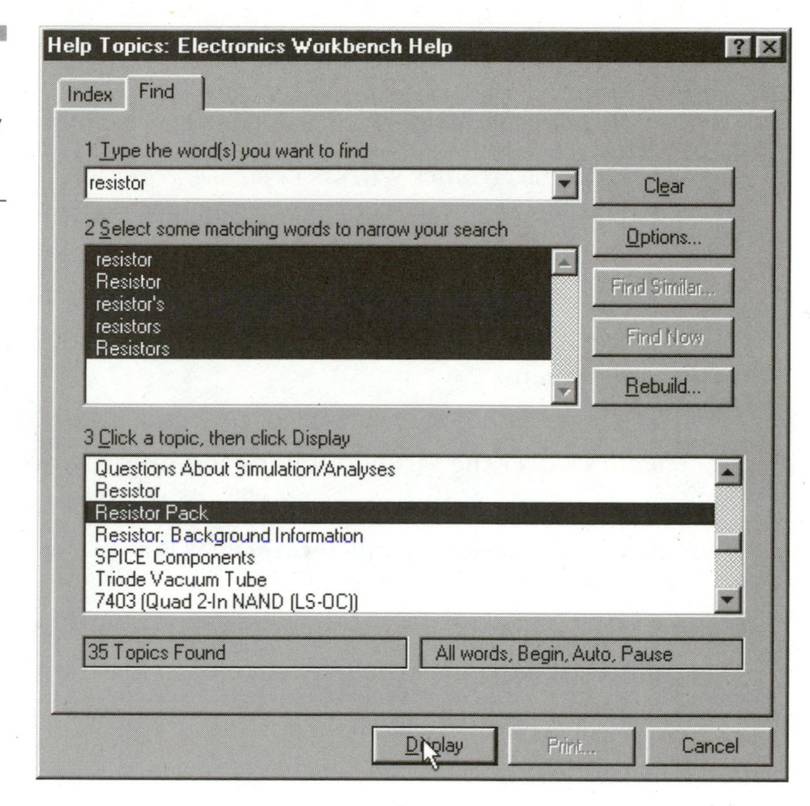

Figure 8.41 The Find tab allows you to search for any incidences of a specific word.

1. With the Help Index open, Choose **Bookmark** > **Define...**
2. A window pops up, prompting you to name the page. A title is automatically inserted into the field, but you can add whatever word(s) will help identify it later.
3. Hit **OK**.

By choosing **Print topic** from the **File** menu, you can make a hardcopy of that page.

Release Notes

Under **Help** > **Release Notes**, you can find tons of useful information that is not available in the EWB 5 manual. It includes updates, hints, and notes the Electronics Workbench Company has supplied. Read through it one day when you have circuit-design block.

Interface Exercises

To get used to the interface, let's try a few exercises.

1. Open Electronics Workbench.
2. Have a good long look at the screen. See what each item is, where it is placed, and how it looks in relation to other items.
3. Open a few menus just to see what they contain.
4. Right-click in the middle of the circuit window and have a peek at that menu.
5. Move the Zoom icon over to the right to see how this works.
6. Click over the various toolbins and note their contents.
7. Try some of the hotkeys as described in the menu section.
8. Create a directory called **C:\MyEWB** and save a file to it. Close the file.
9. Open some of the files in **C:\EWB5\SAMPLES**.
10. Relax! Now start the next chapter to learn how to draw the EWB way.

Summary

The user interface uses simple procedures that any Windows user can master in no time. If you are in doubt about a certain procedure, read through all the chapters for further details. The help menus that are provided with EWB may shed some light on otherwise difficult tasks. Just enter the Find feature in the Help Index and type a few words you think relate to what you are trying to do. If you are still having problems, you are always only a phone call away from Electronics Workbench's tech support line. See Appendix C for more information on contacting them.

The next four chapters delve deeper into specific interface procedures as they relate to issues such as drawing circuits, components, instruments, simulation and analysis, and building circuits.

CHAPTER 9

Drawing with EWB: Schematic Capture

Actually, Electronics Workbench is more like an electronic drawing pad used to doodle schematics.

Schematic capture is an electronic design term that describes drawing schematics with a computer. This is the ability to use Electronics Workbench to create schematic representations of circuits in a high-quality format. It may include lists of parts needed to build the circuit and all other pertinent information. Circuit diagrams can then be printed or saved for archive purposes or to serve team members in the development of circuits. It is also the starting point of simulation and analysis. Think of Electronics Workbench's schematic capture feature as a computer-aided drafting (CAD) program used for electronics, complete with premade symbols, and the ability to connect terminals together and make notes of circuit features. Instead of using a real electronics workbench, you are presented with a virtual workbench. The following sections teach you everything there is to know about producing complete schematics for either output or electronic simulation.

Also see Chapter 14, Unleashing Multisim/EWB, for plenty of valuable circuit drawing tips.

Think of the circuit window as a sheet of paper upon which components and wiring are drawn.

Getting Ready to Draw

To draw with EWB it is best to set up certain items. The following sections show you how to set up your monitor, adjust the toolbars and desktop to your liking (see Chapters 8 and 14 for tips), adjust the drawing's or User preferences, and name and save the drawing.

Setting Up Your Monitor

You monitor must be set to a minimum resolution of 640×480 pixels to operate Electronics Workbench. I recommend using a higher resolution, such as 800×600. To do this, go to **Start** > **Settings** > **Control Panel**. Double-click the **Display** icon. Click the **Settings** tab. Under Desktop Area move the slide until the 800×600 setting appears (see Figure 9.1). A higher resolution is fine. You may be prompted to confirm this change. Do so if it is safe (some monitors won't support higher resolutions). If you are having difficulties with this procedure, refer to the Windows 95/98/NT manual or help file, the monitor's manual, or the PC video card's readme file.

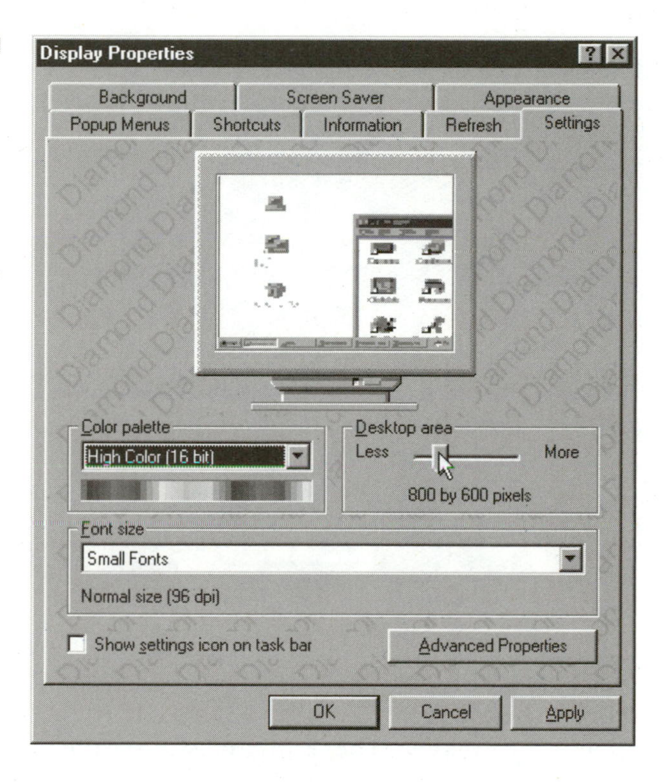

Figure 9.1 Your monitor's resolution must be set to 640×480 or greater to run Electronics Workbench.

Setting Up the Drawing

First things first: When you open Electronics Workbench, it is best to set the schematic options: what elements are present on the drawing, how it will appear when printed, and such. You may relate this to getting your pencils and drawing tools together. This section covers these procedures.

SCHEMATIC OPTIONS

Open this dialog box by going to **Circuit** > **Schematic Options**; you can also right-click in an empty area of the circuit window and select

Schematic Options. Fill out each page of data and hit the **OK** button. You will see the following pages:

- **Grid Tab**—Controls the grid, which is used as a guide to help you place components along specific increments. (See Figure 9.2).
  - **Show Grid**—Checkmark = grid visible.
  - **Use Grid**—Checkmark = use grid. I recommend leaving the grid on, because it can be difficult aligning parts otherwise.

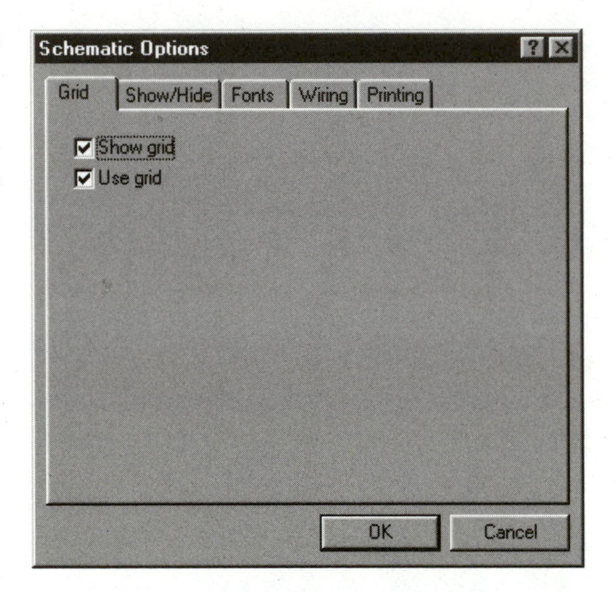

Figure 9.2
To control the grid, use the Schematic Options, Grid Tab.

- **Show/Hide Tab**—Sets circuit items you want displayed. Also adjusts the Parts Bin Toolbar settings. If the item has a check in it, then it is displayed. The items are: Show Labels, Show Reference ID, Show Models, Show Values, Show Nodes (see Figure 9.3). The parts bin selections give you the option to hide the toolbins (the toolbin disappears after you have selected a part to place). You can also select "Keep parts bin position," which opens the toolbar in the same position each time it is accessed; otherwise it defaults to a set location.
- **Fonts Tab**—Sets label and value fonts. Click the **Set label** button to open the Font dialog box. Once the font, font style, and size are selected, hit the **OK** button. The Set Value font button operates in the same fashion (see Figure 9.4).

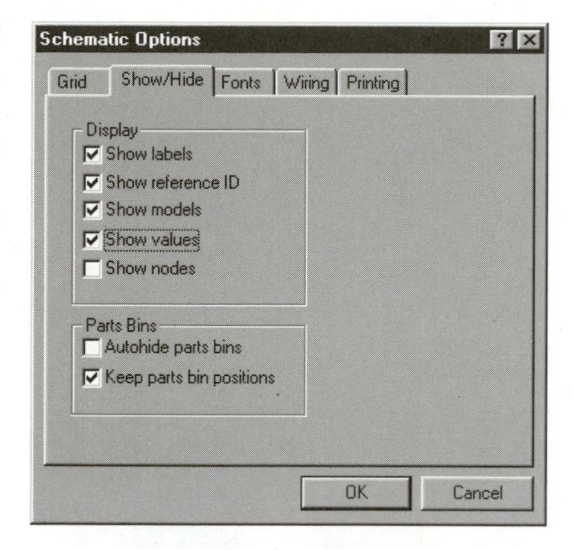

Figure 9.3 The Show/Hide tab determines what labeling elements will be displayed with a component or circuit.

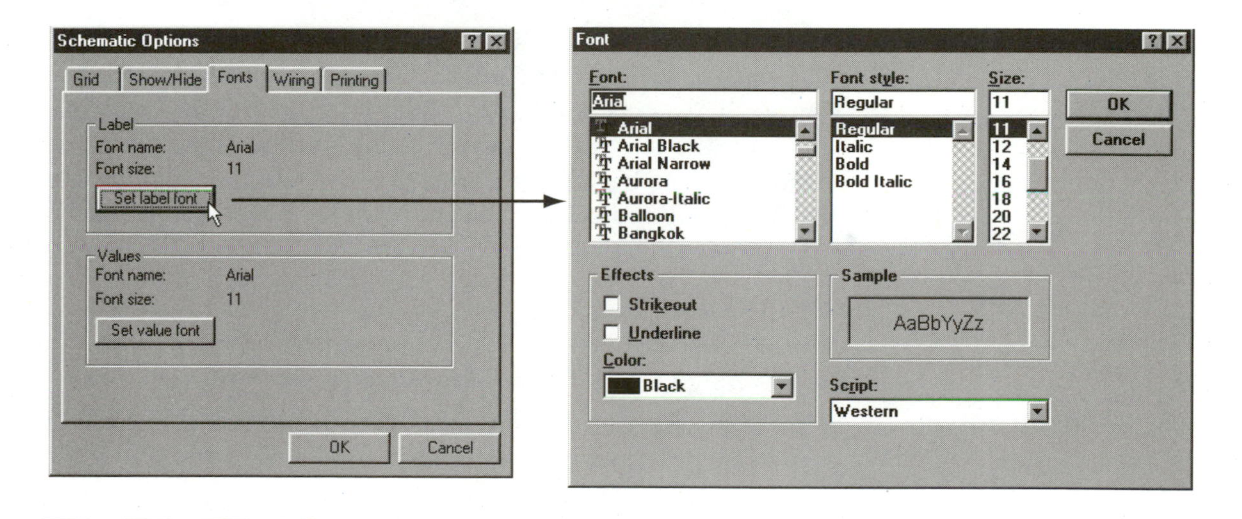

Figure 9.4 The Font Tab lets you adjust how labeling appears.

- **Wiring Tab**—Gives you full control over wiring options (see Figure 9.5).
  - **Routing Options**
    - **Manually route wires**—Lets you run wires one corner at a time. Left-click and release on a component terminal. Move the mouse in the direction you wish to go. Click the mouse for each corner you wish to place.

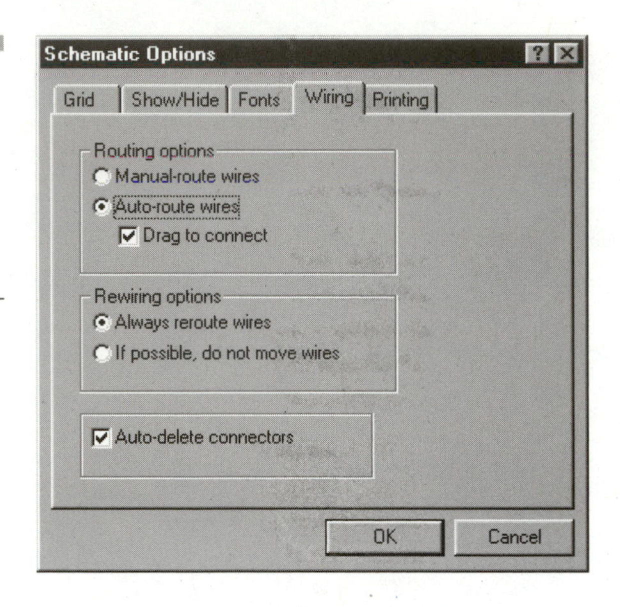

Figure 9.5 The Wiring Tab allows pseudo-control over how Electronics Workbench strings the components together.

— **Autoroute wires**—Automates the wiring task. If Drag to Connect has a checkmark in it, then click on the first terminal and hold the left mouse button. Drag over to the new terminal and release. If Drag to Connect is not selected, click on the first terminal and release the button. Go to the second terminal and left-click once more.

— **Rewiring Options**

— **Always reroute wires**—If a component is moved, this option lets EWB move the wires into new configurations.

— **If possible, do not move wires**—If a component is moved, it tries to keep a semblance of the original wiring routes.

— **Autodelete connectors**—When a wire or part is deleted, EWB removes all unnecessary connectors, if this option is selected.

- **Printing Tab**—Adjusts various printer settings, including zoom level and display of page breaks (see Figure 9.6).

— **For printing, zoom to**—Lets you select the zoom level of the printout. I recommend using the Fit to Page option at the bottom of the selection, or 200 percent.

— **Use visual breaks for main workspace**—If a circuit is too large to fit on one page, it prints to multiple pages. Selecting this item lets you see where the breaks occur. Grey lines on the schematic represent the breaks.

— **Use visual breaks for subcircuits**—Same as previous, but used for subcircuits.

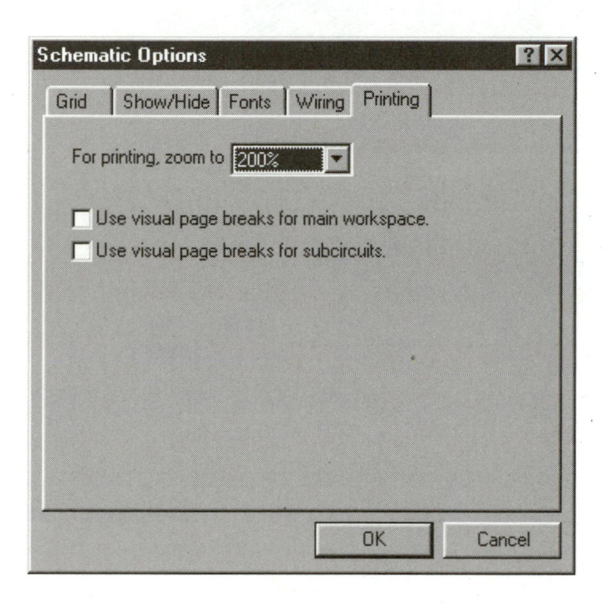

Figure 9.6 The Printing tab controls hardcopy preferences.

Schematic Option settings are set for each circuit and return to default when you start a new circuit or restart the program. To keep these setting for all drawings, open the file called: ***C:\EWB5\default.ewb****. Adjust your schematic options and save this file. Each new drawing created will use the settings. You may want to back up default.ewb to a temporary directory first.*

Open Electronics Workbench. Go to ***Circuit*** *>* ***Schematic Options*** *and set up a drawing to your taste. Turn the grid on/off. Make the drawing show all items. Set the Label font to System, bold 10. Adjust the wiring options and printer. Do you want these to be your master default drawing settings now? Open* ***Default.ewb*** *and make the changes again, saving to this file.*

Title Block

Use this feature to name, number, and date your document. Do this at the beginning of a session to avoid forgetting it later on (refer to Figure 9.7).

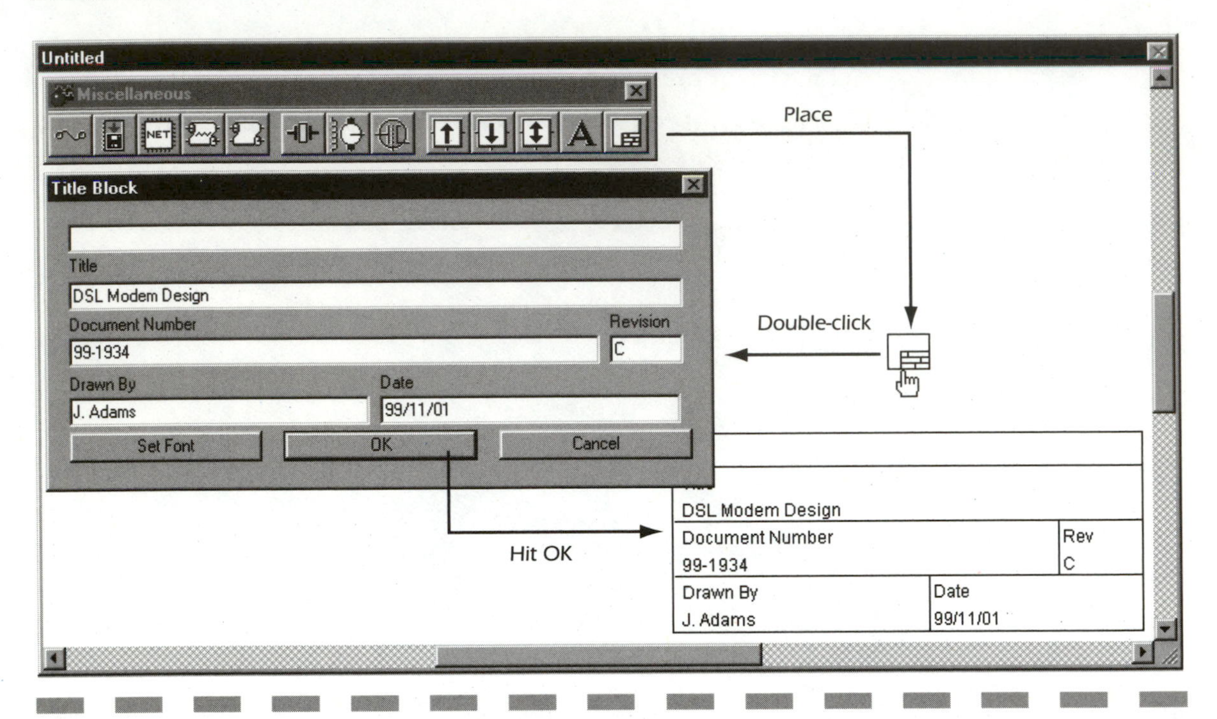

Figure 9.7 A title block lets you add information such as date and designer to help you later identify the drawing.

1. Open the Miscellaneous parts bin.
2. Click and drag into the drawing area a copy of the last icon to the right. Position doesn't matter yet.
3. Double-click the small icon that was created.
4. Fill in the relevant information.
5. Adjust the font, font size, and style by pressing the Set Font button at the bottom.
6. Hit **OK**, **OK.**
7. Move the Title Block into position.

Place the Title Block in the lower-right corner for a professional drawing.

Placing Components

Subsequent chapters explore component placement in more detail, but for now, let's see how the Parts Bin toolbars operate. When you left-click an icon on the Parts Bin toolbar, another component bin appears. The Parts Bin toolbars contain sets of icons that represent individual components from which you can choose to lay into your drawing. Select an item by moving the mouse pointer over it and clicking the left button. Hold the button down and drag the component onto the drawing. Releasing the left button places the component. By double-clicking the component, you can change its properties.

For example, if you want to place 100-ohm resistor on the drawing, click the mouse pointer over the Basic icon on the Parts Bin toolbar (Figure 9.8). The Basic toolbar opens. Left-click and hold the resistor icon and drag the pointer to the screen. Release. Double-click the resistor and select the value tab. In the Resistance Box, type **100** and spin the multiplier to Ω. Hit **OK**.

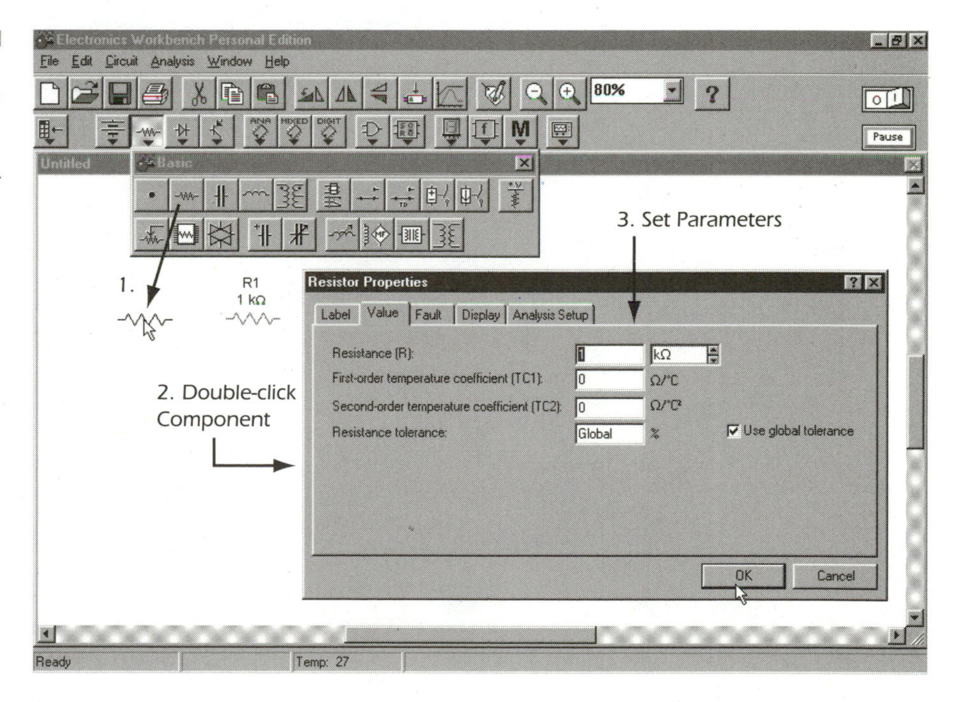

Figure 9.8
Placing a component is as easy as one, two, three!

TIP

You can close the Parts bins by clicking the icon once more or hitting the X in the top-right corner of the toolbar. If you want them to close automatically, select ***Autohide parts bins in the Circuit > Schematic Options > Show/Hide*** *tab page.*

Placing Instruments

The Instruments toolbar is on the Parts Bin toolbar; last right icon. Click it. Once the instruments toolbar is open, left-click on an Instrument icon and hold it. Move the pointer over to the circuit window and release. Wires can then be run to connections on the instrument. You can open the instrument controls and options by double-clicking the instrument on the drawing (see Figure 9.9).

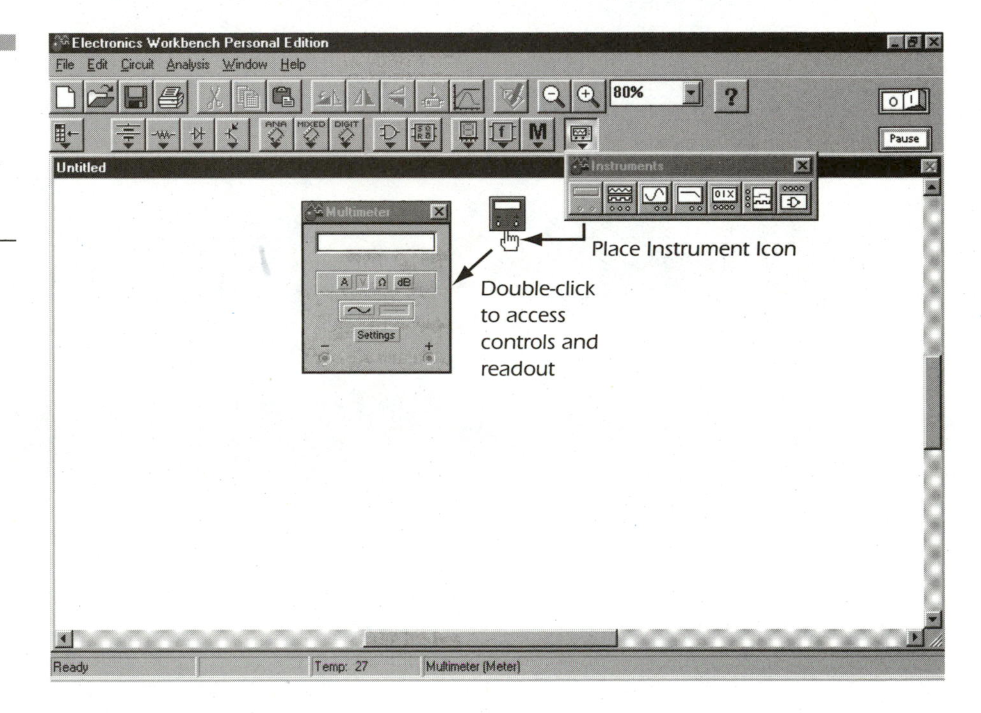

Figure 9.9 Placing a virtual instrument is followed by double-clicking it to access its controls.

Quick Parts

Electronics Workbench keeps track of the last five parts placed. You can right-click anywhere on the drawing and bring up a menu with those five parts listed at the bottom. Scroll down and select a part to place (see Figure 9.10).

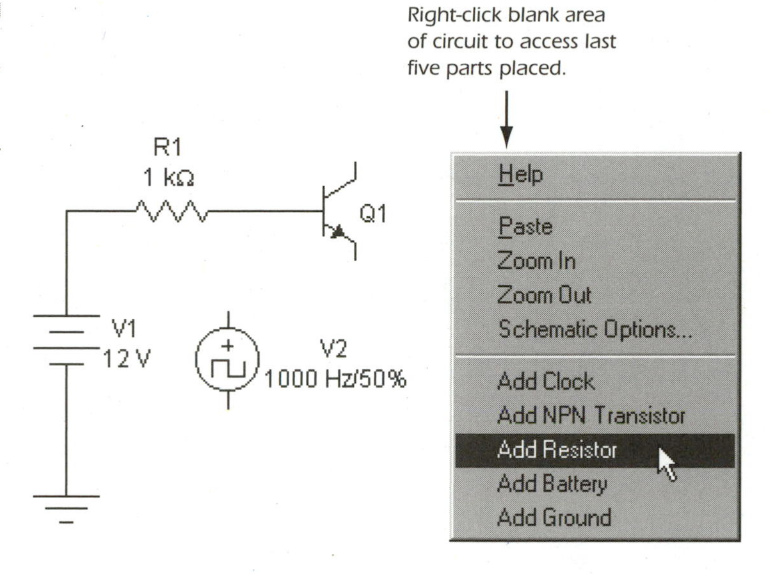

Figure 9.10
A lightning-quick way to place commonly used parts.

Add a ground to the drawing first. This makes grounds available faster by using the Quick Parts method.

Place a few components onto a blank drawing. Throw in some instruments. Try the Quick Part method once you have a few parts placed. Don't worry about wiring or moving them around. Once you have too many parts on the screen and can't tell what is what, start a new drawing. Repeat until you are comfortable with placing parts and instruments.

Selecting Components and Labels

To move or alter a part or instrument, you must first select it. There are several ways to select one part or a group of components.

To select one component, place the mouse pointer over the schematic symbol of the part. Click the left mouse button once. The component is now red, indicating it is selected (see Figure 9.11). If the mouse

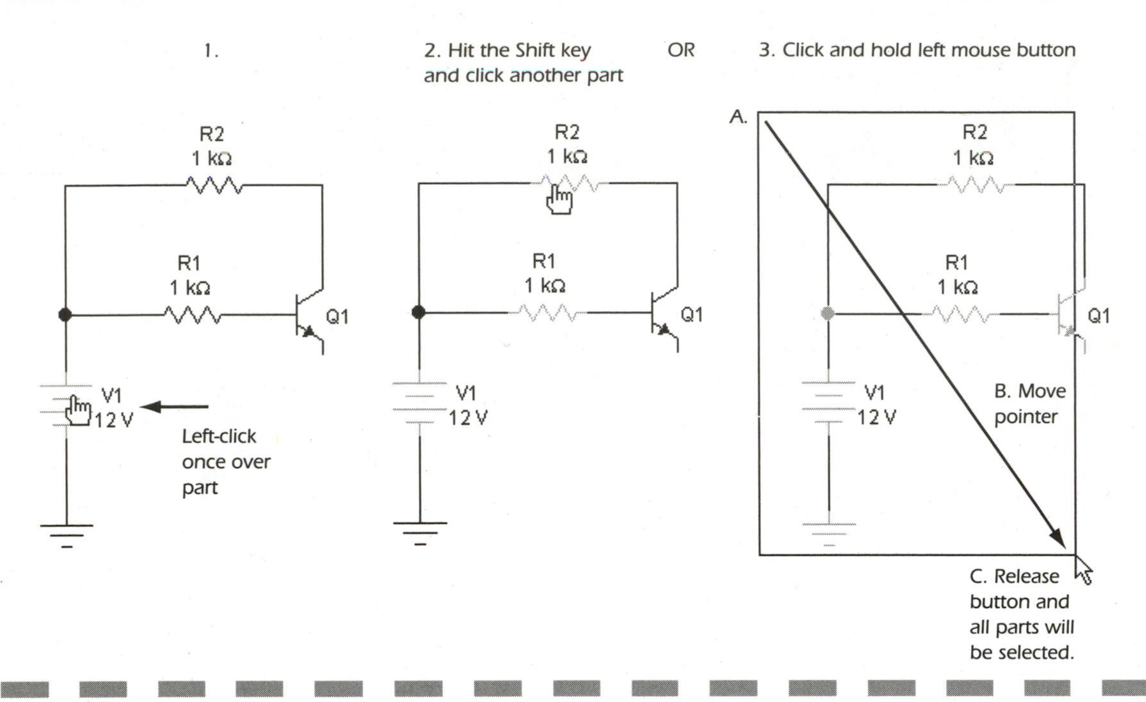

Figure 9.11 Working with components usually requires that you first select them. Here are three methods of doing this.

pointer is placed directly over the selected component, its specs are displayed in the right side of the status bar at the bottom of the screen.

You only have to partially surround a component to select it.

To select multiple components, hold the Shift key down and left-click on those components you wish to add to that group (see Figure 9.11). Each selected component turns red. If you mistakenly select the wrong component and don't want to start over, just click the component again to deselect.

To select a whole area of components, place the mouse pointer in an empty area at the top left of the section you want to select. Click and hold the left mouse button. A dotted-box appears when you move the mouse. Move it to the bottom right corner of the area you are selecting and release the button. Each selected component turns red (see Figure 9.11). If you want to add parts to the selection, hit the Shift key and click the component you wish to add. To remove parts, hold the Shift key and left-click.

TIP

If you don't know what a part is that you already placed on the screen, move the mouse pointer over it and read the contents of the Status Bar (bottom of screen).

Deselecting Components

Left-click a blank area in the circuit window to deselect everything. If you want to deselect only one component in a group, press the Shift key, then left-click that component.

Moving Components

You can move a selected component or just grab the component and move it, skipping the selection step.

To move a single component, place the mouse over the schematic symbol itself (not the labels). Click the left mouse button and hold. Slide the part to where you wish to place it. Release the button (see Figure 9.12).

Figure 9.12 Moving a component: click, drag, release!

To move multiple components, select components you wish to move. Place the pointer over any of the schematic symbols selected. Left-click and hold. Drag and drop.

You can "bump" a selected component one grid length one way or the other. Just hit the appropriate arrow key after the component is selected (see Figure 9.13).

NOTE

All connections remain hooked up while parts are moved.

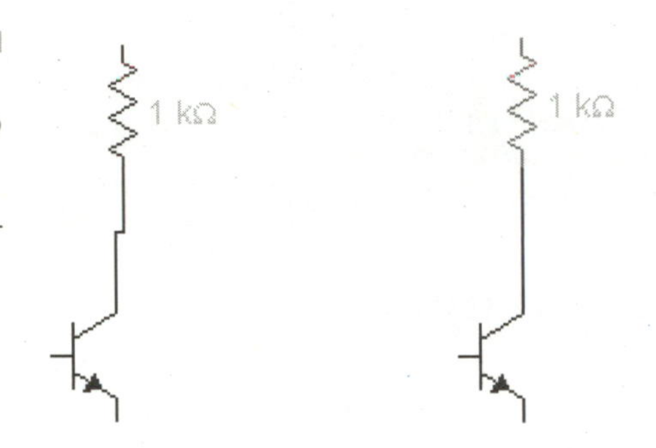

Figure 9.13
Use the arrow keys to "bump" a selected component slightly.

EXERCISE

Place six components on the screen. Select and move one of the components. Now select three of the six components. Move those around. Try the Shift key method of selecting an additional component.

Making Copies of Components

Sometimes it is necessary to make a component facsimile to use in another section of the drawing. It eliminates the hassle of laying a new component with identical settings. There are two commands used. Cut erases the item from the drawing but leaves a copy of the component and settings on the Windows clipboard. Copy places a carbon in the clipboard, leaving the original on the drawing. (Figure 9.14).

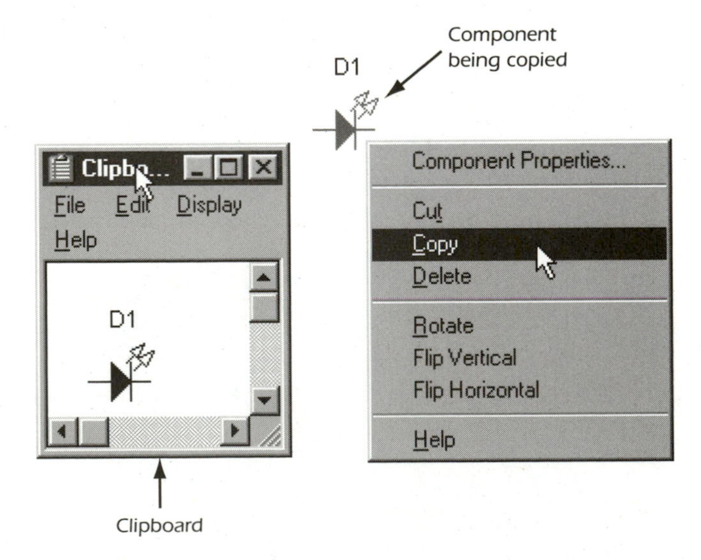

Figure 9.14
Copying or cutting a component places an exact copy in the Clipboard to paste somewhere else.

Select the component(s) you wish to copy. Go to the **Edit** menu. Choose **Copy** or **Cut** (or hit **Ctrl+X** to cut, **Ctrl+C** to copy). Choose **Edit** > **Paste** (or hit Ctrl-V). A copy of that part appears in the center of the drawing for you to move into place. You can also use the toolbar icons that execute the cut/copy and paste commands.

Eliminating the selection step speeds this procedure. Right-click the component. A menu opens. Choose **Copy** or **Cut**. Place the cursor where you want the new component. Right-click and choose **Paste**.

Viewing the clipboard is as easy as going to ***Edit*** *>* ***Show Clipboard****.*

Deleting Components

Now that you have placed 20 parts to test out the interface (four batteries, ten grounds, an oscilloscope, and parts for which you have no idea of their functions) it is time to learn how to delete them. To delete a single component, select the component. Hit the **Delete** key. You can also go to the menu item **Edit** > **Delete** or choose **Delete** after right-clicking the component. A window prompts you to confirm the deletion.

To delete a group of components, use the above procedure after you have selected all of the chosen components.

An undelete feature is not available on EWB V5. I recommend using cut and paste instead of Delete. This at least keeps a copy of the parts' settings.

Place a couple of components onto the screen. Make a copy of one part and paste it somewhere else. Try cutting a component from the drawing. Now delete a few of the other components.

Rotating a Part

It is often necessary to turn a part to make a drawing more aesthetically pleasing and readable. This can be performed with or without a wire connected to it (refer to Figure 9.15).

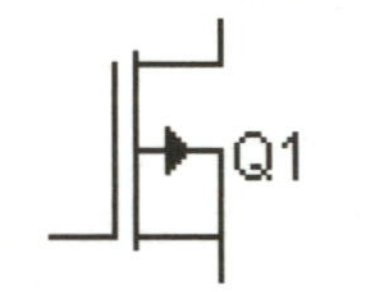

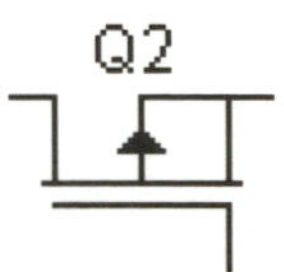

Figure 9.15 Rotating a component.

NOTE

Some items, such as some ICs, cannot be rotated.

To use the toolbar to rotate an item, select the item and hit the **Rotate** icon on the Circuit toolbar.

To use a menu, select the part. Choose **Circuit** > **Rotate** from the menu bar. The part turns 90 degrees counter-clockwise. You can also hit **Ctrl+R** instead of going through the menus. The component must first be selected.

To perform a quick rotate, right-click on top of the component and choose **Rotate**.

Flipping a Part

Mirroring a part along the x- or y-axis is simple (see Figure 9.16).

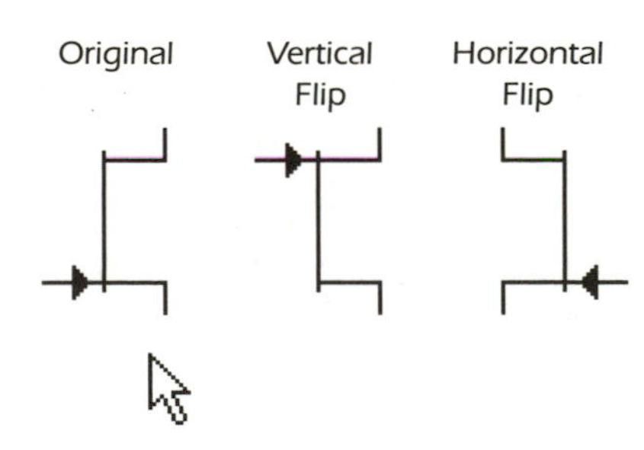

Figure 9.16 Flipping a component.

- **Vertically**—This mirrors the part along a horizontal axis.
  - **Toolbar**—Select the part. Hit the Flip Vertical icon on the circuit toolbar.
  - **Menu or Keys**—Select the part. **Circuit** > **Flip Vertical**.
  - **Quick Vertical Flip**—Right-click the component and choose **Flip Vertical.**
- **Horizontally**—This mirrors the part along a vertical axis.
  - **Menu or Keys**—Select component. Choose **Circuit** > **Flip Horizontal**.
  - **Quick Horizontal Flip**—Right-click on the component and choose **Flip Horizontal**.

Rotating or flipping a group of selected components makes them flip/rotate around their own axis, not the whole selected area.

Place a stack of components onto the screen. Rotate and flip them.

Wiring

Electronics Workbench's various components and instruments must be wired together to create electrical schematics and simulations. Each component contains connection(s) to which a wire can be attached and from which it can be stretched to another connection. Fast wiring techniques make for speedy diagram creation. Here is how to pencil in conductors (see Figure 9.17):

- **Auto Wiring Component**

  1. Move the pointer over the part's terminal until a black dot appears.
  2. Click the left mouse button and hold it. A black line appears.
  3. Move the pointer to the other component terminal and release the left button when on that connection.

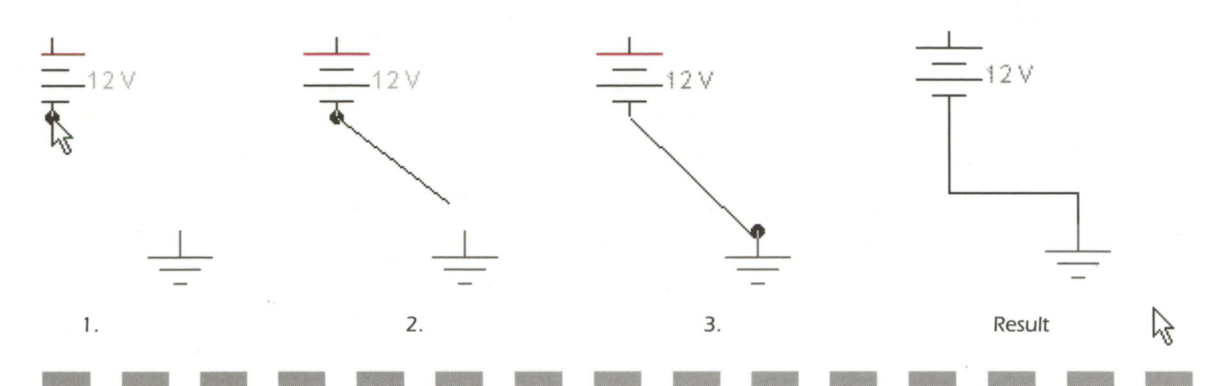

Figure 9.17 Autowiring a component is another three-step process.

- **Another Way to Autowire**

1. Go to **Circuit** > **Schematic Options** and choose the **Wiring** tab.
2. Deselect **Drag to Connect**. Click **OK**. Now you do not have to hold the mouse button down and drag the wire at the same time.
3. Click and release while over the part's terminal.
4. Move the pointer over the new component's terminal and click once more.

Place two resistors onto the screen. Make a connection between two of the connectors.

- **Manual Wiring**—Autowiring can be a less than perfect way to create visually pleasing designs. Solving the spaghetti of wires in a complex circuit can be trying on the eyes and mind. You have the ability to lay wires exactly where you want them. You can still autowire but now you can also draw a line with as many corners as you want (see Figure 9.18).

1. Go to **Circuit** > **Schematic Options** and choose the **Wiring** tab.
2. Click **Manual-route wires.** Hit **OK**.
3. Move the pointer over the part's terminal until a black dot appears.

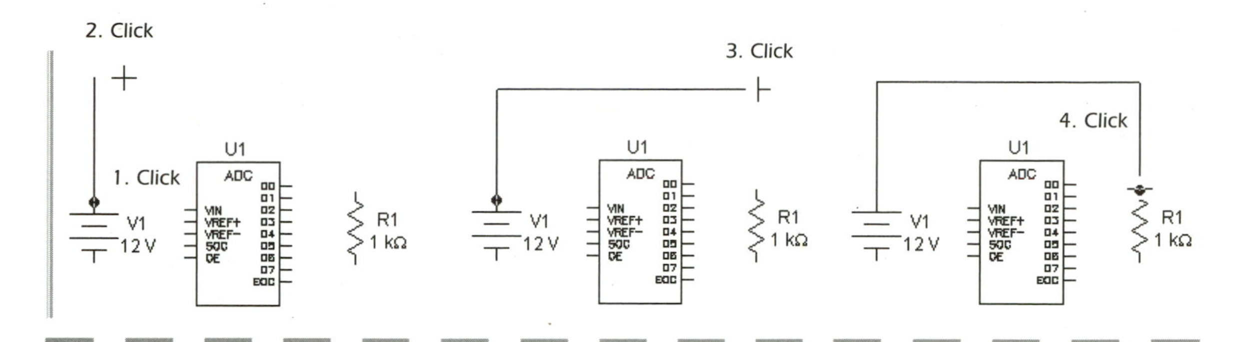

Figure 9.18 Manually wiring components together gives a bit more customization.

4. Click the mouse pointer, then release. A black line and crosshair (+) appears.
5. Move the pointer in the direction you wish the wire to run (horizontal or vertical axis only) and click the mouse button. The wire is still connected to the mouse pointer.
6. Move the mouse to the next point. Do this for each new corner you wish to create.
7. Finish off by clicking on the other component's terminal.

Right-click the mouse button to release the wire from the mouse pointer if you make a mistake while running the wire. Start over.

- **Moving Wires Already Placed**—Electronics Workbench lets you perform minor post operations on the wire; you can move sections of the line slightly (see Figure 9.19).

1. Place the pointer over the section of the wire you wish to move.
2. Left-click and hold. A small double arrow appears.
3. Moving the mouse moves the wire along a perpendicular axis.
4. Release the button when in place.

Changing the Wire's Color

You can change the color of the wire for instrumentation signal identification or color coding. Double-click the wire and the Wire Properties dialog box opens. Click one of six colors then hit **OK**.

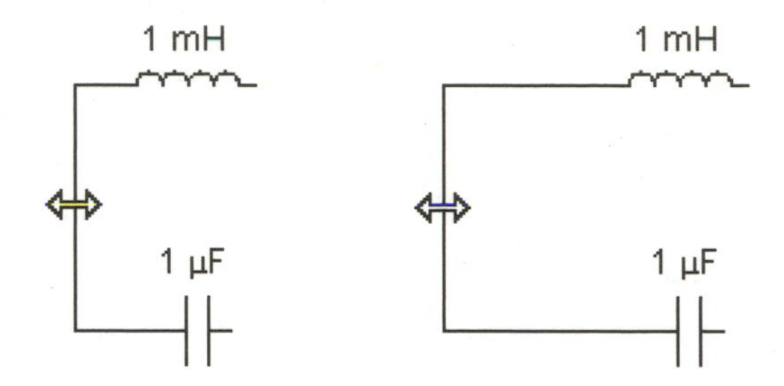

Figure 9.19 Sliding already-placed wires.

TIP

When wires connecting to instruments are colored, this same color appears on the instrument's graph to help identify which signal is coming from which wire. For example, if you color an oscilloscope's signal wire red, the waveform seen on the oscilloscope appears as red.

Deleting a Wire

If a trace is incorrect, you can delete it. Either select the wire then choose **Edit** > **Delete** or hit the **Delete** key. The quickest way to delete a wire is to just right-click over the wire and choose **Delete**.

Inserting a Component into a Prelaid Wire

It is possible to place some components, such as resistors and capacitors, in the middle of a wire and make the terminals connect automatically. Select the part and drag it over the wire. Make sure it is oriented along the same axis. If not, rotate it; the wiring is run automatically (see Figures 9.20 and 9.21).

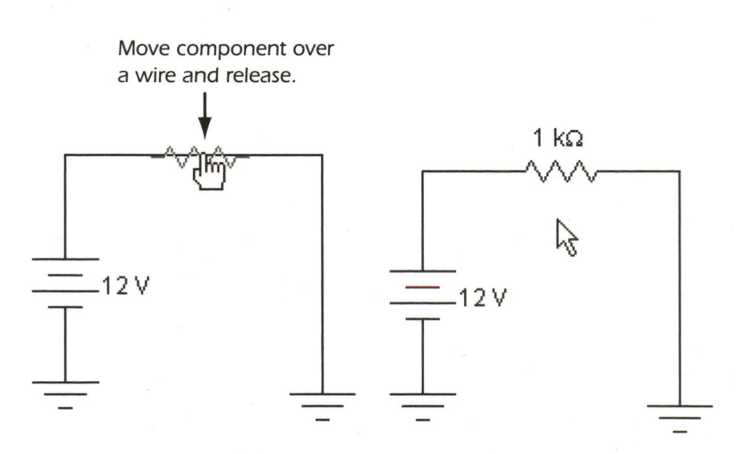

Figure 9.20 Forgot to place a component along a wire that is already stretched? No problem. Just move it over the wire and it will snap into place.

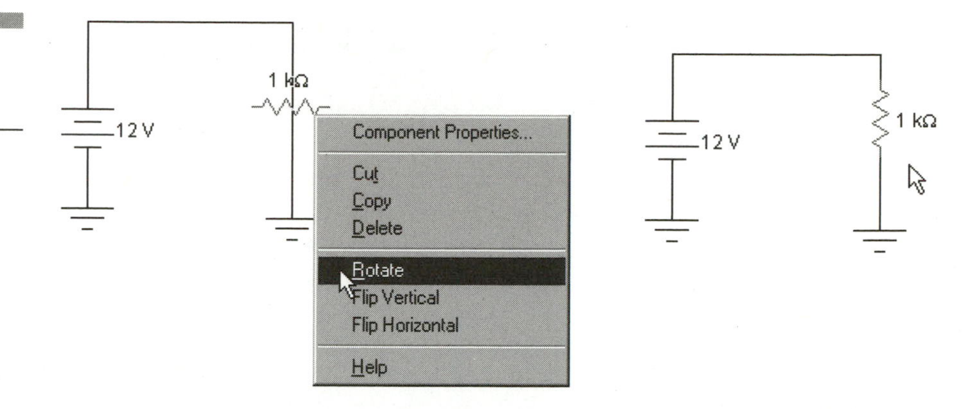

Figure 9.21
Rotating it into the wire works also.

Creating a Connector Anywhere Along a Wire

A connector (junction) can be created anywhere along a wire without having to drag it from the tool bin. If you are already running a wire and want to connect it to the center of another wire, move the tip of the mouse pointer onto the line and release. See Figure 9.22 for an example.

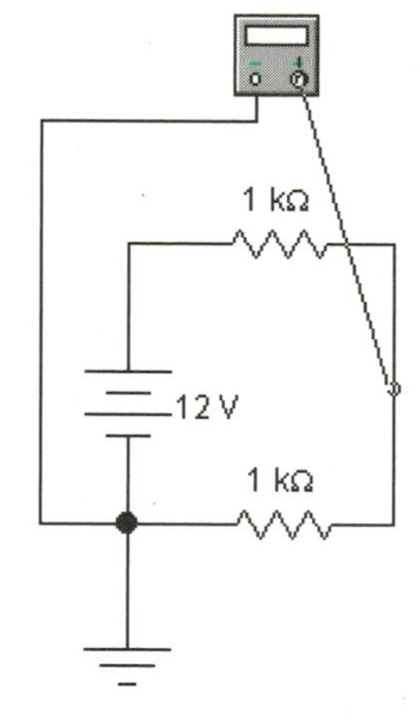

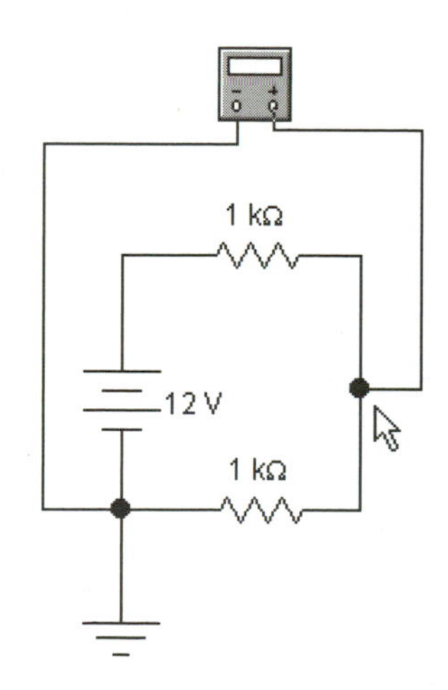

Figure 9.22
Connectors are automatically created when joining a terminal to a wire.

EXERCISE

Place a string of components. Begin wiring them together using the various methods just learned. Make wires connect to each other. Change the color of some of the wiring.

Nodes and Connectors

A node is a common reference point in the circuit. If four component terminals are connected to each other, each wire running to the parts has a common node (see Figure 9.23). Electronics Workbench automatically names the node with a number, but you can change the node name by double-clicking any of the wires that make up the node. This opens the Wire Properties dialog box, Node tab. You will see in Chapter 12, Simulation and Analysis, why these nodes are essential to your drawing.

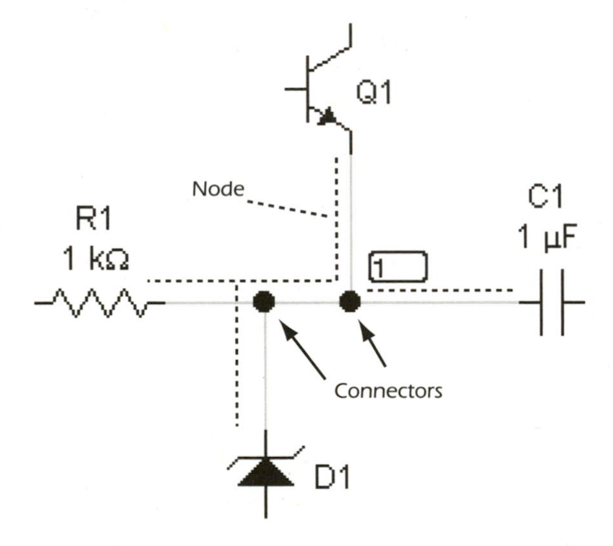

Figure 9.23 Do not confuse a node with a connector. Many connectors can exist on the same node.

If you want to know which wires are part of which node, you can choose to color the node, and thus color each wire along that node. For example, if you want to color the ground node green, double-click any of the wires on the ground node (0) and select the Node tab. Pick the color green from the Set Node Color box. Hit **OK**.

NOTE

The ground node (0) does not display its node number.

Connectors and Junctions

EWB 5's official term for a connection point is *connector*. Do not mistake a connector for a node, because the node can contain many connectors and wires. The connector is merely a drawing feature to let you connect up to four wires cleanly. Also see Figure 9.23.

You cannot hook a component's terminal to itself, as when connecting one pin of an IC to another. Instead, use a connector and place a wire from the component to the terminal and connect back to the component's other pin (see Figure 9.24).

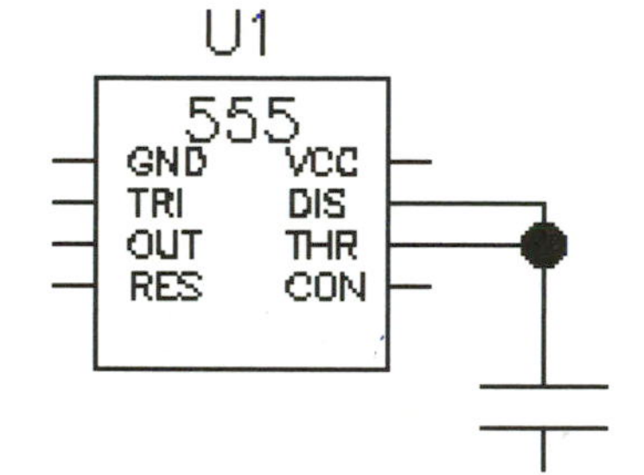

Figure 9.24 Hooking a component to itself requires a connector.

Labels, Values, Node Names and Reference IDs

Electronics Workbench can display a Label, Model Name, Value, or Reference ID for each component, node, and instrument in a drawing (see Figure 9.25). The label is the name you wish to give that part, such as "Load." The alphanumeric Reference ID names the part for schematic reference, such as R1 or Q1. The Model displays the name of the model used for that part such as MMBR941 for a transistor. The Value is the component's electrical value, such as 1kΩ. The node number is given to the entire conductor-path in a circuit, such as the number 1. We can add to this the keyboard letter used to control an electromechanical component such as a potentiometer.

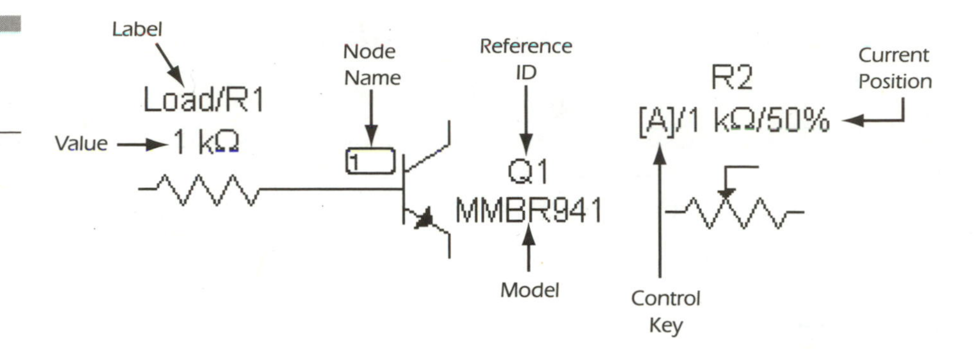

Figure 9.25 Component and circuit labels.

NOTE

Components are described in detail in the next chapter.

To change a component label, double-click the component and select the Label tab. Type in the name you wish to give the component inside the Label box. Hit **OK**.

CAUTION

If the label is not displayed, you must either set labels to display in the Display tab of the Properties box or set the Schematic Option Global Settings correctly.

Once a component is placed, you can change the electrical value by double-clicking it and choosing the Value tab. Type the value into the box and hit **OK**.

The reference ID gives a generic alphanumeric name to each component. For example, if there are five resistors and two capacitors in a circuit, then the first resistor would have a reference ID of R1. The second resistor would be R2, etc. Capacitors would likely be C1, C2, etc. EWB assigns a Reference ID to each component automatically but they can be changed. You can make this reference ID anything you want.

Double-click the component and select the Label tab. Type in the new Reference ID and hit **OK**. If you use an ID name that is already in the circuit, EWB refuses the name.

TIP

If, for example, you want to rename C2 to C1 and C1 to C2 you must first rename C1 to something else, like C-T. Then rename C2 to C1. Finally, rename C-T to C2. Whew!

A number identifies each node. This is sometimes helpful when debugging a circuit and identifying connection points inside simulations or analyses you are running. Electronics Workbench automatically assigns node numbers, but these can be changed. Double-click anywhere along the wire that the node represents and the Node tab on the Wire Properties dialog box opens (see Figure 9.26). Type in the new number and hit **OK**.

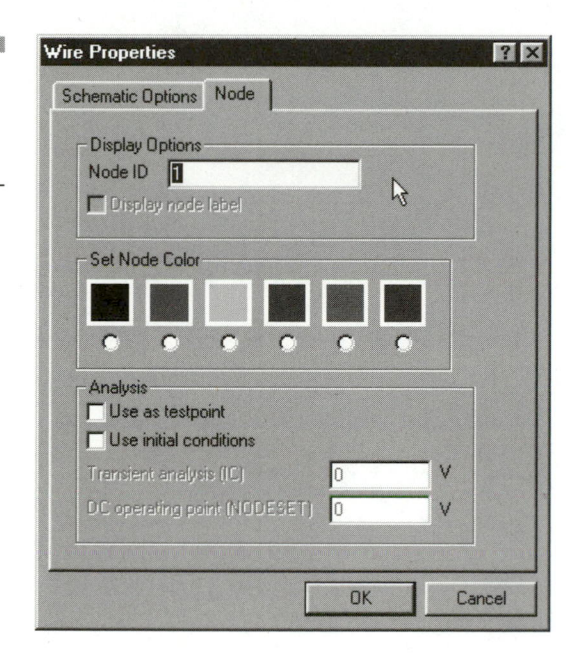

Figure 9.26 Node information can be controlled from this dialog box.

NOTE

Nodes can only be numbered, not named with letters.

Using Global Settings

Global settings are settings that are applied to the entire drawing on which you are currently working. You can choose to display all the Reference IDs, Values, Labels, and Node Names or any combination throughout the entire drawing (refer to Figure 9.27).

1. Right-click in a blank area of the drawing and choose **Schematic Options.**
2. Hit the Show/Hide tab.

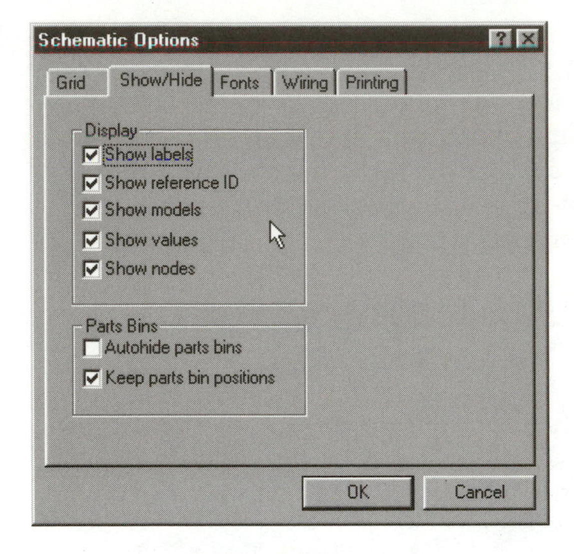

Figure 9.27 Global settings tell EWB to place each of the chosen labels onto this drawing.

3. Place a check in those items you want displayed with the components.
4. Hit **OK**. The setting is applied to each component in the circuit.

Overriding Global Settings

You can override the global settings for any separate component in the circuit. To display only the information you want for that specific component while leaving the rest of the component labels intact (refer to Figure 9.28).

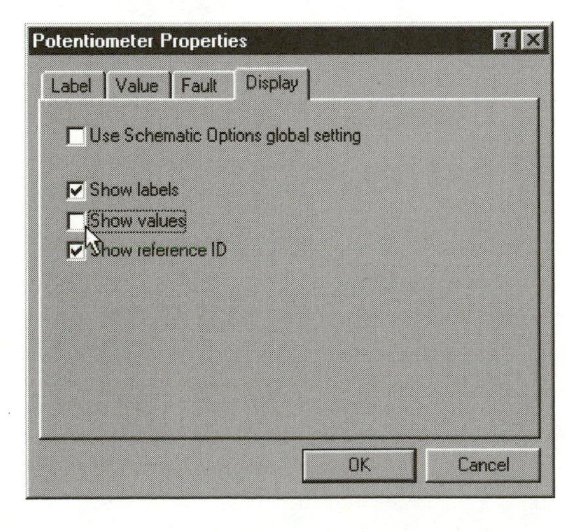

Figure 9.28 If you don't want a specific part in a circuit to use the global display settings, you can override it.

1. Double-click the component you wish to change.
2. Choose the Display tab.
3. Deselect the **Use Schematic Option global setting**.
4. Now check on those elements you want displayed Show labels, Show Values, Show Reference ID.

Text

Adding miscellaneous text is sometimes required to customize a drawing for printouts. You can pencil in personal notes, further labeling, or messages (see Figure 9.29).

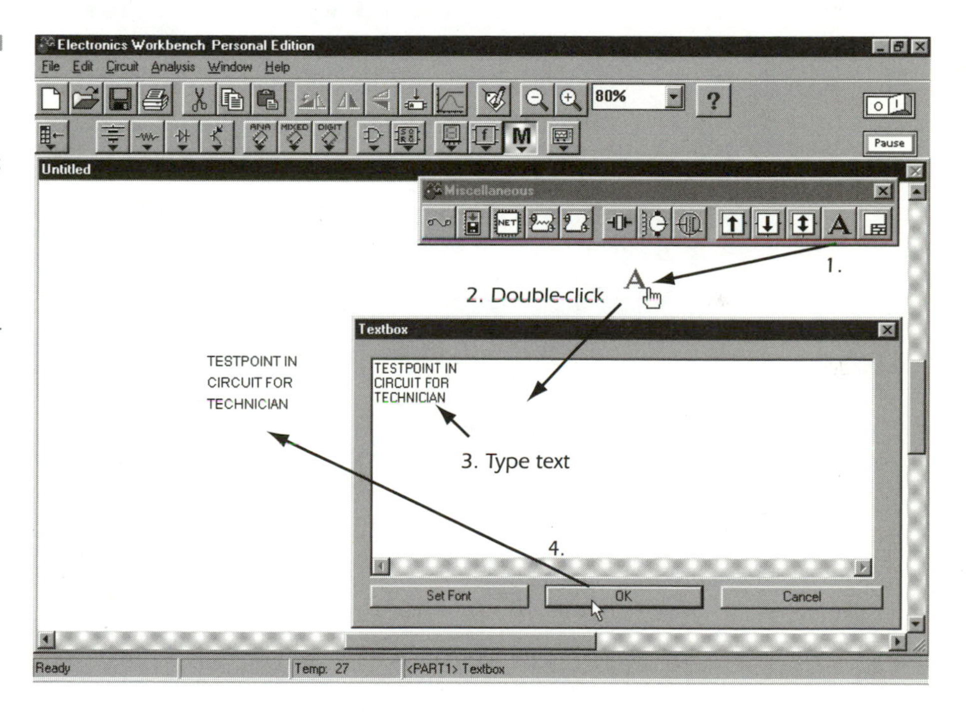

Figure 9.29
Placing text into a drawing helps to further clarify sections of the circuit or add your comments. You also have the ability to change the text color and font.

To place text:

1. Open the Miscellaneous parts bin. The second to last icon on the right is the Textbox. Drag a copy of it to your drawing.
2. A large, bold capital '**A**' appears. Move it to where you want the text placed and double-click it.

3. The Textbox window opens to let you type in the block of text.
4. To set the font, press the Set Font button and choose the appropriate text type.
5. Hit **OK**, **OK**.

Description Window

In addition to placing miscellaneous text on a drawing, you can also add a separate window that describes the circuit in detail (see Figure 9.30). This is a separate movable and sizable window that is typically used to place larger blocks of text such as description of circuit function. To open this window, choose **Window** > **Description** or hit **Ctrl** + **D**. Once the window is open, you can change its size as you would any other window. Click in the middle of the window and begin typing.

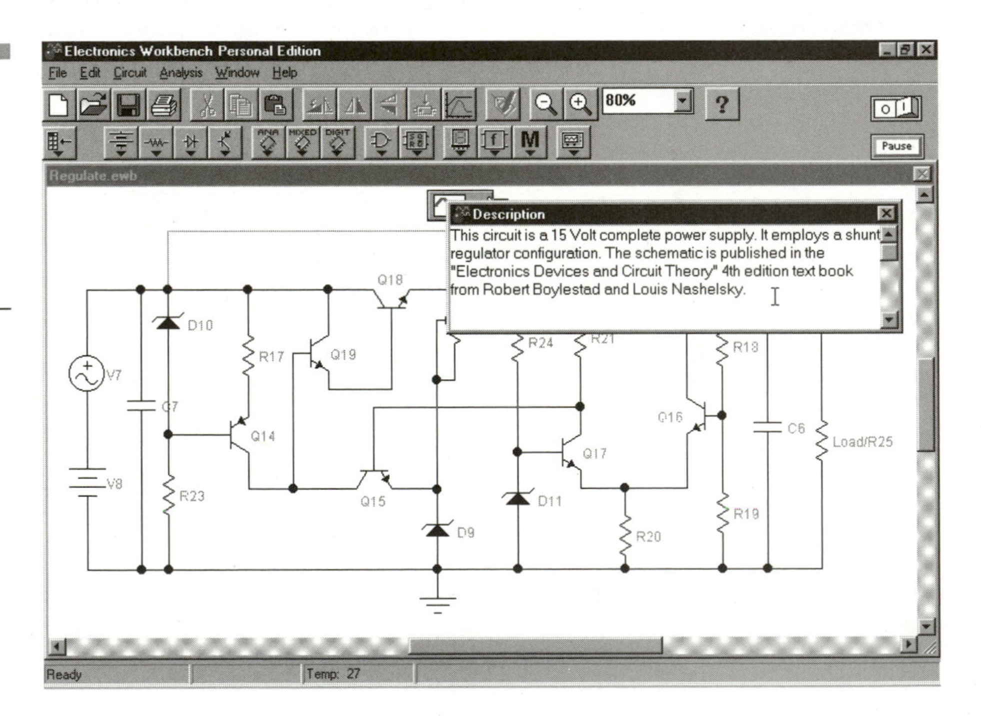

Figure 9.30 The Description window lets you scribble an explanation of the circuit's workings or any other important information.

Subcircuits

"Circuits within circuits within..." This embodies the concept of subcircuits. They are timesavers when you are making designs that use similar circuitry. Simply recycle saved circuits by making them into subcircuit components. These components will include all the modeling, schematics, and values set by you. You merely hook up the wiring of new circuit and let it fly.

To create a subcircuit from an existing circuit:

1. Create the circuit you wish to make a subcircuit. If you already have the components open in an existing drawing, proceed to next step.
2. Select all the components to be part of the subcircuit.
3. Choose **Circuit** > **Create Subcircuit** or hit **Ctrl+B**.
4. A window opens, prompting you to name the subcircuit (eight letters/numbers only). Type in the name and hit **Copy from Circuit** (see Figure 9.31).

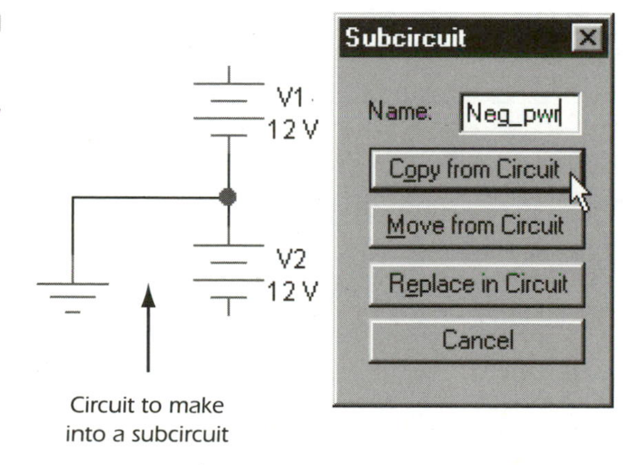

Figure 9.31 Creating subcircuits.

5. The subcircuit opens into a new window. You must create input/output terminals. Drag a wire from each input/output terminal to the edge of the sheet (see Figure 9.32).
6. Your subcircuit is stored in the Favorites toolbar.

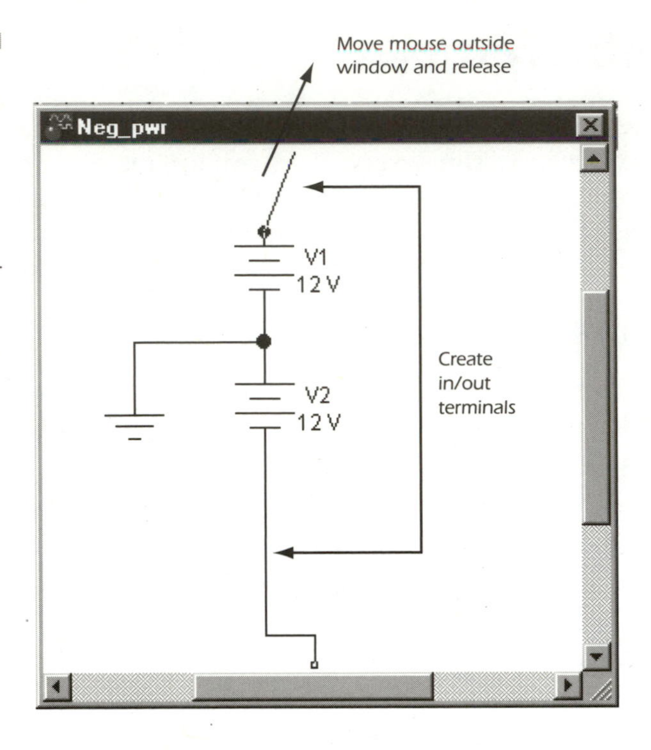

Figure 9.32
Once the subcircuit is in its own window, you must drag connections out to the perimeter of the window.

Choose ***Replace in Circuit*** *to have the components you selected turn into the subcircuit icon.*

To place a subcircuit into the current circuit, go to the Favorites toolbar and move a copy of the Sub icon to the drawing (Figure 9.33). A window opens, allowing you to select the subcircuit name. Once this is chosen, hit **Accept**. The new subcircuit icon is now on your schematic.

To edit or view a subcircuit, double-click its icon.

Printing Electronics Workbench Files

Have computers created a paperless society? Are you kidding? If anything, they have increased the amount of paper output from organi-

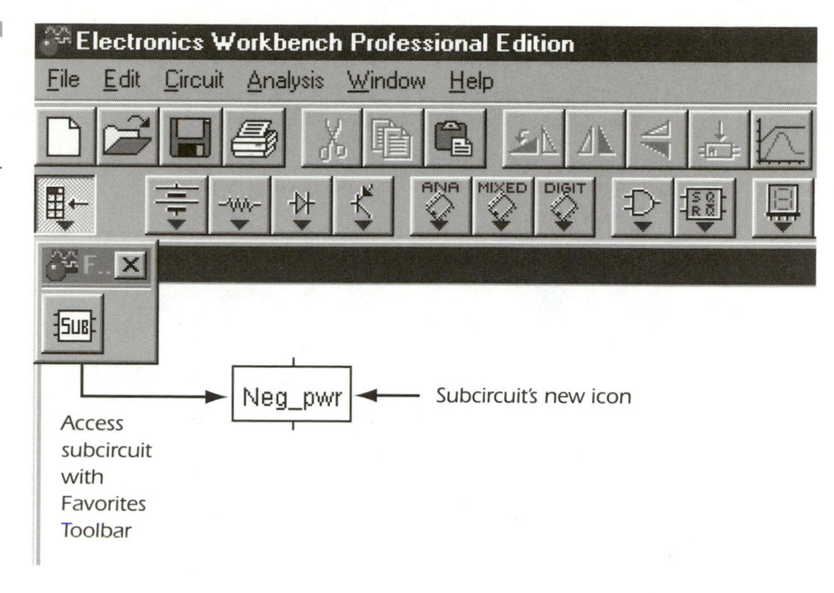

Figure 9.33 Placing a subcircuit is accomplished with the Favorites toolbar.

zations. We still require circuit printouts for fast reference and a more reliable record than the typical computer hard drive file. Electronics Workbench allows you to print the schematic drawing, description window, parts list, model list, subcircuits, and any of the instrument's readouts.

NOTE

The EWB V5 demo restricts you from printing circuits.

Printer Setup

When you select **File** > **Printer Setup**, the Printer Setup dialog box appears. This may be different from that shown in Figure 9.34, because each printer's setup is unique. To select a printer, choose it from the List box on the top. If you wish to adjust the printer's properties, hit that button and make the appropriate settings according to your printer's manual or help file.

You can select the size of the paper, its orientation and, in the case of my laser printer, the tray from which the circuit will print.

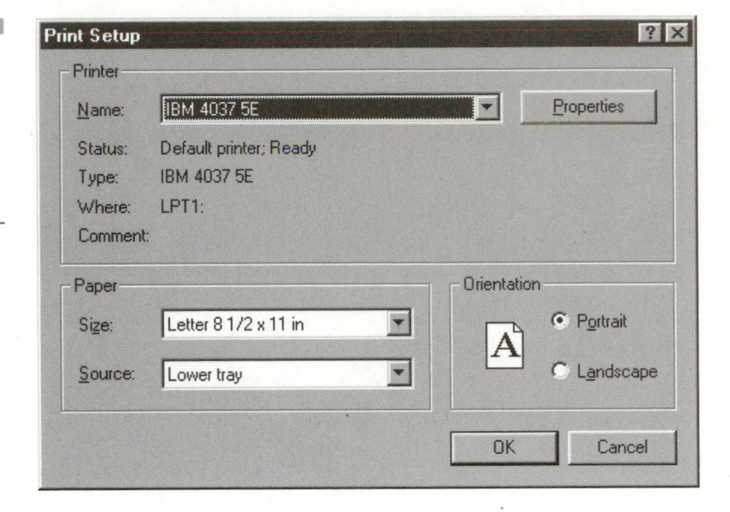

Figure 9.34
The Printer Setup dialog box tells the Windows printer exactly what to print and how it will look.

Printing a Circuit

EWB has the ability to print out a variety of items besides the circuit itself. **File > Print**, or hitting **Ctrl+P**, opens the EWB Print dialog box (see Figure 9.35). From here, you can check on which items you wish to print. Hitting **Print** sends the results to the printer. Items are printed one after the other, in the order you checked them.

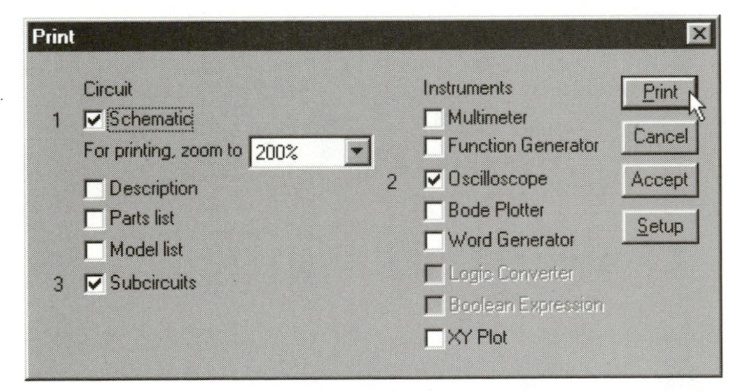

Figure 9.35
What do you want to print today?

TIP

EWB prints items in the order you checked them. See the numbers next to the checkboxes in Figure 9.35.

NOTE

If an item is greyed-out, it means there is nothing to print in that category.

- **Schematic**—Prints the schematic drawing currently on your screen.
- **For printing, zoom to**—Adjusts the zoom level of the printout.
- **Description**—Prints the contents of the Description window.
- **Parts List**—Prints a summary of all the components used in your circuit.
- **Model List**—Prints a complete breakdown of each component model used in a circuit.
- **Subcircuits**—Prints the subcircuit as though it were a regular schematic circuit.

NOTE

The printout uses as many pages as needed to list each item you asked to print.

- **Instruments**—If you used the multimeter, function generator, oscilloscope, Bode plotter, word generator, logic converter, Boolean expression, or XY plot in your circuit, their settings and results can be printed, too.

TIP

Hit the **Print** *button on the toolbar if you wish to print the circuit only.*

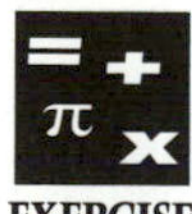
EXERCISE

Print a few circuits. Run the simulation and stop the instruments. Print their results.

NOTE

See Chapter 12, "Simulation and Analysis," for more information on printing graphs and simulation results.

Zooming In and Out of the Circuit Window

Sometimes it is necessary to zoom in or out of a section of your drawing. If you are making a rather large schematic you may need to zoom out. Packing a lot of components into a small area may require a zoom in. Electronics Workbench can zoom into a screen-filling 500 percent or out to a puny 20 percent. There are many ways to activate zoom in and out:

- **Menu**—Choose **Circuit** > **Zoom In/Out**: This zooms in or out one level. The levels are 20, 40, 60, 80, 100, 120, 150, 200, 250, 300, 350, 400, 450, and 500 percent.
- **Tool Bar**—Click the icon on the circuit tool bar that looks like a magnifying glass with a plus or minus sign inside. You can also select the pull-down box to adjust to a pre-set zoom level.
- **Hot Keys**—Hit the Ctrl key, then the + or − sign on the right side of the keyboard (the keypad). Only the − above the keyboard letters will work.

Open a circuit and try each method of zooming. With which one are you most comfortable?

Summary

Drawing with Electronics Workbench can be simple for some people and difficult for others. Practice the procedures until you feel confident and able to whip out passable designs. After that, check out Chapter 14, Unleashing Multisim/EWB, to learn cleaner drawing techniques and tricks to speed you along. I still think drawing with a pencil and paper is faster. But then again, you can't simulate a hand-drawn graphite circuit.

CHAPTER 10

EWB Building Blocks: Components

Components are the building blocks required to assemble our electronic toys.

Building virtual circuits requires an understanding of Electronics Workbench V5's building blocks. This chapter is a guide that describes the components that make an EWB circuit. To use the software efficiently, it's important to have an understanding of the various virtual components and how they function, which includes a working knowledge of everything from simple resistors, to MOSFETs, to complex digital ICs. Each component has values, simulation factors, electrical properties, symbols, and the like that you must understand to run your simulations more efficiently. This chapter also gives you the foresight to model more realistic components of your own design as well as use the pre-existing components shipped with EWB V5. This chapter will be a valuable component reference for your designs, and will allow you to visualize just how your computer sees these virtual parts.

Overview of EWB's Virtual Components

Resistors, capacitors, inductors, transistors, and hundreds of other electrical components come together to make modern electrical devices. Each electronic component performs a set task in relation to current, voltage, or signal. For example, a resistor can be used to adjust voltage or current in a circuit. A capacitor can temporarily store a charge and later release it, or block DC and pass AC. Electronics Workbench contains a virtual representation of thousands of these real-world electronics components; instead of being made of metals, plastic, carbon, ceramic, and silicon, they are calculations and schematic symbols that perform the same function inside the computer. A group of virtual components wired together gives you an idea of how a real circuit performs and the results it produces. So when I talk about components in relation to EWB V5, I am referring to all the information that lets your computer simulate a real-world component.

A symbol does not an Electronics Workbench component make. An EWB component is a model used in simulations, a set of electronic properties (Figure 10.1) that are used in specialty scenarios, package information of the real-world part and, of course, a way to label the component. Together this information can be used in millions of simulations and analyses.

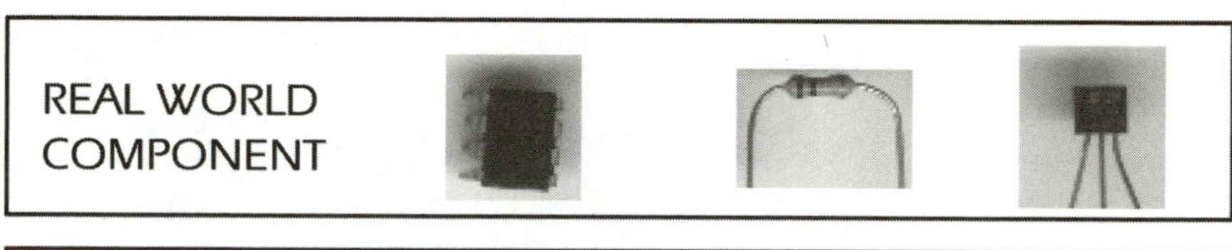

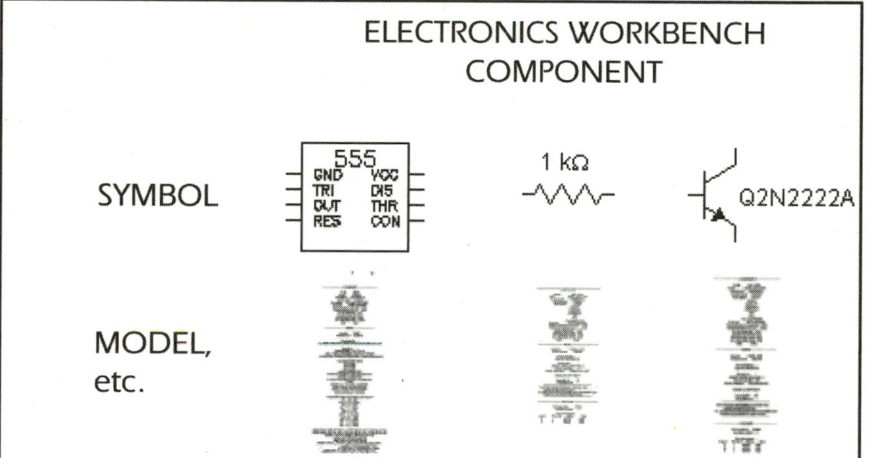

Figure 10.1 An EWB component is a complex mix of mathematics, symbols, labels, and PCB information; not just a schematic symbol.

EWB Component Library Structure and Access

Electronics Workbench components are stored in a database and each component is accessible through its toolbin. You open toolbins by selecting a component group icon from the Parts Bin toolbar and clicking it. Once you select a component, you can drag-and-drop it onto your circuit drawing. Once the part is placed, you can further access its variables and models by double-clicking it (see Figure 10.2). To further complicate things, model libraries (described later) are also available.

Schematic Symbol

Electronics Workbench uses a schematic symbol to represent each electrical component, which includes the value, reference ID in the circuit, model name, and a personalized label (see Figure 10.3). The daunting task of drawing schematics by hand is now eliminated

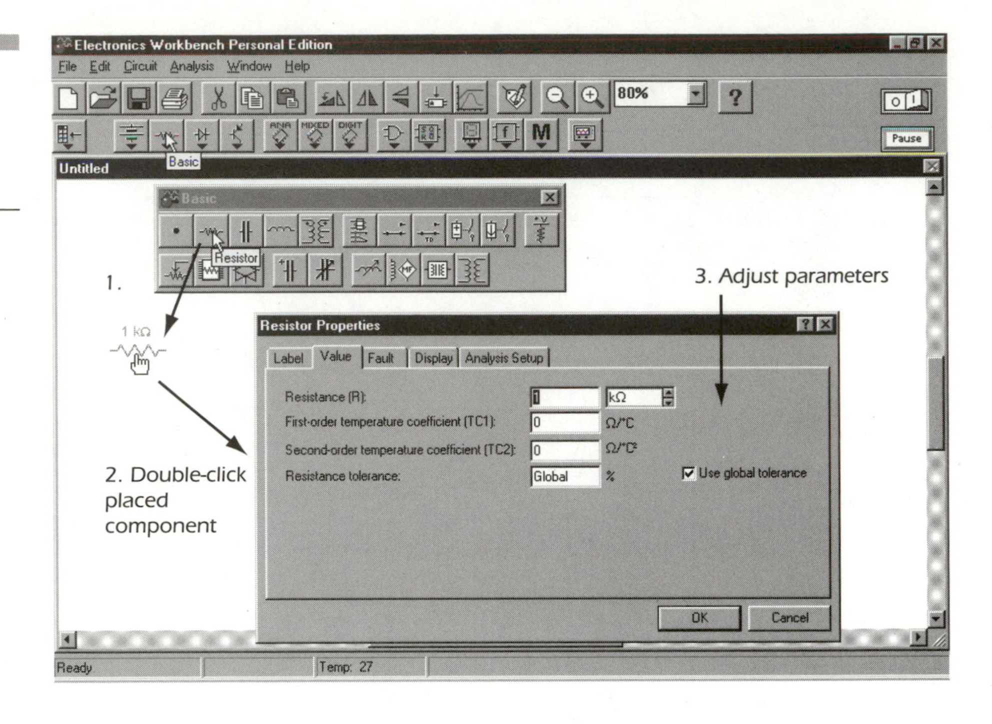

Figure 10.2 Accessing EWB components and their parameters is easy as 1-2-3.

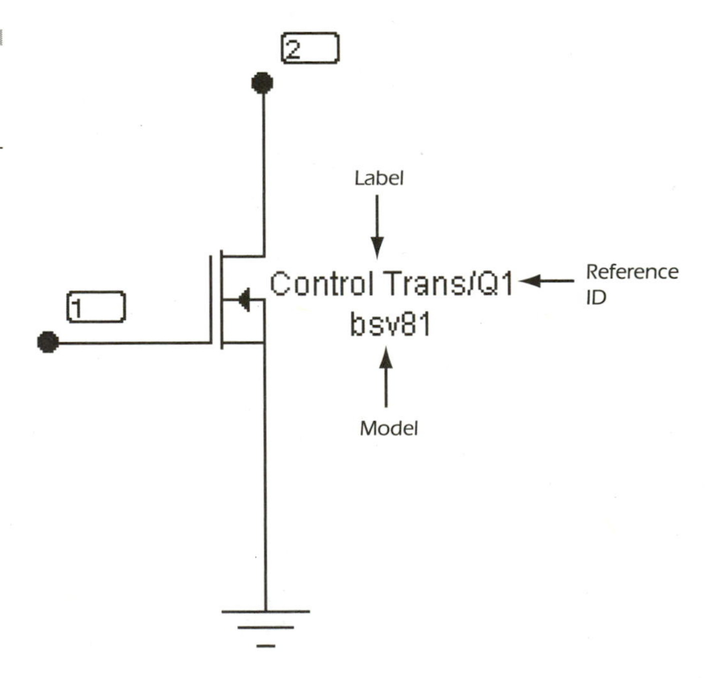

Figure 10.3 Labels of a component.

through *schematic capture.* Each symbol has terminals to connect wires (conductive paths) to other symbols. In this way, the schematic drawing can be created.

NOTE

Another advantage to drawing schematics by computer is the ability of the diagram to be used as a pseudo-breadboard. With it you can quickly "test" a circuit.

Component Models

A component model is a computer representation of the real-world component, which gives the computer the information it needs to simulate that part within a circuit. The component model is pooled with the schematic symbol to create an Electronics Workbench component. For example, a BJT transistor symbol may be combined with a mathematical model of a 2N2222A to create a virtual National transistor.

Component models are creating with several languages or conventions, but Berkeley SPICE is by far the favored; with this language, thousands of real-world components can be modeled.

How Parts Are Simulated

Place a part onto your drawing—begin with a resistor (see Figure 10.4). You will notice that it has two terminals that connect to additional components using a wire as a conductor. The various terminals in the circuits, along with their paths, are called *nodes.*

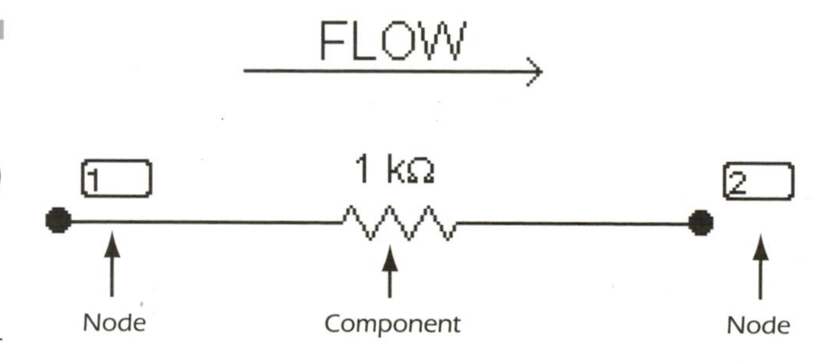

Figure 10.4 The component's information (mathematical model) tells EWB what it is doing when placed between nodes in a circuit.

Call the left side of the resistor node 1 and the right side node 2, and assume that electricity is flowing from node 1 to node 2. While it goes through the resistor, it experiences a change in voltage, amperage, frequency, resistance, or capacitance, etc. In this case, a resistance is created in the flow of electricity. A component model tells the computer exactly what is taking place electronically between the nodes.

This is done using a modeling language such as Berkeley SPICE (which is explained in Chapter 15, SPICE & Netlists).

NOTE

Other modeling languages can be used, but SPICE is by far the most popular.

Each component used in Electronics Workbench 5 uses a mathematical model with adjustable values and input characteristics that closely approximate a real component's properties and the changes that occur in the electrical flow between nodes. EWB contains a powerful database (library) of preset models that you can use for your own custom parts.

Depending on which edition of EWB Version 5 you own, there can be up to 4,000 premade models to use in your designs. Model expansion packs are also available from the Electronics Workbench company (for a premium); these add another 2,000 models.

Accessing Existing Models

EWB 5 contains up to 4,000 premade models of various components such as diodes, transistors, various analog and digital integrated circuits, transmission lines, crystals, and motors. These are accessible by double-clicking a component to open its Component Properties dialog box. From there, a Model tab is available (see Figure 10.5). Each similar model is stored within a library, which can contain one or a thousand different real-world models available to use in your designs.

Component Model Libraries

Each EWB component library contains several models within the same domain. For example, under the Crystal Properties window you

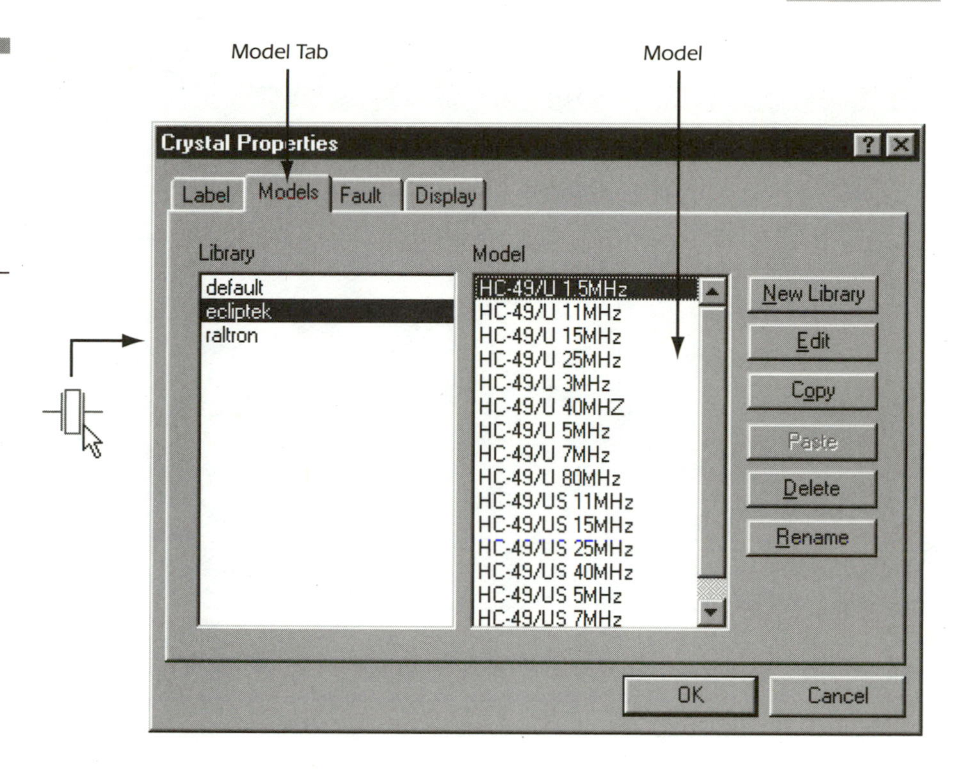

Figure 10.5 Double-click the place component and choose the Model tab to access the component's model window.

see a library called ECLIPTEK. (Figure 10.5). This company makes crystal-based components and has provided several models for that library. Clicking on a library highlights it, and shows each component model available in the library, in the field next to it.

Model Field

The model field contains all the models in a specific library. Want to see or edit specs? Select the specific model, hit the **Edit** button and the Component's Model dialog box opens (see Figure 10.6). Each field can be adjusted to your tastes. Some components contain several pages of specs that can be filled in or modified. Most component models are accessed in this way.

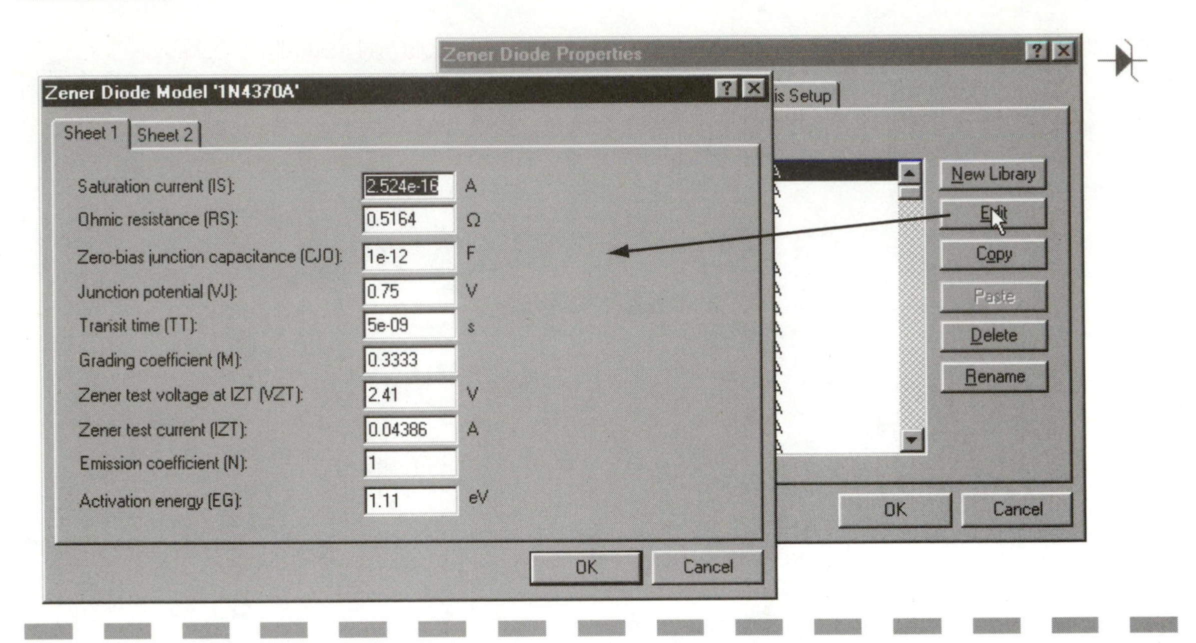

Figure 10.6 Use the Edit button to access the meat of the component's model.

How to Model Your Own Components

Not every component has a complex model, but parts such as transistors require a lot of information to simulate and thus contain a complex model. Electronics Workbench does not contain a database of every manufacturer's components in the entire world. If a component you are looking for has no existing model, you must create your own from a datasheet or databook. Most companies provide a datasheet with a real-world component's electrical characteristics listed (see Figure 10.7). Datasheets are usually downloadable free from the Internet (on the manufacturer's website) or offered in a databook that contains many datasheets. Databooks can be ordered, sometimes free of charge; contact the manufacturer of the component for further information about datasheets. Once you have a datasheet, you can double-click the component and choose the Model tab to input the information and create the component's computer model.

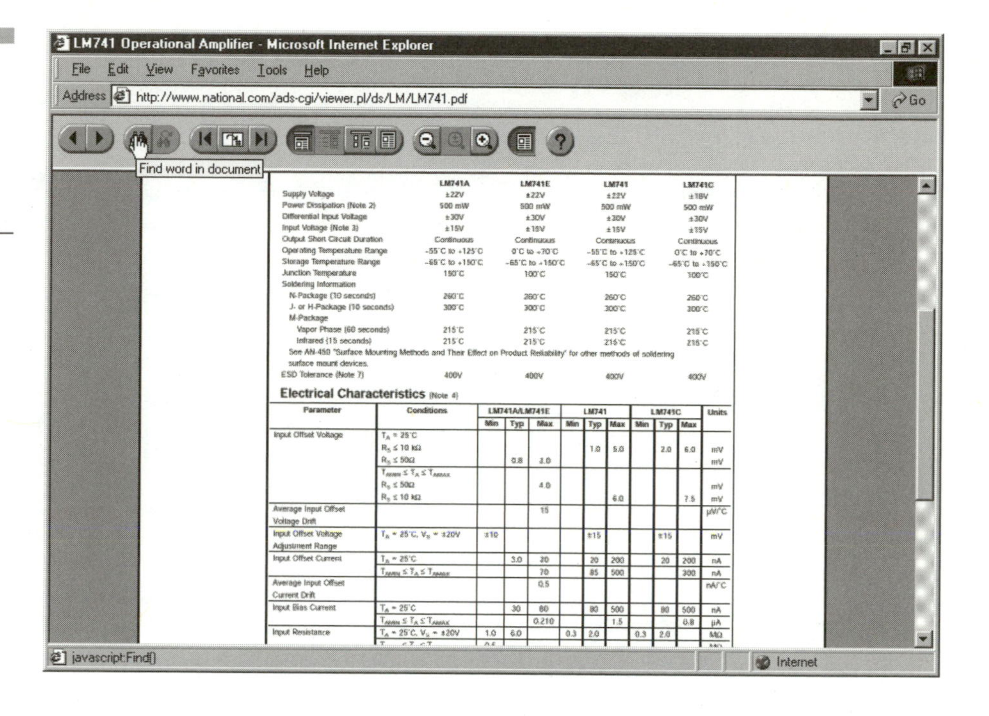

	LM741A	LM741E	LM741	LM741C
Supply Voltage	±22V	±22V	±22V	±18V
Power Dissipation (Note 2)	500 mW	500 mW	500 mW	500 mW
Differential Input Voltage	±30V	±30V	±30V	±30V
Input Voltage (Note 3)	±15V	±15V	±15V	±15V
Output Short Circuit Duration	Continuous	Continuous	Continuous	Continuous
Operating Temperature Range	−55°C to +125°C	0°C to +70°C	−55°C to +125°C	0°C to +70°C
Storage Temperature Range	−65°C to +150°C	−65°C to +150°C	−65°C to +150°C	−65°C to +150°C
Junction Temperature	150°C	100°C	150°C	100°C
Soldering Information				
N-Package (10 seconds)	260°C	260°C	260°C	260°C
J- or H-Package (10 seconds)	300°C	300°C	300°C	300°C
M-Package				
Vapor Phase (60 seconds)	215°C	215°C	215°C	215°C
Infrared (15 seconds)	215°C	215°C	215°C	215°C

See AN-450 "Surface Mounting Methods and Their Effect on Product Reliability" for other methods of soldering surface mount devices.

	LM741A	LM741E	LM741	LM741C
ESD Tolerance (Note 7)	400V	400V	400V	400V

Electrical Characteristics (Note 4)

Parameter	Conditions	LM741A/LM741E			LM741			LM741C			Units
		Min	Typ	Max	Min	Typ	Max	Min	Typ	Max	
Input Offset Voltage	$T_A = 25°C$, $R_S \le 10\ k\Omega$					1.0	5.0		2.0	6.0	mV
	$R_S \le 50\Omega$		0.8	3.0							mV
	$T_{AMIN} \le T_A \le T_{AMAX}$, $R_S \le 50\Omega$			4.0							mV
	$R_S \le 10\ k\Omega$						6.0			7.5	mV
Average Input Offset Voltage Drift				15							μV/°C
Input Offset Voltage Adjustment Range	$T_A = 25°C$, $V_S = \pm 20V$	±10				±15			±15		mV
Input Offset Current	$T_A = 25°C$		3.0	30		20	200		20	200	nA
	$T_{AMIN} \le T_A \le T_{AMAX}$			70		85	500			300	nA
Average Input Offset Current Drift				0.5							nA/°C
Input Bias Current	$T_A = 25°C$		30	80		80	500		80	500	nA
	$T_{AMIN} \le T_A \le T_{AMAX}$			0.210			1.5			0.8	μA
Input Resistance	$T_A = 25°C$, $V_S = \pm 20V$	1.0	6.0		0.3	2.0		0.3	2.0		MΩ

Figure 10.7
You can download datasheets from the component manufacturer's website.

Adjustable Variables

Each component has a specific model with values adjustable by the user or preset to a default value. For example, if we are dealing with a capacitor, the capacitance is adjustable as well as the tolerance (see Figure 10.8). If a component such as a transistor, has many values, a model is created. It contains each variable that is changeable by the user (see Figure 10.9). Other variables include the temperature of the part while in simulation, timing factors, response items, and many others. To see exactly what is adjustable for any part, double-click that component and choose the Value tab. If the component has several variables in the model, click the Models tab (a value tab will not be available). Select the appropriate model and click the **Edit** button. There may be several sheets of variables to complete.

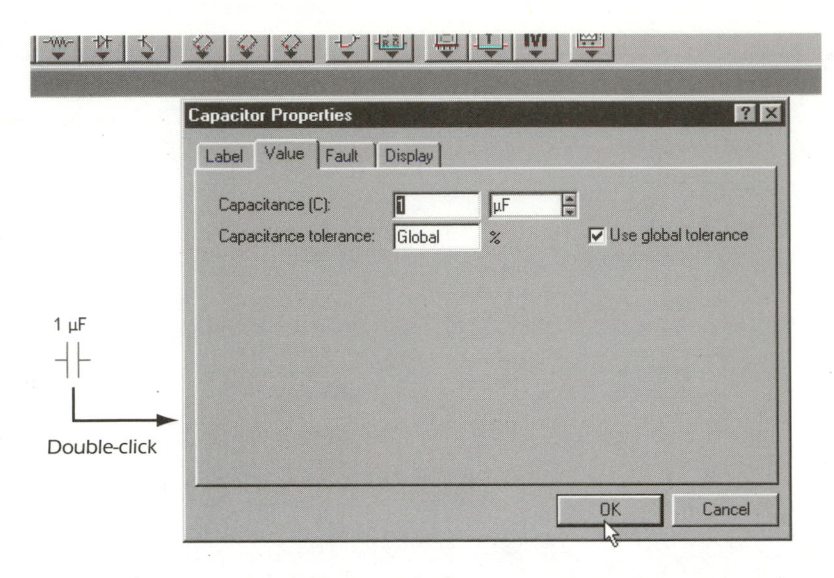

Figure 10.8 You can adjust the value of a component by double-clicking it and using the Value tab.

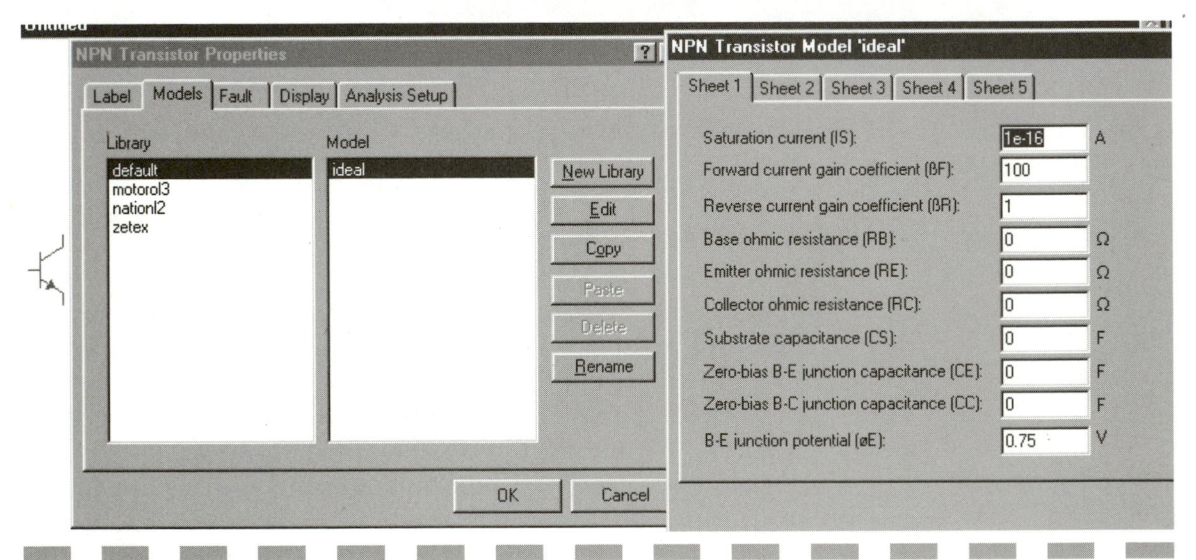

Figure 10.9 Other components require you to select a specific model using the Model tab.

Creating Your Own Parts from Scratch

Not every component is available to use within an EWB simulation, even though there are thousands of common electronic components that are fully modeled and available to you. EWB allows you to make minor modifications to these existing components, thus giving you the ability to create a custom component. This is done by copying an existing component, making modifications, and putting the model into your own personal library of component models.

Creating a Library

Let's use a Zener diode as an example of creating a personal library (see Figure 10.10).

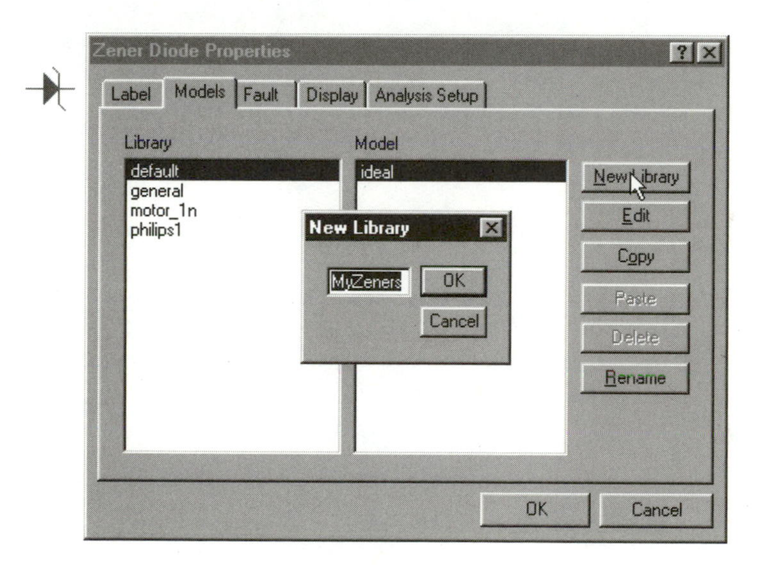

Figure 10.10 Creating component libraries.

1. Place a Zener diode onto a fresh drawing.
2. Double-click it or hit the Component Properties icon. The Zener diode properties box opens.
3. Choose **Models**.
4. Hit the **New Library** button. Name your new library something like MyZeners. The name must be eight characters or less.
5. Choose **OK**. The library is now added to the list.

Copying a Component to Your Library

Now that a custom library is created for you, it is time to add the models (see Figure 10.11).

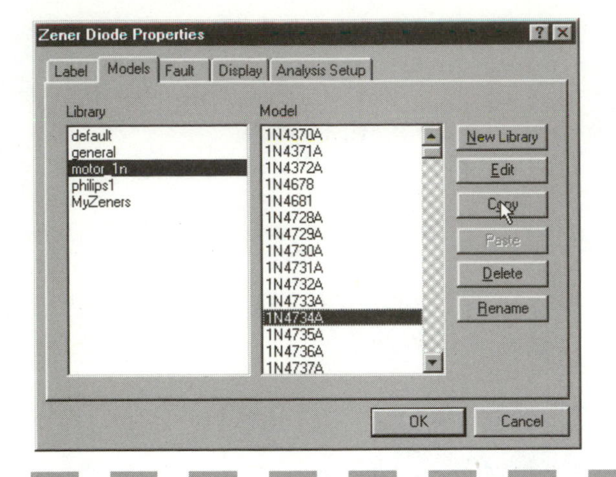

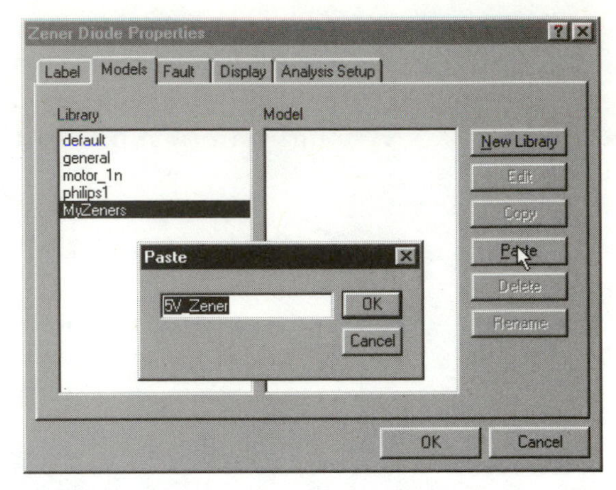

Figure 10.11 Copying a component's model to another library.

1. Click onto the **motor_1n** library and select **1N4734**.
2. Hit the **Copy** button.
3. Click onto **MyZeners** library once more and choose **Paste**.
4. A box appears, asking you to rename it. Lets call it **5V_Zener**. Hit **OK**.
5. Choose the **Edit** button.
6. Change the 5.624 in the Zener test voltage at IZT (VZT) to 5.0.
7. Hit **OK** and **OK** again.

NOTE

When you exit EWB, you will be asked if you wish to save the changes to the Zener diode library MyZeners. Choose yes to save your hard modeling work.

Identifying Parts

If you dragged a part onto the circuit window, only the value of that component is visible. However, it is nameless. You discern it from all the other similar parts by further identifying it. Look at Figure 10.12. Which diagram is more understandable: A or B? There are two ways to perform this task, each with multiple procedures.

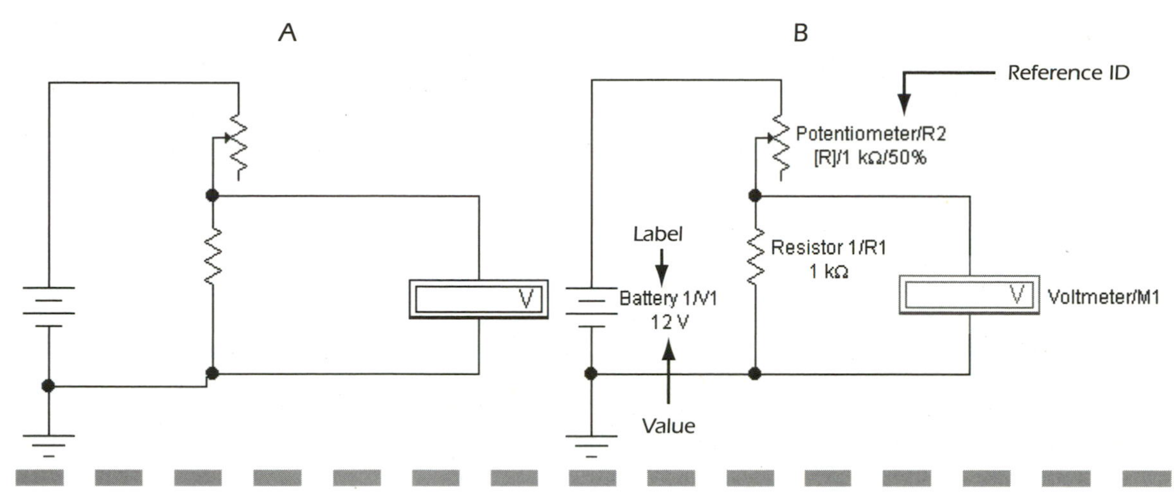

Figure 10.12 Labeling a component helps to identify it clearly. Which would you prefer, A or B?

One is to use a label and the other is to make EWB display the component's *reference ID*.

Labels

If you wish to call the part by a name that suits your tastes, use the following procedure (see Figure 10.13):

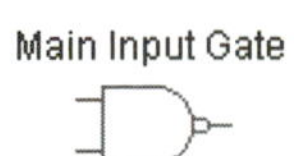

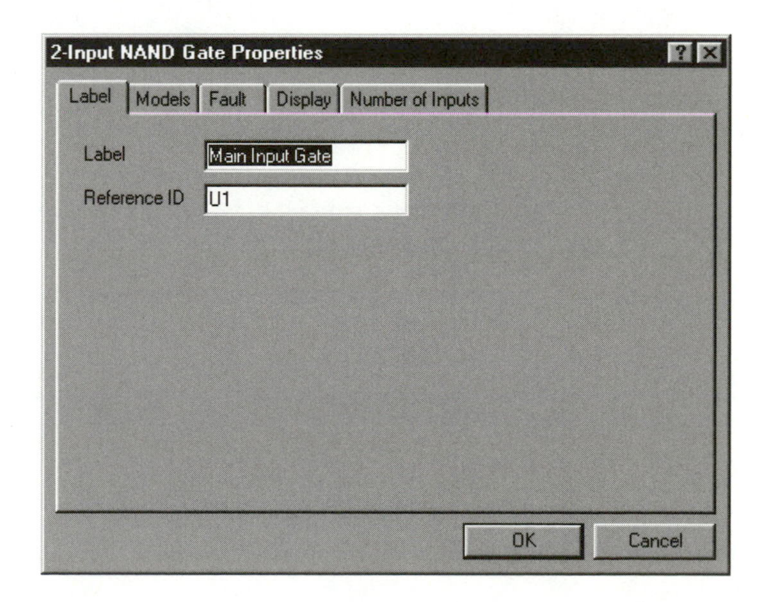

Figure 10.13 Label a component by double-clicking it and selecting the Label tab.

1. Select the component.
2. Open the Component Properties for that part by double-clicking it.
3. Click the **Label** tab.
4. Type in the name or number by which you wish to call this component.
5. Click **OK**.

NOTE

*If you do not see the label, go to **Circuit > Schematic Options > Show/Hide Tab** and make sure **Show labels** is checked on.*

Reference IDs

You can let EWB name the part for you with a generic tag like R1, C3, or Q2 (see Figure 10.14). This is perfectly acceptable for most applications useless you wish to further customize the information with a label. There are two methods to make Electronics Workbench display a components' reference ID using its schematic options.

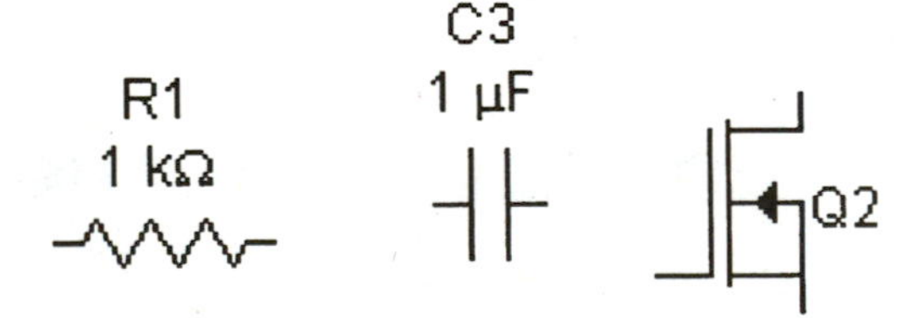

Figure 10.14 Using reference IDs makes identifying parts easier.

CHANGE THE SCHEMATIC OPTIONS

Using the Show/Hide Tab in Schematic Options, you can tell EWB what you want the component to display. In this case, we want the reference ID to show automatically for every part (see Figure 10.15). Here's how:

1. Open the **Circuit** menu and choose **Schematic Options** (or right-click on a blank area in the circuit window and choose Schematic Options from the pop-up menu).
2. Click the **Show/Hide** tab.
3. Check on the Show Reference ID option under display.

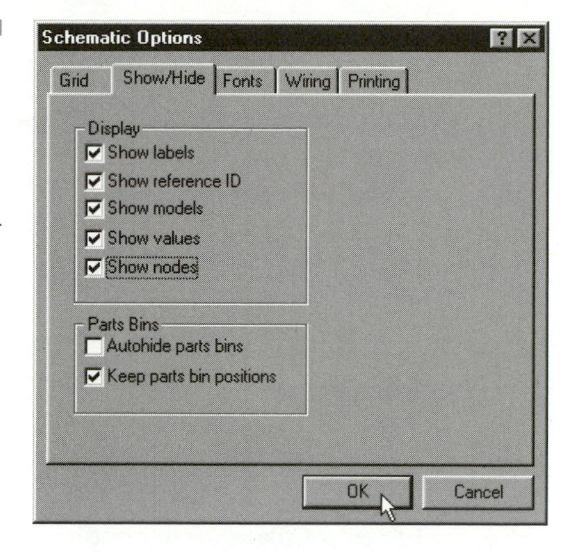

Figure 10.15 Make sure the drawing is set to display reference IDs using the Schematic Options dialog box.

4. While you are at it, check each item you want displayed globally.
5. Click **OK**.

This displays the reference ID of every part in your circuit.

PARTICULAR PART DISPLAY

If you want EWB to display only the reference ID of a component on which you are working, use the following procedure (see Figure 10.16).

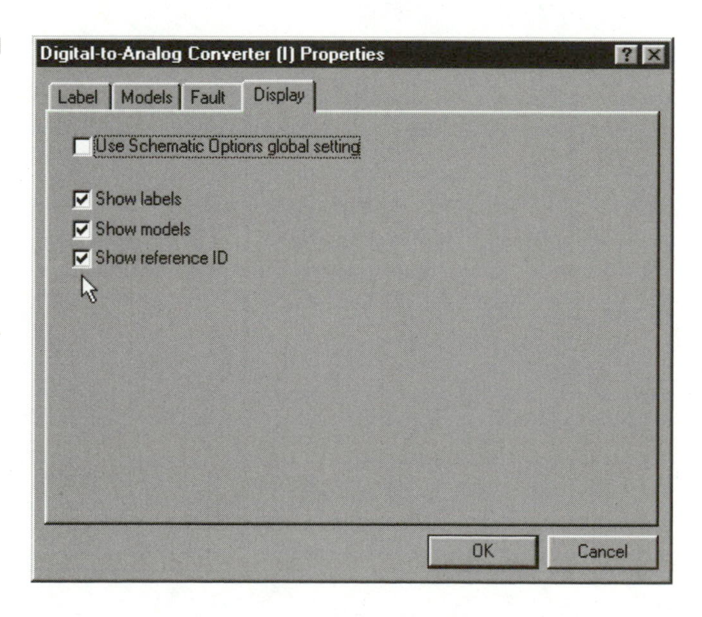

Figure 10.16 You can make EWB display only a specific component's ID and label items by double-clicking the placed item and selecting the Display tab.

1. Double-click the component.
2. Click the **Display** tab.
3. Uncheck the **Use Schematic Options global setting** box and only place a check in those that apply.
4. Click **OK**.

Creating a Fault in a Component

Electronics Workbench Version 5.x contains a powerful feature that deals with more realistic training and simulation. It lets you create a flaw in a circuit either to simulate those conditions for design reasons (torture testing) or to help train a technician. For example, if you are simulating how a circuit would react to a deadly short, you can view its results with this feature. Or if there is a short in a circuit somewhere (embedded into the simulation by the instructor), you can tell a student technician to use the instruments to locate the fault.

In the Component Properties of each part is a tab titled Fault, which opens up the component's fault options and allows you to control that part's glitches (see Figure 10.17).

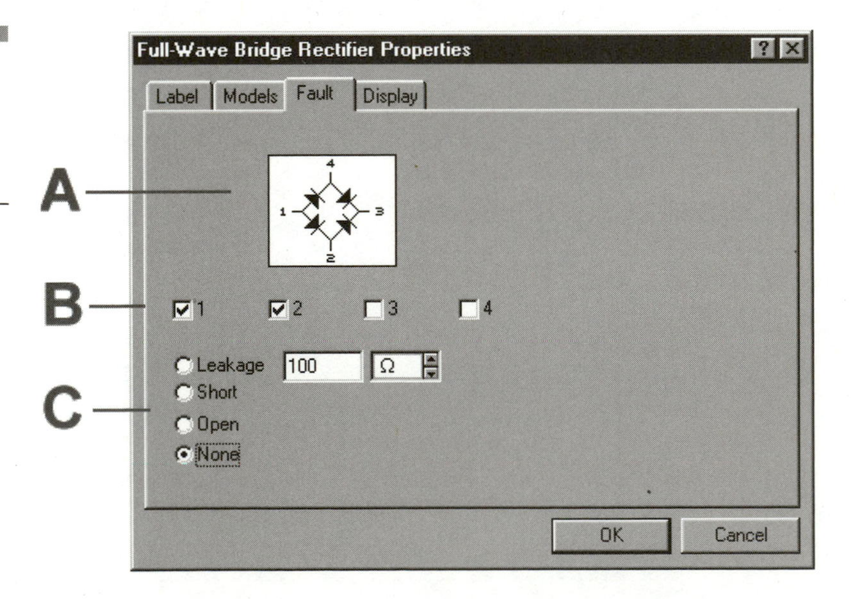

Figure 10.17 The Fault tab allows you to simulate problems within a component.

Each fault can be placed between any of the component's terminals.

- **Component Picture**—At the top is a depiction of the part. Each of its terminals are labeled with numbers (see A in Figure 10.17). This shows between which points the fault(s) is/are being simulated.
- **Terminal Check Boxes**—Check boxes are below the component's image so that you can select the terminals you wish to have a fault between (see B in Figure 10.17). For example, if you want to create a fault between terminals 1 and 2 of a full-wave bridge rectifier, check off 1 and 2. If you want it between 2 and 3, then choose the 2 and 3 boxes.
- **Type of Fault**—Under the terminal check boxes is the selection for the actual fault you want to create (see C in Figure 10.17). The choices are:
  - **Leakage**—Leaks a set amount of resistance between the terminals you have indicated with the Terminal Check Boxes.
  - **Short**—Creates a 0-ohm short between the terminals.
  - **Open**—Simulates an open circuit condition between the points you have set.
  - **None**—(Default) No simulated fault in the component.

How to Engineer a Fault

Let's use a few examples to learn how to set up a faulty component.

1. Create the circuit shown in Figure 10.18.
2. Run the simulation.
3. Take note of the transistor's gain (# = 100) and the voltage when the potentiometer is set to 50 percent.
4. Double-click the **NPN Transistor**.
5. Click the **Fault** tab.
6. Check on the 1 and 3 terminal boxes to create a fault between the base and emitter of the NPN transistor (see Figure 10.19).

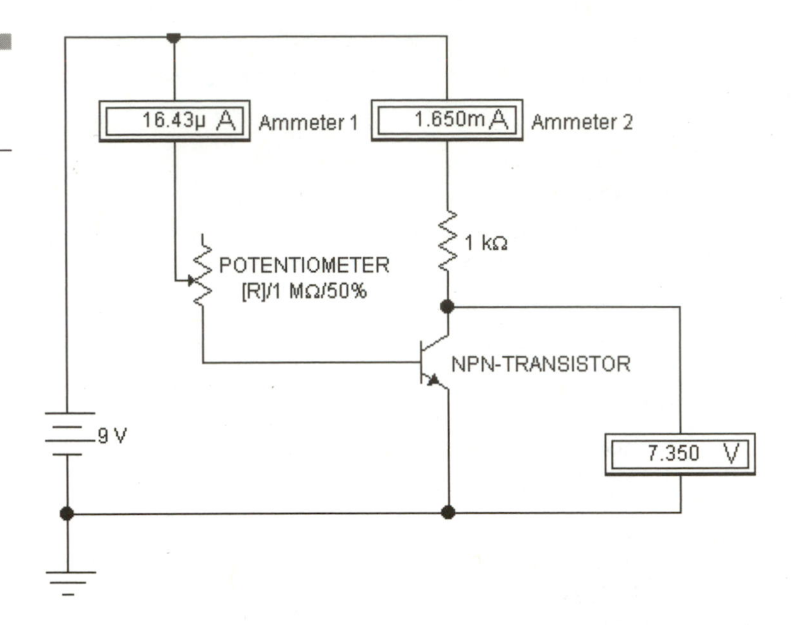

Figure 10.18 Testing the fault feature.

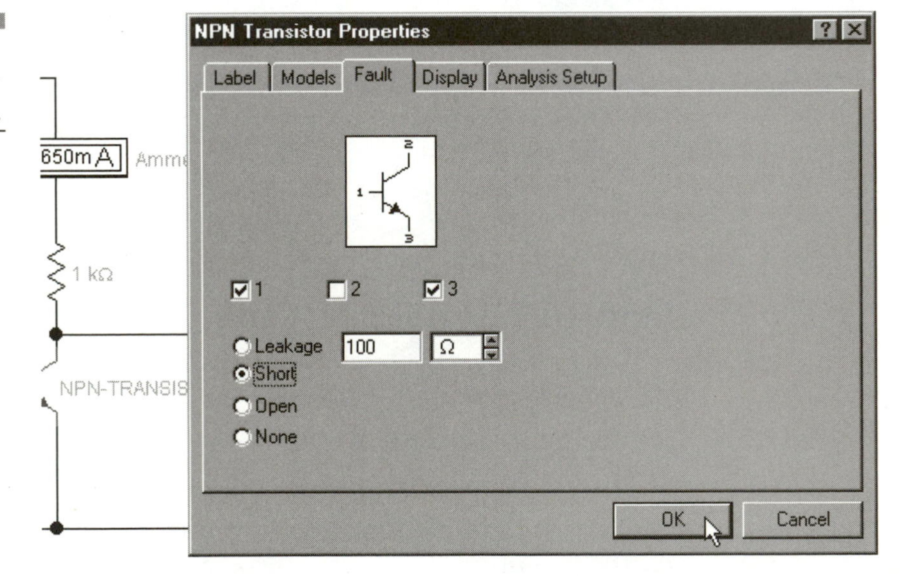

Figure 10.19 The base and emitter of the NPN transistor.

7. Select **Short**.
8. Click **OK** at the bottom.
9. Run the simulation again.

What are the results now? Try different types of faults and note the results.

Parts Descriptions and Examples

The next section briefly explains the use of each component, and also contains examples of some component's uses for easier understanding. If you want a complete engineering description of the parts, refer to Electronics Workbench Technical Reference available from Electronics Workbench (1-800-263-5552).

You will notice some of these components have real-world equivalents (such as resistors, LEDs, etc.), but others are considered subcircuitry (such as volt- or current-controlled voltage and current sources). If you wish to transfer your designs to a PCB layout program in the future, I suggest you use them only in test simulations. Instead, build the specific circuitry yourself, and store it in subcircuit libraries for retrieval into your designs.

NOTE

All toolbar items are described from top (left to right) then bottom (left to right).

Sources

The Sources toolbar is shown in Figure 10.20.

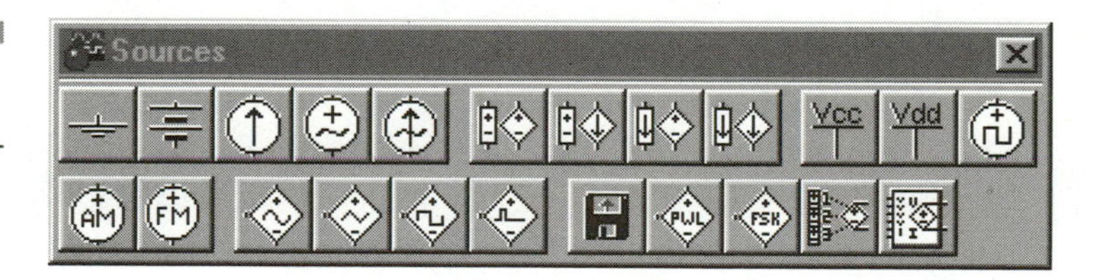

Figure 10.20
Sources toolbar.

Ground—The Ground symbol represents a common point of reference for positive or negative measurements; it represents zero volts. It is a return point in a circuit (usually hooked to the negative side of a battery or voltage source) and each circuit must contain this point of reference—every circuit must contain at least one ground.

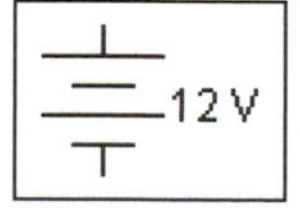

DC Voltage Source (Battery)—The battery functions as a DC voltage source. It can operate as low as the µV range, all the way to a killer kV range. This lets you simulate anything from a Duracell battery to a power station.

When using a battery in a circuit, keep in mind that EWB does not add an internal resistance such as a real-world battery. If you want to use a battery in parallel with another battery or switch, insert a 1 milliohm resistor in series with it.

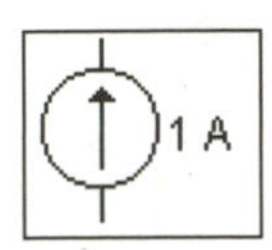

DC Current Source—Adds a DC current to your design. It can be adjusted from a negligible µA range to a deadly house-melting kA range. It is a good component with which to learn current analysis.

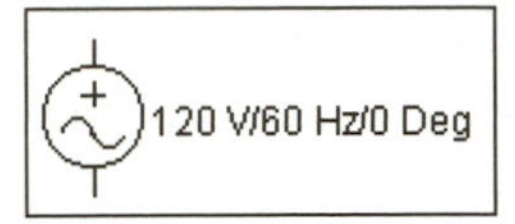

AC Voltage Source—Injects an AC voltage sine wave signal into a circuit. This component is measured in RMS voltage (.707 times the peak voltage). Double-click the component to set its voltage, frequency, and phase angle.

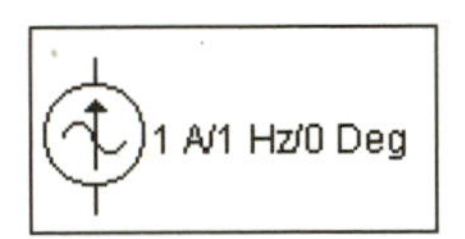

AC Current Source—Lets you pump out an AC current. Like the AC voltage source, it is also measured in RMS not *peak*. The frequency and phase angle are also adjustable. Double-click the component to set these parameters.

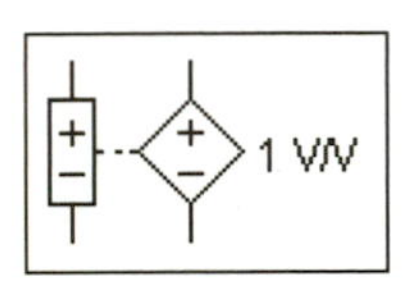

Voltage-Controlled Voltage Source (VCVS)—Outputs a set voltage determined by the input voltage and gain (E). The input side is the rectangle with the + and − terminals. Output is the diamond with the + and − terminals. If, for example, you have 12 volts hooked to the input and the gain set to 1kV, then the output voltage would increase to 12 kV.

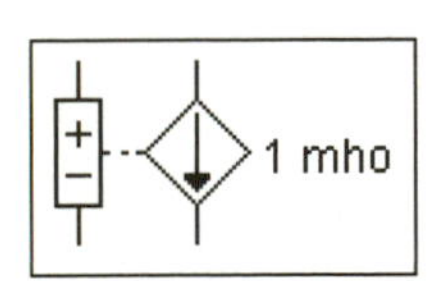

Voltage-Controlled Current Source (VCCS)—Outputs a current value set by the input voltage level. Double-click the component to open parameters. Current and voltage are related by a parameter called *transconductance* (G), which is the ratio of the output current to the input voltage. It is measured in *mhos* (also known as *siemens*) and can have any value from mmhos to kmhos. Use the formula: G = Iout / Vin to change a voltage reading into a current reading.

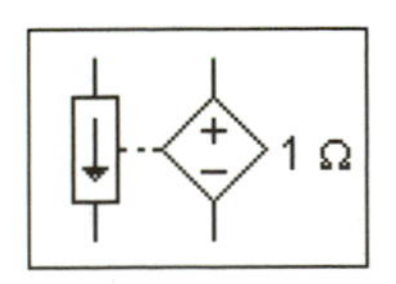

Current-Controlled Voltage Source (CCVS)—Outputs a set voltage determined by the current at its input. The rectangle with the arrow is the current input and the triangle is the associated output voltage. Double-click the component to access the adjustments. The two are related by a parameter called *transresistance* (H), which is the ratio of the output voltage to the input current. It can have any value from mW to kW. Use the formula: H = Vout / Iin to change a current reading into a voltage reading.

Use the CCVS to read current on the oscilloscope. Just wire it into the circuit, as shown in Figure 11.15 in Chapter 11.

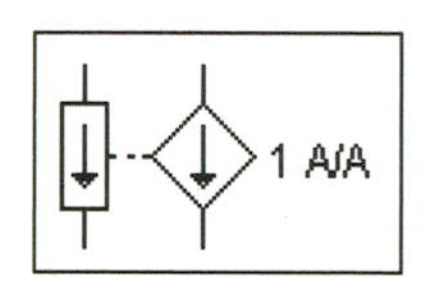

Current-Controlled Current Source—Current output is adjusted by current input. The two are related by a parameter called *current gain* (F), which is the ratio of the output current to the input current. The current gain can have any value from pA/A to TA/A. The formula is: F = Iout/Iin, to change the level of current.

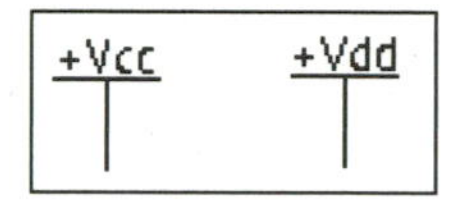

Vcc and Vdd Voltage Sources—Provide a voltage point: 5V for Vcc and 15V for Vdd. Vcc is typically used to connect to TTL ICs and Vdd to CMOS ICs. They are mainly use to supply voltage to the Vcc/Vdd pin of an IC or connected to a gate to throw it high (logical 1).

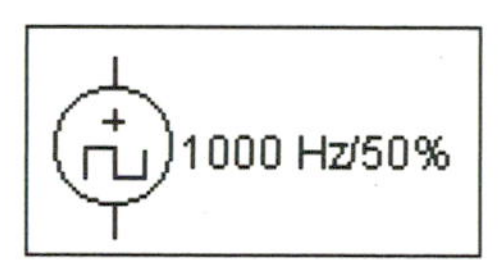

Clock Source—Many ICs require a square wave to "clock" their circuitry; this source lets you input an adjustable square wave into a circuit (i.e., a square-wave generator). You can adjust the amplitude (voltage), duty cycle (on-to-off ratio), and frequency by double-clicking the placed component. If you need to run more than one square-wave clock signal in a circuit and the function generator is already used, this makes a great replacement.

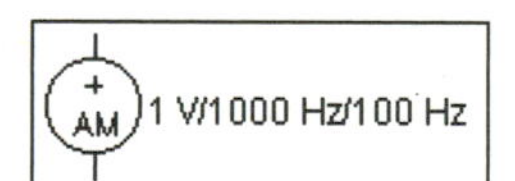

AM Source—This single-frequency amplitude modulation source generates an amplitude-modulated wave. It is used to build and analyze communications circuits. Double-click the placed components to adjust settings: carrier amplitude, carrier frequency, modulation index, and modulation frequency.

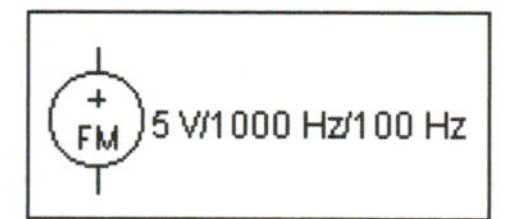

FM Voltage Source—This single-frequency frequency modulation source generates a frequency-modulated wave. It is used to build and analyze communications circuits. Double-click the placed components to adjust settings: voltage amplitude, voltage offset, carrier frequency, modulation index, and signal frequency.

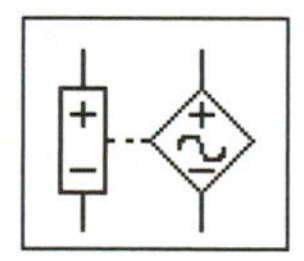

Voltage-Controlled Sine Wave Oscillator (VCO)—Creates a waveform with a frequency determined by the level of inputted voltage. You must adjust the control and frequency arrays to fit your circuit by double-clicking the placed component.

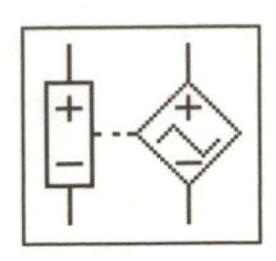

Voltage-Controlled Triangle Wave Oscillator—Functions the same as the VCO but outputs a triangle wave. Requires more waveform tweaking.

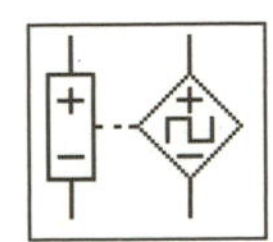

Voltage-Controlled Square Wave Oscillator—Functions the same as the VCO but outputs a square wave. Requires more waveform tweaking.

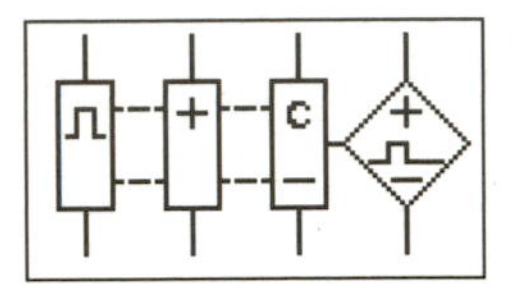

Controlled One-Shot—This oscillator takes an AC or DC input voltage, which it uses as the independent variable in the piecewise linear curve described by the control pulse width pairs. From the curve, a pulse width value is determined, and the oscillator outputs a pulse of that width. You can change clock trigger value, output delay from trigger, output delay from pulse width, output rise and fall times, and output high and low values. When only two coordinate pairs are used, the oscillator outputs a linear variation of the pulse with respect to the control input. When the number of coordinate pairs is greater than two, the output is piecewise linear.

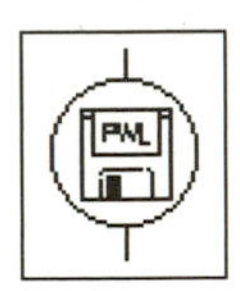

Piecewise Linear Source—This device is one of EWB's customization tools, which lets you output a voltage waveform that has been described in a text file. The file combines pairs of time and voltage level information about the wave to be outputted on the component's terminals. A previously recorded *.scp (oscilloscope file) or Write Data file (as seen in Figure 10.21) will also works. To connect and run:

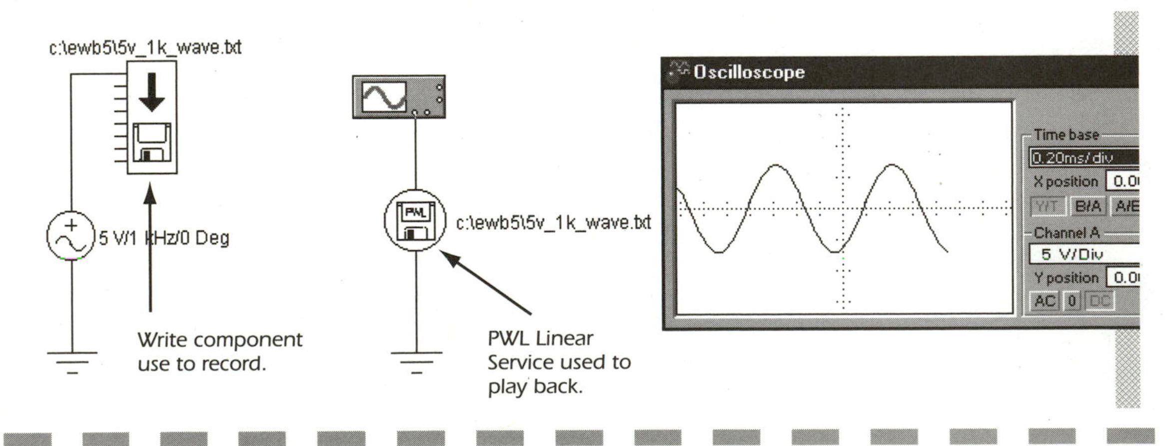

Figure 10.21 Piecewise Linear Source plays back recorded voltage data.

1. Place the component onto the screen and double-click it, choosing the **Value** tab.
2. Hit the **Browse** button to open a text file with *data pairs*. This can also be the data that was recorded earlier with the Write Data component. This file will contain the wave information.
3. Hit **OK**. Connect one of the terminals to the circuit and run. Note that the bottom terminal is 180 degrees out of phase from the top terminal and is typically connected to ground.

There are two ways to build a text file to run this circuit. The difficult one is to hand-type two columns in a text file: the left column is the time points and the right is the voltage level at that time (Figure 10.22). A much simpler method is to hook the circuit up to one of the channels on the Write Data component and save the data as a text file. The waveform that was recorded is output by the piecewise component into your circuit (see Figures 10.23 and 10.24).

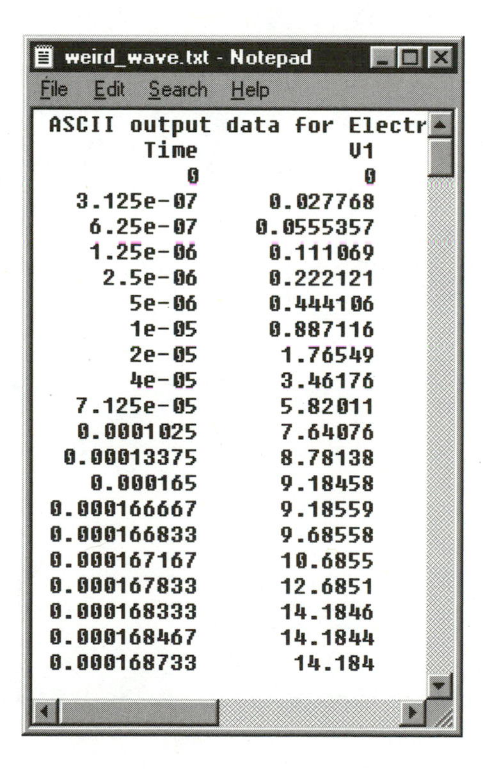
weird_wave.txt - Notepad

File Edit Search Help

ASCII output data for Electr

Time	V1
0	0
3.125e-07	0.027768
6.25e-07	0.0555357
1.25e-06	0.111069
2.5e-06	0.222121
5e-06	0.444106
1e-05	0.887116
2e-05	1.76549
4e-05	3.46176
7.125e-05	5.82011
0.0001025	7.64076
0.00013375	8.78138
0.000165	9.18458
0.000166667	9.18559
0.000166833	9.68558
0.000167167	10.6855
0.000167833	12.6851
0.000168333	14.1846
0.000168467	14.1844
0.000168733	14.184

Figure 10.22 Use Windows Notepad to hand make PWL files.

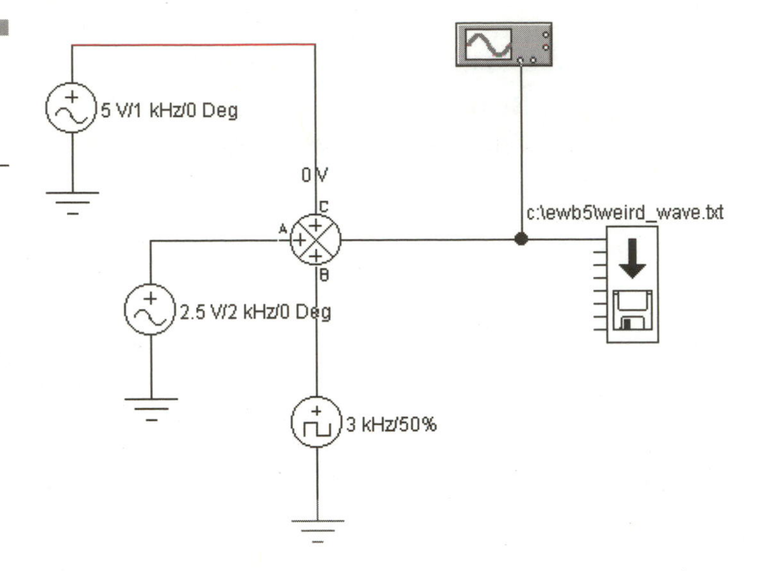

Figure 10.23 Write component used to record PWL file.

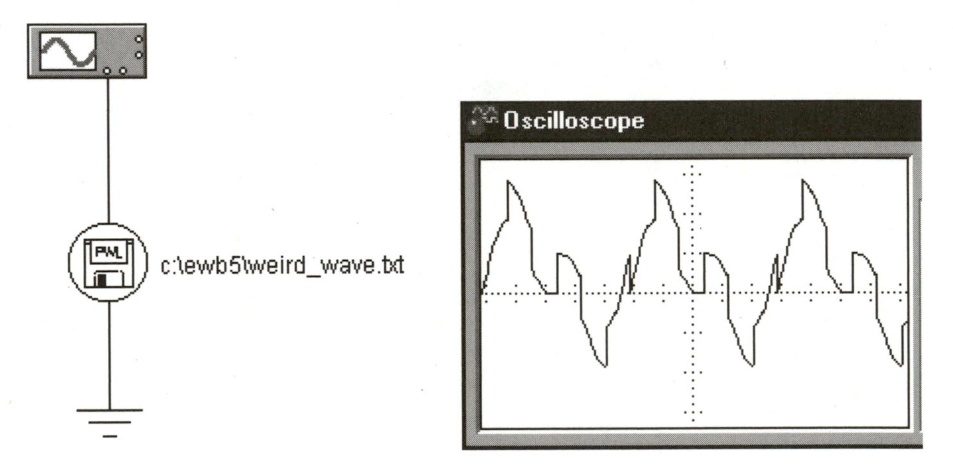

Figure 10.24 PWL source plays back file.

In just about any application that needs a custom waveform, this component really releases the power of Electronics Workbench. You will see some interesting examples of this component's capabilities in later chapters.

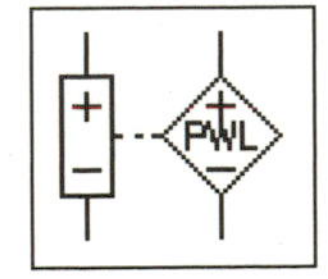

Voltage-Controlled Piecewise Linear Source—Allows you to control the shape of the output waveform by entering up to five (input/output) pairs. To access this, double-click the component, select the Model tab, and edit the default component, making sure to save the new model under a different name. The X values are input coordinate points and the associated Y values represent the outputs of those points. If you use only two pairs, the output voltage is linear.

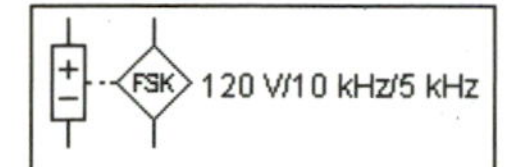

Frequency-Shift-Keying (FSK) Source—Used for keying a transmitter for telegraph or teletype communications by shifting the carrier frequency over a range of a few hundred hertz. The frequency shift key modulated source generates the mark transmission frequency, f1, when a binary 1 is sensed at the input, and the space transmission frequency, f2, when a 0 is sensed. FSK is used in low speed modems. Double-click the placed component to set the mark and space frequency as well as the amplitude of both signals.

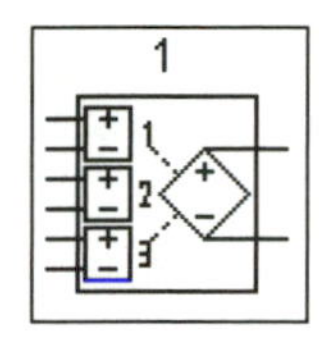

Polynomial Source—This voltage-controlled voltage source is set up using a *polynomial transfer* function. It is a specific case of the more general nonlinear dependent source. Use it for analog behavioral modeling. In EWB5, the polynomial source has three controlling voltage inputs: V1, V2, and V3.

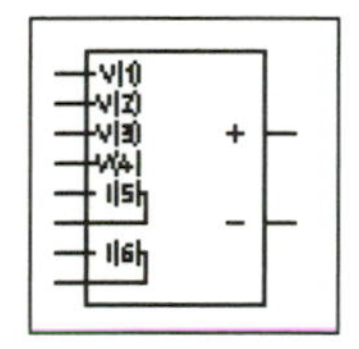

Nonlinear Dependent Source—This generic source allows you to create a sophisticated behavioral model by entering a mathematical expression. Expressions may contain the following operators: +, −, *, /, ^, unary-, and the predefined functions abs, asin, atanh, exp, sin, and tan. Double-click the component to enter the Source Expression in the Value tab.

Basic Toolbar

The Basic toolbar is shown in Figure 10.25.

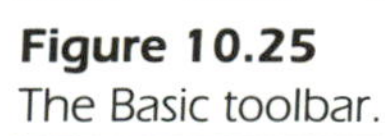

Figure 10.25
The Basic toolbar.

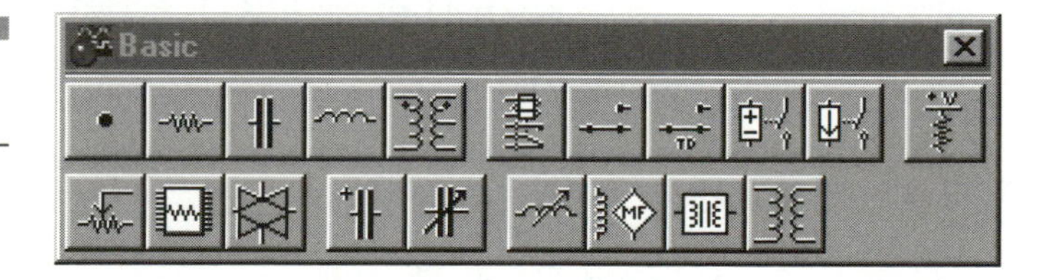

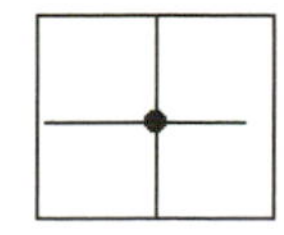

Connector—A junction to connect four wires.

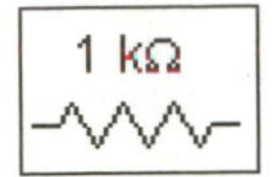

Resistor—These are perhaps the most versatile electronic components in existence. They can be used to control voltage and current, pull up or down a logic gates state, provide a test load, and in countless other tasks. Resistors come in a variety of sizes and power ratings depending on the application. Electronics Workbench simulates many resistor-related variables including power dissipation, tolerance, minimum operating temperature, maximum temp, and others. To access these parameters, double-click the component.

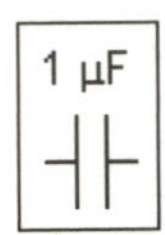

Capacitor—The capacitor is perhaps the second most used electronics component after the resistor. Capacitors store energy in the form of an electrostatic field. They are typically used to filter or remove AC signals from a variety of circuits. In a DC circuit, they can be used to block the flow of direct current while at the same time allowing AC signals to pass. DC circuits also use capacitors as *resistor/capacitor (RC) time constant* circuits. The resistor allows the capacitor to charge at a rate set by the RC combination, then drain at a set rate. Capacitance is measured in *farads*. As with the resistors, the capacitor's values are adjusted by double-clicking the component.

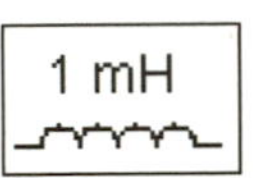

Inductor—The inductor is similar to a capacitor in that it stores electricity, but in the form of an electromagnetic field. It does this when the current flowing through it changes. This characteristic is useless in DC circuits, but in an AC circuit, an inductor opposes a change in current flow. This is called *inductance*. Inductance (L) is measured in *henrys* (H). Double-click the inductor to access its value tab.

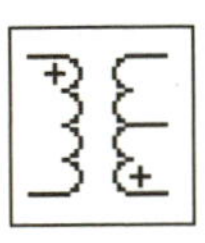

Transformer—Used to step down or step up voltage in an AC circuit. Once the component is placed, double-click it to select between several models. These include default, audio and miscellaneous. You can also create your own transformers by adjusting model item parameters like primary-to-secondary turns ratio and primary-to-secondary resistance. Transformers are typically used in appliances to change step-down or step-up voltage; say 120 volts to 12 volts.

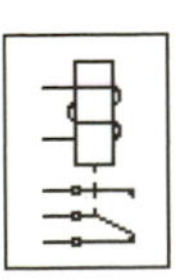

Relay—Modeled after a magnetic relay such as those used in automatic car door locks. A small current charges a magnetic core and causes a switch to turn on or off. You are controlling a higher current circuit with a low power source. Relays are often used as an interface between a low-power IC and high voltage/current devices. The relay component has five connections: the two top connections are the magnetic coil, the three bottom terminals are the contacts. When the coil has a set amount of current going through it, the contacts close (or open, depending on the hookup configuration). Figure 10.26 shows a low-power 5-volt clock used to energize the coil. Two high-wattage light bulbs are hooked to the switching side of the relay.

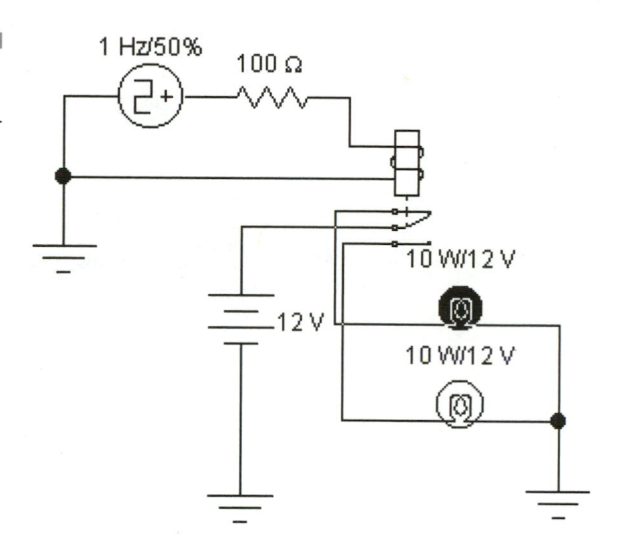

Figure 10.26
Example relay circuit.

[Space]

Switch—Allows you to control your circuit on the fly. The switch contains three terminals: common, position 1, and position 2. It conducts from common to one position until a control key is struck, changing it to the other position. It is used to control a circuit while it is active, using your keyboard. Figure 10.27 is a single pole, double throw (SPDT) switch; let's assume the terminal on the left side is number 1, the top-right terminal is A, and the bottom-right is B.

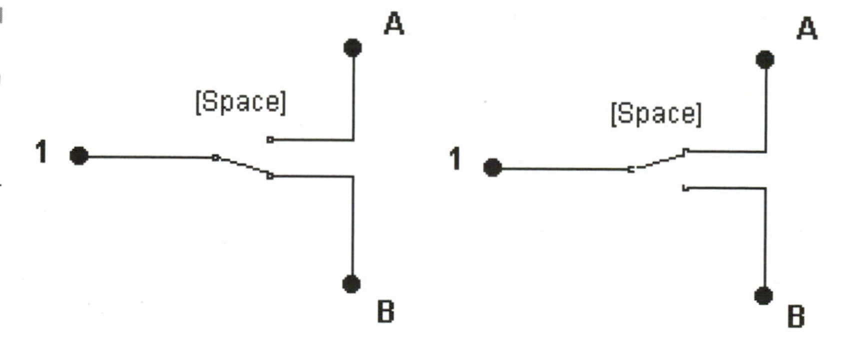

Figure 10.27
Switch conducts from 1-to-B or when the spacebar is hit, from 1-to-A.

- 1 to B conducts upon activation.
- Hitting the Space key makes 1 to A conduct.
- Hitting the Space key again makes 1 to B conduct once more.

Here is how to set the key to operate as a switch in your circuit (see Figure 10.28).

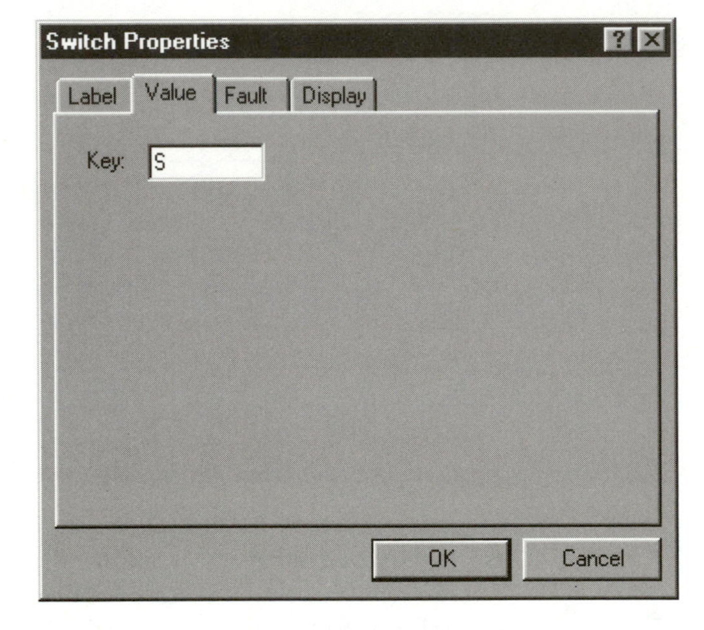

Figure 10.28 Double-click the placed switch to set the control key.

1. Place a switch on the drawing board and connect the appropriate terminals.
2. Double-click the switch or use the menu option to get to its properties.
3. Choose the **Value** tab.
4. Enter the character or number you want to use to control the switch during simulation, for example, 'S' for switch.
5. Hit **OK**.
6. Activate the circuit.

Pushing the **S** key once flips the switch to the other terminal; hitting it again places it back to the original terminal.

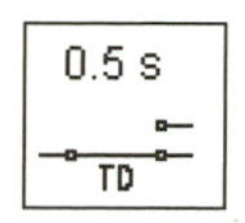

Time-Delay Switch—Opens and closes at a specific time interval. Double-click the placed component to adjust time on/off.

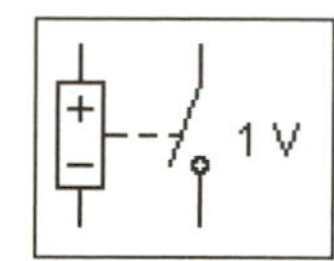

Voltage-Controlled Switch—Opens/closes at specific voltage levels set by the user. Double-click the placed component to access variables.

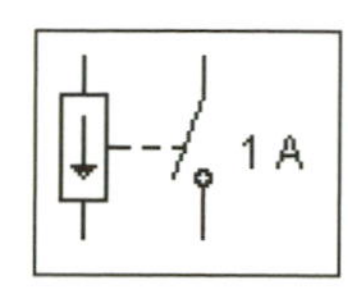

Current-Controlled Switch—Opens/closes at specific current levels set by the user. Double-click the placed component to access variables.

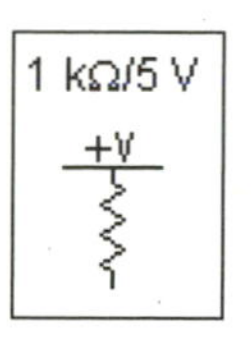

Pull-Up Resistor—Great when you are working on digital or mixed circuits. Sometimes one of the IC pins has to be raised to 5 volts (or whatever voltage you wish) through a current-limiting resistor. This component saves having to run all sorts of irritating spaghetti lines and multiple components. The resistor has a default value of 1Kohms and 5 volts. Double-click the placed component to change these values.

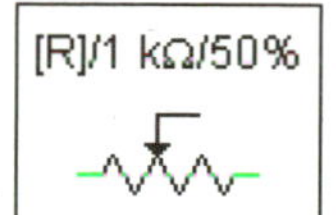

Potentiometer (Variable Resistor or POT)—A user-controlled component that works in much the same way as does the volume control on a stereo (most volume controls are pots). It provides adjustable resistance paths for current to flow; by varying the amount of resistance from one side to the other, you can control the voltage or current. First, identify a control key; when you hit that key on the keyboard, it decreases the value by a preset percentage. If you hit the Shift key and the letter you chose again, the value increments. Its not a quick-turning knob but it will do for simulations.

Once the potentiometer is placed, double-clicking the component opens the window in Figure 10.29. You can set the keyboard letters you wish to use to operate the pot, and also adjust the initial setting and by how much a key-press will increase or decrease the pot.

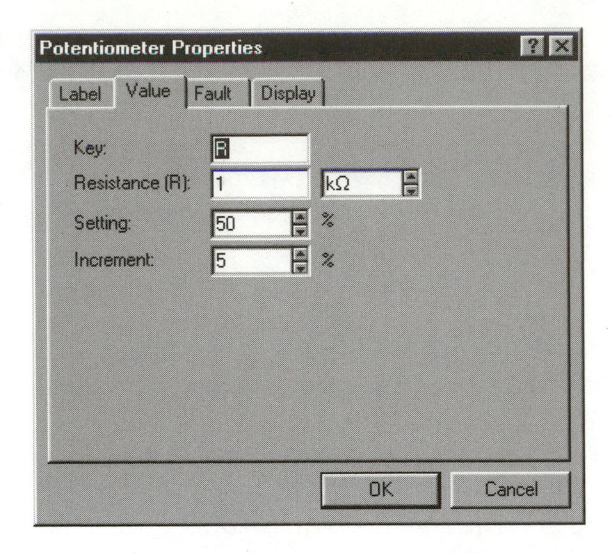

Figure 10.29 Double-click the placed POT to set control keys and other parameters.

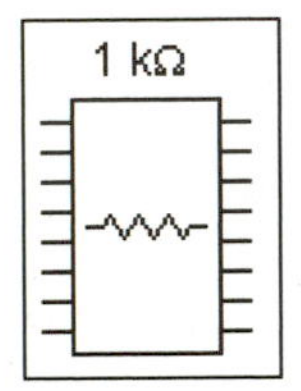

Resistor Pack—Several resistors are placed within one package, each terminal with a separate pin (DIP). It's a cleaner way to design circuits that use a lot of resistors. Double-click the component to set the resistor values. A typical use would be to run each line in an 8-bit bus through a 1K resistor (see Figure 10.30).

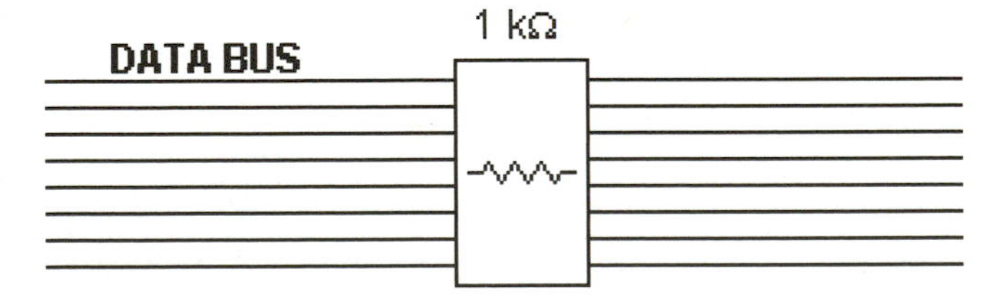

Figure 10.30 Resistor pack used inline with a databus.

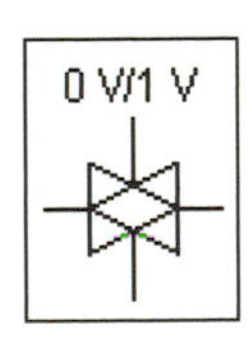

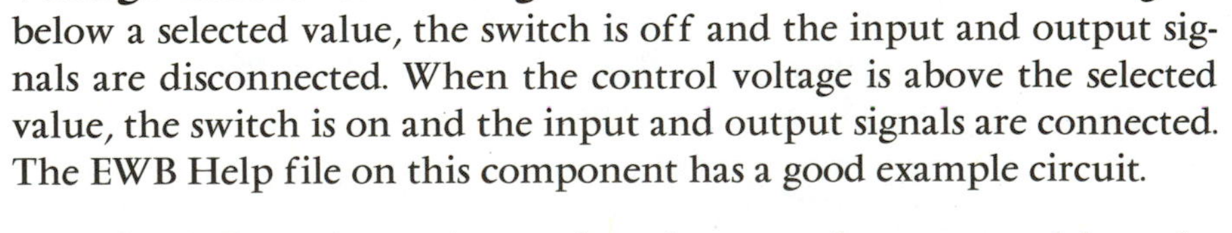

Voltage-Controlled Analog Switch—When the control voltage is below a selected value, the switch is off and the input and output signals are disconnected. When the control voltage is above the selected value, the switch is on and the input and output signals are connected. The EWB Help file on this component has a good example circuit.

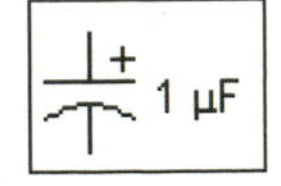

Polarized Capacitor—A capacitor that must be connected into the circuit, observing its polarity (±).

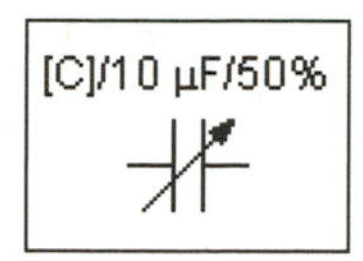

Variable Capacitor—Allows you to adjust the capacitor's farads while the circuit is running. It is done by assigning a keyboard letter as a control; hitting the key lowers the value by a set percentage and pushing Shift + the key increases. It operates the same way as the potentiometer. It is used in radio frequency circuits (tank circuits) to dial in a certain frequency. In fact, on old radios, the frequency dial was connected to a variable air core capacitor.

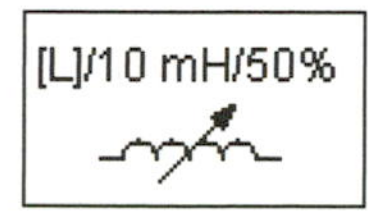

Variable Inductor—Same as variable capacitor, but offers changing inductance. Double-click and choose the Value tab to set the control key and other variables.

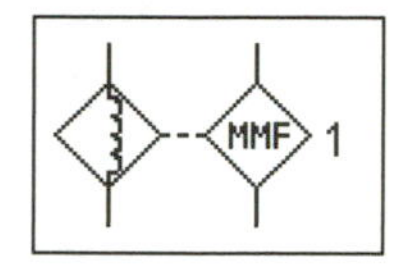

Coreless Coil—A conceptual model that you can use as a building block to create a wide variety of inductive and magnetic circuit models. Typically, you would use the coreless coil together with the magnetic core to build up systems that mimic the behavior of linear and nonlinear magnetic components. It takes a current input and produces a voltage. The output voltage behaves like a magnetomotive force in a magnetic circuit; that is, when the coreless coil is connected to the magnetic core or some other resistive device, a current flows.

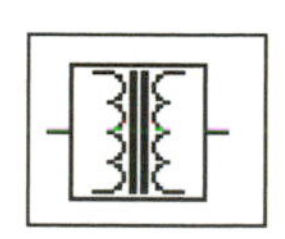

Magnetic Core—Allows you to accurately model an inductive/magnetic component. See the Help file for more information.

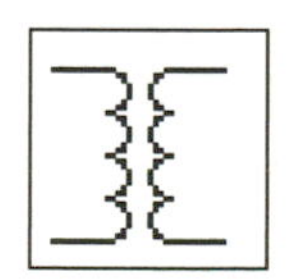

Nonlinear Transformer—Using this transformer, you can model physical effects such as nonlinear magnetic saturation, primary and secondary winding losses, primary and secondary leakage inductance, and core geometric size. See the Help file for detailed information on this component's use.

Diodes Toolbar

Diodes are components that allow current to flow in only one direction. They are used in applications such as simple solid-state switches in AC circuits, to help change AC into DC (rectification), and in the case of an LED, to provide a light show. The Diode toolbar is shown in Figure 10.31.

Figure 10.31
The Diode toolbar.

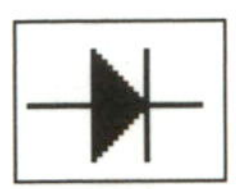

Diode—Select among hundreds of signal and general-purpose diodes. Double-click the component to access the models available to you.

Zener Diode—Operates in *reverse breakdown region* (called the *Zener region*). It is used to provide voltage regulation. Think of it as being equal to a battery that provides the amount of voltage as set in the zener voltage (Vz) (see Figure 10.32).

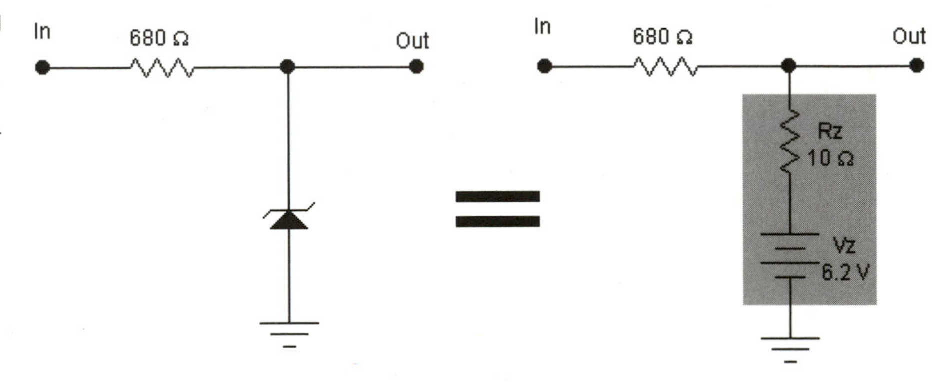

Figure 10.32
A Zener diode and its equivalent circuit.

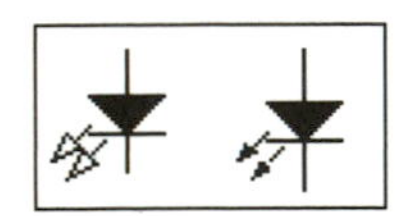

Light-Emitting Diode (LED)—A special diode that emits light when current is flowing through it; typically used as an indicator. For example, it can signal the fact that a circuit is indeed getting current. EWB's schematic representation of an LED is not too exciting but it does the job. When functioning, the center of the two arrows becomes filled with color when activated, indicating a lit state. Several LEDs are selectable from the Component properties, Model tab, including various colors.

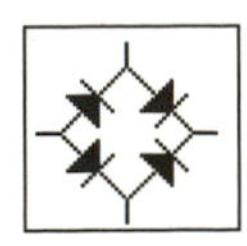

Full-Wave Bridge Rectifier (FWB)—A combination of four diodes that are configured to perform full-wave rectification. It's a great tool to simulate power supplies. By using this component you can save the time it would have taken to draw four diodes. There are gazillions of models to choose from for accurate simulations, analyses and export to PCB layout.

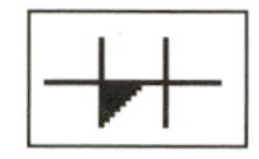

Shockley Diode—Similar to a standard diode except the Shockley diode remains off (or in *forward blocking region*) even if it is forward biased. The Shockley diode conducts current in one direction when the forward voltage exceeds the *forward breakover voltage* (also called the *switching voltage*) and the current is above the holding current.

The thyristors are a family of devices containing four layers of semiconductor material. They are diodes that conduct once a certain voltage level is reached. In most cases, the cathode-to-anode current flow is controlled with a gate. When the gate is activated, there is a low resistance between the anode and cathode. When the gate is shut off, the low resistance continues until the level drops below a certain point or shuts off.

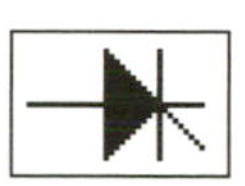

Silicon-Controlled Rectifier—This thyristor can be thought of as a one-way valve with a switch that activates the flow. It consists of three connections: anode, cathode, and gate. When a pulse is applied to the gate (switch), the diode section conducts in one direction. When the gate is off, the diode continues to conduct until it drops below a certain level. An SCR is typically used when interfacing a low-power digital circuit to a high-voltage AC line, such as in your home. A lamp dimmer and motor control are two examples.

Diac—This thyristor is a bidirectional device without a gate to switch it off/on. It conducts current in both directions, once a certain voltage is reached. There are only two terminals on this component.

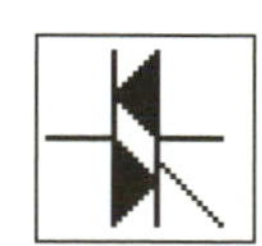

Triac—This thyristor is a bidirectional device with a gate.

Transistors Toolbar

The transistor is the hands-down winner for the Invention of the 20th Century Award. It has created a whole electronics industry that thrives on its ability to miniaturize. A transistor is a valve used for controlling electric current. It is typically used for amplification and switching in billions of circuits. A transistor usually contains three terminals. A *bipolar junction transistor* (BJT) uses a collector, base, and emitter. A *field-effect transistor* (FET) uses a drain, gate, and source. Each transistor has its own applications and form of operation (see Figure 10.33).

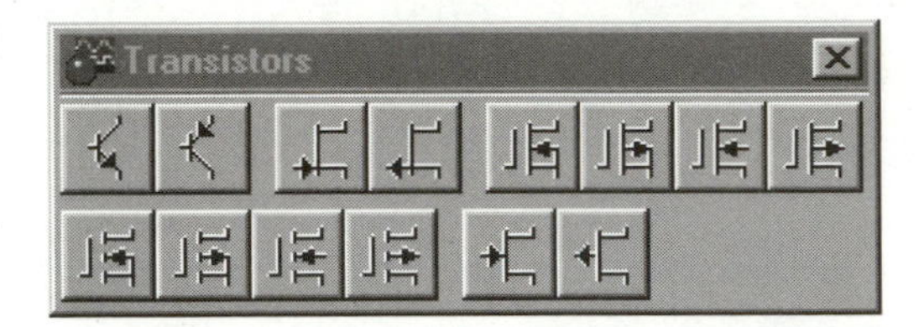

Figure 10.33 The Transitors toolbar.

BIPOLAR JUNCTION TRANSISTORS (BJTS)

Simply stated, a small change in current flow between the base-to-emitter of the transistor creates a large change in current flow in the collector-to-emitter current. A BJT is made with three layers of doped semiconductor regions with two PN junctions between the layers. A BJT transistor operates on the principle that the current involves carriers of both polarities—*holes* and *electrons*.

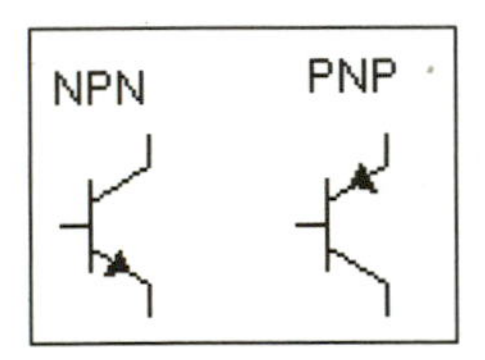

BJT_NPN (Bipolar Junction Transistors, Negative-Positive-Negative)—An npn transistor consists of two negative layers separated by a positive junction layer. The npn or negative-positive-negative transistor is the most commonly used model. It can be used for switching and amplification purposes. The base is positive and the collector/emitter is negative. A small base-to-emitter current flow causes a greatly amplified collector-to-emitter current flow.

BJT_PNP: Same as BJT_NPN except there are two positive layers and a negative one.

JUNCTION FIELD-EFFECT TRANSISTORS (JFETS)

Junction field-effect transistors use either a negative or positive channel with two opposite polarity gates on either side of the channel. Think of one as a garden hose with a hand wrapped around it. The top of the channel (hose) is the drain and the bottom of the hose is the source. The gate is the hand wrapped around the hose. In an n-channel FET, if the gate-to-source junction is reverse-biased (negative voltage applied), a field is created that lowers the drain-to-source current. It's like squeezing the hose with a hand (refer to Figure 10.34). These transistors are used in applications that require a high-input impedance and high-frequency response.

METAL-OXIDE-SILICON FIELD-EFFECT TRANSISTORS (MOSFETS)

MOSFETs are a category of FETs but differ from JFETs in that they have no PN junction structures. The gate is instead insulated from the channel with a layer of silicon dioxide.

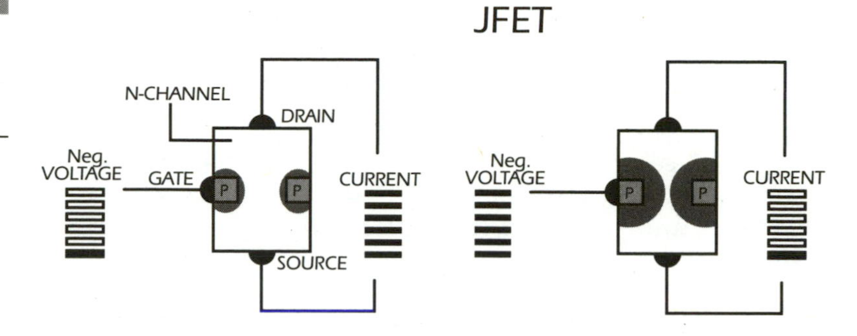

Figure 10.34
Operation of a JFET. See text.

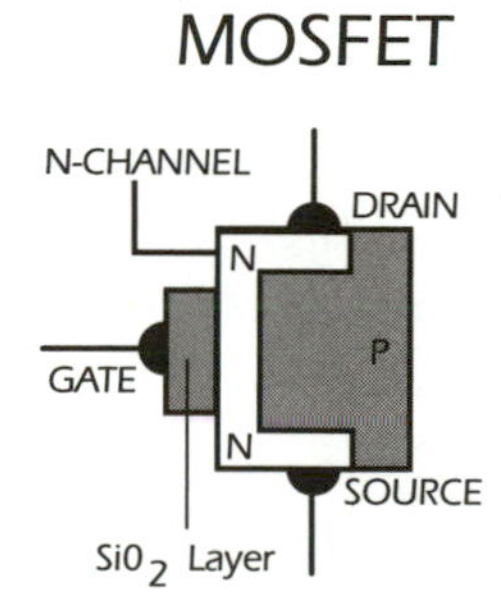

Figure 10.35
An N-Channel DE MOSFET.

Figure 10.36
(JFET_N) N-Channel JFET.

Figure 10.37
(JFET_P) P-Channel JFET.

MOSFETs come in two basic types: depletion-enhancement (DE) and enhancement only (E). Each can further be broken into positive or negative channel. The DE MOSFET (Figure 10.35) can operate in either depletion or enhancement mode, whereas the E MOSFET can only operate in the enhancement mode. Depletion mode (in an n-channel MOSFET) is when the n channel is depleted of some of its electrons, thus decreasing the channel's conductivity; therefore a higher negative gate-to-source voltage lowers the drain-to-source's current, until ultimately it is nil. Enhanced mode (in an n-channel MOSFET) is when more electrons are conducted into the n channel with a positive voltage, thus enhancing the conductivity of the channel; therefore a greater positive gate-to-source voltage increases the channel's drain-to-source current (see Figure 10.38).

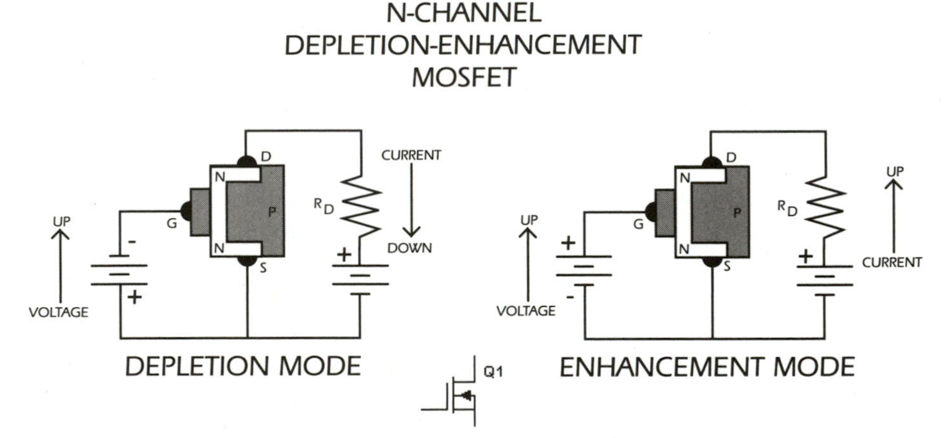

Figure 10.38 An n-channel DE MOSFET can be used in depletion and enhancement mode.

An enhanced MOSFET does not have a physical channel. Instead (on an n-channel enhanced MOSFET), a positive voltage on the gate induces a channel. The higher the voltage, the greater the conductivity of the induced channel (gate-to-source). If the voltage is below a certain threshold, there is no induced channel (see Figure 10.39).

Figure 10.39
An n-channel E Mosfet on works in enhancement mode.

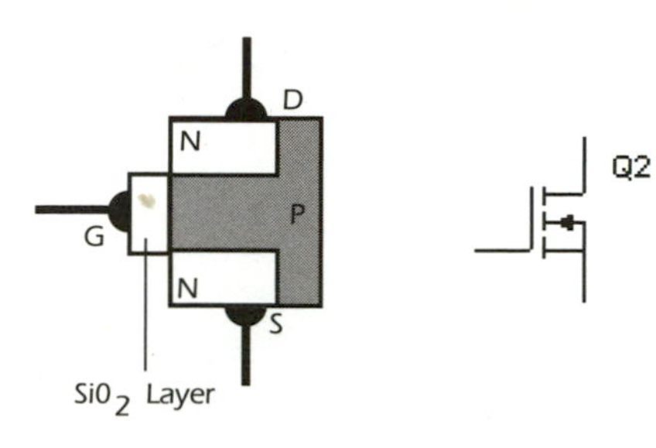

Figure 10.40
(MOS_3TDN) 3-Terminal Depletion N-MOSFET.

Figure 10.41
(MOS_3TDP) 3-Terminal Depletion P-MOSFET.

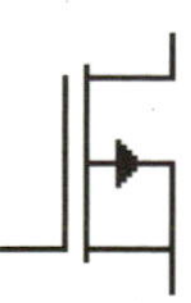

Figure 10.42
(MOS_4TDN) 4-Terminal Depletion N-MOSFET.

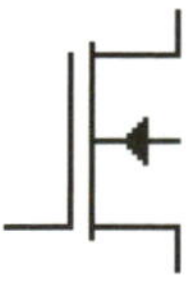

Figure 10.43
(MOS_4TDP) 4-Terminal Depletion P-MOSFET.

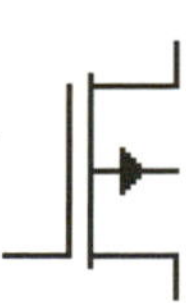

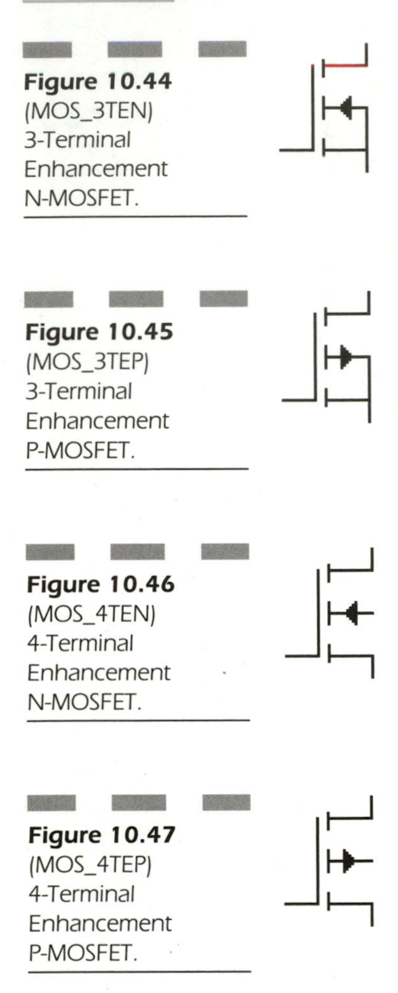

Figure 10.44 (MOS_3TEN) 3-Terminal Enhancement N-MOSFET.

Figure 10.45 (MOS_3TEP) 3-Terminal Enhancement P-MOSFET.

Figure 10.46 (MOS_4TEN) 4-Terminal Enhancement N-MOSFET.

Figure 10.47 (MOS_4TEP) 4-Terminal Enhancement P-MOSFET.

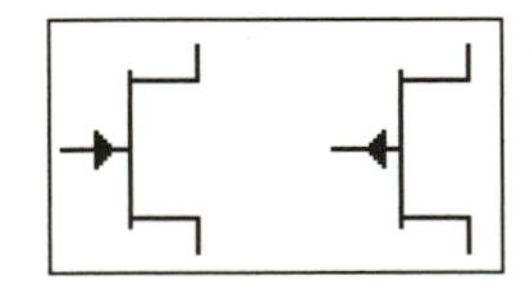

GALLIUM ARSENIDE FIELD-EFFECT TRANSISTORS (GAASFETS)

GaAsFETs are high-speed field-effect transistors that use gallium arsenide (GaAs) rather than silicon as the semiconductor material. They are generally used in very high frequency amplifiers (into the gigahertz range), such as those found in satellite applications and cell phones.

Analog ICs Toolbar

Figure 10.48 shows the Analog IC toolbar.

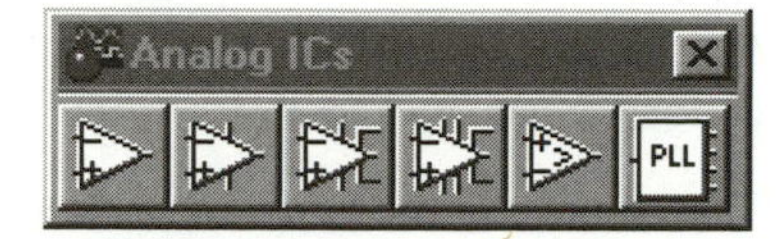

Figure 10.48 The Analog ICs toolbar.

OPERATIONAL AMPLIFIERS

Operational amplifiers are practical amplifier ICs used in millions of electronic applications. The op amp, in its simplest form, consists of three terminals: non-inverting input (A), inverting input (B), and output. It is basically a differential amplifier, meaning it will subtract the voltage of B from the voltage of A, then multiply it by the amount of gain. The results are placed at the output terminal. It can also be used as an inverting amplifier and non-inverting amplifier such as those you see in stereo circuitry.

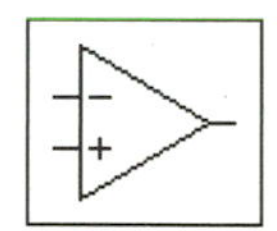

Two-Terminal Op Amp—The three-terminal op amp simulates much faster, but because the model is not as complex, it does not model all the characteristics of an op amp—for example, positive feedback.

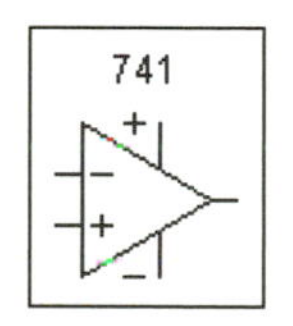

Five-Terminal Op Amp—The five-terminal op amp has two extra terminals (i.e., the positive and negative power supply terminals) at top and bottom, respectively. It models some second-order effects like common-mode rejection, output voltage, and current limiting.

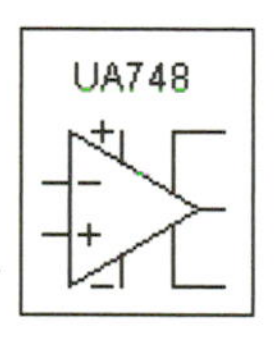

Seven-Terminal Op Amp—See the EWB5 Help file on this component for a great example and explanation.

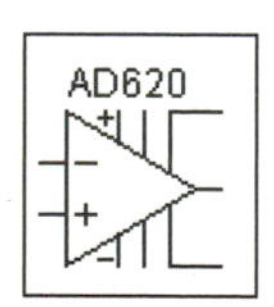

Nine-Terminal Op Amp—See the EWB5 Help file on this component for great example and explanation.

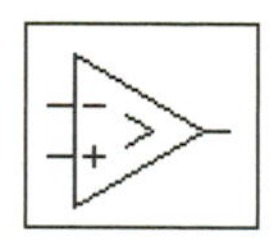

Comparator—Takes two voltage levels and compares them, then sets the output in one of two states: On or Off, according to which input has a greater level. For example, if there were 5 volts on the inverted input, and 4 volts on the non-inverting input, the output would be low. However, if we change the non-inverting input to 6 volts, the output goes high (see Figure 10.49). It's commonly used as an interface between analog and digital circuits.

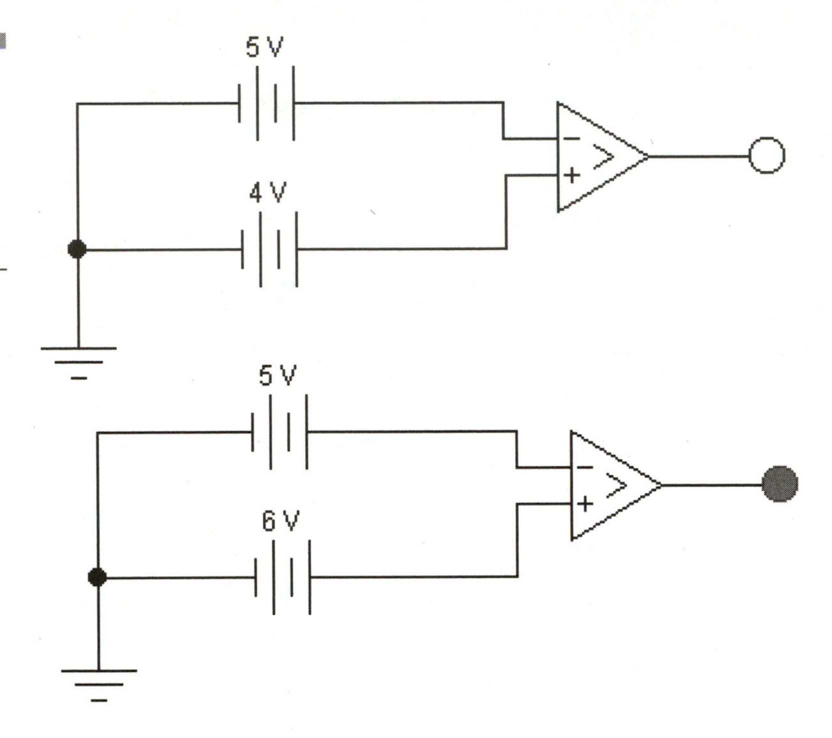

Figure 10.49 Comparator compares two voltage levels and outputs a given logic state according to the results.

Phase-Locked Loop (PLL)—Has a built-in oscillator whose output phase and frequency are steered to keep it synchronized with an input reference signal. It's used in modern radio frequency applications to "lock" onto a certain frequency. See the help file for more information.

Mixed IC Toolbar

The Mixed IC toolbar is shown in Figure 10.50.

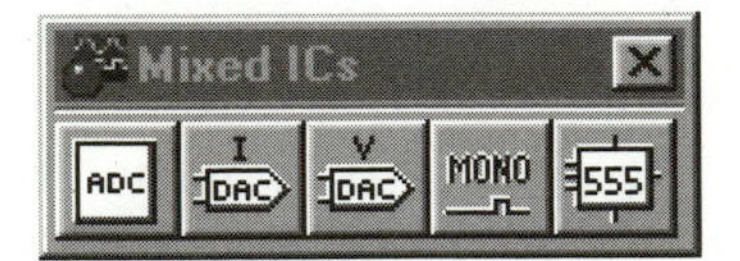

Figure 10.50 The Mixed IC toolbar.

Figure 10.51
ADC_DAC (Analog-To-Digital Converter).

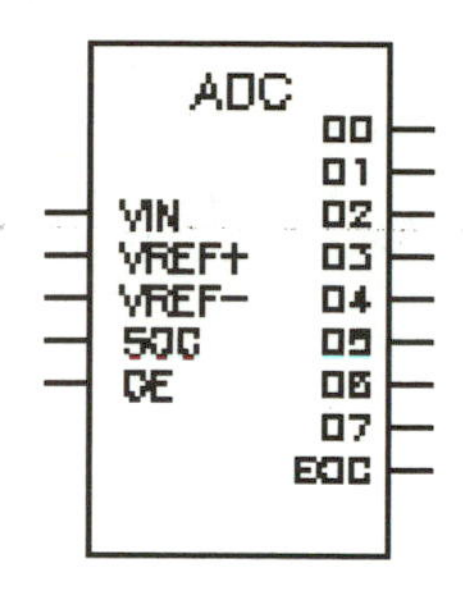

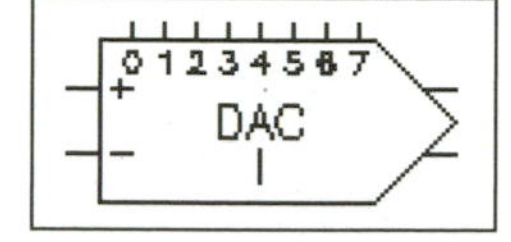

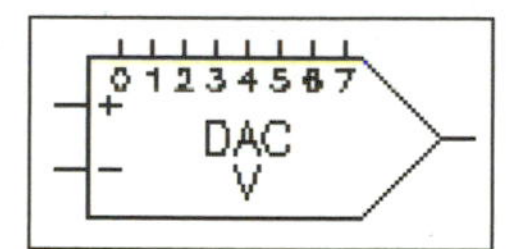

Digital-To-Analog Converters (Current and Voltage)—An **ADC** is an encoder that converts an analog voltage into a digital word. There are five inputs and nine outputs. The **DAC** takes a digital signal and changes it to a voltage or current level. These chips are used in millions of devices to convert analog signals to digital, process the digital information (or even store it), then output the converted data through another converter as an analog signal. The finest example of ADC_DAC chips is in modern compact disc (CD) technology: an analog signal and (through a microphone) is put an ADC; the digital information is stored on the CD; and the CD player uses a DAC to turn the digital bits of a CD back into the sound waves (sent to a speaker).

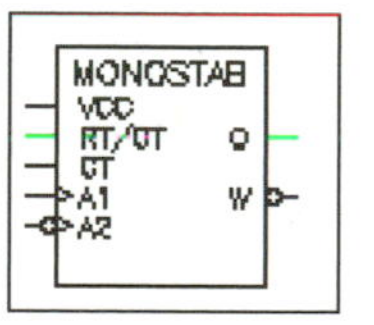

Monostable Multivibrator—This component produces an output pulse of a fixed duration in response to an edge trigger at its input. The length of the output pulse is controlled by the timing RC circuit connected to the monostable multivibrator.

Timer—This is one of my all-time favorites. I can remember building hundreds of projects with 555s bought from the local electronics store. With Electronics Workbench, I now have unlimited timers. The timer can be configured as a monostable or astable multivibrator, an ADC, orused in thousands of other applications, using only minimum components. The timer can be used for almost anything from a simple wave generator, to a short or long timer, to a missing pulse detector—a good example is an automobile's intermittent windshield wipers. Timers are even used in some solid-state ignition systems. Chapter 13 contains several circuits that make use of this gem.

Digital ICs Toolbar

The next components utilize what Interactive calls *IC templates*. By selecting an IC family from the toolbin (Figure 10.52), you are assigning it to a generic IC template. When you go to place the component, a window pops up to ask you to assign a specific IC to the template (see Figure 10.53). The component is now placed onto the drawing as an IC with each input/output and power pin usually labeled.

Figure 10.52
The Digital ICs toolbar.

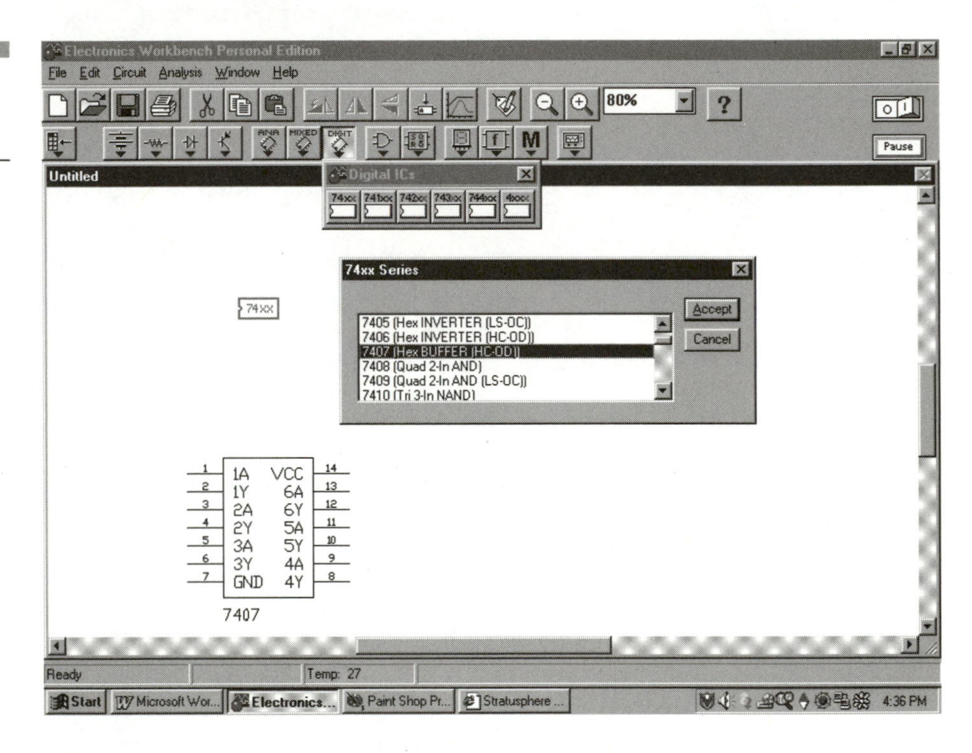

Figure 10.53
Assigning an IC to a template.

TRANSISTOR-TRANSISTOR LOGIC (TTL) ICS

A TTL IC (Figure 10.54) is a digital integrated circuit (such as a gate, flip-flop, inverter, etc.) that operates at a greater speed than CMOS circuitry, but uses much more power. They are typically used in digital circuits that don't require battery power such as cable boxes, televisions, computer boards, etc.

Figure 10.54
TTL ICs.

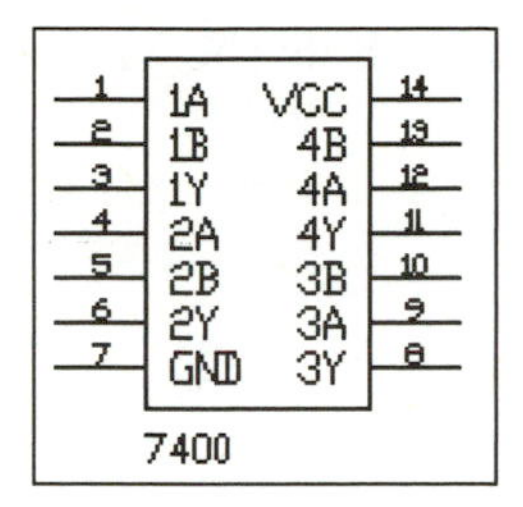

74xx Series—This is a standard 74xx series IC; the 7400 is a quad two-input NAND gate that contains a large repository of digital gates and other logic.

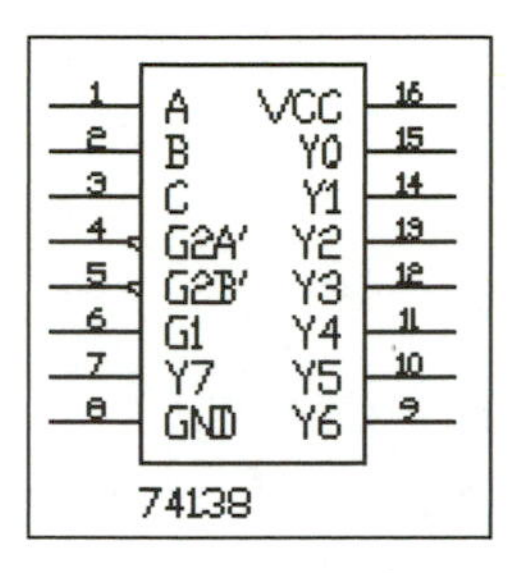

741xx Series—Contains TTL ICs in the 741xx series, such as the 74138, a three-to-eight decoder/multiplexer.

Remember to connect a + Vcc voltage source or 5V power supply to the chip's Vcc pin and a ground to the GND pin.

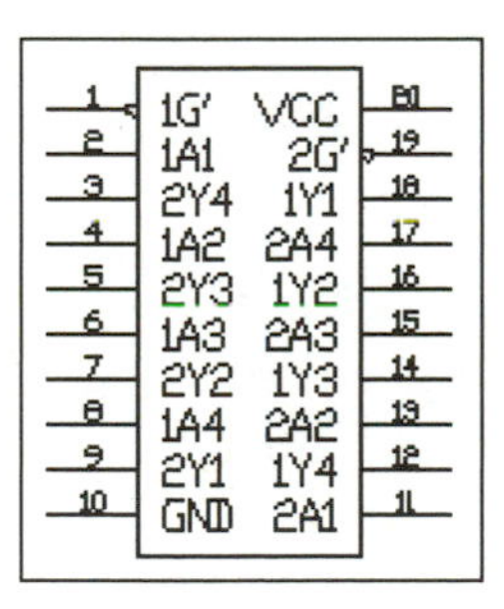

742xx Series—Contains TTL ICs in the 742xx series, such as the 74240, an octal buffer with three-state output.

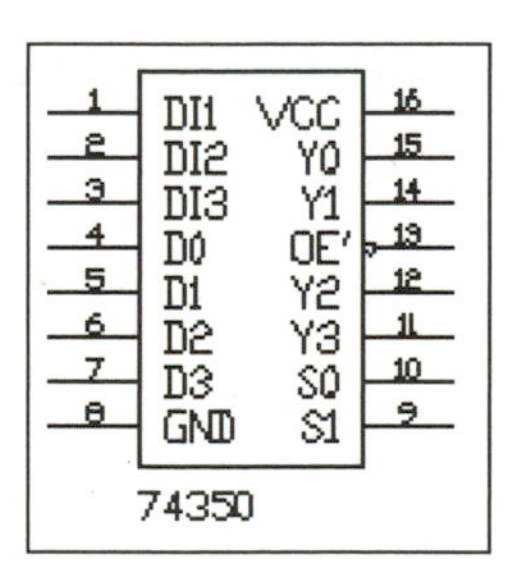

743xx Series—Contains TTL ICs in the 741xx series, such as the 74350, a four-bit shifter with three-state output.

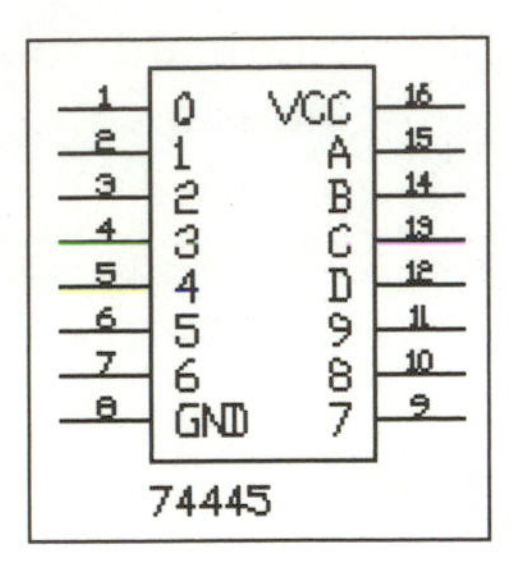

744xx Series—Contains TTL ICs in the 741xx series, such as the 74445, a BCD-to-decimal dec.

TIP

If you do not quite know what a TTL or CMOS IC is, place a copy of it, select it, and hit ***F1***.

COMPLEMENTARY METAL-OXIDE SILICON ICs (CMOS)
CMOS ICs (Figure 10.55) are digital circuits (such as gates, flip-flops, inverters, etc.) that operate at a slower speed than TTL chips, but use a fraction of their power. CMOS ICs are able to operate over a wider voltage range and are based on both p- and n-channel MOSFETS. Typically, CMOS chips are used in portable applications that require battery power; an example is a laptop computer or cell phone.

Figure 10.55
CMOS toolbar.

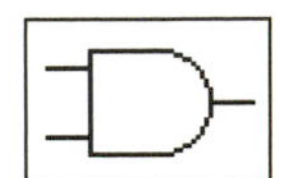

CMOS 4xxx series—These CMOS-based ICs are plugged into digital templates for simulation; an example is the 4009 hex inverter.

TIP

Remember to connect a + Vdd voltage source or power supply to the chip's Vdd pin and a ground to the Vss pin.

Logic Gates Toolbar

The components in this toolbar (Figure 10.56) contain various digital gates. They are mainly used in testing concepts as opposed to design (which should use digital components with IC templates). To see the *truth table* of each component, place it and hit **F1**.

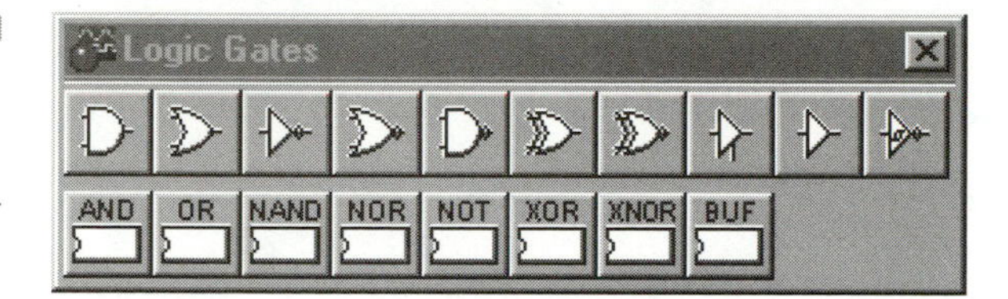

Figure 10.56 The Logic Gates toolbar.

You can select between a CMOS, TTL, or default gate model for logic gate components. Double-click the placed component and select the Models tab. You can also add inputs to the gate with the Number of Inputs tab (see Figure 10.57).

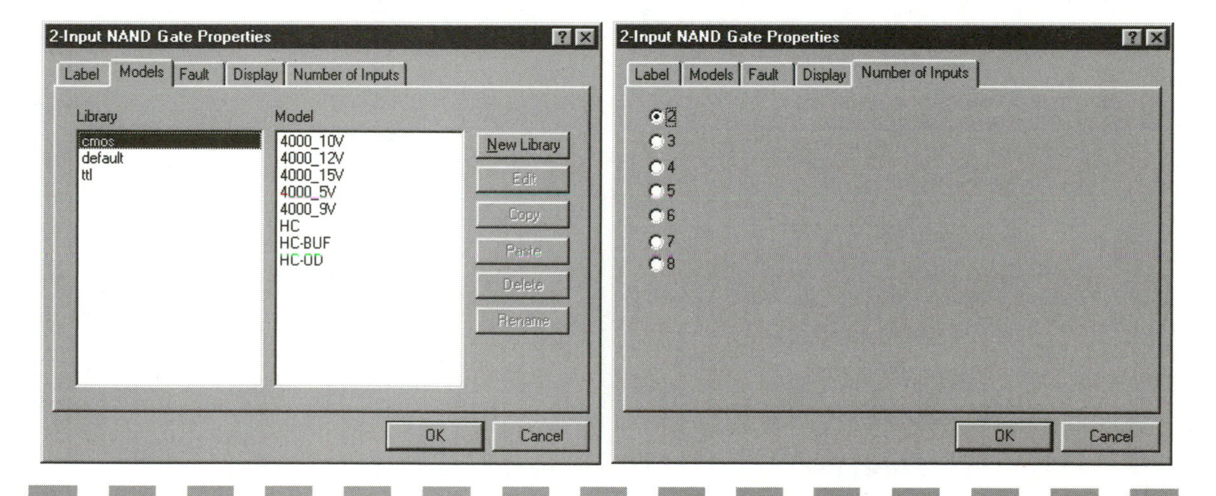

Figure 10.57 Adding inputs to the gate with the Number of Inputs tab.

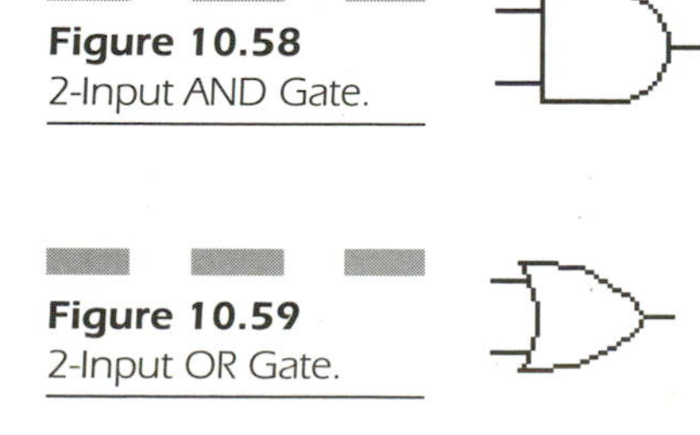

Figure 10.58 2-Input AND Gate.

Figure 10.59 2-Input OR Gate.

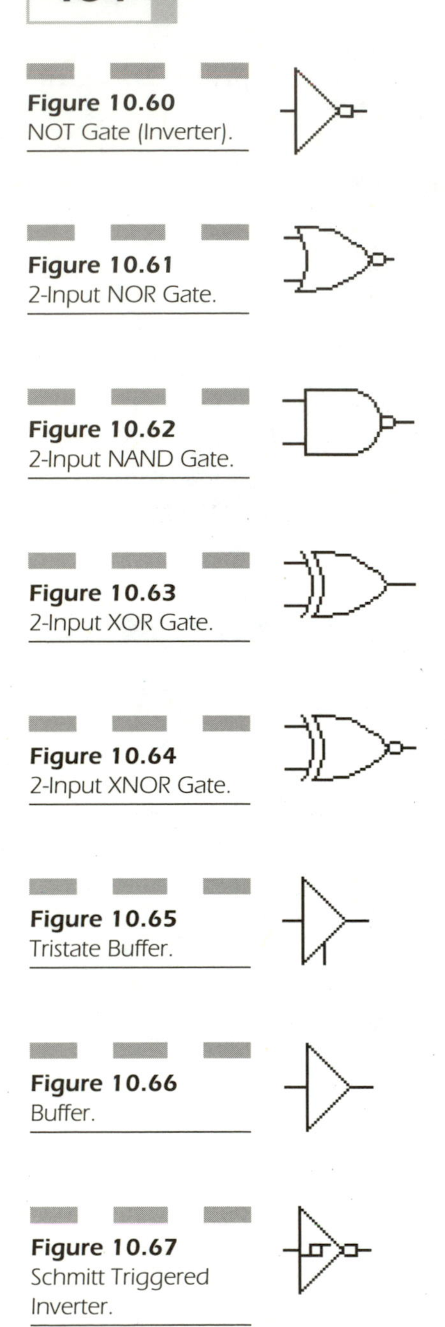

Figure 10.60
NOT Gate (Inverter).

Figure 10.61
2-Input NOR Gate.

Figure 10.62
2-Input NAND Gate.

Figure 10.63
2-Input XOR Gate.

Figure 10.64
2-Input XNOR Gate.

Figure 10.65
Tristate Buffer.

Figure 10.66
Buffer.

Figure 10.67
Schmitt Triggered Inverter.

The logic gates are simply an easier way to access the templates; for example, if you are looking for a TTL IC that contains four 2-input XOR gates, select the XOR icon and it gives you a choice of three ICs that fit that description—one being a 7486. You have a template choice among AND, OR, NAND, NOR, NOT, XOR, XNOR, and Buffers (see Figure 10.68).

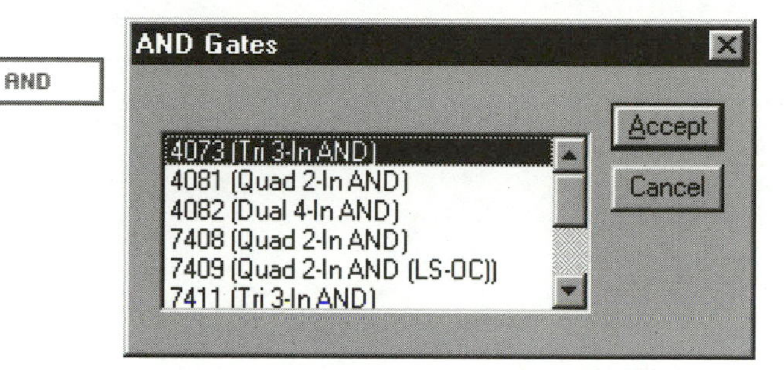

Figure 10.68 Use the bottom section of the Logic Gates toolbar for quick access to templates.

Digital Toolbar

The Digital toolbar, shown in Figure 10.69, is a mixture of digital components not listed elsewhere.

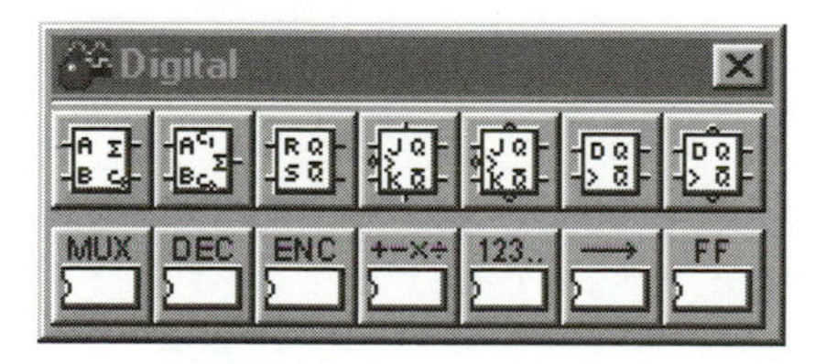

Figure 10.69 The Digital toolbar.

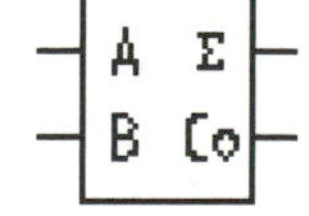

Figure 10.70 Half-Adder.

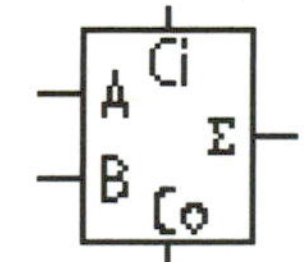

Figure 10.71 Full-Adder.

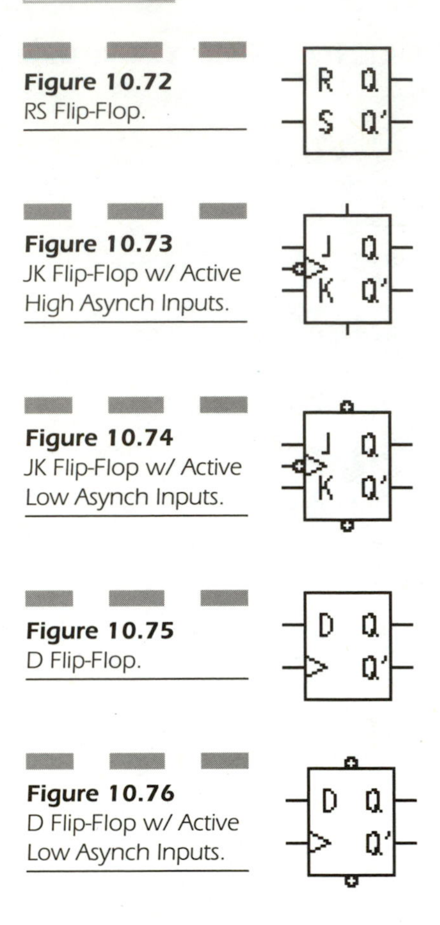

Figure 10.72
RS Flip-Flop.

Figure 10.73
JK Flip-Flop w/ Active High Asynch Inputs.

Figure 10.74
JK Flip-Flop w/ Active Low Asynch Inputs.

Figure 10.75
D Flip-Flop.

Figure 10.76
D Flip-Flop w/ Active Low Asynch Inputs.

Digital items are simply an easier way to access the templates; for example, if you are looking for a TTL IC that contains a decade counter, select the Counters icon and it gives you a choice of 15 ICs that fit that description—one being a 7490. You have a template choice among multiplexer, demultiplexer, encoders, arithmetic, counters, shift registers, and flip-flops (see Figure 10.77).

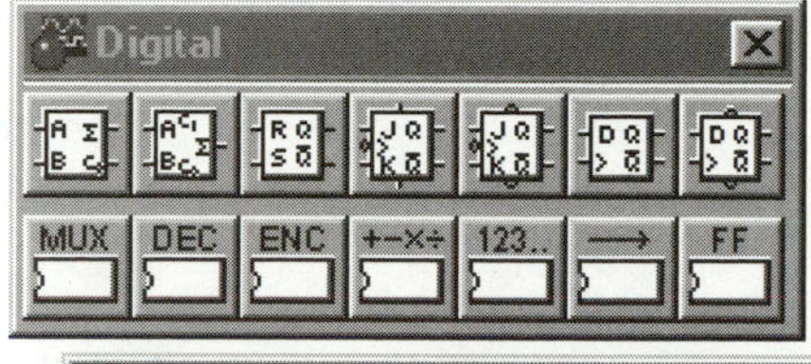

Figure 10.77 Use the lower part of the Digital toolbar for quick access to templates.

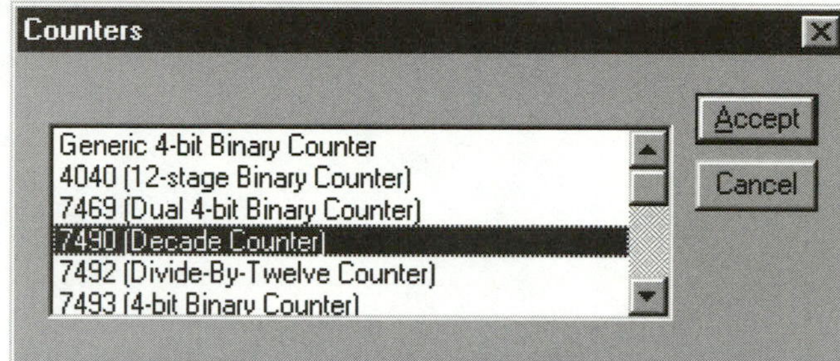

Indicators Toolbar

The Indicators toolbar is shown in Figure 10.80.

Figure 10.78 The Indicators toolbar.

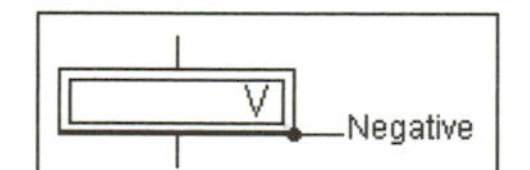

Voltmeter—Measures voltage in a circuit, giving a visual numeric readout. If you need to place multiple indicators to show voltage in a circuit, it is easier to use many voltmeters instead of multiple multimeters set to voltmeter. Be sure to obey the polarity and direction when placing voltmeters in a circuit; the bold line on one side of the meter indicates the negative terminal.

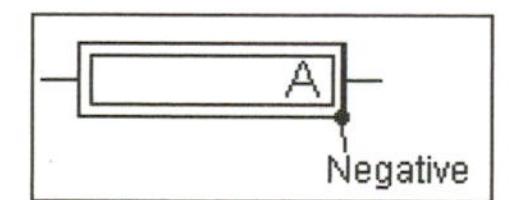

Ammeter—Measures amperage (current) in a circuit, giving a visual numeric readout. Observe polarity with this component. The ammeter is used in place of multiple multimeters set to ammeter.

You must double-click the placed voltmeter or ammeter and adjust it to read either DC (default) or AC (in RMS) (see Figure 10.79). You can also adjust the meter's resistance for a greater degree of simulation accuracy.

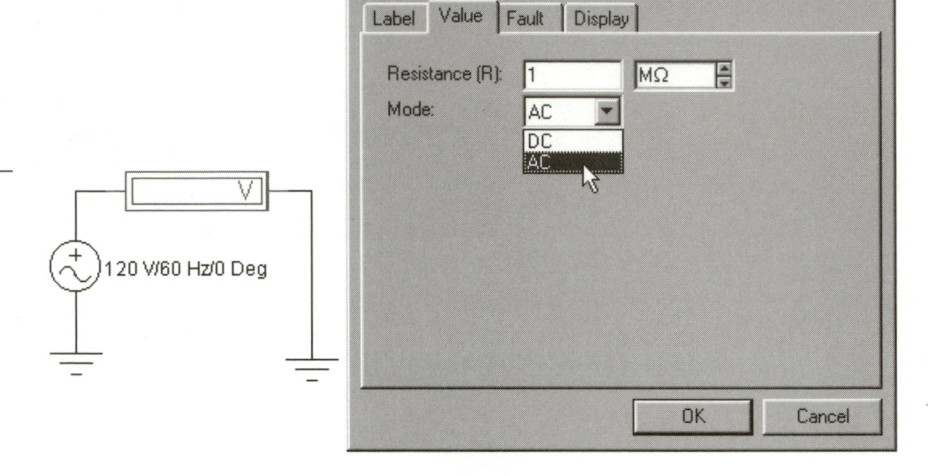

Figure 10.79 Double-click the placed voltmeter or ammeter and adjust it to read either DC or AC.

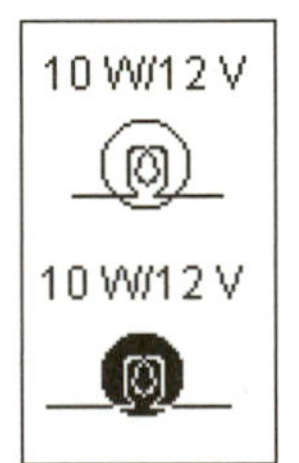

Lamp—A visual indicator that animates when the correct voltage or wattage is applied. If this is exceeded, the bulb burns out—think of it as a light bulb. You can adjust the wattage and voltage by double-clicking the placed component.

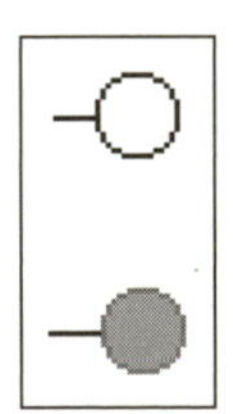

Probe—A visual indicator that animates when the threshold voltage is reached. The circle fills with color (set by the user) when lit—think of it as a logic probe. It is used as a quick visual method of reading the outputs of digital circuits. You can change its color by double-clicking the placed component and selecting the Choose Probe tab. You have a choice between red, green, or blue. You can also use the Models tab to adjust among TTL, CMOS, or default threshold voltages.

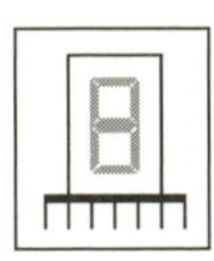

Seven-Segment Display—These seven elongated LEDs, when lit in a specific order, create the numbers 0 through 9 or the letters A through F. Each of the seven terminals controls one of the seven LEDs. You can set the model among CMOS, TTL, and default by double-clicking the placed component. The component is used in any simulation that requires an alphanumeric character.

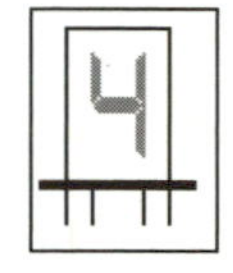

Decoded Seven-Segment Display—A seven-segment display that is controlled with a binary 4-bit nibble. The four input lines are hooked to a 4-bit signal.

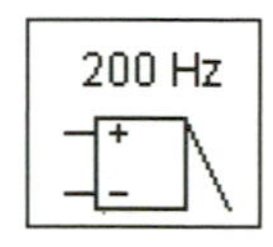

Buzzer—When a specific current is flowing through this device, your PC speaker will "sound" at the frequency you set. Double-click the component to adjust its frequency, voltage, and turn-on amperage.

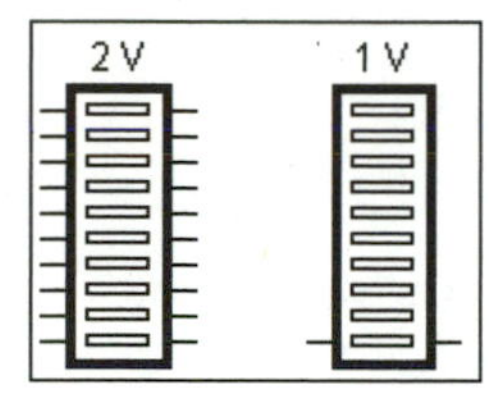

Bargraph and Decoded Bargraph—A series of elongated LEDs stacked on each other. It is typically used as a power level indicator; the more LEDs lit, the higher the level. The decoded bargraph has built-in voltage conversion circuitry and only requires you to hook a voltage level to it. There is also a version where each LED has its own terminal connection. Double-click the placed decoded bargraph to set what power level will activate the LEDs.

Controls Toolbar

The Controls toolbar is shown in Figure 10.80.

Figure 10.80
The Controls toolbar.

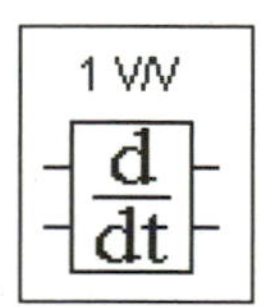

Voltage Differentiator—Calculates the derivative of the input voltage (the transfer function, 1/s) and delivers it to the output. It is used in control systems and analog computing applications. *Differentiation* may be described as a rate of change function and defines the slope of a curve. Rate of change = dV/dT. Double-click the placed component to adjust its settings.

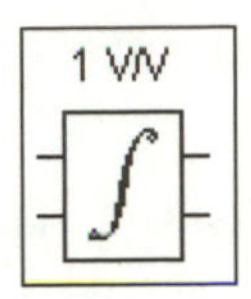

Voltage Integrator—Calculates the integral of the input voltage (the transfer function, *1/s*) and delivers it to the output. It is used in control systems and analog computing applications. Double-click the placed component to adjust its settings.

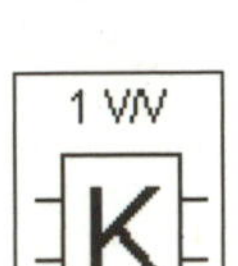

Voltage Gain Block—Amplifies the input by a gain level (K) set by you. Double-click the placed component to adjust its settings.

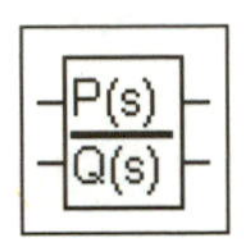

Transfer Function Block—This component models the transfer characteristic of a device, circuit, or system in the *s* domain. The transfer function block is specified as a fraction with polynomial numerators and denominators. A transfer function up to the third order can be directly modeled. This component may be used in DC, AC, and transient analyses.

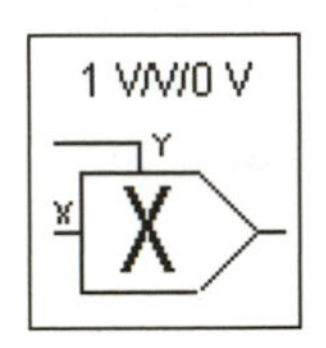

Multiplier—Multiplies two input voltages (X and Y) and outputs the result to A. By double-clicking the placed component, you can adjust its offset and gain.

TIP

The Electronics Workbench Help files contain in-depth explanations and example circuits of the Control items.

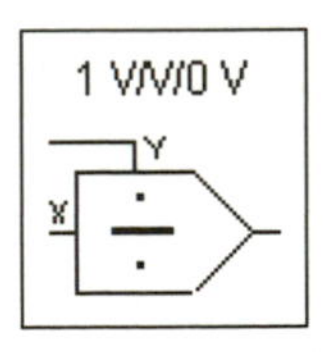

Divider—Divides two input voltages (X and Y) and outputs the result to A. Offset, gain, lower limit, and smoothing domain are accessed by double-clicking a placed component.

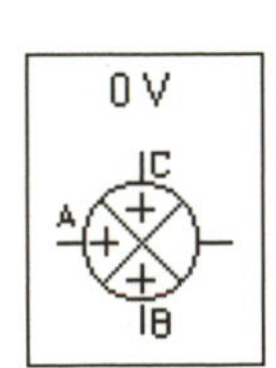

Voltage Summer—Adds three input voltages and outputs the results.

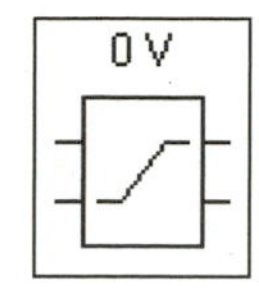

Voltage Limiter—A voltage *clipper;* the output voltage excursions are limited, or clipped, at predetermined upper and lower voltage levels while input-signal amplitude varies widely. See the Help file for examples and further explanation.

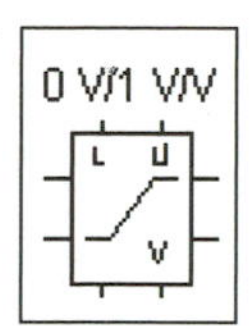

Voltage-Controlled Limiter—A voltage *clipper;* a single input, single output function. The output is restricted to the range specified by the output lower and upper limits. Output smoothing occurs within the specified range. The voltage-controlled limiter operates in DC, AC and transient analysis modes.

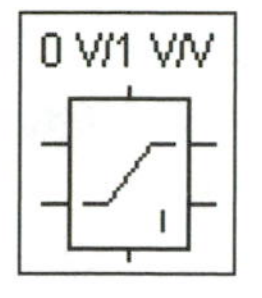

Current Limiter Block—Models the behavior of an operational amplifier or comparator at a high level of abstraction. All of its pins act as inputs; three of them also act as outputs. The component takes as input a voltage value from the "in" connector. It then applies the offset and gain, and derives from it an equivalent internal voltage, Veq, which it limits to fall between the positive and negative power supply inputs. If Veq is greater than the output voltage seen on the "out" connector, a sourcing current flows from the output pin. Otherwise, if Veq is less than the output voltage, a sinking current flows into the output pin.

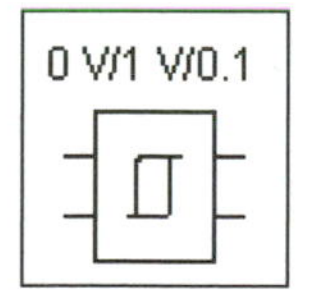

Voltage Hysteresis Block—A simple buffer stage that provides hysteresis of the output with respect to the input. See the Help file for further details.

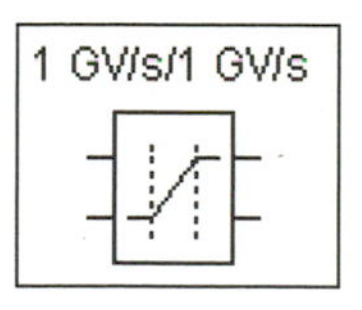

Voltage Slew Rate Block—Limits the absolute slope of the output, with respect to time, to some maximum, or value. You can accurately model the actual slew rate effects of over-driving an amplifier circuit by cascading the amplifier with this component. Maximum rising and falling slope values are expressed in volts per second.

Miscellaneous Toolbar

The Miscellaneous toolbar is shown in Figure 10.81.

Figure 10.81 The Miscellaneous toolbar.

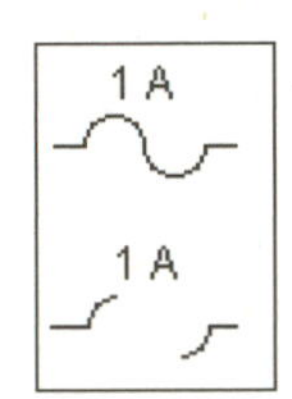

Fuse—A device that protects a circuit from excessive current. You can set the maximum current by double-clicking the placed component. Once the fuse is blown in the circuit, you must reset the simulation to replace it.

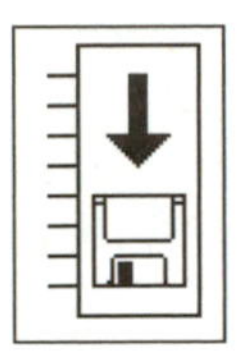

Write Data—This component records up to eight waveforms as text files. You hook a terminal to each signal output and double-click the component to name a file to which the waveform information will be recorded. The text file contains two or more columns: the left column is the time and the right column(s) are the voltage readings at that time point (see Figure 10.82). The data can be used with the piecewise linear source to pump the text recorded waveform into a circuit or output it to a spreadsheet program such as Microsoft Excel.

Figure 10.82 The Write Data component creates a text file you can open with Windows Notepad or Miscrosoft Excel.

Test.txt - Notepad

File Edit Search Help

ASCII output data for Electronics Workbench circuit untitled

Time	V1	V2	V3	V4
0	0	0	0	0
5.20833e-06	0.166608	-0.166608	0.333215	-0.333215
1.04167e-05	0.333215	-0.333215	0.666426	-0.666426
2.08333e-05	0.666426	-0.666426	1.33281	-1.33281
4.16667e-05	1.33281	-1.33281	2.66529	-2.66529
8.33333e-05	2.66529	-2.66529	5.32795	-5.32795
0.000166667	5.32795	-5.32795	10.6349	-10.6349
0.000333333	10.6349	-10.6349	21.102	-21.102
0.000666667	21.102	-21.102	40.8782	-40.8782
0.0011875	36.7305	-36.7305	66.2217	-66.2217
0.00170833	50.9473	-50.9473	81.4836	-81.4836
0.00222917	63.2064	-63.2064	84.3404	-84.3404
0.00275	73.0364	-73.0364	74.3571	-74.3571
0.00327083	80.0597	-80.0597	53.0536	-53.0536
0.00379167	84.0063	-84.0063	23.6732	-23.6732
0.00400101	84.6874	-84.6874	10.5708	-10.5708
0.00452184	84.0933	-84.0933	-22.4527	22.4527
0.00504268	80.2675	-80.2675	-52.0579	52.0579
0.00556351	73.3571	-73.3571	-73.7378	73.7378
0.00608434	63.6277	-63.6277	-84.1918	84.1918

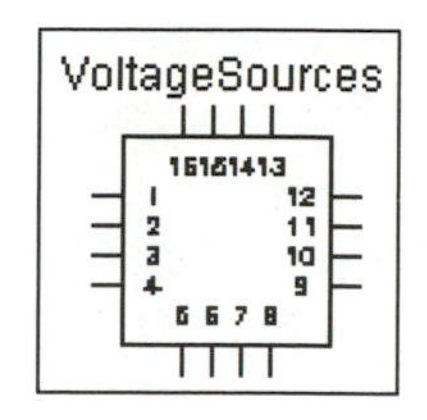

Netlist Component—Pre-made Netlist components (SPICE subcircuits) that you can place into a circuit. They contain voltage regulators, voltage references and more.

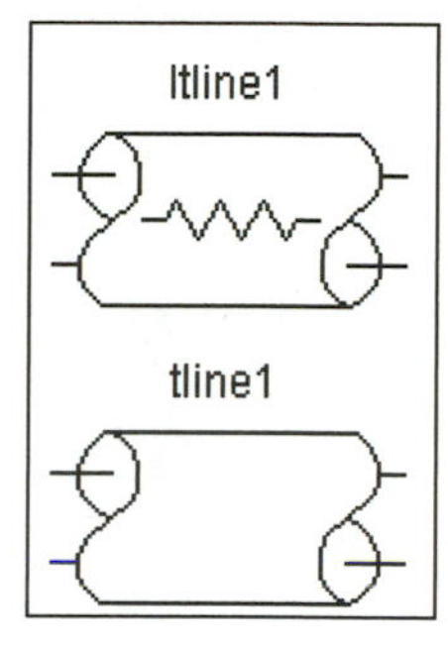

Lossy Transmission Line and Lossless Line—Wire and cable are not perfect conductors; they add increasing amounts of resistance, inductance, capacitance, and conductance for each meter of cable or wire. These components allow you to simulate the results of these cables and wires by inputting their characteristics. Double-click the placed component and select a model of cable to use. You can also create your own model by copying an existing version and editing the parameters. Sometimes, these numbers are available in datasheets or catalogs from electronic distributors. The *lossless model* is an ideal cable or wire that simulates only the *characteristic impedance* and *propagation delay* properties of a transmission line. The characteristic impedance is resistive and is equal to the square-root of L/C. These values are also accessed by double-clicking the placed component and choosing to edit the model.

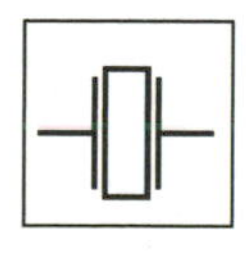

Crystal—Used in oscillator circuits to give a very stable and specific frequency output, this is a quartz crystal that resonates at a specific frequency determined by its physical size. Because of its accuracy, it is used in timing applications that require an exacting measurement of time. Your computer's microprocessor makes use of one such XTAL.

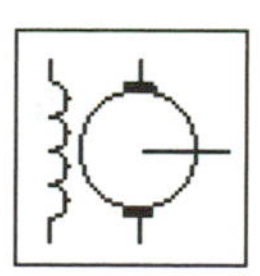

Motor—This is a virtual DC motor with connections that allow its stator to create a magnetic field, and terminals for the brushes, which send power to the rotor. See the Electronics Workbench Help file for more component information.

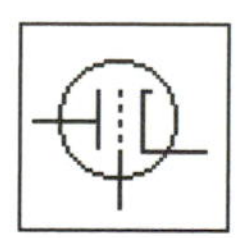

Vacuum Tube—This is the transistor's ancestor; even though silicon wonders have all but replaced the old tubes, some exotic stereo and musical equipment still use them in their designs. Double-click the placed component to select a model.

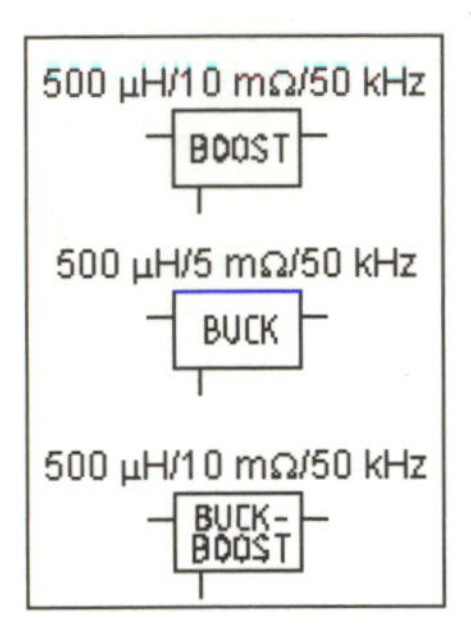

Boost Converter, Buck Converter, and Buck-Boost Converters—These components are used to simulate DC, AC, and large-signal transient responses of switched-mode power supplies operating in both the continuous and discontinuous inductor current conduction modes (CCM and DCM, respectively). To adjust the settings, double-click the placed component.

Textbox—Used to place a block of miscellaneous text onto the screen. Move a copy of the "A" onto the drawing and double-click it to set the text and font size (see Figure 10.83).

Figure 10.83
Use the textbox icon to place text onto drawings.

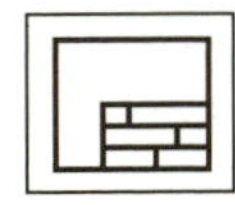

Titleblock—Places miscellaneous information about the drawing into a neat package that includes the drawing's name, author, date, etc. Place a copy of the block and double-click it to set the information (see Figure 10.84).

Title Block
EWB Test Circuit
Title
EWB Test Circuit
Document Number
1999-202
Revision
b
Drawn By
J. Adams
Date
8-25-1999
Set Font
OK
Cancel

Figure 10.84
Use the Title Block to document your drawings.

Component Details

Electronics Workbench provides extensive data on each component. This includes how the component is modeled, its electrical characteristics, and, in some cases, its applications. Under **Help > Help Index** you will see a list of each Parts Bin. Using the Help files saves having to print out a technical tome. You can also hit the **Index** tab and simply type in the name of the component, and EWB will locate it. Or you can follow this procedure:

1. Place the component for which you require more information.
2. Hit **F1**.

I have a section of component examples on my website, at **http://www.basicelectronics.com/ewb/**. If you wish to add a circuit, visit this site.

TIP

To find out more information about a component quickly, right click and choose help.

TIP

To learn exactly what an IC contains (gates, resistors, etc.), you may want to purchase a manufacturer's databooks or find datasheets on the company's website.

Models and Libraries

The information in this chapter should give you a good idea of what is involved in creating custom components. The following information is a more detailed description of how to download and add models or libraries to EWB.

To avoid confusion, a model contains all the information needed to simulate a component and is stored under a file with the *.mod extension. A library contains one or several model files grouped into one file; it has a *.lib extension. Under the directory **C:\EWB5\MODELS** are subdirectories that contain the actual library files. For example, there is a subdirectory called **BJT_NPN**. Notice in that directory that there are several *.lib files for each company's NPN models. You can download *.lib files and place them directly into the appropriate subdirectory and they will become available for use in the EWB Program.

The Mod and Lib files are available from manufacturer's websites. The Electronics Workbench company maintains a list of some of these sites at **http://www.electronicsworkbench.com/html/spice_models.html**.

I also maintain a list at **http://www.basicelectronics.com/ewb/**.

Once in a manufacturer's site, look for SPICE models or libraries (see Figure 10.85) and download them into a temporary directory or, if they are *.lib files, directly into the appropriate Model subdirectory.

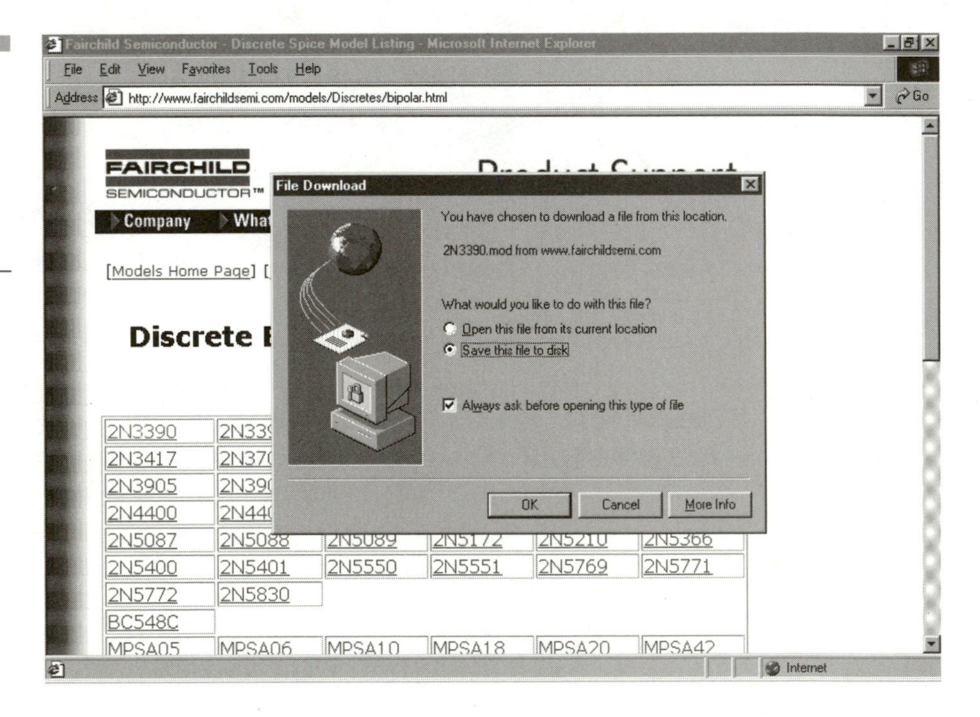

Figure 10.85 Look for *.lib or *.mod files on manufacturer's websites to expand EWB5's component libraries.

You may run across companies that only put out model files (*.mod). You can either group these into one large library file, using a word processor, and save the file as a *.lib file or simply rename the *.mod extension as *.lib; it will be a library of one model then. Once inside the EWB5 program you can cut and paste the individual models wherever you want.

Summary

Electronics Workbench uses complex algorithms to simulate most types of real-world components, allowing us to model them accurately in our circuit designs. Armed with the knowledge you have learned in this chapter, you should be able to create just about any virtual component not included in the software.

CHAPTER 11

EWB Instruments

An Electronics Workbench instrument is your window to the electronic world.

You no longer have to pay thousands of dollars for electronic instrumentation and equipment—there's a cheaper alternative. One of Electronics Workbench's most powerful features is *virtual instruments*. They are just as powerful and versatile as the real thing. (Look out FLUKE™!) They include simple devices such as the multimeter, function generator, oscilloscope, and logic analyzer and also include instruments that have no real-world equivalent but are nevertheless, powerful tools: the Bode plotter, logic converter, and word generator.

The main function of these instruments is to provide an electrical signal or to read an output signal. This chapter breaks down each piece of equipment's functions and uses.

CAUTION

You cannot rotate instruments or run more than one copy of an instrument with EWB5.

Accessing EWB Instruments Toolbar

Electronics Workbench's Instrument toolbar is located on the far right of the Parts Bin toolbar. To open it, move the mouse over the icon and click it. I recommend leaving it open while designing for faster access to instruments (see Figure 11.1). The components of the Instrument toolbar are:

- **Multimeter**—Ohmmeter (measures resistance), voltmeter (measures voltage), and ammeter (measures current). Can also measure decibel loss between two points.
- **Function Generator**—Used to output various electrical waves to a circuit.
- **Oscilloscope**—Allows you to visualize an electrical wave.
- **Bode Plotter**—Graphs a circuit's response to various frequencies; helpful to analyze filter circuits.

- **Word Generator**—Outputs a custom digital waveforms to a circuit.
- **Logic Analyzer**—Views up to 16 digital waveforms.
- **Logic Converter**—Converts a digital signal or circuit into a truth table or truth table into a circuit. No real-world equivalent.

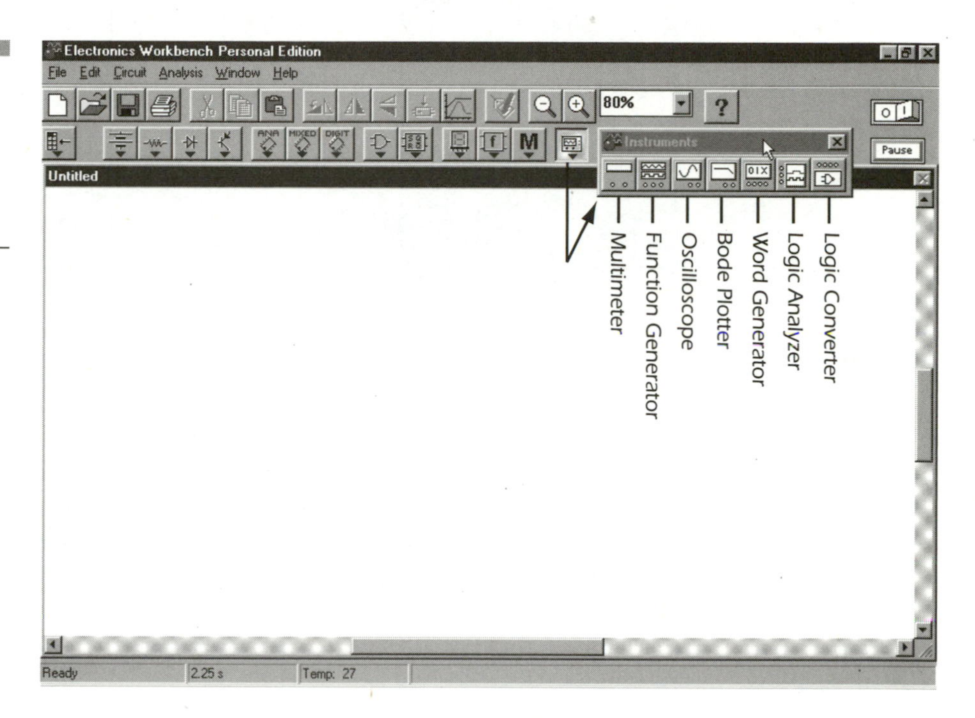

Figure 11.1
The Instrument toolbar is opened with the last icon to the right of the Parts Bin toolbar.

Placing Instruments

An instrument icon must first be placed into the circuit in order to attach and run it. First open the Instrument Toolbar. Place the mouse pointer over the instrument's icon on the toolbar. Click and hold the left mouse button. Move it into the circuit window and place the instrument icon by releasing the button.

Each instrument contains terminals that must be connected to the circuit. Wire each applicable terminal into the circuit. Once a terminal is connected, double-click the instrument to open it. For example, if you are measuring the voltage across a resister, hook up the circuit as seen in Figure 11.2.

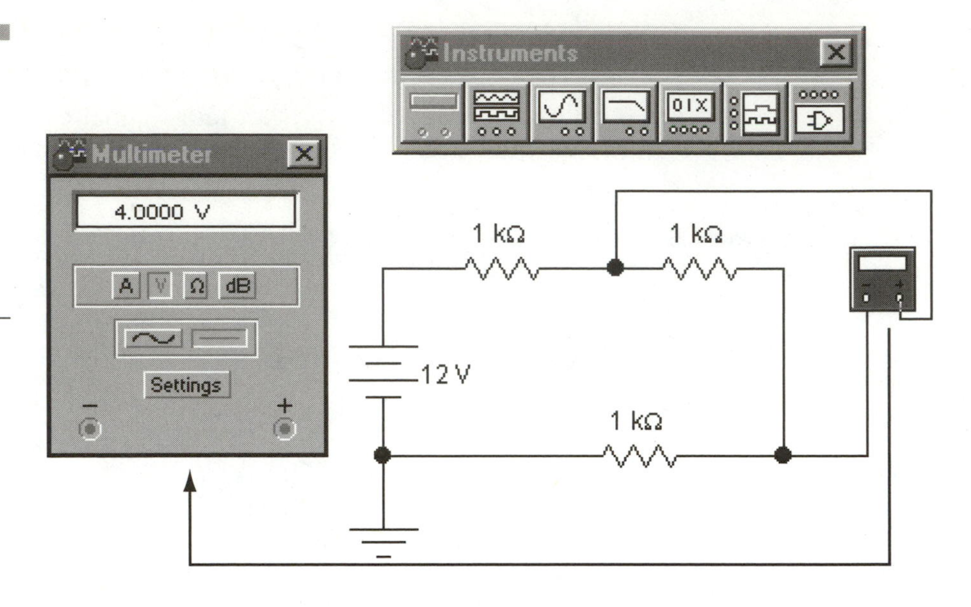

Figure 11.2 Placing, connecting, and opening instruments. In this example, we are measuring the voltage across a resistor using the multimeter.

TIP

Open up the instrument if you are having trouble seeing where the terminals are. Go back to the Instrument icon and move the pointer over that general area until a blue circle appears. If you can't quite figure out what the terminal is, open the instrument to see the label for that point.

Adjusting an Instrument's Settings

After the instrument is wired in, it must be adjusted to measure exactly what you want it to, or output a signal according to your specs. Double-click the instrument's icon to open it; each instrument has a window with various buttons, boxes, and spin controls (see Figure 11.3). Some of these can be adjusted while the simulation is in progress (this is sometimes referred to as on-the-fly adjustment). The controls for each instrument are described in detail later in this chapter.

Analysis Options: Instruments Tab

Under the **Analysis** menu is the command to open the Analysis Options dialog box (or hit **Ctrl+Y**). Find the tab labeled **Instruments**. From here, you can adjust instrumentation settings and control the

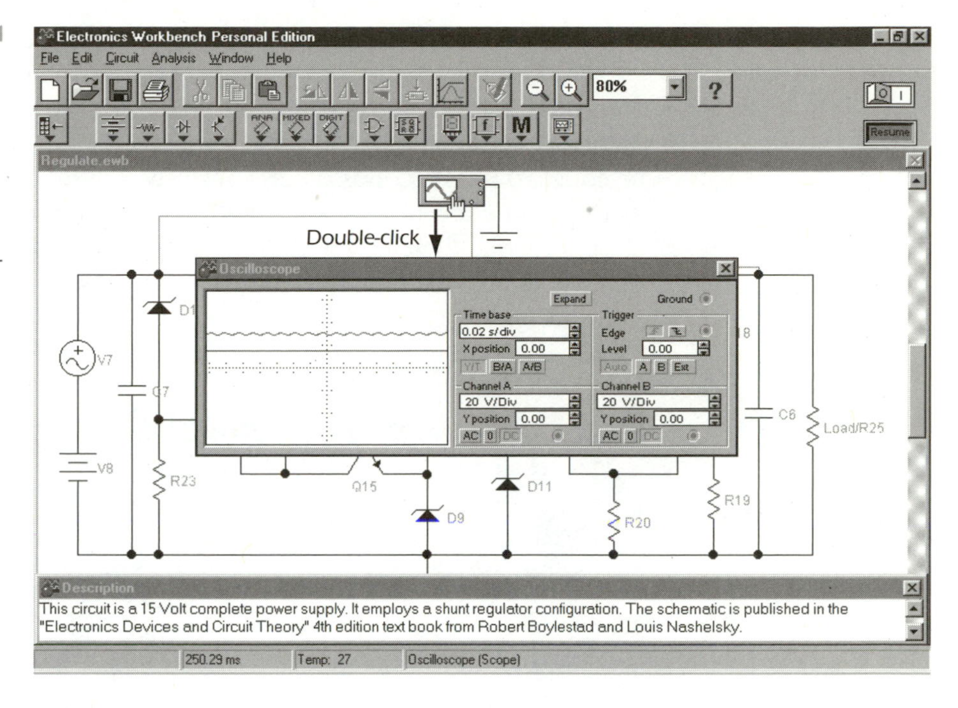

Figure 11.3 By double-clicking the instrument's icon, you can access its controls. The oscilloscope is pictured here.

efficiency of the simulation. If you would like more complex explanations, open up EWB Help and enter the Instruments Analysis Options section:

- **Oscilloscope**—Each item (Figure 11.4) effects the oscilloscope's settings.
  - **Pause after each screen**—Temporarily halts the simulation when the waveform reaches the end of the scope's screen for you to read the oscilloscope's graph.
  - **Generate time steps automatically**—Lets EWB select the next two settings for you.
  - **Minimum number of time points**—Lets you manually adjust how many times EWB will calculate the voltage level or your circuit. For example, if you want the oscilloscope to read the voltage level 100 times for each screen width (14.4 grid lines on the scope's window), then you would set this number to 100. If you want to speed a simulation, reduce the value. However, the accuracy will suffer and the waveform will appear choppy.
  - **Maximum Time Step**—Lets you manually adjust the time between voltage readings the scope will make (every tenth of a second or whatever). You may notice that timepoints adjusts along

with this number. For example, if the oscilloscope is set to 1.00s/Div Timebase and you open the Instrument options (timepoints are set to 100), this number will be set to 0.144. This is because 100 divided by 14.4 seconds (width of scope's screen) is equal to 0.144 seconds. Therefore, EWB will take a voltage reading every one-seventh of a second. Are you confused yet? Don't be. Just think of each oscilloscope's screen width as one cycle. So if the scope is set to 0.02s/Div Timebase and the timepoints are set to 100 (as in Figure 11.4), then each screen's width is 0..288 seconds long. If you divide that by 100 timepoints we get 0.00288 for TMAX. See?

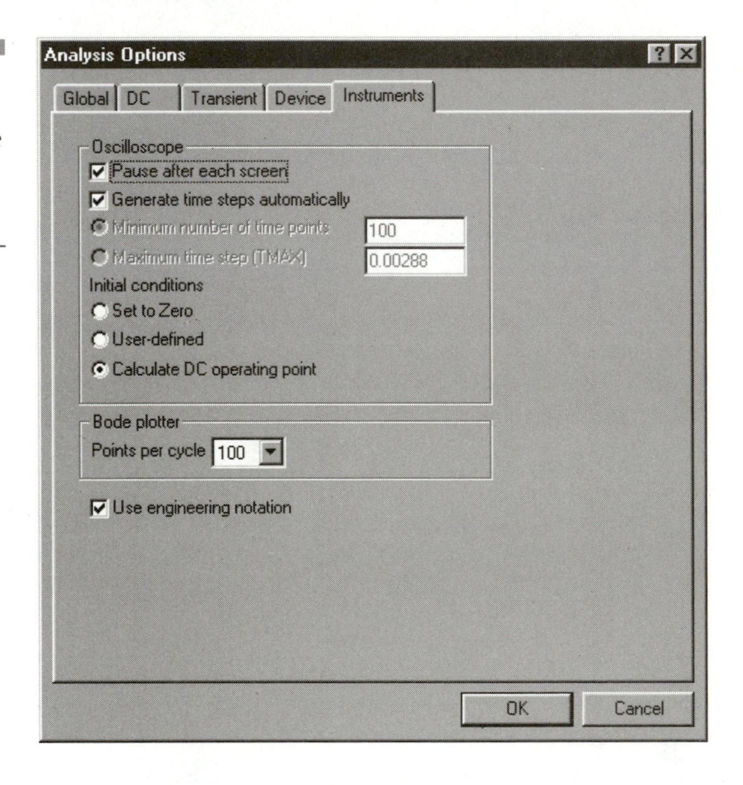

Figure 11.4 Some instruments require defaults to be set under Analysis > Analysis Options > Instruments Tab.

— **Initial Conditions:** Determines how the simulation starts up:

- **Set to Zero**—If using this setting, all conditions are set to zero. I recommend using this if you are having problems reading such things as a changing DC signal.
- **User-Defined**—Tells EWB to use your preset conditions, which are usually set up on each node. To access these, double-click anywhere along the node on the drawing and choose the

Node tab. Under Analysis you will see the **Use initial conditions** checkbox. Check it on and fill in the information below.

- **Calculate DC operating point**—Starts at the voltage levels determined by the DC operating point. EWB defaults to this but I don't like to use it because the scope does not always pick up valuable information at the beginning of a simulation; set to zero is more revealing

- **Bode Plotter Points per Cycle**—Determines how many points will be analyzed for each cycle of the simulation. It is best to leave the setting alone, but if there are problems with the simulation, lower the value to 50 and reactivate.
- **Use Engineering Notation**—Tells EWB to display all values in engineering notation. For example, instead of displaying 5.10937e-03 (scientific notation), the scope will read 5.1094 ms. I recommend leaving engineering notation on (checked-on).

Printing Instrument Readings and Settings

You can make a hard copy of each instrument's settings and its current readout. Go to **File** > **Print** or hit **Ctrl+P** to open the Print dialog box (see Figure 11.5). Check on which instruments you wish a printout for and hit the **Print** button. Note that the order you click them is the order they will print. If you want a printout of specific readouts, such as the oscilloscope, use the Grapher. Open this by choosing **Analysis** > **Display Graphs**.

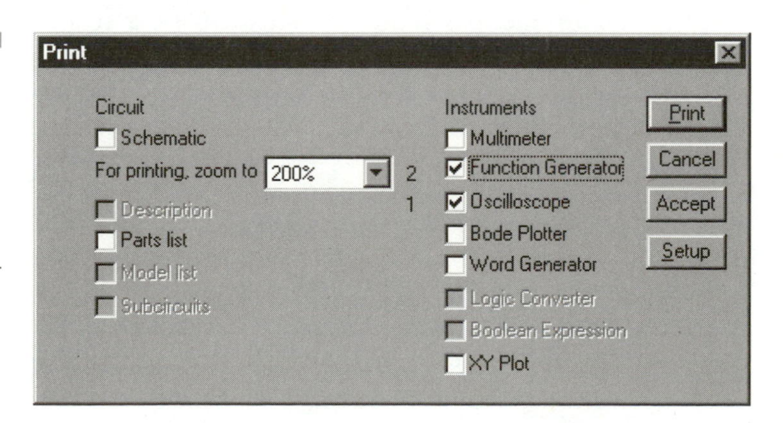

Figure 11.5 Electronics Workbench lets you print a picture of the opened instruments and later refer to its settings.

For tangle-free lines, allow a bit of room when laying instruments.

Instrument User Guide

The remainder of this chapter describes each instrument, along with details of how to hook up the instrument, examples, a few hints, and tricks. The instruments are presented in logical order.

Multimeter

The multimeter is the most valuable instrument I own on my real workbench. It measures resistance, AC and DC voltage/current, and decibel loss. It replaces an ammeter, voltmeter, and ohmmeter. The Electronics Workbench virtual multimeter is no different. Hook up your leads, activate the circuit, and a reading magically appears.

Multimeter Connections

The negative terminal is the left connector at the bottom of the meter's icon. The positive terminal is the right connection. Think of the negative side as a black probe and the positive as a red probe. If you are reading a positive signal in reference to ground, then connect the negative to ground and the positive to a point in the circuit where a measurement is required. This can be a connector, a component's terminal, or a pin anywhere along the path of a wire (this creates a connector from the wire to the meter). The next few sections and examples illustrate how to hook up the multimeter.

Multimeter Operation

Once the multimeter is connected to a circuit or component group, it is time to adjust its measurement settings. Double-click the multimeter symbol. It defaults to measure DC voltage (see Figure 11.6). The top set of buttons lets you select a mode:

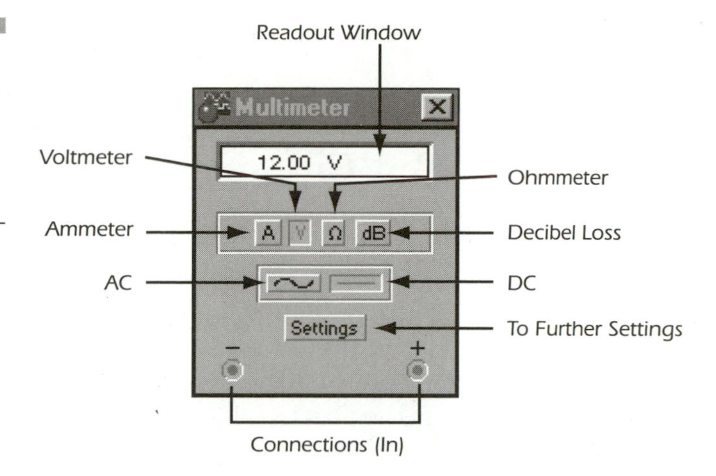

Figure 11.6 Controls and readouts on Electronics Workbench's multimeter.

- **A** for ammeter
- **V** for voltmeter
- Ω for ohmmeter
- **dB** for decibel loss between two points.

The lower two buttons are to choose between AC or DC.

- The squiggly line is AC.
- The straight horizontal line is DC.

The last button adjusts the meter's internal settings. The **Set** button opens the Electronics Workbench Settings dialog box.

A real multimeter is not an ideal model—it can induce resistance that may result in slightly inaccurate measurements. By adjusting the virtual multimeter's internal settings, you can approximate these imperfections to make for a realistic multimeter. To change the settings, either use the spinner or type in the new value. Make sure to set the multiplier as well (n, p, M, G, etc.).

NOTE

You don't have to select a range, because the meter is autoranging—if you're measuring a 1V source and suddenly have to measure 24 volts, you don't have to change any range buttons or dials.

Once your preferred settings are adjusted, it is time to turn the circuit on and take the readings. Leave the multimeter open and activate your circuit. A reading should appear in the top white box; this may vary as time goes on.

Multimeter Test Circuits

DC READINGS

Build the circuit pictured on the left in Figure 11.7, Measuring Voltage. Open the multimeter and activate the circuit. The settings are on DC voltmeter so you should get a reading of 8 volts. Hit the 'r' key and watch the results on the meter. Hit the **Shift** and 'r' key and note the reading now. Hit the **Shift** and 'r' again.

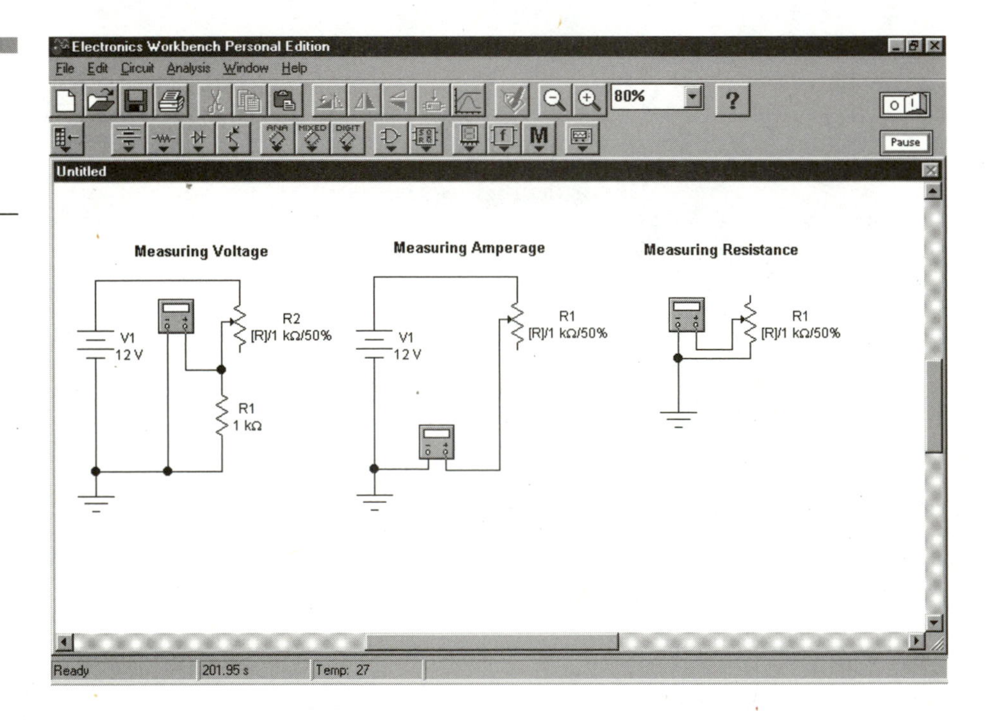

Figure 11.7 Measuring voltage, amperage, and resistance with the multimeter.

Modify the circuit to appear as in Figure 11.7, Measuring Amperage. Hit the ammeter button (A) on the meter and reactivate the circuit. Adjust the potentiometer and note the readings.

Again, rework the circuit to appear as in Figure 11.7, Measuring Resistance. Select the ohmmeter by hitting the Ω symbol. Activate the circuit and play around with potentiometer once more.

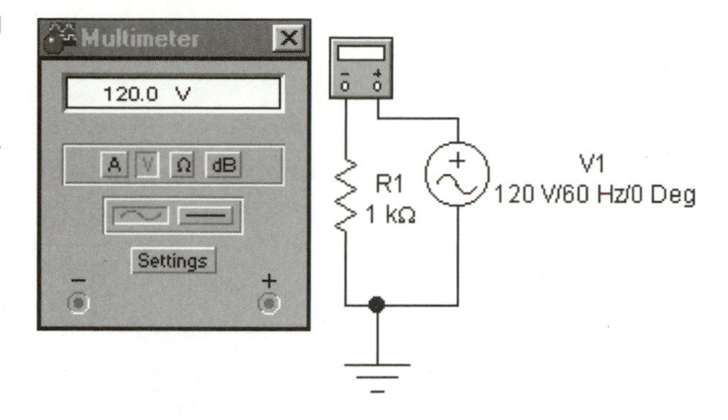

Figure 11.8
Measuring AC with the multimeter.

AC READINGS

Alternating current and voltage readings require that you set the meter to AC. Build the circuit pictured in Figure 11.8 and make the appropriate settings to the multimeter to read AC voltage. Activate the circuit. What is the voltage reading? Select the AC ammeter now. What are its readings?

NOTE

EWB5's multimeter measures the RMS of a voltage/current AC wave (.707 of peak).

Multimeter Tips

Here are two tips to help you use the multimeter more effectively:

- **Swapping Polarity**—If the probes are on the wrong side of the meter, create a criss-cross subcircuit as pictured in Figure 11.9. This makes for cleaner designs.
- **Multiple Readings**—Use voltmeters and ammeters instead of the multimeter.

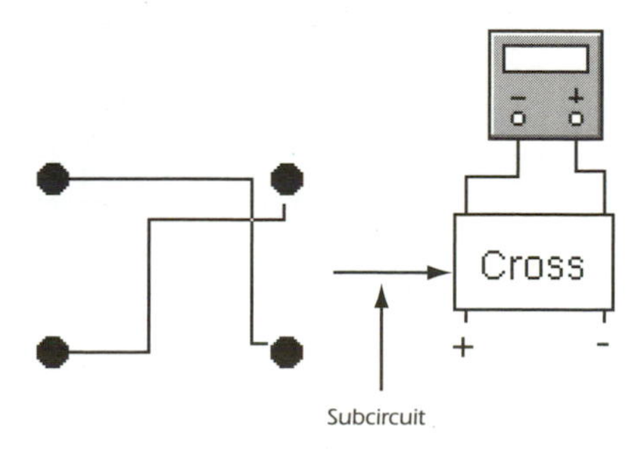

Figure 11.9
Create a "crossing" subcircuit to avoid tangled lines with the multimeter.

Oscilloscope

This complex instrument sometimes scares away newcomers to electronics, but don't fret. It's just another instrument to learn. An oscilloscope measures and displays the magnitude and frequency variations of electronic signals. It is used to test a functioning circuit. The "scope" gives technicians, students, and engineers a visual idea of what is happening inside a circuit. It displays waveform properties that are otherwise undetectable (because they are simply too fast). I like to think of the oscilloscope as a machine that slows time to let us see the intricate details (voltage, current, waveform and frequency) of a circuit's operation.

Electronics Workbench comes with a virtual dual-channel oscilloscope. On your real-life workbench a scope is a major expense; aside from the hundreds or thousands of dollars for a decent scope, you also need an oscillator to test your scope or learn its functions. This requires a function generator or hand-built circuits; in other words, more time and money. EWB replaces this expense with its own savable, printable, hi-tech scope.

Connecting the Oscilloscope

Electronics Workbench's oscilloscope contains four connections (see Figure 11.10).

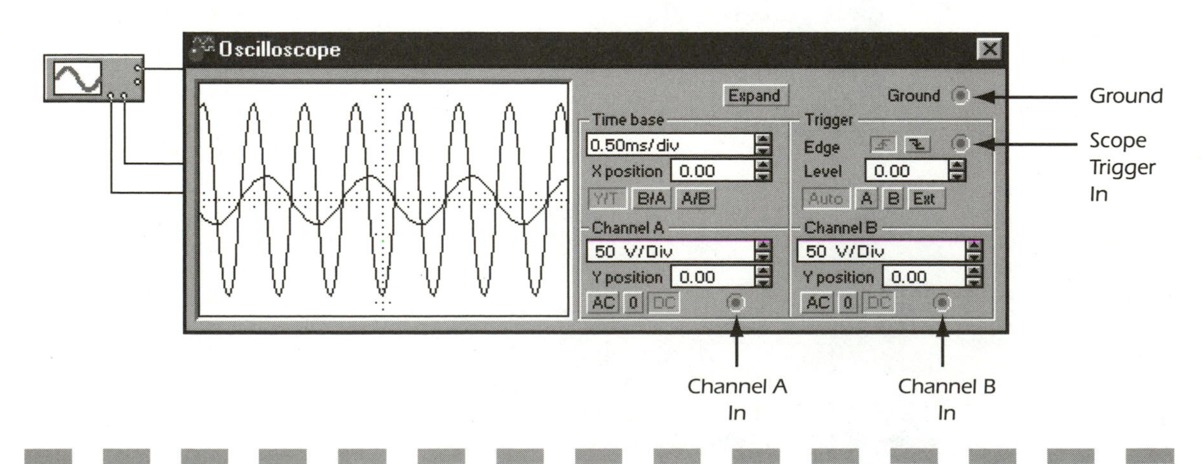

Figure 11.10 Oscilloscope connections.

- Channel A (Probe)
- Channel B (Probe)
- Trigger (external)
- Ground

1. It is not necessary to ground the scope if the circuit being tested already contains a ground. I make it a practice to run a ground to it, however, just in case I forget to ground the circuit.
2. If you want to 'scope out' a node in the circuit, place a wire from either the channel A or channel B terminal to the node you want to measure.
3. Because the oscilloscope is a dual-channel virtual model, you can connect another probe and read it at the same time. This is handy if you want to see what a signal is doing before and after a section of a circuit.
4. The external trigger is the last connection. It allows an electrical input to trigger the scope's readings.

Oscilloscope Operation

Once the oscilloscope is wired into a circuit, you must set its controls. This can also be done while the circuit is in operation, although this sometimes causes all sorts of strange and quirky (but cool) effects. The following list of control settings are applicable to the *expanded* scope; to access this, hit the **Expand** button at the top of the opened scope.

- **Graphic Display**—The readout appears here as a graphical waveform (see Figure 11.11).
- **Scrolling Bar**—Electronics Workbench records the waveform as the simulation runs. If you wish to scroll back to the beginning, use the bar below the Graphic Display window.
- **Cursor Readouts**—A set of vertical markers slide across the Graphic Display. Cursor 1 is red and cursor 2 is blue. These can move back and forth to give you an instantaneous measurement of the wave at that point in the x-axis plot. The windows below the scroll bar give those figures as follows:

  — T1 is the time position of cursor 1's point.

  — VA1 is channel A's voltage at cursor 1's point.

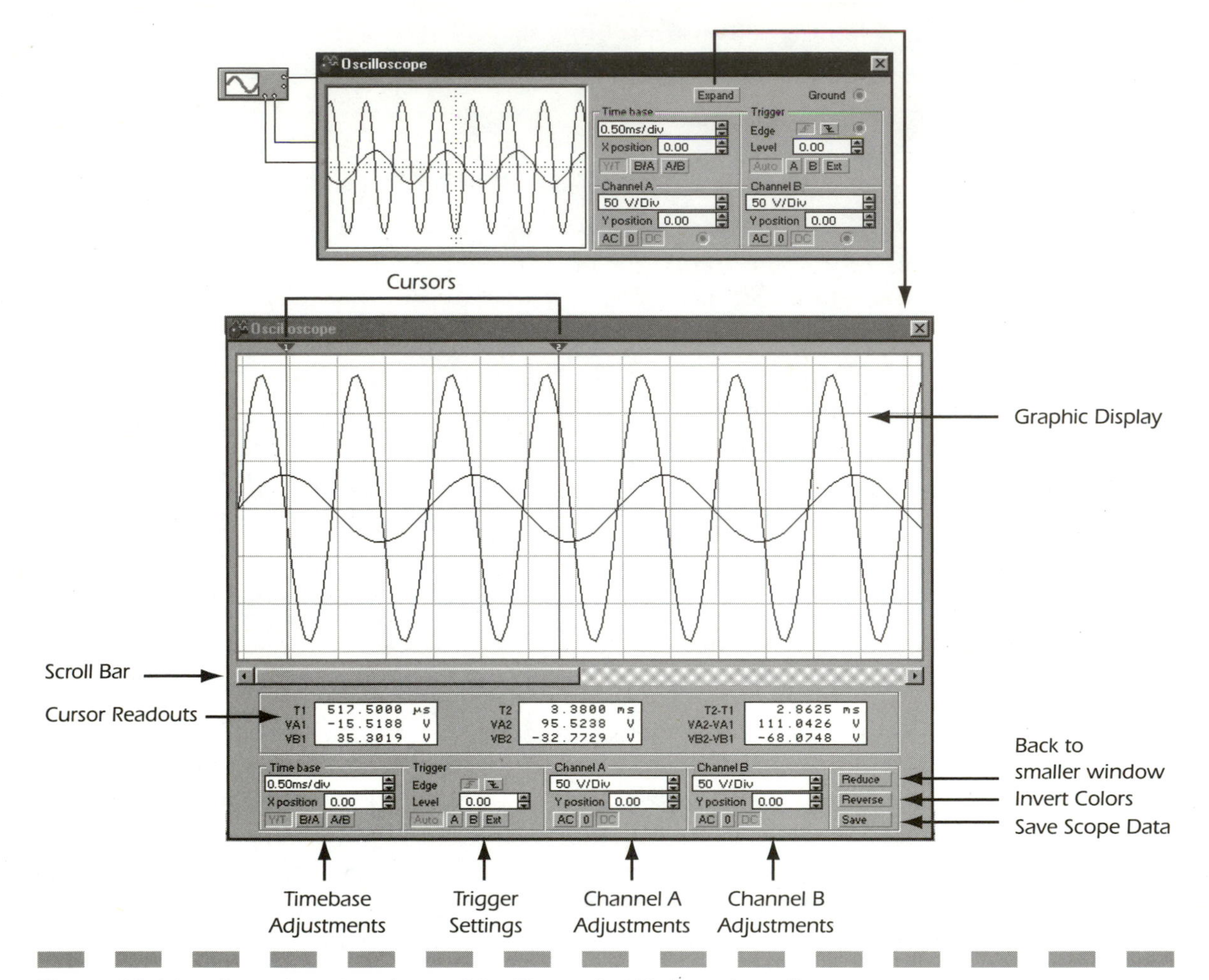

Figure 11.11 Controls and readouts of the Electronics Workbench oscilloscope.

— VB2 is channel B's voltage.

— T2 is the time position of cursor 2's point.

— VA2 and VB2 are cursor 1 and 2's voltages.

— T2-T1 is the time between cursor 1 and cursor 2.

— VA2-VA1 is the difference in voltage between those cursor points.

— VB2-VB1 is the voltage difference between those cursors.

- **Time Base**—You will notice that the graphic screen of the oscilloscope has a vertical and horizontal grid. The time base settings control the scale of the oscilloscope's X position (vertical grid lines) when comparing the amplitude against time (Y/T).

— **Scale**—Adjust the time units of the each grid line (the vertical grid lines).

— **X-Position**—This setting controls the signal's starting point on the x-axis. When X position is 0, the signal starts at the extreme left edge of the display. A positive value shifts the starting point to the right. A negative value shifts the starting point to the left.

— **Axes**—The axes of the oscilloscope display can be switched from showing waveform magnitude against time (Y/T), to showing one input channel against the other (A/B or B/A). The latter settings display frequency and phase shifts, known as *Lissajous patterns,* or they can display a *hysteresis loop.* When comparing channel A's input against channel B's (A/B), the scale of the x-axis is determined by the volts-per-division setting for channel B (and vice versa).

NOTE

Most readings use the Y/T setting.

- **Channel A and B Settings**

— **Volts-per-Division**—Used to set the scale of the Y-position. One horizontal grid line equals one division. If it is set to 1V, then each grid line represents 1 volt. The reading shoots to the top of the axis if the scale is too low. Roll the number up until you are able to view the reading. Conversely, if you can barely make out the waveform, roll the volts/div down (see Figure 11.12). Channel A is set too low and Channel B is set too high.

— **Y Position**—This offsets the reading to make it clearer. For example, if you are measuring two 0- to 5-volt square waves that are exactly opposite to each other, it is best to offset one slightly higher or lower (see Figure 11.13).

— **Input Coupling**—With AC coupling, only the AC component of a signal is displayed. AC coupling has the effect of placing a capacitor in series with the oscilloscope's probe. As on a real oscilloscope using AC coupling, the first cycle displayed is inaccurate. Once the signal's DC component has been calculated and eliminated during the first cycle, the waveforms will be accurate.

With DC coupling, the sum of the AC and DC components of the signal is displayed. Selecting 0 displays a reference flat-line at the point of origin set by Y position.

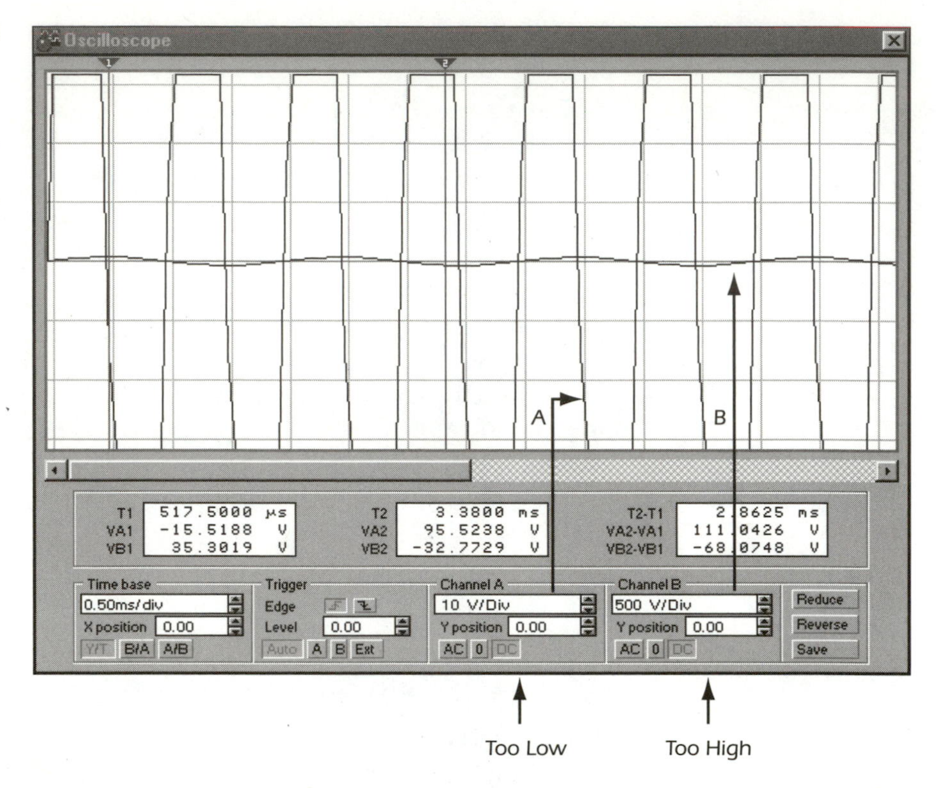

Figure 11.12 Setting the volts/div too high or too low kills a good picture of the waveform.

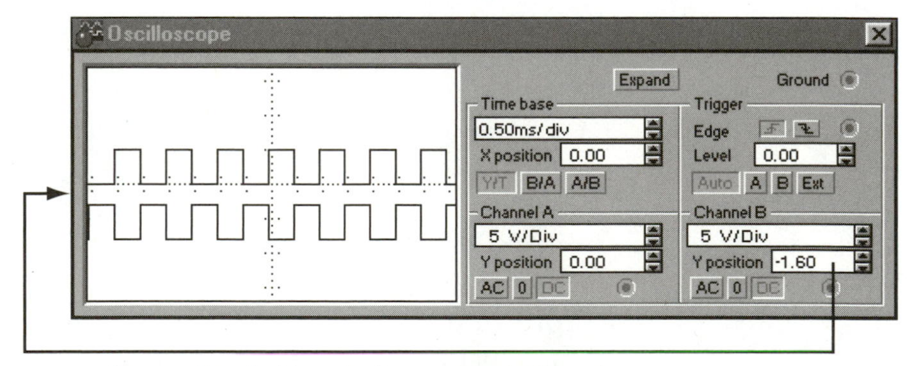

Figure 11.13 Use the Y position for a better view of signals that are too similar to make out if they are on the same line.

Do not place a coupling capacitor in series with an oscilloscope probe. The oscilloscope will not provide a path for current, and the analysis will consider the capacitor improperly connected. Instead, choose AC coupling.

- **Trigger**—These settings determine the conditions under which a waveform is first displayed on the oscilloscope.
  - **Edge**—On an ascending or descending signal.
  - **Level**—When the trigger reaches a preset level.
  - **Trigger Signal Location**—Determines from where the trigger will be received: Auto determines this itself, A comes from channel A, B from channel B, and Ext means external trigger signal.
- **Reduce Button**—Reduces the enlarged scope back down to the original view.
- **Reverse Button**—Makes a negative display (white on black instead of black on white).
- **Save Button**—Save the scope's data as a *.scp file for later use. This can be opened with Windows Wordpad or any word processor.

TIP

To pause the screen at the end of each display screen to read the display, choose ***Analysis > Analysis Options > Instruments*** *tab and place a check in the* ***Pause after each screen*** *box. Hit* ***OK.*** *Rerun the simulation.*

Readouts

Electronics Workbench's Analysis Graph keeps track of the oscilloscope's readings and convert them to a more pliable format. Go to **Analysis > Display Graphs** after the simulation is stopped. This opens the Analysis Graph, which has a tab for each oscilloscope reading.

Oscilloscope Test Circuits

The oscilloscope will be one of your most used instruments. Learning a few basic hookup principles will help you follow along in the rest of the book.

BASIC VOLTAGE WAVE READING

Build the circuit as shown in Figure 11.14. Make sure to adjust the AC sources as shown. Open the oscilloscope and adjust its settings to match Figure 11.14. Activate the circuit. Hit the pause button on the scope and take a look at the wave. Stop the simulation.

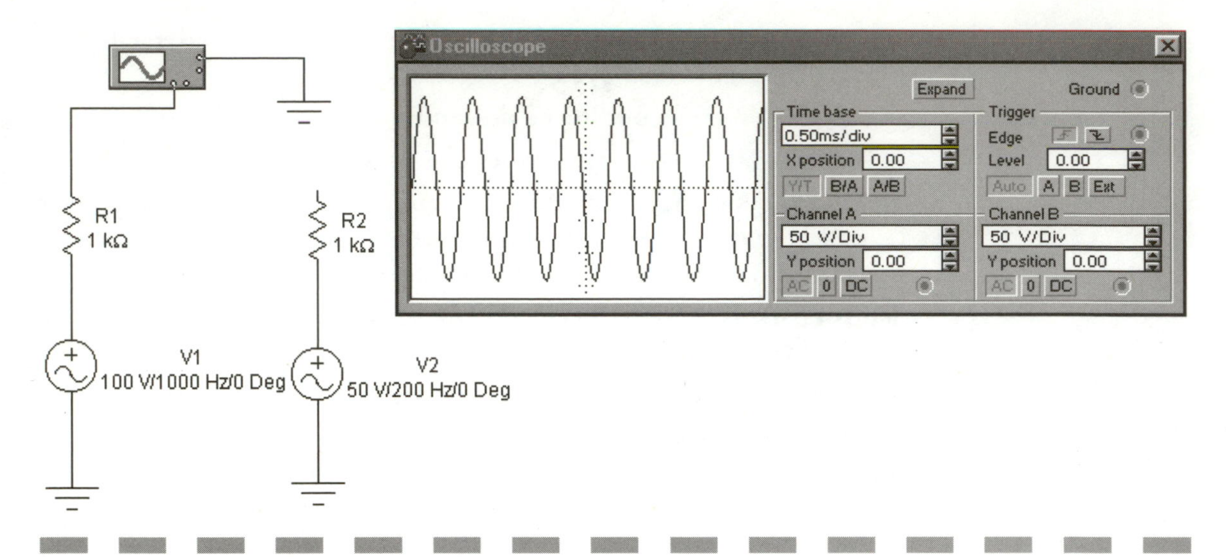

Figure 11.14 Reading a voltage wave with the oscilloscope.

READING TWO WAVES AT ONCE

Referring to the last circuit (Figure 11.14), add a wire from R2 to channel B on the scope. Make sure the AC source is adjusted to 50V, 2000 Hz. Color the wire red. Activate the circuit again.

Oscilloscope Tips

Here are a few tips to help you use the oscilloscope more effectively:

- **Use Colors**—Make the wires connecting two channels different colors. This lets you see which signal is for which channel. I typically use blue for channel A and red for channel B. I also make it a point to use channel A as the "before" signal and B as the "after". This lets you keep better track of what the circuit is doing to the signal.
- **WINDOWS 98**—With Windows 98, you can place the instrument onto a separate screen. providing you are running a two-monitor setup.
- **Reading Current with the Scope**—The scope will not normally read a current wave, but with the addition of a current-controlled voltage source, this is possible. Hook up the circuit in Figure 11.15 and run it with the scope open.

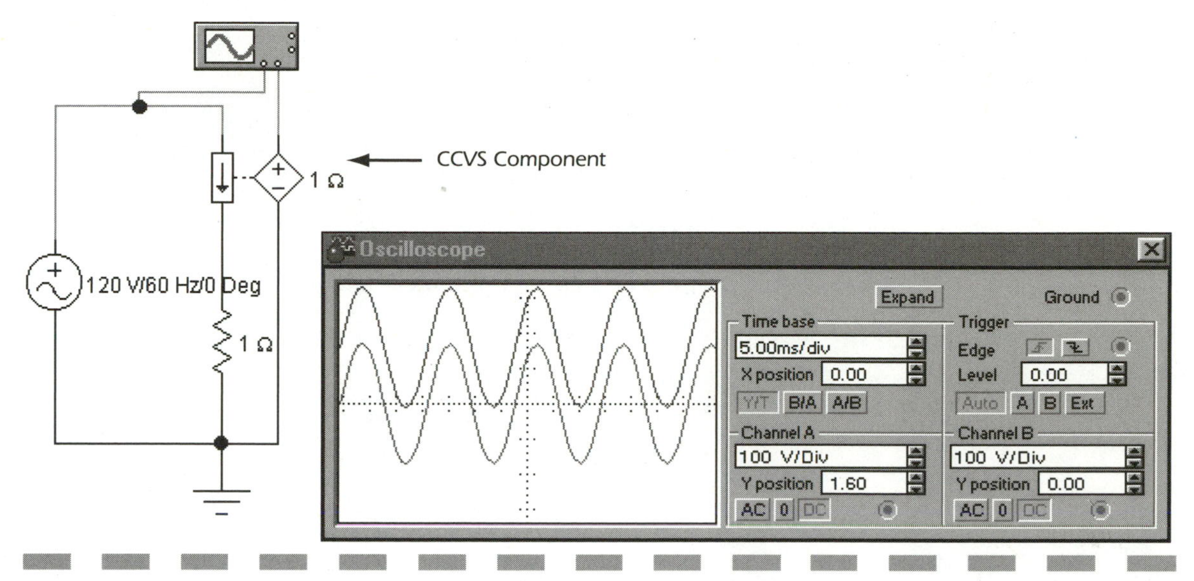

Figure 11.15 You can measure current waves with the oscilloscope by first changing the current into voltage.

Function Generator

It is sometimes necessary to inject a signal into a circuit, and this is the job of a function generator. This voltage source provides a sine, triangle, or square wave at a frequency set by you. In reality, this is quite a costly piece of equipment. But with Electronics Workbench a virtual model is included free for the price of the software. Here is how to make use of it.

Connecting the Function Generator

There are only three connections to worry about with this instrument: the center is the common and is usually connected to a ground; the '+' is the positive output of the function generator; the negative (−) is the mirror of the positive output. If you want to inject a signal into a circuit, connect a ground to the common lead and the positive to the point at which the signal is going to be introduced to the circuit.

Function Generator Operation

After connecting the function generator to the circuit, double-click the icon to open the Adjustments window for the instrument (refer to Figure 11.16).

WAVEFORMS SELECTION BUTTONS

Select one of three waveforms to output (Figure 11.17):

- **Sine wave**—Gives a pure sinusoidal wave. If the amplitude's voltage setting is at 10 volts, then the wave swings from zero to +10 volts to zero to −10 volts and starts over.

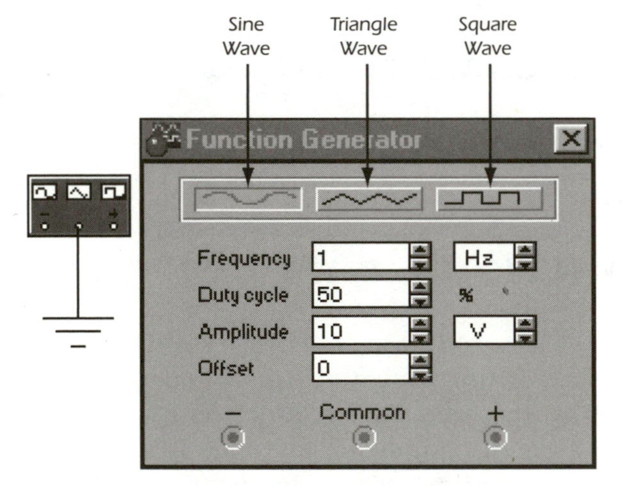

Figure 11.16 The Electronics Workbench function generator.

- **Triangle wave**—This wave goes from zero to 10 volts on a steady rise, peaks sharply, then declines to 0 and −10 volts at a steady descent.
- **Square wave**—This wave starts at 0, almost instantly rises to +10 volts, and sustains that level until there is a sudden drop to −10 volts. Then it stays at −10 volts for a set period of time before skyrocketing back to +10 volts.

FREQUENCY

You can adjust the wave's frequency (cycles per second) between 1 hertz (1 Hz) and 999 megahertz (999 MHz). Choose the variable (1 through 999) and the multiplier (Hz, kHz, MHz).

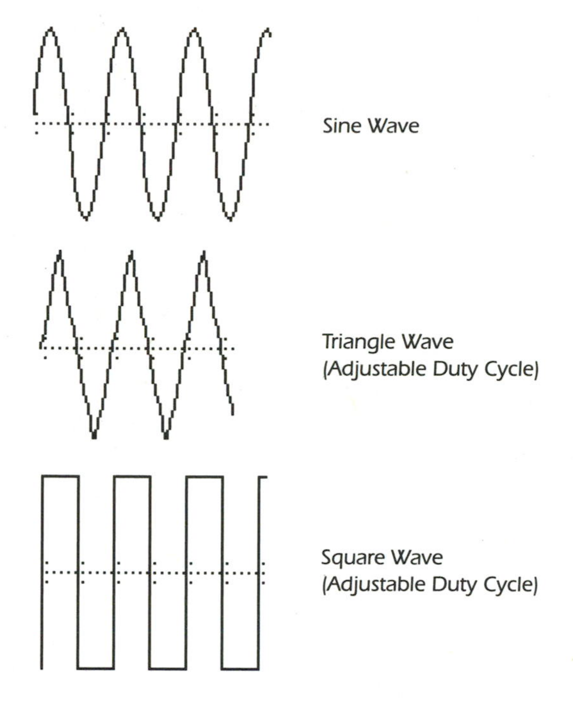

Figure 11.17 Waveforms explained.

DUTY CYCLE

You can adjust the duty cycle of the triangle or square wave. The duty cycle is the ratio of time on to time off. You can adjust it anywhere between 1 and 99 percent (see Figure 11.18). A is 20 percent, B is 50 percent and C is 90 percent.

AMPLITUDE

With the amplitude adjustment you can change the amplitude (height) of the wave from 1 microvolt (1μV) to 999 kilovolts (999kV). Choose the number and the multiplier.

OFFSET

The offset option controls the DC level about which the alternating signal varies. An offset of 0 positions the waveform along the oscilloscope's x-axis (provided its Y pos setting is 0.0). A positive value shifts the DC level upward, while a negative value shifts it downward. Offset uses the units set for amplitude.

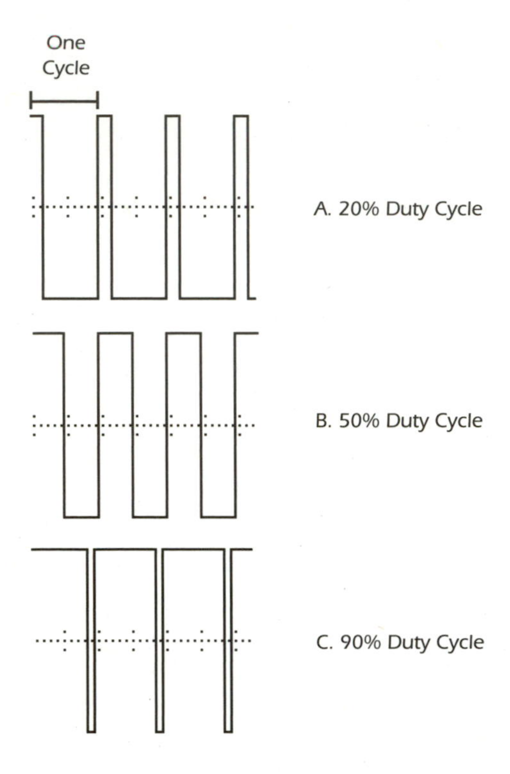

Figure 11.18
Duty cycle explained.

SET RISE/FALL TIME

If a square wave is being produced, you can add a slight delay in the time the signal takes to go high and low. This more accurately mimics true digital signals.

Function Generator Test Circuit

The fastest way to learn about the function generator is to connect it to the oscilloscope. If you are unfamiliar with the scope, backtrack to that section.

1. Build the circuit as shown in Figure 11.19.
2. Make sure all the settings on the function generator and oscilloscope match the ones shown in Figure 11.19.
3. Activate the circuit while the instruments are open.
4. Begin adjusting a few of the settings. Change the waveforms and observe the results on the oscilloscope. Alter the frequency slightly and then the duty cycle.

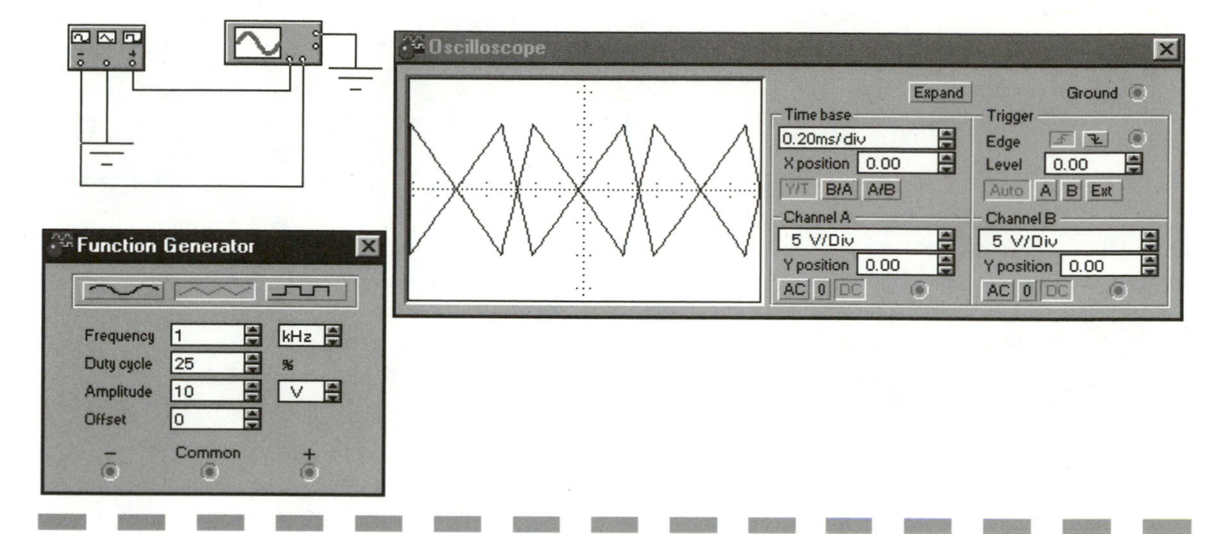

Figure 11.19 Using the oscilloscope to test the function generator.

5. Try the amplitude settings now. Change the offset to 5, then 10, and see where the wave positions itself on the scope.

Function Generator Tips

Here are two tips to help you use the function generator more effectively:

- **Creating a 0V to 5V clock signal**—The function generator's normal square wave swings from positive to zero to negative. However, most digital applications require a 0 to 5V clock signal, which requires that you set the amplitude to 2.5 and the offset to 2.5 (see Figure 11.20).
- **More than one waveform needed**—If you want to inject multiple waveforms into the circuit, I suggest using an AC voltage source for a sine wave, or a clock for square waves. Both are found in the Sources tool bin.

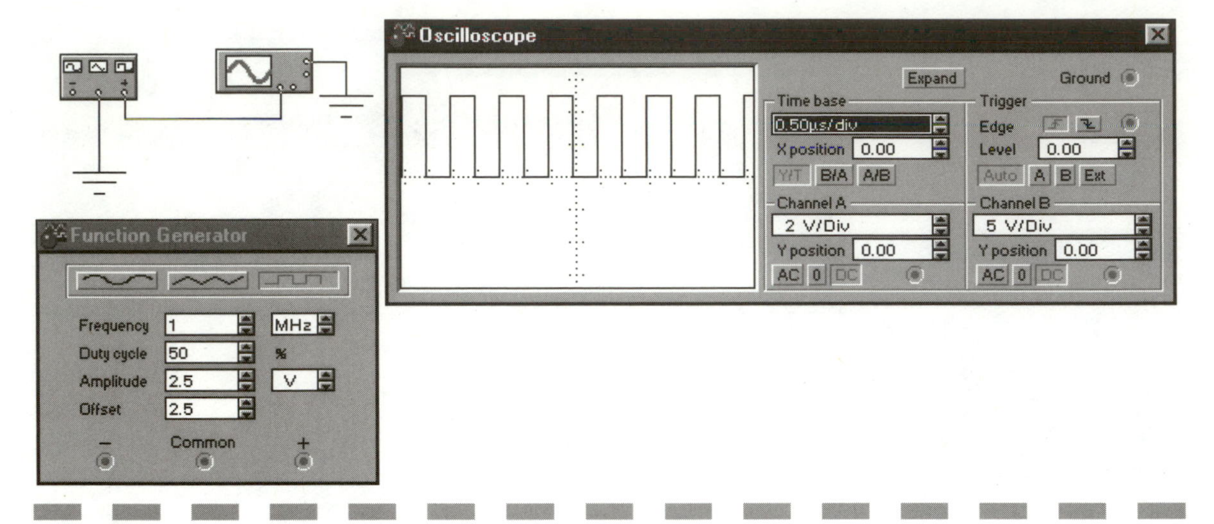

Figure 11.20 Using the function generator to create a clock signal.

Bode Plotter

The Bode plotter produces a graph of a circuit's frequency response and is useful for analyzing filter circuits. It is used to measure a signal's voltage gain or phase shift. It performs a spectrum analysis by injecting a range of frequencies into a circuit and then charting the circuit's reaction to those numbers.

Connecting the Bode Plotter

The plotter has two sets of connections V+In, V-In, V+Out, V-Out. The magnitude mode of the Bode plotter measures the ratio of magnitudes (voltage gain, in decibels) between two points, V+ and V−. Phase measures the phase shift (in degrees) between two points. Both gain and phase shift are plotted against frequency (in hertz). Here's how to connect the Bode plotter in these modes, if V+ and V− are single points in a circuit (see Figure 11.21):

1. Attach the positive In terminal and the positive Out terminal to connectors at V+ and V−.
2. Attach the negative In and Out terminals to a ground component.

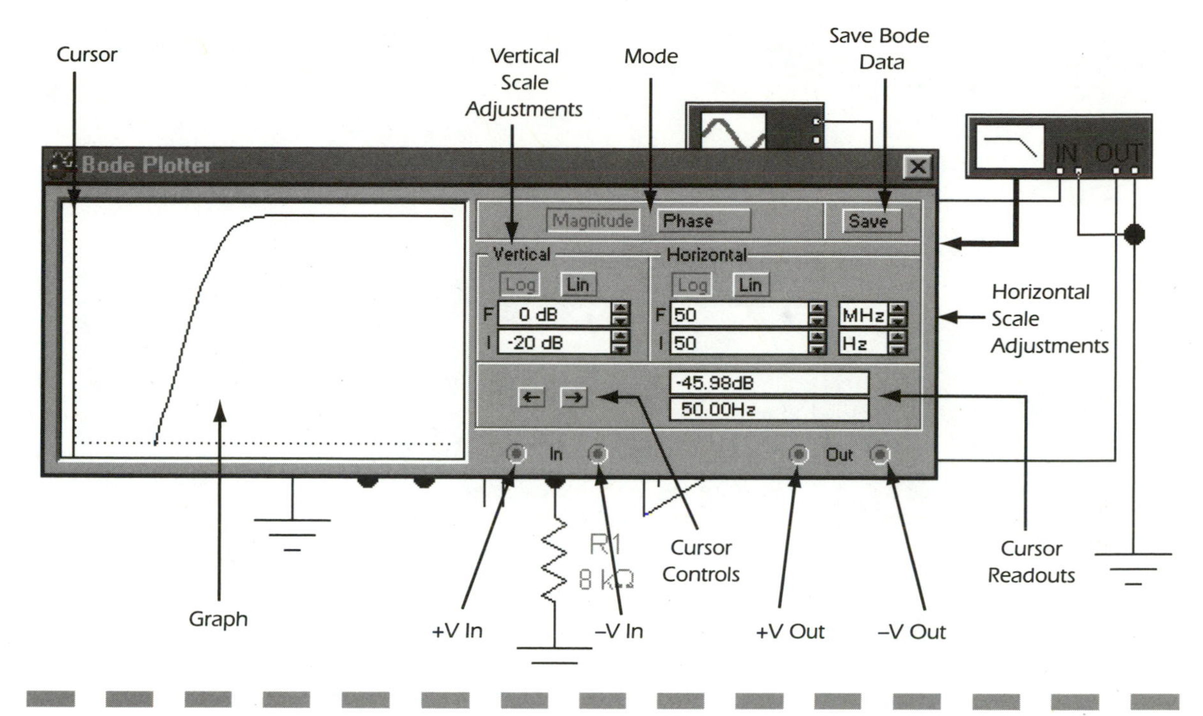

Figure 11.21 Connecting a Bode plotter.

3. If V+ (or V−) is the magnitude or phase across a component, attach both In terminals (or both Out terminals) on either side of the component.
4. In addition to connecting the Bode plotter's terminals, the circuit must contain some kind of AC source although this does not really affect the operation of the plotter. In Figure 11.22 we are using a function generator.

Bode Plotter Operation

After placing the Bode plotter into a circuit, set it to output the readings you choose. The settings are described in the following sections.

MODE SELECT BUTTONS

The two top left buttons marked *magnitude* and *phase*. Press either to select the type of graphs you wish to create.

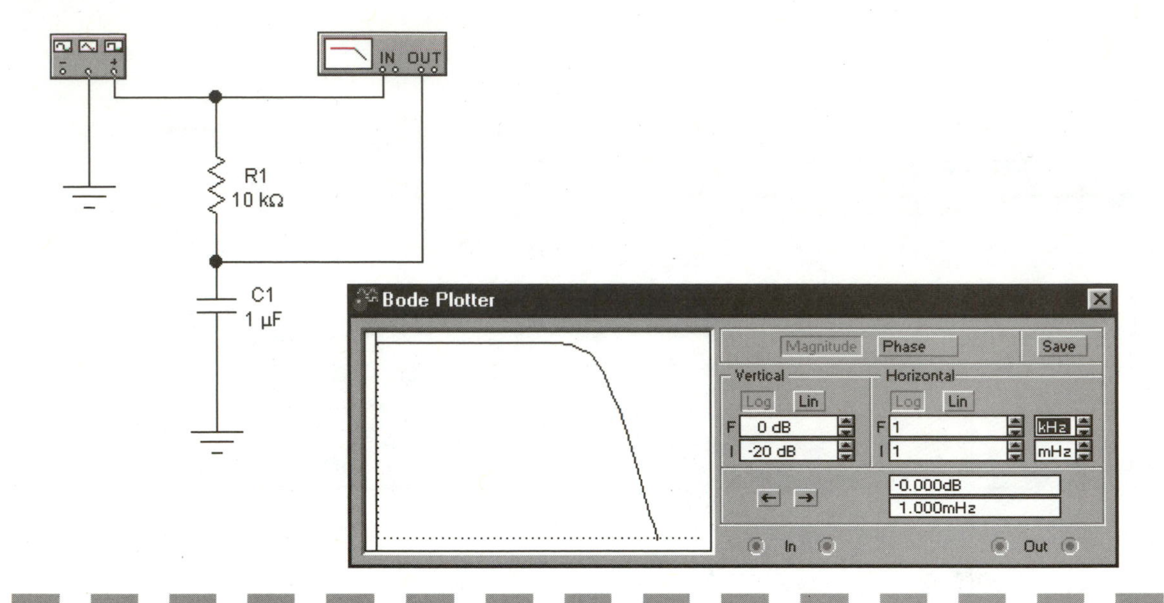

Figure 11.22 You can use the Bode plotter in filter applications such as determining the cutoff frequency of this RC circuit.

SAVE BUTTON

The top-right button saves the data to a *.bod file for later reference. This produces a text file, which charts all the instrument's data into columns. It can later be imported into Excel for a useful hard copy.

VERTICAL AXIS SETTINGS

Sets the plotting device's vertical axis:

- LOG = Logarithmic—Used when the values being compared have a large range.
- LIN = Linear—Graphs non-logarithmically (1,2,3,4...).
- F = Final value—Sets the final graphed value.
- I = Initial value—Sets the first graphed value.

HORIZONTAL AXIS SETTING

Sets the plotting device's horizontal axis. Settings are similar to vertical settings except that they adjust the frequency ranges used for the plot.

ARROWS

The two arrows on the bottom left move a vertical cursor across the screen. Where that cursor lies determines where the values are read out.

READOUTS
These two windows show the exact values at the point where the vertical cursor lies on the graph. The top field is the vertical and the bottom is the horizontal readout.

Bode Plotter Test Circuit

It is easiest to learn how to operate the Bode plotter with an example of how to graph magnitude. Assemble the circuit pictured in Figure 11.22, which is a simple RC filter circuit. The Bode plotter is being used to find the frequency cutoff. Adjust the settings to match the plotter in the screen capture. Activate the circuit. The Bode plotter begins to generate a series of frequencies starting at the initial value of 1 megahertz and ending with the final value of 1 kilohertz. The graph is a plot of the ratio of output voltage to input voltage as a function of frequency. Move the cursor to approximately −3dB. The frequency cutoff should be around 16 Hz. Try changing the value of the capacitor to 10mF and see the new frequency cutoff.

Bode Plotter Tips

Here are two tips to help you use the Bode plotter more efficiently.

- **Saving Data**—You can hit the **Save** button and store the Bode data as a *.bod file. It can be opened with a text editor or imported into Excel or MathCAD.
- **Use AC Analysis Instead**—If you want to have a printable graph and more control over the simulation, use AC analysis instead. Go to **Analysis > AC Frequency** to open the dialog box. Make sure to fill out each tab or page of information.

Digital Instruments

The next set of instruments is digitally based, meaning the instruments either read or output a digital (0 or 1) signal. These include the logic analyzer, word generator and logic converter.

LOGIC ANALYZER

The logic analyzer instrument lets you examine 16 digital signals running at once. It is used to visualize the output of digital logic circuits in much the same way as the oscilloscope illustrates an analog wave. If you are working on digital circuits and need to view more that two outputs (all that is possible with an oscilloscope), you can hook each output to the logic analyzer and take a peek at up to 16 waveforms.

PLACING THE LOGIC ANALYZER

1. Open the **Instruments** tool bin.
2. Move the mouse pointer over the logic analyzer's icon.
3. Click the left mouse button and hold it.
4. Drag the logic analyzer's icon onto the drawing board and place it approximately where you want it.
5. Release the left mouse button.

LOGIC ANALYZER CONNECTIONS

Each of the 16 inputs on the left side is a data input channel. The bottom left terminal is the *external clock* hook up. The bottom center is the *clock qualifier.* The last terminal is the *trigger qualifier.*

- **Data Inputs**—Any digital signal can be hooked to these 16 inputs directly.
- **Clock (External and Qualifier)**—Lets you use an external clocking source for the logic analyzer. Also has a clock qualifier that filters the clock signal.
- **Trigger Qualifier**—Input signal that filters the triggering signal.

LOGIC ANALYZER OPERATION

Open the logic analyzer by double-clicking its icon. The following is an explanation of the conventions used (see Figure 11.23):

- **Graphic Display**—The digital waveforms appear here, along with clock and trigger readings.
- **Scroll Bar**—Lets you scroll back and forth along the graphic window.
- **Stop Button**—Stops the instrument's reading but continues the simulation.

- **Reset**—Starts the instrument's readings from scratch.
- **Cursor Windows T1, T2, and T1—T2**—Cursor 1 is red and cursor 2 is blue. The windows show the time position of each cursor. T1 and T2 shows the time distance between cursors.
- **Clock**—To adjust the clock settings, see Figure 11.24:

1. Click **Set** in the clock area of the logic analyzer. The Clock Setup dialog appears.
2. Select external or internal clock mode.

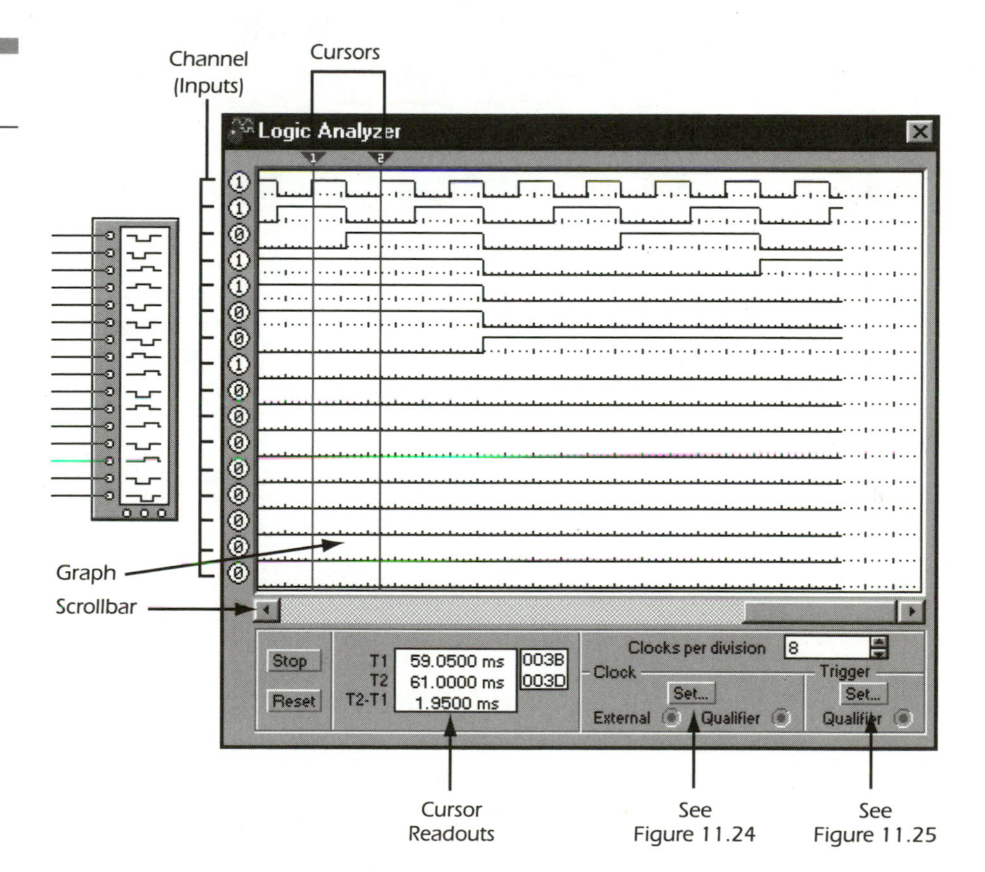

Figure 11.23
The logic analyzer.

3. Select clock rate.
4. Set clock qualifier if set to external.
5. Set Sampling Settings: pre-trigger samples, post-trigger samples, threshold voltage.
6. Click **Accept**.

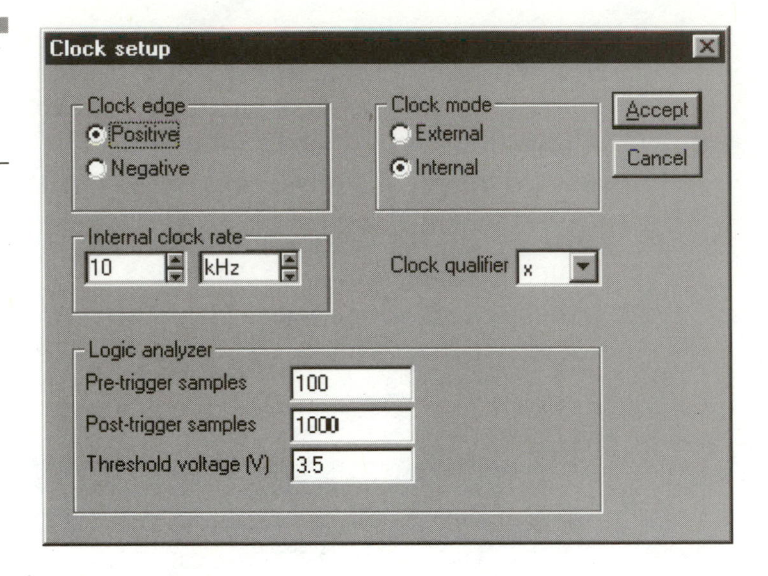

Figure 11.24
Adjusting the clock settings of the logic analyzer.

- **Clock Qualifier**—An input signal that filters the clock signal. If it is set to x, then the qualifier is disabled and the clock signal determines when samples are read. If it is set to 1 or 0, the samples are read only when the clock signal matches the selected qualifier signal.
- **Trigger**
  1. Hit the **Set** button (see Figure 11.25).
  2. The Trigger Settings dialog box opens. Select which trigger clock edge to use.
  3. Select the trigger qualifier. This is an input signal that filters the triggering signal. If it is set to x, then the qualifier is disabled and the trigger signal determines when the logic analyzer is triggered. If it is set to 1 or 0, the logic analyzer is triggered only when the triggering signal matches the selected trigger qualifier.
  4. Set the trigger patterns (A, B, C) and their combinations. An x means the variable can be either a 1 or 0. You can also enter a binary word.
  5. Hit the **Accept** button.

RUNNING THE SIMULATION TO TAKE READINGS To begin the simulation, hit the **Start** button on the Simulation switch. You can stop the logic analyzer graph by hitting the **Stop** button on logic analyzer. Hit the **Reset** button to clear the display, but continue the simulation.

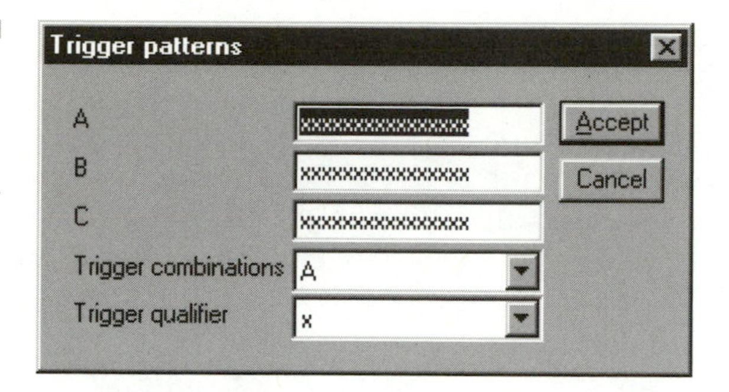

Figure 11.25 Adjusting the trigger settings of the logic analyzer.

LOGIC ANALYZER TEST CIRCUIT It's easiest to use the word generator for logic analyzer tests. Build the circuit as shown in Figure 11.26 and adjust the settings as seen in both instruments (load the up-counter in the word generator). Run the simulation and watch the readout.

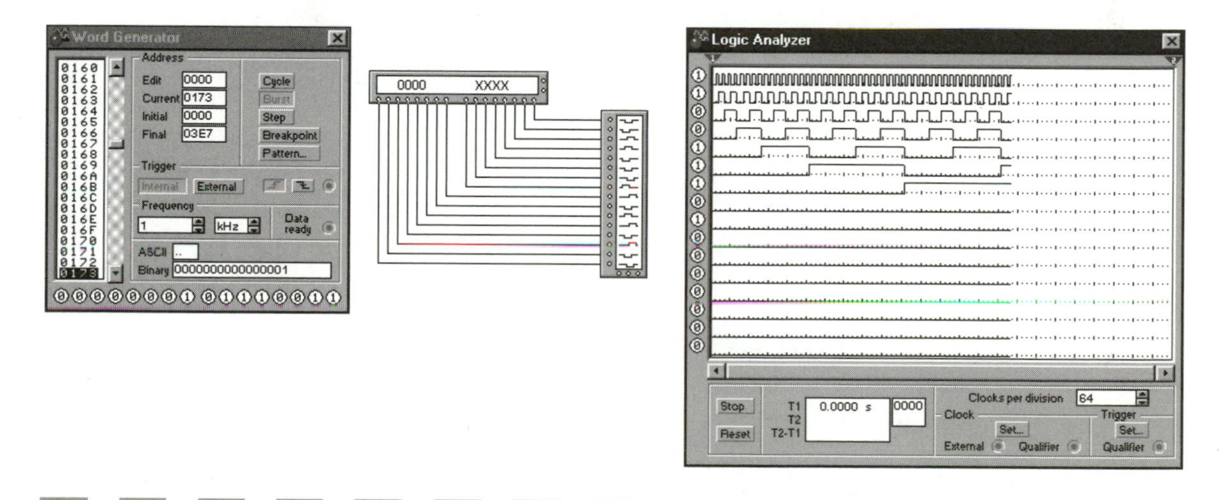

Figure 11.26 Hooking the word generator to the logic analyzer provides a quick learning circuit.

LOGIC ANALYZER TIPS

- **Use Wire Colors**—When you color the input wires to the logic analyzer, the readout displays that color as well. This makes for a quick visual check instead of your having to trace a finger across the screen for each input.
- **Use Instead of an Oscilloscope**—If you have multiple digital waves to read, use the logic analyzer instead of multiple oscilloscope channels and reconnects. The simulation runs much faster as well.

Word Generator

The word generator creates digital words or patterns of bits to feed into a digital circuit. It can be used for a variety of applications that require a specific digital input in your designs. I personally use it as a shortcut to mimic a complex circuit or IC output like a microcontroller. Think of it as a way to communicate digital words directly to a simulated device.

There are patterns built into Electronics Workbench, or you can create your own and save them for later retrieval. Creating patterns requires setting up many parameters, but there is no other way to input these digital bits into a circuit.

Connecting the Word Generator

The word generator has 16 output channels (pins) as well as an external trigger and data ready output.

DIGITAL OUTPUT PINS

The bottom row has 16 digital output channels. These are the output terminals that send the 5-volt digital signals. The rightmost pin is the *Least Significant Bit* (LSB) and the leftmost pin is the *Most Significant Bit* (MSB). To connect the outputs to a circuit, run wires accordingly from these terminals to the circuit's connections, keeping in mind to start with the right-side pin (LSB) (see Figure 11.27).

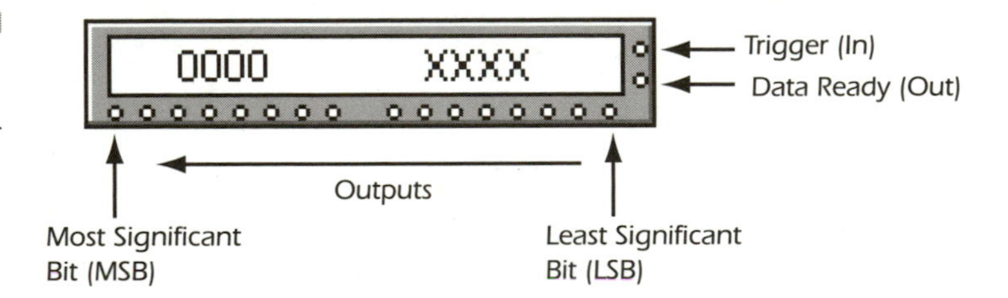

Figure 11.27
The word generator.

EXTERNAL TRIGGER

This is an Input terminal that allows you to control the word generator by an external means such as a clock or other circuit. Hook up the incoming signal to the terminal and select whether you want it to cycle on a negative-going or positive-going pulse.

DATA READY

This is an Output terminal that lets the circuit know that data from the word generator is ready to transmit. This can be used in conjunction with the external trigger to allow an external circuit to control the word generator.

Word Generator Operation

Apply the word generator to the circuit window. Open it by double-clicking the icon. Take a good look at the controls (see Figure 11.28).

Figure 11.28 Using the word generator.

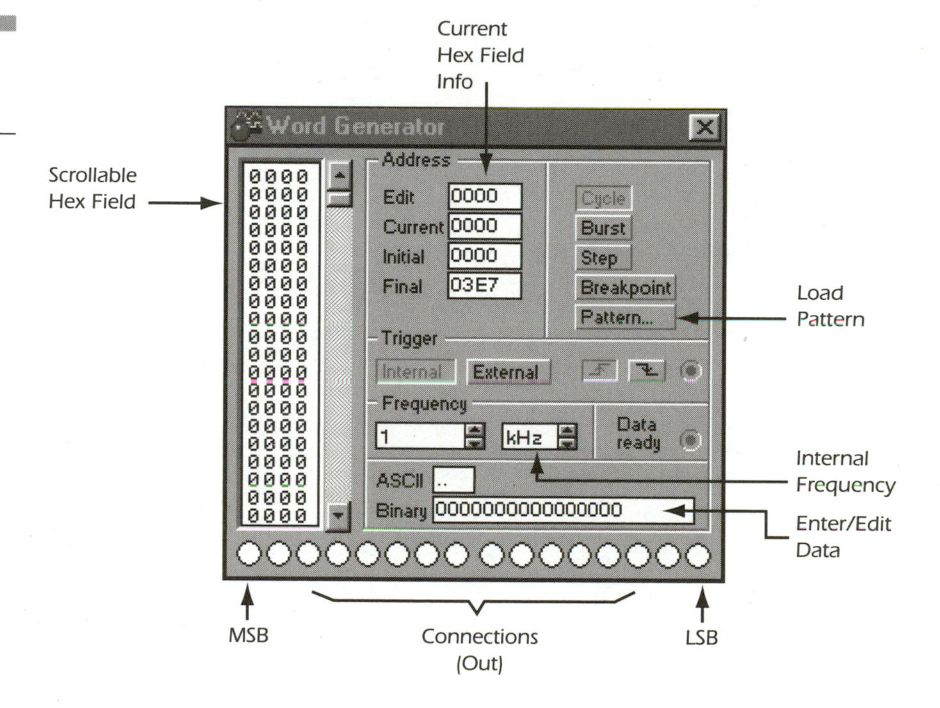

- **Scrollable Hex Field**—Controls the output of the word generator using hexadecimal numbers (0 to 9 and A to F). This is the hex equivalent of the 16-bit binary word, as shown in the binary window to the right.
- **Address**—Sets the variables related to the hex field addresses. Includes **Edit**, **Current**, **Initial**, and **Final**.
- **Trigger**—Selects either internal or external trigger as well as a signal trigger direction (rise or fall).

- **Frequency**—Sets the frequency at which words are sent out of the generator.
- **Edit Area**—Lets you edit the current word in either ASCII or binary modes.
- **Control Buttons**—Activates the word generator and simulation with your variable set to the outputs. These preset ways to run the generator include:
  - **Cycle**—Starts at the initiate address; goes sequentially to the end. The address is reset to the beginning, and the sequence is played again.
  - **Burst**—Runs through the complete list of addresses once and then pauses the simulation.
  - **Step**—Goes forward one address at a time.
  - **Breakpoint**—Lets you set the current address as a breakpoint; the address line appears as a grey box where selected. When you run the simulation, it automatically pauses at that point.
  - **Pattern**—Opens the Pre-Setting Patterns dialog box (see Figure 11.29). You can select to clear the buffer, open an archived pattern, save a pattern, create an up-counter (1, 2, 3...), create a down-counter (4, 3, 2, 1), create a shift-right (bit shifts to the right, one significant bit), or a shift-left pattern.

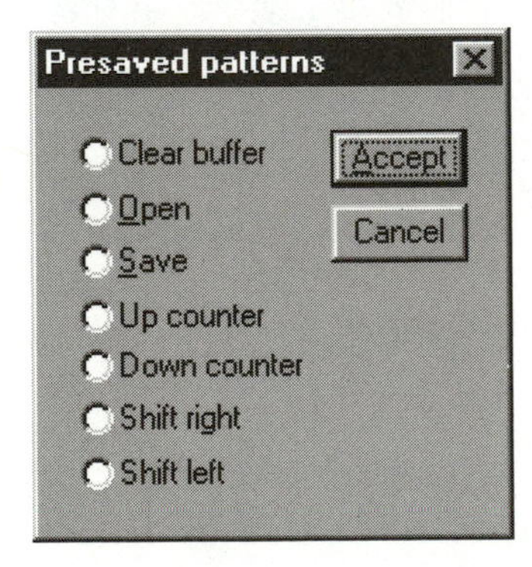

Figure 11.29 Word generator digital patterns can be loaded or saved.

- **Loading a File**—A previously stored file can be opened with this button. It is stored in a *.dp file, which is a text file with four hexadecimal characters for each entry.
- **Entering Words**—To enter words in the generator follow this procedure.

1. Highlight the word that exists in that address already.
2. Enter the hex data into that address or go down to the ASCII or binary field to change the data.
3. Click the next word/address you wish to edit; repeat steps 1 through 3 until complete.
4. Store your pattern by pressing **Pattern** > **Save** and name the *.dp file.

- **Clearing the Buffer**—To write zeros to each address, hit the **Pattern** > **Clear buffer** > **Accept** button.
- **Copy and Pasting HEX**—Take sections of hex from the field and use them in other areas to speed the process of making complex digital waveforms (not currently available on Multisim V6). Place the mouse pointer inside the hex field to the left of the first address you wish to select, left-click, and hold. Drag the mouse to where you wish the selection to run and release the button. The selected addresses and hex are highlighted in blue. Hit **Ctrl+C** to copy. Move the pointer and click where you want the copied section to be placed and hit **Ctrl+V** to place it. See Figure 11.30.

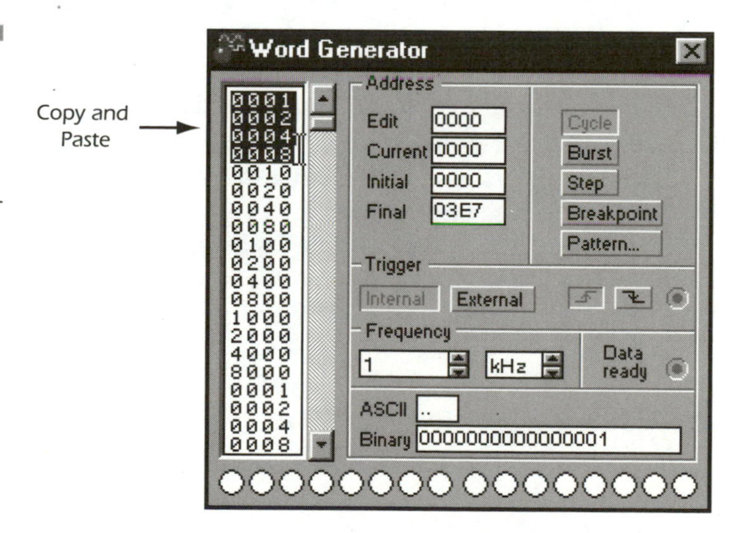

Figure 11.30 Copy and paste sections of hex to save pattern creation time.

Word Generator Test Circuit

Build the test circuit in Figure 11.31. Open the word generator. Set the frequency to around 10 Hz. Hit the **Pattern** button and select the

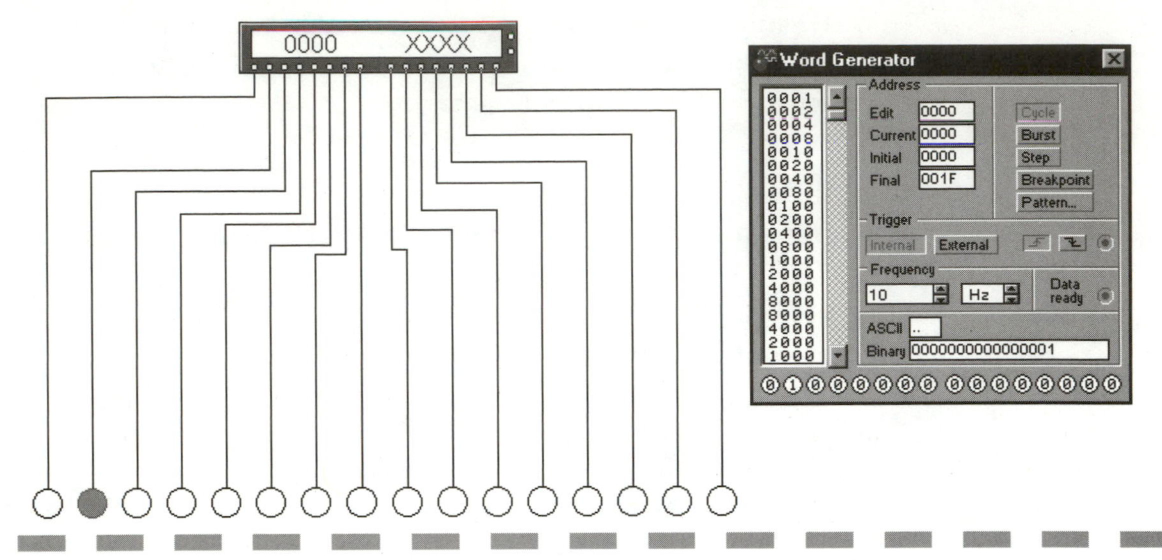

Figure 11.31 Knight Rider circuit to test the word generator.

Shift left pattern. Activate the circuit, making sure it is set to cycle mode.

Highlight the first 16 lines (0000 to 000F) and hit **Ctrl+C**. This copies those numbers into the clipboard. Hit the **Pattern** button again and load the **Shift right** pattern. Move the cursor into the hex field in the very top left corner of line 0000. Hit **Ctrl+V** to paste the shift-left pattern into the shift-right. You must tell the word generator to stop at the last word; set the initial address to 0000 and the final address to 001F. Go to line 0020 and set it as a breakpoint. Save the pattern by hitting **Pattern** > **Save** > name it KITT.DP and hit **Accept**. Place the pointer onto the left side of Address 0001. Activate the circuit. Remind of you an old TV show?

TIP

Create a directory in C:\EWB5, called Patterns to save your word generator files.

Word Generator Tips

Here are a few tips to help maximize your use of the word generator.

- **Copy and paste**—Cut and paste sections of familiar patterns to create your own. Try to build up a library of commonly used patterns and save them as *.dp files in a directory called C:\EWB5\PATTERNS\.

- **Mimic microcontroller output**—I use microcontrollers in many of my projects; because Electronics Workbench doesn't have this component, I make do by using word generator to emulate microcontroller outputs. For example, if I am simulating a device that turns on four relays one second apart, I adjust the generator to 1 Hz and use four of its terminals. If the controller needs an input before performing this task, I can use the external trigger on the word generator to activate the outputs.
- **Create your own patterns with a text editor**—You can make a pattern by typing columns of hex into Notepad, saving to a DP file, and loading the file into the word generator. Make sure the file starts with **Data**: and each four hex number has a break at the end:

```
Data:
0001
0002
0004
0008
Initial: 0000
Final: 0003
```

Logic Converter

The logic converter is a timesaving digital design tool. Although there is no such creature in the real world of electronics, the logic converter can take a digital circuit and derive a truth table or Boolean expression from it, or conversely creating a circuit from your truth table—a tremendously tedious task when done by hand. The logic converter saves your brain and pencil from digital design burnout.

Connecting the Logic Converter

There are eight input connections that are hooked to the inputs of the digital circuit (see Figure 11.32). You can use one or all of the eight connections. The connection on the far right is the output, which reads the final outcome of the digital circuit. The best way to describe the function and operation of this instrument is by giving you two examples of how to use it.

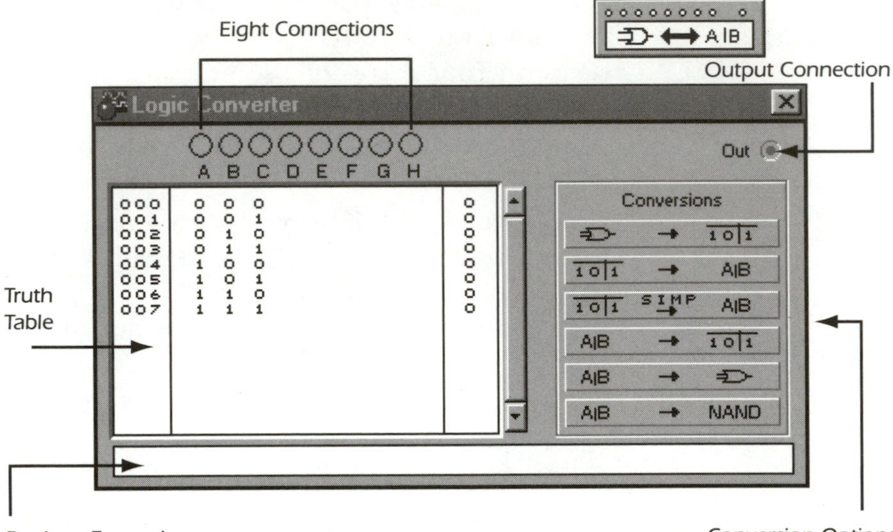

Figure 11.32 Connections on the logic converter.

DERIVING A TRUTH TABLE FROM A CIRCUIT

Figure 11.33 shows a simple digital circuit made of two NAND gates and one NOR gate. There are 16 possible input combinations that can be input into the A, B, C, and D inputs (which are connected to A, B, C, and D of the logic converter). There are also 16 possible outcomes presented at the Output of the circuit (hooked to the logic converter's Output terminal). Open (double-click) the logic converter by pressing the **Circuit to Truth Table** button. The computer converts the signals into an Output table; this may take some time. You see the results of the conversion in the far right column. As you can see, this circuit has only one combination that activates a high output.

CREATING A CIRCUIT FROM A TRUTH TABLE

This is the coolest feature of this instrument. Say, for example, that you want to build a circuit that tells when any two out of three parking spots are filled. A sensor is placed on each parking spot. These sensors are connected to a digital circuit to make an LED illuminate when any two spots are taken. We can use the logic converter to figure out the logic circuit. See Figure 11.34:

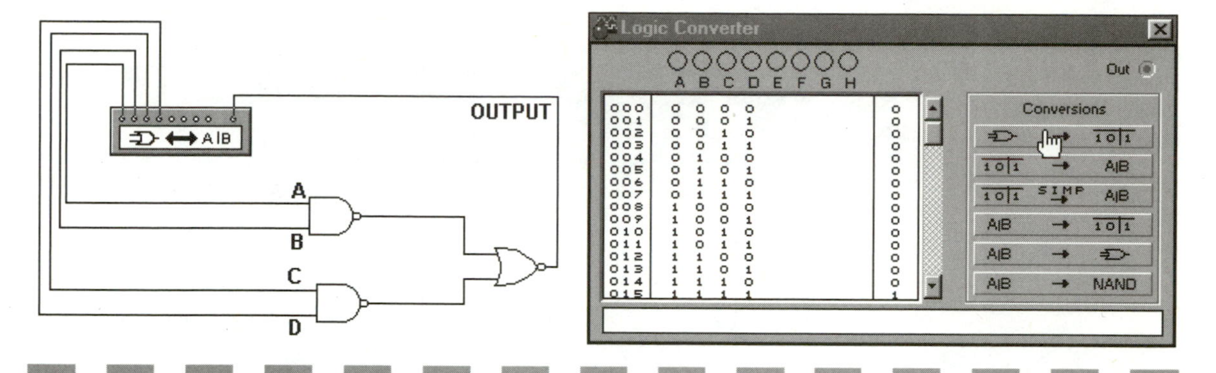

Figure 11.33 The fastest way to make a truth table!

1. Open a new circuit and name it Park.ewb.
2. Place a Logic Converter icon onto the drawing and double-click it.
3. Place the mouse pointer over the A connection (top-left terminal) and left-click it once. Do the same for the B and C connections. You should see a series of consecutive numbers in the left column, zeros and ones in the middle column, and zeros in the right column. To create a circuit, first create the truth table. The zeros in the rightmost column will be the result of each row and must be adjusted to either a **1**, **0**, or **X** (X means 1 or 0 is acceptable). In the case of the parking circuit, we want each row with two 1s to equate to a logic 1. There are three of these: **0 1 1**, **1 0 1**, and **1 1 0**. For each of these lines (003, 005, and 006) we make the result a logic 1.
4. Move the mouse pointer over 0 in the right column, row 003. Click once. A small box appears around the number. Hit the numeral 1 on the keyboard. The number changes. Note that you can either hit **0**, **1**, or **x** on the keyboard to change the value.
5. Change lines 005 and 006 to a logic **1** as well.

   There is no direct way to convert a truth table into a circuit, so we must first create a Boolean expression.
6. Press the **Truth Table to Boolean Expression** button. You see the Boolean expression in the bottom field.
7. Convert the **Boolean Expression to a Circuit** using that button. Click it and allow it to draw the circuit for you. The resulting circuit is shown in Figure 11.35. That's it. You can now use that circuit in your parking circuit.

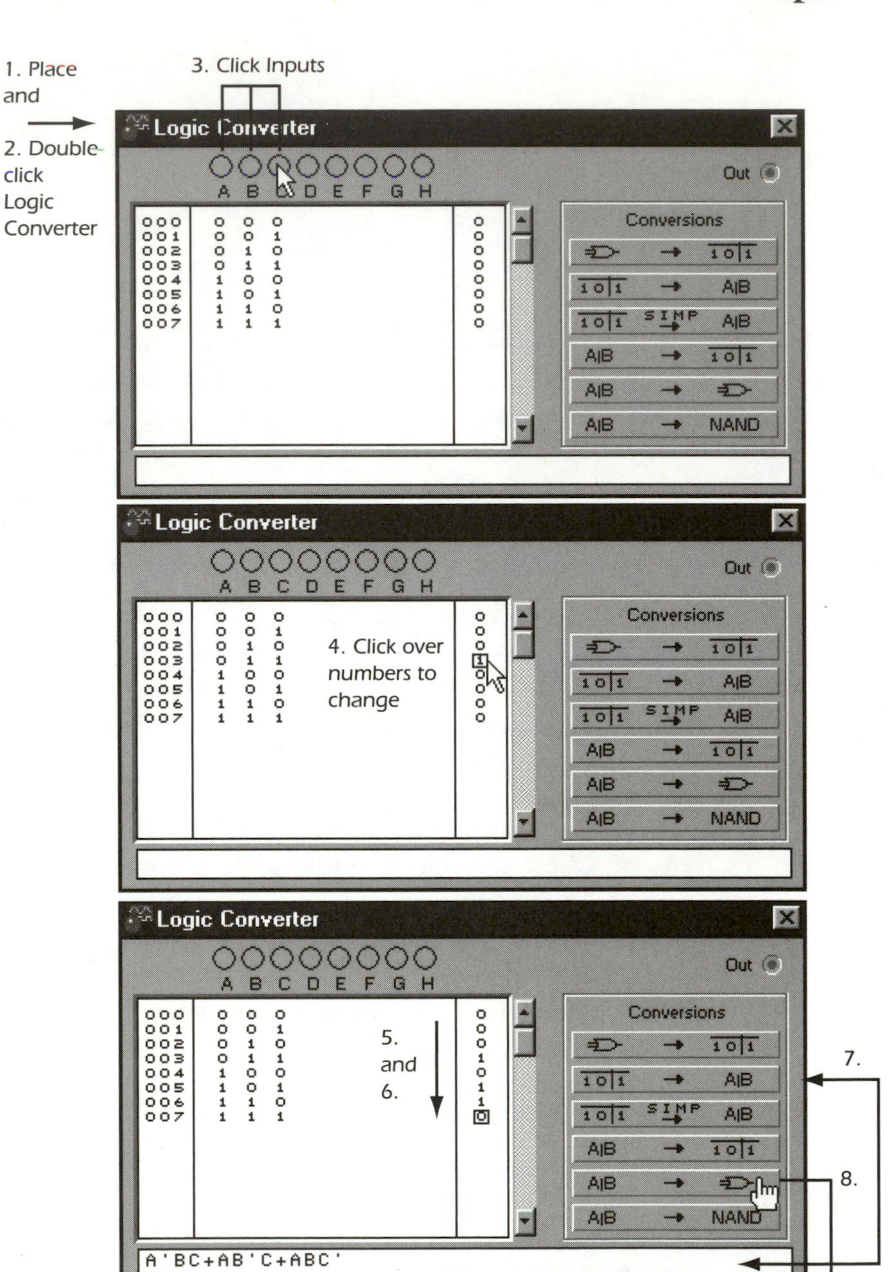

Figure 11.34 Letting Electronics Workbench do your digital design deeds.

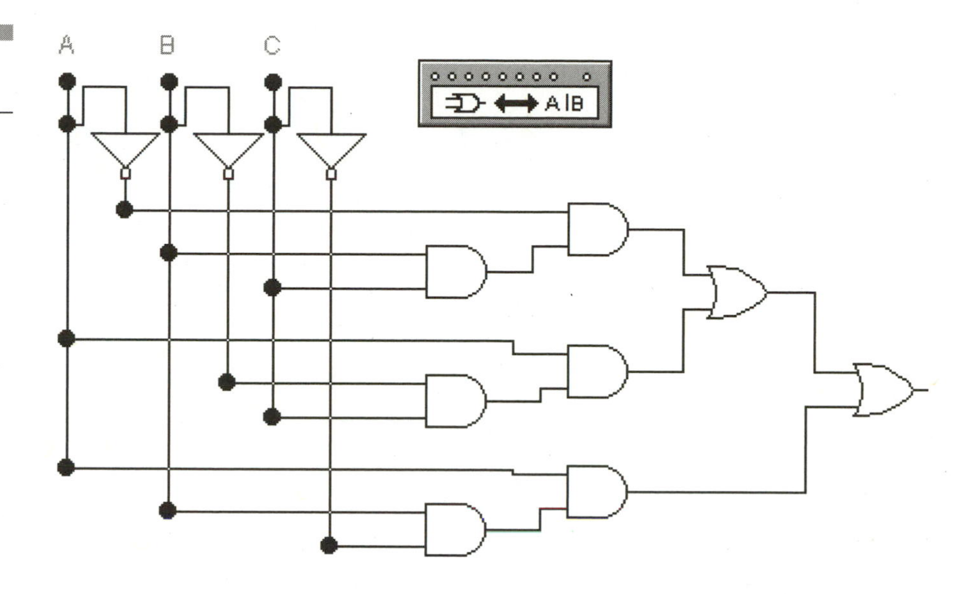

Figure 11.35
The resulting circuit.

TIP

If you wish to make a circuit entirely out of NAND gates, click the ***Boolean Expression to NAND gate Circuit*** *button instead.*

Summary

Each instrument is complex and requires some fiddling and experimenting to learn. Open some of the sample circuits included in the C:\EWB5\Samples directory and locate the different instruments, then think of ways to read your own circuits using virtual equipment. Pretty soon, you will be turning and adjusting settings as quickly and as easily as you do your home theater equipment or microwave.

CHAPTER 12

EWB V5: Simulation and Analysis

Investigating components and circuits and how they interact under different situations is the work of analysis and simulation.

It's a waste of resources to use a computer only to draw circuits, when the same thing can be achieved with a pencil and paper. Our aim is to use circuit drawings for other rewarding purposes, like simulation and analysis, which lets you apply the hard work of drawing the circuit to perform tests. It's as if you drew a circuit on paper, built the circuit in real life, and hooked up countless instruments to it. With Electronics Workbench, all you have to do is draw and flip a virtual switch to get results.

NOTE

Simulation *uses the computer to simulate a real-world circuit using software.* Analysis *analyzes a circuit, circuit segment, or component and submits it to many tests or scenarios.*

Simulation in EWB

EWB simulates the electronic flow through a conductor, piece of carbon, semiconductor, or other electronic component material. The software pretends real-world electronic components are strung together and activated using two methods:

- **Mixed-Mode Simulation**—As discussed earlier, Electronics Workbench is capable of simulating a mixture of analog and digital parts. Mixed-mode simulation means mixing resistors with AND fates, capacitors with inverters, etc.
- **B SPICE**—Electronics Workbench's simulation engine is based on Berkeley SPICE3F5 algorithms, which describe each component in a specific circuit, as well as how they are interconnected. For example, a transistor may be described as **.MODEL 2N2222A NPN (IS=1.16e-14 BF=200 BR= 4 RB=1.69 RE=0.423 RC=0.169 CJS=0...)**. Each parameter describes the properties of that component. The more parameters, the more detailed the model, and thus the more realistic the part in the circuit simulation.

You've been sweating for hours creating a perfectly drawn circuit with components, wires, labels, etc. Now what? How about activating the electrons to see if the circuit would actually work in the real world. To simulate something we first need a complete circuit (Figure 12.1) that would otherwise operate in the real world. We also require a way to visualize the output; otherwise we won't know what the circuit is doing. You could use an indicator or instrument for this purpose.

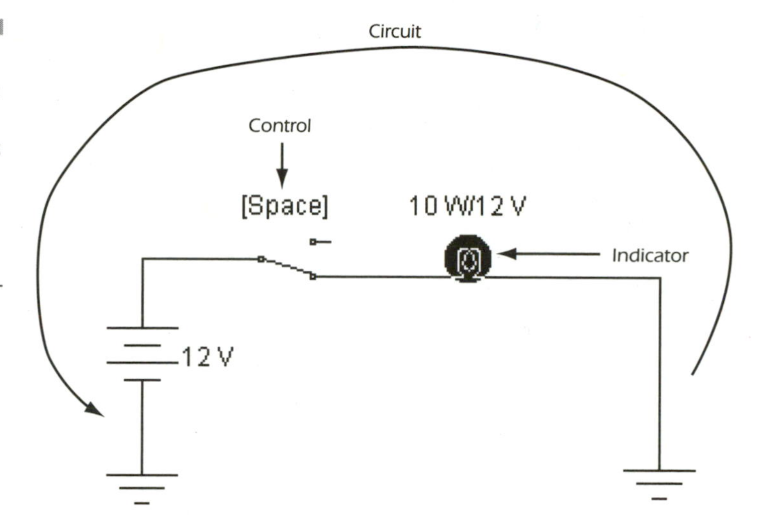

Figure 12.1
Electronics simulation requires a "visual" to see what the circuit is doing. This can be as simple as a bulb or even a readout from an instrument.

In our example, the circuit contains a battery hooked to a switch and a light bulb (grounded). The bulb is the visual indicator we need to view the results. If we look at this from the viewpoint of a computer, all that is seen is a list of mathematical formulas to calculate at the speed of light. That's the actual simulation.

NOTE

An Electronics Workbench simulation is a transient analysis that continues to run until you stop it with the simulation switch.

Simulation Switch

The simulation switch activates the circuit and the simulation. It is located in the top-right corner of the desktop when Electronics Workbench is first opened (see Figure 12.2). 1 means activate and 0 means deactivate; a pause button is also available to halt a simulation temporarily.

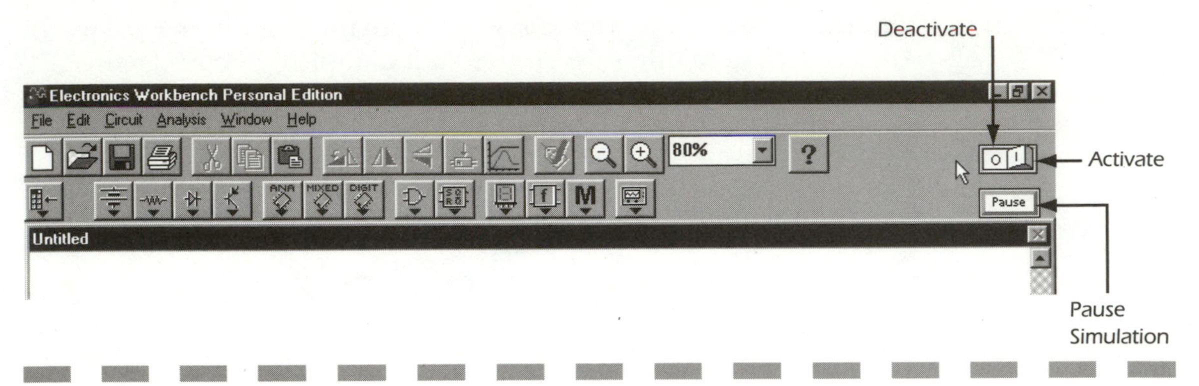

Figure 12.2 The Simulation Switch activates, deactivates, or pauses the circuit.

Statistics Log

If you open the Grapher (**Analysis** > **Display Graphs**) after the simulation has run, you will see an *analog statistics chart* which lists information about the simulation as well as any errors that occurred. (Figure 12.3). This is handy when debugging a troublesome simulation. The window updates each time the simulation is run.

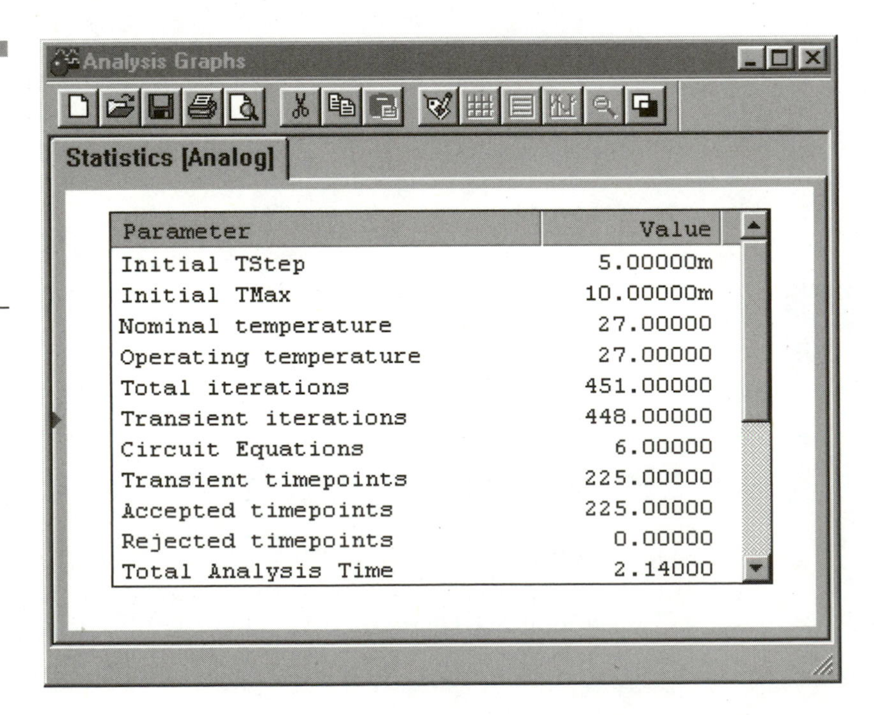

Parameter	Value
Initial TStep	5.00000m
Initial TMax	10.00000m
Nominal temperature	27.00000
Operating temperature	27.00000
Total iterations	451.00000
Transient iterations	448.00000
Circuit Equations	6.00000
Transient timepoints	225.00000
Accepted timepoints	225.00000
Rejected timepoints	0.00000
Total Analysis Time	2.14000

Figure 12.3 The Statistics Log keeps track of the analyses, along with any errors that may occur. It pops up inside the Grapher after each analysis.

Electronics Workbench Analysis

Electronics Workbench's most powerful feature is its ability to analyze circuits inside and out. You can analyze a circuit, a component, how a component reacts to a certain circuit, how it reacts to worst-case scenarios, how its readings change with temperature—the list goes on and on.

In an EWB analysis, the user defines areas in a circuit (nodes) from which he or she wishes to take readings. These can be instantaneous readings, values that change with time or frequency, or even reactions to a varying temperature, voltage, etc. Electronics Workbench then calculates the appropriate variables and outputs them in the form of a table, graph, or a series of both. The results can be recorded, printed, or sent to other programs for further processing.

Electronics Workbench performs three major types of analysis: DC, AC, and transient. Each major type can be performed in conjunction with several other analyses. DC analysis ignores all AC elements in a circuit and calculates the steady-state voltage or current for each node in the activated circuit. It is also used to determine certain transistor and diode variables to be used in other analyses. AC analysis takes the results from the DC analysis and calculates AC circuit response as a function of frequency. Transient analysis lets you view what a circuit is doing over time. This may become your most used feature in EWB V5, because it offers greater control over what you want to read, as opposed to the virtual instruments. Most simulations running in Electronics Workbench are of the transient variety.

Use DC analysis instead of positioning multiple multimeters around a circuit.

Why Analyze?

You may wonder why you would need such powerful analysis capabilities. The answer? To better visualize *exactly* what a circuit is doing while active: A student can learn the characteristics of circuits and components; engineers and designers can fine-tune circuits or components they are designing or test "what-if" scenarios in seconds; technicians can simulate problems they encounter with real-world devices on the bench. The good thing about analysis is that a record can be kept of the results and graphs can be printed for later examination or presentation.

Steps of Analysis

It is easiest to break down the analysis processes into steps:

1. Draw circuit. Make a schematic drawing of the circuit you wish to analyze (covered in Chapter 9).
2. Setup up analysis. Select the correct analysis and input the variables required.
3. Analyze. Activate the analysis.
4. Read results. View the Grapher's results.
5. Make adjustments to circuit and reanalyze.
6. Save, print, or export results. Keep your results, make a hard copy, or pump them over to other software.

NOTE

Analyses is the plural form of analysis.

Setting up the Analysis

Assuming your circuit is assembled and ready, first set up the options to be used for a specific analysis by going to the Analysis menu and selecting the appropriate choice (example, **Analysis** > **Transient...**). A window opens such as in Figure 12.4. Each analysis presents you with a different window, but the idea is the same. (The Analysis Examples section shows how to set each Analysis dialog box.)

TIP

*If you don't know which nodes you want to have graphed or simulated, go back to the circuit window, set Show Nodes (**Schematic Options** > **Show/Hide** tab), and note them on a piece of paper. Go back to the Analysis dialog box and set the variables.*

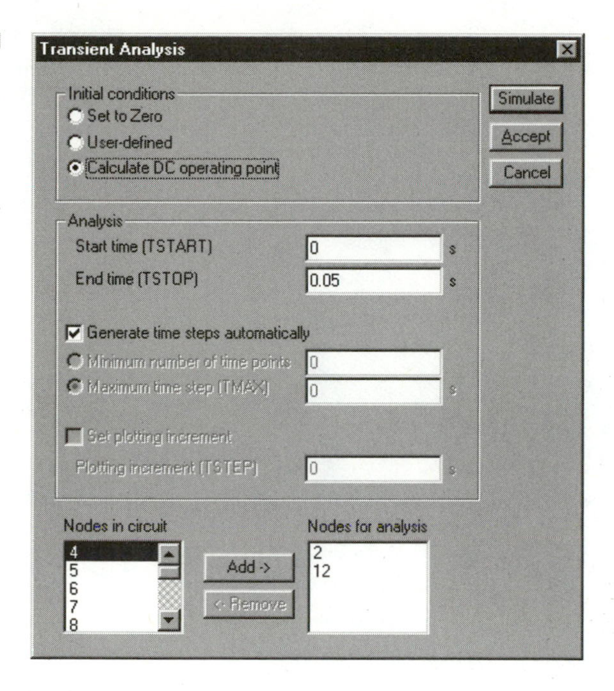

Figure 12.4
The Transient Analysis dialog box best exemplifies how an analysis is set up.

Initial Conditions

This section of the Analysis dialog box lets you tell EWB how to begin a simulation. There are three choices:

- **Set to Zero**—Sets each variable to 0. I use this option to determine RC time circuits and when I want an exact reading at the beginning of the simulation. If you do not use it, the analysis or simulation will not pick up the charge or discharge curve.
- **User-Defined**—You can set the initial voltage levels of any point (node) i the circuit drawing. Under the **Connector Properties** > **Node** tab dialog box (double-click connectors) you set the voltage level at which the node starts off a simulation (see Figure 12.5). For example, some oscillator simulations require a sort of "jump start voltage". You would set the node's initial voltage level to, say, .000001V and run the simulation with the user-defined initial condition set.
- **Calculate DC Operating Point**—This determines the steady voltage levels around the circuit before it begins the simulation. This is set by default.

Use the Set to Zero initial condition to debug troublesome simulations or analyses.

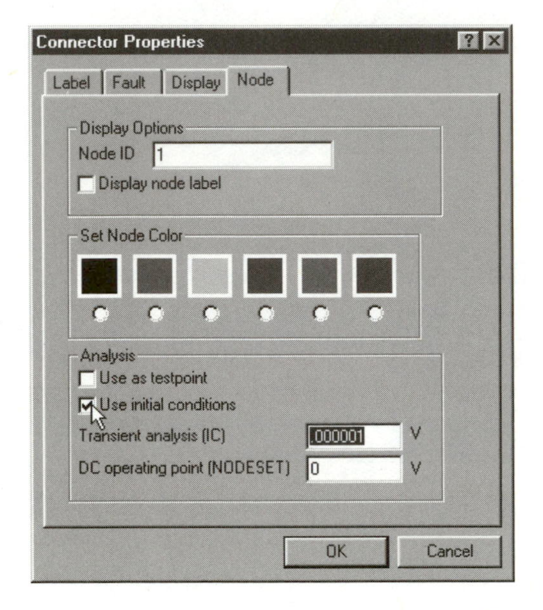

Figure 12.5 When you use User-Defined Initial Conditions, the node's properties have to define the exact voltage level at which you wish the simulation to begin.

Nodes and the Art of Choosing Variables

Each analysis requires you to select one or multiple test point(s) at which the readings will take place. This information is placed inside the Nodes for Analysis field in each of the Analysis dialog boxes. It tells the software what exactly you want to look at in the circuit. Most of the time we are comparing an input signal with an output signal; therefore the two nodes that represent these can be used. Other times you may be *sweeping* (constantly changing) the parameters of a component to see the circuit's reaction. For example, you may choose to sweep the value of a resistor as it relates to its tolerance. This requires you to set not only an output node but also the component to sweep. When you are first beginning analyses you may merely want to select all the variables and use the Toggle Legend on the Grapher to sort out what is what. You can then run the analysis again after you have narrowed the variables. With practice you will become familiar with which variables to select.

If you check Use as Test Point under the Connector's Properties window (Figure 12.5), that node is automatically included in the Nodes for Analysis box in an analysis.

SELECTING NODES FOR ANALYSIS

In Figure 12.6, you can see the Nodes in circuit and Nodes for analysis field. This is present on all analysis setup dialog boxes. You must highlight a node or multiple nodes on the left and hit the **Add** button to move them to the right. If you wish to remove them from the Nodes for analysis list, highlight them again and hit the **Remove** button.

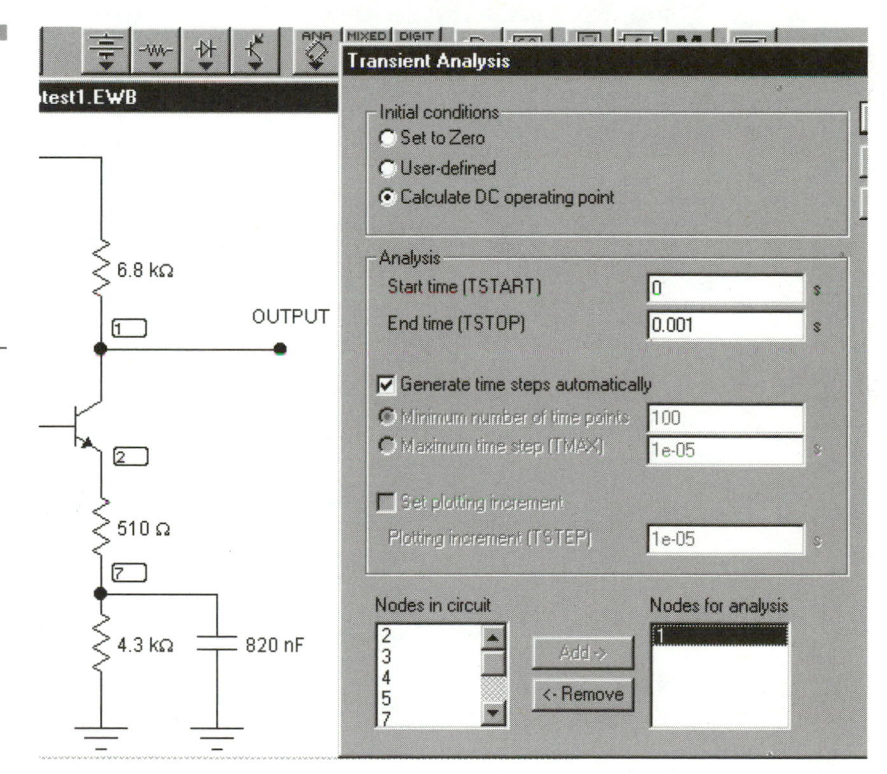

Figure 12.6 To take readings from around the circuit, you must first tell EWB which to use. This is done using the Nodes for Analysis field at the bottom of the Analysis dialog box.

To select a series of nodes, highlight one, hold the Shift key down and hit the Up or Down arrow keys. To select specific nodes, hold down the Ctrl key and click those you wish to select/deselect. Release the Ctrl key when all are selected.

Adjusting Analysis Options

Customizable options give you greater control of the analyses, but I do not recommend changing values unless you absolutely know the results that will occur. **Analysis** > **Analysis Options... (Ctrl+Y)** opens the dialog box that contains five pages of variables. They are **Global** tab (Figure 12.7), **DC** tab (Figure 12.8), **Transient** tab (Figure 12.9), **Device** tab (Figure 12.10), and the **Instruments** tab, which was described in the previous chapter. For a complete list of each item and how to adjust (or not adjust), see the EWB Help file under Analysis Options. Open the main help menu, go to the Index tab, and type in "analysis options" to access each item. For notes on how they related to speed and accuracy, see that section in Chapter 14.

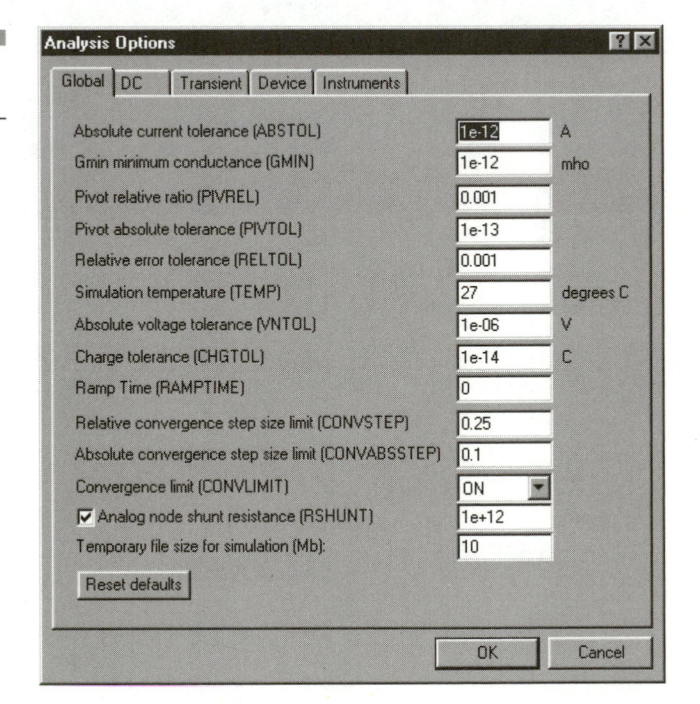

Figure 12.7
The Global tab.

If you can't quite remember what the analysis options were originally set to, hit ***Restore defaults*** *on each page.*

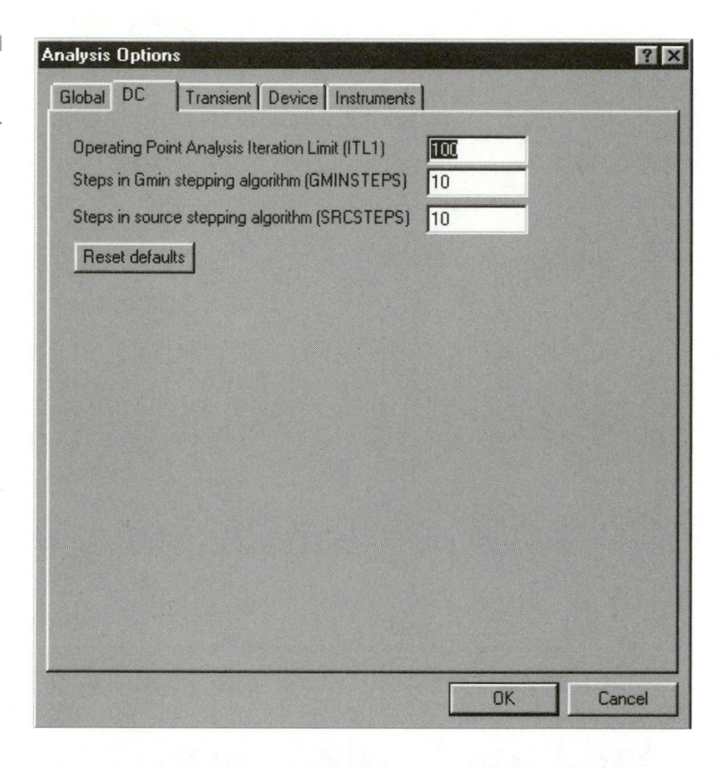

Figure 12.8
The DC tab.

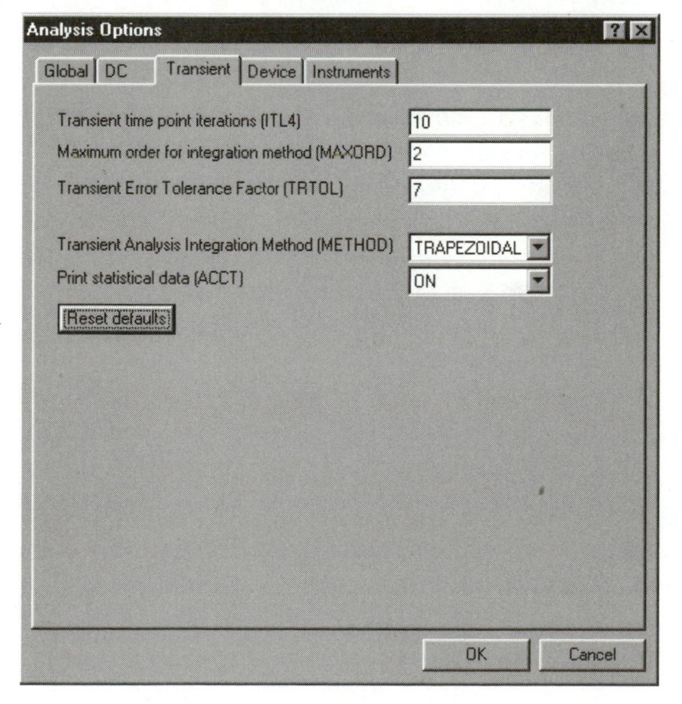

Figure 12.9
The Transient tab.

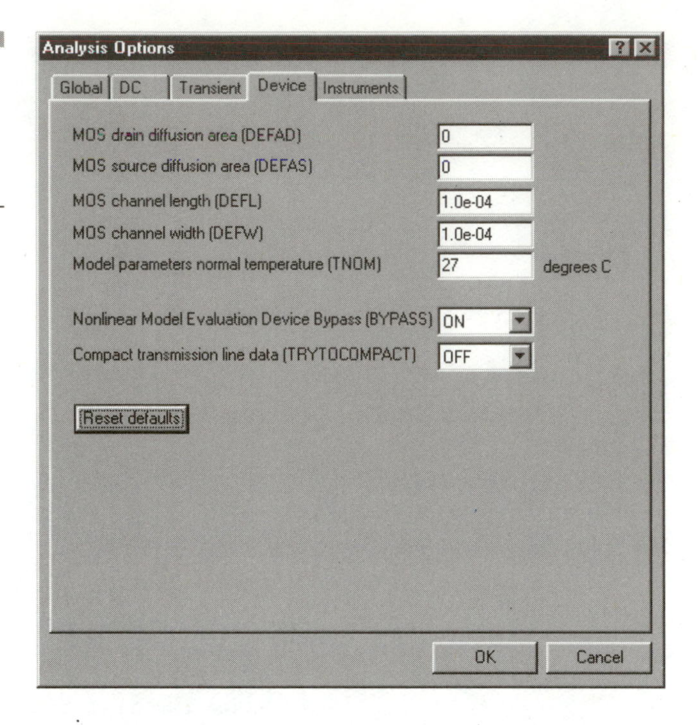

Figure 12.10 The Devices tab allows you to adjust a graph after it has been created.

Analysis Explanations, Setup, and Grapher Results

The Analysis menu items can become quite complicated, so realize I'm presenting as simple an explanation as possible. For in-depth explanations of each analysis you may want to refer to the *Technical Reference* that came with EWB or a good circuit analysis book, such as *Electronic Circuit and Simulation Methods* (McGraw-Hill), *Introductory Circuit Analysis* (Prentice Hall), or *Electronic Devices and Circuit Theory* (Prentice Hall).

NOTE

The Graphs window is called Grapher throughout this chapter for easier understanding. This is a Multisim term but seems to describe it well.

NOTE

Use of the Grapher is explained later in this chapter.

DC Operating Point

Each analysis begins with a DC operating point calculation, when Electronics Workbench calculates the steady-state voltage of each node in an active circuit. The current from each voltage source is also calculated. It's like placing a multimeter at each node in the circuit. During the DC analysis, AC sources are nullified, capacitors are open, and inductors are shorted. Electronics Workbench figures out the *quiescent operating point,* which is the steady-state operating conditions of a valve or transistor in its working circuit but in the absence of any input signal. This takes the task of complex circuit calculation out of your hands.

Setting Up a DC Operating Point Analysis

Let's use two examples to explain this better. First look at Figure 12.11. It's a multiple voltage-source circuit; we will use DC operating point to figure out the voltage at each node. The circuit consists of two batteries and three resistors.

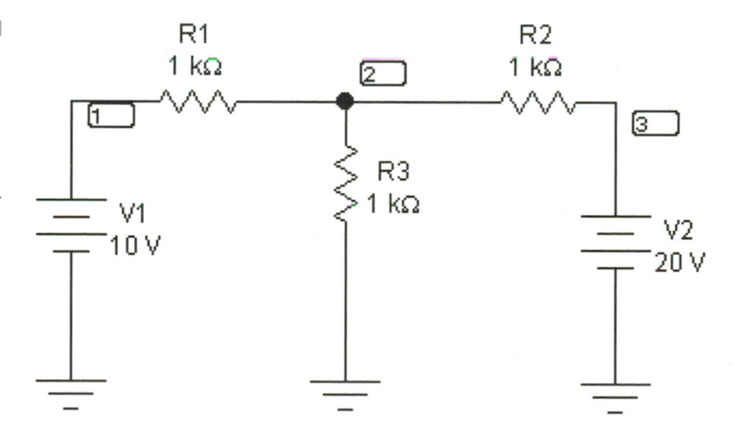

Figure 12.11
A dual DC source circuit is used to test the DC operating point analysis.

1. Build the circuit as pictured in Figure 12.11 and adjust the component values to match.
2. Choose **Analysis** > **DC operating point**. The analysis is performed automatically and no dialog box opens to select nodes. Instead, every node is plotted.

3. The Grapher opens and indicates the voltage level of every node in the circuit. Note the voltage of node 2 (see Figure 12.12). You may have to widen the window to see the readings (depending on which resolution your monitor is set). Close the Grapher and save the circuit as DC_OP1.EWB.

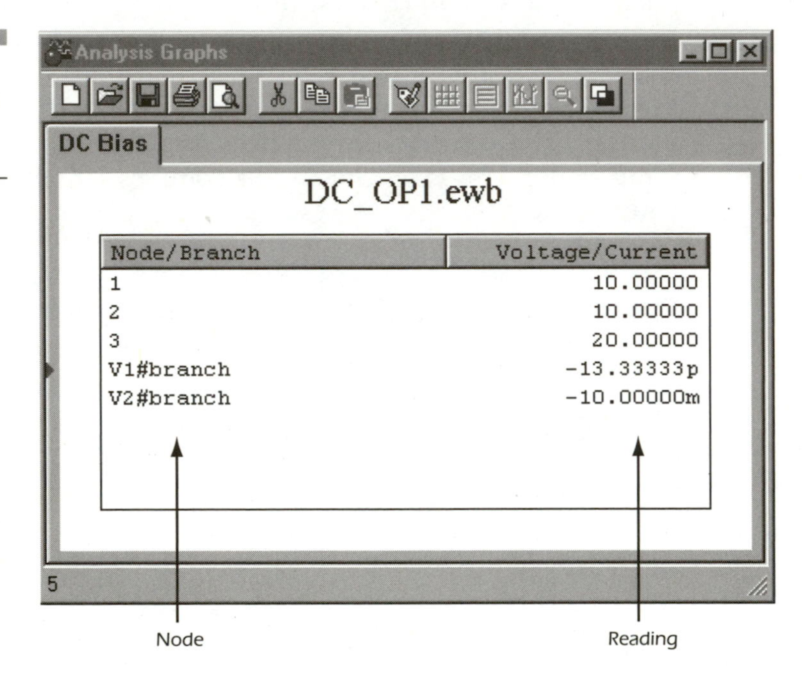

Figure 12.12
The Grapher displays the results of the DC op analysis.

NOTE

Nodes are numbered in the order that you connect their terminals.

Figure 12.13 is a bit more complicated. It is a DC-biased transistor used as a simple amplifier. Say, for example, you want to determine the DC bias point at the transistor's collector (node 1). By using DC operating point, we can figure this out in seconds:

1. Build the circuit pictured in Figure 12.13 and save it as AMPTEST1.EWB for later use.
2. Choose **Analysis** > **DC operating point**.
3. The Grapher pops up with each node's voltage. Node 1 is the DC bias point of the amplifier. Save the circuit once again.

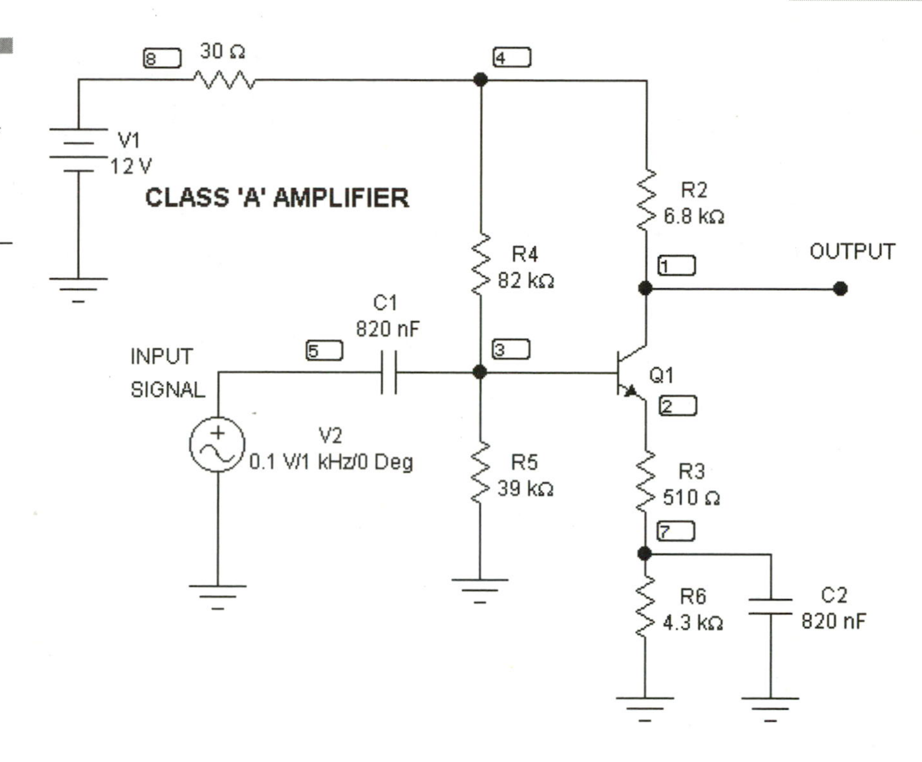

Figure 12.13
A Class A amplifier is used to test many of the analyses, including the DC operating point.

The only readout sent to the Grapher after the simulation is a list of voltage or current values for every node. There are two columns: left is the variable (node); the right is the reading in voltage or amperage.

When you need to know the voltage readings of each point in a DC circuit fast, the DC operating point analysis is much quicker than placing a multimeter at several locations and running the simulation. It is also handy in finding the bias point of amplifiers or any other scenario that requires complex voltage calculations.

AC Analysis

AC analysis serves the same function as the Bode plotter: It plots the circuit's response to a series of frequencies. However, AC analysis offers more flexibility. All DC sources are given zero values and AC components, such as capacitors and inductors, are modeled for the analysis. Remember, AC sources do not particularly matter because Electronics Workbench inputs its own sinusoidal waveform, graphing the circuit node responses. Also, note that digital components are treated as large resistances to ground.

Setting Up an AC Analysis

Have a look at the two filter circuits in Figure 12.14. The top is a band-pass filter that will only let an exact frequency band through the filter. The bottom circuit is a band-stop filter that will inhibit a certain frequency band. To perform this analysis:

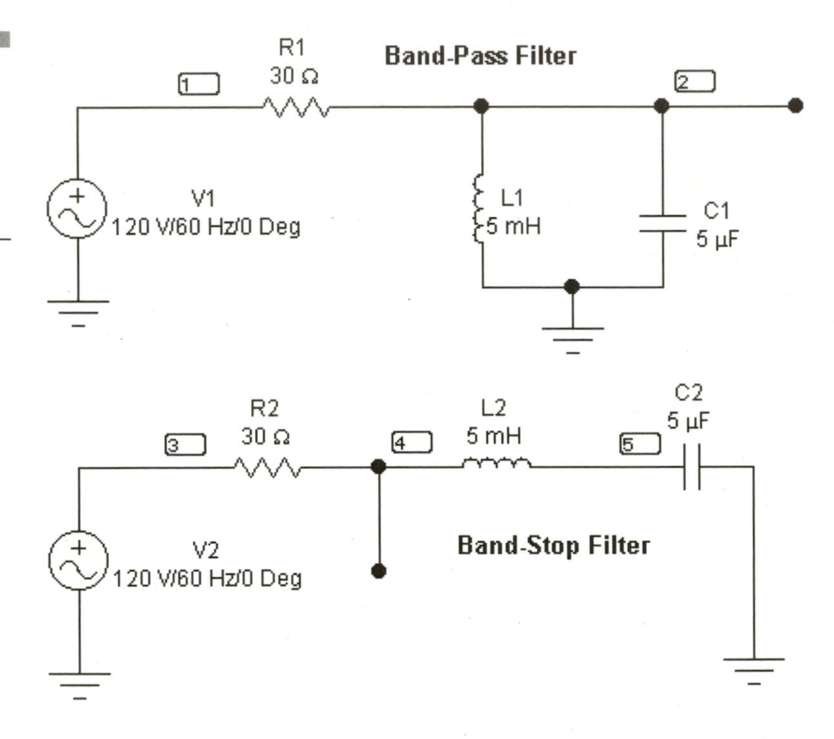

Figure 12.14 The band-pass and band-stop filters make perfect examples for an AC analysis.

1. Build both circuits and save the file as ACA_TEST.EWB.
2. Choose **Analysis** > **AC Analysis**. The AC Analysis dialog box opens.
3. Adjust the frequency parameters as seen in Figure 12.15. Make sure to set the vertical scale to linear (this makes the information easier to read).
4. Choose **2** and **4** as the nodes for analysis.
5. Hit the **Simulate** button. The Grapher opens and begins to plot the solution.
6. If you want to check the bandwidth, turn the cursors on, position them until y1 and y2 are around 707.0000m. The bandwidth is in the cursor readout window (Figure 12.16).

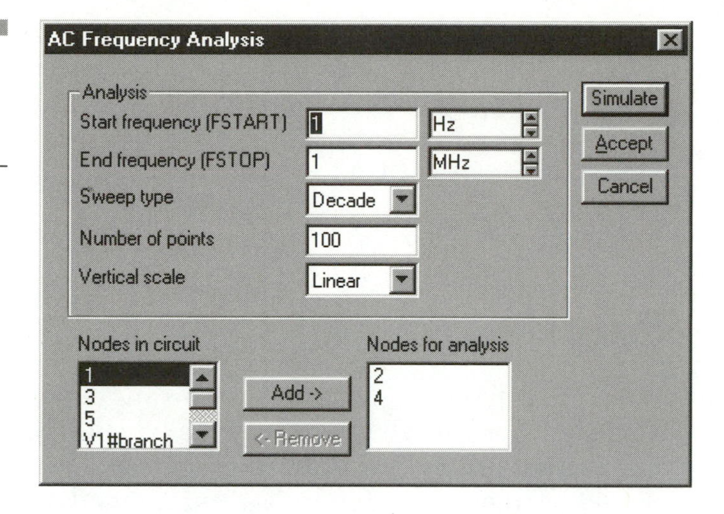

Figure 12.15 Setup the dialog box as shown for this AC analysis.

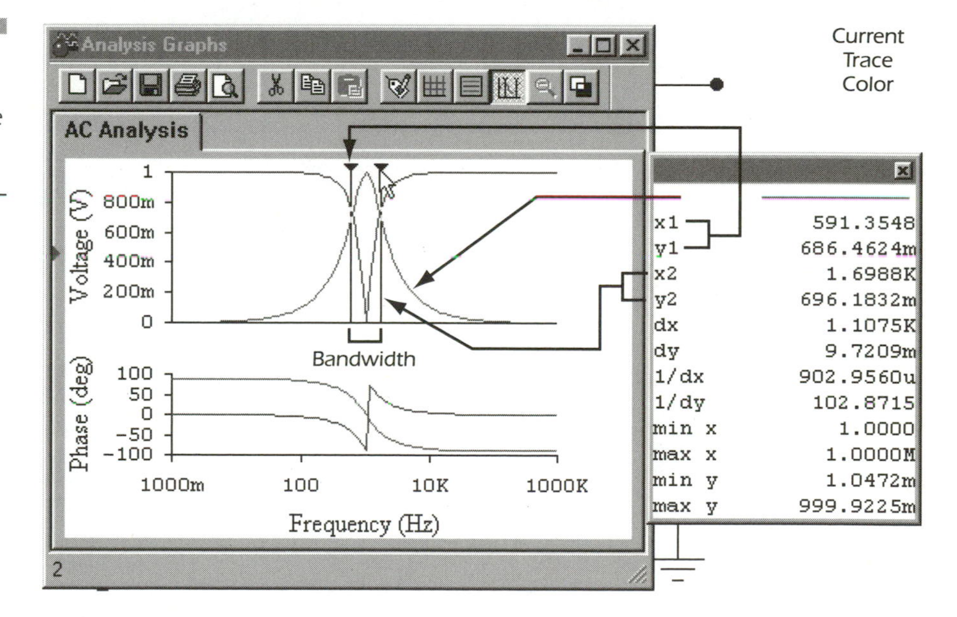

Figure 12.16 Once the Grapher appears, you can use cursors to determine the bandwidth.

TIP

To select another graphed trace for cursor readouts, left-click anywhere along the trace. The currently selected line will be on the top of the cursor readout window and also indicated at the bottom-right of the Grapher (Figure 12.16). More information on the Grapher is presented later in this chapter.

Electronics Workbench has a voltage versus frequency graph, as well as a phase versus frequency graph. I recommend turning on the grid when using the cursors.

AC analysis is handy when designing filter circuits. It's much faster to set up AC analysis than to use the Bode plotter, and certainly faster than pencil, paper, and calculator marathons.

Transient Analysis

Transient analysis adds a time factor to your circuit's readings. DC and AC analysis are basically steady, "one moment in time" analyses; transient analysis breaks time down into segments and calculates the voltage and current levels for each given interval. For example, a node may be .1 volt at 1 millisecond, .2 volts at 2 milliseconds, .3 volts at 3 milliseconds, etc. Electronics Workbench strings them together to give you graphs that analyze the ever-changing circuit (see Figure 12.17).

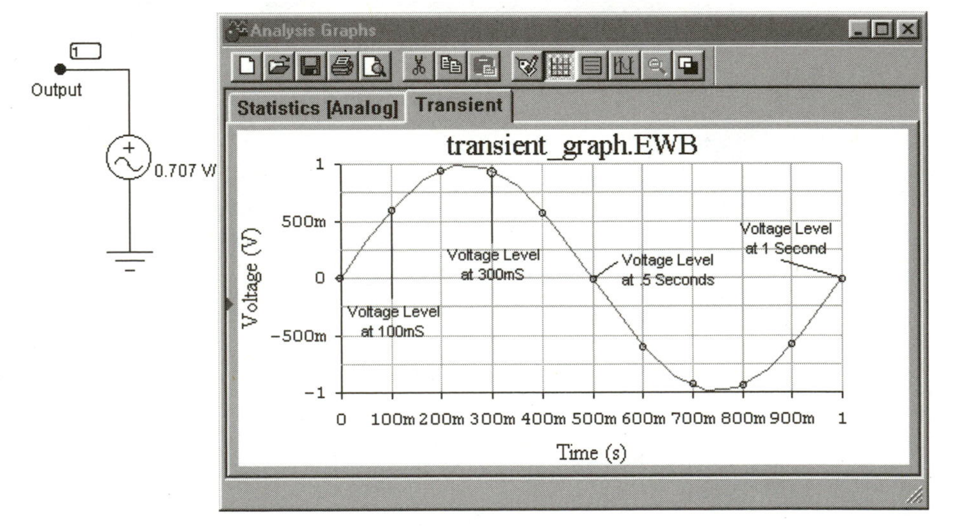

Figure 12.17 A transient analysis adds a factor of time to a circuit in order to create waveforms, etc.

NOTE

Transient analysis is also called time-domain transient analysis.

Setting Up Transient Analysis

Let's go back to the transistor amplifier circuit we made with the DC operating point analysis example (see Figure 12.18). Set up the circuit again or open the AMPTEST1.EWB file you saved.

1. Once the circuit is opened, save the file as AMPTEST2.EWB.
2. Go to **Analysis** > **Transient Analysis**. The dialog box opens.
3. Adjust the analysis parameters to match Figure 12.19.
4. Select the Output node (in this case, **1**) and the input node (**5**) as the Output variables under that tab.
5. Hit the **Simulate** button.
6. The Grapher opens, as pictured in Figure 12.20. The blue signal is the Input AC signal; the red is the amplified Output signal. You will notice the blue is larger than the red signal. This is because the input signal uses a different voltage scale on the right axis. Right-click over the graph and adjust the right and left axis to match each other for a better graph (see Figure 12.21).
7. Resave the circuit.

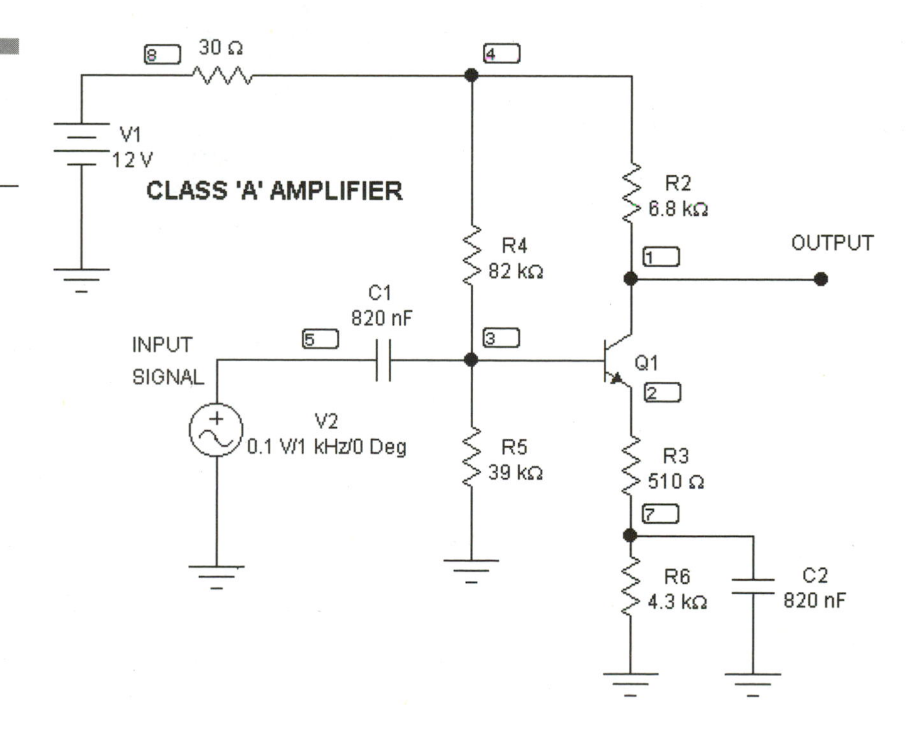

Figure 12.18
Reopen the Class A amplifier to test the transient analysis.

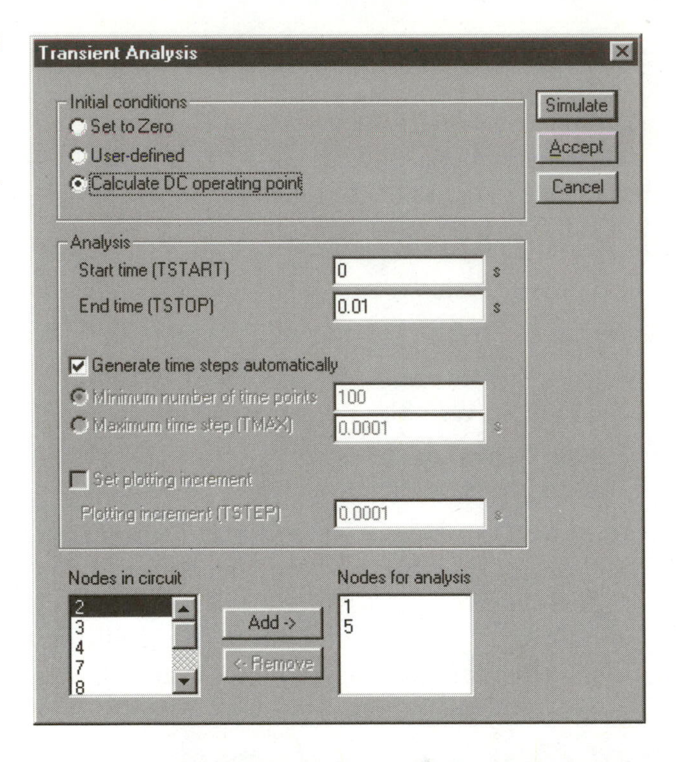

Figure 12.19 Adjust the transient analysis parameters as seen here.

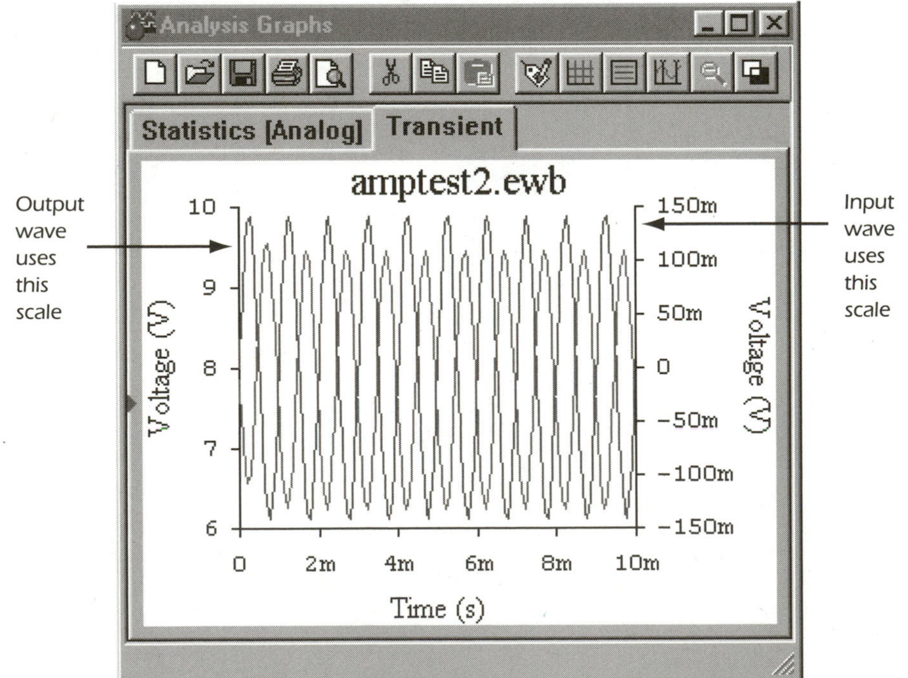

Figure 12.20 The Grapher opens with a waveform for the input and output of the amplifier.

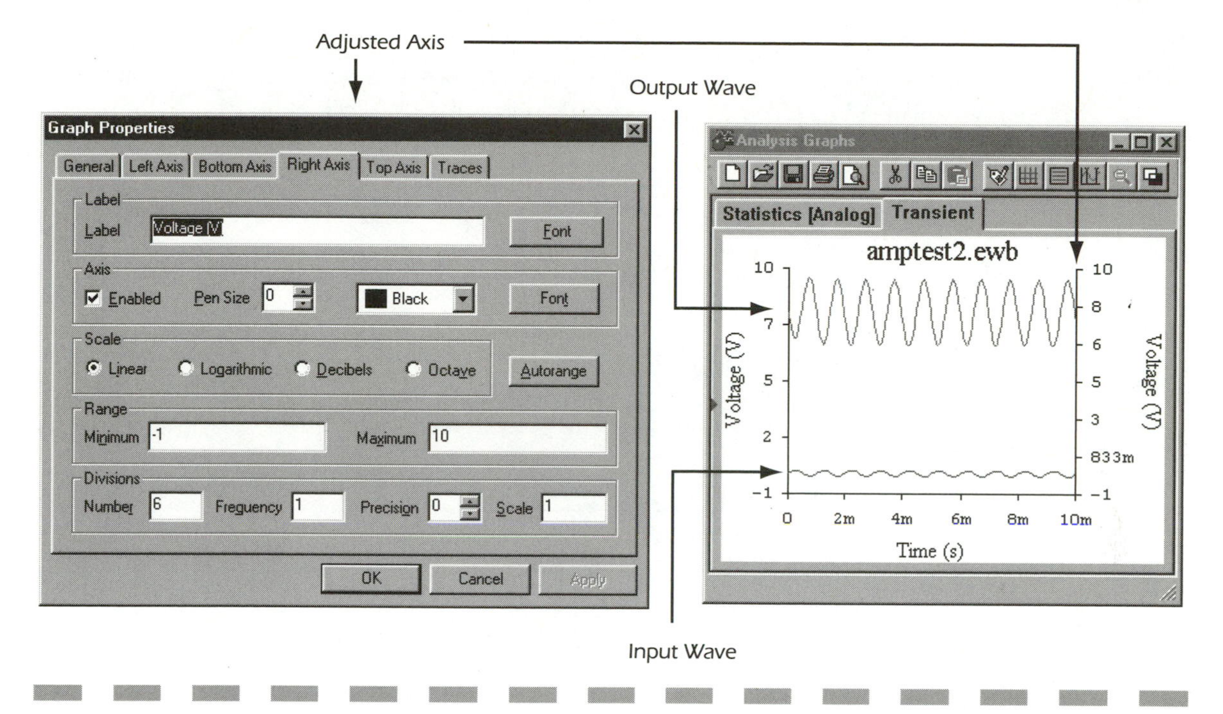

Figure 12.21 *Adjusting the graphs' properties can make the waveforms more readable.*

The Grapher plots a waveform for each node you have selected. The x-axis is time and the y-axis is voltage/current. Each waveform is color-coded; if you can't identify waveforms, toggle the legend on. You can also use the cursors to determine specific readings along the wave.

Use transient analysis for just about any circuit that requires you to view the voltage/current level over time. It replaces having to set up 20 oscilloscopes.

DC Sweep

Have you ever wanted to see how a circuit responds to varying DC voltages? DC sweep automatically adjusts (sweeps) the DC voltage of one to two voltage sources in the circuit (according to your settings), then uses this information to graph the circuit's response. Technically, it is performing a DC operating point for each voltage level(s), which saves having to perform a DC operating point, change the DC voltage levels, and do another DC operating point on each source. It's also great for visualizing non-linear components such as transistors and diodes or just to see how a DC circuit will respond to various DC voltage levels.

Setting Up a Single-Source DC Sweep

We will use a simple Zener diode to test this circuit.

1. Build the circuit as seen in Figure 12.22 and save it as DCSWEEPTEST.EWB.

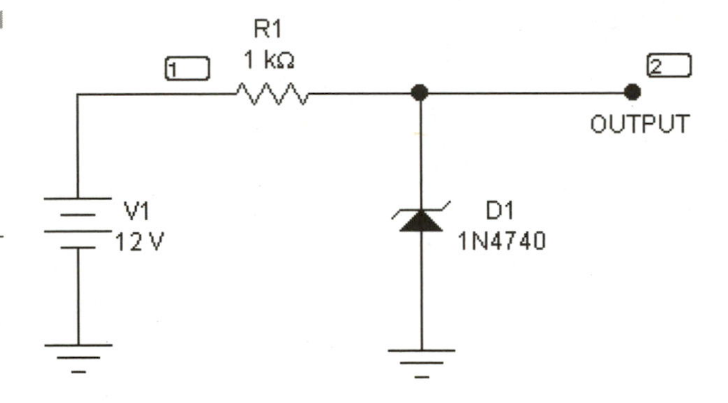

Figure 12.22 A simple Zener-control voltage regulator circuit is used to test the DC sweep analysis.

2. Choose **Analysis** > **DC Sweep**.
3. Adjust the analysis parameters page as seen in Figure 12.23.
4. Choose node 2 (Output) as the Output variable.
5. Hit **Simulate**.
6. The graph shows the output voltage (y-axis) as it relates to the sweeping DC input voltage (x-axis). You will see that the voltage steadies at around 10 volts (see Figure 12.24).

NOTE

The next chapter gives an example of a dual DC sweep.

If you are only reading one voltage source, then the x-axis is varying voltage source and y-axis is the output voltage. If you are using two voltage sources, the second voltage source is represented incrementally by alternating color lines. To see what each line represents, toggle the legend.

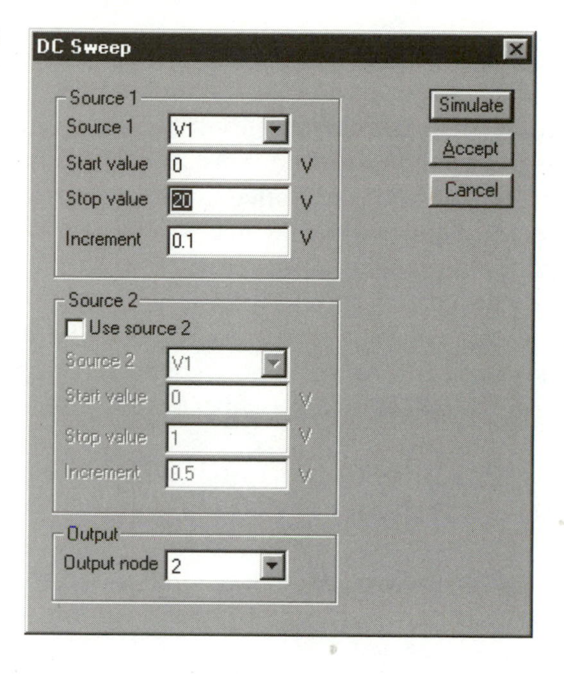

Figure 12.23 Adjust the DC sweep analysis parameters as shown.

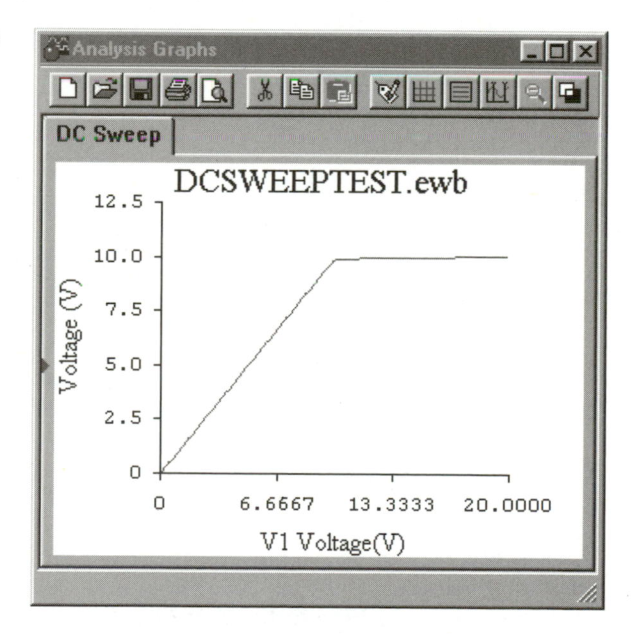

Figure 12.24 The resulting graph shows the output voltage climbs with the input until it reaches a level of 10 volts, at which point it steadies.

DC sweep can be used for any circuit you wish to map the results of varying DC voltages. It's also a fast way to make transistor curve graphs (discussed in the next chapter).

Fourier Analysis

Fourier analysis breaks a complex waveform down into its component frequencies and amplitudes. Any complex waveform consists of *fundamental frequencies* combined with *harmonics* of those frequencies. For example, a complex voice wave may consist of only three fundamental frequencies along with many harmonics. By using this analysis we can get a picture of exactly what frequencies and harmonics are involved and their relative strengths. The mathematics of how this works would fill up this whole chapter, so I will simply give you the results with the following example.

Setting Up a Fourier Analysis

In Figure 12.25, the three AC sources are added together to produce a complex waveform. What elements make up this waveform?

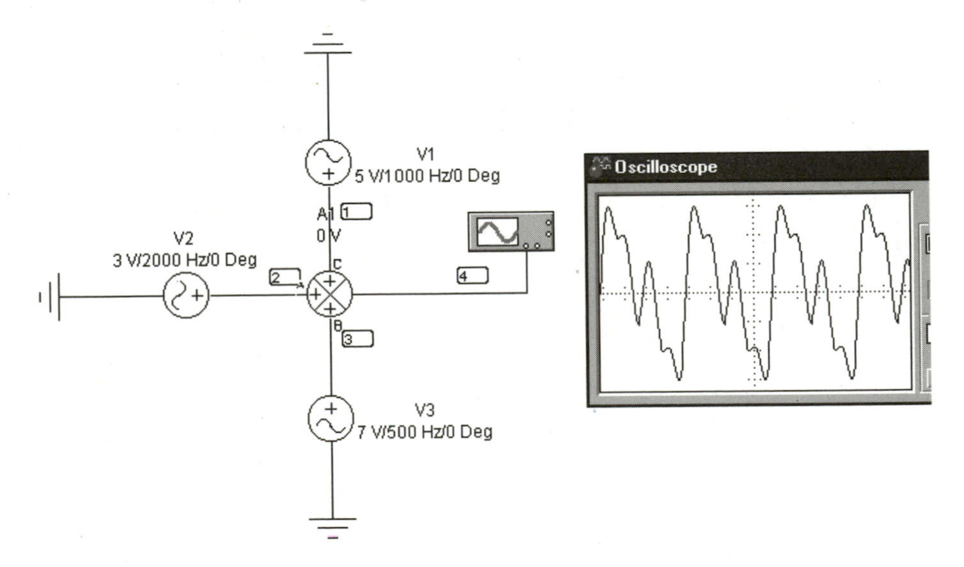

Figure 12.25 Build this three-wave adder circuit to test the Fourier analysis. It will create the non-sinusoidal waveform, shown on the scope.

1. Build the circuit in Figure 12.25 and save it as Fourier.EWB.
2. Select **Analysis** > **Fourier Analysis**.
3. Set up the parameters as shown in Figure 12.26. The fundamental frequency is usually the lowest frequency or lowest common denominator of multiple frequencies—in this case, 500 Hz.

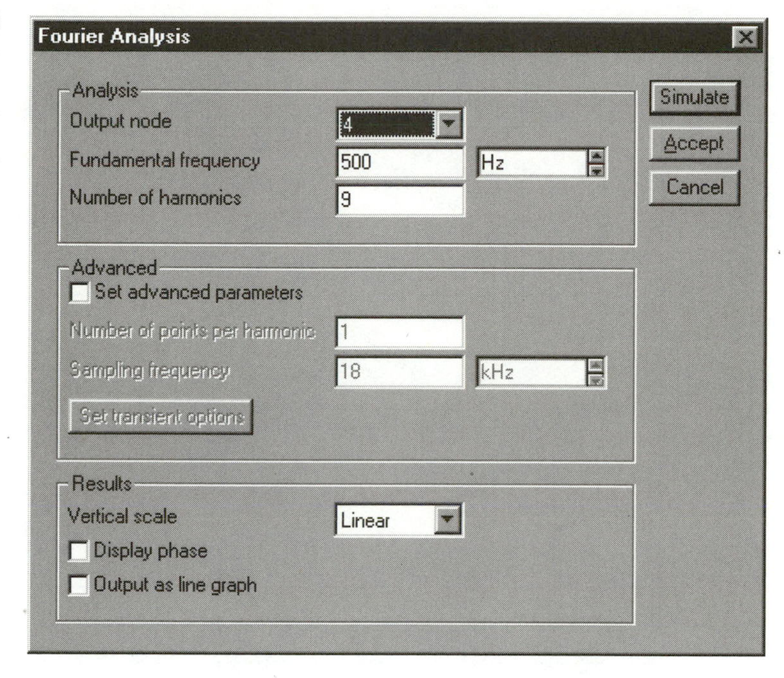

Figure 12.26
Set up the Fourier Analysis dialog box as shown.

4. Hit the **Simulate** button. The Grapher may take a moment to come up for this simulation (Electronics Workbench is calculating).
5. Figure 12.27 appears. The top graph shows the frequencies and strengths of those frequencies; the bottom section shows the *total harmonic distortion* (THD).

You can also make a line graph by checking on the **Output as line graph** option in the Analysis setup dialog box.

Use Fourier analysis for anything that involves a complex waveform, such as speech recognition technology.

Noise Analysis

Noise can be defined as any unwanted voltage or current that ends up in a circuit's output and dilutes the originally intended signal. For example, a crackle may creep into an audio amplifier and cause annoyance to your ears and muddying of a recording's playback; or noise-prone components can cause snow on a television picture tube.

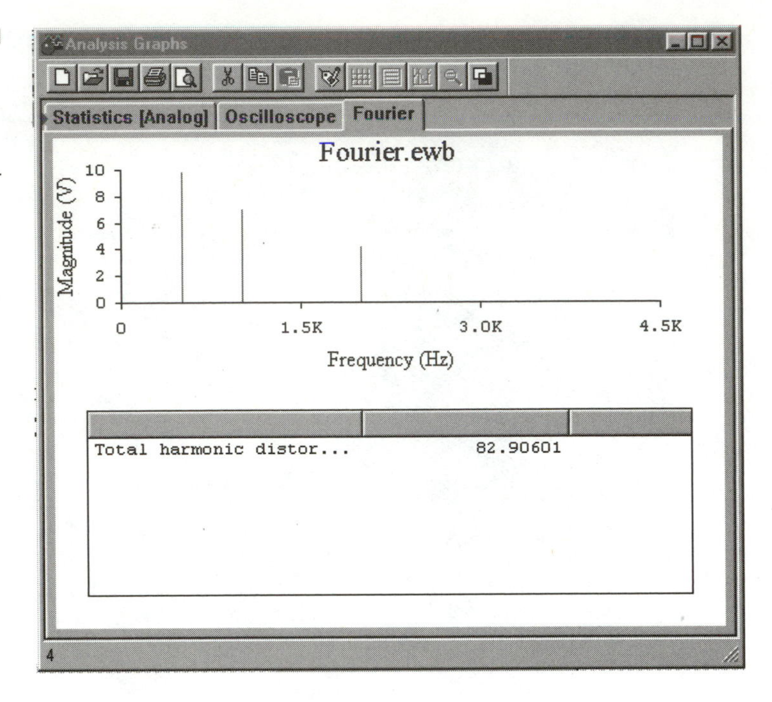

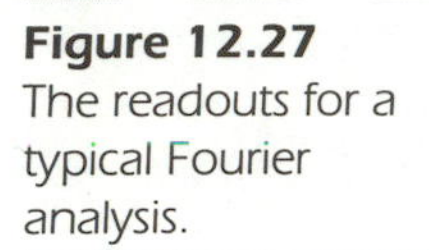

Figure 12.27 The readouts for a typical Fourier analysis.

Each resistor and semiconductor in a circuit contributes to this noise. This is due to a combination of *thermal* noise (noise due to temperature factors) in resistors, *shot noise* (found in semiconductors), and *flicker noise* (*pink noise* in BJTs and FETS operating below 1kHz).

Electronics Workbench performs an AC-like analysis on the entire circuit, using special noise models for each component, instead of its regular models. This is used to sum the magnitude of noise power from the input to output of the complete circuit. In other words, it calculates the total noise that the circuit's components are producing or it figures out the noise of individual sections or components in the circuit. It displays this information in the form of a graph of output noise power versus frequency.

Setting Up a Noise Analysis

Let's open up the AMPTEST2.EWB again and follow along.

1. Open or build the circuit, setting each value.
2. Replace the transistor with a 2N3904 by double-clicking the ideal one and choosing a Nationl2 2N3904.
3. Go to **Analysis** > **Noise Analysis**. The dialog box opens.

4. Set up the analysis parameters as seen in Figure 12.28.

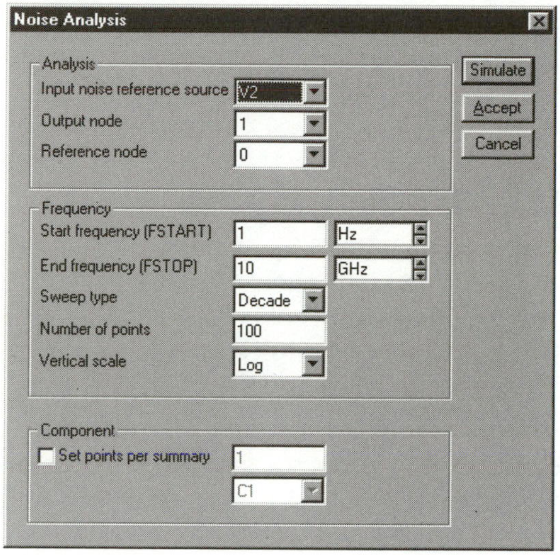

Noise Analysis
Analysis
Input noise reference source
V2
Output node
1
Reference node
0
Simulate
Accept
Cancel
Frequency
Start frequency (FSTART)
1
Hz
End frequency (FSTOP)
10
GHz
Sweep type
Decade
Number of points
100
Vertical scale
Log
Component
Set points per summary
1
C1

Figure 12.28 Setup the Analysis parameters as shown for the noise analysis.

5. Run the simulation by hitting the **Simulate** button.
6. The Grapher appears with a graph of input-to-output noise. The blue trace (lower one) is input and the red is output noise (see Figure 12.29).
7. Save the circuit as NoiseTest.EWB.

Analysis Graphs
Noise
NoiseTest.ewb
Output Noise Power (V^2/Hz)
1e-014
1e-015
1e-016
1e-017
1e-018
1e-019
Input Noise Power (V^2/Hz)
1e-012
1e-013
1e-014
1e-015
1e-016
1e-017
1e-018
1e-019
1000m
100
10K
1M
100M
10G
Frequency (Hz)

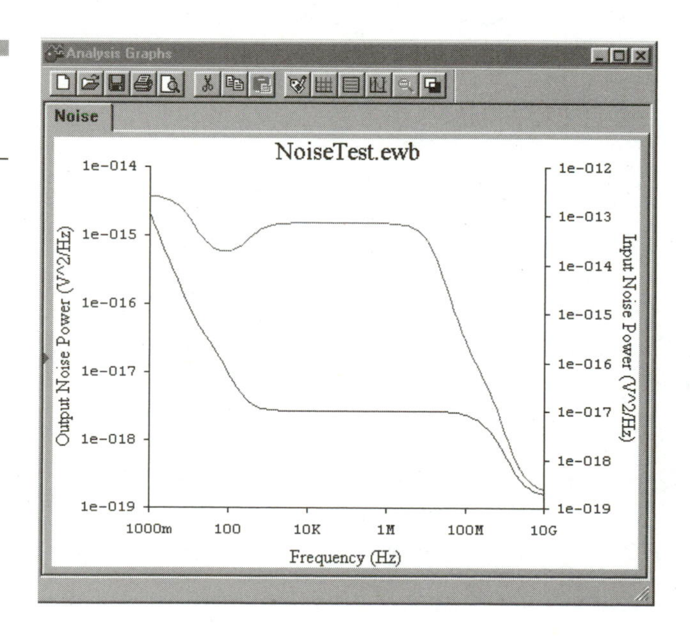

Figure 12.29 The results of the noise analysis.

The graph is a comparison of input noise to output noise or the level of noise for a specific component at a varying frequency. If you are trying to create a circuit with the lowest noise possible, utilize noise analysis; it is often used in telecommunications applications for this purpose.

Distortion Analysis

Distortion is any change in the original waveform (see Figure 12.30 for a few examples). Distortion analysis is used to locate small amounts of distortion that are not picked up by a transient analysis. These include *harmonic distortion* (created by harmonic energy) and *intermodulation* (IM) *distortion* (created by mixing signals at different frequencies). Electronics Workbench does a small-signal distortion analysis of the circuit, sweeping one or two frequencies. The results are graphed depending on whether you used one or two frequency sweeps.

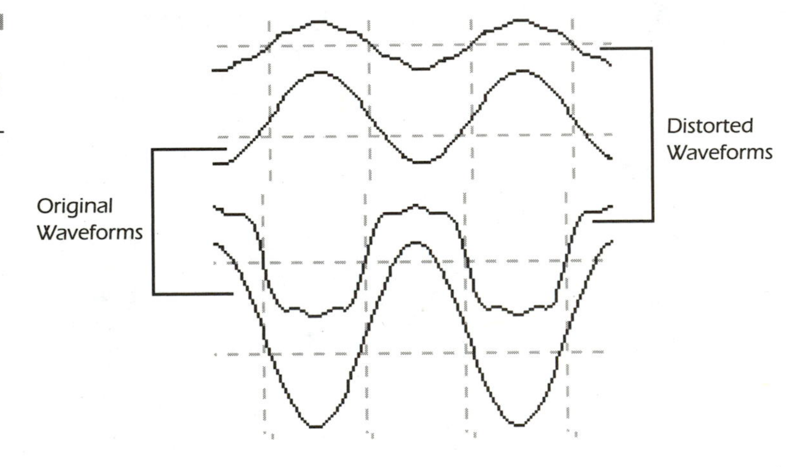

Figure 12.30
Examples of distorted waveforms.

NOTE

If F2/F1 ratio is checked, then an IM distortion analysis is performed; otherwise a harmonic distortion analysis is done.

Setting Up Harmonic Distortion Analysis

1. Open the Noisetest.EWB file and save it as DistortionTest.EWB.
2. Choose **Analysis** > **Distortion Analysis**.

3. Leave the analysis parameters as they are and select node 1 as the node for analysis.
4. Hit the **Simulate** button.
5. A page with two graphs opens. (Figure 12.31). Two graphs are created: a voltage/frequency graph for the second harmonic and another for the third harmonic.
6. Leave the circuit open.

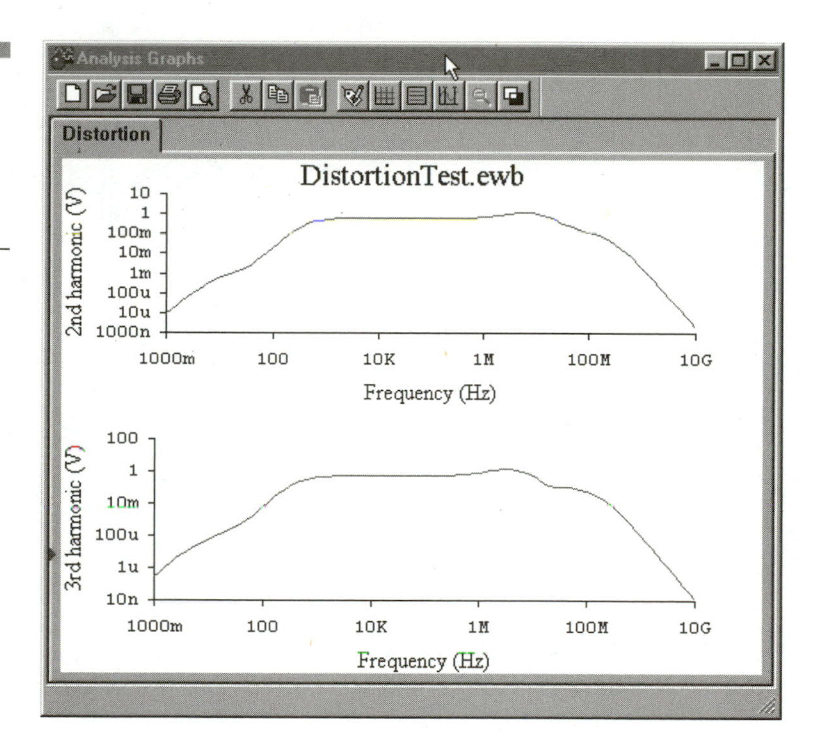

Figure 12.31 The harmonic distortion analysis creates two graphs: 2nd and 3rd harmonic distortion.

Setting Up IM Distortion Analysis

1. With DistortionTest.EWB still open, choose **Analysis > Distortion Analysis.**
2. Check on the F2/F1 ratio and hit the **Simulate** button.
3. Three graphs are created: Frequency(F) 1 + Frequency(F) 2, F1−F2, and 2(F1)−F2 (see Figure 12.32), each comparing voltage/frequency.

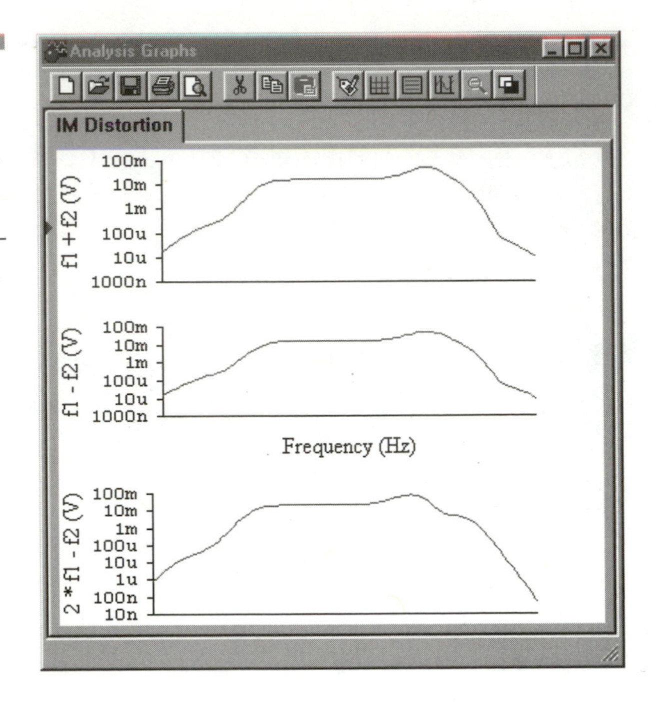

Figure 12.32 The intermodulation (IM) distortion analysis creates three graphs: f1+f2, f1-f2, and 2*f1-f2.

TIP

Right-click on a graphed line to determine its label.

Additional Analyses

The remaining analyses are only available with the Professional edition of Electronics Workbench Version 5. I've provided a brief explanation of each, but refer to the EWB Technical Reference or Help files for more details.

- **Parameter Sweep**—Automatically changes preset parameters so that you avoid having to run a simulation, change a value, and run the simulation again. It is a quick way of fine-tuning a component value or other circuit parameter.
- **Temperature Sweep**—Each component responds differently to varying temperatures; this analysis simulates varying temperatures and the circuit's reaction. The results are output according to your settings.

- **Pole Zero**—Helps to determine the stability of a circuit by calculating whether the output signal remains bounded or increases indefinitely, possibly damaging components.
- **Transfer Function**—Calculates the DC small-signal transfer function between an input source and two output nodes (for voltage) or an output variable (for current). It also calculates the input and output resistances.
- **Sensitivity**—Takes the guesswork out of determining which components must be high-end and which can come from the bargain bin at Joe's Electronics Shop. Calculates which components will most effect the DC bias point so that you can see which components are most sensitive to variation so you can refine a design.
- **Worse Case**—A statistical analysis that gives you an idea of the circuit's response to the worse possible changes in certain component parameters. For example, if you were designing a power supply and wanted to know how certain parameters, taken to the edge, would effect power output, you would run this analysis.
- **Monte Carlo**—Want to "roll the dice" with your circuit? Monte Carlo randomly changes component variables and notes the circuit's response.

Grapher

Now that you know how to start and run an analysis, it's time to learn how to interpret exactly what Electronics Workbench is trying to tell. The Grapher is a visual way to read results. After an analysis has been run, the Grapher opens automatically or you can go to **Analysis** > **Display Graphs.**

NOTE

The Grapher is the name given to the Display Graphs window. It is not an official EWB5 term; it actually comes from Multisim 6. But for better clarification, I have chosen to use it here also.

The window remains open and visible until you either close the window or choose **Analysis** > **Display Graphs** from the menu again (this removes the check from the menu item).

If you are running dual monitors in Windows 98, place the Grapher on the other screen.

Analysis Graphs Window and Icons

The Analysis Graphs popup window contains each analysis graph (see Figure 12.33). There are no menu items: toolbar icons and right-click menus are the only way to access commands. Below the icons is a page, which in turn may contain one or more graphs. First, let's look at the icons (Figure 12.34):

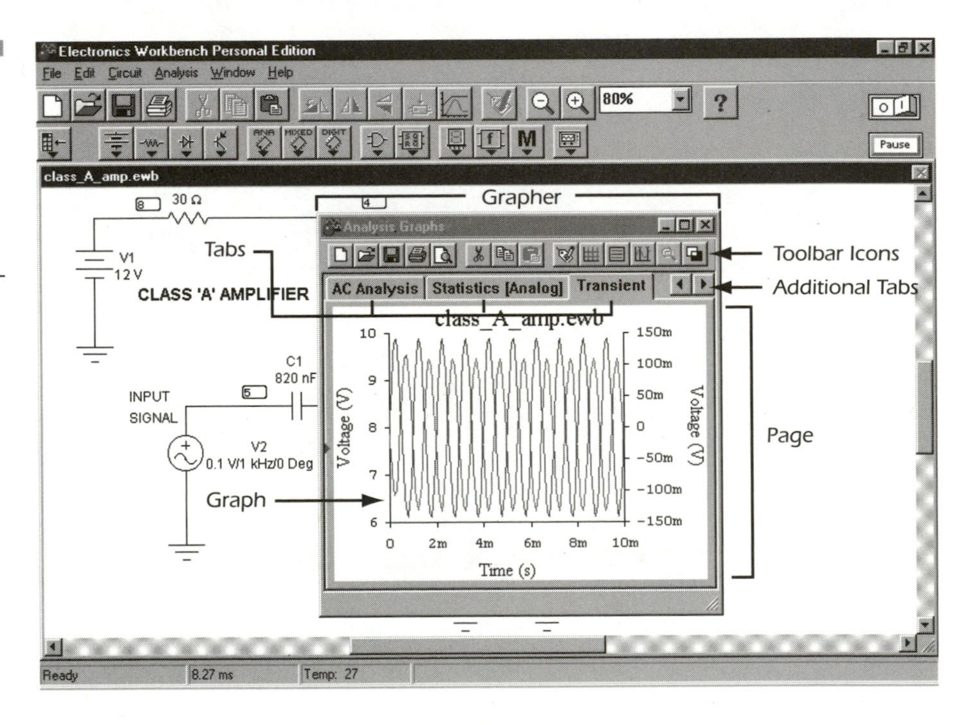

Figure 12.33 The Grapher displays graphs and readouts for each analysis performed. They can then be saved or printed.

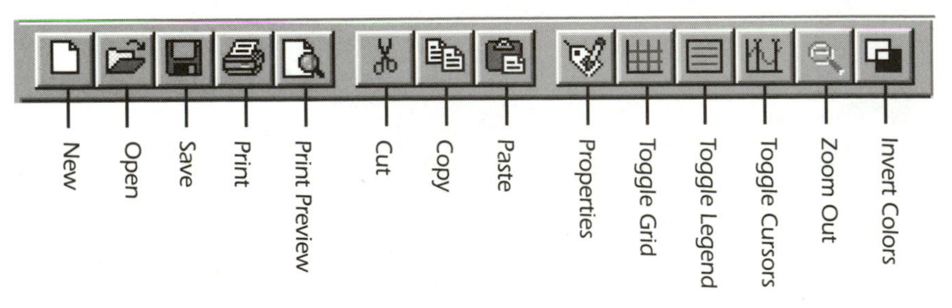

Figure 12.34 Grapher icons.

TIP

There are many right-click menus within the Grapher to speed actions. Move the mouse around and check them out.

- **New**—Creates a new page or lets you delete all existing pages.
- **Open**—Opens an existing Grapher file (*.gra), scope data (*.scp), or Bode plotter result (*.bod).
- **Save**—Stores the current Grapher files.
- **Save as**—Name and store the current Grapher files as *.gra files or as text files.
- **Print**—Opens the Print dialog box to make a hard copy of the select graphs.
- **Print Preview**—Previews each page.
- **Cut**—Cuts the current selected graph or page.
- **Copy**—Copies a selection into the current page.
- **Paste**—Pastes a selection into the current page.
- **Properties**—Allows you to modify the page or graph properties.
- **Toggle Grid**—Applies a grid to the graphs for easier reading.
- **Toggle Legend**—Turns on a chart that shows what each color line (trace) represents.
- **Toggle Cursors**—Turns on the cursors and also opens a readout window.
- **Zoom Out to Original Size**—Restores the graph to the original size if you have zoomed into any section.
- **Invert Colors**—Changes white to black, black to white, etc.

NOTE

Help on the Grapher is only accessible from EWB's main Help menu.

Pages and Graphs

Each analysis is on a separate page; the pages are named and numbered and accessible by clicking the tabs at the top (see Figure 12.35). The red arrow to the left of the leftmost tab indicates that a page is selected. If you were to hit the Properties icon on the toolbar, it would bring up that page's properties. Each page may also contain more than one graph (see Figure 12.36). To select a graph, left-click anywhere on that graph and the red arrow shifts to the graph, meaning it is selected. By hitting the Properties icon, the graph's properties are accessible, instead of the page's properties.

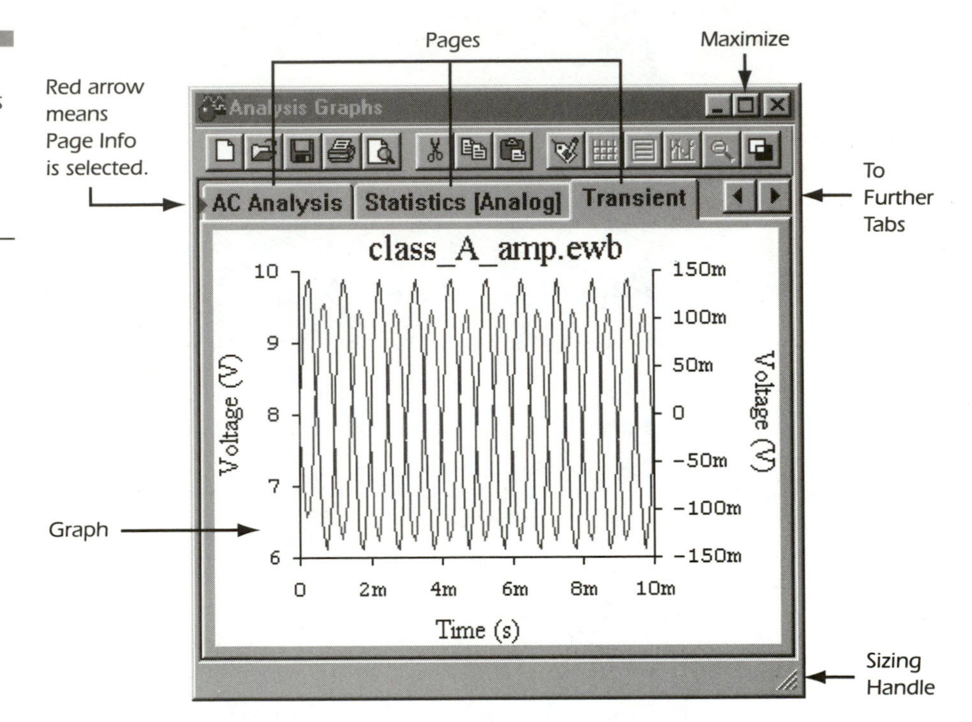

Figure 12.35 One or more graphs are placed on a page. Flip between Grapher pages with the tabs as shown.

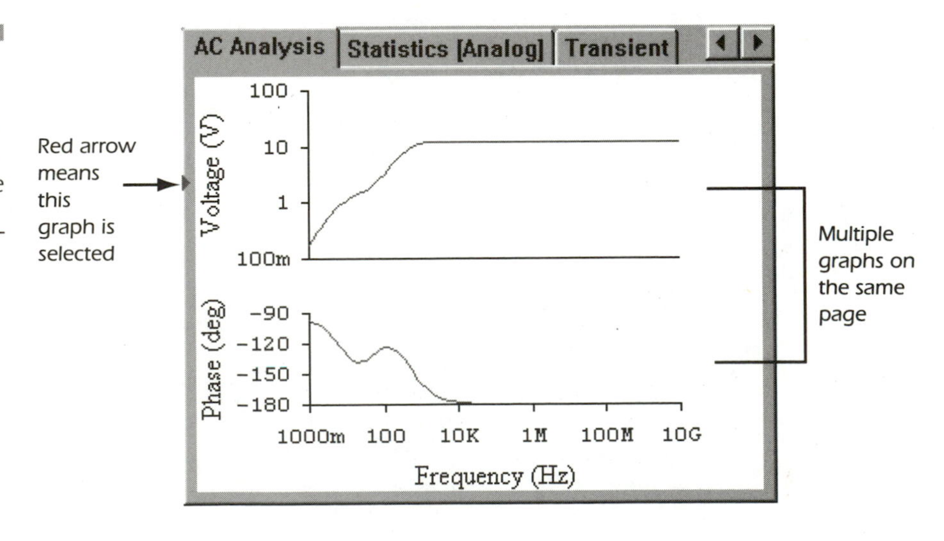

Figure 12.36 There may be more than one graph on a page. Select a graph by clicking the mouse pointer on it.

CAUTION

Sometimes you must widen the Grapher to see all the data, graphs, and text being displayed. Use the handle in the bottom right corner to pull the window open. You can also maximize the window to see as much as possible.

WORKING WITH PAGES

Once a simulation (analysis) has run, the Grapher opens with a page for each analysis (in some cases, multiple pages). Each page may contain one or more graphs or tables. If you have previously run other analyses, they will be available as well. Use the tabs at the top of each page to change between pages. If there are too many pages to list with tabs, then Electronics Workbench creates two arrows to the right of the tabs; hitting them lets you scroll to view more tabs (to the right or left) (Figure 12.35).

NOTE

Electronics Workbench keeps a record of the graphs for each analysis you have performed on a circuit. The only way to delete them is to select that page/graph and hit the cut icon. Note also that if you run the same simulation twice, the first instance disappears. So save graphs before running again.

SELECTING A GRAPH

Each analysis you have selected to run will have a page. Choose the page you want to view by clicking its tab until it appears. The graph(s) should now be visible. Click anywhere on the graph with the left mouse button to make a little red triangle appear on the right-hand side. See it next to the mouse pointer in Figure 12.37? That means that graph is selected.

NOTE

If more than one graph exists on a page, click the mouse button over the one you want to select.

ZOOMING IN AND OUT OF A GRAPH

This is a feature that lets you get a really close look at a section of the graph. It lets you zoom into any area of the graph, which you specify with a box (see Figure 12.38).

Figure 12.37
The red arrow on the left indicates this graph is selected.

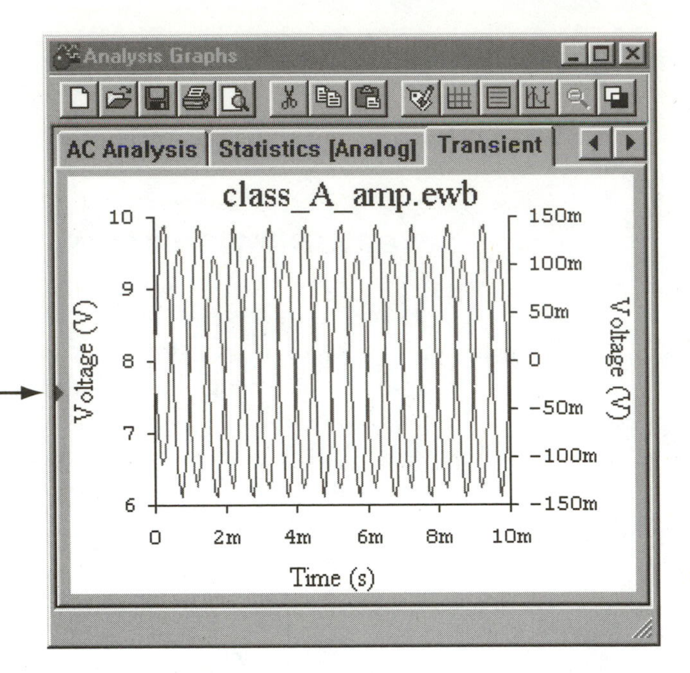

Figure 12.38
Zoom into a section of a graph by clicking and dragging the mouse pointer, surrounding the area you wish to enlarge.

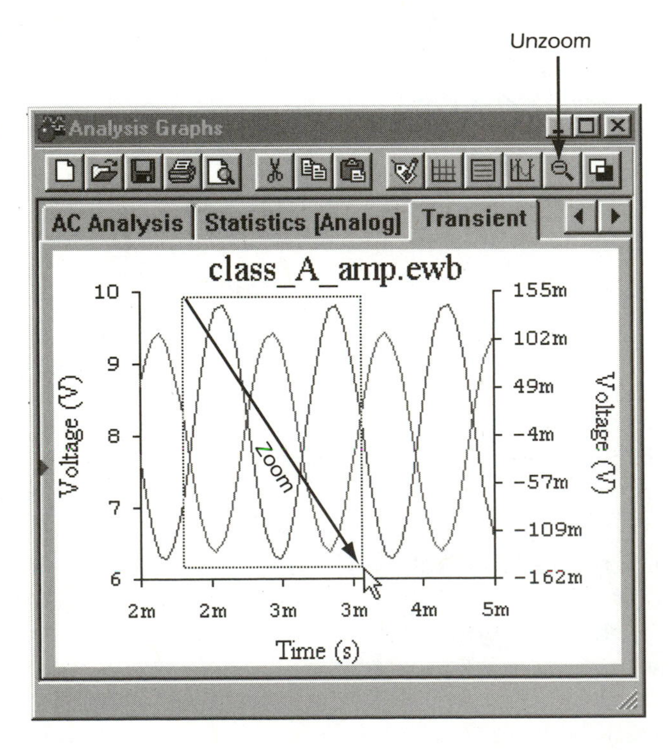

1. Select the graph you want to zoom in on.
2. Click and hold the left mouse button.
3. Drag the mouse down and to the left (enclosing the area you want to zoom). A dotted box appears. Manipulate the box until it surrounds the area of the graph you wish to zoom into.
4. Release the mouse button. A zoomed view of the graph appears.

If you want to zoom in further, repeat steps 2 through 4. When you want to zoom back out of the graph (unless you want to leave it zoomed in) hit the Restore Graph icon or right-click and choose the same.

CHANGING A PAGE'S PROPERTIES

You can open the Properties window of a page by clicking that page's tab and then hitting the Properties icon. (Figure 12.39). You can choose the Tab Name, Title (and font), and Color of the page. Make changes and close the window by choosing the OK button.

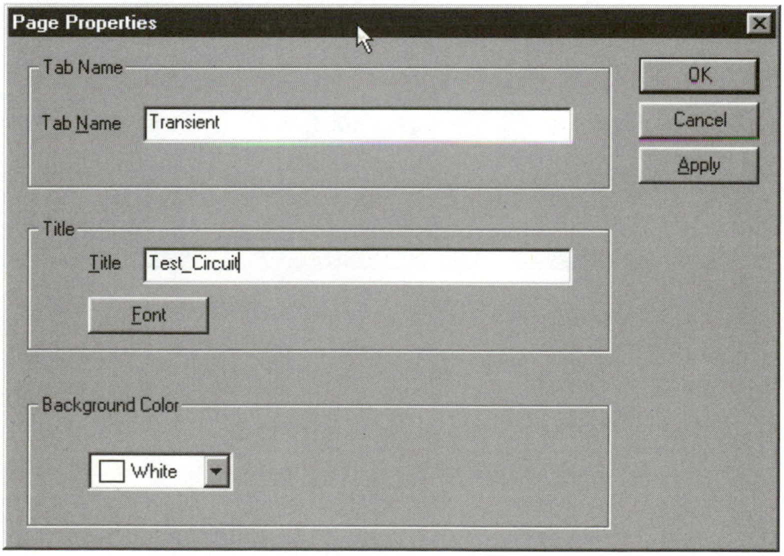

Figure 12.39
The Page Properties dialog box.

INVERTING PAGE COLORS

Want to make the page black and the traces/text white? Hit the Invert Colors icon on the toolbar. If the pages are in color, you may get confused, so try to choose a color combination that will not make your data disappear.

CHANGING A GRAPH'S PROPERTIES

Each item on each tab page can be adjusted in order to customize the currently selected graph (see Figures 12.40 through 12.45).

- **General Page**—*Title* is the name of the graph, *grid* adjusts the grid color and line width, *trace legend* turns on the legend, *cursors* can be turned on/off or used to adjust which traces you want readout. (Figure 12.40).

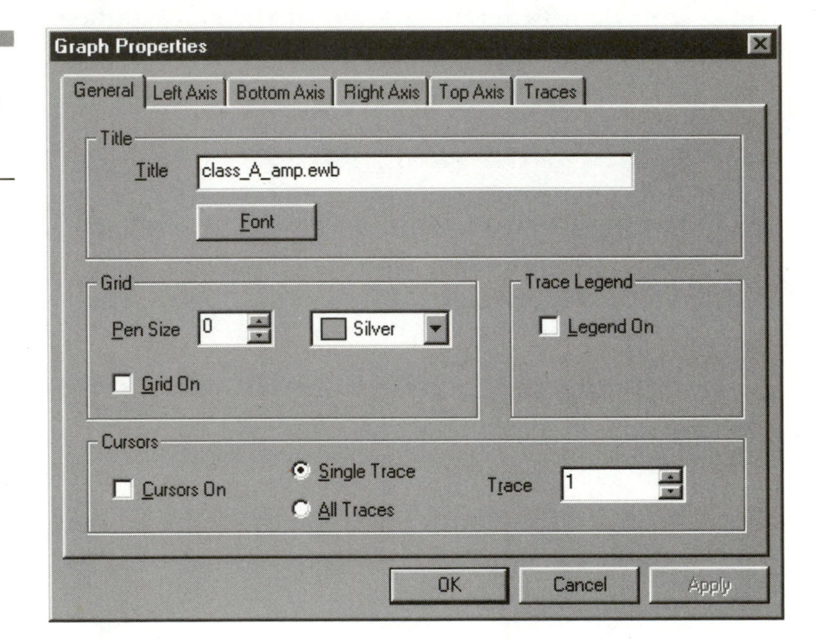

Figure 12.40 General Page of the Graph Properties dialog box.

- **Left Axis, Bottom Axis, Right Axis and Top Axis**—Lets you adjust the axis label and font, axis pen size, color, the scale of that axis, the range of that axis, and the divisions (how many ticks per axis), (Figures 12.41 through 12.44). You can adjust the area you wish to have displayed on the graph; if you wanted to print out the wave from .017ms to .023ms, adjusting the bottom axis range's minimum and maximum readjusts the graph. You can use the same method to shrink the amplitude with the left axis. Play with each setting until you are happy with how the graph looks for printout or storage.
- **Traces**—Adjust the name of the trace, the width, the color, the X and Y range, or offset the trace on an X or Y axis (Figure 12.45).

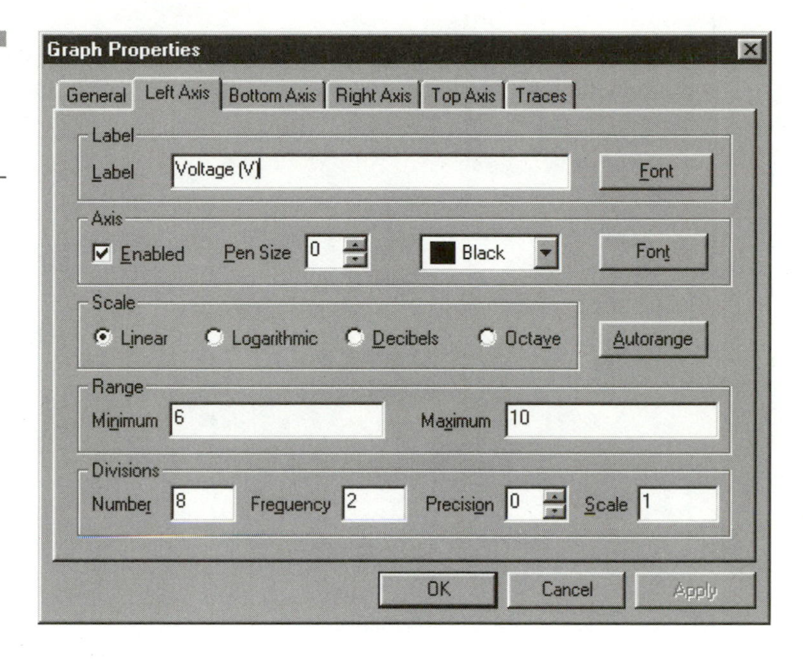

Figure 12.41 Left Axis page of the Graph Properties dialog box.

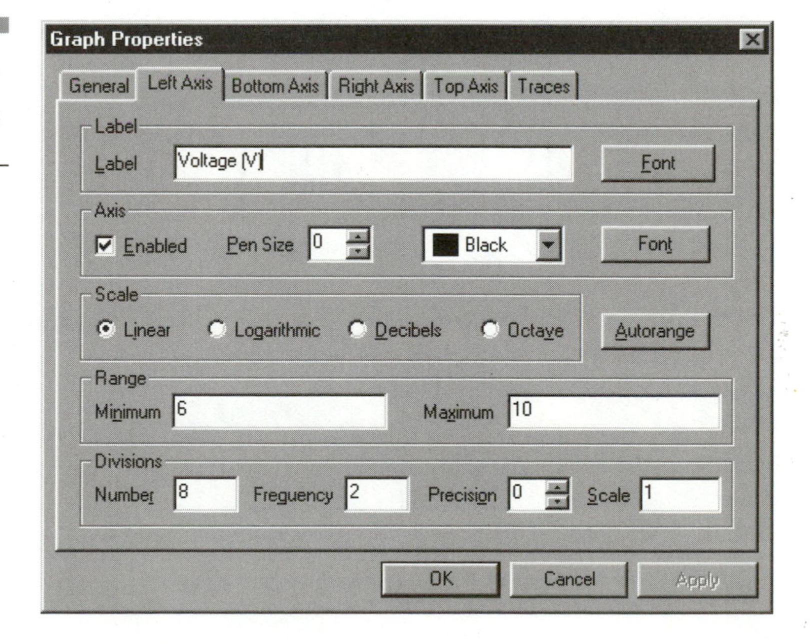

Figure 12.42 Bottom Axis page of the Graph Properties dialog box.

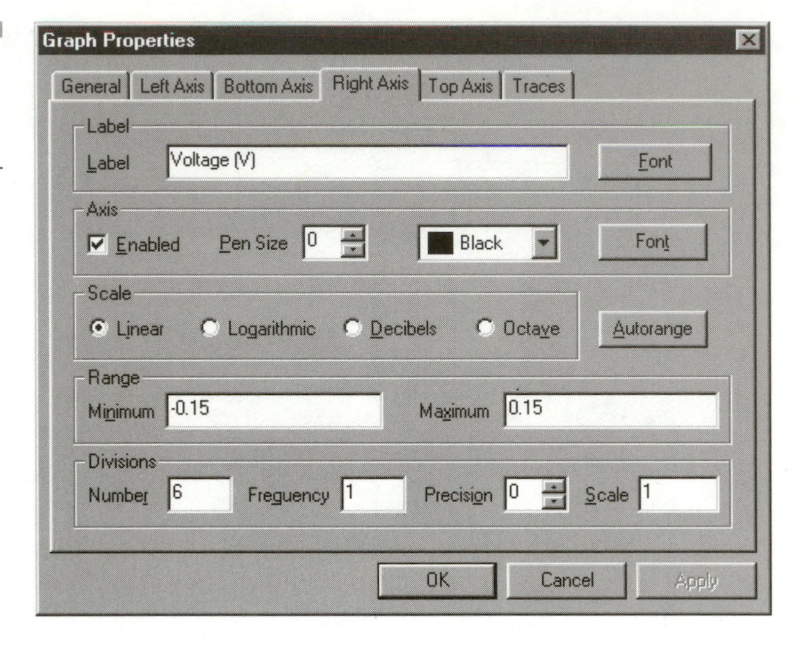

Figure 12.43
Right Axis page of the Graph Properties dialog box.

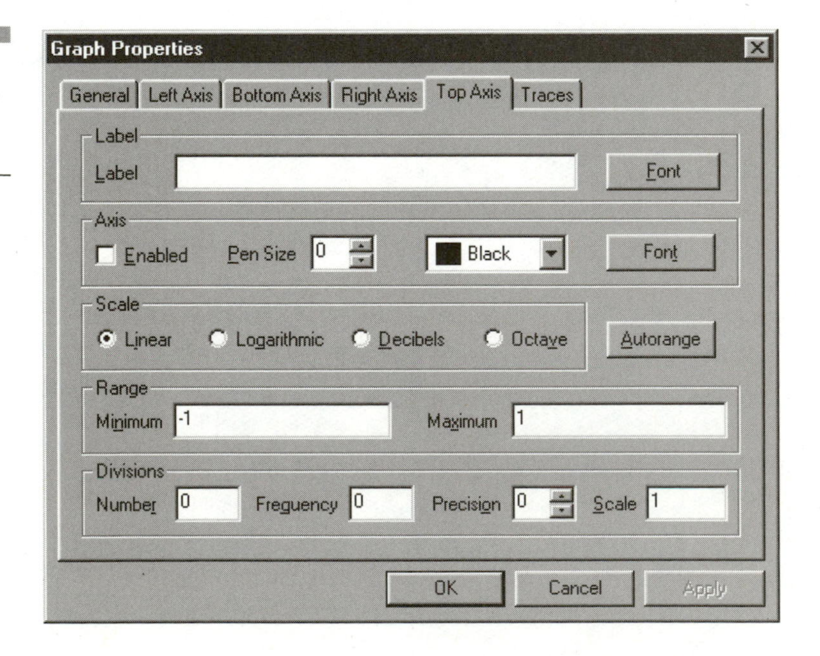

Figure 12.44
Top Axis page of the Graph Properties dialog box.

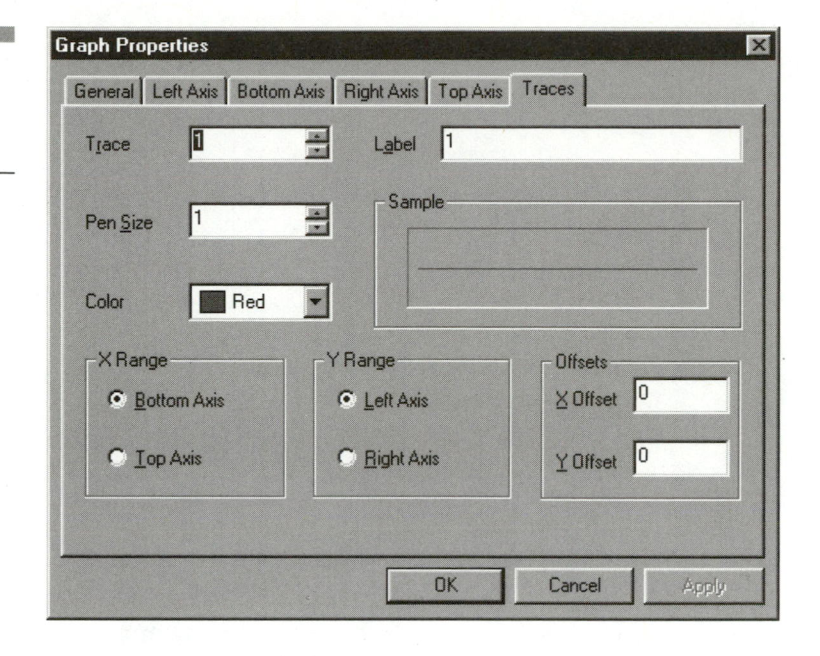

Figure 12.45
Traces Page of the Graph Properties dialog box.

CREATING A NEW PAGE

To create a new page to paste other graphs to, hit the New icon. A window pops up, asking you to name the tab. Type in a name. This creates a new document. To make a title on the page, click once onto that tab; hit the **Properties** icon. Type a name into the title box and select your font. Pick the color you want the page to be. Hit **OK**.

CUT/COPY AND PASTE GRAPHS OR PAGES

You can cut or copy a graph (or a zoomed-in section of a graph) and paste it onto another page. After creating a new page, you can take existing graphs and cut or copy them, then paste them into the new page (see Figure 12.46).

- **Cut**—Select the graph you want to cut, indicated by a red triangle on that graph. Hit the Cut icon or choose **Cut** from the right-click menu.

NOTE

Keyboard shortcuts won't work in the Grapher.

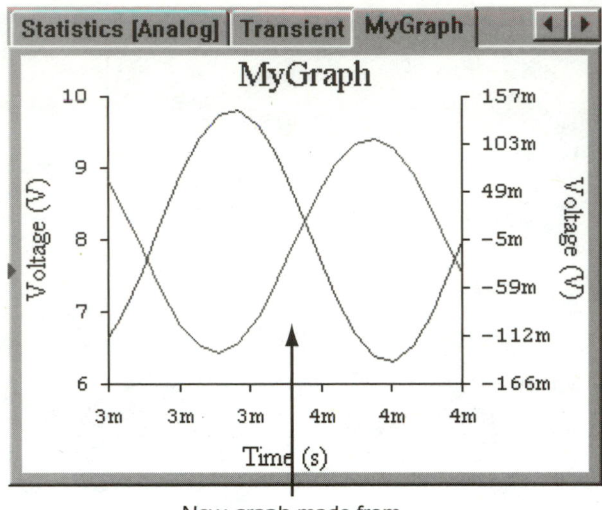

Figure 12.46 Create a new page to paste graphs (or sections of graphs) into.

- **Copy**—Select the graph you want to copy, indicated by a red triangle is on that graph. Hit the Copy icon or right-click and choose **Copy**.
- **Pasting**—Choose the page you want to paste to by hitting its tab. Click the Paste icon or right-click and choose Paste.

PRINTING

You can send one or all graphs to a printer for hardcopy. If you want only one graph printed, find page using the Print Preview and only print that page with the printer control. Hitting the Print icon is the only way to send the graphs to the printer; keyboard shortcuts only access the main EWB program. The Printer dialog box opens. Adjust the settings and choose **OK** to print.

Some menu items are available by placing the mouse over an area and hitting the right mouse button and bringing up the hot menus. Experiment with this.

PHASE GRAPHS

On a phase graph, you may have noticed a steep line in the middle of a waveform (see Figure 12.47). This means the graph is going past a specified amount of degrees (y-axis) and is starting off on the opposite end of the graph. Electronics Workbench chose to do this instead of enlarging a graph to enormous proportions. Try to picture the graph as being one continuous waveform.

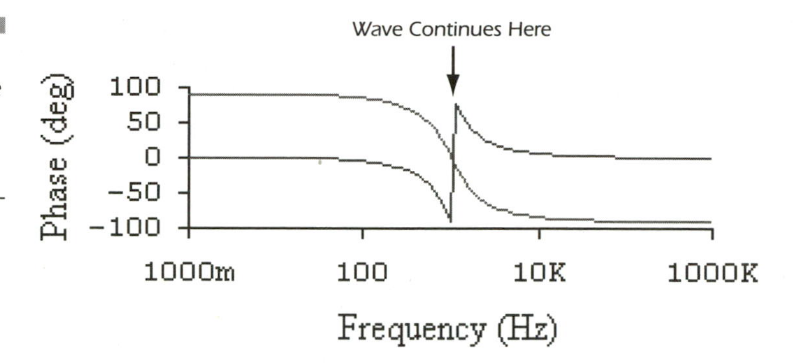

Figure 12.47 The wave on a Phase graph may be split in order the fits. See text.

Working with Cursors

When a graph is selected, a small window opens when you hit the Toggle Cursor icon on the toolbar (as pictured in Figure 12.48). Let's call this the *cursor readout* window. Back at the graph you will notice the two cursors running top to bottom. Cursor 1 and cursor 2 happen to be the same color and are only distinguishable by moving them and observing their results on the cursor readout. You can move these cursors back and forth across the graph and take readings at those points by placing the mouse pointer over the triangle section at the top, left-clicking, and holding it. Drag the mouse left or right to move it.

To see the readouts of each cursor's position, first select a trace. Click over the trace you wish to view and the readings appear at the top of the cursor readout window. To view the numbers on other traces, alternately click the pointer over those. Here is a list of what each item in the window means:

- **x1 and y1** = coordinates for cursor 1 (red).
- **x2 and y2** = coordinates for cursor 2 (blue).
- **dx** = x-axis delta between the two cursors (i.e., difference between cursor readouts 1 and 2 on the x-axis).
- **dy** = y-axis delta between the two cursors. (i.e., difference between cursor readouts 1 and 2 on the x-axis).
- **1/dx** = reciprocal of the x-axis delta.
- **1/dy** = reciprocal of the y-axis delta.

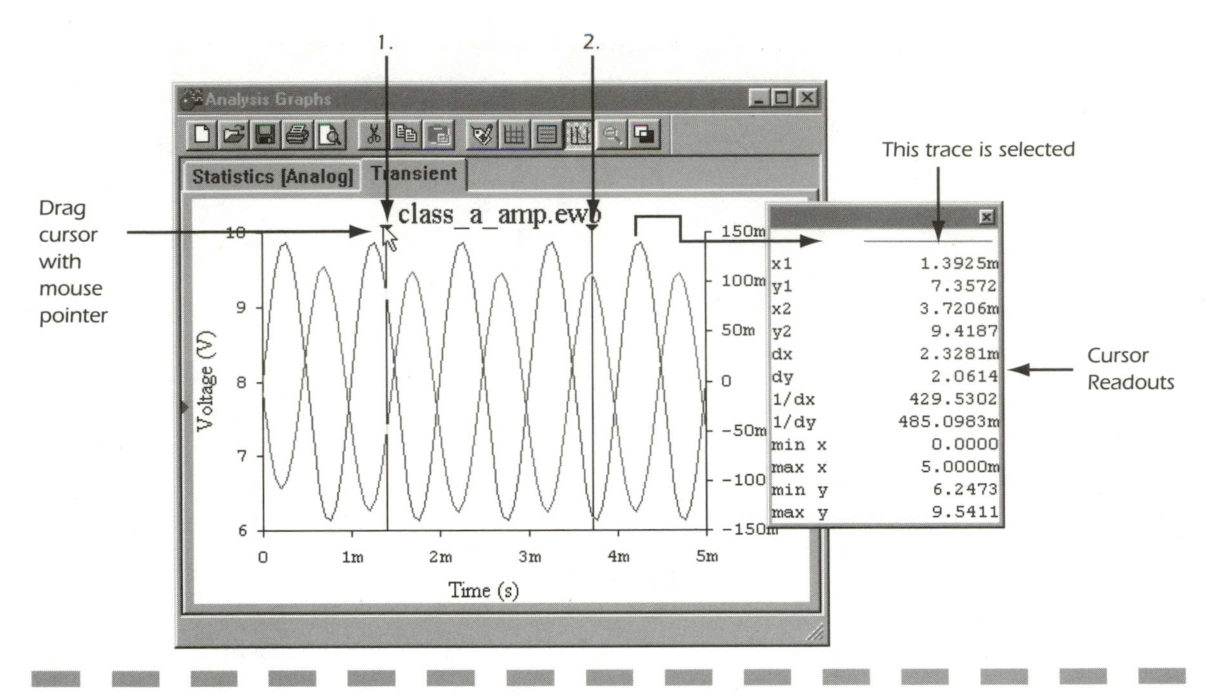

Figure 12.48 The cursors and cursor readout window are used to take numeric readings of a graph.

- **min x, min y** = x and y minima within the graph ranges (i.e., the lowest reading on the x or y axis).
- **max x, max y** = x and y maxima within the graph ranges (i.e., the highest reading on the x or y axis).

Say, for example, you want to know the exact voltage of the waveform at 1ms (see Figure 12.49). Grab cursor 1 and move it until **x1** reads approximately 1.0000m. What does **y1** show as the voltage? Another cool thing about the cursors is that you can figure out the peak voltage in a split second. See it as **min/max y**?

***1/dx** can be used as a simple frequency counter. Place cursor 1 at the start of a cycle and cursor 2 at the end of a cycle and take the reading. In our example (Figure 12.50) the frequency is 1,447 Hz or 1.4470 KHz.*

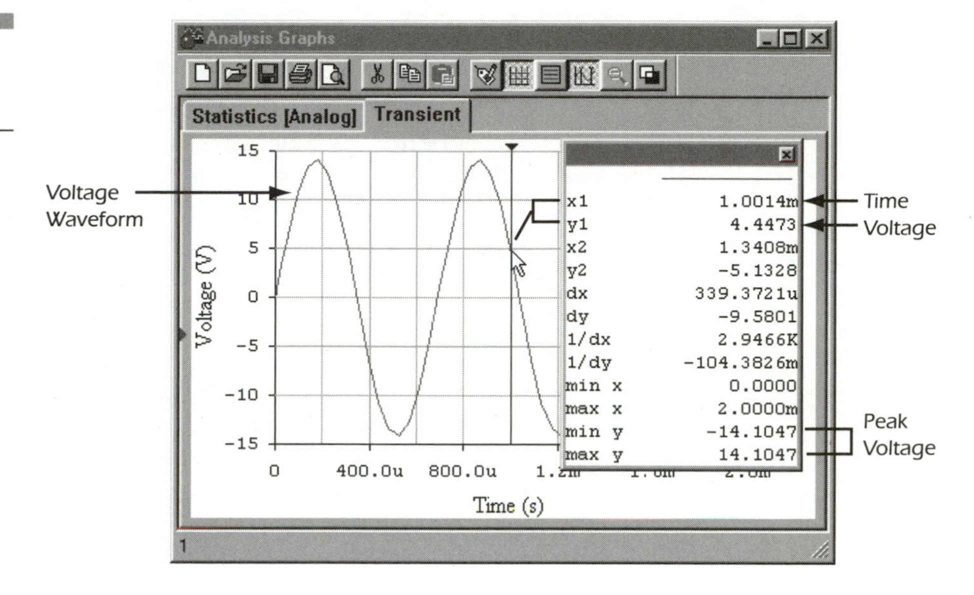

Figure 12.49 An example of how to use cursors.

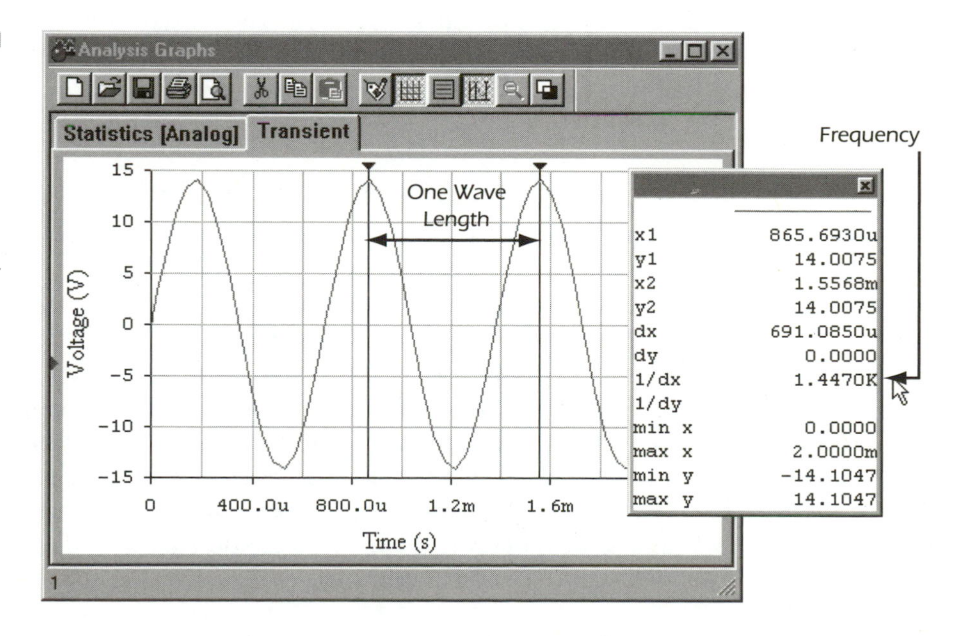

Figure 12.50 Place the cursors one wavelength away from each other and use the 1/dx reading as a frequency counter.

Tweaking Simulations and Analysis

GETTING DIFFICULT SIMULATIONS TO RUN

Electronics Workbench can be a "bang-the-keyboard" program at times, usually because you are not able to get a simulation or analysis to actually work the way you think it should. This often results in annoying, cryptic error messages. I would estimate that 50 percent of the simulations I create with EWB 5 don't work the first time the switch is flipped. In fact, when working on any mildly complex simulation, I've come to expect virtual electrical sparks to fly. This is not always due to poor design but rather a matter of telling Electronics Workbench exactly what is happening or tweaking the simulation's parameters until all is working correctly. The following are a few observations and hints I have come across. (Also see the "Debugging Circuits" section in Chapter 14 for more suggestions.)

GROUNDS

Make sure a ground is placed somewhere in the circuit to help EWB reference the circuit's readings.

CONNECT ALL LINES

Make sure all lines that are supposed to be connected actually are. Sometimes EWB5's autorouting feature makes it appear that a line is connected when it really isn't. Also, if you have been editing connectors, an entire line may have disappeared without you even knowing it. Go over each wire, remembering that a junction (connector) is always present only when an intersecting line is actually connected.

FAILURE TO CONVERGE IN DC OPERATING POINT

Electronics Workbench uses a certain method to analyze circuits, called the Newton-Raphson algorithm. I won't tell you the gory details, but it basically guesses at the solution to a circuit according to all its voltage and current information and then tries to solve the circuit. If it doesn't quite come out right, it repeats this procedure again and again (*iterations*). If the voltages and currents finally come together within a certain level, convergence occurs. If it doesn't match the user-defined convergence parameters within a set number of repetitions, then the circuit "Fails to Converge." Electronics Workbench usually gives a cryptic error message such as "singular matrix," "Gmin stepping failed," "Source stepping failed" and "Iteration limit reached."

To correct these errors, EWB recommends the following:

1. Check the circuit topology and connectivity. Make sure that:
   - The circuit is correctly wired and includes no dangling nodes or stray parts.
   - You haven't confused zeros with the letter O.
   - Your circuit has a ground node, and every node in the circuit has a DC path to ground. Make sure no sections of your circuit are completely isolated from ground by transformers, capacitors, etc.
   - Capacitors and voltage sources are not in parallel.
   - Inductors and current sources are not in series.
   - All devices and sources are set to their proper values.
   - All dependent source gains are correct.
   - Your models and subcircuits have been correctly entered.
2. Increase operating point analysis iteration limit to 200—300. This allows the analysis to go through more iterations before giving up.
3. Reduce the RSHUNT value by a factor of 100.
4. Increase the Gmin minimum conductance by a factor of 10.
5. Enable the option **Use zero initial conditions**.

TRANSIENT ANALYSIS TROUBLES

Timestep too small: this is the most common error message I get. Here is what IIT recommends:

"If transient analysis is being performed (time is being stepped) and the simulator cannot converge on a solution using the initial time step, the time step is automatically reduced, and the cycle is repeated. If the time step is reduced too far, an error message ("Timestep too small") is generated and the simulation is aborted. If this occurs, try one or more of the following:

1. Check the circuit topology and connectivity.
2. Set relative error tolerance to 0.01. By increasing the tolerance from 0.001 (0.1 percent accuracy), fewer iterations are required to converge on a solution and the simulation finishes much more quickly.
3. Increase transient time point iterations to 100. This allows the transient analysis to go through more iterations for each time step before giving up.

4. Reduce the absolute current tolerance, if current levels allow. Your particular circuit may not require resolutions down to 1 mV or 1 pA. You should allow at least an order of magnitude below the lowest expected voltage or current levels of your circuit.
5. Realistically model your circuit. Add realistic parasitics, especially junction capacitances. Use RC snubbers around diodes. Replace device models with subcircuits, especially for RF and power devices.
6. If you have a controlled one-shot source in your circuit, increase its rise and fall times.
7. Change the integration method to gear. *Gear* integration requires longer simulation time, but is generally more stable than the trapezoid method."

BREAKING THE CIRCUIT INTO SUBCIRCUITS

If a rather complex circuit refuses to run, begin by dividing it into subcircuits and chart their results. You can cut out the subcircuit and create a new document until that section runs correctly. Then paste each of the working sections into a new document.

RESTARTING ELECTRONICS WORKBENCH

Electronics Workbench will often outright refuse to perform a procedure. If all else fails, try restarting the program. When certain memory errors occur, I also recommend saving all your work and resetting the entire system.

EWB SUPPORT AND HELP

http://www.electronicsworkbench.com/ contains a Technical Support section, which contains a list of common simulation and analysis problems. Look under the Analysis and Simulation sections. Bookmark those pages and visit often for new information.

No software program is perfect. Minor improvements must be added and bugs squashed. EWB periodically sends an updated version to registered customers. For example, Electronics Workbench Version 5.12 was the last produced before Multisim 6 came out. Contact EWB for more version number information. As a last resort, you can have an EWB tech staff member look at your circuit. Allow time to answer your e-mail, because they get busy at times. There is an e-mail form at **http://www.electronicsworkbench.com/html/support_form_product.html**.

NOTE

EWB5 is a very stable program. I recommend that Multisim owners use it when having problems with version 6. I mention this so that you are aware that glitches in the setup of the simulation/analysis and not software bugs may be causing your woes.

Making the Most of Your Analyses

Electronics Workbench's Analysis is only as useful as you make it. Learning the different analyses is simple; just open circuits you have made or found in books and practice on them. EWB's website provides primers and test circuits to help with this also. (I post test circuits at **http://www.basicelectronics.com/ewb/**.)

Which Analysis Should I Use?

Most newcomers to EDA are intimidated by the prospect of actually using the Analysis features in Electronics Workbench 5. It's no wonder, with the complex engineering explanations the Electronics Workbench company and other sources give. The main analyses (DC op, AC, transient, and DC sweep) are actually quite easy to use once you get the hang of them. After learning a bit about them, it's simply a matter of choosing which instruments you would have hooked into the circuit and replacing them with the analyses. This makes for quicker, more accurate, customizable results. Here are a few replacements:

- **Multimeter**—Use DC operating point to determine steady voltage levels as well as current.
- **Oscilloscope**—A transient analysis replaces an infinite number of scopes when set up correctly. This allows more traces to be plotted, and permits you to narrow the readings down to only those sections you really need.
- **Bode Plotter**—Although the Bode plotter is a great addition to the EWB V5 software, it can be a pain to set up. Running an AC analysis instead speeds the process.

Analyses can also replace certain components or subcircuits. For example, DC sweep can be used in place of a potentiometer, battery, and resistors in a variable voltage source configuration.

Which analysis do I personally make the most use of and why? By far, the winner is transient analysis because it is much faster than hooking in an oscilloscope and futzing with its settings; I also make use of DC operating point, DC sweep, and AC analysis (in that order).

Summary

Analysis can be one of the most complex sections of Electronics Workbench to learn. Once you figure it out, however, the Analysis menus will become the most used. The next chapter makes use of everything you have learned by helping you build some test circuits.

CHAPTER 13

Twenty-five Circuits in EWB V5

Building a circuit piece by piece reinforces your understanding of the whole.

There is no substitute for hands-on knowledge. Constructing an object with your own hands (be it with tools or a mouse and monitor) helps to seat concepts in the mind. Building the circuits in this chapter will progressively give you a better understanding of the workings and capabilities of Electronics Workbench 5. The circuits are designed to take you from simple "get used to the software" projects, to more complex simulations. Follow the text as well as possible but don't be afraid to experiment with your own alterations, tests, or designs. Most of the following circuits come from my twisted mechanical mind but some are compiled through other various sources. Open up EWB and let's build some imaginary gadgetry.

Building Them Yourself

Instead of just opening these circuits from the companion CD-ROM, I recommend building each circuit as you see it in the diagrams, then running it. The whole point of this chapter is to learn the workings of the program; each circuit contains custom settings that must be adjusted or certain analysis that must be performed (see each circuit for details). If you are having trouble making out a certain feature of the circuit, open it up from the CD-ROM and take a peek, then go back to building it yourself.

Where to Get More Circuits to Build

Just about any circuit that can be built in real life can be constructed in a digital domain. Electronics Workbench 5 can be used to test circuits you may find in books or magazines, so that you can carry them over into your own projects. I recommend subscribing to a few electronic periodicals and collecting circuits that way. Also, Tab Electronics puts out several great series of books that contain circuit cookbooks. There is the *Encyclopedia of Electronic Circuits* (Volume 1—6), *The Master Handbook of IC Circuit Applications, The McGraw-Hill Circuit*

Encyclopedia and Troubleshooting Guide (Volume 1—4), and many more. You can also search the Internet with the keyword, "circuits."

NOTE

Visit the new Circuit Knowledge Center at ***http://www.electronicsworkbench.com*** *for more EWS5 circuits.*

CD-ROM

The included CD contains all of Chapter 13's circuits. They can be opened from the CH13 folder once the EWB5 demo is loaded. See the CD's readme.txt file under CD_DRIVE_LETTER:\readme.txt for further information or any changes in the current edition. When loading the CD, there will be further instructions as well.

Ohm's Law (OHMS1.EWB and OHMS2.EWB)

For the first circuit, I have chosen a build-it-in-two-minutes lesson. Walking through this first circuit helps you bring a simple drawing together.

1. Open EWB 5 and follow along. You should know how to place components and wire them together, as we saw in the previous chapters on Electronics Workbench.
2. Set up the document to display every label and node available by right-clicking the empty drawing and choosing **Schematic Options** > **Show Hide** tab; check-on every option under Display.
3. Place a DC voltage source (battery) in the middle of the screen.
4. Place a ground below the battery, a few grid segments below.
5. Place a resistor above and to the right of the battery. It will default to 1kOhm.
6. Place an ammeter to the right of the resistor. All the components are placed (see Figure 13.1). Now it is time to connect them.
7. Connect the negative terminal of the battery to the ground.
8. Connect the positive terminal to the left terminal of the resistor.
9. Connect the right terminal of the resistor to the positive terminal of the ammeter.

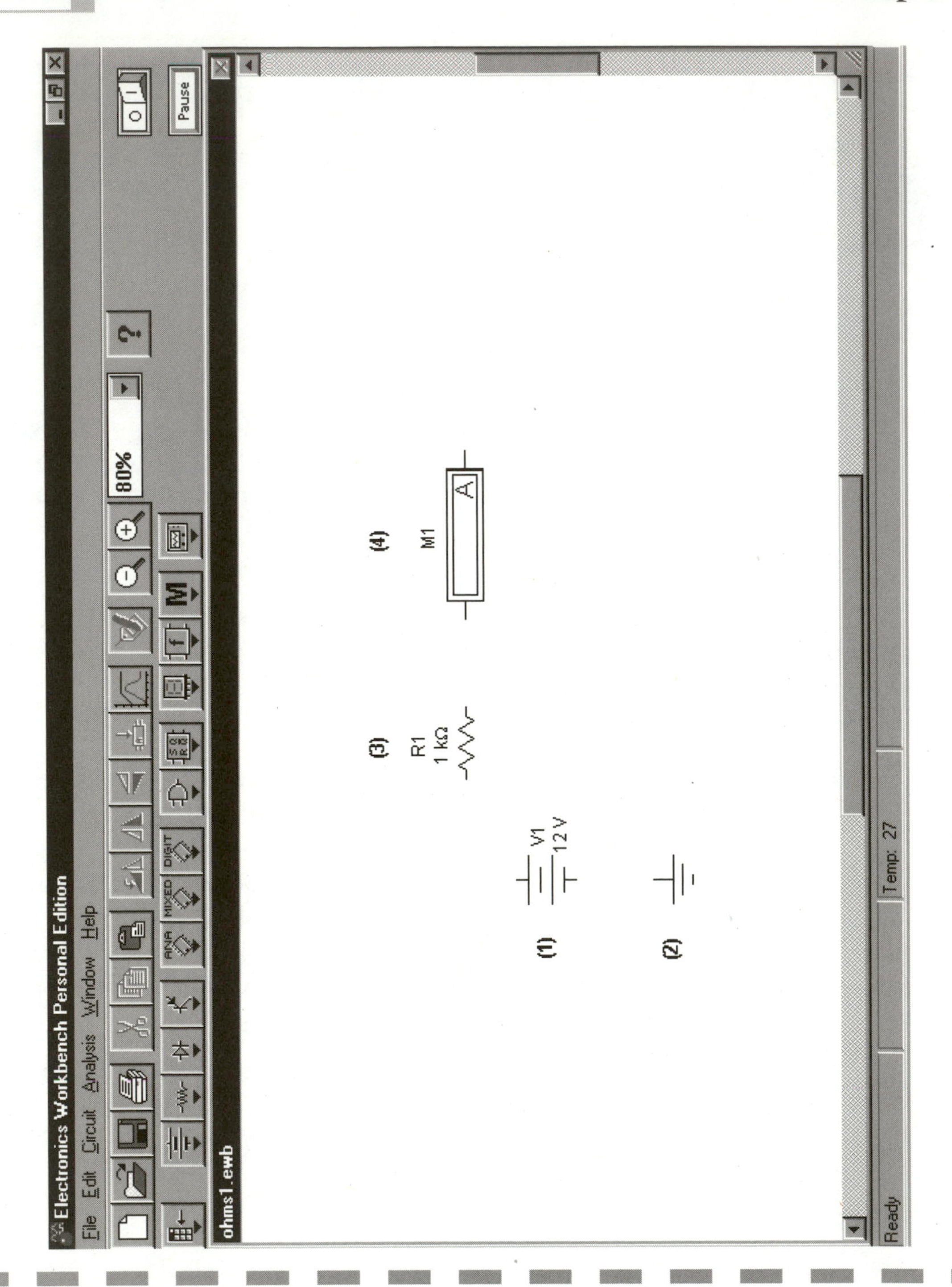

Figure 13.1 Building OHMS1.EWB.

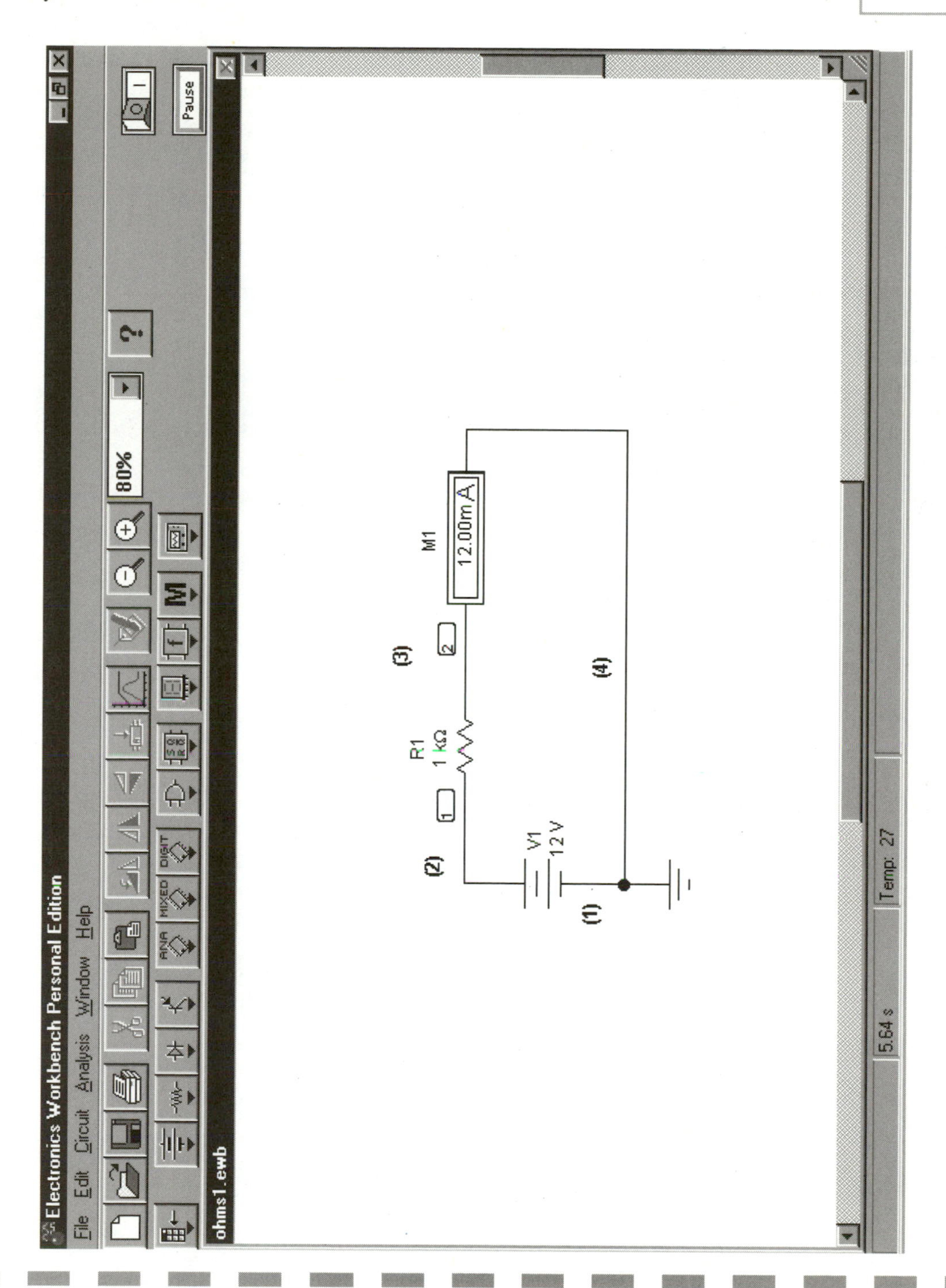

Figure 13.2 OHM'S LAW, PART I—OHMS1.EWB.

10. Connect the negative terminal of the ammeter to the center of the ground wire, thus creating a connector there. The circuit should look like Figure 13.2.
11. Activate the circuit and note the amperage going through the circuit. Is this a correct reading? Remember I = V/R. Deactivate the circuit.

NOTE

The nodes were named in the order you wired the circuit.

12. Save the circuit as Ohms1.ewb.
13. Now let's change the resistor in the circuit to a potentiometer. Delete the resistor.
14. Place a potentiometer into the circuit, as shown in Figure 13.3.

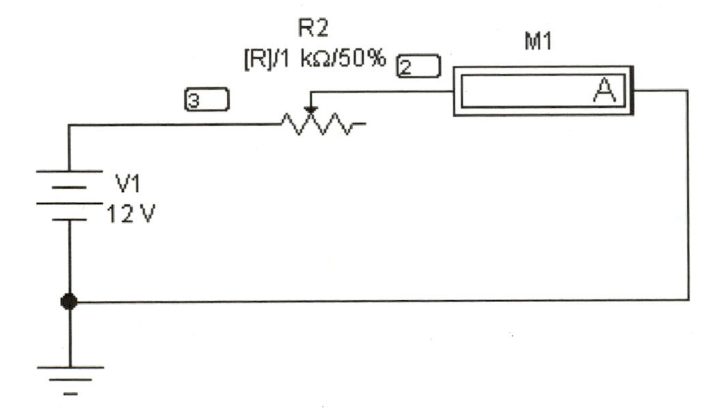

Figure 13.3
OHM'S LAW, PART II—OHMS2.EWB.

15. Activate the circuit and note the ammeter reading.
16. Hit the 'r' key on the keyboard once and watch the readings. Hit it again and again and note the readings. Hit **Shift+R** now.
17. Deactivate the circuit and save it as Ohms2.ewb.

That's it.

Voltage Divider (VOLT_DIV.EWB and VOLT_DIV2.EWB)

Resistors can be used to derive various voltages from a single source. If the ground is offset, you can even produce a negative voltage. In the circuit in Figure 13.4, we are able to create four voltage levels using three resistors in series. The circuit in Figure 13.5 uses two resistors to produce a +6V and −6V level.

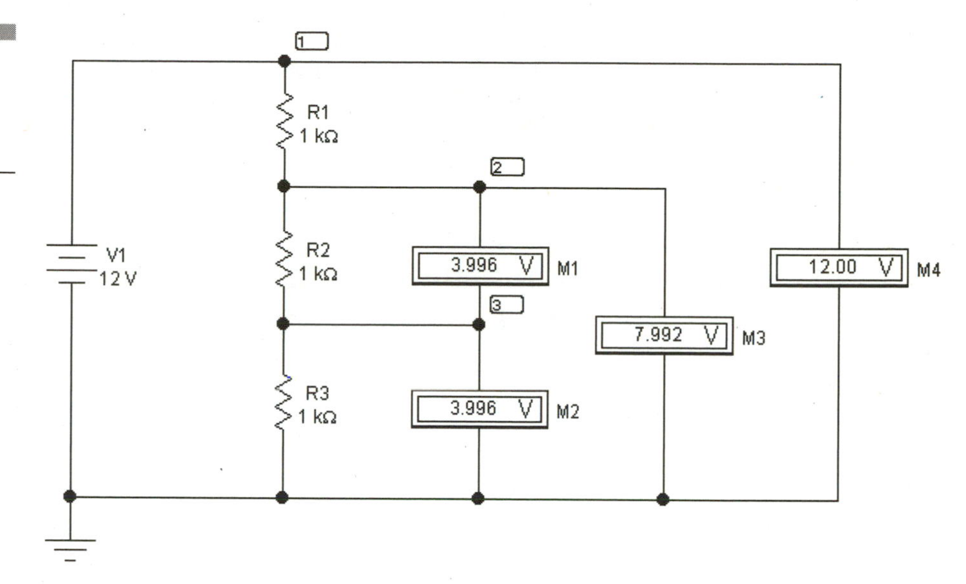

Figure 13.4 VOLTAGE DIVIDER, PART I—VOLT_DIV.EWB

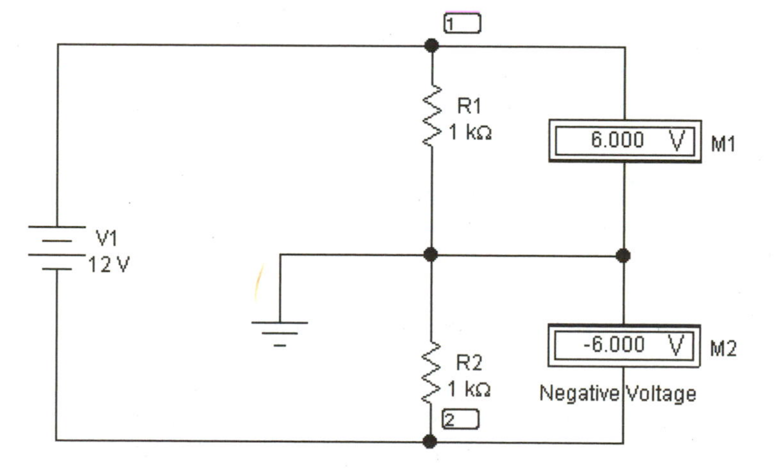

Figure 13.5 VOLTAGE DIVIDER, PART II—VOLT_DIV2.EWB

This next circuit's instructions are more condensed, now that you know the simple procedure of building and activating the circuit. Let's start with the circuit in Fiugre 13.4.

1. Place all components as seen in Figure 13.4. Make sure the voltmeters are in the correct direction (bold line is negative side).

2. Wire the components together as shown, in the order the nodes are shown in on the schematic. Remember node 0 is ground.
3. Activate the circuit and note the readings on the four voltmeters.
4. Deactivate the circuit and save it as Volt_Div.ewb.
5. Perform a DC analysis by selecting all of the nodes in the circuit. Do the voltages match the ammeter readings?

Now build the circuit as seen on the left in Figure 13.5 and run the simulation and DC analysis again. Save the circuit as Volt_Div2.ewb.

- **Custom Settings**—Make sure **Show Nodes** is set in the **Schematic Options**.
- **Analysis**—Perform a DC operating point analysis.

Kirchhoff's Current Law (KIRCHHOFF1.EWB)

The remaining circuits will be simply described. You should have a basic idea of how to build them now.

Kirchhoff's Current Law (Figure 13.6) states "The sum of all the currents flowing to a point in a circuit must be equal to the sum of all the currents flowing away from that point." We have input a 1-ampere current source into three resistors in parallel. Each branch's ammeters give you the amount current flowing through each path. When they are recombined, another ammeter is used to prove this law.

- **Custom Settings**—None.
- **Analysis**—None.

Kirchhoff's Voltage Law (KIRCHHOFF2.EWB)

Kirchhoff's Voltage Law states "the sum of all the voltage sources acting in a complete circuit must be equal to the sum of all voltage drops in that circuit." We have provided a 10-volt battery as voltage source. It is connected to four resistors in series and the voltage drops are recorded with voltmeters hooked to each. Do the voltage drops add up to the voltage source? See the circuit in Figure 13.7.

- **Custom Settings**—Label the nodes A, B, C, D, E, F or place text in proximity.
- **Analysis**—Perform a DC operating point analysis to confirm readings.

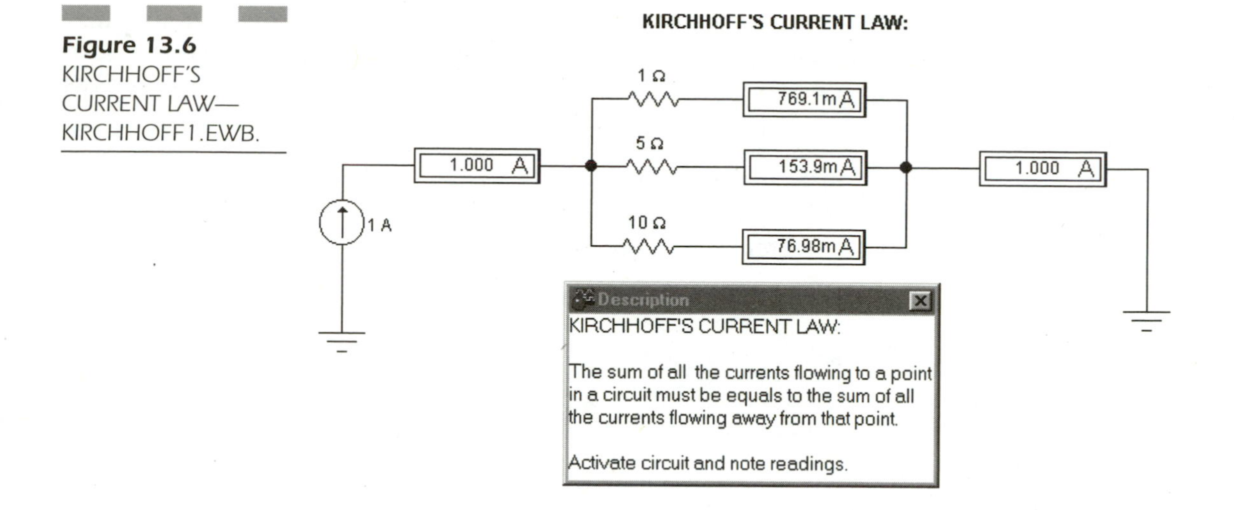

Figure 13.6 KIRCHHOFF'S CURRENT LAW—KIRCHHOFF1.EWB.

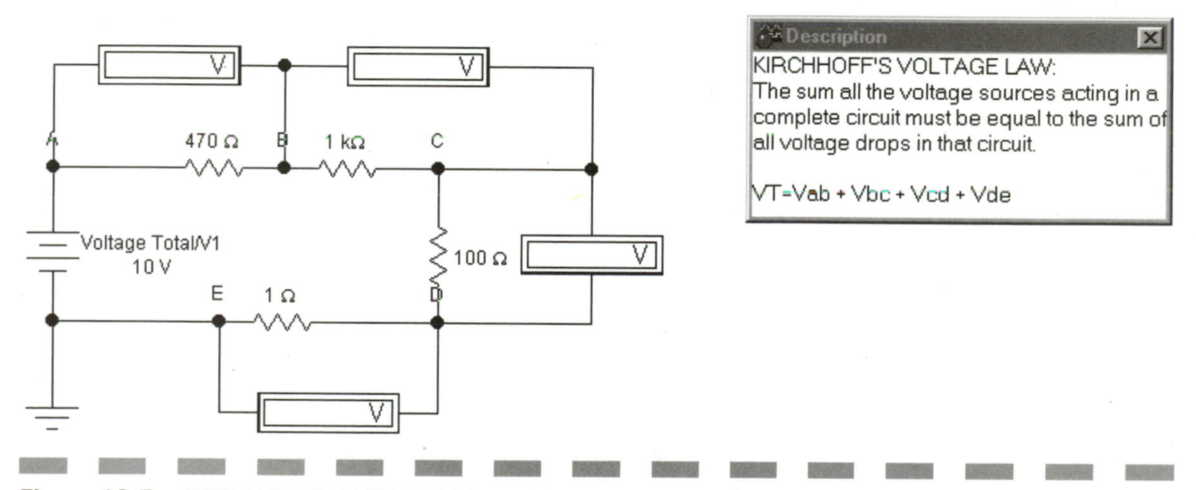

Figure 13.7 KIRCHHOFF'S VOLTAGE LAW—KIRCHHOFF2.EWB.

Wheatstone Bridge (Wheatstone.EWB)

Build the circuit as pictured in Figure 13.8. This circuit consists of three resistors and a potentiometer in a bridge configuration. The Wheatstone bridge is often used to find an unknown resistance. When the bridge is balanced (V1/V3 = V2/V4) then the voltage on the multimeter should read near zero volts. If R1 = R3 * (R2/R4). At what percentage does R1 need to be set to balance the bridge? Remember each 5 percent change of the POT is 50 ohms. Figure it out and set it to that figure using the 'r' or 'R' key. Does the multimeter read near zero volts now?

- **Custom Settings**—Set the POT to use the R key. Adjust the multimeter to read DC voltage. Try the voltmeter instead of the mulimeter; it's a faster way to take readings.
- **Analysis**—None.

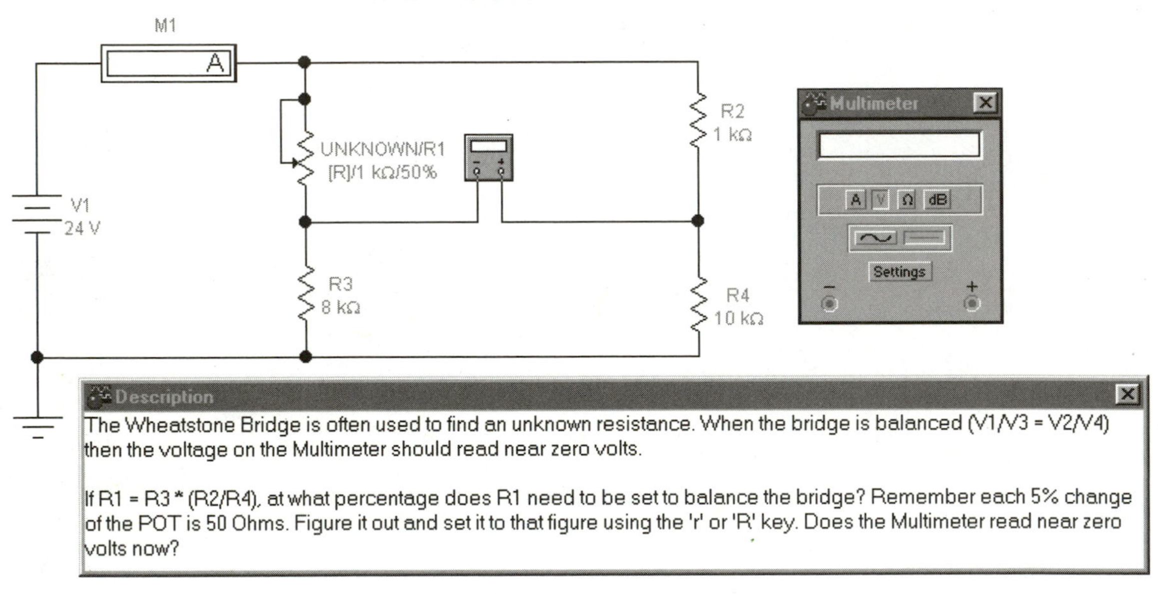

Figure 13.8 WHEATSTONE BRIDGE—WHEATSTONE.EWB.

Resistor/Capacitor (RC.EWB)

Build the circuit as pictured in Figure 13.9. A resistor and capacitor are often used in conjunction with each other in a DC circuit as a timing device. Go to **Analysis** > **Analysis Options** > **Instruments** tab and set the simulation to **Pause after each screen**. When the circuit is built, open the oscilloscope, and activate the circuit with the simulation switch. You will see the 10V battery begin to charge the capacitor through R1. The scope will pause at the end of the screen. Click the pointer into the circuit window somewhere and hit the **Space** bar; this switches the RC circuit to a discharge cycle. Click the **Resume** button on the simulation switch and watch the results of the capacitor discharging through R2.

- **Custom Settings**—Under **Analysis** > **Analysis Options** > **Instruments** tab set initial conditions to zero. Adjust the oscilloscope's settings to match the diagram, making sure to have **Pause after each screen** checked on before activating the simulation.
- **Analysis**—Run a transient analysis, setting initial conditions to zero and selecting node 4 (or that node to which the scope's channel A is connected). Set TSTOP to .005 seconds. Simulate it. Toggle the switch and run the same transient analysis.

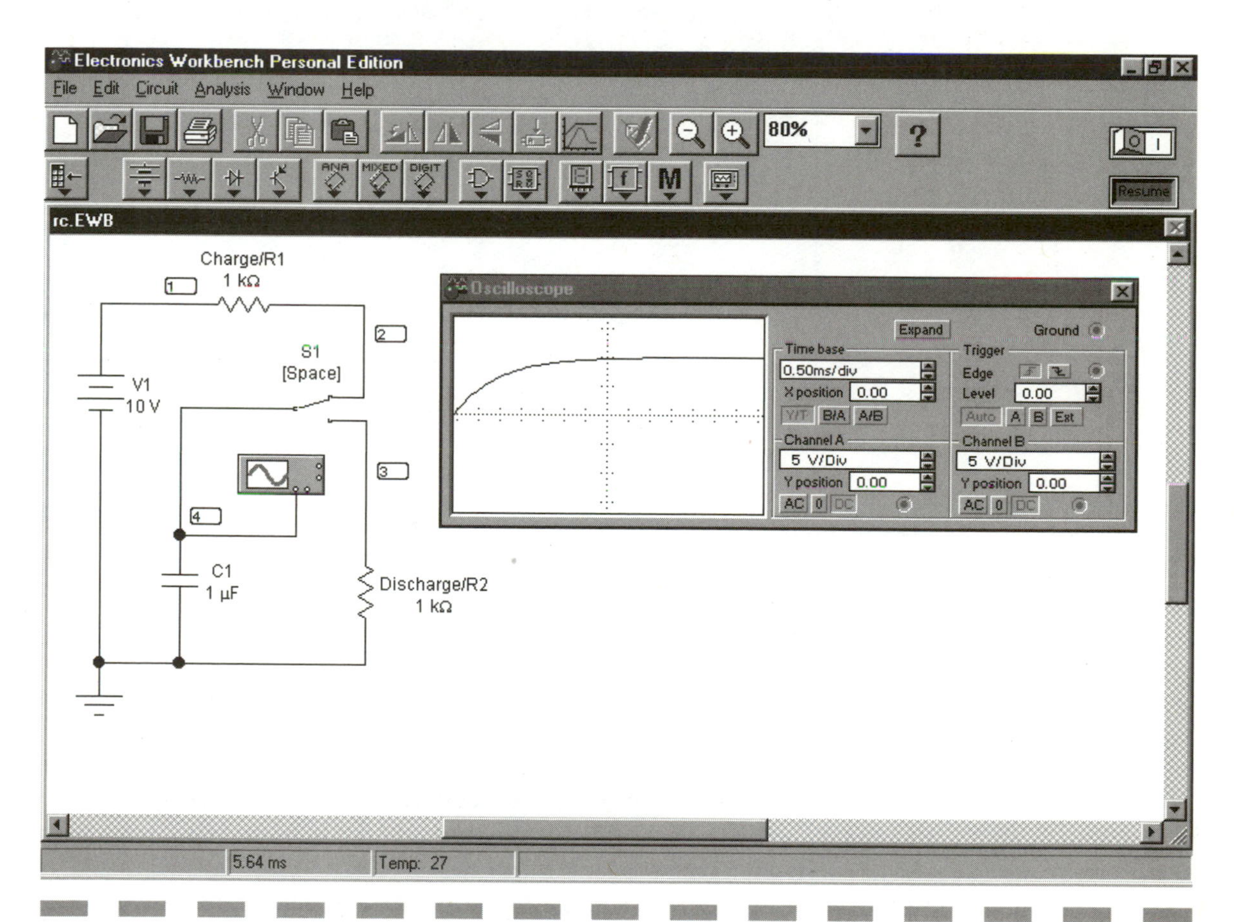

Figure 13.9 RESISTOR/CAPACITOR CIRCUIT—RC.EWB.

Automotive Lights-on Reminder (REMINDER.EWB)

This circuit (Figure 13.10) demonstrates uses for the buzzer, bulb, diodes, and switches all in one. It models an automobile lights-on-reminder. If your headlights are turned on and the ignition key switch is turned off, the warning buzzer will sound to warn you that if you walk away from the car, a dead battery will be waiting in the morning.

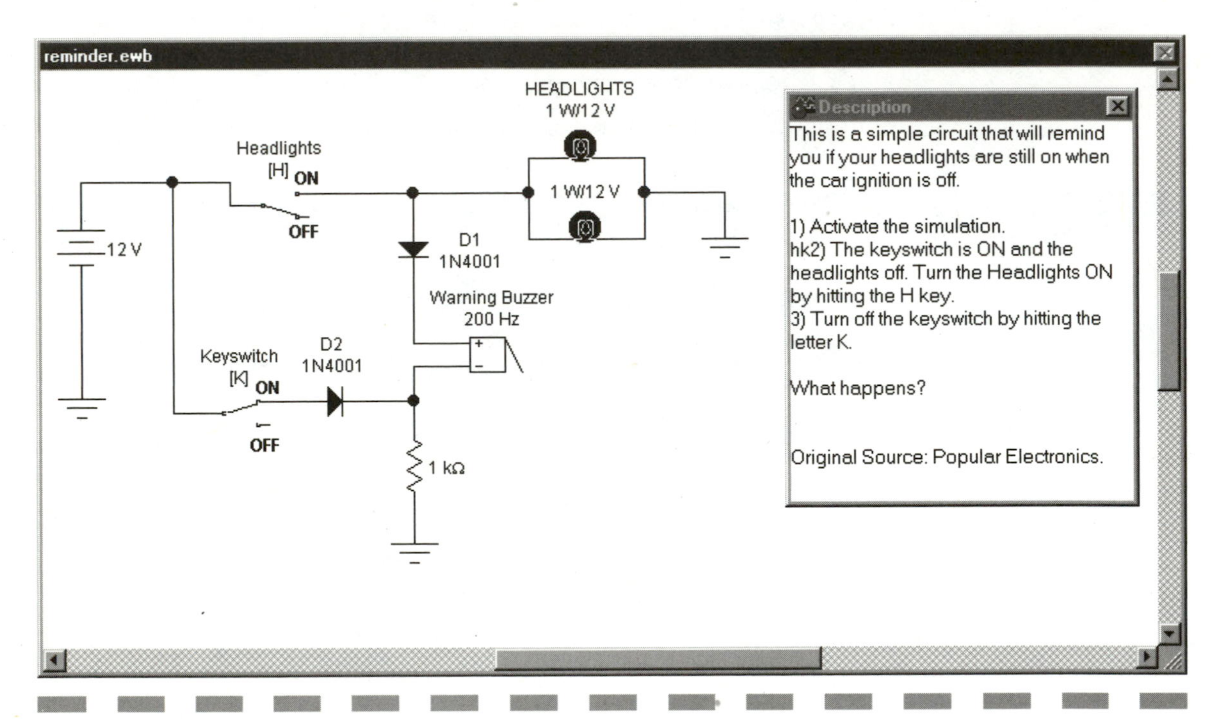

Figure 13.10 AUTOMOTIVE LIGHTS-ON REMINDER—REMINDER.EWB.

The circuit uses two diodes. When both the ignition key switch (letter K) and the headlights (letter H) are on, the buzzer draws no current and remains silent. With only the key switch on, diode 1 (D1) is reverse-bias and will prevent current from flowing through the buzzer. When the headlights are on and the key switch is off, the buzzer will energize and alert the driver to turn the headlights off. With the key switch off and lights on, D2 is reverse-biased, stopping current from flowing to the key switch. The 1K resistor prevents a short circuit when the key switch is on.

To use the circuit, activate it and use the K key to control the ignition key switch and the H key to control the lights. Turn the ignition on and off with the headlight in the off position (switch in the up position is off). Now turn the headlights on, making sure the key switch is also on. Now turn off the headlights. What happens?

- **Custom Settings**—Assign the headlight switch to the H key and the key switch to the K key. Set the buzzer to 9V and 1 mA, 200 Hz.
- **Analysis**—None.

This circuit first appeared in *Popular Electronics*.

Two-transistor Multivibrator (MV.EWB)

This is a demonstration of a classic *astable multivibrator* (oscillator) circuit built with two transistors, two capacitors, and four resistors (see Figure 13.11). Electronics Workbench's oscilloscope needs some fiddling to make this circuit work, so I have included instructions on both the scope and a transient analysis.

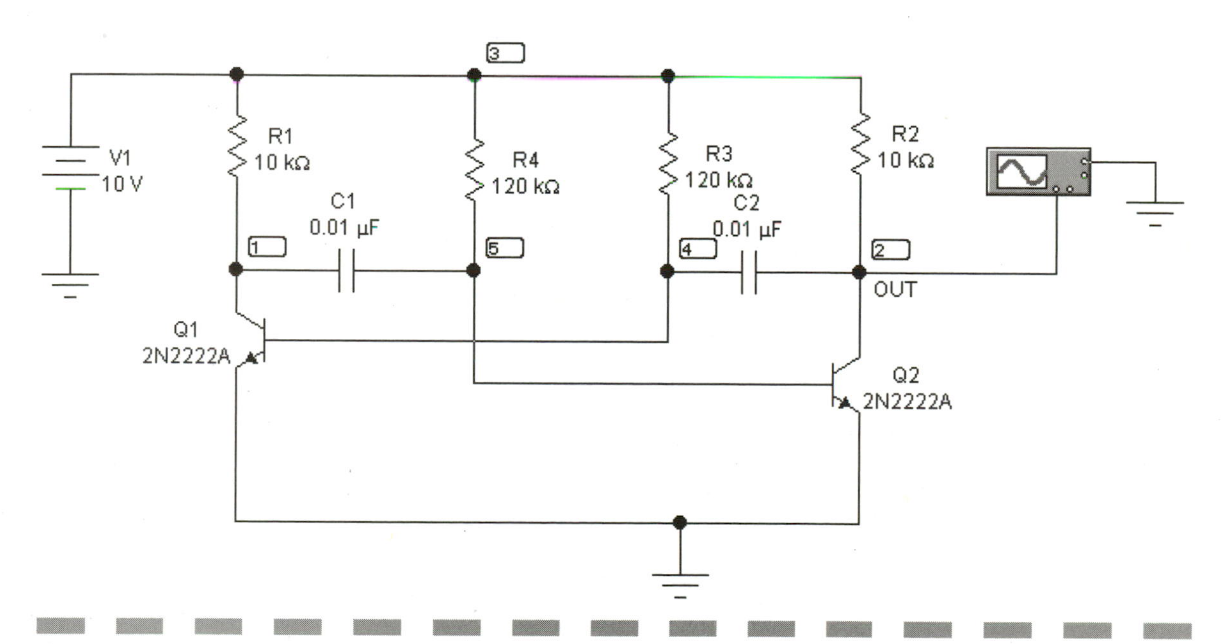

Figure 13.11 TWO-TRANSISTOR MULTIVIBRATOR—MV.EWB.

Build the circuit and adjust the component values as shown. For the oscilloscope:

1. Electronics Workbench will not automatically handle the time steps correctly with certain oscillator circuits so it's time to switch to manual drive. Go to **Analysis** > **Analysis Options** > **Instrument** tab and check on **Minimum number of time points** and set it to **1000**.
2. Open the oscilloscope and set the timebase to 1ms/Div and channel A to 5 volts.
3. Activate the simulation.

You can also use transient analysis to run this circuit (preferable):

1. Go to **Analysis** > **Transient Analysis**.
2. Set the following settings: TSTOP = .01sec, turn off **generate time steps automatically**, and set the maximum time step TMAX to 1e-05 seconds.
3. Choose the Out node as the output variable; in this case it is node 2.
4. Hit the **Simulate** button. The multivibrator will take a moment to stabilize. Use the cursors to determine the frequency (this is the 1/dx reading when the cursors span one wavelength).

- **Custom Settings**—As discussed above.
- **Analysis**—As discussed above.

Simple Push-pull Amplifier (PUSHPULL.EWB)

This circuit (Figure 13.12) uses two BJT transistors to create a simple Class B push-pull amplifier: an npn on the top and pnp on the bottom. The function generator simulates signal input. The 1K-ohm resistor (RL) simulates a load. A switch is used to simulate a resistance, or you can change over to two crossover distortion-eliminating diodes. Figure 13.13 shows the two resulting waveforms. Open the scope and run the simulation, noting the crossover distortion. Hit the spacebar and watch the distortion disappear.

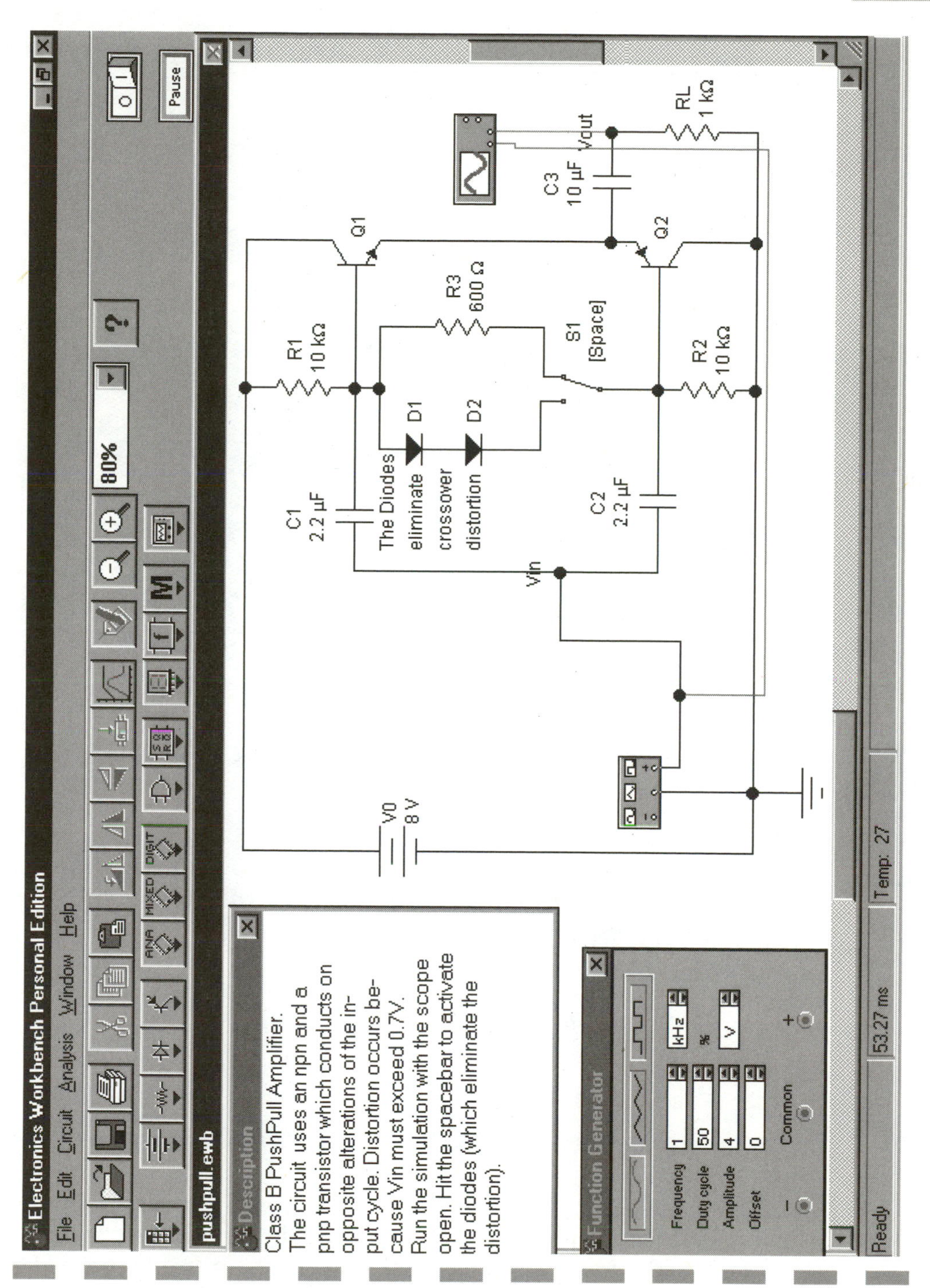

Figure 13.12 SIMPLE PUSH-PULL AMPLIFIER—PUSHPULL.EWB.

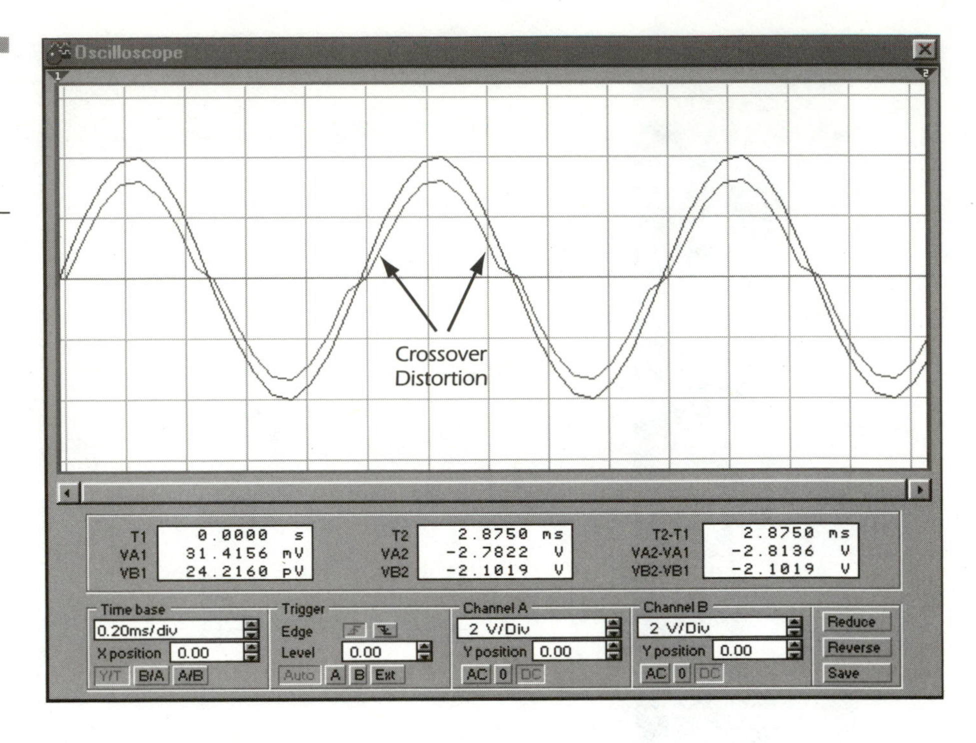

Figure 13.13 Distortion is corrected using two diodes in the push-pull amplifier.

NOTE

*You must change the **Analysis Options** slightly to make this circuit work. Change the **ITL4** setting to **100**, under **Analysis** > **Analysis Options** > **Transient** tab.*

- **Custom Settings**—Make the spacebar control the switch. Set the oscilloscope as seen in Figure 13.13. Set the function generator to 1kHz, 4V, sine-wave.
- **Analysis**—None.

Based on a circuit from *150 Circuits,* 1993 Interactive Image Technologies Ltd.

FET Curves (FETCURVE.EWB)

Have you ever wanted to create current-voltage curves (drain-source characteristic curves) to visualize a certain transistor's properties? This could take hours using a pencil and calculator (see Figure 13.14). With this circuit, we can do it in seconds. It is an excellent example of a two-

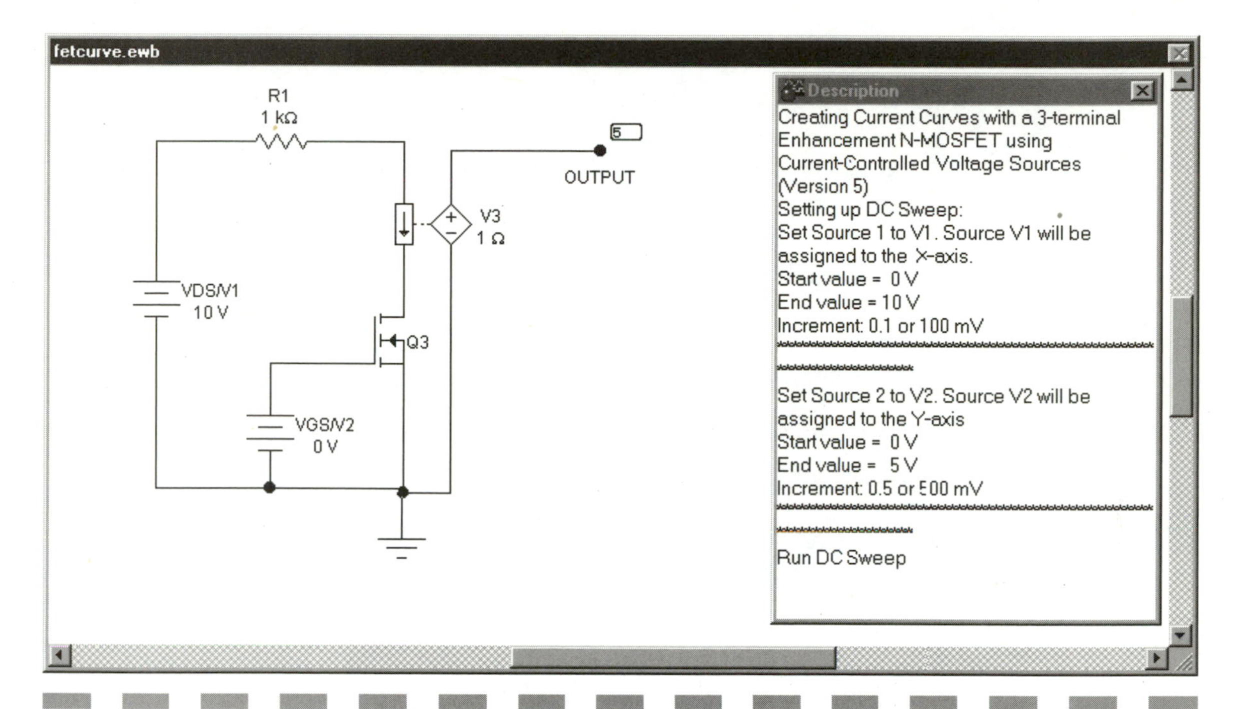

Figure 13.14 FET CURVES—FETCURVE.EWB.

voltage source DC sweep. We are essentially turning V1 (voltage supply) into a current reading output with the current-controlled voltage source (CCVS) then sweeping it. Build the circuit as seen in Figure 13.14. Open the DC sweep analysis and set the parameters described under **Analysis**. Simulate. The Grapher will open, and each colored line will represent a half-volt increase on the gate; this may take a minute or so. Replace the MOSFET with a similar model and rerun the DC sweep. Notice the characteristic changes? Remember you can save the results as graph data.

- **Custom Settings**—None.
- **Analysis**—Under DC sweep, set source 1 as V1, start value as 0V, stop value as 10V, and increment as 0.1V. Check on **Use Source 2** and set source 2 as V2, start value as 0V, stop value as 5V, and increment .5V. Under the Output variable tab, choose 5 as the node to read (or whichever node is the output).

From the Interactive Image Technologies, Ltd. (EWB company) FTP site (**ftp.interactive.com**).

12-Volt Simple Power Supply (POWERSUP.EWB)

Figure 13.15 is an easy-to-build power supply, which will convert 120-volt AC into approximately 12-volt DC. The alternating current is stepped down with the transformer, put through a bridge rectifier (four diodes) and then filtered with capacitors. The DC voltage output will settle at around 16.5 volts. A simple Zener diode steadies the resulting voltage to just below 12V. Beware! This is by no means a design for a real power supply; it is merely to give you an idea how to use the components involved.

- **Custom Settings**—None.
- **Analysis**—None.

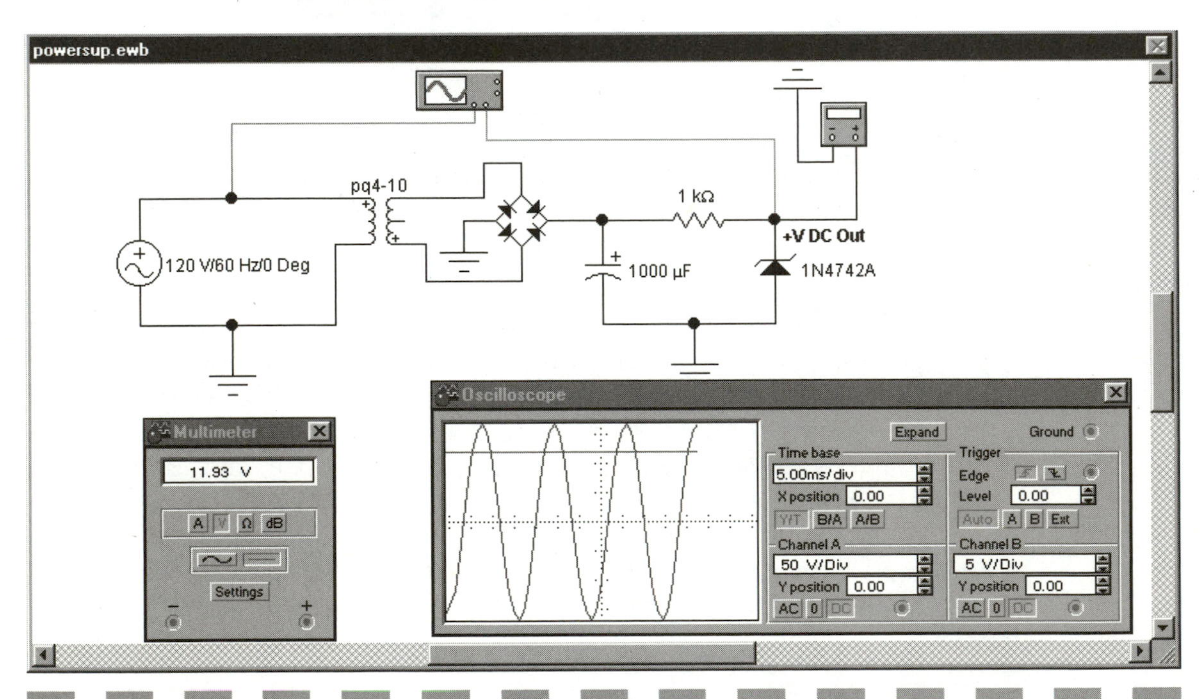

Figure 13.15 12-VOLT SIMPLE POWER SUPPLY—POWERSUP.EWB.

Astable Multivibrator (ASTABLE.EWB)

The circuit in Figure 13.16 demonstrates one of the thousands of uses of the 555 timer. In this case, it is used as a basic oscillator operating at a low frequency. The design is of a .5 Hz oscillator used to light an

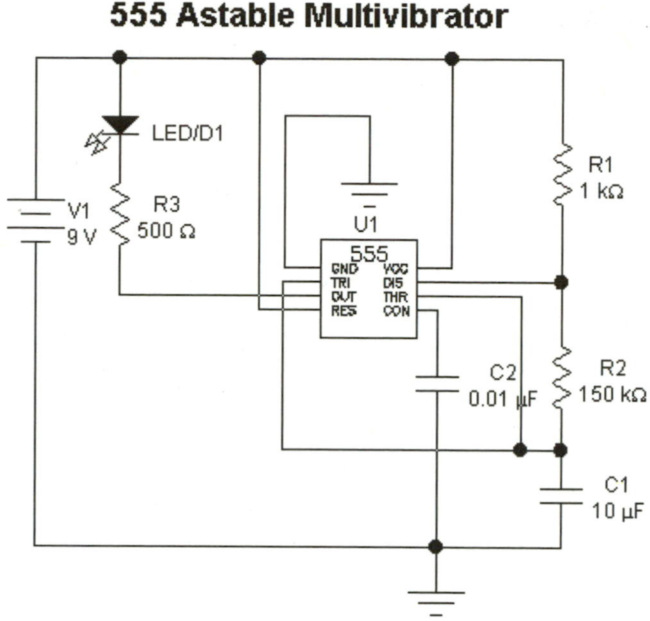

Figure 13.16 ASTABLE MULTIVIRATOR—ASTABLE.EWB.

LED. R2 and C1 control how long the LED will be lit and R1 + R2 and C1 controls the time the LED will be extinguished. Time High (LED OFF) = 0.693 × C1 (R1 + R2). Time Low (LED ON) = .693 × C1 × R2. Build the circuit as shown and activate it. Try changing the values for R1, R2, and C1 and note the results. You may need to hook up the oscilloscope if the frequency becomes too high to view with the LED.

- **Custom Settings**—None.
- **Analysis**—None.

From my website (**http://www.basicelectronics.com**). For more information on 555 timers in general, see the website.

Car Battery Level Indicator (BAT_LEV.EWB)

Have you ever had a dead car battery in the middle of winter in 30 below weather while dressed in light clothes? Let me tell you, it's not a pleasant experience. The circuit in Figure 13.17 demonstrates several components to simulate this condition. The car battery is a subcircuit of sorts, which simply uses a potentiometer to control voltage from

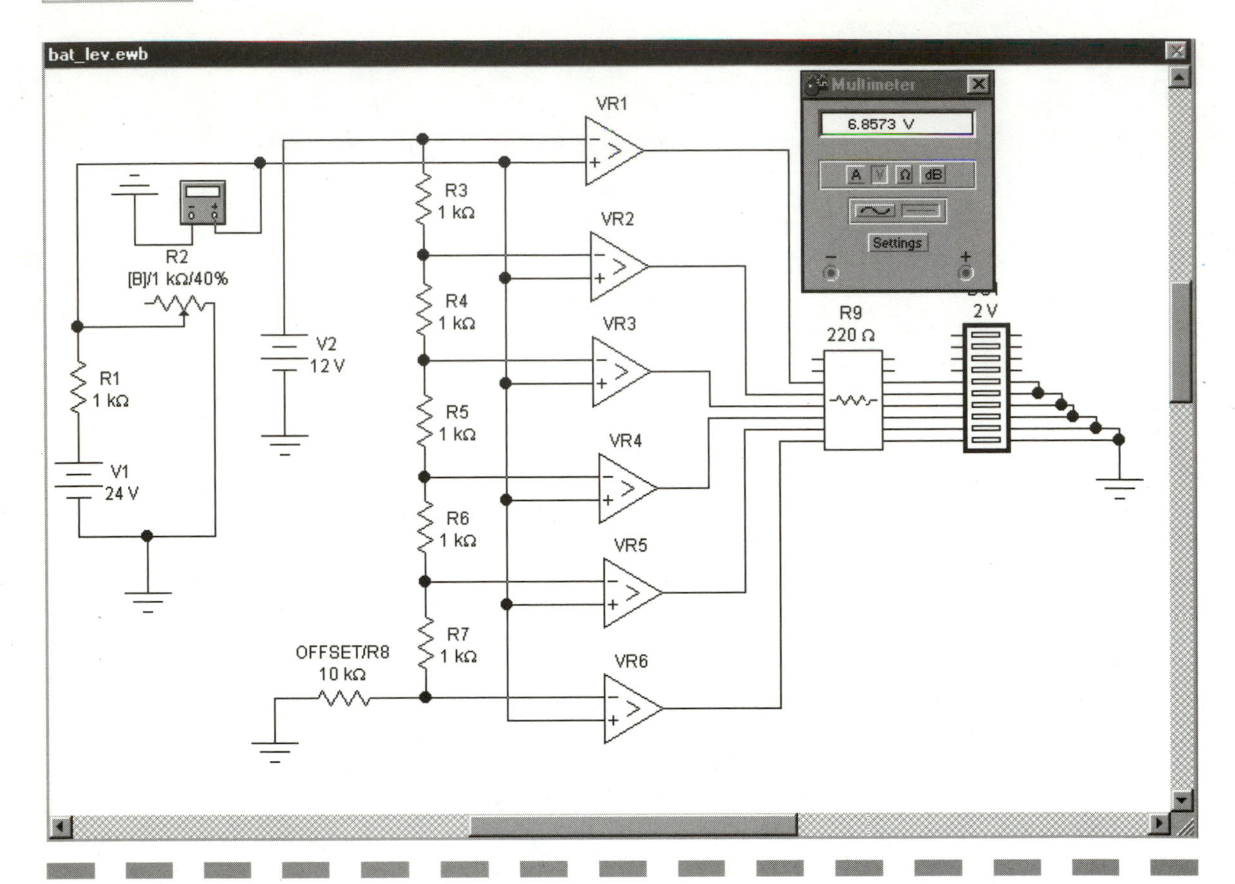

Figure 13.17 CAR BATTERY LEVEL INDICATOR—BAT_LEV.EWB.

12V to 0V (remember the voltage divider circuit earlier?). This is connected to a series of comparators. Another voltage divider is used to derive reference voltages for the other side of each comparator. R8 is used to offset the voltage ladder so any voltage under 7 volts will not register on the bargraph.

Run the simulation and hit the letter **b**, watching the voltage come down on the multimeter. The LEDs on the bargraph will alternately extinguish as the voltage hits a preset level.

- **Custom Settings**—Set the potentiometer to 100 percent to begin the simulation and assign the letter **B** to control it. You must go to **Analysis** > **Analysis Options** > **Transient** tab and change **ITL4** to **50** or you will get a time step error.
- **Analysis**—Select the nodes of each point on the voltage reference resistor ladder and perform a DC operating point analysis.

NOTE

*If you are getting an error message such as "Timestep too small" in a simulation, change the **ITL4** setting to **50** under **Analysis** > **Analysis Options** > **Transient** tab. You may also have to go to the Global tab and change **RELTOL** to **.01** or lower.*

Inverting Operational Amplifier (PGM_OPAMP.EWB)

The operational amplifier or op amp is one of the most widely used ICs in electronics. The most basic configuration used is the *inverting amplifier*. The example circuit in Figure 13.18 controls the gain by altering the resistance on the input. By hitting combinations of A, S, D, and F you can control the gain of the amplifier from −15 to 0. Open the scope and run the simulation. Change over to the main screen and begin hitting combinations of A, S, D, and F and note the results on the scope. Remember, each resistor added in parallel lowers the total resistance and thus raises the gain on the amplifier (Gain = −R5/Rtotal of R1, R2, R3, and R4).

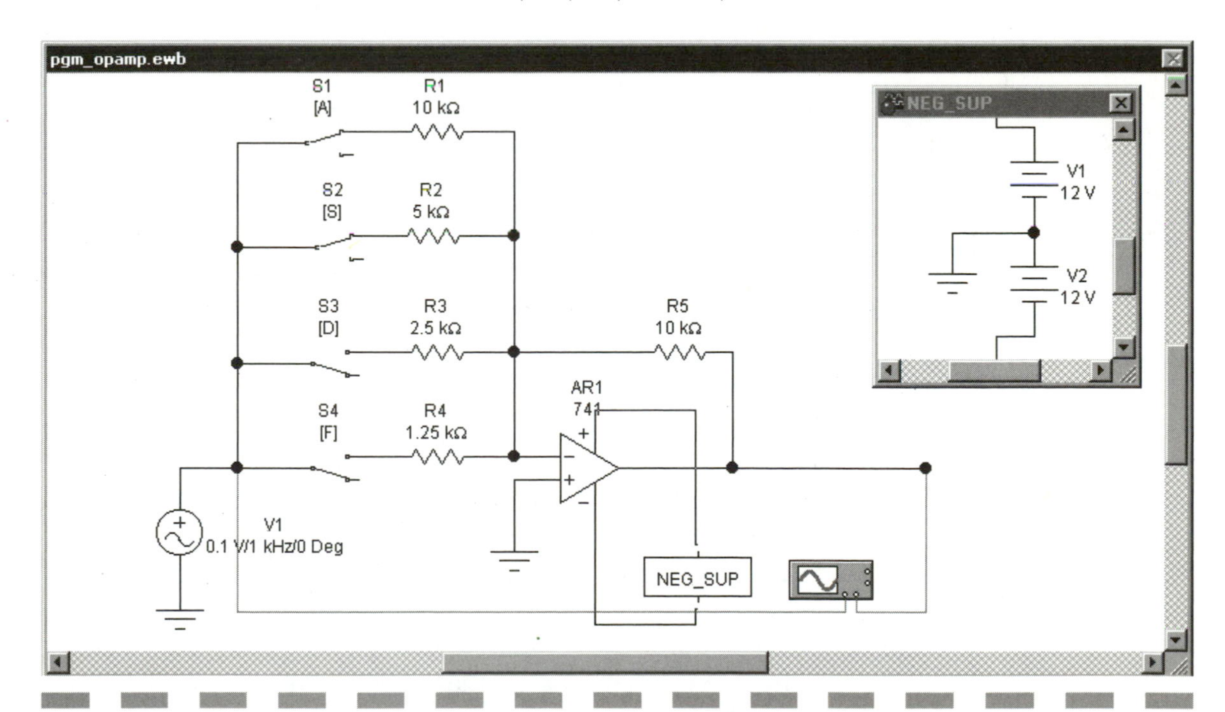

Figure 13.18 INVERTING OPERATIONAL AMPLIFIER—PGM_OPAMP.EWB.

- **Custom Settings**—Switches set to A, S, D, and F. Set the oscilloscope to a timebase of .5ms/Div and each channel to 500mV/Div.
- **Analysis**—None.

TIP

The NEG_SUP subcircuit in the op amp example can be used to supply negative voltage to the appropriate terminals on many ICs. Use it with other circuits to shorten design time.

Analog-to-digital Converter (ADC_DAC.EWB)

Analog-to-digital converter and digital-to-analog converter ICs are finding their way into millions of modern applications. These two components are in heavy use because the world is generally analog and most processing is done in the digital domain, then sent out as an analog signal again. A great example of this is compact disc (CD) technology: It uses an analog-to-digital conversion (ADC) technology to change the analog sound waves into digital words (record), which are stored as digital bits onto a plastic disc. When the disc is placed into a player, it converts the digital bits back into analog waves with a digital-to-analog converter (DAC).

In our example circuit in Figure 13.19, we are taking an analog wave, sending it through an ADC, and then back to a DAC. Hook up the circuit as seen and open the scope. Run the simulation and notice the stepped wave output. Note that I have flipped the DAC for clarity; the numbers had to be reversed. Modern DACs will smooth this wave out before it is used as an output signal.

- **Custom Settings**—Once you run the circuit as shown, change the Start of Count to 40 kHz and note the results. This setting changes how many samples of voltage are taken per second.
- **Analysis**—None.

Simple Digital Circuit (ADDR_DEC.EWB)

Figure 13.20 shows a simple decoder that uses four NOT gates and four AND gates to turn a two-bit binary number into one of four decimal numbers. The truth table is pictured in Figure 13.20 as well. Build the circuit as shown. Activate and use the **A** and **B** keys to display the four possible outputs.

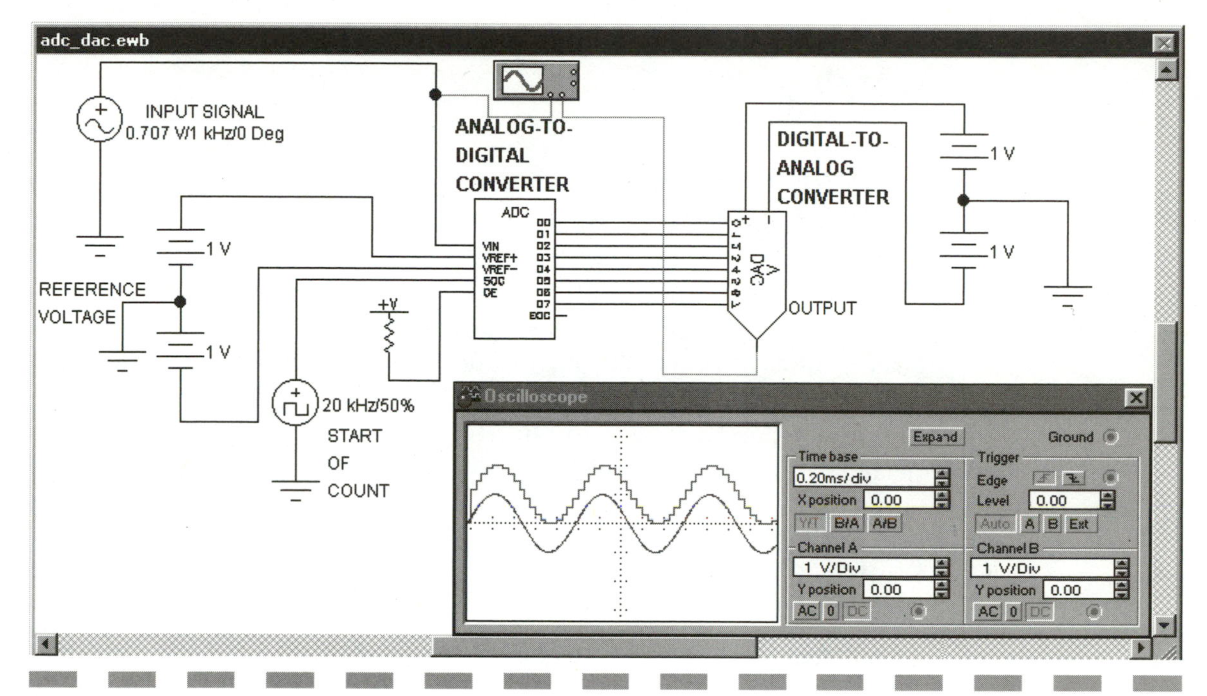

Figure 13.19 ANALOG-TO-DIGITAL CONVERTER—ADC_DAC.EWB.

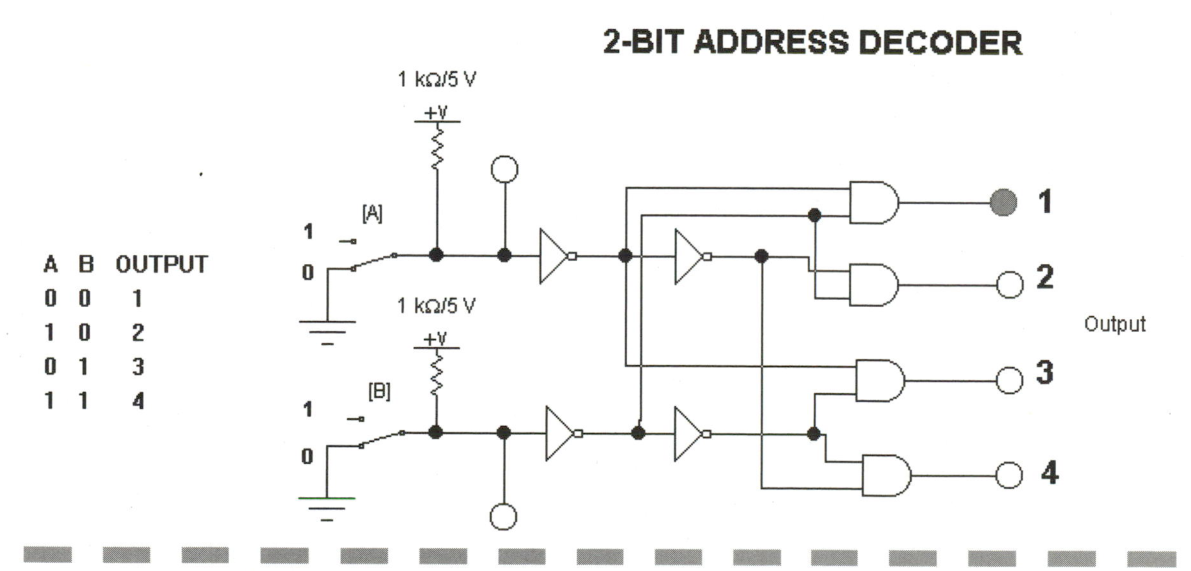

A	B	OUTPUT
0	0	1
1	0	2
0	1	3
1	1	4

Figure 13.20 SIMPLE DIGITAL CIRCUIT—ADDR_DEC.EWB.

- **Custom Settings**—Make the top switch A and the bottom switch B.
- **Analysis**—None.

Decimal-to-BCD Seven-segment Display (DECTOBCD.EWB)

The circuit in Figure 13.21 helps you learn the function of digital gates as well as the seven-segment display. It's a rather complex looking circuit but will build fast if you use cut/paste for sections such as the switches. Each switch is hooked to ground and Vcc component (see note). For example, when you hit the number 1, the signal wire is connected to the Vcc (High). When you hit 1 again, it is hooked to ground (Low). This network of switches gives a signal to the logic gates (consisting of six NOR gates, two NAND gates, one inverter, and a triple-input NAND gate [7410]). When the circuit is activated, the logic decodes the decimal and turns it into a 4-bit binary number (BCD). The BCD signal is fed to the seven-segment decoder, which changes it into a visual digit between 0 and 9. So if you hit the number 4 on the keyboard, 4 is changed into 0100 and fed to the display, which shows the number 4.

- **Custom Settings**—Build it as you see it. Set the switches to 0, 1, 2,...
- **Analysis**—None.

NOTE

An actual signal switching circuit that eliminates floating gates (i.e., the gate is getting no definite signal while being changed from 0 to 1) may contain a pull-up or pull-down resistor always connected to ground or Vcc (see Figure 13.20 for an example). The pull-up creates a high signal until a momentary button closes a circuit to ground. Conversely, a pull-down creates a low signal until a pushbutton completes a circuit to Vcc. However EWB5 does not contain a momentary pushbutton component. Dectobcd.ewb's switches were created another way for schematic clarity. This may be used on other circuits in this book, but if transferred to the real world, it can create weird problems with a circuit due to a temporary floating condition. If you build this circuit on a breadboard or PCB, be sure to use the pull-up or pull-down design instead.

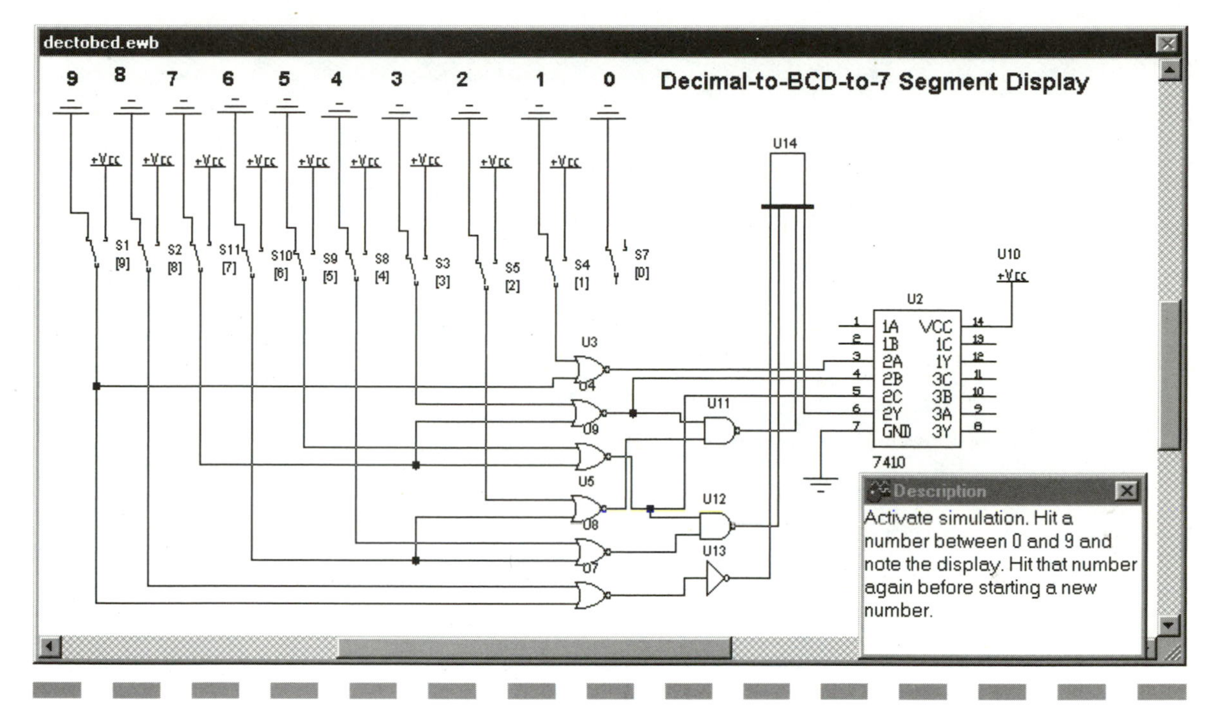

Figure 13.21 DECIMAL-TO-BCD-7 SEGMENT DISPLAY—DECTOBCD.EWB.

Flip-flop Circuit (FLIPFLOP.EWB)

I decided to use a game to demonstrate the use of the RS flip-flop and the D flip-flop (see Figure 13.22). The D flip-flops are used as a counter activated by a High signal on the D terminal and then clocked with a clock source. This in turn lights up four lights one after the other. The inverted output of the "Go" D flip-flop and an AND gate are used to prevent players from hitting their keys too soon. It will only make a ground signal available to the RS flip-flop when the countdown is done. The RS flip-flop determines which signal was switched over to ground (Low) first. The results of the contest are sent to either the Q or inverted Q outputs (Q with line over it means Not Q or simply the opposite of Q).

This is a two-player game. Choose a letter (A for Left player) and (L for right player). Hit the Spacebar to begin the countdown. When the Go light is red, see who can hit their letter fastest. If you hit the key too soon, the high-pitch buzzer sounds. You must reset the game in this case. After the contest, hit the Spacebar again until the countdown is reset and replace the player switches to the top positions.

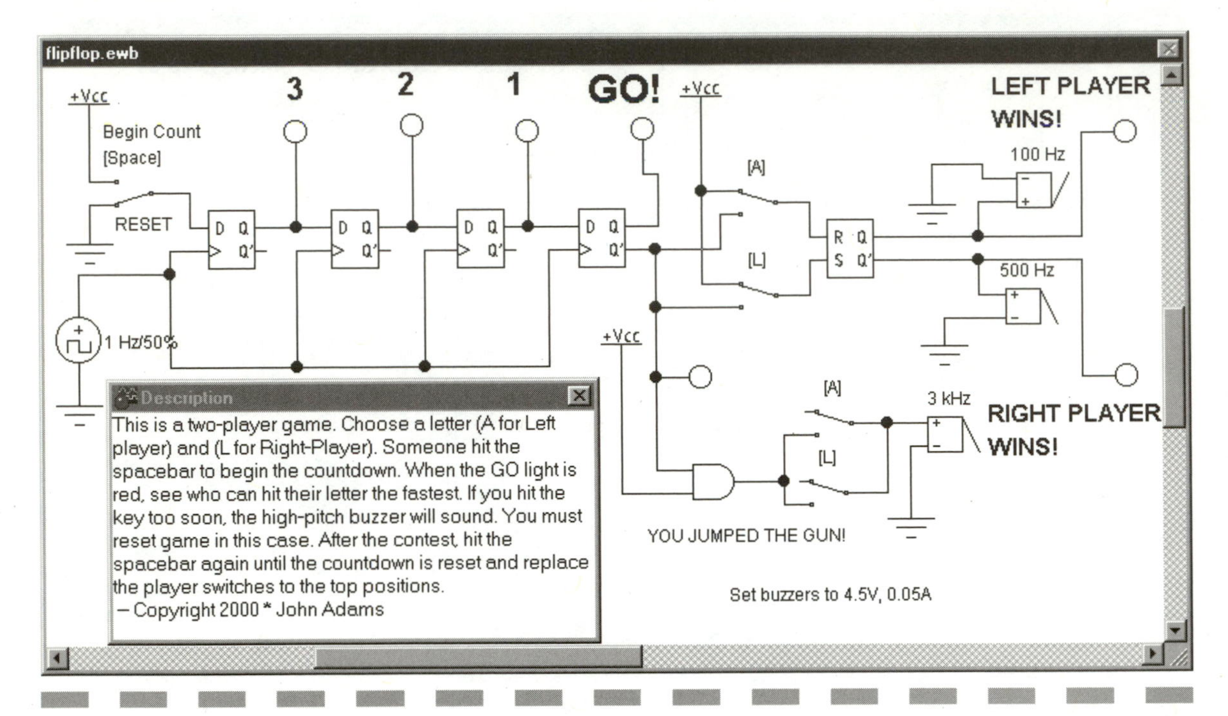

Figure 13.22 FLIP-FLOP CIRCUIT—FLIPFLOP.EWB.

- **Custom Settings**—Set the **Begin Count/Reset** switch to the Spacebar. Make the top switch (left player) the A key and the bottom switch (right player) the L key. Set all buzzers to 4.5V, 1 mA and set the frequencies as shown in Figure 13.22.
- **Analysis**—None. Have fun instead!

Parallel-to-serial-to-parallel Circuit (PAR-SER.EWB)

This is a rather complex circuit (Figure 13.23) that demonstrates many digital components, including the use of IC templates. Make sure to adjust the two clock sources as shown, paying attention to the duty cycle on the top one. If you are having trouble with the circuit, open it from the CD and try to figure it out. When running the circuit, give it time, because this is a slow simulation.

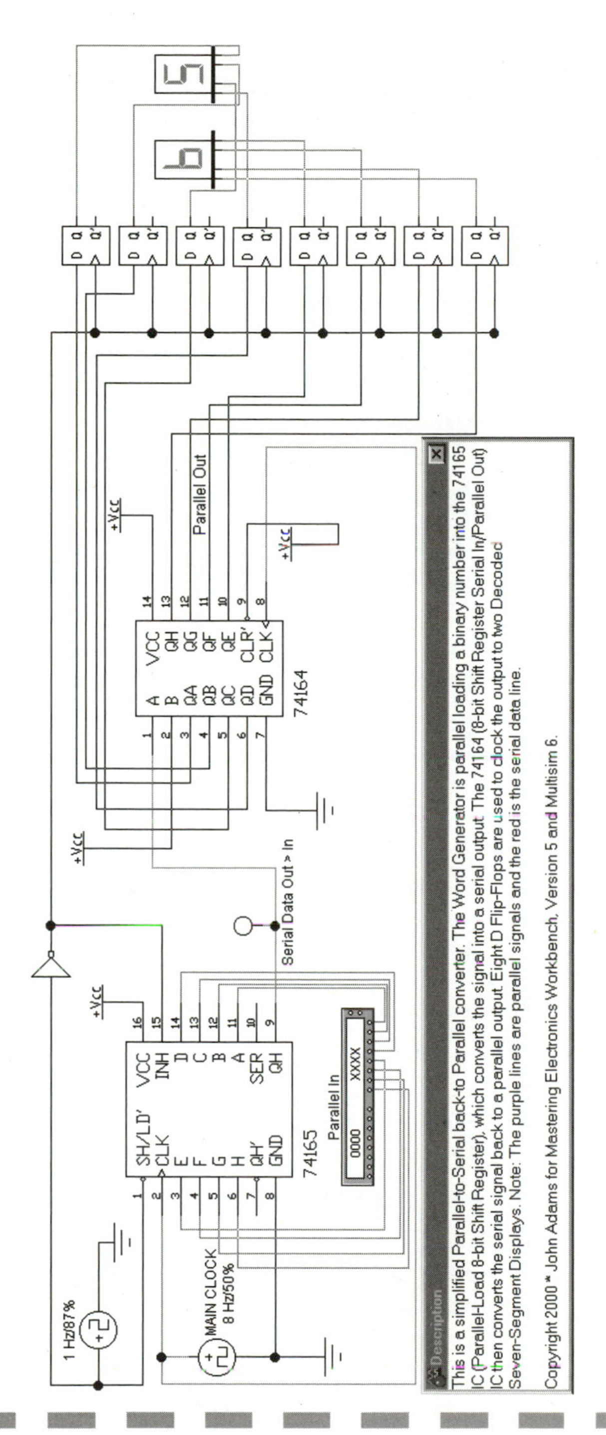

Figure 13.23 PARALLEL-TO-SERIAL-TO-PARALLEL CIRCUIT—PAR-SER.EWB.

This is a simplified parallel-to-serial back to parallel converter. The word generator is parallel loading a binary number into the 74165 IC (parallel-load 8-bit shift register), which converts the signal into a serial output. The 74164 (8-bit shift register serial in/parallel out) IC then converts the serial signal back to a parallel output. Eight D fip-flops are used to clock the output to two decoded seven-segment displays. (The purple lines are parallel signals and the red is the serial data line).

- **Custom Settings**—The top clock source must be adjusted to 1 Hz and 87 percent duty cycle. The main clock is simply set to 8 Hz. The word generator should be loaded with the up-counter pattern and the frequency set to 2 Hz.
- **Analysis**—None.

Control Circuit (CONVERT_TEMP.EWB)

Moving from the United States to Canada, I was forced to learn the metric method of reading temperature. I still have difficulty, (along with most Canadians) converting the weather report's Celsius readings into the familiar Fahrenheit scale. The circuit in Figure 13.24 uses a few of the control components to do just that.

This simple control circuit converts degrees Celsius to degrees Fahrenheit. Start the circuit and hit the **d** or **Shift+D** keys to change the Celsius reading. Each push of the key is one degree Celsius. The scale goes from +50 to −50 degrees Celsius.

- **Custom Settings**—Set the potentiometer key to D and make its increments 1 percent. Set ITL4 to 25 (under **Analysis** > **Analysis Options** > **Transient** tab).
- **Analysis**—None.

Netlist Example (NETLIST.EWB)

Netlist components can be downloaded from semiconductor websites, which lets you add components not included with EWB5. The *.cir files can be placed into the directory **C:\EWB5\MODELS\SUBCKTS** and renamed to be a *.lib file. In this example, I visited **http://www.national.com/models/spice/ComLinear/** and downloaded the **clc449.cir** file to the Netlist directory (1.1 GHz ultra-wide-

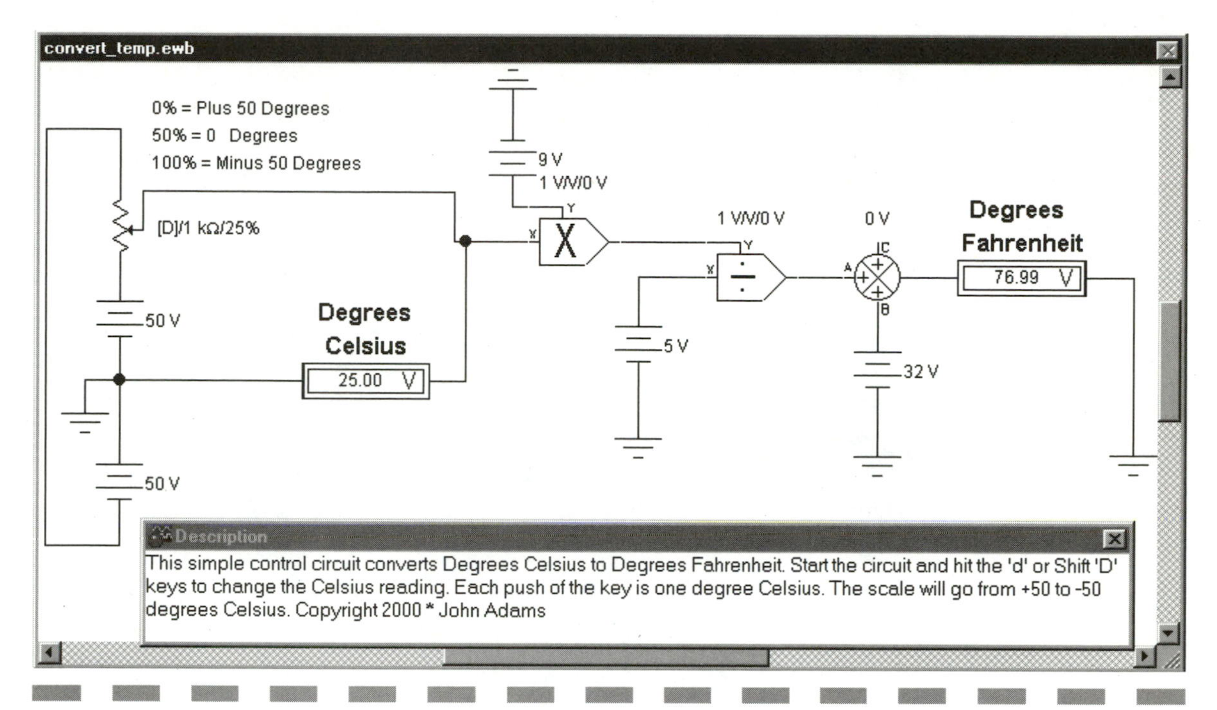

Figure 13.24 CONTROL CIRCUIT—CONVERT_TEMP.EWB.

band op amp; note that the website lists this as a 1.2 GHz model). I then changed the file extension to read **clc449.lib**. You could choose any file name as long as the extension is *.lib or you can even combine multiple SPICE model files and rename the whole file *.lib. Start EWB5. Open the Miscellaneous toolbin and place a Netlist component. Double-click the IC and the model window appears. Choose the CLC449.lib and then the CLC449 from the model field. Luckily, National Semiconductor Corporation includes a graphic of the subcircuit (Figure 13.25). Unfortunately, EWB5 does not use the same node or pin numbers as the figure shown. Instead we must open the text file in a word processor and see in what order the pins appear. In this case, the order is non-inverting, inverting, output, +Vcc, and −Vee. EWB assigns numbers in that order. So pin 1 on the Netlist component is now the non-inverting input, pin 2 is the inverting input, etc. Hook up the circuit as seen in Figure 13.26. Note that the op amp symbol in the top right corner is merely a visual aid and not necessary to the drawing. Make the adjustments as noted in the custom settings section. Open the oscilloscope and activate it. Run an AC frequency analysis of the op amp as described in the analysis section.

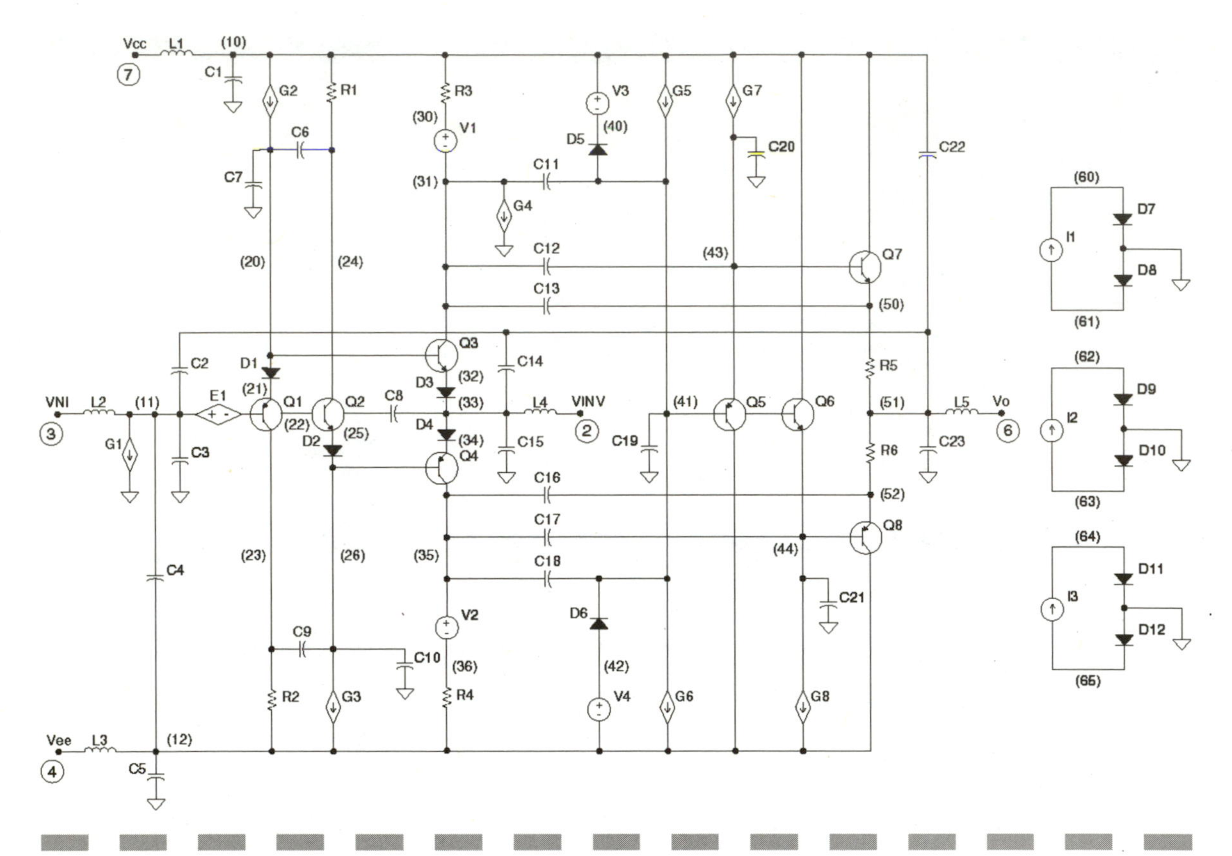

Figure 13.25 This is a diagram of the CLC449 SPICE subcircuit (from National Semiconductor Corporation website).

*Importing the *.cir file into a blank circuit page can also show you the Netlist subcircuit. Beware! Some of these can be enormous and take minutes to convert.*

- **Custom Setting**—Set the function generator to 100 MHz, 1V. Adjust the oscilloscope's timebase to 5.00ns/Div and channel A and B to 1V/Div. Display the output's node.
- **Analysis**—Go to **Analysis** > **AC Frequency**. Set FSTART to 1Hz, FSTOP to 10GHz, decade sweep, 100 Points, Log scale, and select the Output node as the node for analysis; in this case it is node 6. Hit the **Simulate** button. Toggle the cursors and see if the amplifier's bandwidth does indeed drop at around 1.1 GHz.

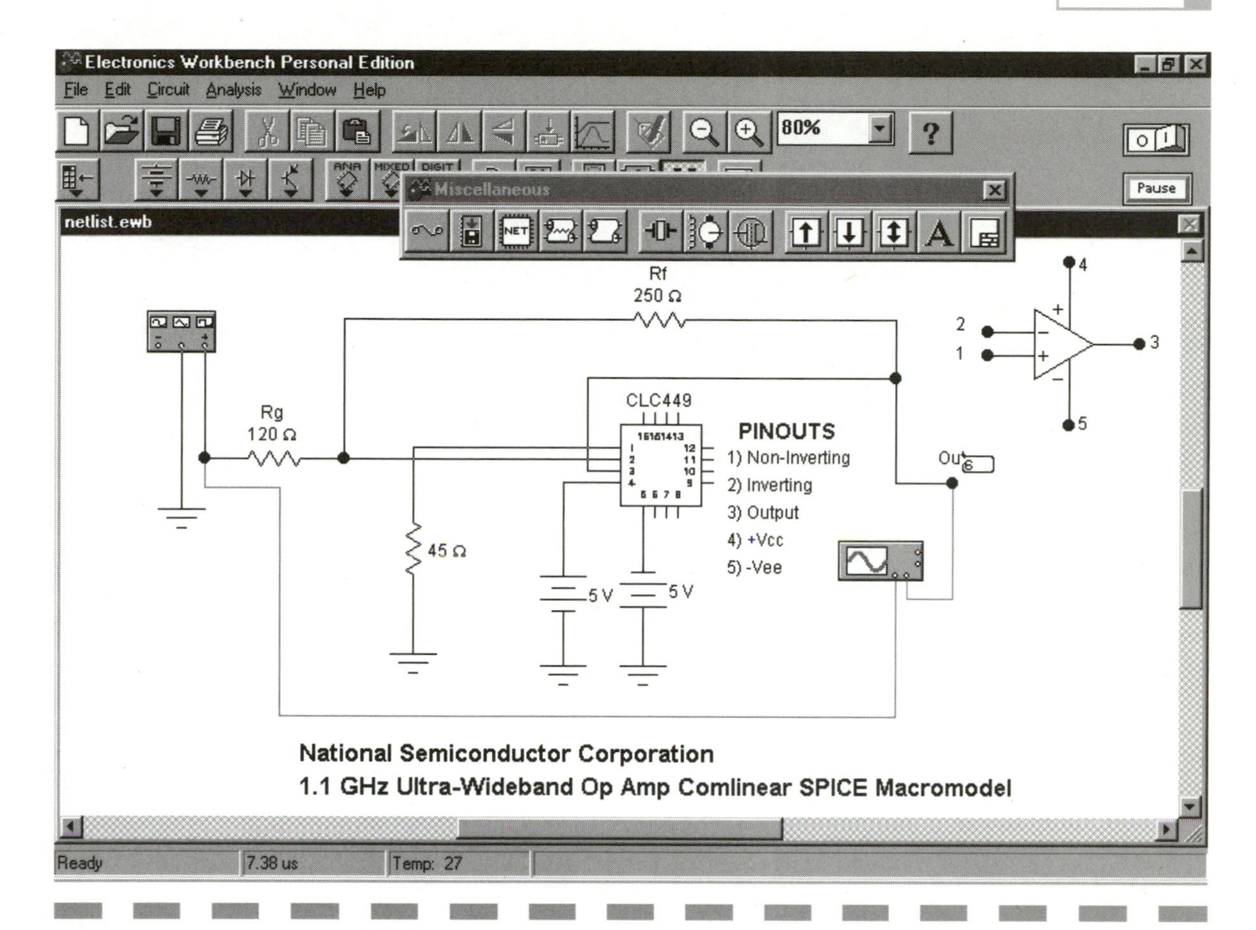

Figure 13.26 NETLIST EXAMPLE—NETLIST.EWB.

You can typically download a datasheet for a SPICE model component to get an idea of its pinouts and applications.

PWM Voltage Controller (CHEAP_PWM.EWB)

By varying the duty cycle of a square wave, you can control the voltage level of many DC-based devices such as LEDs, transistors, DC motors, etc. In this quick and dirty pulsed-width modulation circuit, I used a 555 timer to control the on to off time of a 1 kHz square wave.

The resulting signal is put through a capacitor to even out the voltage level. If you were using this to control the brightness of an LED, the capacitor could be eliminated because the human eye cannot see the LED going on and off at even lower frequencies. Build the circuit in Figure 13.27 and adjust the custom settings. Open the oscilloscope and run the circuit. Hit the **w** or **Shift+W** keys to control the duty cycle of the output and thus the voltage level.

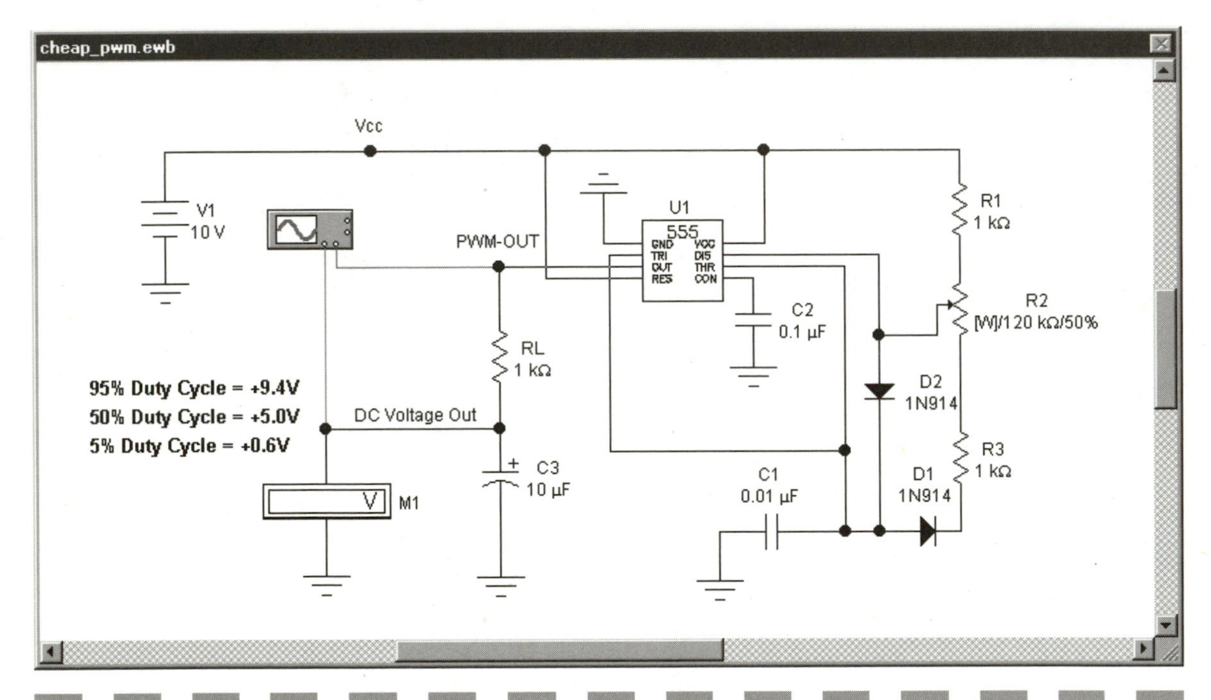

Figure 13.27 PULSE WIDTH MODULATION VOLTAGE CONTROLLER—CHEAP_PWM.EWB.

- **Custom Settings**—Set ITL4 to 25 or higher. Assign the W key to the potentiometer. On the oscilloscope, set the timebase to 0.50ms/Div, channel A and B to 5V/Div.
- **Analysis**—None.

DC Motor (DCMOTOR.EWB)

The simple circuit in Figure 13.28 demonstrates the use of the DC motor component. You can investigate the speed characteristics by

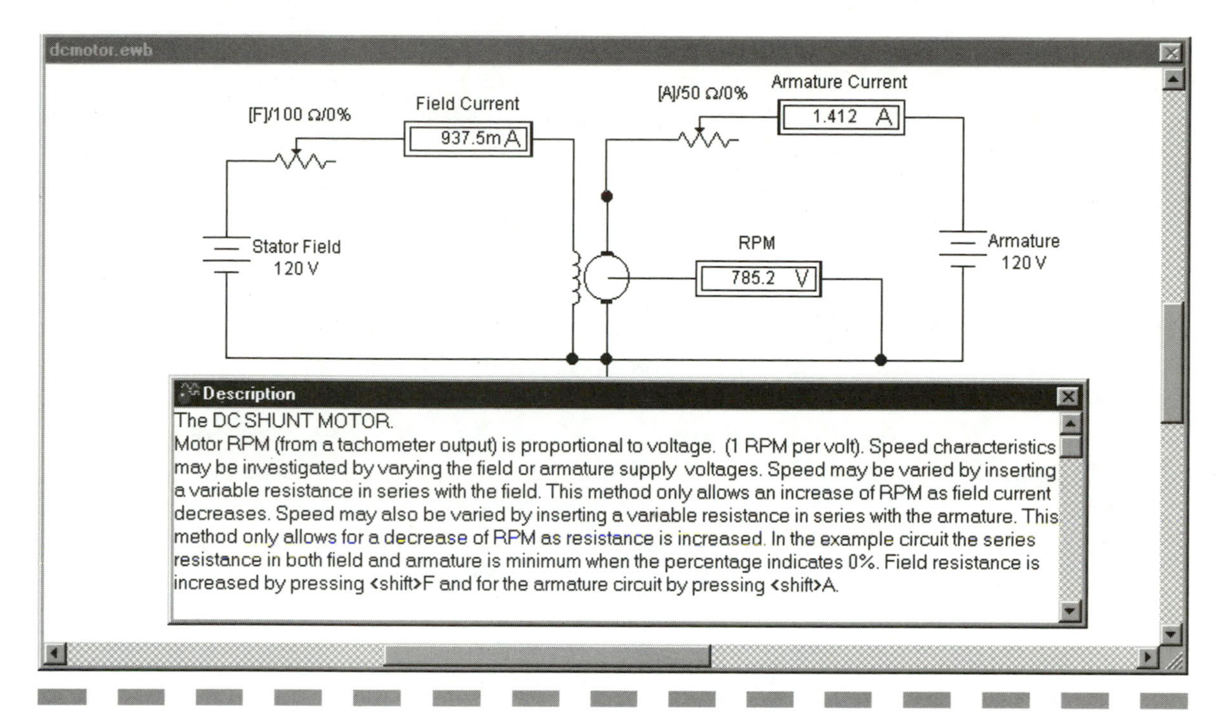

Figure 13.28 DC MOTOR—DCMOTOR.EWB.

adjusting the current in the stator and armature. You may want to experiment with a feedback system that will adjust each field to receive optimum speed of the rotor.

In the DC shunt motor, motor RPM (from a tachometer output) is proportional to voltage (1 RPM per volt). Speed characteristics may be investigated by varying the field or armature supply voltages. Speed may be varied by inserting a variable resistance in series with the field, which only allows an increase of RPM as field current decreases. Speed may also be varied by inserting a variable resistance in series with the armature, which only allows for a decrease of RPM as resistance is increased. In the example circuit, the series resistance in both field and armature is minimum when the percentage indicates 0 percent. Field resistance is increased by pressing **Shift+F** and for the armature circuit by pressing **Shift+A**.

- **Custom Settings**—Assign the field potentiometer to F and the armature to A.
- **Analysis**—None.

From the Interactive Image Technologies, Ltd. ftp site.

Phototransistor Test (PHOTOTRANS.EWB)

This circuit (Figure 13.29) will help reinforce two concepts: emulation and using the quirks of EWB to your advantage. With the circuit in Figure 13.29 you will realize that even though EWB has its limitations, you can still bypass these to emulate what you are trying to simulate. In other words, if it's not built into EWB5, build it in! In this case we are using an ideal BJT as a pseudo-phototransistor. We are creating a phototransistor symbol (schematic capture) by utilizing one of EWB's bugs; the ability to make an invisible wire or part.

This demonstration circuit emulates a phototransistor used to count objects passed between the IR LEDs and the phototransistor. What's the secret to this circuit? The "object" is actually a subcircuit with a 90 percent duty cycle clock source and a 4.7 Mohm resistor, which is connected to the transistor. You cannot see the connection because there is a copy of the two connectors and wire placed over the original, making one cancel the other. See the left side of the figure for an example; the L with the connectors is placed over the object-to-transistor connection. This quirk in EWB can be used to create many different symbols and connections that are otherwise not possible. The circuit is not a real-world circuit but can be easily converted to one. The RA would be replaced by a potentiometer to adjust the sensitivity of the phototransistor. The NOT gate may need to be replaced with a transistor or Schmitt Triggered device as well.

Figure 13.30 shows the object subcircuit as well as the procedure in making the invisible connection to the emulated phototransistor.

- **Custom Settings**—Use an ideal BJT transistor and label it phototransistor. Set the Object subcircuit's clock source to 5V, 1 Hz, 90 percent duty cycle.
- **Analysis**—None.

Piecewise (PIECEWISE1.EWB and PIECEWISE2.EWB)

In my opinion, the piecewise linear source and the write component are the most valuable components included with EWB5. They let you record a waveform (digital or analog) onto your hard drive (as a text file), then play it back through a different circuit. The simple circuits

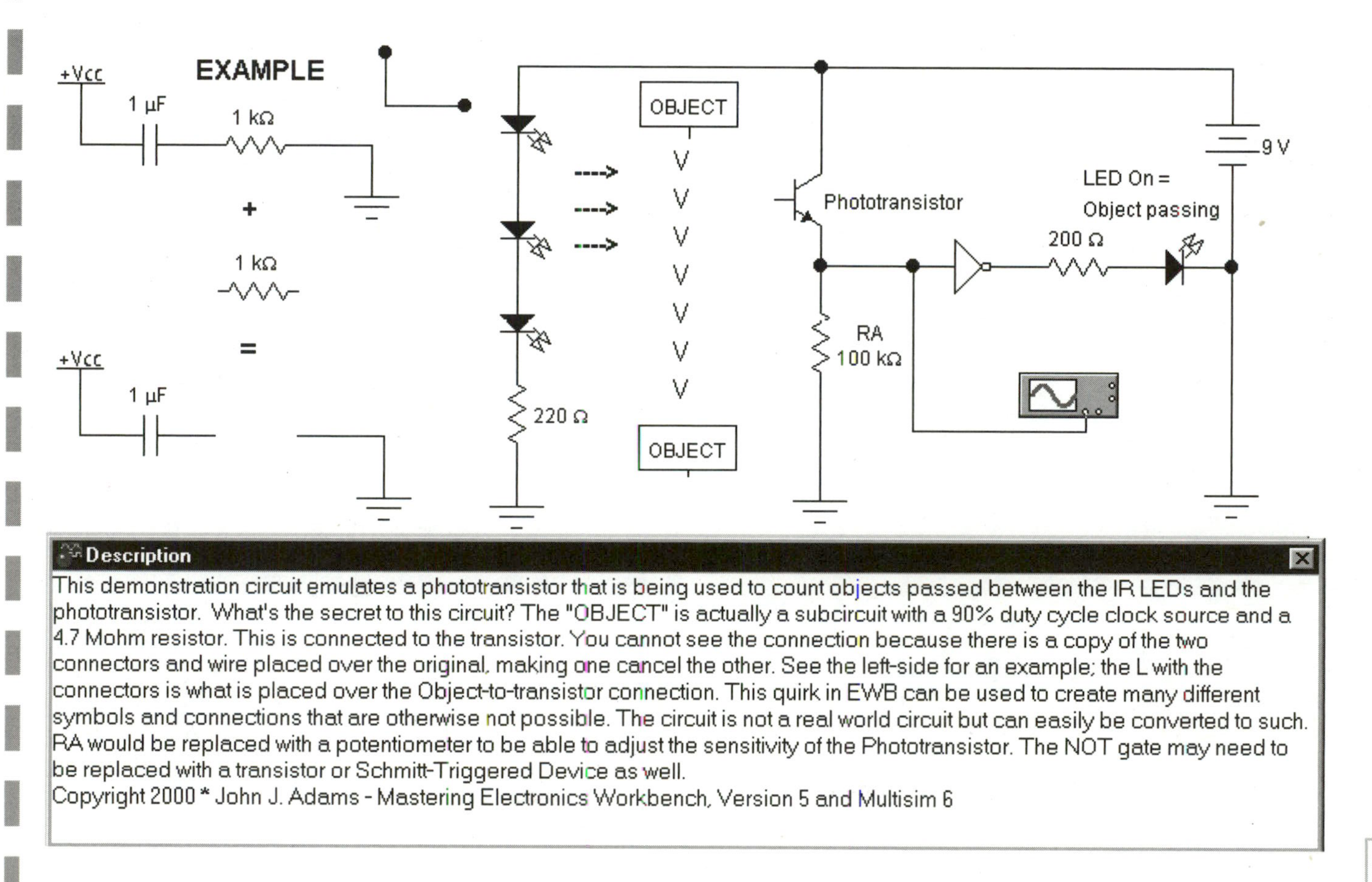

Figure 13.29 PHOTOTRANSISTOR TEST—PHOTOTRANS.EWB.

in Figures 13.31 and 13.32 demonstrate this ability. In **Piecewise1.ewb**, three waveforms are mixed to create a simulated voice wave and recorded to **piecewise1.txt**. Build this file and activate it for a minute or so. In **Piecewise2.ewb** we are playing back this file through two inverting amplifiers. Build the circuit, open the oscilloscope, and activate the circuit.

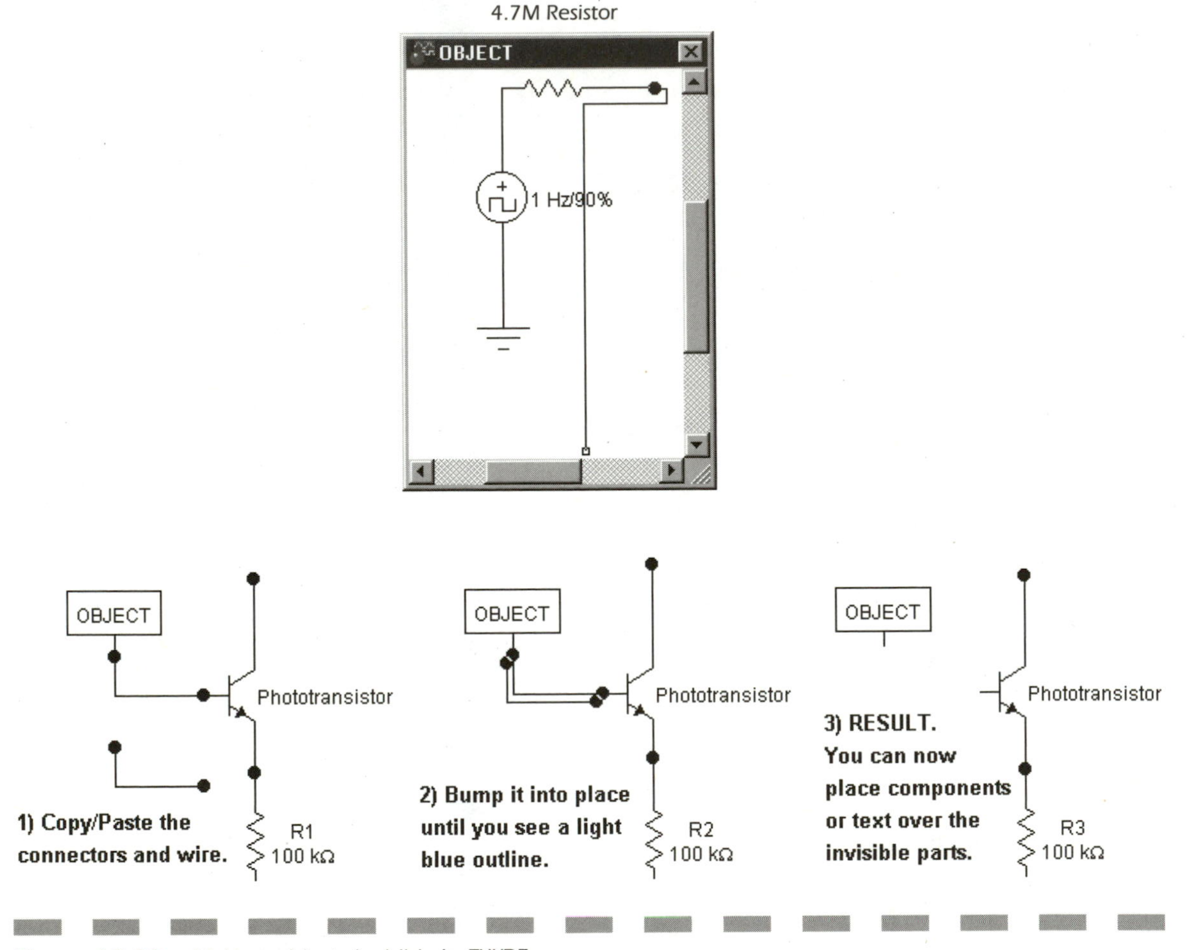

Figure 13.30 Making objects invisible in EWB5.

I am developing advanced circuits and special utilities that will let you take these EWB waveform text files and convert them into sound files (to play back through your sound card). A more advanced version will also be developed to let you connect a digital voltmeter to a real-

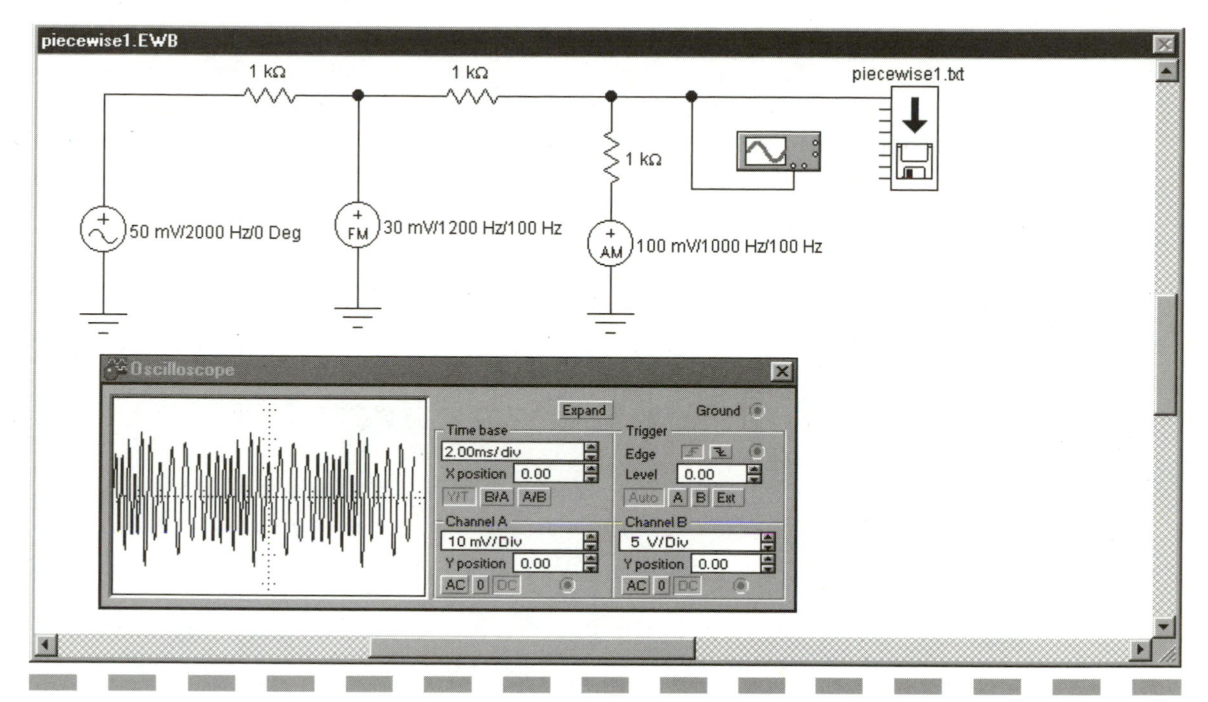

Figure 13.31 PIECEWISE LINEAR SOURCE EXAMPLE CIRCUIT, PART I—PIECEWISE1.EWB.

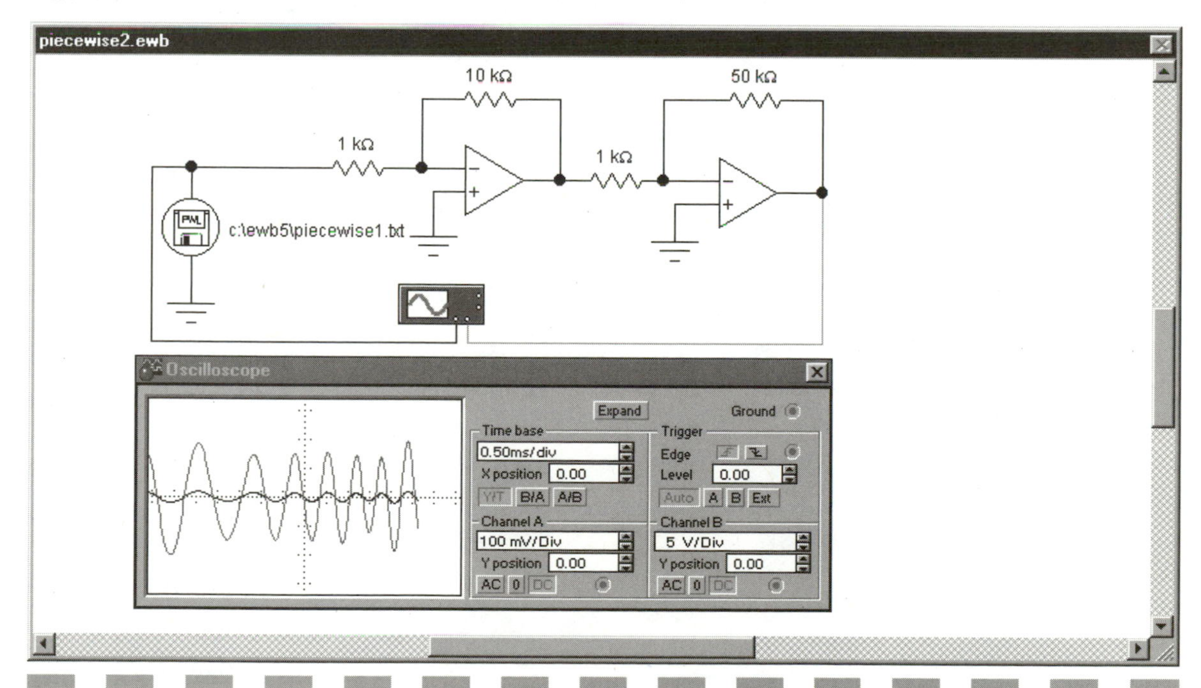

Figure 13.32 USING THE PIECEWISE FILE WITH THE WRITE COMPONENT, PART II—PIECEWISE2.EWB.

world circuit and read data to an EWB piecewise file. This can be used in your EWB simulations and output to a black box device connected to your computer. Watch my website (**basicelectronics.com**) for further details; this is going to be one awesome hobbyist application.

- **Custom Settings**—Save the piecewise file to **C:\EWB5\PIECEWISE1.TXT**.
- **Analysis**—None.

Making Your Own Circuits

This chapter was merely an introduction to teach you the various EWB components and their functions. The software is actually capable of extremely detailed simulations that can help your hobby or education immensely. Just think of ways to overcome the limitations of EWB as we saw in the Phototransistor circuit (which, by the way, I developed to use with an HO slot car lap counter). Find projects that interest you and simulate them. Once they are developed, you have the option of transferring them to layout and creating PC boards for your own use. There's nothing like holding that PCB in your hand, knowing it was created from scratch.

CHAPTER 14

Unleashing Multisim/EWB—Customization, Tricks, and Power Techniques

Releasing the power of sany oftware lies in its everyday use.

The power of a tool is determined by its value to you. How will you use it? Can you apply it to every facet of your hobby or profession? This chapter shows you all sorts of technical treats and tricks to empower you to squeeze every last feature out of Multisim or Electronics Workbench. It is jam-packed with helpful data, so read it over as thoroughly as possible and try out the techniques.

Customizing Multisim 6

Multisim is highly customizable and can be adjusted according to taste. For example, a user can adjust the toolbar items, a hobbyist can create a master default document that opens with his or her settings already plugged in, or a teacher can disable certain functions on some editions to train students. This section outlines each customization trick.

Default Settings

Multisim does away with the default.ewb scheme used in EWB5. It is replaced with **Edit** > **User Preferences** (or **Ctrl+U**). This contains all the information about the various default settings that are enacted when a new circuit is created. For example, you may set the circuit to Show Nodes, but if this is not set in User Preferences then the setting is lost next time you open a fresh file. Here is how to use User Preferences better.

1. When Multisim is loaded and first opened, go to **Edit** > **User Preferences**.
2. Adjust each page of settings for future documents and hit **OK**.
3. Choose **File** > **New**. It's that simple. Each new document created will contain those settings and components. Try it.

*Place common components, such as resistors, grounds, and caps and save the circuit as something like **Master.EWB** or **MSM**. Open this file instead of using **File** > **New**. This makes for quicker designs. Remember that you can always choose Select All and delete for a clean board.*

Moving Toolbars

Multisim has a highly adjustable workbench. Toolbars can be placed just about anywhere around the screen or made to float. Just grab an area between the icons and the edge of the toolbar and drag it to a new location. Older versions of Multisim place moved toolbars into their original position upon resetting because the toolbar positions could not be saved. Newer Multisim versions (6.11+) can now save toolbar positions.

'Favorites' have been eliminated in Multisim.

Component Databases

Instead of "Favorites", Multisim uses a user or corporate database that contains components of your choice. To turn on the **Component Group Toolbars**, go to **View** > **Toolbars** > **Databases** and check on the appropriate toolbar.

To add components to these toolbars, you must first use the Component Editor or chose the **Edit** button while in the Component Browser. Be sure to choose **View** > **Refresh component toolbars** once a new component category is added or you will not see it available.

Making Multisim Feel Like EWB5

I am partial to the simplified interface that Electronics Workbench 5 offered. In fact, I must confess to reverting to EWB when in a hurry. However, I'm getting used to Multisim more and more since I have made a few customized settings. Some of them must be set each time you restart Multisim.

- **Adjusting the Colors and Shutting Off the Show Grid**—In my default settings (**Edit** > **User Preferences**) I set the colors to white background with black components and labels; I find the colored component schemes distracting and not well suited for laser printing. The grid dots are also unnecessary at times, so I turn off the Show Grid option.

- **Moving the Toolbars to the Top**—The second thing I do is move all the toolbars to the top section, as shown in Figure 14.1. This makes it easier to remember where everything is.

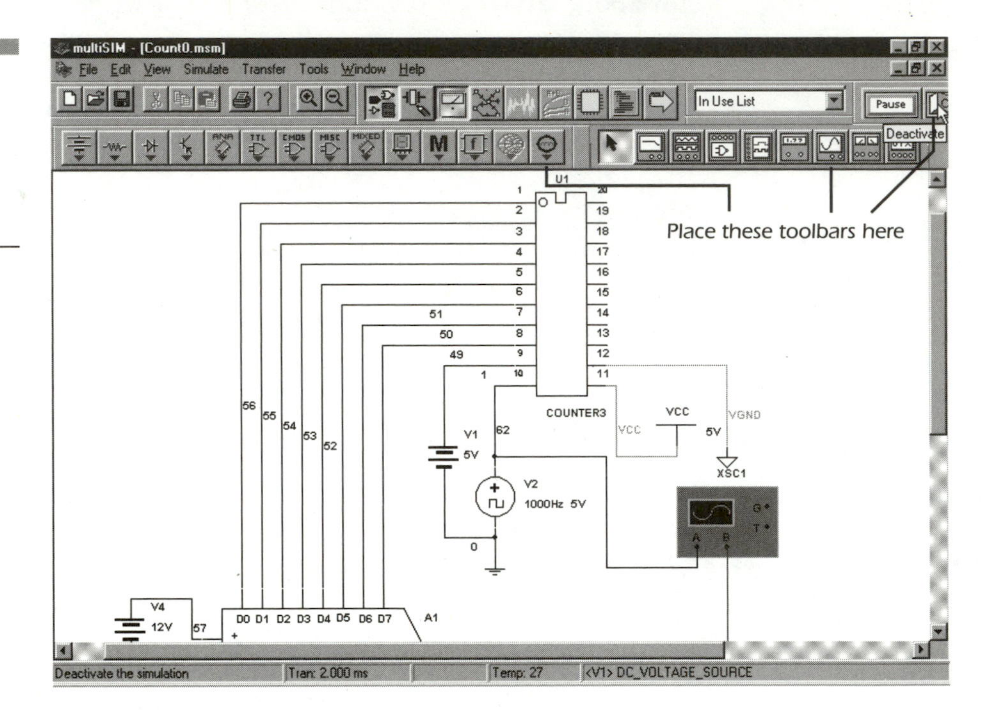

Figure 14.1
With a few movements of the toolbars, you can have Multisim look like EWB5.

- **Always-On Instrument Toolbar**—I find it handy to leave the instrument toolbar open and displayed at the top as well. Otherwise you are opening and closing or moving the toolbar around too much.
- **Use Virtual Components When Doodling**—If you are only testing concepts or ideas, use virtual components (such as resistors, caps, inductors) in place of the Component Browser. This saves a few steps and your valuable time. You can always redesign the circuit with non-virtual components later in the development stage.

CAUTION

If you move toolbars in 6.11 or above, I recommend first backing up the file called ***C:\Multisim\Users\User.ini****. I have noticed a severe intermittent bug that causes the toolbars to go totally screwy next time Multisim is opened. If this error occurs, replace the User.ini with your backed-up file. You may lose some customization info, but it's better than having to reload Multisim.*

Customizing EWB

EWB can be adjusted according to taste as well. For example, a user can adjust the toolbar items, a hobbyist can create a master default document that opens with his or her settings already plugged in, or a teacher can disable certain functions on some editions to train students.

Default.EWB

The master setting file, **DEFAULT.EWB**, is in the **C:\EWB5** directory. It contains all the information about the various default settings that are enacted when a new circuit is created. For example, you may set the **Schematic Options** to **Show Nodes.** If this is not saved to the default.ewb file, then the settings are lost next time you open a fresh file. Here is how to work with the Default.EWB file:

TIP

*Back up DEFAULT.EWB into another directory, such as **C:\EWB5\BACKUP**, before making any changes.*

Back up DEFAULT.EWB somewhere, just in case you make inadvertent changes. (On the CD-ROM, the original default.ewb file is under **(CD_DRIVE_LTR):\MISC\DEFAULT.EWB**. I've also posted this file at: http://www.basicelectronics.com/ewb/default.ewb.)

Now that you have backed up the file, changes can be made and saved:

1. Start Electronics Workbench V5.
2. Open the **DEFAULT.EWB** file in the **C:\EWB5** directory.
3. Each setting or part placed in this document will dictate all future new documents. Make each change and place those parts that you want automatically placed. when a new document is started.
4. Choose **File** > **Save**. Overwrite? Yes.
5. Choose **File** > **New**.

It's that simple. Each new document created will contain these settings and components.

ITEMS THAT CAN BE ADJUSTED WITHIN DEFAULT.EWB

- **Circuit** > **Schematic Options**. Every item in this dialog box can be changed and saved for the DEFAULT.EWB file.
- All settings in the **Analysis** > **Analysis Options** dialog box will be saved.
- If there are components placed in your drawing when you save the DEFAULT.EWB file, they will appear each time you create a new drawing. I usually place a ground symbol.
- Any toolbars open will be open for each new circuit.
- Any subcircuits created and saved under "Favorites" will be available.
- The zoom level will be preset to your tastes.
- Title blocks and miscellaneous text can be permanently affixed.

*Place common components, such as resistors, grounds, and caps onto the DEFAULT.EWB board and save. This makes for quicker designs. Remember, you can always choose **Select All** and delete them for a clean board.*

Moving Icons

With EWB Version 5.12 you are able to reposition individual icons. Hit the Shift key while the mouse pointer is over an icon and left click, hold. Drag the icon to a new location on the toolbar and release both the mouse button and shift key. Remember that icons cannot be placed anywhere other than on a toolbar. You may have learned this the hard way and had certain shortcuts disappear. If so, restart EWB and they will be in their unspoiled position. Note that these icon positions cannot be saved.

"Favorites" Component

By placing a popular component into the Favorites bin, you are making it available that much faster. Simply right-click over the component icon and choose **Add to Favorites**. The component can be placed by opening Favorites and grabbing it. Just a little tip: open the

DEFAULT.EWB and place the most-used components into the Favorites bin, then save. Those components will then be quickly accessible in all future documents.

Clean Designs

I have a propensity to be a perfectionist when it comes to circuit design: I'm a stickler for ultra-legible designs. This means:

- Being able to understand what each component is.
- Knowing where it is connected.
- The ability to move around the drawing easily without heavy scrolling.
- All components are clearly labeled and any other text-based information important to the circuit is correctly placed.

I've always made an effort to meet these criteria when designing circuits, because I loathe coming back to a circuit and scratching my head trying to figure out what it is or what is happening. The solution is to use the spic-and-span circuit design techniques listed.

NOTE

These hints apply to both EWB and Multisim, unless specified.

Rotation or Flipping of Parts

A part must often be rotated or flipped to orient it better. This makes connecting wires cleaner, better routed, and easier to understand (see Figure 14.2). A ground must be run from pin 1 of a 555 timer. The standard way (A) would be to run the line below the chip and connect the ground at the bottom. A better way (B) would be to place the ground on top of the chip, so that wires are not run together, creating confusion. The best method (C) would be to rotate the ground 180 degrees and place it above the IC: there is no crossing or unnecessary turns in the wire. I know some designers would scoff at this practice, but which design is easiest to read? Flipping the component can also reorient parts and terminals for better placement and wiring.

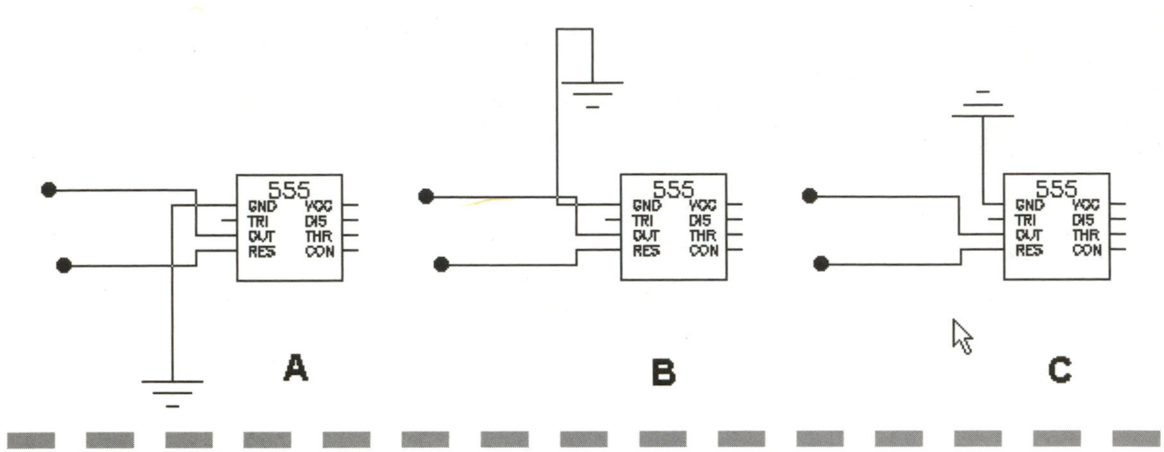

Figure 14.2 Rotate or flip components to clean the drawing up. Which example is easiest to read?

CAUTION

Many components that were rotatable in EWB5 are currently not rotatable in Multisim 6.11 or lower. In this case, either design your own rotated components or wait for the next version to come out.

Create or Modify Symbols in Multisim

Some symbols in Multisim are less than perfect. You may want to correct the incongruity of parts. An example is the op amps, which have connectors one or two pixels off. Irritating! You can correct this by using the Symbol Editor and turning off the snap-to-grid feature, then moving the pins one pixel over (see Figure 14.3). Other symbol corrections may include making a rotated symbol that is not rotatable in the current Multisim; an example is a potentiometer. I also recommend making some of your own component symbols and sending them to the User database for use in your drawings. This may include custom symbols that are simply used for schematic capture only. Chapter 4 touched on this subject, showing a speaker symbol we created and Chapters 7 and 13 showed the phototransistor trick.

Power-Bus Lines or Rails

Another common practice in schematic capture is to use **power-bus lines** or **rails**. This means drawing circuits in between two horizon-

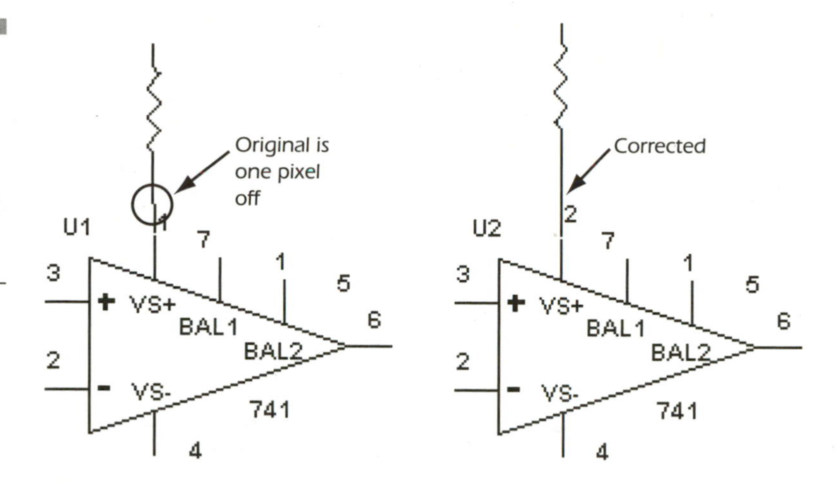

Figure 14.3 Make your own library of symbols in Multisim. In this example, I have corrected a one-pixel error in Multisim's 741 op amp.

tally placed parallel lines; like building the circuit inside a railroad track. The top rail is Vcc or V+; it is typically hooked to the positive terminal of the battery or power supply. The bottom is the ground line and usually connected to the negative terminal of a battery or power supply. Sometimes there is a third rail representing a negative voltage. In between the lines, I weave parts together from left to right. A signal enters from the left side; is conditioned by an IC, transistor, or other component and is output on the right. That output becomes the input to another component placed to the right of the first. This is a standard practice in schematic capture and electronics design. It just so happens that it is very conducive to EWB's graphical user interface as well.

In Figure 14.4, you will notice the V+ bus at the top and the ground bus on the bottom. In between are signal-conditioning transistors and an IC. The right connection is the output.

CONNECTORS

You can never have enough connectors in a drawing. They help to verify and clarify junctions at a glance. Don't hesitate to add a connection in the middle of a line and name it if it helps you to understand that point in the circuit. I often use them as test points or input/output points. In fact, if you use Multisim you can use the In/Out connector (right-click in an empty area of the screen to place one) to identify nodes and testpoints easier. In Figure 14.5, I have used eight connectors as a data bus and marked the output for later reference.

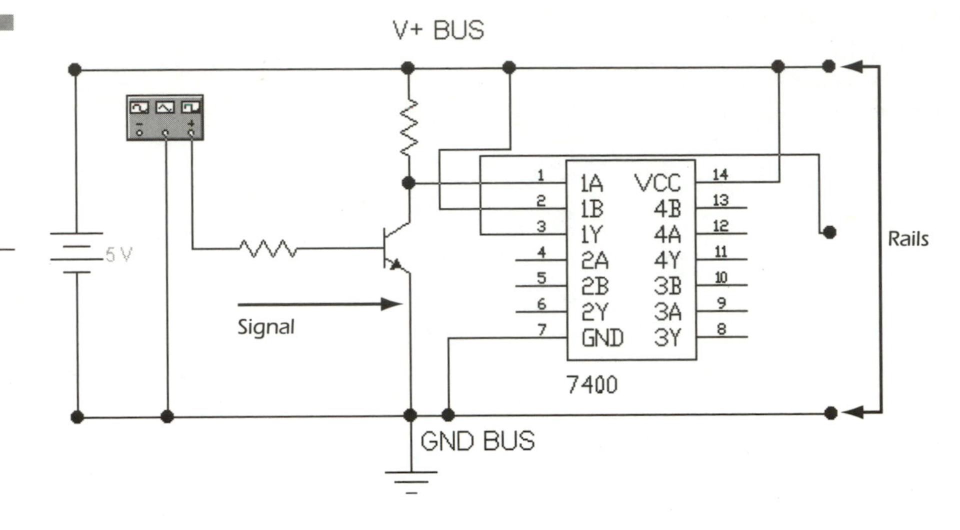

Figure 14.4 A power bus (rails) make it easier to understand schematics. It's like reading left to right, within the rails.

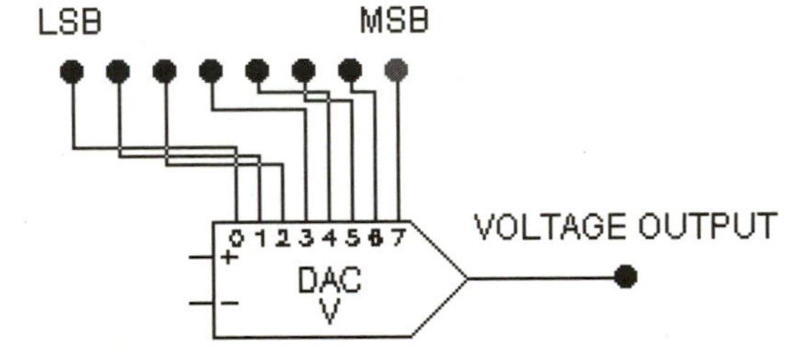

Figure 14.5 Use as many connectors as possible to make a drawing more understandable.

TIP

If you have trouble identifying a particular wire in a jumble, temporarily color that line. A faster method is to highlight it with a single mouse click over the wire. In EWB5 this bolds the wire. In Multisim it places handles along the length of the wire instead.

Rerouting Lines, Moving Wires, and Spacing Parts

Electronics Workbench does a good job of running lines cleanly (most of the time). But there are times it gets finicky. This requires either moving parts or adjusting and positioning the wire lengths. As you can see in Figure 14.6, the drawing on the left is not very clear, because the parts are too close together and the wires going to the connector where the mouse pointer's arrow is pointing are not connected to the proper terminals. Now look at the drawing on the right side. To solve

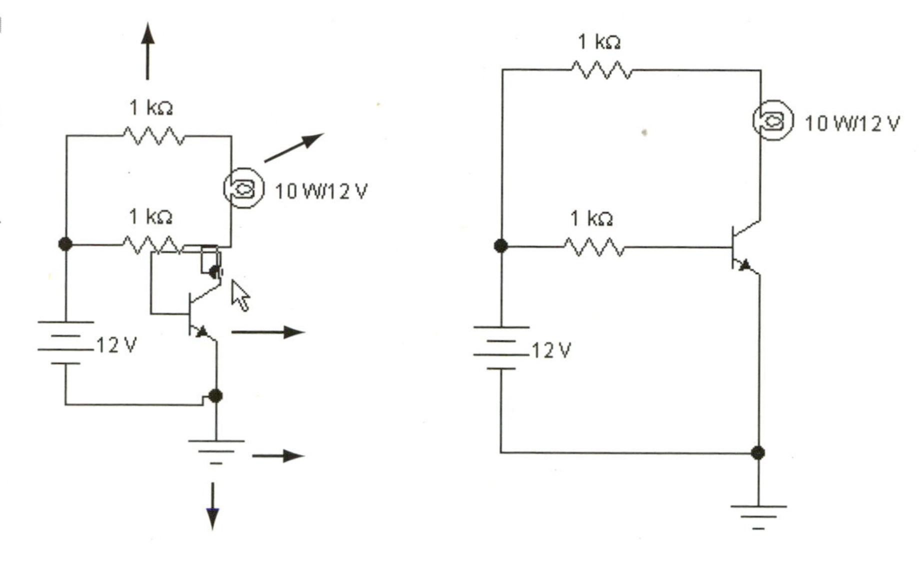

Figure 14.6
Spacing components just a bit farther apart creates a more aesthetically pleasing drawing.

this, I merely moved the parts away from each other and rerouted the wires to the various terminals (if necessary). Use this same technique in Multisim.

CAUTION

Moving a component once wires have been connected will almost always mess up your work. If possible, try not to move the component once placed.

Data Buses

Multisim includes a *bus drawing feature* (see Chapter 3, Drawing with Multisim, for more details).

In EWB5, working with the number of lines, digital ICs, and circuits required to make a complex circuit can make you dizzy. A 4-, 8-, or 16-line data bus that carries bidirectional signals is the answer. But this can sometimes create oodles of wires that confuse even the most experienced. The trick with EWB is to make them as easily readable as possible, by placing them parallel to each other and not crisscrossing. Run a trace, then manipulate the wire to run parallel (see Figure 14.7). Be consistent. You can also create a data bus along the top or bottom of the screen, similar to the power bus, and run each IC to it (see Figure 14.7. Label the lines with the text feature.

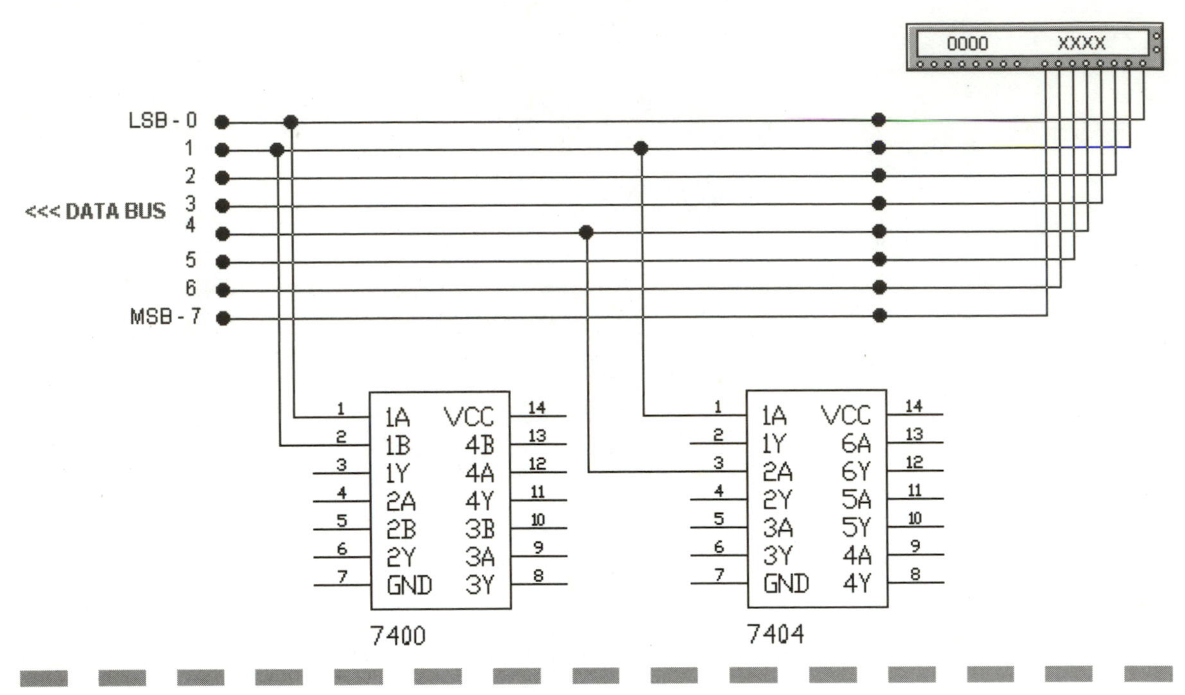

Figure 14.7 A data bus can organize a ton of wiring at once.

One good solution is to take two 16-pin ICs and place them horizontally parallel on either side of the screen and run the lines between them. Paste a bunch of connectors along the wires, then delete the ICs. This leaves a tidy-looking data bus. You can also use the resistor pack if you want the lines closer together.

Labels

Labeling a part ensures understanding in the future. You may remember what the part is now, but when you open that circuit in five months, you are likely to be scratching your head and kicking yourself for not identifying it before. Get in the habit of naming parts and displaying their reference IDs (R1, C2, Q5...). Make sure to set the Show Labels option under **Schematic options** > **Show/Hide**. In Multisim, right-click in a blank area of the screen and choose **Show**.

Notes and Text

As well as drawing parts on the screen, you have the ability to place a variety of text onto your drawing (see Figure 14.8). Most people don't even realize this feature is available on EWB5. If you need to add weird symbols, formulas, or notes, this is the feature to use.

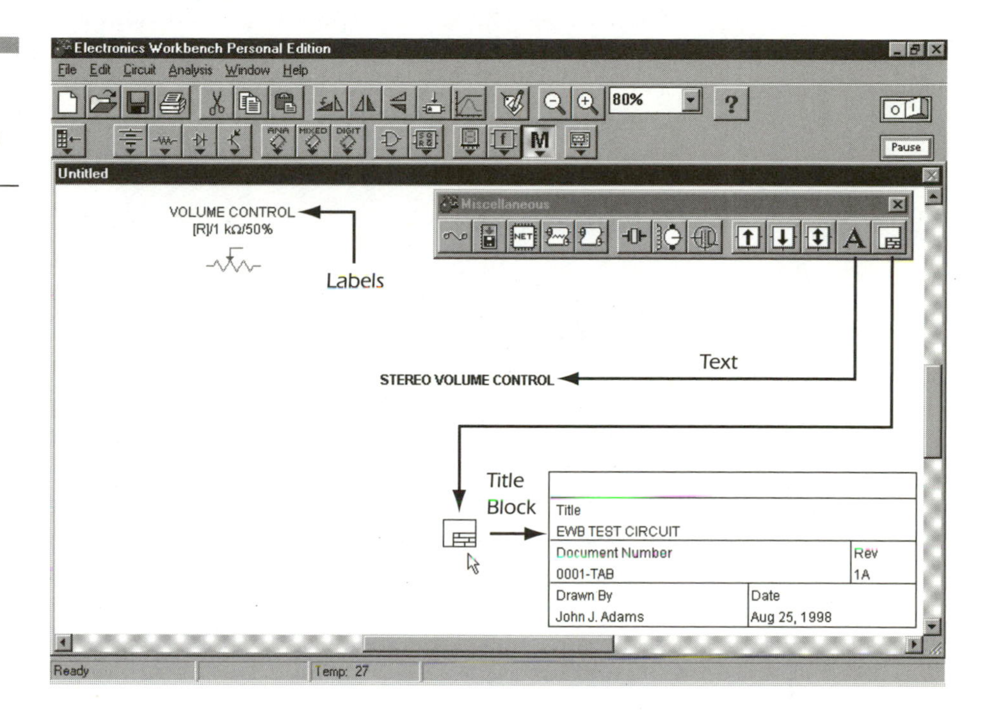

Figure 14.8 Text and title blocks help to document a drawing.

To place text in Multisim, right-click in a blank area of the screen and choose **Place Text**. Click the pointer where you want the text placed. Type in the text. Click outside the text somewhere to place the text. Unlike EWB's adjustable fonts, Multisim can only use one set font, but you can change the color by right-clicking the text.

Title Blocks

If you are using Electronic Workbench 5 or Multisim 6 in an engineering capacity, you may want to use the title block feature. This places a box where all pertinent information about the drawing can be added to help further document the circuit.

Descriptions

The Description box is an excellent way to document a drawing. I usually use it to remind myself of an analysis to run or settings in the circuit. This separate window can contain text and symbols. The box is sizable, so you can either display tons of text on the screen at once or shrink the window and display only a few choice words. To access the Description box in Multisim, choose **Edit** > **Description** or hit **Ctrl+D**. To access the Descriptions box in EWB5, choose **Windows**, **Description** or hit **Ctrl+D**.

Subcircuits

If you are building a rather complex circuit, it is sometimes best to divide it into sections and convert those into subcircuits. For example, if you require a power supply to be inserted into a design on which you are working, you can build that supply, convert it to a subcircuit, and insert an icon with one or two output terminals onto the main schematic.

See Chapter 3 for more information on subcircuits used in Multisim. With EWB5, If you want to change a section on the subcircuit, just double-click it and a window will open with that circuit (see Chapter 9).

Crossovers

If a drawing becomes a wiring nightmare, with lines crossing over each other, maybe you want to create a *crossover subcircuit*. You can eliminate the spaghetti by making the line crosses inside a subcircuit (see Figure 14.9). In Multisim you can also use *virtual wiring* to achieve the same purpose or flip the instrument to changes its terminals direction. See Chapter 3 for more information on this.

Colors

If you are trying to read signals from various sources, it is best to color the wires. For example, if you are trying to see what the first data channel on the logic analyzer is reading, color the wire red. For the second, use blue. Keep adding lines and different colors. Now

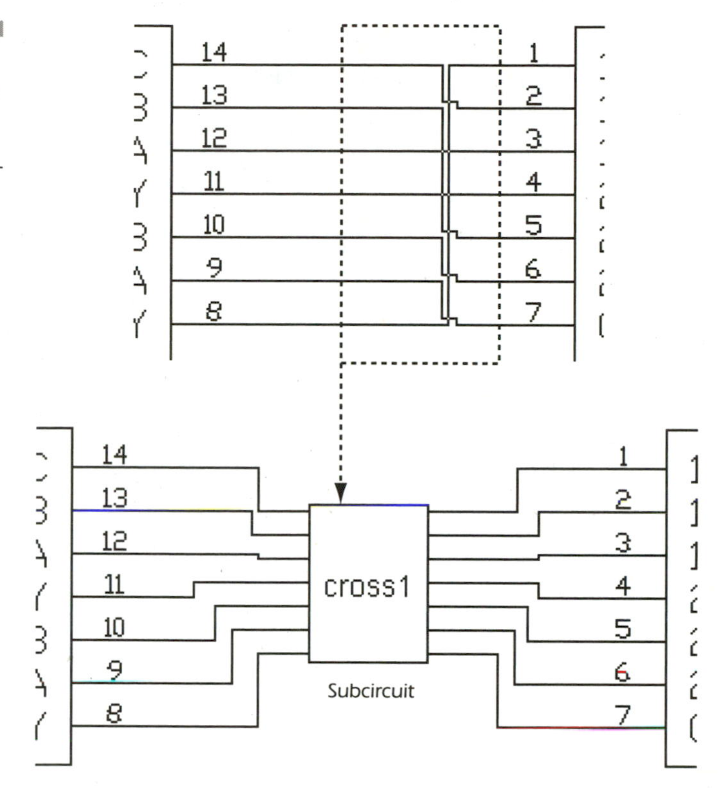

Figure 14.9 Create a subcircuit to eliminate crossed wires.

open the instrument and run the simulation. Each terminal's trace will be in a different color.

Grounds, Vcc, and Vpp

It is not always necessary to run a positive and ground bus. If you want to make things clearer and use less wiring (less is more), then use the ground symbol instead of running lines to the bottom, the Vcc of a 5V power supply, or the Vpp of a 15V supply (see Figure 14.10). There is also a pull-up resistor in the Sources Toolbin that pulls the pin up to Vcc level. This is used quite often in digital circuitry.

Zoom-in/out

Make liberal use of zoom in and out. If you have spaced out parts, it is sometimes hard to take it all in with one screen (not everyone has a

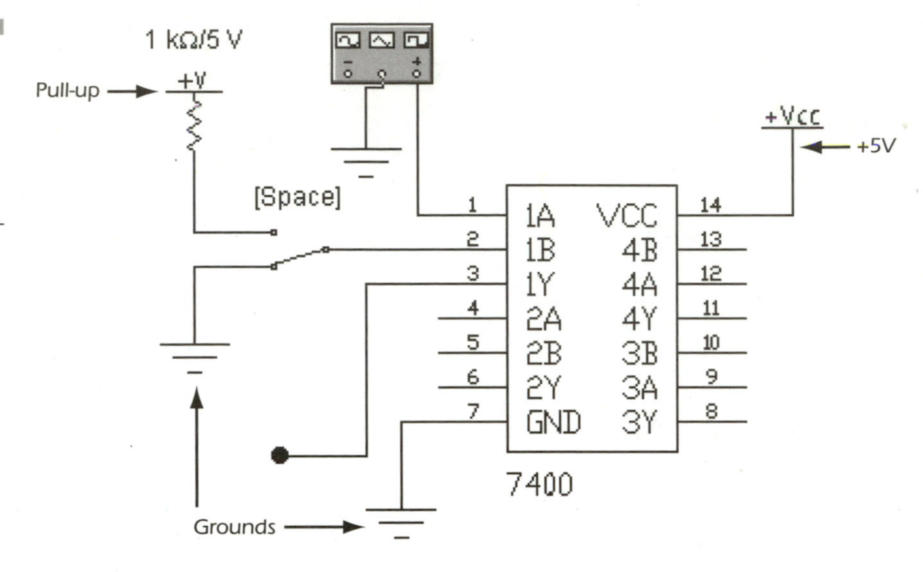

Figure 14.10 Vcc, ground, and pullup resistors eliminate unnecessarily long wiring.

21-inch monitor!). So make use of the zoom-in command to tweak the drawing then back out and take an overview of it. (Of course, you can always try to convince the Powers That Be that a 21-inch monitor is the *only* solution.)

Fast Design

Speed is the key to fighting design boredom. Time slows to a snail's pace when you are building a circuit that you know could be done faster if only you knew a few hints. Learning these tricks and tips will save you design and simulation time:

1. The best way to speed up schematic design and simulation is by learning the program inside and out. This means knowing where each command resides, how it is used, and if there is a faster way to perform this command.
2. Use the copy/paste command on common components.
3. Build the DEFAULT.EWB file up to include commonly used components and subcircuits. When you start drawing a new circuit, delete each component you don't think you will need.
4. Place a copy of each component you think you will need for a new circuit onto the screen at the start, making them available for quick copy/paste.

5. Use a combination of keyboard shortcuts with one hand and the mouse with the other.
6. Get a scroll/zoom mouse. My mouse has a wheel to let me scroll a window. I zoom in and just scroll with the mouse instead of using the scroll bars. You can also program an extra button to perform whatever command you wish.
7. Add memory and a faster processor. You can never have too much Megabytes or Megahertz.
8. Doodle your designs on a pad while waiting for Multisim or EWB to open, or while long simulations are being performed. This gets you into the design frame of mind.
9. Surf the 'Net to see if there are pre-made circuits or Netlists available.
10. Learn how to debug designs faster.

Debugging Circuits

Having patience is truly a virtue. Having patience to deal with software anomalies is nearly impossible and has driven many a human to the edge. When I began this book, almost every mildly complex circuit I built *would not run*. What was I doing so wrong to make those annoying error windows keep coming up? After taking a few seconds to think (smacking or kicking a monitor is expensive, painful, and doesn't help) I would do one of the following:

1. Scream! OK, now that I am slightly calmer...
2. Break a circuit up into several parts to debug it. Copy/paste the sections into a new circuit and run it there. Once each section is debugged, work on piecing together the master design.
3. Pay attention to the simulation error log/audit trail. Read through it and see what Multisim thinks is the problem. Access it by selecting **View > Show/Hide Simulation Error Log/Audit Trail**. Older versions of Multisim automatically open this at the end of every simulation. In EWB, check out the Statistics section after running a simulation. Access it by going to **Analysis > Display Graphs**.
4. Go to **http://www.electronicsworkbench.com/html/technical_support.html**. My site at **http://www.basicelectronics.com/ewb** also has a tech section that lets EWB users help each other.

5. Contact EWB's technical support team at 416-977-5550, (by fax 416-977-1818) or fill out the Technical Support Form on their website.
6. Adding small resistors in line with certain components can correct quirks in EWB and Multisim. An example is adding a 0.1-ohm resistor in series with multiple switches in parallel. Setting IT4 from 10 to 100 also helps switch and POT problems.

Additional debugging tips are described in Chapters 6 and 12. Once you become familiar with the programming quirks, you can leave out step 1. (If only there were a way for Multisim or EWB to explain to you the exact problem and solve it for you automatically.)

Increasing Simulation Speed or Accuracy

One of the most-asked questions the EWB Company receives is "How do I speed up simulations?" Electronics Workbench cannot simulate a circuit in real time, and we are likely to be years away from this. If it *were* possible, we could have an interface card that lets you go from real-world to EWB/Multisim and back to the real world; the ultimate breadboarding system.

There is a Catch-22 involved in simulation. You can speed up the simulation but lose accuracy or slow the simulation and raise accuracy; speed and accuracy just don't mix with EWB/Multisim. The trick is to start with a higher degree of precision and begin to work your way down until there is a happy medium between speed and accuracy.

Electronics Workbench V5

The accelerator pedal for speed and accuracy in EWB is the dialog boxes under **Analysis** > **Analysis Options**. There are five tabs to deal with: Global, DC, Transient, Device, and Instruments.

GLOBAL TAB

The only change EWB recommends is to RELTOL (Relative Tolerance Error) (Figure 14.11). It is a universal accuracy control. Changing the value between 0 and 1 significantly effects the speed or accuracy. Raising the value lowers the accuracy but speeds the simulation. I recommend first taking a zero from the default 0.001 and making it 0.01 for greater speed; add a zero for greater accuracy.

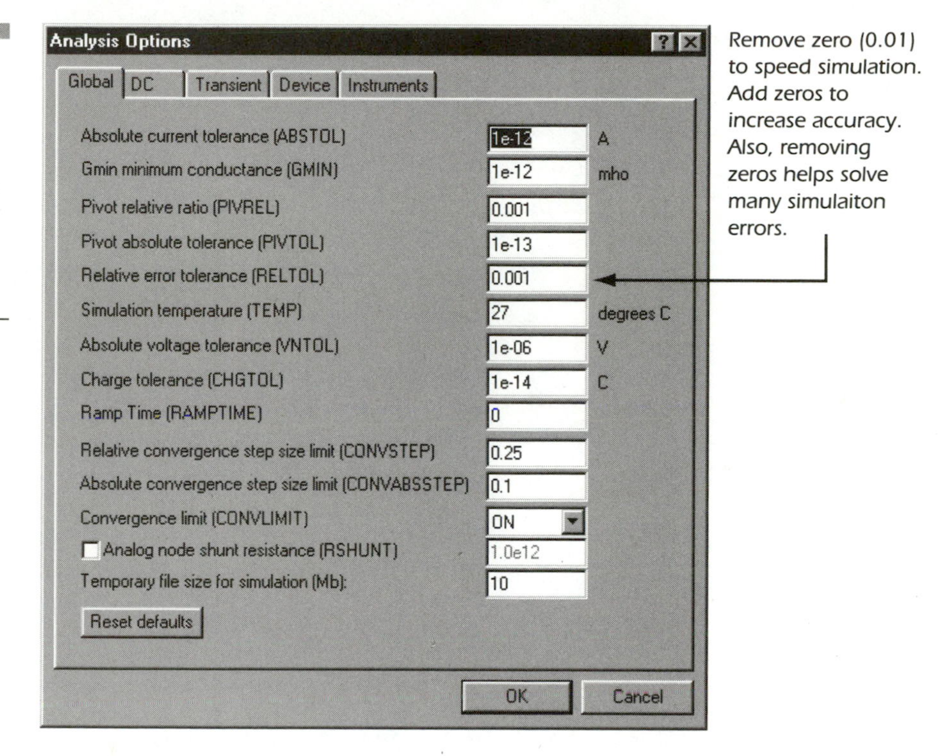

Figure 14.11 The Analysis Option tabs are used to adjust simulation speed and accuracy. On the Global tab, I do not recommend changing anything but RELTOL.

DC TAB

The DC operating point (Figure 14.12) goes through x calculations, estimating and attempting to figure out the solution to each circuit node. You can speed this up (but reduce accuracy) by lowering the number of iterations (repetitions) it takes to come to a solution. This is done by lowering ITL1. This goes the same for GMINSTEPS under the same dialog box. Try reducing each by half to speed up. Conversely, if you are trying to raise the accuracy of the circuit, these numbers can be increased. For greater accuracy, increase ITL1 to 1000 and increase GMINSTEPS by a factor of 10 each time.

DEVICE TAB

Leave these settings alone unless you are experienced in FET modeling.

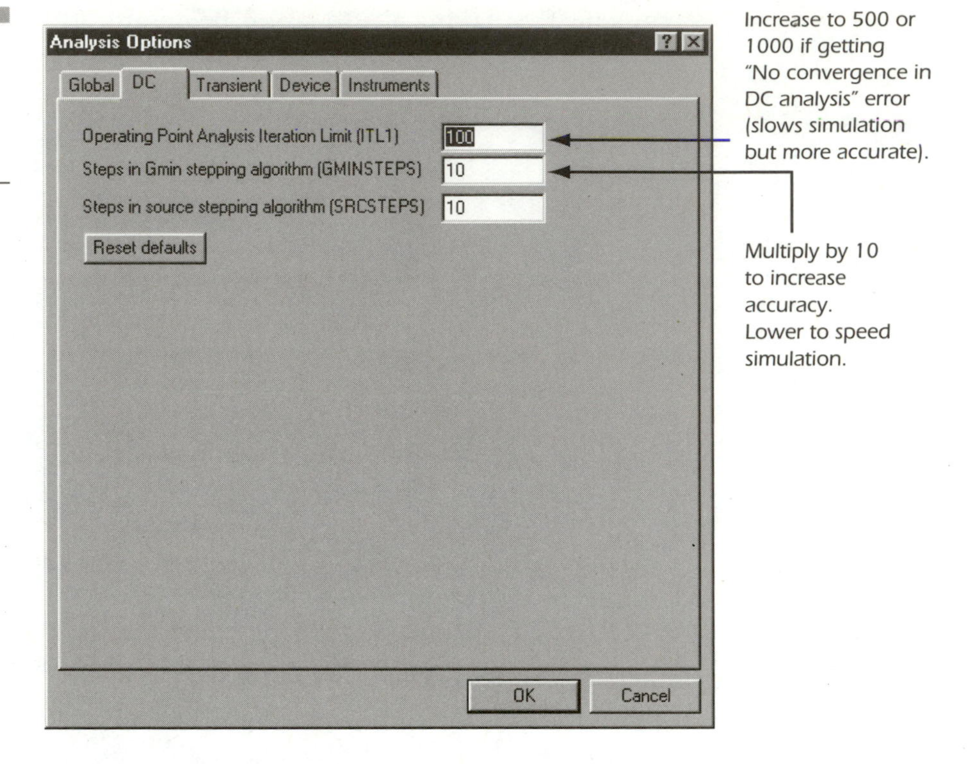

Figure 14.12 On the DC tab, you can change ITL1 or GMINSTEPS as shown.

TRANSIENT TAB

If you are performing a transient analysis, this dialog box (Figure 14.13) can be used to increase speed or accuracy. Increasing the ITL4 setting slows simulation but increases accuracy; 10 to 25 is about normal. MAXORD should not really be touched, but you can experiment by raising up to 6 to increase accuracy. The next key point is the method. The TRAPEZOIDAL method provides the fastest simulation but becomes unstable if used for oscillation circuits. In this case, use the Gear method. Also, ideal switch or SCR problems can be solved by using Gear.

INSTRUMENT TAB

The Instrument Settings Dialog Box under Analysis Options can also be used to speed simulations that use the oscilloscope or Bode plotter (see Figure 14.14). The idea is to force EWB to calculate fewer numbers in a given time. This is done by lowering the **Minimum number of time points** or raising the **Maximum time step** (time between calculations). After unchecking **Generate time steps automatically**, check on either **Minimum number of time points** or **Maximum**

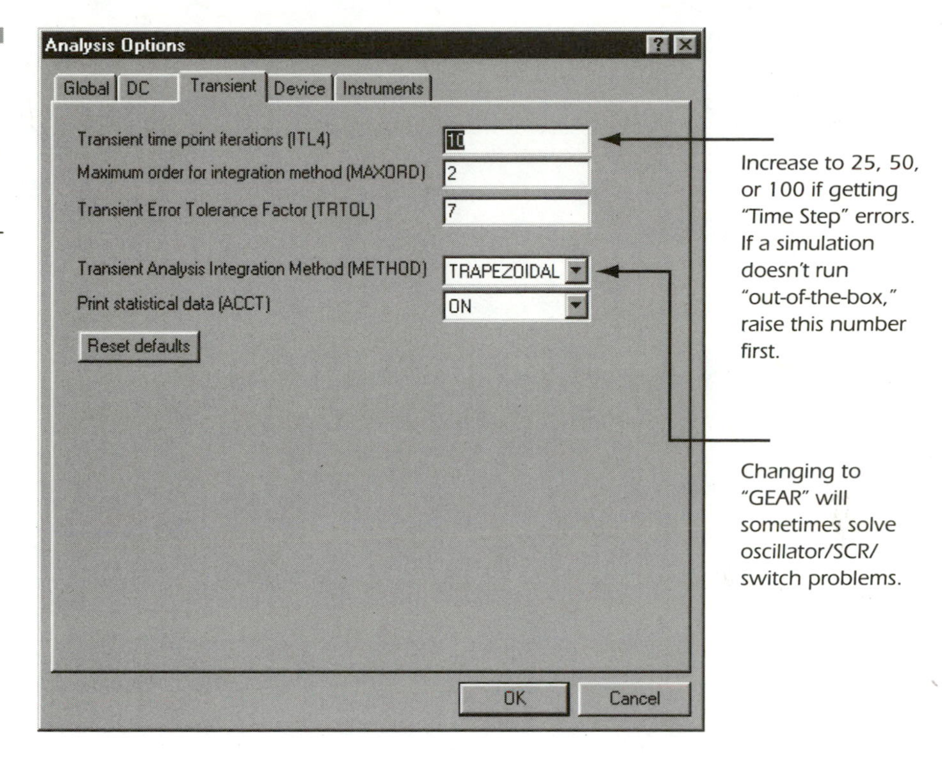

Figure 14.13
On the Transient tab, raising ITL4 and using GEAR will solve many problems (but slow the simulation).

time steps (TMAX). Lower or raise the numbers to speed the simulation. For example, if the time points are set to 100, lower them to 50. If TMAX is set to .0072, raise it to .072. Be aware that this will create a more jagged waveform (Figure 14.15). However, if you are just trying to get a quick idea of what a circuit is doing, this is fine; you can always change the numbers back later. The same goes for the Bode plotter time points: lower for greater speed, less accuracy.

Multisim 6

Multisim uses very similar settings to adjust speed and accuracy, but they are accessed with greater difficulty. Most of the settings are under **Simulate** > **Default Instrument Settings** > **Miscellaneous Options** tab > **Analysis Options** button. The Analysis Options window will open as pictured in Figure 14.16. The description of a highlighted option can be seen at the bottom. By unchecking **Use Default Value**, you can adjust the settings. The options use the same abbreviations or acronyms as EWB; for example, RELTOL is listed under the

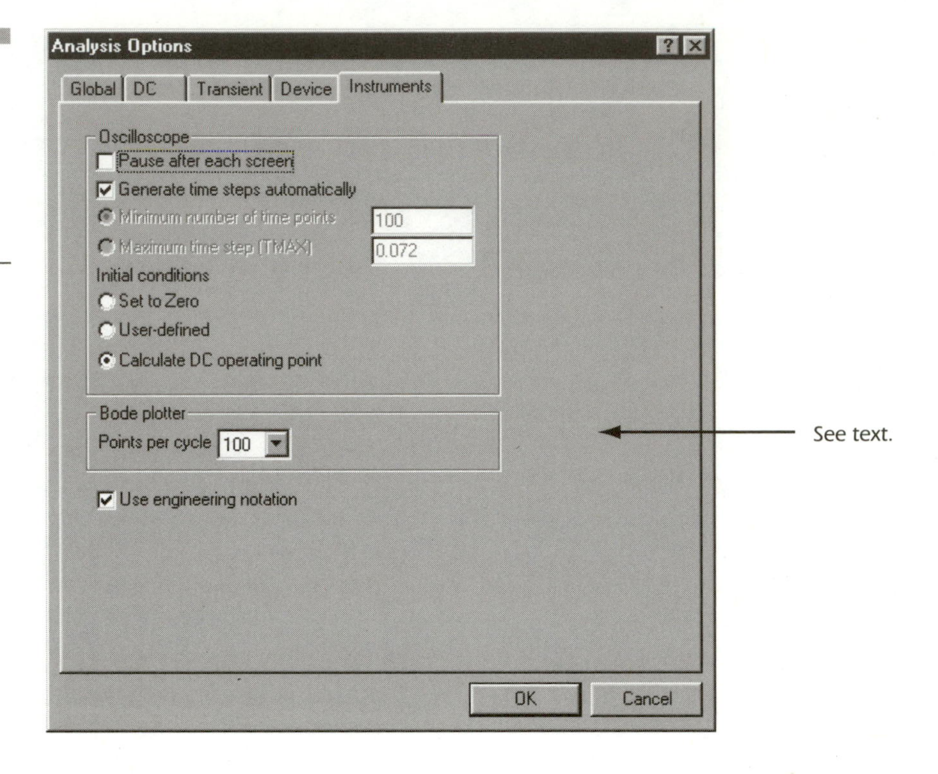

Figure 4.14
Adjusting the time step settings on the Instruments tab give a great deal of control over speed and accuracy.

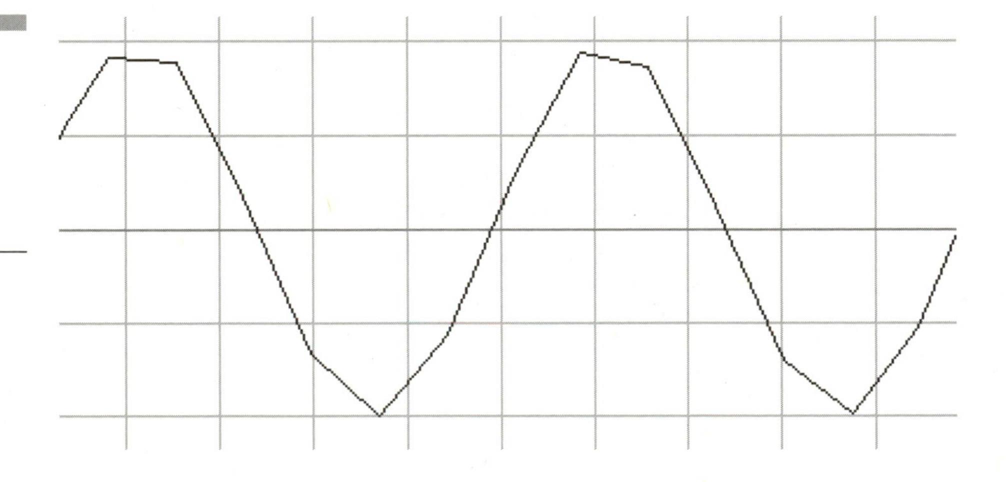

Figure 4.15
Adjusting the time steps or TMAX can create a jagged waveform (but speeds the simulation.

Global tab as RELTOL in EWB5. See the previous section for details on each setting and how to adjust them.

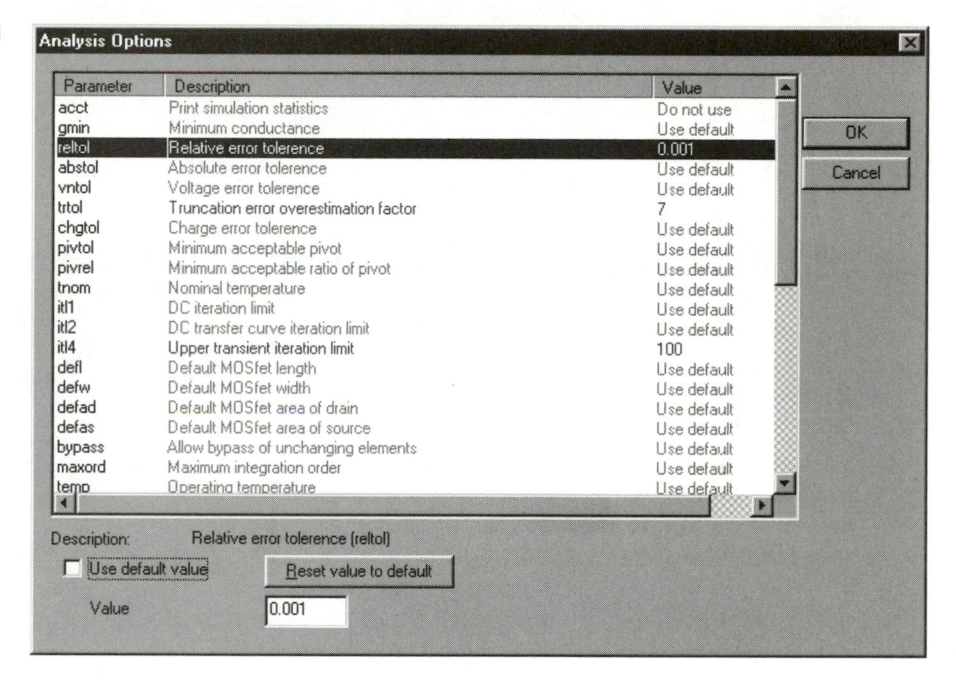

Figure 14.16 In Multisim, the Analysis Options are more difficult to access and change. You must go to Simulate > Default Instruments Settings > Miscellaneous tab > Analysis Options button.

You can also access the Analysis Options from the Miscellaneous tab under each Analysis dialog box.

Aside from adjusting the previous settings in either EWB5 or Multisim, you can simply break a circuit into smaller pieces and test each section this way. The smaller the circuit (fewer components), the faster the simulation. The last tip I have is the obvious one: speed up your computer with a higher-speed CPU or additional RAM. For more tips as they become available, see EWB's website as well as my own (**www.basicelectronics.com/ewb**).

Having Fun with EWB/MULTISIM

Now it's time to make a few interesting circuits using Multisim and Electronics Workbench.

Making Noise

Life is too serious. These two circuits came to me when I was told EWB could not output a signal to a sound card (and why not?) Build the circuit shown in Figure 14.17, making generous use of cut and paste for the buzzers. Set each buzzer's voltage to 4.5 volts and adjust the frequencies as shown. Load different patterns into the word generator's buffer and run the circuit.

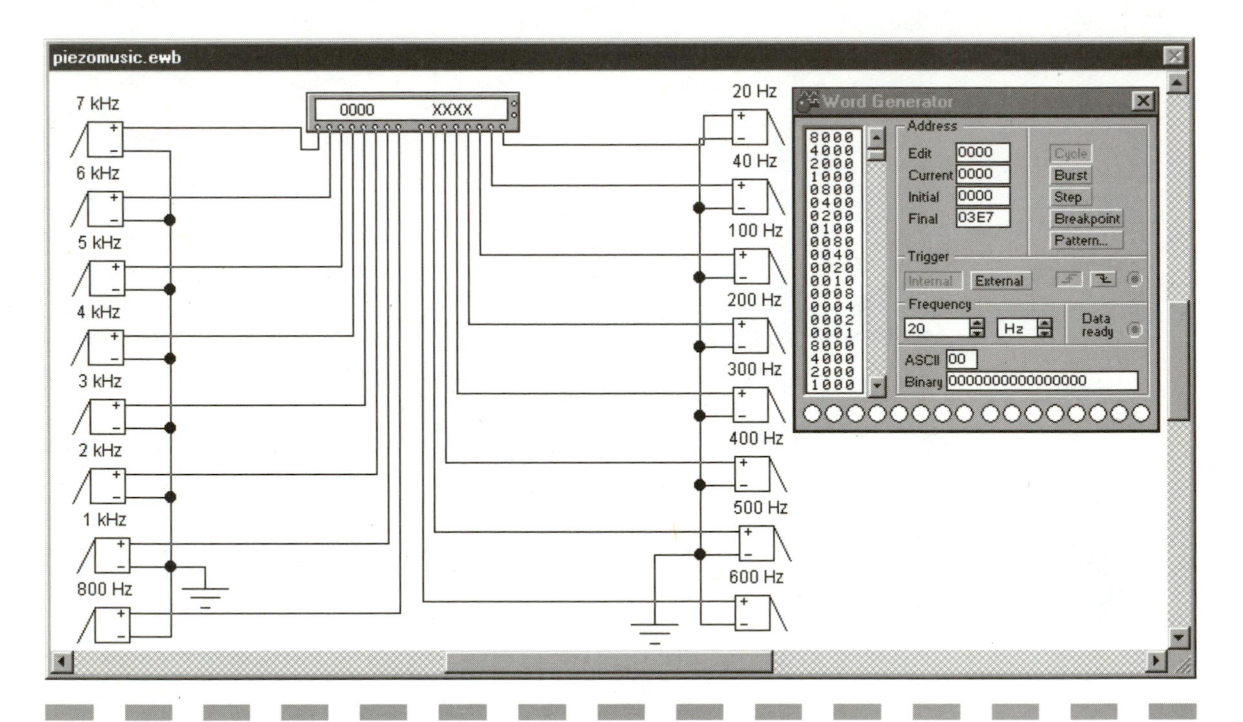

Figure 14.17 Let's make some noise!

I have tried unsuccessfully to get this circuit to work with Multisim 6.11. (In fact, I cannot get any circuit with more than one buzzer to work with Multisim.) This may be an anomaly in my computer setup or something that will be fixed in 6.2 or higher. I have noted it just in case this happens.

Heads or Tails?

Can't make a decision? Let Electronics Workbench do it for you with this interesting circuit. It's a classic project many of you may have built when beginning electronics, and I've decided to create a digital version here. There are many configurations, but I chose this one to be compatible with EWB/Multisim. Build the circuit as shown in Figure 14.18 and set the flipping key to F. Start the switch in the Reset position. Hit the **F** key once to flip and then again to reset the circuit.

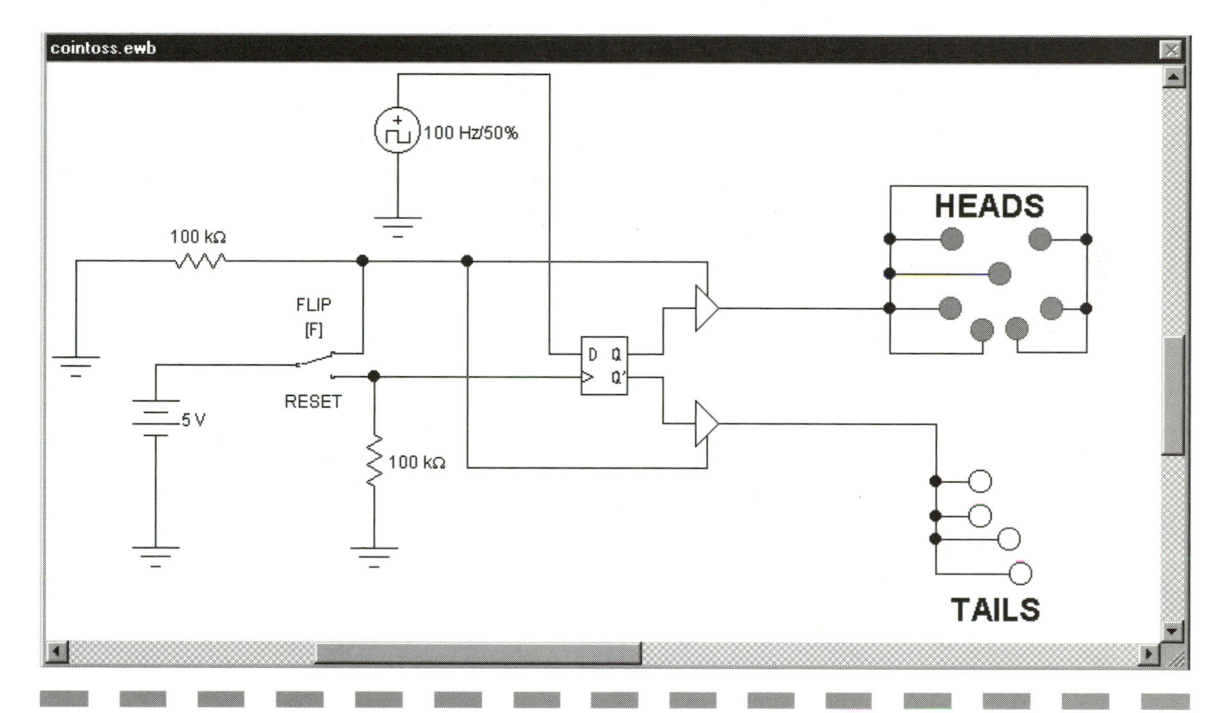

Figure 14.18 My take on the classic Electronic Coin project.

Lissajous Figures

Hooking two AC voltage sources into the oscilloscope's X and Y readings can create some interesting patterns. Build the circuit as shown in Figure 14.19 in either EWB5 or Multisim 6. Set the AC source to 120V, 3kHz and the function generator to 120V, 2.7kHz to start. Set the oscilloscope as shown, making sure to click the A/B axes. Start the circuit with the scope open and let it run for a moment. Expand the scope if you wish. Now adjust the frequency and amplitude of the function generator slightly and rerun the circuit.

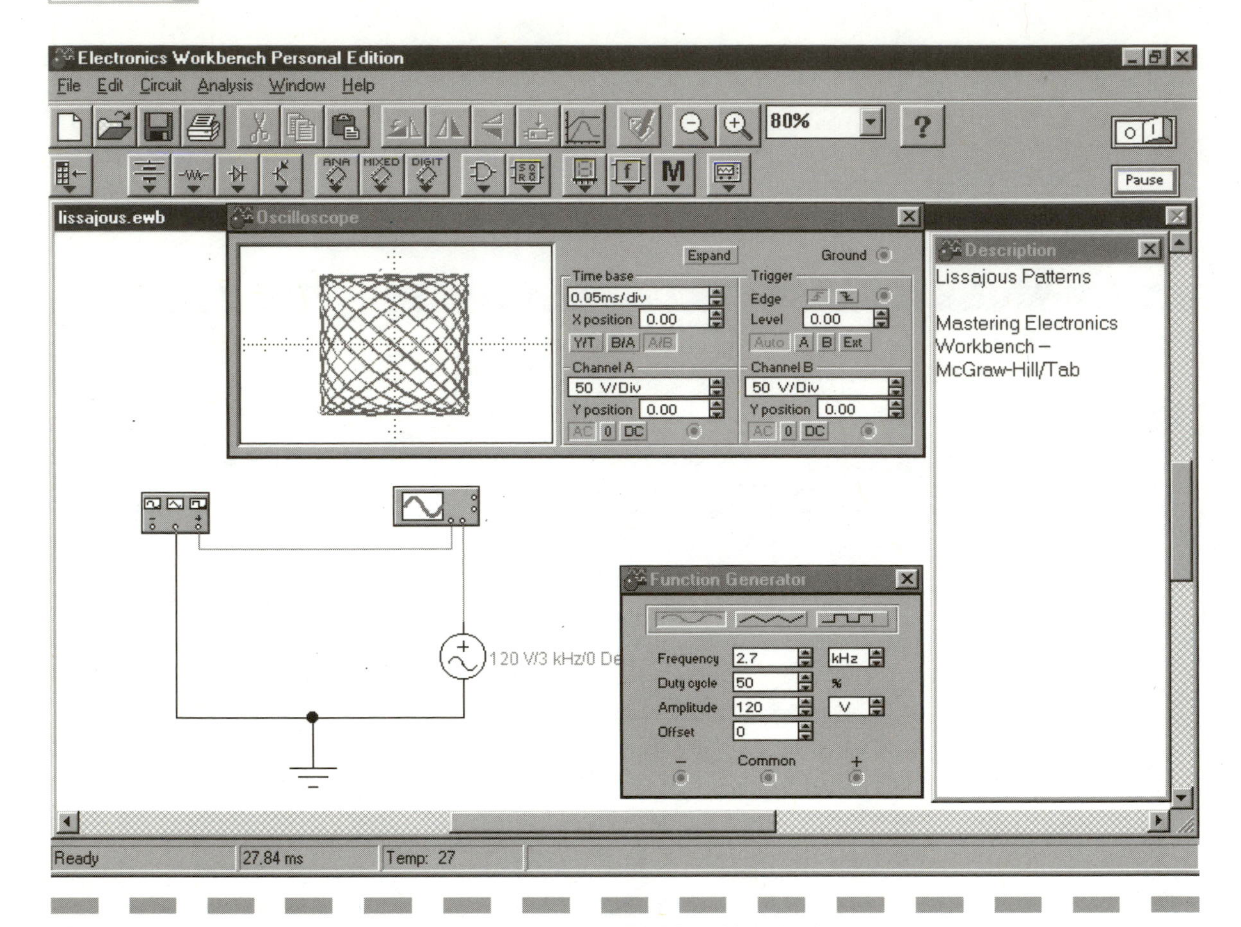

Figure 14.19 Lissajous figures can be created using EWB5 or Multisim 6.

Missing Pulse Detector

Build the circuit as shown in Figure 14.20. This circuit uses a 555 timer to watch for pulses and warn if they are not present. Think of it as a button that resets a bomb countdown. Every once in a while you have to push it to reset the counter. However, if the countdown runs out, BOOM! The Space key is used. Start in the off position. When the circuit is started, hit the **Spacebar** to move it to the on position and then hit it again to the off position. Do this every other second or so. Stop for a bit and see what happens.

If you build this circuit in Multisim 6.11 or lower, you must fix the 555 first because it doesn't work. (The instructions are in Chapter 7, "The Astable Miltivibrator Circuit.")

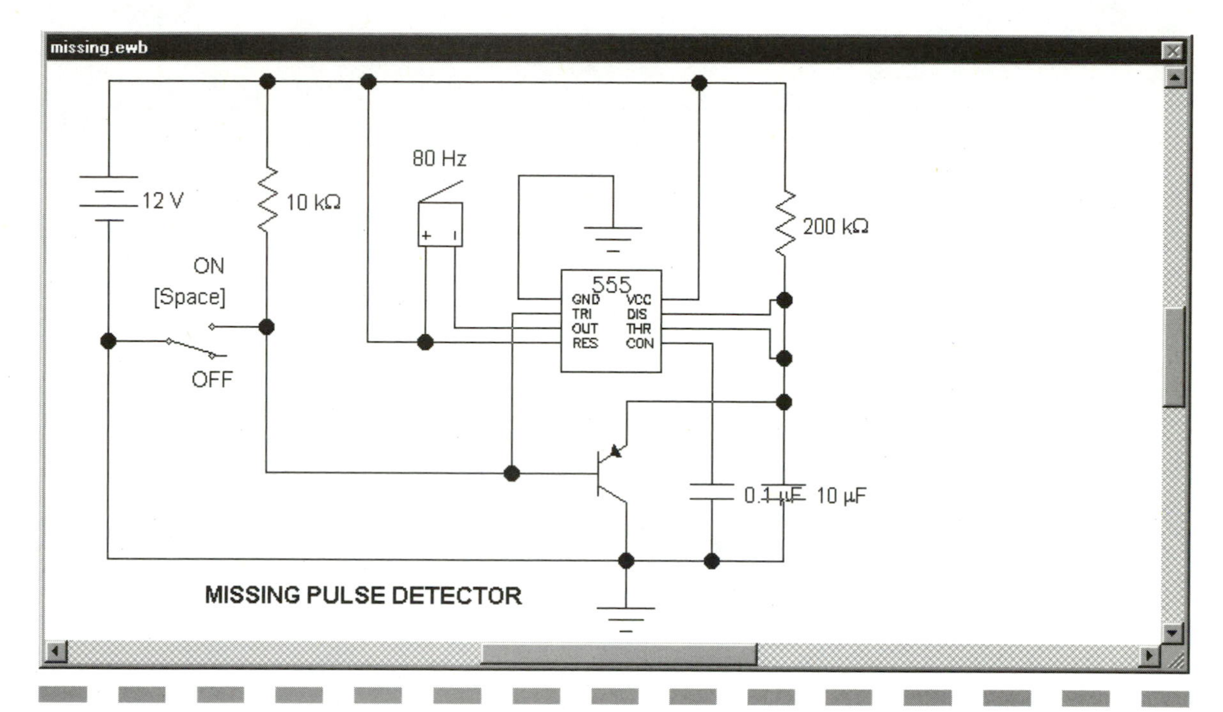

Figure 14.20 Keep hitting the space key or BOOM!

Three-way Switch

I often am asked how a three-way switch works (like those in your hall). This simple demonstration clears that up. Build the circuit as shown in Figure 14.21. Set the upstairs switch to A and the downstairs one to L. Use them as you would a real three-way switch in your house and note the results. Color the wires to show how the wiring is actually done in a house.

Momentary Switch with EWB 5

It's a real pain to hit a switch key twice to change it from Off to On to Off. Most digital applications only require a momentary change in state to trigger a signal. EWB does not contain a momentary pushbutton, so I created one (Figure 14.22) with a capacitor, AND gate, and two switches set to the same key. The only limitation is that it can only supply a 5-volt signal to a digital circuit (from the output of the AND gate). You can build the circuit into a subcircuit and use the tip at the end of this chapter to make it available for every digital drawing you do.

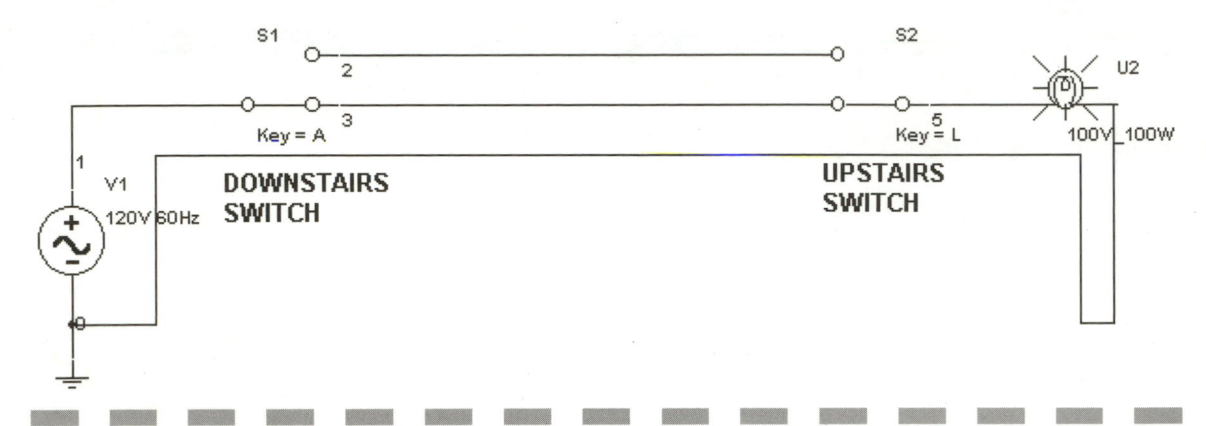

Figure 14.21 Want to know how a three-way hall switch works?

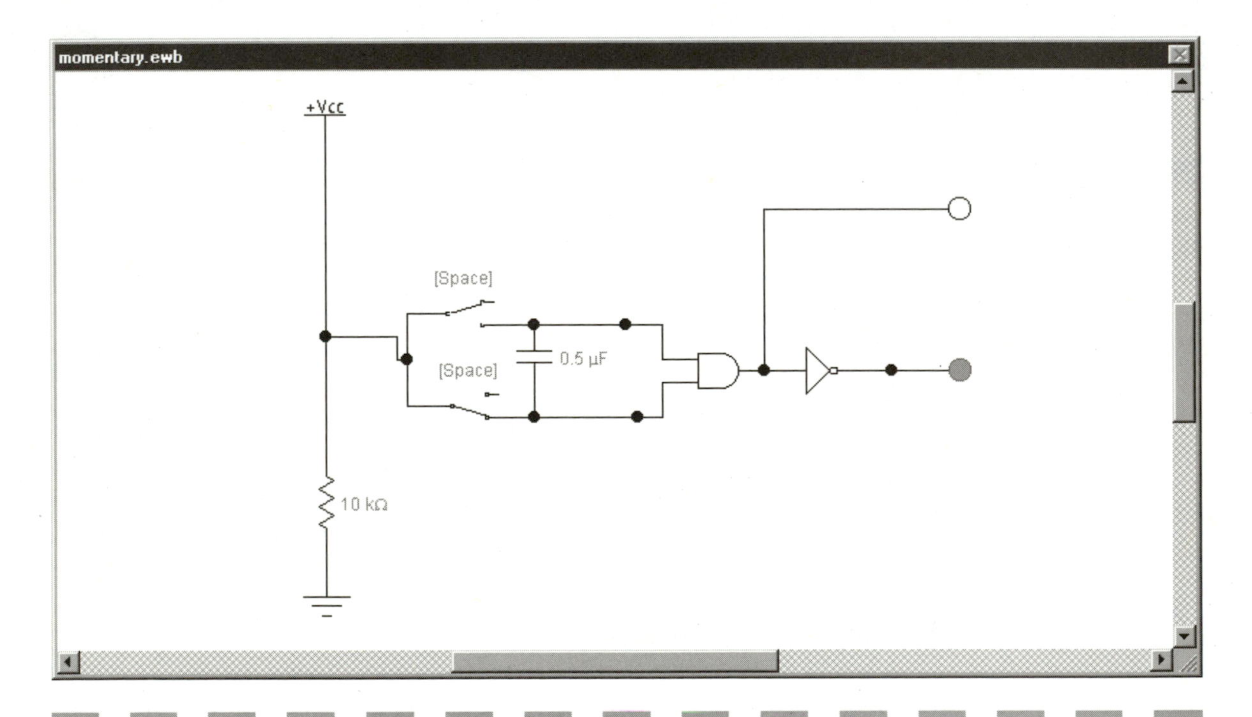

Figure 14.22 Create this momentary switch with EWB5.

EWB5 Wattmeter

Electronics Workbench 5 does not contain a wattmeter, as Multisim 6 does. The only solution is to build the circuit shown in Figure 14.23,

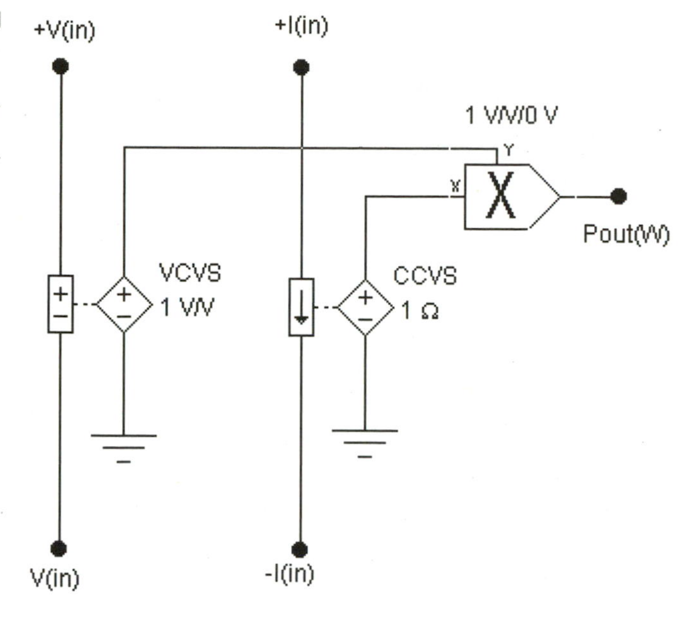

Figure 14.23
This subcircuit emulates a wattmeter in EWB5.

make it into a subcircuit, and use a voltmeter as a readout device. An example circuit is shown in Figure 14.24.

Making Your Own Circuits

Building fun electronics projects is the goal of just about any hobbyist. The same can hold true using EWB/Multisim. Try to think of different ways to apply this software to keep your interest in it strong. Assemble virtual circuits from magazines that put out new interesting circuits each month. After using the software for some time, you will learn the tricks to get around EWB's and Multisim's limitations. When you do, send me your fun circuits. Also visit **http://www.basicelectronics.com/ewb** for more circuits as they become available.

Learning with Your Own Applications and Circuits

I own a lot of books about electronics—too many! Some of them are for electronic fundamentals, but my favorites are those with circuits

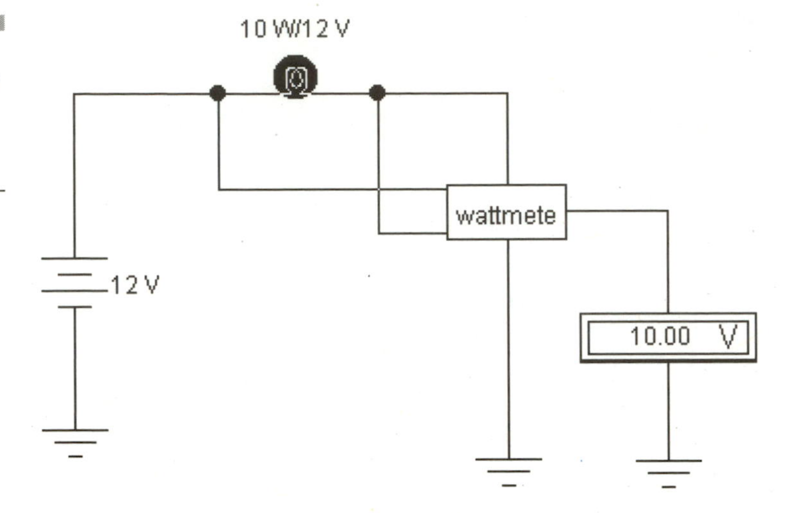

Figure 14.24 An example circuit of the wattmeter created in Figure 14.23.

to build. Take for example the Rudolf F. Graf *Encyclopedia of Electronic Circuits* series published by Tab Electronics or Delton T. Horn's project books (also by Tab). I've learned more about electronics with these circuit catalogs and Electronics Workbench than with any other method. It also gives me a stronger knowledge of EWB and its capabilities as I build the circuits. Just throwing together a circuit in minutes easily tests concepts, and I can save the circuit for later importing into my projects. Quite a timesaver.

But don't limit yourself to building circuits from books. I also tend to look at real-world electronic items and try to figure them out using EWB—sort of reverse engineering. After studying a PCB, I build it as a schematic. From there, I can get an idea what is happening and why.

My own circuit ideas are enhanced using EWB as well. There are many designs that pop into my head that I would love to test quickly somehow. I try to figure out how EWB can simulate the design and build it. Debugging can solve a lot of idiosyncrasies I may not have considered. Either way, using EWB in all aspects of electronics strengthens your knowledge and speeds up your design or test time, which leads us into the next section.

EWB for Me

Each person will use Electronics Workbench and Multisim for different purposes. For example, I use it to design projects as well as to reverse-engineer devices (I always want to know how something works). It's helped me learn complex electronic theory and teach others basic principles. I also use it to do schematic capture, placing the images on my website as well as sending people ideas or explanations. Most of my uses are geared toward a hobbyist, but these are not the only uses for this powerful software package. The following is a suggested list of what you may want to use EWB or Multisim for in your electronic endeavors.

- **Students**—Many students around the world use Electronics Workbench to learn. It helps to teach basic electronics principles in a fast and interesting manner. Instead of wasting hours building each circuit to test a concept, they can assemble circuits digitally in minutes.
- **Teachers**—By using faulty components or locking certain circuit elements, an educator can teach students important principles. This helps to teach troubleshooting and basic electronic concepts as well.
- **Corporate Trainers**—EWB is used by various organizations to educate employees who require a basic knowledge of electronics. This may include technicians, computer or network specialists, repair persons, installers, and many others. In fact, some electronics boutiques are requiring their salespersons have a basic knowledge of electronics to further assist their "knowledgeable customers."
- **Hobbyist**—As a hobbyist, I use EWB to test project ideas. This includes little tidbits that sit in the back of my mind awaiting a quick method of escaping. With EWB I am able to hit a few keystrokes, make a few swipes of the mouse, and have a proof-of-concept circuit up and running—quite a bit faster than penciling the design and then breadboarding, possibly burning out a few expensive ICs in the interim. I can use the analyses to fine-tune the design, and once I have a completely functioning design, I can route it over to Layout or Ultiboard to create a PCB speedily. This saves another few hours that would have been spent in sit-and-think mode.
- **Repair Technicians**—It's sometimes hard or impossible to find a schematic for a device you may be repairing. Using EWB or Multisim, you can, to a degree, simulate the circuit to find trouble points. For example, a subcircuit for a television may be blowing transistors. Using the program to test the subcircuit by creating shorts may produce the same results. You can then take this infor-

mation to the real circuit. This is much cheaper than blowing expensive transistors or ICs.

- **Medical Electronics**—Many medical devices today are electronics-based. Technicians who use the equipment are being required more and more to have a basic knowledge of the electronics involved inside this high-tech equipment. EWB can help train students in this field.
- **Circuit Designer/Engineers**—Engineers and designers use EWB in a virtual development cycle consisting of design, testing, debugging, retesting, and PCB layout. In theory, the engineer doesn't even have to touch a soldering iron or resistor anymore.
- **Component Designer/Engineers**—Because component parameters can be adjusted and then applied to a test circuit so fast, EWB makes a perfect component development platform. Component manufacturers can go from idea to silicon in a much shorter time.
- **Automotive Technicians**—Cars today are rolling microprocessors. Electronics control everything from the engine to the safety systems. A mechanic can no longer have a simple understanding of volts, current, and resistance; he must now know modern digital electronics and electrical engine-management systems, or risk being outmoded. There are discs and books available that use EWB to train the automotive technician quickly.

Miscellaneous Gems and Little-Known Tricks

Many tips and tricks have been shown throughout this book, and here are a few more I have collected:

1. While a component is selected, hit the **F1** key to access information about that part.
2. In EWB5, you can changing the number of inputs to a gate by double-clicking it and selecting the **Number of Inputs** tab.
3. Rename a *.mod to *.lib and place it in the **C:\EWB5\Models** and the appropriate subdirectory (example BJT_NPN).
4. Build a database of rotated components in Multisim.

5. If you own EWB V5, visit **http://www.electronicsworkbench.com/html/version_5.html** for tons of simulation, component, and bug-fixing tips.
6. If you own Multisim 6.xx, visit **http://www.electronicsworkbench.com/html/technical_support.html** for tons of simulation, component, and bug-fixing tips.
7. To have EWB5 display statistical data in the Grapher at the end of a simulation, choose On at **Analysis Options** > **Transient** > **Print statistical data (ACCT)**.
8. Study the help files that come with EWB5 and Multisim 6.
9. Call the EWB Company to make sure you have the most current version of Multisim.
10. The battery in EWB has no resistance. If you want to use a battery in parallel with another battery or a switch, insert a milliohm resistor in series with it.
11. The 555 on Multisim version 6.11 or earlier has a bug in it. To correct this you must create your own inside the User database. See Chapter 7, "The Adjustable Multivibrator Circuit."
12. To make a subcircuit available to each new drawing, open DEFAULT.EWB and create the subcircuit. Choose Replace in the circuit button when naming it. Run the connections to the ends of the subcircuit window. Delete the subcircuit from the drawing. Save the file. Now when you create a new circuit, you can access that subcircuit with the Favorites icon on the top left of the screen.
13. Make your own databases of subcircuits and component models to use in day-to-day designing.
14. Locate component application notes from manufacturers' websites to get circuit ideas.
15. Create a negative power supply to use with op amps and other circuits. Make it a subcircuit and then make it available in the DEFAULT.EWB as previously described.
16. Make a certain color of wire represent a radio wave or light beam while doing schematic capture.
17. The piecewise components are very good customization tools. Create variable voltage files to test your circuits under varying conditions.
18. Use the potentiometer component as a thermistor or photocell.

19. If you need more than two channels on the oscilloscope, use transient analysis instead.

Summary

Tricks and tips come to me each day while I'm making use of EWB and Multisim. Some of them derive from "fooling" the software and others are based on electronic principles that I ported over to the simulator. Collect as many of these as you can and pass them on to me to post on the EWB section of my website for others to use. Maybe we can eventually have our features and fixes incorporated into newer versions of EWB and Multisim.

CHAPTER 15

SPICE and Netlists

Think of each electronics component as an input/output device. A signal is taken in, changed and sent out.

Electronics Workbench and Multisim make use of Berkeley (SPICE) F3. Simulation Program With Integrated Circuit Emphasis. It's a mouthful, but what is it exactly? We briefly explained SPICE in chapter 1. This chapter expounds upon it further.

SPICE is a language used to mimic electronic circuitry using a personal computer. It is a programming language that the computer uses to simulate a circuit, much like Basic or C. The SPICE file is a text file that allows simulators, such as Electronics Workbench and Multisim, to simulate and analyze a circuit and its components in great depth. Every circuit is broken down into node points (the point between each component's terminal) and the SPICE language determines voltage, current, etc., at these laid out connections. In this way, you can simulate nearly any electronics component in a circuit of your design.

Why Is It Important to Learn SPICE?

Electronics Workbench & Multisim are basically graphical ways to input SPICE. Who wants to sit with a pencil, design a circuit, and then spend hours writing a complex SPICE Netlist to simulate it? Not only is this a waste of valuable time, but it is just one more subject you would have to pile onto your curriculum.

Learning SPICE can be difficult, but because SPICE allows you to customize a design and gives a more accurate simulation or analysis, it is vital. Not every electronics component or circuit in existence is modeled yet. And even if they were, new ICs are coming out daily. This is where your knowledge of SPICE will give you designing power. It's similar to learning to draw with a pencil and drafting instruments; even though you use a computer and mouse. SPICE gives you the inside skills to squeeze extra horsepower out of Multisim and Electronics Workbench software.

The subject of SPICE can fill an entire book. In this chapter we merely present an overview of this language—just enough to be of use in basic design and theory. There are several books on the market that cover the subject of SPICE in great detail, including *Inside Spice* (McGraw-Hil) and *The Spice Book* (John Wiley & Sons). There are also many websites devoted to this, available to the diligent search hound.

SPICE Basics

SPICE, in the most general terms, breaks down a circuit into points where voltage or current can be measured or simulated under various conditions. To explain further, let's look at how we would "SPICE up" the circuit in Figure 15.1.

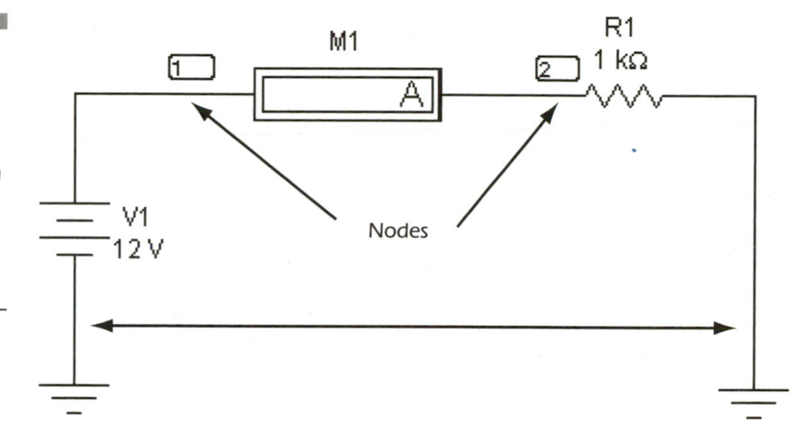

Figure 15.1 With SPICE, circuits are broken at the nodes. Each component between them is defined so that the computer understands what's happening.

We've broken down this circuit into its elementary parts. Between each part is a node point, or simply a node. Think of a node as a connecting wire, or any point along a connecting wire, between two or more adjoining components. It is merely a reference point that is used in SPICE. A circuit can contain any number of nodes.

You can see from the diagram that there are three nodes. I have marked them with EWB's **Show Nodes** feature. The ground is node **0**. Between the positive lead of the battery and the left side of the ammeter is node **1**. The connection from the right-side of the ammeter and the right terminal of the resistor is node **2.** Keep this figure in mind as we continue.

Netlists

A SPICE Netlist is a text file that uses SPICE syntax and is used to input SPICE-based simulators (such as EWB and Multisim).

Create the simple circuit in Figure 15.1. Go to **File > Export** and save the file as **C:\EWB5\My-EWB\SPICE_IT.CIR**. In Multisim, go to **Transfer > Export Netlist**. Open a text editor such as Windows Notepad and go to the directory in which you saved the .cir file (in this case **C:\EWB5\My-EWB**) and open it (see Figure 15.2). This is the circuit's *Netlist*.

Figure 15.2
A SPICE Netlist of the circuit in Figure 15.1.

```
1  ***********************  ***********************
2  *  Interactive Image Technologies               *
3  *                                               *
4  *  This File was created by:                    *
5  *    Electronics Workbench to SPICE netlist     *
6  *    conversion DLL                             *
7  *                                               *
8  *  Wed Jan 27 02:04:15 1999                     *
9  ************************************************
10
11 * Ammeter(s)
12 *
13 X_ammeter_0 1 2 ammeter_1mohm
14
15 * Battery(s)
16
17 V1 1 0 DC 12
18
19 * Resistor(s)
20 *
21 R1 2 0 1K
22
23 * Misc
24 .OP
25 .SUBCKT ammeter_1mohm 1 2
26     R 1 3 1mohm
27     V 3 2 DC 0V
28 .ENDS
29
30 .END
```

Once you have the file open, you see the following code (I have added line numbers to the text file to make the explanation easier):

NOTE

Any line that begins with an asterisk (*) is commented out by the program. It is merely used to make pertinent notes. Electronics Workbench SPICE export function produces comments for each text file. Multisim creates a similar Netlist file.

```
1 *********************** ***********************
2 * Interactive Image Technologies        *
3 *                         *
4 * This File was created by:           *
5 *  Electronics Workbench to SPICE netlist    *
6 *  conversion DLL             *
7 *                         *
8 * Wed Jan 27 02:04:15 1999           *
9 ************************************************
```

The next section of the SPICE Netlist is the list of components—their name, value, and the placement in the circuit.

```
11 * Ammeter(s)
12 *
13 X_ammeter_0 1 2 ammeter_1mohm
14
15 * Battery(s)
16 *
17 V1 1 0 DC 12
18
19 * Resistor(s)
20 *
21 R1 2 0 1K
```

Line 15 is a comment stating that the following code describes a battery in the circuit. In this case, we used a 12-volt model.

Line 17 is the actual data that the computer uses to model the battery inside the circuit. The V1 indicates the parts reference ID. The next two numbers are the battery's placement in the circuit. In this case, it is between node 1 and node 0. DC means that it is a direct current voltage source. The last number is the battery's value, in this case 12 volts.

In line 21, the resistor's code, R1 is the reference ID. The position is between node 0 and 2. The value is 1K or 1000 ohms.

The next section of the SPICE Netlist is the miscellaneous section, which can include subcircuits and models.

```
23 * Misc
24 .OP
25 .SUBCKT ammeter_1mohm 1 2
26   R 1 3 1mohm
27   V 3 2 DC 0V
28 .ENDS
```

Line 23 contains the comment that this section is for miscellaneous code.

Line 24 shows the operating point analysis.

In line 25 `.subckt` means the following code is a subcircuit. In line 13, the ammeter was described; the last section of that line "ammeter_1mohm" referred to this subcircuit. The 1 and 2 are its position in the circuit (between node 1 and 2).

Lines 26 and 27 are the description of the subciruit. It is very similar to the component's code used in lines 11 to 21. This creates a circuit as pictured in Figure 15.3.

Line 28 ends the subcircuit routine and returns back to component section to continue. Line 30 ends the Netlist.

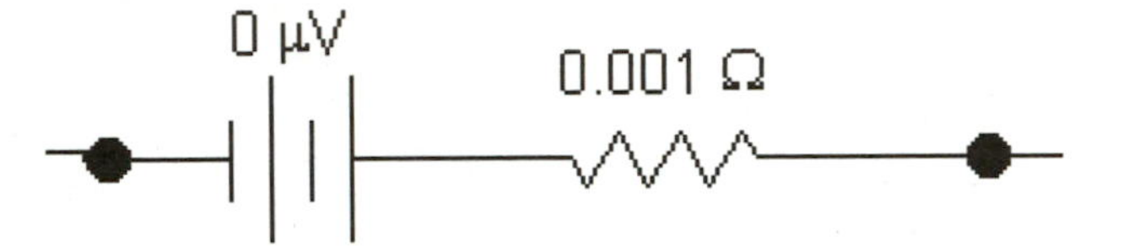

Figure 15.3
A subcircuit is created for the ammeter.

Models

In the next example we look at how to model a component using SPICE. Construct the circuit in Figure 15.4. Now export the circuit and save it as **C:\EWB5\MY-EWB\SP_TRANS.CIR**. Open this file in Notepad. You will see the following:

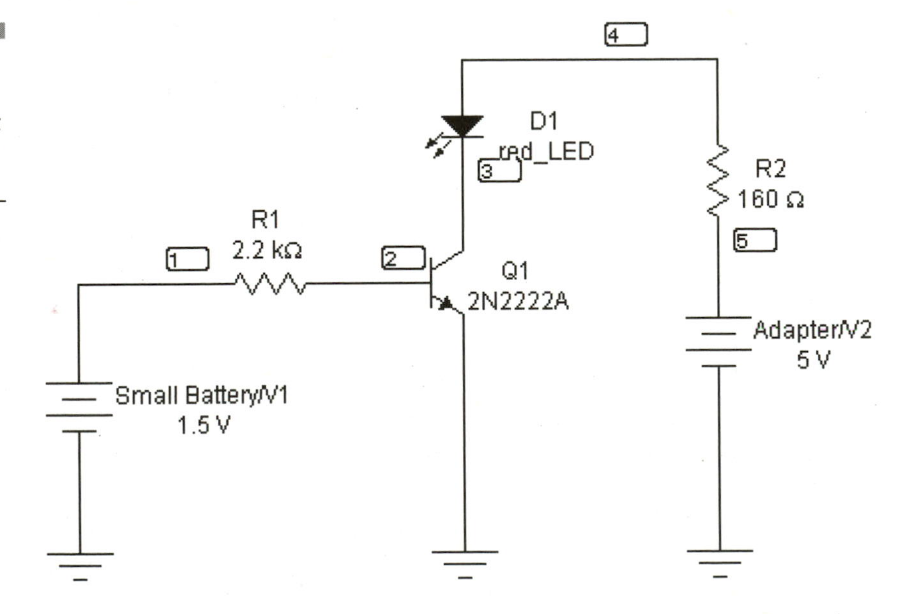

Figure 15.4
This transistor circuit contains two models: one for the transistor and one for the LED.

```
********** c:\ewb5\my-ewb\sp_trans.ewb ***********
* Interactive Image Technologies     *
*            *
* This File was created by:       *
* Electronics Workbench to SPICE netlist  *
* conversion DLL        *
*            *
* Sat Nov 14 10:14:33 1999      *
**************************************************
* Battery(s)
* Adapter
V2 5 0 DC 5
* Small Battery
V1 1 0 DC 1.5

* Resistor(s)
*
R2 5 4 160
*
R1 1 2 2.2K

* NPN Transistor(s)
*
Q1 3 2 0 Qn2n_2N2222A

* LED(s)
*
D1 4 3 LED_red_LED

* Misc
.MODEL Qn2n_2N2222A NPN(Is=11.6f BF=200 BR=4 Rb=1.69 Re=423m
Rc=169m Cjs=0 +Cje=19.5p Cjc=9.63p Vje=750m Vjc=750m Tf=454p
Tr=102n mje=333m mjc=333m +VA=113 ISE=170f IKF=410m Ne=2 NF=1
NR=1 VAR=1e+30 IKR=1e+30 ISC=0 NC=2 +IRB=1e+30 RBM=0 XTF=0
VTF=1e+30 ITF=0 PTF=0 XCJC=1 VJS=750m MJS=0 XTB=0 +EG=1.11 XTI=3
KF=0 AF=1 FC=500m TNOM=27)

.MODEL LED_red_LED D(Is=10f Rs=37.5 Cjo=0 Vj=750m Tt=0 M=0)

.OPTIONS ITL4=25
.END
```

TRANSISTOR MODELS

Let's look at the transistor model now. Under the component description you will see the following:

```
* NPN Transistor(s)
*
Q1 3 2 0 Qn2n_2N2222A
```

The first two lines are comments. The third line is the basic description of the component. Q1 is the Reference ID. The transistor is placed between nodes 3, 2, and 0. Node 3 is the collector, node 2 is the base, and node 0 is the emitter terminal.

The Qn2n_2N2222A refers to the model statements later in the Netlist.

```
.MODEL Qn2n_2N2222A NPN(Is=11.6f BF=200 BR=4 Rb=1.69 Re=423m
Rc=169m Cjs=0 +Cje=19.5p Cjc=9.63p Vje=750m Vjc=750m Tf=454p
Tr=102n mje=333m mjc=333m +VA=113 ISE=170f IKF=410m Ne=2 NF=1
NR=1 VAR=1e+30 IKR=1e+30 ISC=0 NC=2 +IRB=1e+30 RBM=0 XTF=0
VTF=1e+30 ITF=0 PTF=0 XCJC=1 VJS=750m MJS=0 XTB=0 +EG=1.11 XTI=3
KF=0 AF=1 FC=500m TNOM=27)
```

This is the meat of the component. It contains each of the parameters you would see when editing the model inside EWB (Figure 15.5). For example, the saturation current is the first number (Is=11.6f), and the forward gain (amplification) of the transistor is BF=200.

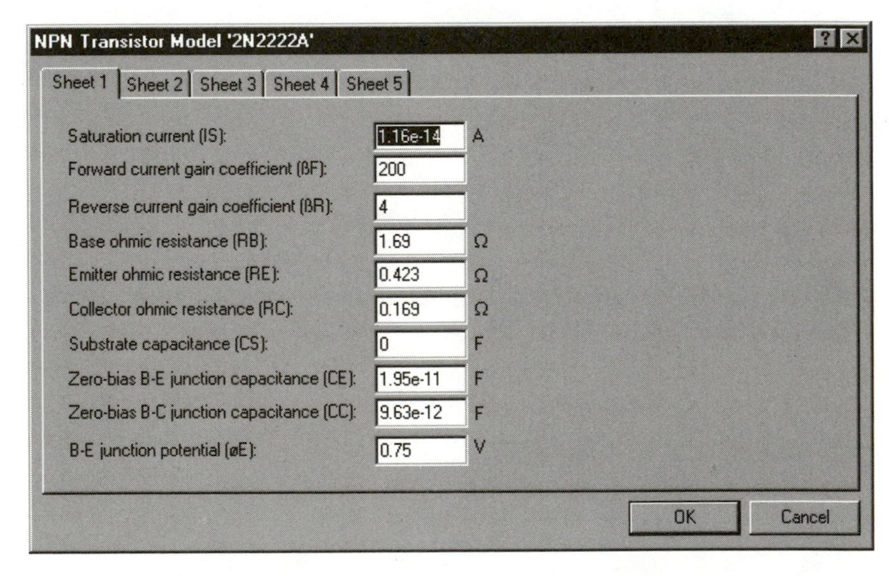

Figure 15.5 In EWB5, you can see what each component model parameter is by double-clicking the component, selecting Model tab, and clicking the Edit button. Unfortunately, Multisim did away with this great feature; you must hand-enter variables as text now.

Each component that is modeled inside EWB will contain these lines in the SPICE Netlist. The LED is another example:

```
.MODEL LED_red_LED D(Is=10f Rs=37.5 Cjo=0 Vj=750m Tt=0 M=0)
```

NOTE

For more information on SPICE model parameters, see the EWB5 or Multisim 6 documentation or purchase a specific SPICE book.

Using EWB5/Multisim Files in Another Simulator (Exporting)

EWB5 and Multisim 6 files are usually not compatible with other SPICE simulators. The only way to transfer these files is to turn them into a Netlist file and import them into other SPICE software. Do this by using the **Export** command in EWB under **File** > **Export**. In Multisim, use **Transfer** > **Export Netlist**. The file is saved as a *.cir file that can be used in just about any SPICE simulator or PCB package.

Importing SPICE

If you have a SPICE Netlist and want to use it in EWB or Multisim, you can import. EWB5 requires you to go to **File** > **Import**. With Multisim 6, go to **File** > **Open** and select Spice Netlist Files (*.cir) as the **Files of type**. Locate the *.cir file and choose the Open button. In EWB5 you will be asked several questions before the circuit "builds itself". With Multisim, you will be asked to scale the drawing. The circuit opens. Have patience, because this may take some time. Also, just as a note, the circuit will look nothing like the one you created with the other software program; EWB and Multisim use a specific drawing algorithm that can make it look like spaghetti. If this is the case, you may need to rearrange things and then save it as an EWB or Multisim file.

Finding SPICE Netlist Files

Many component manufacturers post SPICE Netlists and SPICE models on their websites. Electronics Workbench maintains a list of these at **http://www.electronicsworkbench.com/html/spice_models.html**.

You can also use a search engine, such as **www.google.com** with the terms SPICE and/or NETLIST. Also, keep an eye on the

sci.electronics.* newsgroups as well as **alt.binaries.schematics.electronic**.

Multisim and SPICE

If you have Multisim Personal Edition, the only way to make your own models is by using the SPICE Netlist importation button under the Component Editor dialog box, Model tab (Power Pro contains a Model Maker). By learning a bit of the SPICE language, you can customize the variables right inside the model data window once a premade Netlist is open. The component can then be saved into the User database for use in your simulations.

Summary

The main function of graphical electronics simulators, like Multisim and EWB5, is to make SPICE transparent. Learn the SPICE language bit by bit from sources on the Internet as well as SPICE-specific books; build circuits and export the Netlists, then open the *.cir files with your text editor to see what is happening. Learning SPICE helps you to increase the power you can squeeze out of EWB5 or Multisim 6.

CHAPTER 16

Layout and Other EWB(IIT) Software

Being able to create a PCB right from an EWB simulation makes this software my greatest hobbyist-related asset.

Electronics Workbench has a new product that further exemplifies the practical side of EWB—*Electronics Workbench Layout* (the newest version is called *Ultiroute and Ultiboard,* which is described later). The software is a PCB layout program that integrates nicely with Electronics Workbench Version 5, Personal or Professional Edition; Ultiboard and Ultiroute are used with Multisim 6, but can still be used with EWB5 if necessary. You are able to design and test a circuit with EWB or Multisim and then export it the layout programs. From there, a printed circuit board (PCB) can be designed and saved, printed, or sent to a PCB manufacturer. You can also use special material in a laser printer that allows you to take the PCB trace pattern just created and etch it onto a copper-clad board. The combination of EWB5/Multisim and Layout/Ultiboard/Ultiroute can take you from idea to PCB in a matter of minutes.

This chapter introduces you to the software and gives you an example of how to take a circuit from EWB to PCB. It is mostly written with EWB Layout 5.3; the same procedures can be used with minimal change in Ultiboard.

Goal of Layout/Ultiboard/Ultiroute

Have you ever tried to design a PCB by hand? Ack! It's next to impossible with even mildly complex designs. Layout, Ultiboard, and Ultiroute automate this process; the PCB software moves the footprint information of the components from your EWB5/Multisim schematic captures into Layout/Ultiboard. From there, traces (conduction lines) can be automatically and manually routed; this is called *autorouting.* It saves hours by eliminating the impossible task of figuring out criss-crossed traces by hand.

Electronics Workbench Layout 5.3xx

This was EWB's original PCB layout offering which is a powerful board layout package for producing high-quality, multi-layer printed circuit boards. It has tight and seamless integration with the schematic capture feature, and allows you to incorporate board layout and design to bring well-designed boards to production quickly. It integrates with EWB5.11, allowing you to either export schematics or to create new PCBs of your own design and specification. It's a super deal at $399, but requires a steep learning curve. Once you see an example, things start to click into place. This chapter uses Layout in its example, but Ultiboard is very similar and only requires a few changes in procedures. See Figure 16.1.

Figure 16.1 Electronics Workbench Layout integrates with EWB5.

Ultiboard

Ultiboard is a powerful, easy-to-use PCB layout and routing program that delivers the features and flexibility you need to build reliable boards effectively, coupled with a unique combination of advanced functionality and exceptionally easy use. It integrates with Multisim, or you can open previous Layout files to edit and reroute. As with Layout, it presents a challenge in mastering the ins and outs. See Figure 16.2.

Ultiroute

Essentially this is an add-on to Ultiboard that performs advanced autoplacement and autorouting. Autoplacement places and rotates component footprints in the position and direction in which the software thinks it should be; the autorouting uses a more advanced method that can optimize costs and other important factors. See Figure 16.3.

Figure 16.2
Ultiboard helps you create PCBs from your Multisim files.

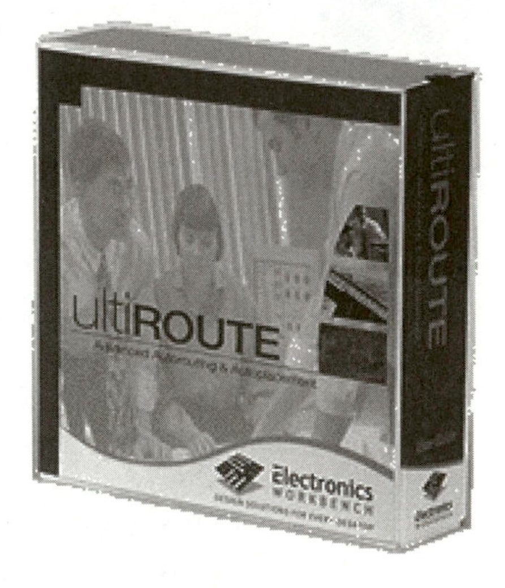

Figure 16.3
Ultiroute is an advanced autorouter and autoplacement product sold separately from Ultiboard and Multisim.

CAUTION

The personal edition of Ultiroute cannot perform autoplacement.

Which Program Should I Use and Where Do I Get It?

This advice is based on my personal experience as a hobbyist. If your budget is limited, try to locate the older version of Electronics

Workbench Layout and EWB5 as a package. If you want some more advanced features and software that is more integrated with Multisim's schematic capture features, choose a package of Multisim 6 and Ultiboard Personal Edition; it will handle most hobbyist applications. You can also ask EWB about their current package specials and save $100 or so if you buy the two together. Unless you are doing a lot of advanced PCB boards that are being sent to a manufacturer, I would forgo the cost of Ultiroute unless you have unlimited funds. It's useless to buy the cheaper Personal Edition, because there is no autoplacement feature. If you do purchase it, definitely order the much more expensive Professional Edition. What do I use? I hate to say it, but the old tried and true EWB5/EWB Layout is my first choice for quick projects (although I am turning to the EWB5/Ultiboard and Multisim/Ultiboard more and more). I can always import my Layout files to Ultiboard later. If I am working on a PCB that will possibly go into production, I use Multisim/Ultiboard and Ultiroute Professional.

If you want more information and current pricing/package specials, visit **http://www.electronicsworkbench** and choose the product information link you wish to read about. You can also call the EWB sales department at 800-263-5552.

Loading the Software

Loading procedures change with each update. Follow EWB's recommendations. As a guide, load Multisim, then Ultiboard, then Ultiroute. Placing the CD in the CD-ROM bay typically brings up the loader program. If not, locate the **setup.exe** file on the CD and run it. I recommend using the directories that the loader program selects, because some files absolutely must be found in that directory.

Layout/Ultiboard Demo on CD

The CD that accompanies this book contains a demo version of both EWB Layout and Ultiboard. Either one can be used to build the example PCB in this chapter. However, the file cannot be printed or saved. See Appendix D for more information on the demos. To order the Personal or Professional versions of Layout/Ultiboard, see Appendix B. I would like to repeat, Ultiboard can be used with either EWB5 or Multisim 6 and is virtually the same software as Layout.

EWB Layout

Figure 16.4 shows EWB Layout's basic interface. Each significant item is labeled. Please note, my screen captions have a background color of white, to be more compatible with print media. The default color is black. To change the color of the board, go to **Tools** > **Options** and select the **Colors** tab.

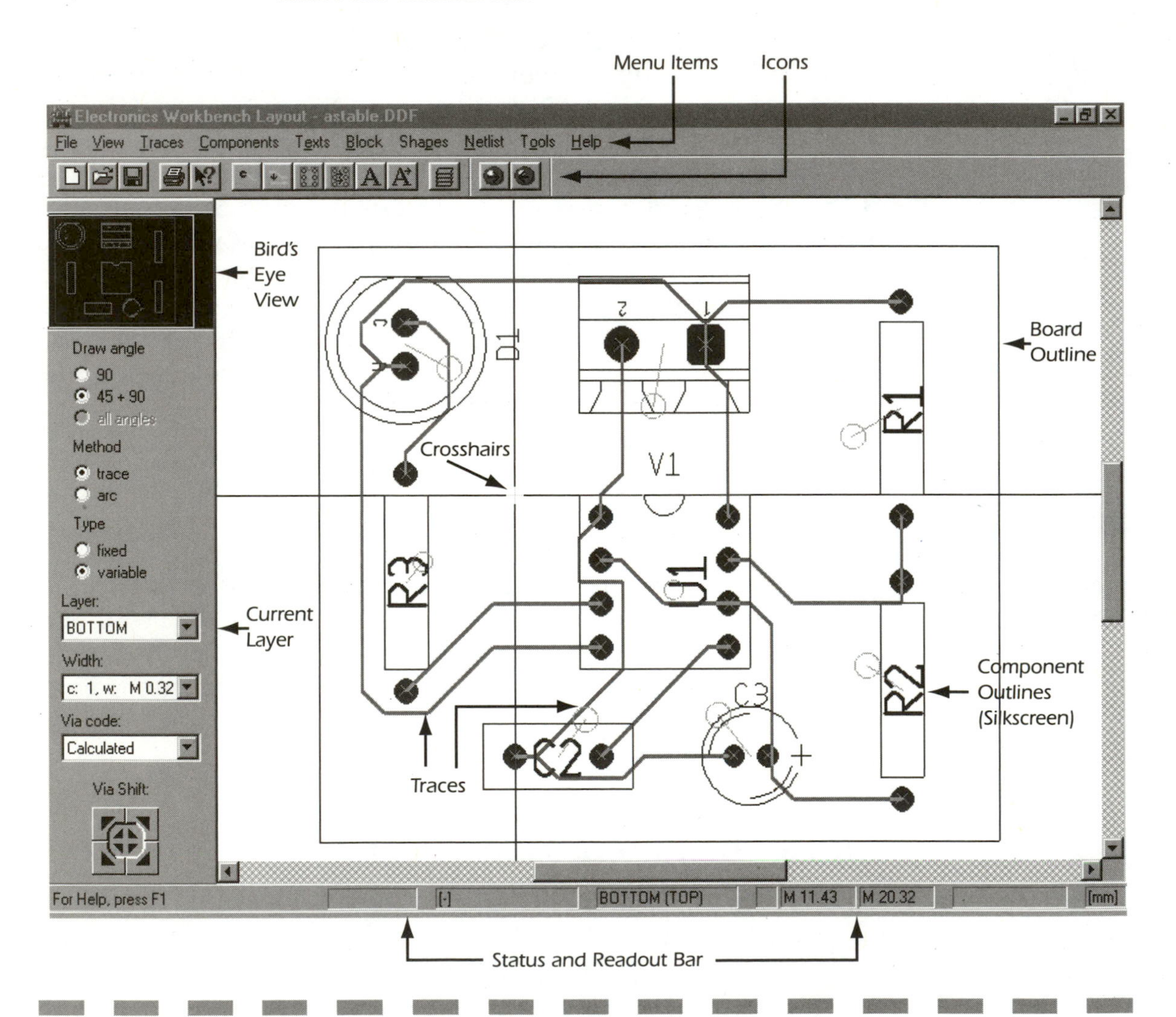

Figure 16.4 The EWB Layout interface.

Getting Started

The following sections are based on using EWB Layout. Ultiboard has nearly the same procedures. An example circuit is used as a training aid.

A 555 timer is used to create an astable multivibrator, which is attached to an LED for output. The circuit turns the LED on for approximately 1 second and off for the same. This produces a rate of about 1/2 of a Hz. Open the file in EWB5 called **555ast.ewb** or build the circuit as shown in Figure 16.5. Use the following procedures to make this schematic into a real-life printed circuit board.

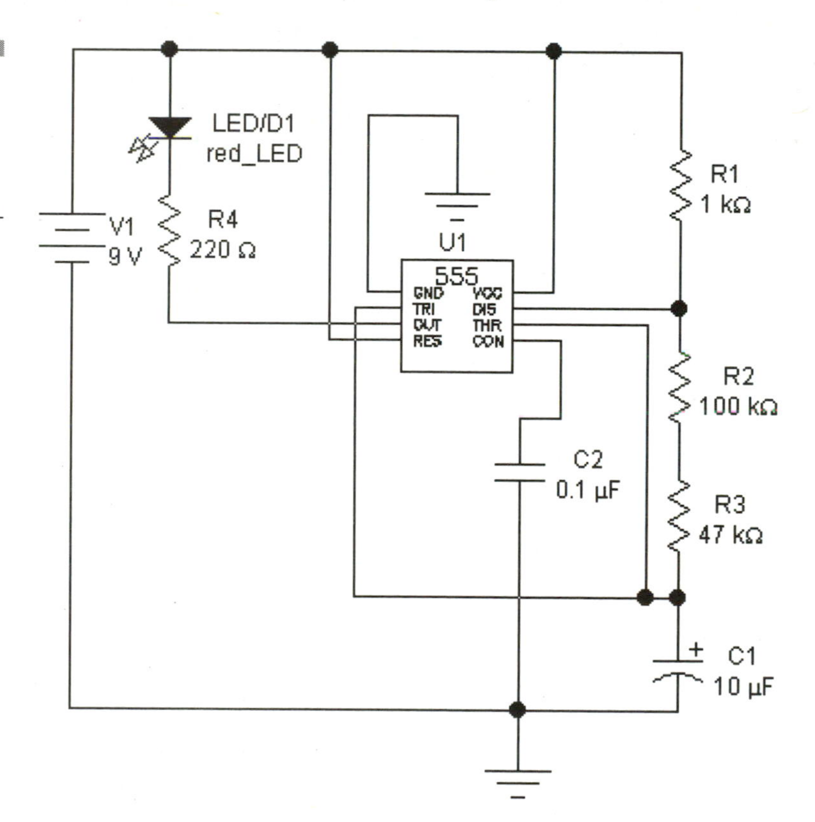

Figure 16.5
An astable 555 timer circuit is used as an example in our EWB Layout design.

NOTE

A 100k and 47k resistor are placed in series to create 147kOhms. Because 150kOhms is not a common value resistor, we must combine two in order to bring it as close to 150k as possible. You could also use two 330k resistors in parallel to produce 165kOhms; the timing will be slightly longer though.

Importing an EWB V5 Design into Layout

Import is the most powerful feature of the EWB software products. It lets you take a design you have been working on in EWB5 and send its information to the Layout software. Each schematic symbol you have used in EWB is replaced with an actual component shape, complete with pin assignments and pads. Layout then lets you place the components onto a PCB and run the conductive traces manually, or more important, by using an autorouting tool. Here is how to import our 555ast.ewb file:

1. Open or create the 555ast.ewb file in Electronics Workbench 5.11.
2. Go to **File** > **Export to EWB Layout**.
3. You will be prompted to name the file in the layout format. The name 555ast should be in this field. Hitting the **Save** button creates two files layout: 555ast.PLC and 555ast.NET. Make sure you save these files into the layout subdirectory or Layout will not be able to send an EWB Layout file back to EWB 5 (back-annotate).
4. Electronics Workbench Layout opens. You will see a 160 × 100 mm PCB appear; each component used in the 555 circuit is placed on the top of the board as shown in Figure 16.6. Go to **File** > **Save As** and save the file as 555ast.DDF. You are now ready to work with the PCB.

NOTE

Saving and printing files is not possible in the demo version of Layout/Ultiboard.

Setting Up the Board

Before we begin the drawing process, we must set the PCB and system settings. This includes the units of measurement used, the design rule settings, the number of board layers and colors, and more.

System Settings

In this example, we will be using metric units, because it is easier to understand component footprints and certain measurements this way

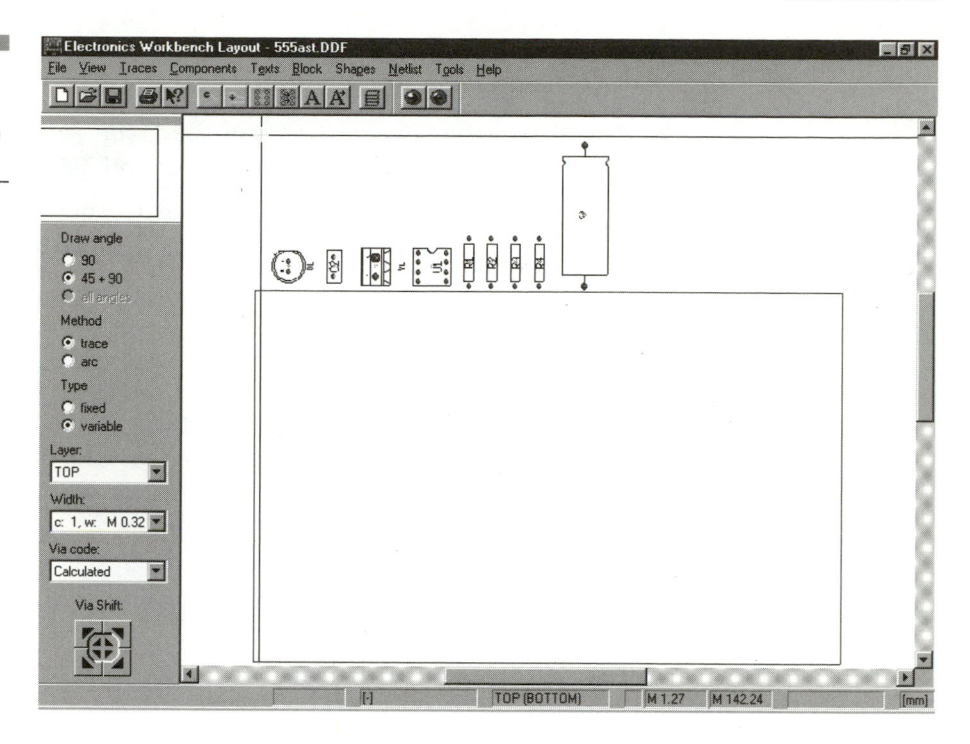

Figure 16.6
Our 555ast.ewb design is exported to Layout.

(remember, most of the world is using the metric system). There is also a setting for Standard (1/200th of an inch per unit) or Inches. To set this, go to **Tools** > **Options** > **System Settings** tab and choose Metric from the Unit box. You may want to turn on and set **Autosave** while you are at it.

Setting the Number of Board Layers

You must set the number of layers the board contains. As a hobbyist, you will rarely use more than a one- or double-sided board. Here is how to select them:

1. Go to **Tools** > **Options** and click the **Board Settings** tab.
2. Under the Layers box you see Max Layers. Choose the maximum number of layers you will use. The default is two, which is plenty for virtually all hobbyist applications. If you will be making a single-sided board, set the layers to two as well (I will explain this in a moment).
3. Click **OK**.

4. Now go to **Tools** > **Autoroute** > **Options**.
5. Select the **Layer Direction** tab.
6. This shows the general direction the layer will autoroute. I usually check on **Horizontal** on the top layer and **Vertical** on the bottom.
7. If you are doing a single-sided board, uncheck the **Top Layer**. This forces Layout to create a single-sided board.

Rule Level

It is best to set the rule level at this time as well. This will tell you if there are any shorts or clearance errors.

1. Go to **Tools** > **Options.**
2. Select the **Rule Level** tab.
3. Select either **Full check** or **Overrule**. I like overrule, because it prompts you to make a correction.
4. Hit **OK.**

Setting the Current Layer

On the left-side of the screen, you will see the pulldown arrow for setting the current layer. Choose bottom layer for this example.

Moving Components onto the Board

When Layout imports your file, it creates a set of components on the top of the working window, as you saw in Figure 16.6. They must be placed and oriented correctly for the program to autoroute the lines. There are several ways to do this, but here is the most efficient:

1. Go to **Block** > **Move**.
2. Move the mouse to the top-left corner, slightly above the components.
3. Click the left mouse button once and drag it to the bottom-right corner and click again.

4. All the components selected will move as a group. Place them somewhere near the middle of the board. The location is unimportant for now.

*To zoom in to a section, place the crosshairs where you wish to zoom in and hit **F8**. To zoom out, use **F9**.*

Now that the components are placed on the board, you can position them. This can be difficult and frustrating for the first-time user, because this is an unnatural interface that requires you to first select the **Move Component** option, then position the part. You *cannot* just grab the part with the mouse pointer and place it as you would in EWB.

You will notice white lines (in Figure 16.7 they are black) running from each component. These are called *force vectors*. In simple terms, they direct you to where the components should be placed on the board. The tiny circles at their ends should all converge in the center of the board. This optimizes the board layout. Now we need to move the components:

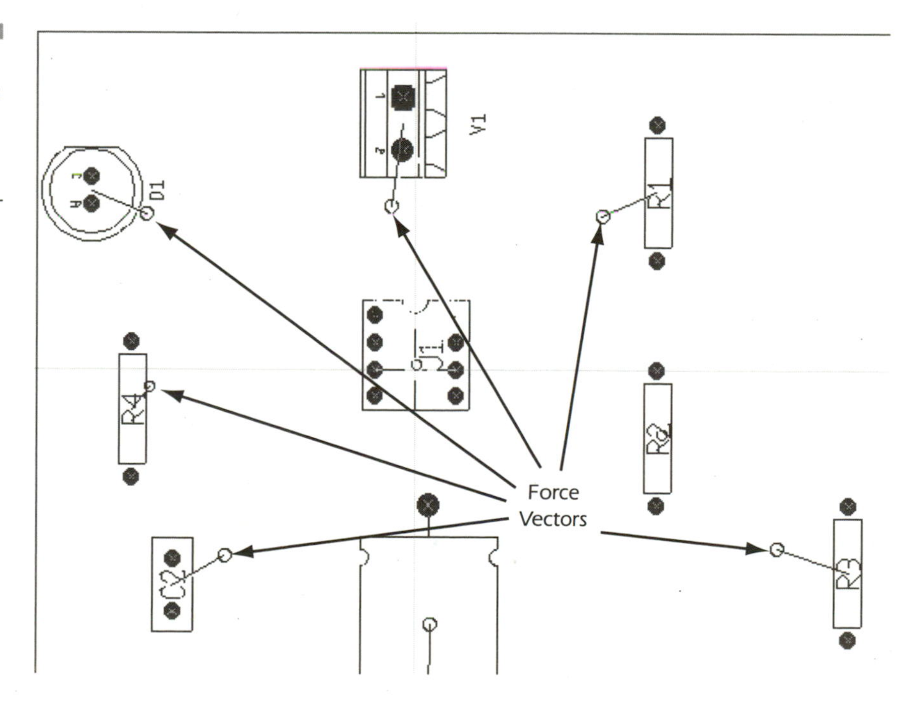

Figure 16.7 Force vectors act as a guide to help you place components onto the board.

1. Go to **Components** > **Move** or select the Move Component icon on the toolbar.
2. Move the mouse over the component you wish to move and left-click once.
3. When you move the mouse, a ghost outline of the component appears. Place that where you wish to lay the component.
4. Click the left mouse button to place it.
5. You can also orient (rotate) the component at this time by using **F2**. You will notice the rotation reading in the very bottom-left of the screen. If you want to move the component between the top and bottom layer, hit **F5**.
6. Right-click the mouse button to release it (if you left-click again the part will be placed somewhere else).
7. While the Move Component command is still on, you can move other components with steps 2 through 6.

The next instructions let you place and orient the components in our test circuit:

1. In our 555 circuit, move the IC socket into place first, as shown in Figure 16.8 (somewhere near the center of the board). Do not change the orientation.
2. Place the LED as shown and rotate to 90 degrees.
3. Place the power supply connection and orient as shown.
4. Place C2 as shown (leave C1 for now, because we will have to change its footprint first—Layout does not import the correct one).
5. Place R1, R2, R3, and R4 and rotate as shown.

TIP

If you get weird artifacts on the screen, choose ***View*** *>* ***Redraw Screen****.*

Changing a Component's Footprint

More often than not, EWB exports the incorrect footprint. In the case of our 555 circuit, the polarized capacitor is way too big. Follow this procedure to correct this:

Figure 16.8
Begin placing and rotating components as shown.

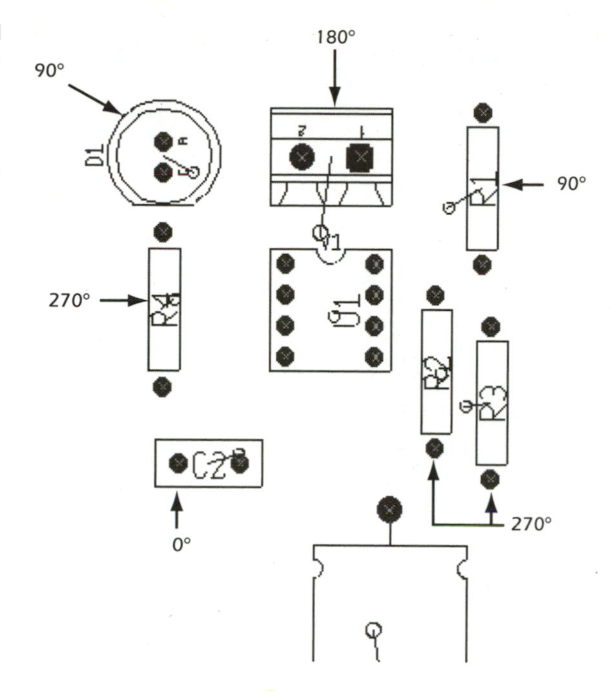

1. Go to **Components** > **Change** and the dialog box in Figure 16.9 opens.
2. Click on **Library** in the Shape source area. You will see files in the Library field open. Scroll down and highlight **l7cap.lib**. In the Shape field, scroll down and select **ELKO5R2**—a 5mm-wide case, with pins 2mm apart.
3. Hit the **OK** button on the bottom of the dialog box. Layout will prompt you, "Do you want all components with this shape to change to the new shape?" Choose **Yes**. The capacitor is now in the correct format.
4. Choose **Components** > **Move** and rotate and place the cap as shown in Figure 16.10.

Defining Board Size

The PCB's shape can be adjusted to just about any shape and to a maximum size of 50 × 50 inches (127cm × 127cm). In this case, we are making a small design, approximate 37mm square. First set the coordinates in the lower left corner of the board to represent 0,0:

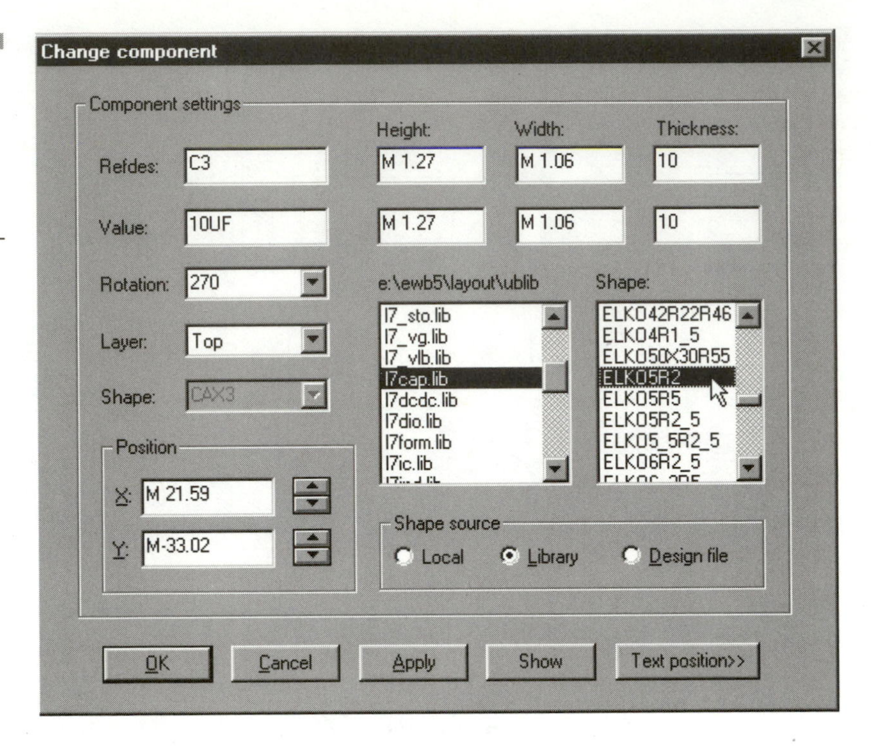

Figure 16.9 The Change Component dialog box is used to fix the polarized capacitor's footprint.

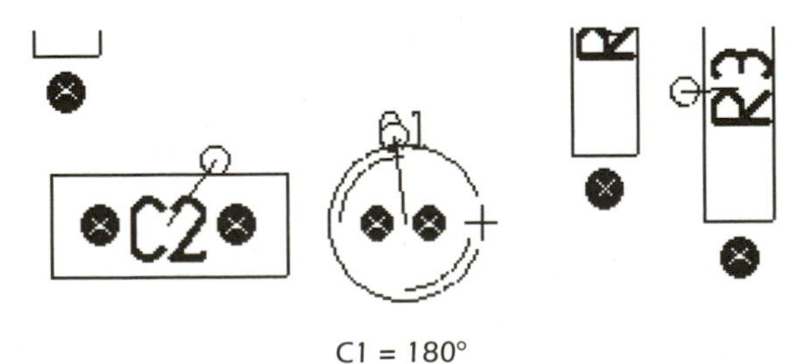

Figure 16.10 Rotate and place the new cap symbol as shown.

1. Use **Tools** > **Reference Point** and selecting **Set by mouse**.
2. Click in the lower-left corner of the board outline shown in Figure 16.11.

That point is now M 0.00, M 0.00 as shown on the status bar at the bottom of the screen. It is now time to outline the PCB. There are two ways to do this:

1. Go to **Tools** > **Board Outline** > **Define by rectangle**.
2. Place the cursor on M 0.00, M 0.00 and click the left mouse button once.

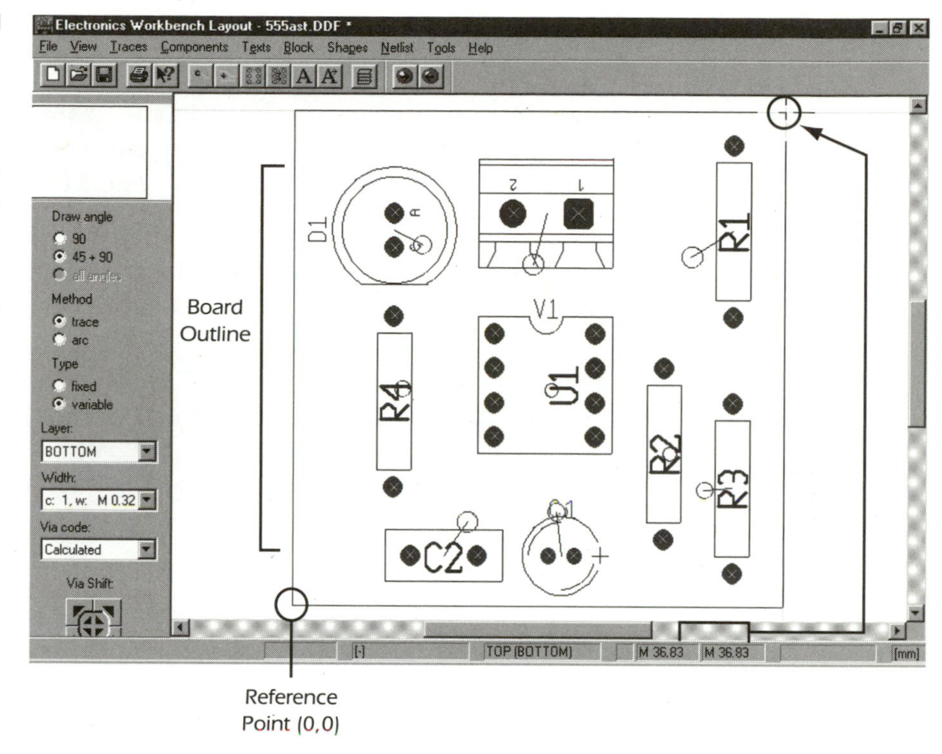

Figure 16.11
Set the reference point of the board as shown. Use the Board Outline feature to correctly adjust the size of the board to approximately 37mm square.

3. Move the mouse to M 36.83, M 36.83 or somewhere near, as shown in Figure 16.11.
4. Click the left mouse button once more.
5. You will be asked "Sure to delete old board shape". Hit **Enter** or **Yes**.

You can also set the board with specific coordinates:

1. Go to **Tools** > **Board Outline** > **Define by rectangle**.
2. Hit the * key on the numeric keypad. A dialog box opens, allowing you to set the lower-left coordinates.
3. Leave the coordinates at 0 and 0. Hit **OK**.
4. Press the * key again and set the upper-right coordinates at 37mm and 37mm.
5. Hit **OK** and the board is set.

Autorouting

Now for the fun part: letting EWB Layout route the traces for you. Because we are using a one-sided board, this could be a difficult task even for a routing program. Follow along and I will show you how to do this:

1. Double-check the layer directions by going to **Tools** > **Autoroute** > **Options.** Select the Layer Direction tab and make sure only the bottom layer is checked on. Set the direction to horizontal. Hit **OK**.
2. Go to **Tools** > **Autoroute** > **Internal** > **Route All**. Let Layout do its thing. Figure 16.12 should appear.

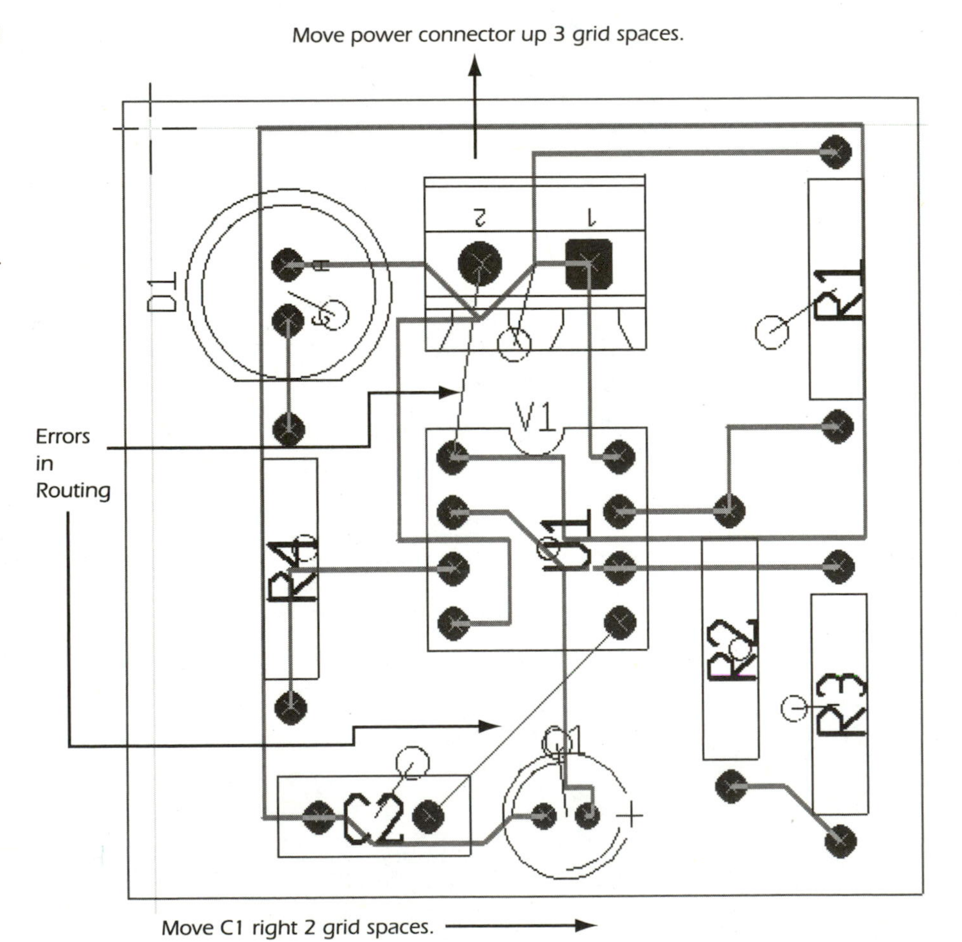

Figure 16.12 The routed circuit. Note that there are corrections that must be made, because certain traces are not run correctly.

3. Next check the connectivity of the circuit to make sure all the pins are actually connected to each other. Go to **Tools** > **Connectivity check**. A dialog box opens. Just hit **OK**.

As you can see, there are errors in the routing. It didn't connect the ground to the IC's pin 1 and one side of C2 is not traced to pin 6. You may have slightly different errors. This is one problem with using Layout's internal routing feature: it requires a lot of futzing around with component placement and orientation to get a decent PCB design out of it. To solve this board's problems do the following:

1. Go to **Traces** > **Delete All Traces**. A dialog box opens. Choose to delete the entire board and hit **OK**.
2. Confirm deletion of traces by hitting **Yes** or **Enter**.
3. Choose **View** > **Redraw Screen** This eliminates any artifacts.
4. Move C1 over to the right by two grids spaces (see Figure 16.13).
5. Move the power connector up three grid spaces, as shown in Figure 16.12.

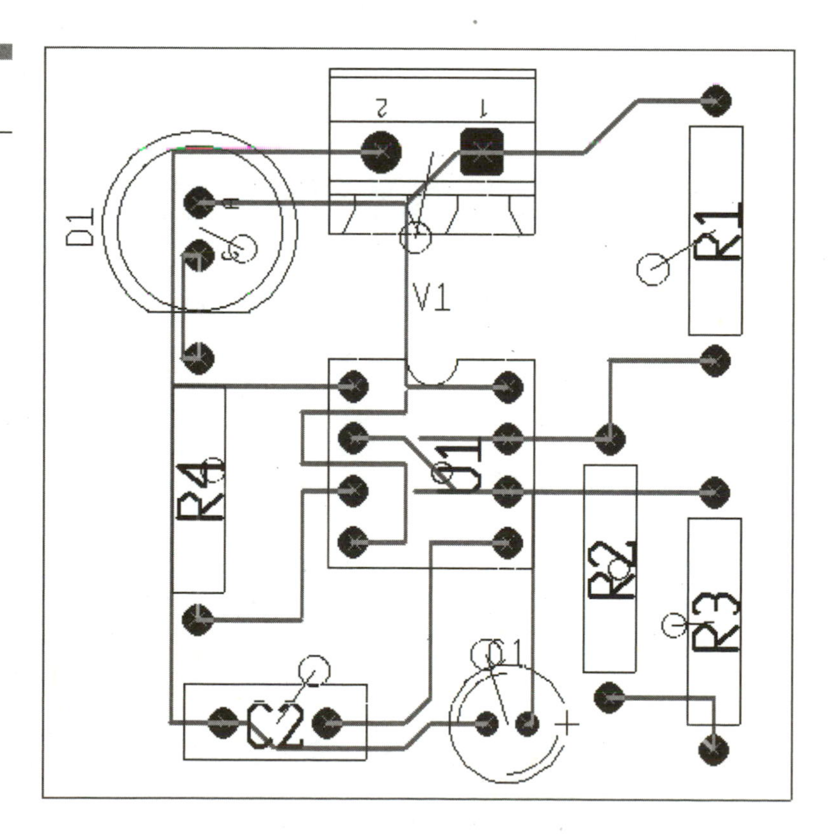

Figure 16.13
Corrected PCB.

6. Redo the routing by going to **Tools** > **Autoroute** > **Internal** > **Route all**. The traces will look as they do in Figure 16.13. If they do not, you need to tear up the traces, move the components around a bit, and reroute.

theRouter!, Ripup/Retry Router

Layout has a built-in advanced routing program that requires less fuss than the internal router. It is a separate program that is launched from Layout. Here is how to use this much simpler method of routing:

1. Using the previous circuit, delete all traces and refresh the screen.
2. Save the file.
3. Choose **Tools** > **Autoroute** > **External** > **To Autorouter**.
4. The router program opens as shown in Figure 16.14 (screen capture colors are inverted).
5. Hit the **Route!** menu item and hit **OK** on the dialog box that opens.
6. A trace design will be created as shown in Figure 16.15. Hit **File** > **Save**.

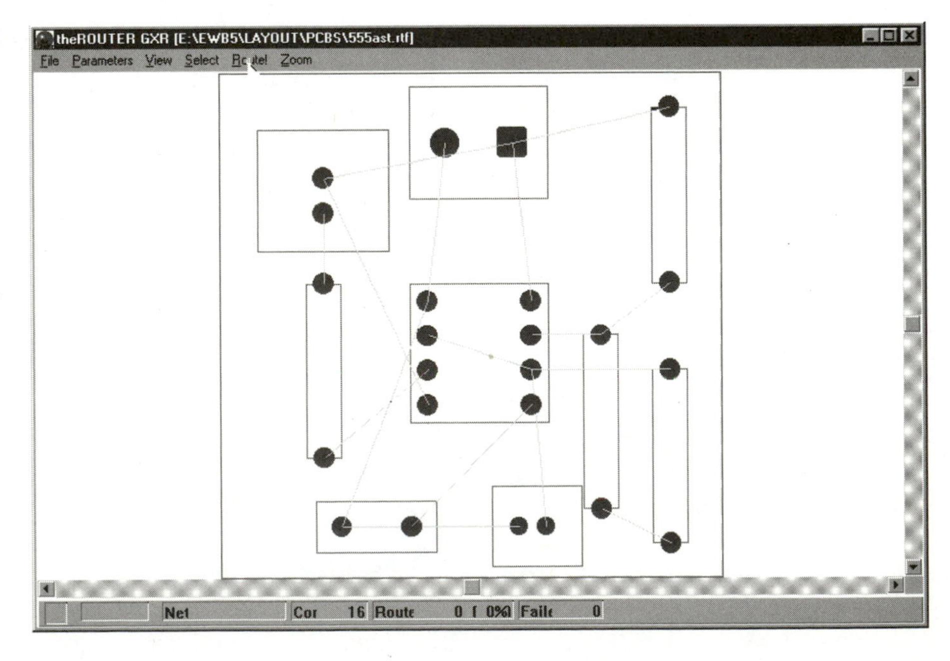

Figure 16.14 theRouter! is a much more advanced, accurate, and simpler autorouter (built into EWB). Lines between pins indicate conduction points.

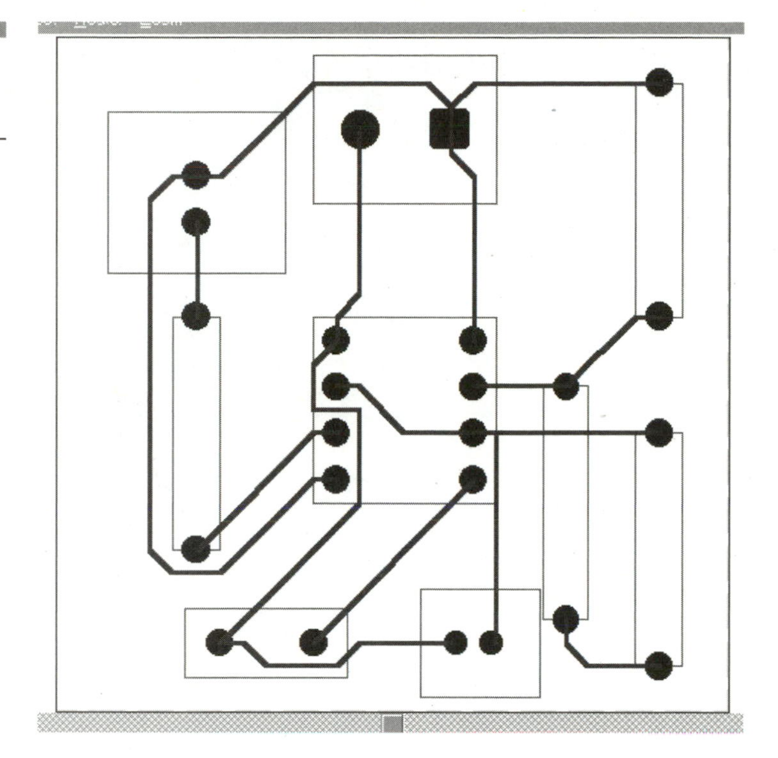

Figure 16.15
The results of this autorouter.

7. Choose **File** > **Exit**.
8. Layout should be open once more. Choose **Tools** > **Autoroute** > **External** > **From Autorouter**. For some reason I have to perform this command twice to send the trace information to Layout.
9. You now have a PCB as shown in Figure 16.16. It's a much cleaner design, isn't it? Save the file.

Outputting Your Design to a Printer

The PCB design is useless unless you can transfer it to a copper-clad board for etching. There is a great product called Press-n-Peel Blue Transfer film offered by Techniks, Inc. (**http://www.techniks.com/index.html**), which is a sheet of a special material that is placed into a

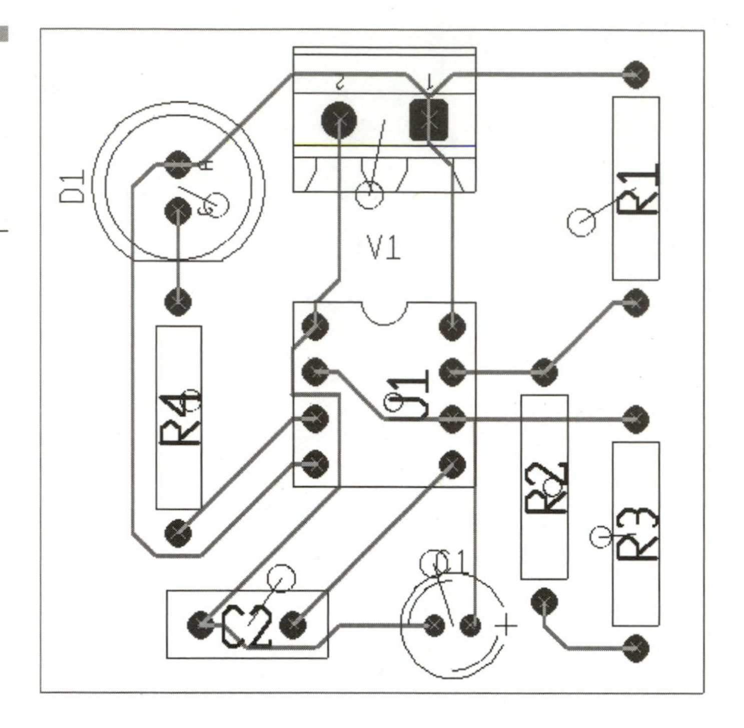

Figure 16.16 The Router information is then sent to Layout and the resulting traces are shown.

laser printer. You print your PCB Layout design onto it, iron the sheet onto a blank copper-clad board, and peel off the sheet. Chemicals are then used to etch away the unused copper. Drilling the holes for each component is required. You can then clean the board and begin populating it with components (soldering time!). The actual instructions are a bit more detailed and are posted at **http://www.techniks.com/press-n-peel.html**. I suggest printing out that page and following the instructions to the letter. The sheets are relatively cheap and you can order copper-clad boards at the same time if you wish (the advantages of on-line shopping).

NOTE

You can also send out your Layout or Ultiboard files to a "one-off" or "prototype" PCB manufacturer. See the back of most electronic magazines for current companies.

Printer Controls

Printing the PCB can be quite a task. There is a printout for each layer that is to be etched, the silkscreen that is printed on top of production PCBs (usually white and containing the component outlines as well as text), a solder mask, and a drill target to show the placement of the drill holes and hole size. To access the controls for the printer, go to **File** > **Print**. (**File** > **Post Processor** in Ultiboard). Select the files you wish printed from the right under the Available setting files and add them to the Settings to Process. Hit GO! For a drill target printout, hit that button on the Post Processing dialog box (see Figure 16.17 for an example of the printouts for the example circuit we created). Unfortunately I do not know a way to save these as .gif files, which are typically used on the Internet. I print out the layers and scan them, then adjust the size of the .gif in a paint program such as Paint Shop Pro.

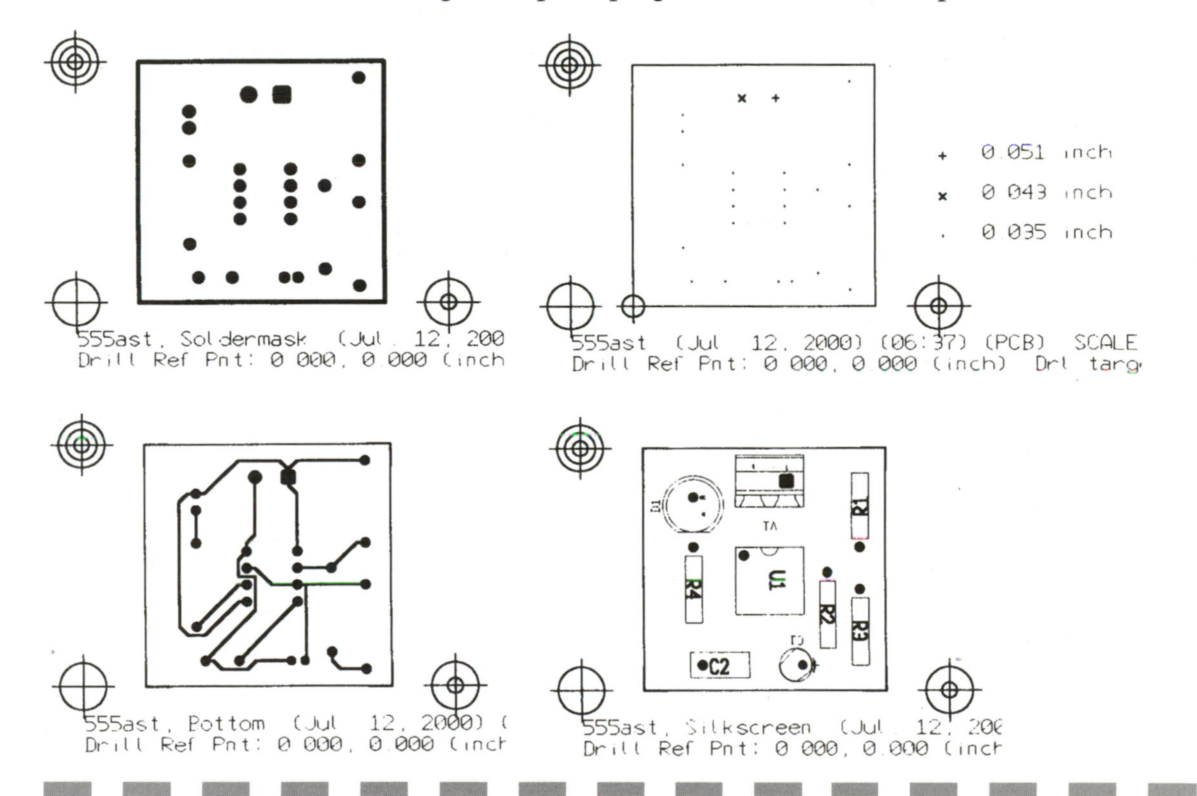

Figure 16.17 Example printouts of our circuit. Set your scanner to the same resolution as your printer to scan a printout. This creates a .bmp or .gif file at the correct scale.

Place the drill targets printout on top of the copper board to drill the appropriate sized holes.

Ultiboard and Ultiroute Notes

This chapter used EWB Layout for its exercises. Ultiboard can be used for just about the same procedures. The problem with the Personal Edition of Multisim is that it does not include the proper DIP package for the 555, which means you are stuck with having to correct this (this can be a complex task). I recommend that if you own EWB5 and Layout, create the design in that, then import it into Ultiboard so the footprint is available.

Here are a few notes about Ultiboard:

- Visit **http://www.electronicsworkbench.com/html/layout_support.html** for up-to-date tech tips and bug reports. The page is currently of mammoth proportions; I recommend printing it out and keeping it handy.
- There are no footprints for sources such as the power connector we used in Layout. Instead, place a 1×2 or 1×3 Header in your Multisim schematic to represent this item. Make the square terminal positive and use the other two for ground and negative supply if applicable.
- Move the component selector field onto the top toolbar as soon as you begin using Ultiboard. It is extremely irritating otherwise.
- Ultiboard has much better documentation than Layout. You may want to read the PDF file included with the software and even try the tutorial.
- Printing is accessed through **File** > **Post Processing**.

Summary

I was recently talking about the difficulty of learning Layout and Ultiboard with one of the tech support personnel at EWB and he hit the nail right on the head. He said that it takes a bit of groping around in the dark with the program until all of the sudden you understand a large percentage of it. So don't feel bad about the difficult learning curve of this software. This chapter gives you a head start, but be sure to check out the Help files and documentation of Layout, Ultiboard, and Ultiroute. It is well worth learning and applying this software to your projects. Great satisfaction comes from holding that completed PCB in your hands, knowing you designed it from scratch; after all that is the ultimate goal of mastering Electronics Workbench.

APPENDIX A

WEBSITE ADDRESSES

TAB ELECTRONICS (McGRAW-HILL)
http://www.tabelectronics.com or
http://www.books.mcgraw-hill.com/tab-electronics/

ELECTRONICS WORKBENCH
(FORMERLY: INTERACTIVE IMAGE TECHNOLOGIES, INC.)
http://www.electronicsworkbench.com/

ELECTRONICS WORKBENCH TECHNICAL SUPPORT
http://www.electronicsworkbench.com/html/technical_support.html

THE AUTHOR'S WEBSITE
http://www.basicelectronics.com/

EWB INFO
http://www.basicelectronics.com/ewb/

EDA.ORG
Tons of EDA information including tools and models.
http://www.eda.org/

EDA TOOLS CAFE
This is a super information site for all EDA areas.
http://www.dacafe.com/DACafe/homepage.html

ABOUT.COMS TAKE ON EDA
http://electronics.about.com/industry/electronics/msub_eda.htm

GOOGLE SEARCH ENGINE
Fantastic Search tool
http://www.google.com/

UNIVERSITY OF CALIFORNIA, BERKELEY
Department of Electrical Engineering and Computer Sciences
SPICE info
http://www.eecs.berkeley.edu/

INTERNET NEWSGROUPS FOR ELECTRONICS AND EDA
sci.electronics.basics—Elementary questions about electronics.
sci.electronics.cad—Schematic drafting, printed circuit layout, simulation.
sci.electronics.components—Integrated circuits, resistors, capacitors.
sci.electronics.design—Electronic circuit design.
sci.electronics.equipment—Test, lab, and industrial electronic products.
sci.electronics.misc—General discussions or the field of electronics.
sci.electronics.repair—Fixing electronic equipment.
misc.industry.electronics.marketplace—Electronics products and services.
alt.binaries.schematics.electronic
alt.engineering.electrical
sci.engr.electrical.compliance
sci.engr.electrical.sys-protection
comp.lang.verilog
comp.lang.vhdl

MAGAZINES
EDN Magazine—http://www.ednmag.com/
EE Times.COM—http://www.eetimes.com/
Poptronics—http://www.gernsback.com/

APPENDIX B

EWB5 AND MULTISIM 6 PRODUCT ORDERING/REGISTERING INFORMATION

All versions of Electronics Workbench and Multisim 6 can be ordered from Electronics Workbench. See Appendix A for contact numbers. Current demos, pricing, and electronic registration are also available on their Website at **http://www.electronicsworkbench.com/**.

Last released version of Electronics Workbench was 5.12. Here are the editions available:

- Educational Edition
- Student Edition
- Personal Edition
- Professional Edition

The current edition for Multisim is 6.20 with 6.5 planned at the time of this writing. Editions of Multisim include:

- Various Education editions are available (some editions are included with training courses)
- Corporate training editions will be available soon.
- Student Edition
- Personal Edition
- Professional Edition
- Power Professional Edition
- Modules for design sharing, VHDL, Verilog and model expansion packages are also available

Layout, Ultiboard, and Ultiroute are currently available. Costs depend on which edition you purchase (Personal, Professional, or Power Pro). See the EWB site for more information.

Costs of these packages are not listed here because they are constantly changing. You can either call EWB, visit their website, or look in the back of electronic magazines (such as *Poptronics*) for specials. EWB has package deals when you order Multisim and Ultiboard/Ultiroute at the same time.

CAUTION

Some EWB products require a hardware device be attached to your computer. This "Dongle" is placed in series with the parallel port and is required to be connected whenever the program is in use.

APPENDIX C

ELECTRONICS WORKBENCH CONTACT INFORMATION

Addresses

ELECTRONICS WORKBENCH USA
908 Niagara Falls Boulevard, Suite #068
North Tonawanda, NY 14120-2060 USA

ELECTRONICS WORKBENCH EUROPE
Energiestraat 36,
1411 AT Naarden,
The Netherlands

ELECTRONICS WORKBENCH CANADA
111 Peter Street, Suite 801
Toronto, Ontario,
Canada M5V 2H1

Contact

Telephone: 416-977-5550
Fax: 416-977-1818

E-Mail

ewb@electronicsworkbench.com

SALES
Sales only: 800-263-5552
Sales@electronicsworkbench.com or isales@electronicsworkbench.com for international inquiries.

Technical Support

Visit the website first to complete the Technical Support form at **http://www.electronicsworkbench.com/html/technical_support.html**. Or phone (416) 977-5550. Fax is (416) 977-1818.

International Customers must contact their local distributor for technical support.

APPENDIX D

ABOUT THE CD

The contents and further instructions for the included CD are listed in a file in the root directory of the CD called CD_INFO.TXT. To run the setup program to install the demo versions of EWB5, MULTISIM, LAYOUT or ULTIBOARD, insert the CD-ROM and follow the on-screen instructions. If the setup program does not automatically run upon inserting the CD-ROM, locate the <CDROM>:\SETUP.EXE file and run it.

CD contains demo versions of:

- EWB 5.12
- EWB Layout
- Multisim 6
- Ultiboard

The circuits in this book can be opened with these demo versions of Electronics Workbench's software.

INDEX

B

C

D

E

H

I

J

K

L

Q

R

S

U

V

W

X

Y

Z